J.M. Trigo-Rodríguez · F.J.M. Rietmeijer · J. Llorca ·
D. Janches
Editors

Advances in Meteoroid and Meteor Science

Foreword by J.M. Trigo-Rodríguez, F.J.M. Rietmeijer, J. Llorca and D. Janches

Previously published in *Earth, Moon, and Planets*, Volume 102, Issues 1–4, 2008

Springer

J.M. Trigo-Rodríguez
Institute of Space Sciences (CSIC-IEEC),
Barcelona, Spain

F.J.M. Rietmeijer
University of New Mexico,
Albuquerque, NM, USA

J. Llorca
Institut de Tècniques Energètiques,
Universitat Politècnica de Catalunya,
Barcelona, Spain

D. Janches
Northwest Research Associates,
Colorado Research Associates
Division (NWRA/CoRA Div.),
Boulder, CO, USA

Cover illustration: South Taurid fireball of magnitude -9 appeared on October 13th, 2007 at 23h48m50 ± 10s UTC. The fireball appears projected over the Pleiades (M45) cluster in this casual picture taken by Mario Ximénez de Embún from Marugán, Segovia, Spain. A Canon 350D camera was used with a 200mm f:2.8 lens plus a Sigma 2X duplicator. The camera was mounted in piggy-back of a telescope.

Backcover illustration: Daylight bolide photographed by Maria M. Robles from Santa Columba de Curueño (León). This magnitude -18 bolide appeared on January 4, 2004, and announced the fall of the Villalbeto de la Peña meteorite studied by the Spanish Meteor and Fireball Network (SPMN). A total mass of more than 3 kg of L6 ordinary chondrites were recovered by researchers of the Spanish Meteor and Fireball Network (SPMN). For comparison, the Moon is clearly visible on the left.

Library of Congress Control Number: 2008923259

ISBN-978-0-387-78418-2 e-ISBN-978-0-387-78419-9

Printed on acid-free paper.

1

springer.com

Contents

Preface

Josep M. Trigo-Rodriguez · Frans J. M. Rietmeijer · Jordi Llorca · Diego Janches

Originally published in the journal Earth, Moon, and Planets, Volume 102, Nos 1–4.
DOI: 10.1007/s11038-008-9228-0 © Springer Science+Business Media B.V. 2008

This volume is a compilation of articles that summarize the most recent results in meteor, meteoroid and related fields presented at the Meteoroids 2007 conference held in the impressive CosmoCaixa Science Museum in Barcelona, Spain. The conference took place between 11 and 15 of June and was organized by the Institute of Space Sciences (*Consejo Superior de Investigaciones Científicas*, CSIC) and the *Institut d'Estudis Espacials de Catalunya* (IEEC). Researchers in meteor science and supporting fields representing more than 20 countries participated at this international conference where 126 presentations were delivered in oral and poster forms. The 69 papers included in this volume represent the work of 154 authors from about 70 different institutions across the globe. The Meteoroids conference is an international meeting that takes place every 3 years since the first one held in Bratislava, Slovakia in 1994. The 2007 meeting was the first one where samples of a comet, 81P/Wild 2, were available from the NASA Stardust mission, and results from laboratory characterizations were presented and discussed. Seemingly aware of the upcoming meeting a bolide was observed over La Mancha, Spain, on May 10. The first five recovered fragments of this event that is known as the "Puerto Lápice" eucrite meteorite fall were shown at the meeting. Eucrites are linked to asteroid 4 Vesta, which is the source of differentiated achondrite meteorites that are igneous rocks formed from basaltic magmas. Puerto Lápice and Wild 2 are at the opposites of the spectrum of

J. M. Trigo-Rodriguez (✉)
Institute of Space Sciences (CSIC) and Institut d'Estudis Espacials de Catalunya (IEEC) Campus UAB, Facultat de Ciències, Torre C-5 parells, 2a planta, Bellaterra, Barcelona 08193, Spain
e-mail: trigo@ieec.uab.es

F. J. M. Rietmeijer
Department of Earth and Planetary Sciences, MSC03-2040, 1-University of New Mexico, Albuquerque, NM 87131-0001, USA

J. Llorca
Institut de Tècniques Energètiques, Universitat Politècnica de Catalunya, Diagonal 647, ed. ETSEIB-C, Barcelona 08028, Spain

D. Janches
NWRA/CoRA Div., 3380 Mitchell Lane, Boulder, CO 80301, USA

DOI: 10.1007/978-0-387-78419-9_1

meteoroid compositions that can interact with the Earth's atmosphere. Laboratory analyses of this meteorite and the comet dust will be critical to elucidating the properties of these known meteoroid-producing sources.

Technological advances in meteor and meteoroid detection, the ever-increasing sophistication of computer modeling, and the proliferation of autonomous monitoring stations continue to create new niches for exiting research in this field. They also allow the built-up of long-term databases providing crucial statistics needed to understand origins and distributions. It was especially gratifying to witness at this meeting the emergence of laboratory-based meteor science.

The conference gave a comprehensive overview on meteoroid and meteor science in two broad-based thematic categories. The first category covered detections, observations and measurements techniques many of which were described in great detail by invited speakers. The contributed presentations in this category focused on the formation of meteoroid streams by active or dormant comets and asteroids, together with dynamical studies of meteoroids moving through the solar system. The study of meteoroids as space hazard is a topic of rapidly increasing interests due to the need of secure the safety and health of manned and unmanned space missions. It is also gaining impetus from the more ambitious initiative to build a human lunar outpost. Papers discussing optical techniques to observe meteor phenomena were prominent and results included the observation of enhanced activities of the 2006 Leonids and 2006 Orionids. The outcomes of years of infrasound and radar detections also showed that these methodologies are no longer stepchildren of meteor science, greatly expanding the mass range of extraterrestrial bodies which can now be studied. Radar meteor detection methodologies have evolved immensely since these instruments were first applied in the 1950s. Greater transmitted power, multi station interferometric techniques and the use of dual frequencies allow meteor radars to provide exciting new data, including the discovery of new meteoroid streams. In addition, in the past decade, the increasing use of high-power and large-aperture radars offer a new look at the meteor phenomena by allowing the routine study of the meteor head-echo, non-specular trails and a particle size range that bridge the historic gap between dust detector on board of satellites and specular meteor radars.

The second category of results included dynamical modeling exemplified by the power of reconstructing past meteor displays and accurate predictions of modern meteor stream activities. Meteor observations are now providing more precise input to fine-tune models, which is an achievement of increasing sophistication in both areas. For example, Comet Wild 2 data were preliminary explored for their relevance to cometary meteoroid properties. With the availability of this comet dust, interplanetary dust particles, micrometeorites and meteorites for laboratory studies, it is but a giant leap to use what we know of these samples as a starting point for experimental meteor science. Results from laboratory simulations of chemical releases during the meteor ablation process are showing that we are closer to understanding how the meteoric mass is deposited in the upper atmosphere. This particular advancement allows linking the meteoric flux with several aeronomical phenomena such as mesospheric metallic layers, noctilucent clouds and meteoric smoke particles embedded in the ionospheric plasma.

The scientific organizing committee (listed below) was responsible for shaping the meeting agenda covering both long-term research directions and objectives while also exploiting opportunities and testing new directions and interactions. These goals were achieved by judicious choices of invited, regular and poster presentations and are reflected in the compilation of articles presented in this book. The meeting also included an invited public lecture by Dr. Clark Chapman entitled "The hazard of asteroids and comets impacting Earth". We would like to take this opportunity to acknowledge and thank the long hours of hard work spent by the members of the local organizing committee (LOC, listed also below). The dedicated work of the LOC along with the tremendous help provided by students from the IEEC, Universitat Politècnica de Catalunya (UPC), Universitat de Barcelona (UB), IEEC, amateur astronomers volunteers and the support received from the CosmoCaixa museum staff resulted in a flawless meeting. This conference highlighted a growing multidisciplinary interest in meteorite, meteoroid and meteor research that we should nurture and also showed that results from this field can provide clues to address unanswered questions in other disciplines (i.e. aeronomy). We look forward to the next Meteoroids conference that will be held in the USA in 2010.

We would like to acknowledge the sponsors for this conference, including the *Ministerio de Educación y Ciencia* (MEC), IEEC-CSIC, and CosmoCaixa. Their financial contributions made it possible to have a successful and exciting scientific meeting and to prepare this tangible record of this proceedings volume.

The papers in this volume underwent the rigorous refereeing process that is applied to other papers in the journal Earth, Moon, and Planets. It could not have been achieved without the time and effort from over 100 referees, who guarded both scientific quality and clarity of the manuscripts. The guest editors of this volume acknowledge the professionalism and diligence of the editorial staff at Springer Science. It really requires all parties to cooperate to turn an idea into a proceedings volume. We also thank the editors and staff of Earth, Moon, and Planets.

Sincerely,
Josep M. Trigo-Rodriguez
Jordi Llorca
Diego Janches
Frans Rietmeijer

1 Scientific Organizing Committee

Peter Brown, University of Western Ontario, Canada
Valeri Dikarev Max Planck Institute for Solar System Research, Germany
Robert Hawkes, Mount Allison University, Canada
Diego Janches, Colorado Research Associates Division, NorthWest Research Associates Inc., USA
Peter Jenniskens, NASA/Ames Research Center, USA
Jordi Llorca, Institut de Tècniques Energètiques, Universitat Politècnica de Catalunya, Spain
Ingrid Mann, University of Münster, Germany
Asta Pellinen-Wannberg, Swedish Institute of Space Physics, Sweden
Olga Popova, Institute for Dynamics of Geospheres, Russian Academy of Science, Russia
Douglas O. ReVelle, Los Alamos National Laboratory, USA
Frans J.M. Rietmeijer, University of New Mexico, USA
Pavel Spurny, Astronomical Institute of the Academy of Sciences, Ondřejov Observatory, Czech Republic
Josep M. Trigo-Rodríguez, Institute of Space Sciences, IEEC-CSIC, Spain
Junichi Watanabe, National Astronomical Observatory of Japan, Japan
Iwan Williams, University of London, U.K.

2 Local Organizing Committee

Josep M. Trigo-Rodríguez, ICE-CSIC (Chair)
Jordi Llorca, UPC (Co-Chair)
Jordi Isern, Director ICE-CSIC
Alberto J. Castro-Tirado, Instituto de Astrofísica de Andalucía (CSIC)
José A. Docobo, Universidad de Santiago de Compostela (USC)
José M. Madiedo, Universidad de Huelva (UHU)
José L. Ortiz, IAA-CSIC
Anna Bertolin and Pilar Montes, ICE-CSIC (secretaries)
Santi Oliveras, ICE-CSIC (webmaster)

Chapter 1. Meteor Shower Activity, Forecasting, Dust Orbits

The IAU Meteor Shower Nomenclature Rules

Peter Jenniskens

Originally published in the journal Earth, Moon, and Planets, Volume 102, Nos 1–4.
DOI: 10.1007/s11038-007-9155-5

Abstract The International Astronomical Union at its 2006 General Assembly in Prague has adopted a set of rules for meteor shower nomenclature, a working list with designated names (with IAU numbers and three-letter codes), and established a *Task Group for Meteor Shower Nomenclature* in Commission 22 (Meteors and Interplanetary Dust) to help define which meteor showers exist from well defined groups of meteoroids from a single parent body.

Keywords Meteor shower · Meteoroid stream · Nomenclature

1 Introduction

Commission 22 of the International Astronomical Union is concerned with all aspects of meteors and with interplanetary dust. It falls under IAU Division III (Planetary Systems Sciences) and is currently chaired by Dr. Pavel Spurny of Ondrejov Observatory.

The International Astronomical Union has the task to define astronomical terms and give names to entities in space whenever needed to further astronomical research. Most recently, it labored over a definition of "planet" and created a category of "dwarf planets" to which Pluto belongs. Until now, meteor showers have not been named officially, as a result of which there is much confusion in the literature. Some showers are well defined but have multiple names (e.g., Draconids, gamma-Draconids, October Draconids, Giacobinids, Giacobini-Zinnerids), sometimes changing name when the radiant moves into another constellation. Many other showers are only ill defined and are given a different name in each new detection, often leaving us confused about whether these proposed showers are indeed groups of meteoroids from the same parent body.

P. Jenniskens
Task Group for Meteor Shower Nomenclature, Commission 22 I.A.U.
http://meteor.asu.cas.cz/IAU/nomenclature.html

P. Jenniskens (✉)
SETI Institute 515 N. Whisman Road, Mountain View, CA 94043, USA
e-mail: pjenniskens@mail.arc.nasa.gov

During the IAU General Assembly in Prague on August 24, 2006, Commission 22 established a new *Task Group for Meteor Shower Nomenclature*, confirmed at the subsequent Division III meeting, with the objective to formulate a descriptive list of established meteor showers that can receive official names during the next IAU General Assembly in Rio in 2009. The objective of this action is, based on our community's work on meteor showers, to uniquely identify all existing showers. This would enable, for example, studies of associations between meteor showers and potential parent bodies among the many Near-Earth Objects that are being discovered.

Current members of the Task Group are Peter Jenniskens (chair), Pavel Spurny (president of C22), Vladimir Porubcan (head of the IAU Meteor Orbit Data Center), Juergen Rendtel (president of the International Meteor Organization), and regional representatives Tadeusz Jopek (Poland), Shinsuke Abe (Japan), Jack Baggaley (New Zealand), and Bob Hawkes (Canada).

To reach this goal, the traditional meteor shower nomenclature practices were formalized (with a few choices made to clean things up) by adopting a set of nomenclature rules, and a two-step approach was taken to uniquely identify meteor showers. First, a Working List of $\sim$230 showers was adopted that gives a summary of showers reported until now from a compilation of past publications (Jenniskens 2006). To facilitate identification, the list is fully cross-referenced by giving the mean orbit and radiant from each prior record, as well as the source of the work.

Each proposed shower was given a name, as well as a unique number and a three-letter code to be used in future publications that discuss the recovery of the stream in orbit surveys and other types of observations. The IAU numbers go back to a system of numbers introduced in the work at the Harvard Smithsonian Center for Astrophysics and now used by the IAU Meteor Orbit Data Center, by simply adding to the numbers given to potential meteor showers in the past. The three-letter code is based on the codes used by IMO, with few exceptions. The designated names are mostly traditional, adhering to a system of nomenclature rules given below, but accepting that it is not always known what is the nearest star to the radiant position at the time of the peak of the shower.

The task ahead is to collect information to add more showers to this Working List, and to collect sufficient information for each shower to establish that the streams of meteoroids responsible are *groups of meteoroids from the same parent body*. The established showers will then be included in an IAU List of Established Meteor Showers, and will be voted on at the Commission 22 meeting in Rio for official recognition.

2 Meteor Shower Nomenclature

The general rule is that a meteor shower (and a meteoroid stream) should be named after the then current constellation that contains the radiant, specifically using the possessive Latin form (Table 1). The possessive Latin name for the constellations end in one of seven declensions:

-ae (e.g., Lyrae),
-is (e.g., Leonis),
-i (e.g., Ophiuchi),
-us (e.g., Doradus),
-ei (e.g., Equulei),
-ium (e.g., Piscium), or
-orum (e.g., Geminorum).

Table 1 Latin possessive names of meteor showers

Constellation	Latin possessive	Shower	Constellation	Latin possessive	Shower
Andromeda	Andromedae	Andromedid	Leo	Leonis	Leonid
Antlia	Antliae	Antliid	Leo Minor	Leonis Minoris	Leonis Minorid
Apus	Apodis	Apodid	Lepus	Leporis	Leporid
Aquarius	Aquarii	Aquariid	Libra	Librae	Librid
Aquila	Aquilae	Aquilid	Lupus	Lupi	Lupid
Ara	Arae	Arid	Lynx	Lyncis	Lyncid
Aries	Arietis	Arietid	Lyra	Lyrae	Lyrid
Auriga	Aurigae	Aurigid	Mensa	Mensae	Mensid
Bootes	Bootis	Bootid	Microscopium	Microscopii	Microscopiid
Caelum	Caeli	Caelid	Monoceros	Monocerotis	Monocerotid
Camelopardalis	Camelopardalis	Camelopardalid	Musca	Muscae	Muscid
Cancer	Cancri	Cancrid	Norma	Normae	Normid
Canes Venatici	Canum Venaticorum	Canum Venaticid	Octans	Octantis	Octantid
Canis Major	Canis Majoris	Canis Majorid	Ophiuchus	Ophiuchi	Ophiuchid
Canis Minor	Canis Minoris	Canis Minorid	Orion	Orionis	Orionid
Capricornus	Capricorni	Capricornid	Pavo	Pavonis	Pavonid
Carina	Carinae	Carinid	Pegasus	Pegasi	Pegasid
Cassiopeia	Cassiopeiae	Cassiopeiid	Perseus	Persei	Perseid
Centaurus	Centauri	Centaurid	Phoenix	Phoenicis	Phoenicid
Cepheus	Cephei	Cepheid	Pictor	Pictoris	Pictorid
Cetus	Ceti	Cetid	Pisces	Piscium	Piscid
Chamaeleon	Chamaeleontis	Chamaeleontid	Piscis Austrinus	Piscis Austrini	Piscis Austrinid
Circinus	Circini	Circinid	Puppis	Puppis	Puppid
Columba	Columbae	Columbid	Pyxis	Pyxidis	Pyxidid
Coma Berenices	Comae Berenices	Comae Berenicid	Reticulum	Reticulii	Rectuliid
Corona Australis	Coronae Australis	Coronae Australid	Sagitta	Sagittae	Sagittid
Corona Borealis	Coronae Borealis	Coronae Borealid	Sagittarius	Sagittarii	Sagittariid
Corvus	Corvi	Corvid	Scorpius	Scorpii	Scorpiid
Crater	Crateris	Craterid	Sculptor	Sculptoris	Sculptorid
Crux	Crucis	Crucid	Scutum	Scuti	Scutid
Cygnus	Cygni	Cygnid	Serpens	Serpentis	Serpentid
Delphinus	Delphini	Delphinid	Sextans	Sextantis	Sextantid
Dorado	Doradus	Doradid	Taurus	Tauri	Taurid
Draco	Draconis	Draconid	Telescopium	Telescopii	Telescopiid
Equuleus	Equulei	Equuleid	Triangulum	Trianguli	Triangulid
Fornax	Fornacis	Fornacid	Triangulum Australe	Trianguli Australis	Trianguli Australid
Gemini	Geminorum	Geminid	Tucana	Tucanae	Tucanid
Grus	Gruis	Gruid	Ursa Major	Ursae Majoris	Ursae Majorid
Hercules	Herculis	Herculid	Ursa Minor	Ursae Minoris	Ursae Minorid
Horologium	Horologii	Horlogiid	Vela	Velorum	Velorid
Hydra	Hydrae	Hydrid	Virgo	Virginis	Virginid
Hydrus	Hydri	Hydrusid	Volans	Volantis	Volantid
Indus	Indi	Indid	Vulpecula	Vulpeculae	Vulpeculid
Lacerta	Lacertae	Lacertid	–	–	–

Custom is to replace the final suffix for "-id", or plural "-ids". Meteors from Aquarius (Aquarii) are Aquariids, not Aquarids. An exception is made for meteors from the constellation of Hydrus, which will be called "Hydrusids", in order not to confuse with meteors from the constellation of Hydra.

When the constellation name has two parts, only the second declension is to be replaced by "id". Hence, meteors from Canes Venatici (Canum Venaticorum) would be "Canum Venaticids".

When two constellations are grouped together, a dash is used and both constellation names will have "id". Hence, Puppids-Velids. As a guideline, the sequence of those constellations are best in the order of which the radiants travel through them (Bootids-Coronae Borealids, not Coronae Borealids-Bootids). "Complex" can be used to indicate groups of meteor showers that may originate from the same (former) parent body, while groups of parent body fragments are usually referred to as a "family" (like the Hirayama families in the asteroid belt). Hence, one could say that the Taurid Complex of meteor showers originated from the Encke family of comets.

If higher precision is needed, then the shower is named after the nearest (if in doubt: brightest) star with a Greek letter assigned, as first introduced in the Uranometria atlas by Johann Bayer (1603), or one with a later introduced Roman letter. If in doubt, the radiant position at the time of the peak of the shower (in the year of discovery) should be taken. Hence, the meteors of comet IRAS-Araki-Alcock would be named "eta Lyrids" (or "eta-Lyrids").

Following existing custom, one may add the name of the month to distinguish among showers from the same constellation. In this case, one could call the shower from comet IRAS-Araki-Alcock the "May Lyrids", in order to differentiate from the more familiar "April Lyrids".

For daytime showers, those with a radiant less than 32 degrees from the Sun, it is custom to add "Daytime", hence the name for the "Daytime Arietids" in June as opposed to the Arietids in October.

South and North refer to "branches" of a shower south and north of the ecliptic plane, resulting from meteoroids of the same (original) parent body. Because they have nearly the same longitude of perihelion at a given solar longitude (the argument of perihelion and longitude of ascending node differing by 180 degrees between South and North), the two branches are active over about the same time period.

If the meteoroid stream is encountered at the other node, it is customary to speak of "twin showers". The Orionids and eta-Aquariids are twin showers, even though each represent dust deposited at different times and are now in quite different orbits. As a matter of custom, twin showers and the north and south branches of a stream carry different names.

Meteor showers are not to be named after their parent bodies (e.g., Giacobinids, IRAS-Araki-Alcockids). The names of comets tend not to be Latin, making the naming not unique. Also, comet names can change when they get lost and are recovered. I like to add that even the proposed association may change, as many Taurids may originate from other parent bodies than 2P/Encke, for example.

In case of confusion, the *Task Group for Meteor Shower Nomenclature* will choose among possible alternative names, in order to establish a unique name for each meteor shower (e.g., eta-Lyrids, not May Lyrids).

3 The Working List

The Working List of Meteor Showers and the nomenclature rules were published in IAU Bulletin 99 (January 2007) and are posted at:

the website of the IAU Meteor Data Center: http://www.astro.sk/~ne/IAUMDC/
the website of IAU Commission 22 (Task Group for Meteor Shower Nomenclature): http://meteor.asu.cas.cz/IAU/nomenclature.html
IAU Information Bulletin January 2007: http://www.iau.org/fileadmin/content/IBs/ib99.pdf

During the Meteoroids 2007 conference in Barcelona, the Task Group convened and worked out the logistics of adding new streams to the Working List and adding new information on streams already in the Working List.

The institute responsible for maintaining the Working List is the *IAU Meteor Data Center*, which is currently managed by Vladimir Porubcan. The person responsible for setting up a website to facilitate the reporting of new streams and new data on existing streams, and give out new IAU numbers, will be *Tadeusz Jan Jopek* of Poznan Astronomical Observatory in Poznan, Poland:

http://vesta.astro.amu.edu.pl/Staff/Jopek/

The International Meteor Organization will take a role in coordinating the reporting of newly discovered streams by amateur meteor observers, mostly to facilitate the inclusion of streams that are only recognized from visual observations of meteor outbursts.

Once a website is in place that can provide updates to the Working List, newly discovered streams should not be reported in the literature without a designated IAU number. Before publication, the IAU MDC (Jopek) should be contacted to obtain a shower number. This will facilitate subsequent discussion in the literature to help confirm the detection. In the near future, it is the intention of the Task Group that a telegram be issued (CBET) with a brief summary of each new find to signify publication of the discovery as part of the process of reporting new streams, and in order to allert the community that new streams have been reported.

4 The List of Established Meteor Showers

In two years from now, in January of 2009, half a year before the next IAU General Assembly in Rio de Janeiro (Brasil), a subset of all showers will be selected for inclusion in the List of Established Meteor Showers. Selection will be based on the work in our community up to that point. The proposed list of established meteor showers will be posted prior to the General Assembly to facilitate discussion on whether there is sufficient evidence to include each shower in this list based on information and sources listed in the Working List. Only those showers that are beyond reproach are expected to pass the vote for official recognition during the Commission 22 meeting at the Assembly, and henceforth be recognized as a unique astronomical entity.

Note added in proofs The website for reporting new meteor showers is now operational at: http://www.astro.amu.edu.pl/~jopek/MDC2007/. An announcement was made on CBET 1088 (Sep. 25, 2007).

References

P. Jenniskens, *Meteor Showers and their Parent Comets*, (Cambridge University Press, Cambridge, UK, 2006), 790 pp
P. Jenniskens, Div. III/Comm. 22/WG Task Group for Meteor Shower Nomenclature. *IAU Information Bulletin* 99, January 2007, 60–62
P. Spurny, J. Borovicka, *Minutes of the Commission 22 Business Meeting*. (IAU General Assembly, Prague, 2006), August 24, 2006

Current Status of the Photographic Meteoroid Orbits Database and a Call for Contributions to a New Version

Jan Svoren · Vladimir Porubcan · Lubos Neslusan

Originally published in the journal Earth, Moon, and Planets, Volume 102, Nos 1–4.
DOI: 10.1007/s11038-007-9167-1

Abstract A central depository for meteor orbits obtained by photographic techniques, as a part of the IAU Meteor Data Center, was moved to the Astronomical Institute of the Slovak Academy of Sciences in Bratislava in 2001. The current version of the catalogue contains data on 4581 meteor orbits obtained by 17 different stations or groups from the period 1936 to 1996. Since 1996 a few huge campaigns were organised including very successful Leonids and Perseids. That is why we would prepare a new more complete version of the database. The main aim of this paper is a call to the observers of meteors having new or recalculated/remeasured data on photographic meteors to send them to the MDC, where after a check and consultations with the observer, the orbits will be included in the database.

Keywords Astronomical databases · Photographic meteor orbits

1 Current Version of Photographic Meteor Orbits Database

The IAU Meteor Data Center in Lund, since it was founded early in the 1980's, has acted as a central depository for meteor orbits obtained by photographic, video and radar techniques. It accumulated a huge number of meteoroid orbits obtained world-wide and is providing them to meteor scientists for various analyses.

In 2001, after Kiruna meteoroids conference, the IAU Meteor Data Center was moved to the Astronomical Institute of the Slovak Academy of Sciences in Bratislava. The database is covering an interval of 60 years—since 1936 when it became possible to determine precise photographic meteor orbits. In Fig. 1 the distribution of 4581 photographic meteors of the database observed over the year is depicted. The majority of well known streams are easily identified. The most populated streams in the database are the Perseids in August and Geminids in December.

J. Svoren (✉) · V. Porubcan · L. Neslusan
Astronomical Institute of the Slovak Academy of Sciences, Tatranska Lomnica 059 60, The Slovak Republic
e-mail: astrsven@ta3.sk

J.M. Trigo-Rodriguez et al. (eds.), *Advances in Meteoroid and Meteor Science.*
DOI: 10.1007/978-0-387-78419-9_3

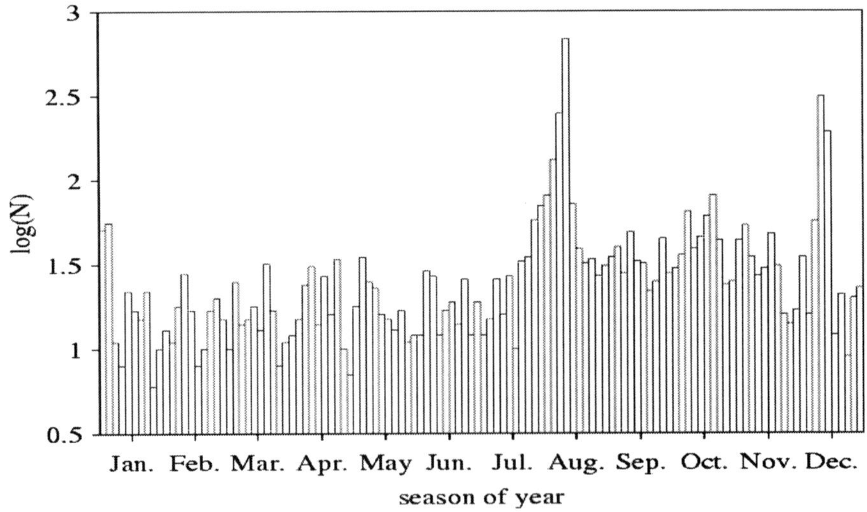

Fig. 1 Distribution of the 4581 photographic meteors of the IAU MDC database

Not only classical photography but also modern optical techniques are used now. It is very pleasant to follow that new catalogues listing meteor orbits detemined by video techniques are published, e.g. catalogue compiled by Koten et al. (2003) containing 817 orbits. The optical meteors cover a wide range of initial particle sizes, from fireballs having masses of 0.1–10 kg to faint TV meteors of the order of 10^{-7} kg. This article deals only with the classical photographic records compiled originally for the IAU Meteor Data Center in Lund (a series of the papers, e.g. Lindblad 1987 and 2001) and completed by additional meteor orbits published mainly by Spurny, Babadzhanov et al. and Halliday et al. The references in detail are published in Lindblad et al. (2005).

The previous versions of the database contained orbital and geophysical data on meteors in two separate files. The separation was mainly due to limitations of computer-memory capacity in the past. Because of a compatibility of the data with the old programs, we still conserved their two-file format in the last version of the database. At the same time, we introduced a new format and wrote the data into a single file named as *all2003.dat.* This merging of the data is not only more comfortable for their reading, but in various studies it is often necessary to utilize the complete information available for each meteor compiled in both original files (orbital—orbital elements; geophysical—radiants, geocentric and heliocentric velocities, etc.). Therefore, the new file contains the merged geophysical and orbital data (in ASCII format) sorted by the date of meteor detection, from January 1 to December 31. A five-line format for each meteor is chosen to provide a comfortable reading of the complete data in one place. A blank line separates the data of two neighbouring meteors. All the values are expressed in full figures. If a given parameter was not published by the original author then zeros are inserted in the file (to enable a formated reading, too). In *all2003.dat* file, all the orbital data are calculated by us by the same procedure, on the basis of the published time of appearance, the radiant position and geocentric velocity. In all of the published data catalogues, except for the MORP (Halliday et al. 1996) and Betlem et al. (1998) orbits, the 1950 equinox was used. In this version we converted the angular elements to J2000.0.

Eccentricities of some meteor orbits in the database considerably exceed unity. A limit of the heliocentric velocity of about 48 km s^{-1} can be regarded as a reasonable limiting value between acceptable and unacceptable heliocentric velocities (Lindblad et al. 2005). We recommend to omit 46 meteors with the heliocentric velocity over this limit from all statistical studies.

The photographic database version 2003, can be downloaded from the IAU MDC at the Astronomical Institute of the SAS from the address: http://www.astro.sk/~ne/IAUMDC/Ph2003/database.html

Available are the geophysical and orbital data on 4581 photographic meteors (ASCII format) sorted as in the original catalogues of the individual authors or stations. The *all2003.dat* file contains the merged geophysical and orbital data.

Besides the three data files listed above, there are at disposal lists of 875 Perseids and 387 Geminids meteoroid streams members selected from the database (Svoren and Kanuchova 2005; Kanuchova and Svoren 2006).

2 Preparation of the Next Version

To detect and resolve any inconsistencies in the orbital data we will recalculate all the obtained orbits based on the position of corrected radiant and geocentric velocity at the time of meteor observation. The IAU meteor database contains geophysical parameters and orbital elements, which are mutually dependent. Therefore, one data set can be used to verify the correctness of the other. To check the consistency of the two data sets, the following two recalculations are made:

(1) Assuming that the published radiant coordinates and geocentric velocity of the meteor at the time of detection were correct, the orbital elements q, e, ω, Ω and i are recalculated.

(2) However, it is obvious that errors sometimes appear also in the published geophysical (encounter) data. Hence we consider the five published orbital elements as the input and recalculate the radiant coordinates α, δ and the geocentric velocity V_g of the meteor. In this recalculation the most optimal method of theoretical radiant prediction for a given orbital geometry (Neslusan et al. 1998) is used.

3 Call for New Observed or Recalculated Observations

The IAU MDC catalogue summarizes photographic meteor orbits observed only until 1996. However, since 1996 more very successful observing campaigns were organised and new meteor orbits were obtained, including very successful observations of the Leonids and Perseids. This is a great motivation to update and prepare a new more complete version of the database. The main aim of this paper is a call to observers of meteors having new or remeasured data of photographic meteors to send them to the MDC. After a check and consultations with the observer, the orbits will be included in the database.

For the future we plan to introduce a new service. Each observer or contributor to the database will be able to perform a preliminary check of the consistency of his own geocentric and orbital data sets before he sends the data to us, by on-line calculator, anonymously.

We plan also a small change in the format of the database. Data, if possible, will be published together with their error bars. Errors could be obtained from the original reports of the observers based on precision of observable techniques and methods used. A comparison of precision of individual groups and stations could be a second way to calculate them. We would like to avoid an inclusion of the errors obtained formally from different statistical processes.

Acknowledgements The authors are indebted to K. Ohtsuka for his collaboration in checking of the published data. This research was supported by VEGA - the Slovak Grant Agency for Science (Grants Nos. 1/3067 and 2/7009).

References

H.C. Betlem, R. Ter Kuile, M. de Lignie, J. van't Leven, K. Jobse, K. Miskotte, P. Jenniskens, Astron Astrophys Suppl Ser **128**, 179–185 (1998)
I.A. Halliday, A. Griffin, A.T. Blackwell, Meteoritics Planet Sci **31**, 185–217 (1996)
Z. Kanuchova, J. Svoren, Contrib Astron Obs Skalnate Pleso **36**, 181–193 (2006)
P. Koten, P. Spurny, J. Borovicka, R. Stork, Publ Astron Inst Sci Czech Rep. **91**, 1–32 (2003)
B.A. Lindblad, in *Interplanetary Matter, Proc 10th ERAM.*, eds. by Z. Ceplecha, P. Pecina (Astron Inst Czechosl Acad Sci, Prague, 1987), pp. 201–204
B.A. Linbdlad, in *Meteoroids 2001 Conf. ESA Publ Div*, ed. by B. Warmbein, (ESTEC, Noordwijk, ESA SP-495), pp. 71–72
B.A. Lindblad, L. Neslusan, V. Porubcan, J. Svoren, Earth Moon Planets **93**, 249–260 (2005)
L. Neslusan, J. Svoren, V. Porubcan, Astron Astrophys **331**, 411–413 (1998)
J. Svoren, Z. Kanuchova, Contrib Astron Obs Skalnate Pleso **35**, 199–220 (2005)

The Dynamics of Low-Perihelion Meteoroid Streams

Paul A. Wiegert

Originally published in the journal Earth, Moon, and Planets, Volume 102, Nos 1–4.
DOI: 10.1007/s11038-007-9182-2 © Springer Science+Business Media B.V. 2007

Abstract The Canadian Meteor Orbit Radar (CMOR) has collected information on a number of weak meteor showers that have not been well characterized in the literature. A subsample of these showers (1) do not show a strong orbital resemblance to any known comets or asteroids, (2) have highly inclined orbits, (3) are at low perihelion distances ($\ll$ 1 AU) and (4) are at small semimajor axes (<2 AU). Though one might conclude that the absence of a parent object could be the result of its disruption, it is unclear how this relatively inaccessible (dynamically speaking) region of phase space might have been populated by parents in the first place. It will be shown that the Kozai secular resonance and/or Poynting–Robertson drag can modify meteor stream orbits rapidly (on time scales comparable to a precession cycle) and may be responsible for placing some of these streams into their current locations. These same effects are also argued to act on these streams so as to contribute to the high-ecliptic latitude north and south toroidal sporadic meteor sources. There remain some differences between the simple model results presented here and observations, but there may be no need to invoke a substantial population of high-inclination parents for the observed high-inclination meteoroid streams with small perihelion distances.

Keywords Meteoroid stream · Poynting–Robertson drag · Secular resonance · Toroidal meteor sources · Meteor shower · Sporadic meteors

We report here on a number of meteor showers that have been recently studied by means of the Canadian Meteor Orbit Radar (CMOR, Jones et al. 2005). These showers are weak to moderate in strength and were either discovered in the CMOR catalogue (Brown et al. 2007) or have only been poorly characterized in previous studies. In Sect. 1, those showers with clear links to parent bodies are discussed. Section 2 deals with links to other better-known showers, and Sect. 3 examines the dynamics of this ensemble of streams and its possible link to the toroidal sporadic meteor sources.

P. A. Wiegert (✉)
Department of Physics and Astronomy, The University of Western Ontario, London, ON, Canada N6A 3K7
e-mail: pwiegert@uwo.ca

DOI: 10.1007/978-0-387-78419-9_4

1 Links with Parent Objects

One new shower has a clear connection to a parent. The Daytime ε Perseids shower has an orbit which bears a similarity to that of comet 96P/Machholz. Table 1 lists their respective orbital elements. The Drummond (1981) D' of this association is 0.14 and the Valsecchi et al. (1999) D is 0.047 though the D of Southworth and Hawkins (1963) is somewhat larger at 0.435. There is a strong resemblance in the perihelion distance q, inclination i and longitude of the ascending node Ω. The match is poorer in the semimajor axis a (which is difficult to measure) and the argument of perihelion ω, possibly due to precession. We conclude that this shower is likely part of the Quadrantid meteor complex, to which 96P has been linked in the past (McIntosh 1990; Babadzhanov and Obrubov 1992; Gonczi et al. 1992; Jones and Jones 1993; Jenniskens 2004; Wiegert and Brown 2005).

2 Links with Known Streams

Some of the other weak showers detected by CMOR are related to the multiple intersections between a meteoroid stream and the Earth's orbit that occur during the stream's precession cycle. For example, the Daytime April Piscids and the South Daytime May Arietids (sometimes called the o Piscids in the literature) are both clearly related to the North and South i Aquariids (see Table 2). Under apsidal precession, the intersection points of this stream with the Earth's orbit can easily be computed to occur near values of the argument of perihelion ω of 50°, 130°, 230° and 310°. We have also verified this by numerical experiment. Thus the Daytime April Piscids and the South Daytime May Arietids, together with the N/S i Aquariids, complete the set of four separate showers produced by the precession of meteoroids released from a single parent.

3 The Remaining Streams

Despite the associations discussed in the two preceding sections, most of the weak showers in the CMOR catalog do not have immediately obvious parent bodies, nor clear links to known streams. In fact, many of these streams have semimajor axes a below 2 AU, perihelia q well inside Mercury's orbit, and high inclinations (Table 3 and Fig. 1), placing them in a region of phase space that is very sparsely populated by comets and asteroids. A search of the asteroid and comet databases turns up no bodies with orbits clearly similar to those of these streams.

One might speculate that the low-perihelion distances of these streams, together with the high activity levels and rapid depletion they would produce in a source comet, might account for the current absence of parent bodies. The parents would simply have disrupted or become inactive or extinct. However this would not explain how the source bodies

Table 1 Comparison of the orbits of 96P/Machholz (Marsden and Williams 2005) and the Daytime ε Perseids

Name	a (AU)	q (AU)	e	i (°)	Ω (°)	ω (°)
D ε Perseids	4.6 ± 1	0.13 ± 0.01	0.97 ± 0.01	63 ± 2	96 ± 0.3	40 ± 2
96P/Machholz	3.01	0.123	0.959	59.9	94.5	14.6

Errors for the shower elements are approximate

Table 2 The elements of the Daytime April Piscids and South Daytime May Arietids, together with those of the better-known North and South *i* Aquariids

Name	a (AU)	q (AU)	e	i (°)	Ω (°)	ω (°)
Daytime April Piscids	1.51	0.26	0.83	4.7	25	50
S Daytime May Arietids	1.51	0.27	0.82	5.1	227	232
N *i* Aquariids	1.52	0.27	0.83	5.7	159	309
S *i* Aquariids	1.55	0.22	0.86	5.3	309	134

The orbits are from the CMOR catalogue

Table 3 A selection of the new or previously little-studied meteor showers in the CMOR catalogue

Name	a (AU)	q (AU)	e	i (°)	Ω (°)	ω (°)
N Daytime ω Cetids	1.58	0.12	0.93	34	45	33
S Daytime ω Cetids	1.72	0.14	0.92	36	225	216
S June Aquilids	1.12	0.06	0.94	56	260	159
Daytime γ Taurids	1.57	0.10	0.93	23	266	211
Vulpeculids	0.76	0.17	0.77	55	105	335
N June Aquilids	1.71	0.11	0.94	39	101	328
β Equulids	0.89	0.16	0.82	50	106	330
July σ Cassiopeiids	1.09	1.00	0.08	81	105	217
ψ Cassiopeiids	2.14	0.93	0.56	83	118	141
N δ Aquariids	1.81	0.10	0.95	24	139	329
σ Serpentids	1.92	0.16	0.92	64	276	41
ω Serpentids	1.37	0.16	0.88	56	276	39
θ Coronae Borealids	1.11	0.92	0.17	77	296	125
λ Bootids	1.49	0.96	0.36	79	295	207
ζ Coronae Borealids	2.34	0.82	0.65	80	294	125
α Antilids	2.47	0.14	0.94	64	136	140

reached these orbits in the first place, as the dynamical evolution of bodies into this region is slow.

We report here that Poynting–Robertson (PR) drag is likely responsible for the current orbits of these showers. It will be shown that streams produced by comets at larger a and q can evolve into streams of the type described above (or at least the smaller members of these streams can) on time scales of only thousands of years, short compared to their precession times.

Additionally, we report that many such streams are trapped in the Kozai resonance (Kozai 1962) which causes their eccentricities e and inclinations i to oscillate. Such meteoroids produce radiant distributions with some of the characteristics of the toroidal sporadic meteor sources.

3.1 Investigations

In order to study the dynamics of these streams, the showers in Table 3 were simulated numerically with a symplectic Wisdom and Holman (1991) style integrator able to handle

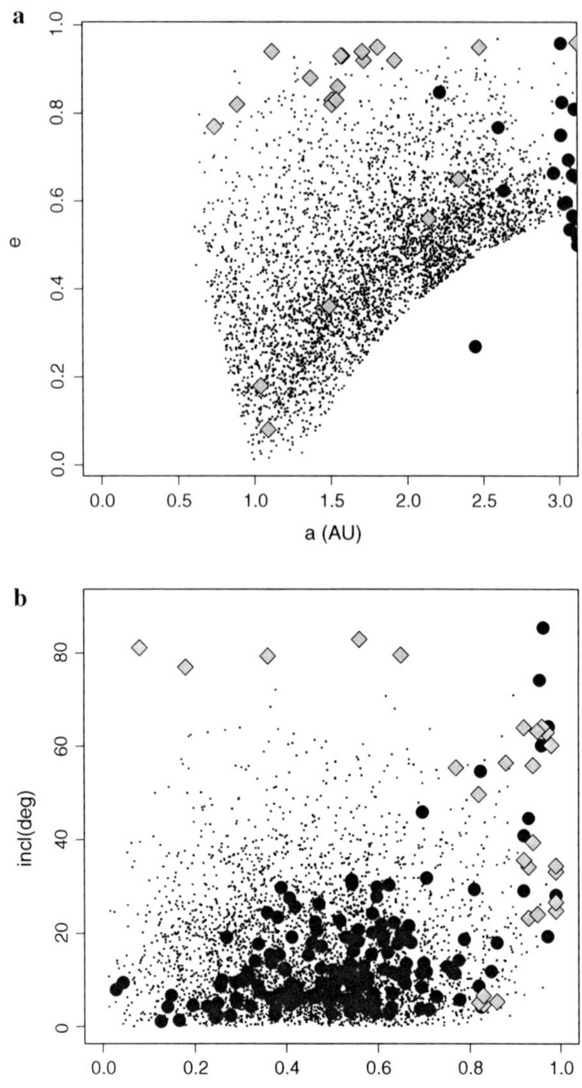

Fig. 1 The orbital distributions of near-Earth asteroids (dots, from the AstDys website http://hamilton.dm.unipi.it/cgi-bin/astdys/astibo), comets (black circles, Marsden and Williams (2005)) and the showers discussed here (grey diamonds) in (**a**) *a*–*e* and (**b**) *e*–*i* space

close encounters by the hybrid method (Chambers 1999). Two sets of ten particles were spread along the orbit of each meteoroid stream at equal intervals of mean anomaly. One set was assigned a beta of zero for comparison purposes. The other set was assigned a β value of 0.0057 to simulate particles of a density of 2,000 kg m^{-3} and a radius of 100 μm (Weidenschilling and Jackson 1993). Each set was integrated backwards for 50,000 years with a time step of one day.

The simulation of multiple particles per stream allows us to better understand the effects of differential perturbations such as planetary encounters. However, these simulations have only a small number of particles and are not of the caliber of those frequently used these days for detailed shower timing and strength predictions, which may involve tens of thousands or more particles. Nevertheless they provide great insight into the dynamical behaviour of these streams.

A common feature of the numerical simulations is a substantial change in the semimajor axes of the stream orbits over time. Some streams can undergo changes in a at rates exceeding 1 AU per 10^3 years, though average rates near 1 AU per 10^4 years are more typical. Thus the stream produced by a Jupiter family comet with $a \approx 3$ AU could become one with $a \sim 1$ AU (like many of the showers in Table 3) in only a few 1,000 years.

An example of the semimajor axis evolution of one such stream, the β Equulids, is shown in Fig. 2. Note how the particles with $\beta = 0.0057$ have rapidly changing semimajor axes while the control particles with $\beta = 0$ remain largely unaffected. This indicates that these changes are indeed the result of radiation forces. If the new showers discussed here are primarily composed of small particles, then they could have been released from comets with larger values of a and q and subsequently transported to their current orbits by PR drag. This might also explain the absence of these showers from visual shower catalogues, as such streams are unlikely to contain many of the larger meteors (with smaller β values) which are more easily observed by optical means.

Figure 3 shows the eccentricity evolution of the β Equulids stream. Notably absent is the monotonic circularization expected for meteoroids experiencing strong PR drag (Wyatt and Whipple 1950), though we note that a careful treatment by Breiter and Jackson (1998) revealed that there were cases where a small increase in e could be expected from PR drag. In the simulations presented here, e is seen to oscillate on time scales of 10^4 years. The reason that an alternation of e occurs rather than a simple reduction in its value is because of the action of the Kozai resonance (Kozai 1962), also known as the secular precession effect discussed by Babadzhanov and Obrubov (1987). This secular effect pumps angular momentum in and out of the meteoroid orbit faster than PR drag removes it, and thus controls the value of e in this dynamical regime.

The secular resonance that affects e also produces an oscillation in the inclination i. Its effect on the β Equulids stream is shown in Fig. 4. Inclination and eccentricity oscillate out of phase with each other, and the meteoroids spend much of their time at high inclination,

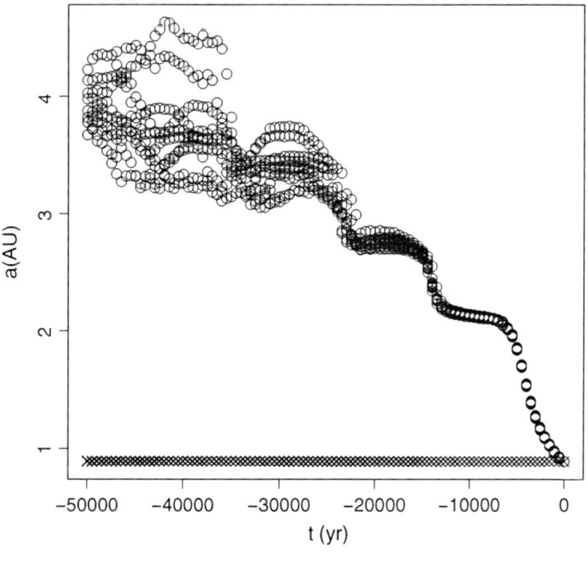

Fig. 2 The evolution of the semimajor axis of the β Equulids meteoroids simulated backwards for 50,000 years. The open circles are 100 μm radius particles ($\beta = 0.0057$), while the crosses are particles with $\beta = 0$

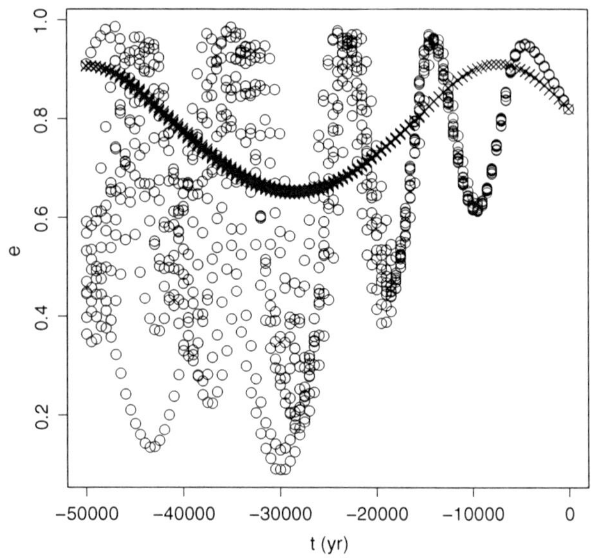

Fig. 3 The evolution of eccentricity of the β Equulids meteoroids simulated backwards for 50,000 years. See Fig. 2 for more details

at a time-average value near 60°. Thus meteoroid streams produced at much lower inclination ($\lesssim 20°$) can be driven up to much higher inclination ($> rsim80°$) by this effect. In fact, these particles spend most of their time at high inclination. This result is relatively insensitive to β, even particles at $\beta = 0$ also have large time-averaged inclinations. Thus, there is no need to invoke a substantial population of high-inclination parents for these streams; they could easily be produced by bodies with a much flatter distribution (e.g. the Jupiter-family comets) pumped up by the secular resonance.

The high time-averaged inclination of these meteoroids also suggests a connection with the north and south toroidal sporadic sources that we investigate next.

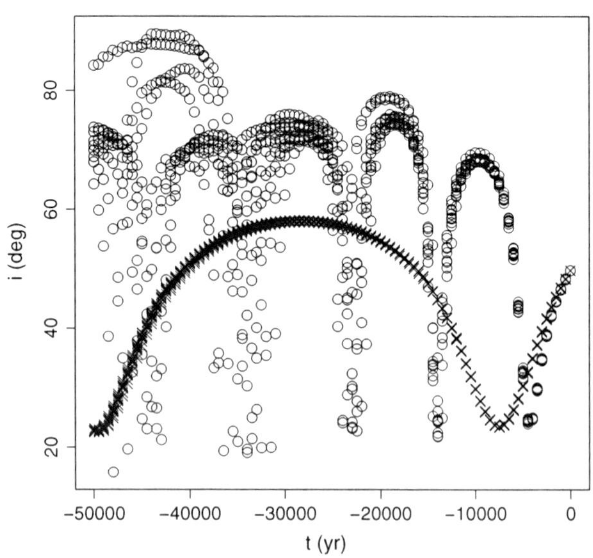

Fig. 4 The evolution of inclination of β Equulids meteoroids simulated backwards for 50,000 years. See Fig 2 for more details

3.2 The Toroidal Sporadic Meteor Sources

The orbital element distributions of the north toroidal sporadic source have been determined (e.g. Jones and Brown 1993), and it is expected that those of the southern toroidal source will be similar. However, the origin of the meteors that produce these sources is not known. The elements presented in Jones and Brown (1993) for the northern source show a peak in a at 1 AU, one in inclination near 60°, and a distribution in e with a preponderance of near-circular orbits. The high-inclination is particularly puzzling owing to the absence of comets or asteroids on such orbits. Could the high inclination showers discussed here be connected to the toroidal sporadic sources? Perhaps as these meteoroids diffuse away from the shower orbits and drift inwards under PR drag, many remain in the secular resonance at high i, ultimately becoming toroidal sporadics?

In order to investigate this possibility, we simulated meteoroid streams originating from hypothetical parents of the high-i streams described above. The difference between these simulations and the ones mentioned earlier are (1) these simulations are run forwards in time, (2) three different particle radii are included: 50, 100 and 200 μm (10 particles each, with appropriate β values) and (3) the meteoroid streams are started with the elements given in Table 3 with the exception that the semimajor axis is set to 3 AU. This provides a proxy for the putative cometary parents of these streams, here assumed to be Jupiter-family comets. By simulating these streams forwards under PR drag, we can make a rough determination of whether or not the meteoroids produced by such parents could produce the toroidal sporadic sources.

Figure 5 shows the resulting density of radiants of the simulated meteoroids with nodes within 0.1 AU of the Earth over 10^5 years (roughly their collisional lifetime (Grun et al. 1985), though their high inclinations are likely to prolong their survival in practice, weighted according to their collision probability with the Earth (from Opik (1951) as given by Galligan and Baggaley (2004)). The radiants are based on the true minimum approach distance between the orbits, not just the distance between the nodes. The radiants are

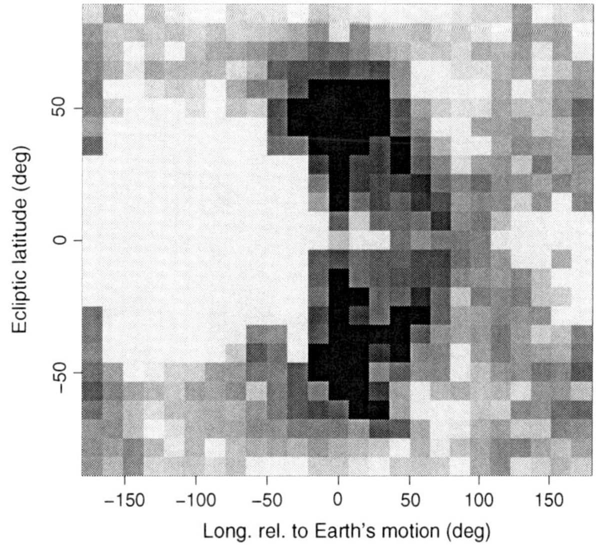

Fig. 5 The radiant distribution of simulated meteoroids weighted according to the collision probability with the Earth. Darker tones indicate a higher density of meteor radiants. The Earth's apex is towards the origin in this plot and the Sun is at a relative longitude of −90°

determined simply from the relative velocity of the meteoroid and the Earth at closest approach.

Both north and south toroidal radiants are reproduced, though they are nearer the ecliptic plane than the observed toroidal radiants which are at ecliptic latitudes of ±60° (Jones and Brown 1993). The orbital distributions of meteors within the toroidal radiant will be examined next. The orbits will be found to bear some resemblance to observed toroidal meteors, but this scenario probably does not provide a complete explanation of the origin of the toroidal sources.

Figure 6 shows the distribution of inclinations within the radiant area defined by a longitude relative to the apex of less than 30° and a latitude (either north or south) between

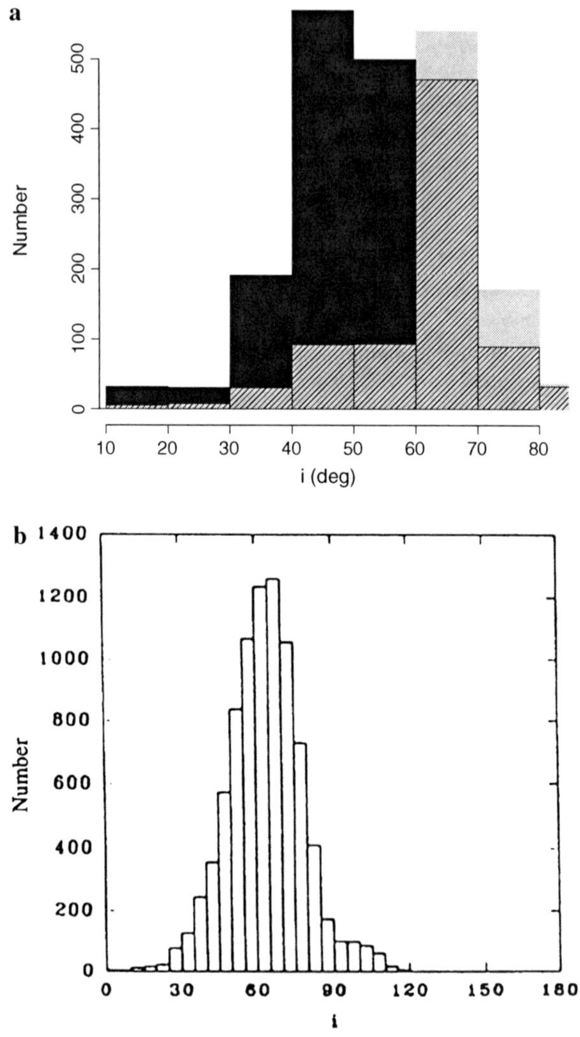

Fig. 6 (**a**) The distribution of the inclinations of simulated meteoroids accumulated over 10^5 years. The histogram in grey is unweighted; the black is weighted according to the collision probability with the Earth, normalized to a similar peak value. Panel (**b**) is the observed distribution of north toroidal source meteors from Jones and Brown (1993)

Fig. 7 The distribution of semimajor axes of (**a**) simulated meteoroids and (**b**) observed north toroidal source meteors from Jones and Brown (1993). See Fig. 6 for more details

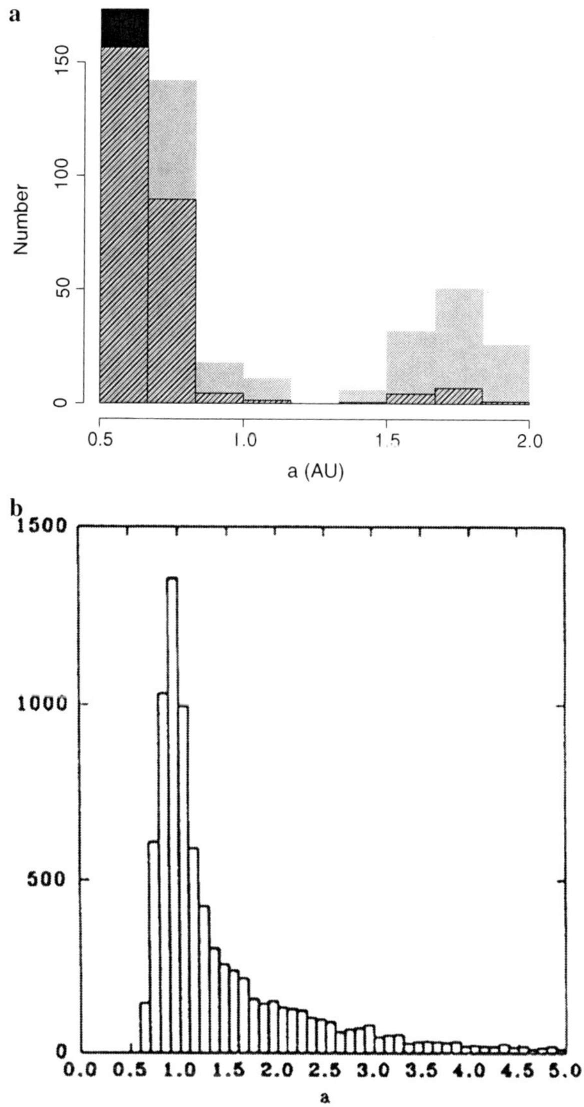

40° and 70°. The distribution shows a peak at high inclination, similar to that observed but not expected given our choice of radiant latitude. Figures 7 and 8 show the orbit element distributions for the semimajor axis and eccentricity for those meteoroids in the above radiants. The weighted distributions bear some resemblance the measured distributions for the north toroidal source, given in Fig. 9 of Jones and Brown (1993), but are not identical. The simulated semimajor axis distributions, both weighted and unweighted, are sharply peaked like the observations, but at values below those of the observed distribution. The simulated and observed eccentricity distributions differ as well. The observed distribution contains a preponderance of near-circular orbits. The unweighted simulated distribution is

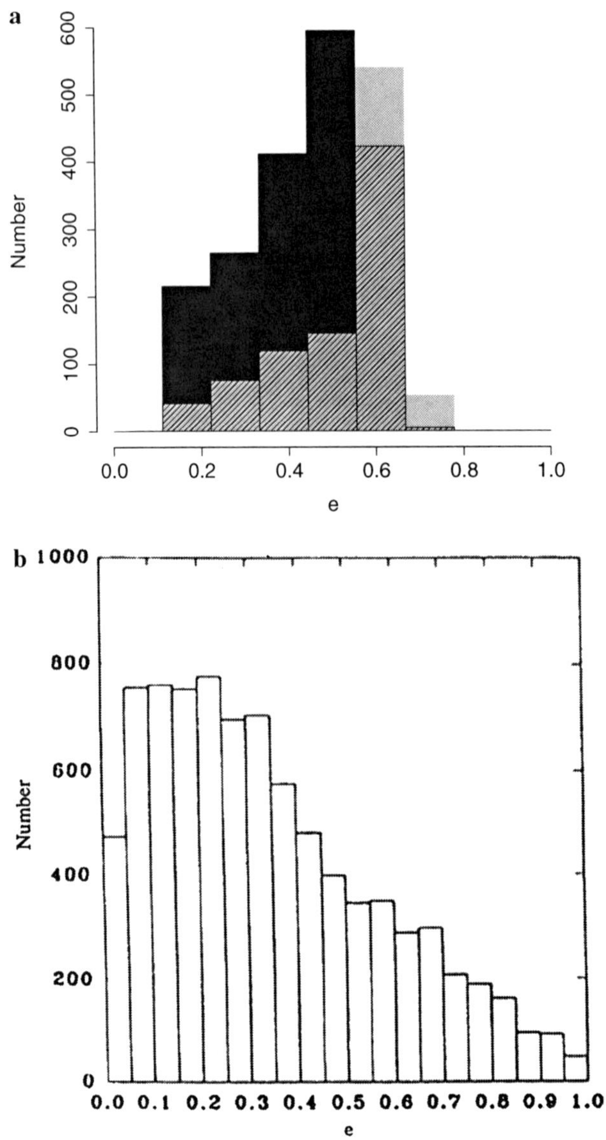

Fig. 8 The distributions of eccentricity of (**a**) simulated meteoroids and (**b**) observed north toroidal source meteors from Jones and Brown (1993). See Fig. 6 for more details

peaked near $e = 0.6$. Though the discrepancy is less pronounced for the weighted distribution (which should more closely match observations), the match is far from perfect. The differences between the experimental and theoretical distributions may simply be due to our coarse modelling of the parent streams. However, it probably also indicates that the crude scenario employed here, despite some intriguing intimations, is insufficient to completely explain the toroidal sporadic sources.

4 Conclusions

The CMOR catalogue of meteor shower orbits provides much new information on the nature of meteoroid streams near the Earth, particularly improved orbits for many weak showers. Using this information, the Daytime ε Perseids can now be connected with comet 96P/Machholz. Two other poorly-characterized showers, the Daytime April Piscids and South Daytime May Arietids, can now be linked to the North and South ι Aquariids. As well, a number of unusual weak showers with small q and a and large values of i were shown to be consistent with small ($r \sim 100$ μm) meteoroids released from comets with much larger a, and evolving under PR drag and the Kozai resonance. Initial indications are that these meteoroids eventually become part of the north and south toroidal sources, though more work is needed.

Acknowledgements PW extends his thanks to J. Vaubaillon for presenting these results at the Meteoroids 2007 conference in Barcelona in the author's absence, as well as to J. M. Trigo-Rodríguez and the LOC for allowing the last-minute replacement. This work was supported by the Natural Sciences and Engineering Research Council of Canada.

References

P.B. Babadzhanov, I.V. Obrubov, Evolution of meteoroid streams, in European Regional Astronomy Meeting of the IAU, eds. by Z. Ceplecha, P. Pecina, vol. 2 (Czechoslovak Academy of Sciences, Ondřejov, Czechoslovakia, 1987), pp. 141–150

P.B. Babadzhanov, Y.V. Obrubov, P/Machholz 1986 VIII and Quadrantid meteoroid stream. Orbital evolution and relationship, in Asteroids, Comets, Meteors 1991, eds. by A. Harris, E. Bowell (University Arizona Press, Tucson, 1992), pp. 27–32

S. Breiter, A.A. Jackson, Unified analytical solutions to two-body problems with drag, MNRAS **299**, 237–243 (1998)

P. Brown, R.J. Weryk, D.K. Wong, J. Jones, The Canadian Meteor Orbit Radar (CMOR) Meteor Shower Catalogue, Earth Moon Planets, this issue (2007). doi:10.1007/s11038-007-9162-6

J.E. Chambers, A hybrid symplectic integrator that permits close encounters between massive bodies, MNRAS **304**, 793–799 (1999)

J.D. Drummond, A test of comet and meteor shower associations. Icarus **45**, 545–553 (1981)

D.P. Galligan, W.J. Baggaley, The orbital distribution of radar-detected meteoroids of the Solar system dust cloud. MNRAS **353**, 422–446 (2004). doi:10.1111/j.1365-2966.2004.08078.x

R. Gonczi, H. Rickman, C. Froeschlé, The connection between Comet P/Machholz and the Quadrantid meteors, MNRAS **254**, 627–634 (1992)

E. Grun, H.A. Zook, H. Fechtig, R.H. Giese, Collisional balance of the meteoritic complex, Icarus **62**, 244–272 (1985). doi:10.1016/0019-1035(85)90121-6

P. Jenniskens, 2003 EH1 is the Quadrantid shower parent comet, Astron. J. **127**, 3018–3022 (2004)

J. Jones, P. Brown, Sporadic meteor radiant distributions-orbital survey results, MNRAS **265**, 524–532 (1993)

J. Jones, W. Jones, Comet Machholz and the Quadrantid meteor stream, MNRAS **261**, 605–611 (1993)

J. Jones, P. Brown, K.J. Ellis, A.R. Webster, M. Campbell-Brown, Z. Krzemenski, R.J. Weryk, The Canadian Meteor Orbit Radar: system overview and preliminary results, Plan Space Sci. **53**, 413–421 (2005)

Y. Kozai, Secular perturbations of asteroids with high inclination and eccentricity, Astron. J. **67**, 591–598 (1962)

B.G. Marsden, G.V. Williams, Catalogue of Cometary Orbits, 16th edn. (IAU Central Bureau for Astronomical Telegrams—Minor Planet Center, Cambridge, 2005)

B.A. McIntosh, Comet P/Machholtz and the Quadrantid meteor stream, Icarus **86**, 299–304 (1990)

E.J. Opik, Collision probability with the planets and the distribution of planetary matter, Proc. R. Ir. Acad. Sect. A **54**, 165–199 (1951)

R.B. Southworth, G.S. Hawkins, Statistics of meteor streams, Smithsonian Contrib. Astrophys. **7**, 261–285 (1963)

G.B. Valsecchi, T.J. Jopek, C. Froeschle, Meteoroid stream identification: a new approach—I. Theory, MNRAS **304**, 743–750 (1999)

S.J. Weidenschilling, A.A. Jackson, Orbital resonances and Poynting–Robertson drag, Icarus **104**, 244–254 (1993)

P. Wiegert, P. Brown, The Quadrantid meteoroid complex, Icarus **179**, 139–157 (2005). doi:10.1016/j.icarus.2005.05.019

J. Wisdom, M. Holman, Symplectic maps for the *n*-body problem, Astron. J. **102**, 1528–1538 (1991)

S.P. Wyatt, F.L. Whipple, The Poynting–Robertson effect on meteor orbits, Astrophys. J. **111**, 134–141 (1950)

Meteor Outburst Profiles and Cometary Ejection Models

D. J. Asher

Originally published in the journal Earth, Moon, and Planets, Volume 102, Nos 1–4.
DOI: 10.1007/s11038-007-9227-6 © Springer Science+Business Media B.V. 2008

Abstract The spatial structure of meteor streams, and the activity profiles of their corresponding meteor showers, depend firstly on the distribution of meteoroid orbits soon after ejection from the parent comet nucleus, and secondly on the subsequent dynamical evolution. The latter increases in importance as more time elapses. For younger structures within streams, notably the dust trails that cause sharp meteor outbursts, it is the cometary ejection model (meteoroid production rate as a function of time through the several months of the comet's perihelion return, and velocity distribution of the meteoroids released) that primarily determines the shape and width of the trail structure. This paper describes how a trail cross section can be calculated once an ejection model has been assumed. Such calculations, if made for a range of ejection model parameters and compared with observed parameters of storms and outbursts, can be used to constrain quantitatively the process of meteoroid ejection from the nucleus, including the mass distribution of ejected meteoroids.

Keywords Celestial mechanics · Comets · Dust trails · Leonids · Meteor outbursts · Meteor streams

1 Introduction: Dust Trail Theories

Dust trail theories have been highly successful in predicting the sharpest storms and outbursts in the Leonids (Kondrat'eva and Reznikov 1985; Kondrat'eva et al. 1997; Asher 1999; Lyytinen 1999; McNaught and Asher 1999) and other streams (e.g. Reznikov 1983, 1993; Watanabe et al. 2005). During a single perihelion return of an active comet, meteoroids are released on to a range of orbits close to, but not identical to, the comet's orbit. The range of orbital periods soon causes the particles to stretch into a trail, which can already be quite long after just a few revolutions.

Even within one revolution, the orbits are subject to gravitational perturbations. The key realisation in the development of dust trail theories is that the perturbations are a function

D. J. Asher (✉)
Armagh Observatory, College Hill, Armagh BT61 9DG, UK
e-mail: dja@arm.ac.uk

only of location along the trail, over short timescales (but at least for a few revolutions). This means that the determination of when outbursts occur reduces to a problem with two end points: the time t_0 when particles are ejected from the comet, and a later time t_1 when the Earth passes through the stream. The trail can produce an outburst at time t_1 if the result of planetary perturbations has been to bring the node of the trail particle orbits precisely to Earth intersection, rather than the node being inside or outside the Earth's orbit. Only one value of the orbital period allows particles to reach their node at time t_1. The idea of dust trail theories is to quickly find (iteratively) this value of the period, calculating the perturbations on just one representative particle with that period. All particles with the same period have comoved between t_0 and t_1, i.e. have continuously been at almost the same point in space as each other, and have therefore been subject to the same perturbing accelerations from the planets.

A set of comoving particles are by definition at a single point along a trail. Although their orbital periods must be the same, their other orbital elements can differ owing to their range of ejection velocities. This leads to the trail having a nonzero width, generally much less than a trail's length but significantly greater than the size of the Earth. A trail is also much narrower than the width of the whole stream derived from the parent comet, as the stream has formed from meteoroids ejected over a long timescale, during which their orbits can diverge to a greater extent.

Over short enough timescales, however, the perturbations on a set of comoving particles are the same and trail cross sections are invariant. The width at a single point along a trail results from ejection velocities, not from differential perturbations. This allows calculations as described below. Such calculations are only applicable to young enough trails. In other cases, more detailed modelling of the stream is necessary (cf. Vaubaillon et al. 2005a, 2005b). For the Leonids, trails whose age is a few revolutions (up to ~ 10) are young enough (see Sect. 3). For other streams, the limiting age depends on the effect of perturbations, and thus especially on factors such as the proximity of the stream orbit to Jupiter. For long period streams (Lyytinen and Jenniskens 2003), a further consideration is the extreme sensitivity of the semi-major axis to perturbing forces; with showers such as the α-Aurigids (Jenniskens and Vaubaillon 2007) this can be relevant.

Some authors have used the parameter Δa_0 to indicate distance along a trail. This is the difference in semi-major axis (i.e. equivalent to period) from the comet at time t_0. Owing to planetary perturbations, the semi-major axis is not constant, even over a few revolutions. As radiation pressure (Sect. 2) affects the semi-major axis, it is easiest to define Δa_0 as being for particles on which radiation pressure is negligible, and to think of all comoving particles as having the same Δa_0, even though their semi-major axis can differ if they are subject to radiation pressure.

2 Method: Calculating Cross Sections

In principle, observations of meteor showers can be used to place strong constraints on where along the parent comet's orbit meteoroids were released, or on the meteoroids' ejection velocities (e.g. Brown and Arlt 2000; Ma and Williams 2001; Ryabova 2001, 2007; Arter and Williams 2002; Asher and Emel'yanenko 2002). The available detailed observations of many storms and outbursts provide excellent opportunities in this regard. The requirement of the modelling is to reliably convert circumstances of ejection into observable quantities such as meteor activity profiles.

Provided that ejection speeds are small relative to the orbital speed (true in practice), calculations of a family of orbits of ejected particles can be linearised. From Lagrange's planetary equations, expressions for changes in orbital elements can be derived as functions of three ejection velocity components (Δv_R, Δv_T, Δv_N), in radial, transverse and normal directions:

$$\begin{aligned} \Delta a &= \mathtt{AR}\,\Delta v_R + \mathtt{AT}\,\Delta v_T \\ \Delta r_D &= \mathtt{RR}\,\Delta v_R + \mathtt{RT}\,\Delta v_T + \mathtt{RN}\,\Delta v_N \\ \Delta\Omega &= \mathtt{ON}\,\Delta v_N \end{aligned} \qquad (1)$$

where a = semi-major axis, r_D = heliocentric distance of descending node (the relevant node for Leonids) and Ω = longitude of ascending node. The coefficients AR etc. are functions of the elements (including the true anomaly ν) given in celestial mechanics books (e.g. Murray and Dermott 1999 Chap. 2; Danby 1988 Chap. 11; Roy 1988 Chap. 6; Bate et al. 1971 Chap. 9). As r_D is not one of the standard elements, Δr_D can be derived from expressions for the changes in eccentricity e and argument of perihelion ω:

$$r_D = \frac{a(1-e^2)}{1 - e\cos\omega}$$

$$\Delta r_D = \frac{\partial r_D}{\partial a}\Delta a + \frac{\partial r_D}{\partial e}\Delta e + \frac{\partial r_D}{\partial \omega}\Delta\omega$$

(see also Pecina and Šimek 1997).

An outburst profile is determined by the ejection velocity distribution and the meteoroid production rate as a function of ν (Ryabova 2001). Any specific Earth encounter with a trail is parametrised by a single value of Δa_0 (Sect. 1). The Earth's passage through the trail occurs at a single value of Δr_D. It therefore suffices to calculate results along just one dimension, i.e. $\Delta\Omega$, which varies as the Earth moves through space and meteors are recorded, although we may calculate results in the (Δr_D, $\Delta\Omega$) plane if we wish to develop a picture of trail cross sections. The parameter Δa_0 can be thought of as measuring the along trail dimension, with Δr_D and $\Delta\Omega$ spanning the across trail dimension.

In meteor outburst calculations, solar radiation pressure, parametrised by β, the ratio of radiation pressure to solar gravity, is less important than planetary perturbations. However, for the majority of particles that produce visual meteors, it is still significant. If $\beta = 0$, the value Δa in (1) equals Δa_0. When $\beta \neq 0$, the effect on the orbital period (critical for determining which particles comove and are therefore part of the same cross section along a given trail) can be calculated (Kondrat'eva and Reznikov 1985; Williams 1997; Asher and Emel'yanenko 2002), and so the correct value of Δa (i.e. that corresponds to the desired Δa_0) can be found, to be used in (1). This correct Δa is a function of ν. In addition, the coefficients AR etc. in (1) can be evaluated using the appropriate value of the central mass, for any given β.

For any ν the Eq. (1) are linear and therefore easily inverted to find (Δv_R, Δv_T, Δv_N) for a given (Δa, Δr_D, $\Delta\Omega$). An assumed distribution of particles in (Δv_R, Δv_T, Δv_N) space therefore yields a density of particles, ejected at that ν, that reach the point (Δa, Δr_D, $\Delta\Omega$) in the trail. An assumed meteoroid production rate then allows the contributions from different ν to be summed, and the overall density for the given ejection model at the given (Δa_0, Δr_D, $\Delta\Omega$) to be evaluated. When integrating over ν, the Jacobian of the transformation from (Δv_R, Δv_T, Δv_N) to (Δa, Δr_D, $\Delta\Omega$) phase space is used for normalisation between different ν values. The Jacobian is a function of ν but not of (Δv_R, Δv_T, Δv_N).

3 Application: Leonids

The above procedure allows density profiles to be quickly generated for a large number of models each with its own ejection parameters. The physics of the cometary mass loss and meteoroid ejection process (Whipple 1951; Jones 1995; Ma et al. 2002) determines the range of models that it is reasonable to consider. Observational data have been obtained for various Leonid storms and outbursts (e.g. Jenniskens et al. 2000). Work will soon be in progress to generate density profiles (for a range of models) relating to Leonid trail encounters for which observational results are available, and to make a careful comparison with observed data. Such an approach allows ejection parameters to be constrained (e.g. Brown and Jones 1998). In the case of the Leonids, it is particularly useful that many outbursts have been observed, as each ejection model can simultaneously yield density profiles for every outburst. This assumes the process of meteoroid production to be the same on each return of the parent comet that has given rise to a trail later encountered by the Earth, indeed the modelling process can be a test of whether this is true.

The procedure is computationally light (thus enabling a wide range of ejection models to be assessed), firstly because it consists of calculations only for those parts of trails encountered by the Earth, and secondly because it does not involve numerical integrations of orbit evolution. Integrations are required only (i) to determine the location in the ecliptic, relative to the Earth's orbit, of a single reference point in a trail cross section, sometimes referred to as the "trail centre" although cross sections are not symmetrical about the trail centre; (ii) to verify that trails being considered have cross sections that are invariant under the planetary perturbations that occur between the ejection epoch and the observed epoch; and (iii) to verify that perturbations occurring during the ejection epoch itself, which may last as long as a year or so, are negligible. Regarding (ii), Leonid trail cross sections seem to remain invariant for a few revolutions, although certainly not for as long as e.g. 20 revolutions, in general (cf. McNaught and Asher 1999; Asher 2005). Regarding (iii), some new test integrations using Everhart's (1985) RADAU integrator as implemented in Chambers' (1999) MERCURY package show that, as a result of perturbations during the ejection arc (taken as $r < 3.4$) alone, the nodal position of Leonid particles is rarely displaced by more than an Earth diameter or so, i.e. a very small distance compared to the entire trail cross section. Exact displacements depend on the relative configuration of 55P/Tempel-Tuttle and Jupiter in their orbits on the given return of 55P.

An example cross section is shown in Fig. 1. For a single value of v and a single ejection speed, the locus of points in the $(\Delta r_D, \Delta\Omega)$ plane is an ellipse (cf. Kondrat'eva and Reznikov 1985; Müller et al. 2001; Welch 2003). The density distribution in Fig. 1 is essentially a sum of (non-concentric) ellipses, ejection occurring over an orbital arc spanning several months before and after perihelion, and with a range of ejection speeds at each v. Future modelling will consider the ejection process in more detail, addressing for example the possibility that meteoroid production from 55P/Tempel-Tuttle concentrates strongly near perihelion, as suggested by observations of the coma (Watanabe et al. 2001).

When the Earth passes through the cross section shown in Fig. 1, it encounters a one dimensional profile as plotted in Fig. 2, i.e. the calculated data in Fig. 2 are basically a subset (for a single Δr_D value) of those in Fig. 1. The width of the profile shown in Fig. 2 is several hours, clearly inconsistent with the accurately determined observed profile (Arlt et al. 1999), immediately showing that the ejection model adopted here for illustrative purposes does not match the real ejection process. A range of ejection models will soon be tested, and indeed as observational Leonid data exist for a different magnitude intervals, it

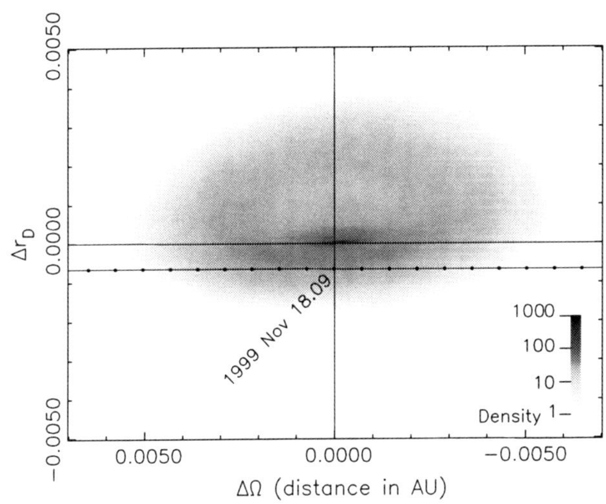

Fig. 1 Density in the ecliptic of the cross section of the 1899 Leonid trail encountered by the Earth in 1999, for one example ejection model. ΔΩ converted from radians to distance in Astronomical Units. Radiation pressure parameter $\beta = 0.001$. Meteoroid production rate uniform in true anomaly (cf. Kresák 1976; Brown and Jones 1998). Mean ejection speed $50/r$ m/s (r in AU), a power law with exponent -1 in heliocentric distance r being assumed for simplicity in this example, with a range around this mean value being allowed at any r. Ejection directions uniform over sunward hemisphere. Earth went from right to left, Earth shown actual size at 1 h intervals

should be possible to constrain the meteoroid production as a function of β (equivalently meteoroid mass) as well as v.

At present it is possible to envisage either of two conclusions. On the one hand, this kind of modelling could provide good fits to sufficiently young (e.g. 1-revolution, or 2-revolution, or more) Leonid trails and thus give strong quantitative constraints on ejection processes. Alternatively this modelling could demonstrate that no ejection model can fit all the observations, suggesting that ejection processes alone do not fully determine

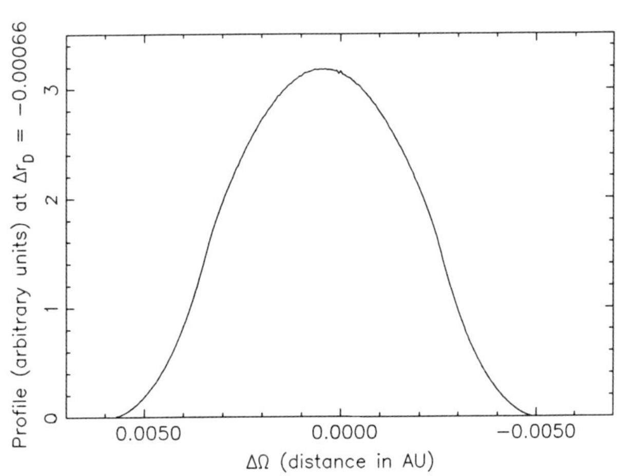

Fig. 2 The density profile encountered by the Earth as it goes through the trail cross section shown in Fig. 1

trail cross sections even on short timescales, and thus verifying that other processes such as radiative forces during the orbital evolution are important even over short times (cf. Lyytinen and van Flandern 2000; Lyytinen et al. 2001). Either of these conclusions would be valuable in the study of dust trails and meteor outbursts.

Acknowledgements Helpful comments from Dr Peter Brown and Dr Jun-ichi Watanabe are greatly appreciated.

References

R. Arlt, L. Bellot-Rubio, P. Brown, M. Gyssens, Bulletin 15 of the International Leonid Watch: first global analysis of the 1999 Leonid storm. WGN **27**, 286–295 (1999)

T.R. Arter, I.P. Williams, Meteoroid ejection velocities deduced from a study of the April Lyrid meteor shower. MNRAS **329**, 175–180 (2002)

D.J. Asher, The Leonid meteor storms of 1833 and 1966. MNRAS **307**, 919–924 (1999)

D.J. Asher, in *Dynamics of Populations of Planetary Systems*, ed. by Z. Knežević, A. Milani. The dynamical structure of meteor streams and meteor shower predictions. Proc. IAU Colloq. vol. 197 (Cambridge University Press, 2005), pp. 375–382

D.J. Asher, V.V. Emel'yanenko, The origin of the June Bootid outburst in 1998 and determination of cometary ejection velocities. MNRAS **331**, 126–132 (2002)

R.R. Bate, D.D. Mueller, J.E. White, *Fundamentals of Astrodynamics*. (Dover Publications, New York, 1971)

P. Brown, R. Arlt, Detailed visual observations and modelling of the 1998 Leonid shower. MNRAS **319**, 419–428 (2000)

P. Brown, J. Jones, Simulation of the formation and evolution of the Perseid meteoroid stream. Icarus **133**, 36–68 (1998)

J.E. Chambers, A hybrid symplectic integrator that permits close encounters between massive bodies. MNRAS **304**, 793–799 (1999)

J.M.A. Danby, *Fundamentals of Celestial Mechanics*, 2nd edn. (Willmann-Bell, Richmond, Virginia, 1988)

E. Everhart, in *Dynamics of Comets: Their Origin and Evolution*, ed. by A. Carusi, G.B. Valsecchi. An efficient integrator that uses Gauss-Radau spacings. Proc. IAU Colloq. vol. 83 (Reidel, Dordrecht, 1985), pp. 185–202

P. Jenniskens, J. Vaubaillon, Aurigid predictions for 2007 September 1. WGN **35**, 30–34 (2007)

P. Jenniskens, C. Crawford, S. Butow, Successful hybrid approach to visual and video observations of the 1999 Leonid storm. WGN **28**, 58–63 (2000)

J. Jones, The ejection of meteoroids from comets. MNRAS **275**, 773–780 (1995)

E.D. Kondrat'eva, E.A. Reznikov, Comet Tempel-Tuttle and the Leonid meteor swarm. Sol. Syst. Res. **19**, 96–101 (1985)

E.D. Kondrat'eva, I.N. Murav'eva, E.A. Reznikov, On the forthcoming return of the Leonid meteoric swarm. Sol. Syst. Res. **31**, 489–492 (1997)

L. Kresák, Orbital evolution of the dust streams released from comets. Bull. Astron. Inst. Czechosl. **27**, 35–46 (1976)

E. Lyytinen, Leonid predictions for the years 1999–2007 with the satellite model of comets. Meta Res. Bull. **8**, 33–40 (1999)

E. Lyytinen, P. Jenniskens, Meteor outbursts from long-period comet dust trails. Icarus **162**, 443–452 (2003)

E.J. Lyytinen, T. van Flandern, Leonid predictions based on the satellite model of comets. Earth Moon Planet. **82**, 149–166 (2000)

E. Lyytinen, M. Nissinen, T. van Flandern, Improved 2001 Leonid storm predictions from a refined model. WGN **29**, 110–118 (2001)

Y. Ma, I.P. Williams, The ejection velocity of meteoroids from cometary nuclei deduced from observations of meteor shower outbursts. MNRAS **325**, 379–384 (2001)

Y. Ma, I.P. Williams, W. Chen, On the ejection velocity of meteoroids from comets. MNRAS **337**, 1081–1086 (2002)

R.H. McNaught, D.J. Asher, Leonid dust trails and meteor storms. WGN **27**, 85–102 (1999)

M. Müller, S.P. Green, N. McBride, in *Proceedings of the Meteoroids 2001 Conference*, ed. by B. Warmbein. Constraining cometary ejection models from meteor storm observations, ESA SP–495 (ESA, Noordwijk, 2001), pp. 47–54

C.D. Murray, S.F. Dermott, *Solar System Dynamics* (Cambridge University Press, 1999)
P. Pecina, M. Šimek, The orbital elements of a meteoroid after its ejection from a comet. A & A **317**, 594–600 (1997)
E.A. Reznikov, The origin of the Bootid meteor stream. Trudy Kazan. Gor. Astron. Obs. **47**, 131–136 (1983) (In Russian)
E.A. Reznikov, The Giacobini-Zinner comet and Giacobinid meteor stream. Trudy Kazan. Gor. Astron. Obs. **53**, 80–101 (1993) (In Russian)
A.E. Roy, *Orbital Motion*, 3rd edn. (Adam Hilger, Bristol, 1988)
G.O. Ryabova, The Geminid meteor stream activity profile. Sol. Syst. Res. **35**, 151–157 (2001)
G.O. Ryabova, Mathematical modelling of the Geminid meteoroid stream. MNRAS **375**, 1371–1380 (2007)
J. Vaubaillon, F. Colas, L. Jorda, A new method to predict meteor showers. I. Description of the model. A & A **439**, 751–760 (2005a)
J. Vaubaillon, F. Colas, L. Jorda, A new method to predict meteor showers. II. Application to the Leonids. A & A **439**, 761–770 (2005b)
J. Watanabe, H. Fukushima, T. Nakamura, in *Proceedings of the Meteoroids 2001 Conference*, ed. by B. Warmbein. The activity profile of comet 55P/Tempel-Tuttle in 1998 return: meteoroid release concentration on the perihelion, ESA SP–495 (ESA, Noordwijk, 2001), pp. 175–178
J. Watanabe, M. Sato, T. Kasuga, Phoenicids in 1956 revisited. PASJ **57**, L45–L49 (2005)
P.G. Welch, Matching cometary ejection processes to the Leonids 1998–2001 using a hybrid numerical model. MNRAS **342**, 971–994 (2003)
F.L. Whipple, A comet model. II. Physical relations for comets and meteors. ApJ **113**, 464–474 (1951)
I.P. Williams, The Leonid meteor shower: why are there storms but no regular annual activity? MNRAS **292**, L37–L40 (1997)

High Inclination Meteorite Streams can Exist

Daniel C. Jones · Iwan P. Williams

Originally published in the journal Earth, Moon, and Planets, Volume 102, Nos 1–4.
DOI: 10.1007/s11038-007-9163-5

Abstract Meteorites represent bodies at the larger end of the meteoroid size spectrum since they are large enough to survive ablation in the Earth's atmosphere. They are thus far less numerous than normal meteoroids that become meteors. A number of meteorites can arrive at around the same time and location and so in some sense represent a stream, but these are just recent fragmentations. Most meteors, according to their cosmic ray exposure age are at least 10 million years old. This is roughly also the timescale for the onset of chaos in the inner Solar System and so conventional wisdom is that meteorites can not survive on such orbits for such a time span and that they certainly cannot survive as coherent streams. We investigate numerically the survival of streams for this time interval.

Keywords Meteorites · Meteoroid streams · Celestial mechanics · Numerical methods: N-body

1 Introduction

Meteorites are defined to be interplanetary bodies that have collided with Earth and survived their journey through its atmosphere. Very small bodies have a high surface to mass ratio and thus slow down very quickly on encountering the atmosphere so that there is virtually no heating for most of their passage. As the mass loss from an ablating object is roughly proportional to the surface area, very large meteors can survive their passage through the atmosphere and it is these meteorites that form the subject of this discussion.

Meteor showers are seen when Earth passes through a meteoroid stream. Such streams form when a comet nucleus ejects dust and gas as seen in normal comet tails (e.g. Wu and Williams 1993) or when the nucleus disintegrates (e.g. Jenniskens 2004; Williams et al. 2004). Such meteors are thus cometary grains ablating. The maximum size of grains

D. C. Jones (✉) · I. P. Williams
Astronomy Unit, Queen Mary, University of London, Mile End Road, London E1 4NS, UK
e-mail: d.c.jones@qmul.ac.uk

ejected by comets is on the order of 10 cm (Williams 1992) which is considerably smaller than many meteorites, also cometary material is relatively weak and is unlikely to survive passage through the atmosphere to be recovered a meteorite. Collisions between asteroids can produce fragments of any size up to the initial sizes of the colliding asteroids, hence, this is the process which produces most of the objects which are recovered as meteorites.

Several meteoroid streams have asteroids associated with them (e.g. Porubčan et al. 2004). Asteroid 2003 EH_1 is believed to be the parent of the Quadrantid stream and a study by Williams et al. (2004) tentatively concluded that the comet C/1490 Y1 could be dynamically related to 2003 EH_1, indicating that in this case the asteroid could be a fragment of the original comet nucleus that is now dormant. By extension, this finding also suggests the possibility that other asteroids which are associated with meteoroid streams could be dormant comets.

Large objects can fragment in the atmosphere causing a meteorite shower to be seen, with the fragments creating fireballs that radiate from the same point. Although the science of such fragmentation is fascinating (e.g. Borovička and Kalenda 2003; Ceplecha and ReVelle 2005) it is not the topic of our investigation. There are also at least 25 meteorites which are known to be fragments blasted off the Moon by impacts and more than 35 which are from Mars, but the vast majority of all meteorites are pieces of asteroids. They have a range of cosmic ray exposure ages (Wetherill 1980) but these are in excess of 10 million years. This means that they have not been in the interior of larger bodies for at least this time interval.

The conventional view is that asteroid collisions, releasing meteorite sized fragments took place within the main Asteroid Belt and that the collisional fragments stayed here until the effects of chaos changed the orbits into Earth crossing orbits (Farinella et al. 1993). This mechanism is unlikely to produce meteorite streams since each potential meteorite will receive different perturbations at different times. It is more likely that the precursor object is already on an Earth-crossing orbit at the time that it is disrupted.

The interest in the existence of meteorite streams arose because of the identification by Halliday et al. (1990) of a number of fireballs in the Meteorite Observation and Recovery Program (MORP) and Prairie network databases with exceedingly similar orbits. The detailed data for 259 fireballs observed with MORP are presented in the paper by Halliday et al. (1996). Recently, Wolf and Lipschutz (1995) and Lipschutz et al. (1997), have argued that there is evidence from meteorite falls for the existence of meteorite streams. A major difficulty in any such discussion is that meteorite falls are rather rare. For example, the flux of meteorites which reach Earth's surface with a mass of 1 kg and over is about 2,000 per year (Bland 2005). Averaging over Earth's surface indicates that an object of this size reaches the surface roughly every 12 months in an area the size of the UK. Further, most meteorites are "finds", that is that they are found on the ground without being observed as meteors and therefore there is no record of the orbital parameters of the progenitor meteoroid.

The dynamical problem we wish to investigate here is not one of finding a mechanism that changes an, essentially stable, orbit into an, intrinsically unstable, Earth-crossing orbit, but rather whether a quasi-stable family of orbits can be found within an essentially chaotic region. This is not impossible, it is known that there are islands of stability in other chaotic regions (e.g. in trans-Neptunian space: Duncan et al. 1995; Jones et al. 2005). If such a region is found, we then ask the further question of not only can such meteorites survive in this chaotic region for millions of years, but also whether a stream (a set of similar orbits) can be detected within the survivors.

2 Identification of Meteorite Streams

Halliday et al. (1978, 1981) observed a fall and recovered nine meteorite fragments with a total mass of 4.58 kg from near Innisfree, Alberta. The orbital elements of this meteorite were also obtained from the MORP fireball data. During a search through the MORP data a second meteor was found which had appeared exactly 3 years after the Innisfree meteorite and which was believed to have fallen as a meteorite near Ridgedale, Saskatchewan, though no meteorite has been actually recovered. This object was found to share almost precisely the same orbit as the Innisfree meteorite (Halliday 1987; Halliday et al. 1990). Thus, the existence of what appeared to be a group of meteorite producing asteroid fragments was established, or, if a small abuse of terminology is permitted, a *meteorite stream*.

After being informed of the finding of another object which may be related to the Innisfree object in the Prairie Network data (McCrosky et al. 1978, 1979), Halliday et al. (1990) conducted a further search comparing this set and the MORP data for more objects related to the conjectural Innisfree-Ridgedale group and also for orbit groupings amongst the MORP events. The search resulted in four meteorite groups being identified and these are shown in Table 1 along with their orbital data and date of observation.

Table 1 Data for four groups of related fireballs.The date is the Julian date of observation of the fireball, and the D' (Drummond 1981) value for each meteor is that relative to the mean for the group to which it belongs. The calculated initial orbital elements and the estimated mass of each object are also shown. The mean orbit for group 4 does not include object 4b. The orbital element symbols represent the following characteristics of each object's orbit: perihelion distance (q), eccentricity (e), inclination (i), argument of perihelion (ω) and longitude of the ascending node (Ω). Data from Halliday et al. (1990)

Designation		Date	q (AU)	e	i (°)	ω (°)	Ω (°)	Mass (kg)	D'
1	Mean 1		0.99	0.47	8	185	319		
1a	MORP 285	2443180.60	0.99	0.47	12	178	317	2.1	0.032
1b	MORP 288	2443191.81	0.99	0.44	1	185	328	6.2	0.052
1c	MORP 545	2444275.59	0.98	0.48	12	187	316	1.8	0.025
1d	PN 40617	2440617.89	0.98	0.52	3	194	311	>0.25	0.058
1e	PN 40996	2440996.84	0.99	0.46	10	181	325	>0.25	0.017
2	Mean 2		0.99	0.59	6	192	123		
2a	MORP 580	2444483.84	1.00	0.59	6	167	159	11.0	0.043
2b	MORP 872	2445527.69	0.99	0.57	8	202	109	10.0	0.027
2c	PN 40405	2440405.77	0.97	0.60	3	208	101	>0.25	0.030
3	Mean 3		0.94	0.52	2	205	170		
3a	MORP 123	2442271.79	0.88	0.56	3	232	139	0.74	0.052
3b	MORP 189	2442669.83	0.93	0.52	2	39	350	0.59	0.047
3c	MORP 498	2444165.65	0.97	0.53	2	158	205	0.36	0.040
3d	MORP 886	2445584.81	0.96	0.49	4	210	164	1.1	0.034
4	Mean 4		0.99	0.59	1	185	241		
4a	MORP 626	2444568.62	0.98	0.57	1	192	242	0.77	0.031
4b	MORP 687	2444985.64	0.89	0.58	11	138	295	1.6	0.086
4c	PN 39815	2439815.65	0.99	0.59	0	183	238	>0.25	0.017
4d	PN 41280	2441280.74	0.99	0.60	2	179	242	>0.25	0.019

Recently it has been found that the Příbram and Neuschwanstein meteorites have exceedingly similar orbits (Kornoš and Tóth 2007). The closeness of these orbits, while suggestive of an association, is not strongly significant in a statistical sense; Pauls and Gladman (2005) found that for a data set as large as the current list of fireballs there is a probability of >70 % of two orbits being this close by chance. However, Kornoš and Tóth (2007) argue that this probability is less than 10%. Pauls and Gladman (2005) further simulated the evolution of the Příbram/Neuschwanstein pair and found that they become decoherent within a short time period; on the order of a few 10^5 years. This shows that orbital similarity at the time of observation alone is not enough to establish an association between objects. A similar conclusion was reached by Jones et al. (2006) in our investigation of the Kappa Cygnid meteoroid complex.

In this paper we investigate the lifetimes of the suggested meteorite streams given in Table 1. Specifically, we investigate whether the given groups of meteorite sized objects can survive as streams at all and if so then if the stream can survive for many millions of years.

3 The Evolution of Objects Back in Time

As well as the individual objects in Table 1 we want to integrate the equations of motion for the mean orbits of each group from a common starting epoch. Since the earliest year of observation for any object in the table is 1967, 1st January of that year at 0 h UT was chosen as the starting epoch. The evolution back in time was calculated numerically using Mercury 6 (Chambers 1999). Mercury 6 differs from other symplectic integrators in the way it deals with close encounters. By creating a hybrid-symplectic program Chambers has allowed the more accurate (but slower) Bulirsh-Stoer integrator (Bulirsh and Stoer 1966) to take over in the event of a close encounter between a test particle and a planet. This differs from, for example, Duncan et al. (1998) who applied a different timestep to each perturbing body, giving stronger perturbers shorter timesteps. Perturbations from the planets Venus, Earth, Mars, Jupiter and Saturn were included. The data for the planetary orbits were obtained from the NASA Horizons System[1]. An object is considered to have been lost from the simulation if its heliocentric distance exceeds 300 AU. Since we want to investigate the behaviour of large particles which are not strongly affected by non-gravitational forces, no such forces were included.

For each mean orbit we placed 10 clone objects at even spacings in true anomaly around these orbits. Thus we have 16 real objects and 40 clone objects in total. The cosmic ray exposure of the recovered Innisfree meteorite was 28 million years (Heusser et al. 1978). For comparison, the corresponding ages for Příbram and Neuschwanstein were 18 million (Lavrukhina et al. 1974) and 48 million years[2], respectively. Hence we will use approximately this timescale for our integration and the orbital elements at the starting epoch were then integrated back for a further 30 million years.

After running the integration for the 30 million year timescale it becomes clear that very few of the objects can survive in the inner Solar System for this length of time. Of the 56 initial objects only seven remain within the limits of the simulation, the rest having collided with a planet or gone beyond the 300 AU outer boundary of the simulated system. Of these, six remain within the orbit of Jupiter for the whole simulation, but the semi-major

[1] http://ssd.jpl.nasa.gov/?horizons

[2] http://tin.er.usgs.gov/meteor/metbull.php?code=16950

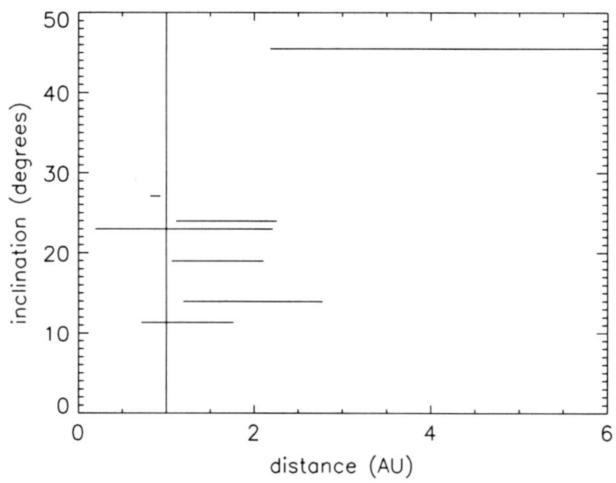

Fig. 1 The inclination of meteorites which survive for the whole length of the integration against their heliocentric distance. The extent of the line shows the range of heliocentric distances that each object covers at the end of the simulation, thus perihelion is the lowest extreme of the line and aphelion the highest. The vertical line indicates the position of Earth and thus any line which crosses this line crosses Earth's orbit

axis of the seventh object only moves beyond 1 AU in the last 50 Myr of the simulation time. Figure 1 shows the inclination against perihelion distance for these seven objects at the end of the simulation. The horizontal lines show the range of heliocentric distances covered by the objects, perihelion is the lowest extreme of the line and aphelion the highest. There are two objects whose orbits cross that of Earth, but since the inclinations are all fairly high a close encounter with Earth would only be possible at the node and then only if the heliocentric distance at the node was close to 1 AU. Amongst the seven surviving objects there was not a complete set from the original four groups. In fact only one of the seven was a real meteorite, the others being clone objects. However, three clones and the one real object (MORP288) that did survive were from group 1. This suggests the possibility that the orbits of the group 1 objects and their clones are more stable than the other groups. We also note that two of these four survivors (MORP 288 and clone 5) have quite similar orbital elements. The fact that so few objects survived may be explained in the following way: The original orbits of the individual objects must have been in a stable region in order for the object to have survived until now, however, the orbits that we currently see have evolved away from their original positions. Producing a group of clone objects around the orbits as they are at present will undoubtedly place some of those objects into an unstable orbit. However, the survival on similar orbits of MORP288 and clone 5 suggests that the original orbit may have been in this locality.

We next looked at the orbits of the six objects which remained within the orbit of Jupiter for the whole length of the simulation. It is possible that these orbits happen to be on small islands of stability in this otherwise unstable region. If this is the case then taking the orbits of these objects at the −30 Myr epoch, creating a group of clone objects around each of them and then integrating them back forwards to the present should result in a relatively high number of objects which survive on an orbit which does not largely change. As before we created a group of ten clone objects around each of these six orbits which survived to the −30 Myr epoch. We have the positions and orbits of the planets at this time as part of the results of the last simulation and these were used for the second simulation.

Table 2 Orbital data for each of the six objects used in the second simulation. The number of clones for each of these objects which survived for the whole length of the simulation back to the present time is also shown

Designation		q (AU)	e	i (°)	ω (°)	Ω (°)	surviving clones
1	MORP288	1.11	0.34	23.95	300.48	356.61	9
2	gp1cl04	0.20	0.84	22.98	270.66	65.23	5
3	gp1cl05	1.06	0.33	18.99	81.01	303.18	9
4	gp1cl07	0.72	0.42	11.34	300.20	143.33	2
5	gp3cl10	1.19	0.40	13.95	346.99	146.76	4
6	gp4cl03	0.82	0.07	27.05	256.08	210.69	8

In this second simulation we find that a much greater proportion of the objects survive. Of the 60 clone objects only 23 are lost. This corresponds to the survival of 62% of the objects, compared to the situation in the first simulation where only 7 of the 56 objects survived, or 13%. Table 2 shows the initial orbital elements of the six long lived objects and the number of clones which survived the simulated 30 Myr back to the present. Of particular interest are the group around MORP 228 and group 1 clone 5, that is objects 1 and 3 in the above table. The combined survival rate for these two objects' clones is 90%, i.e. only two clones are lost. This illustrates again the stability of this locality. The clones of object 6 also deserve investigation as the survival rate is 80%.

Figures 2–4show the evolution of the orbital elements of these three objects respectively. In each figure the light grey lines show the evolutions of the individual clones, the black line shows the mean for all the clones and the dark grey line shows the evolution of the object from Table 2 as calculated in the first simulation.

Figure 2 shows that most of the 10 clones for this group stays at roughly their starting positions. At about −18 Myr the mean diverges drastically away from this stable evolution as the result of the one clone which is lost. As soon as that object is completely lost from the simulation, the mean returns to close to the initial value. At this time one other object starts to diverge from the rest of the remaining group, but is not completely lost by the time the simulation reaches the present. In this case the orbit of the real object remains close to the mean of the clones except in inclination. This family of objects does seem to represent a stream, at least in a loose sense of the word. The orbit of the real object, MORP 288, only changes by a small amount, with the exception of the inclination, which is about 24° at the −30 Myr epoch and is only 1° at the present epoch.

The evolutions of the orbits of the clones for object 3 are shown in Fig. 3. In this case the clones do not all remain close to their starting positions, but they do all remain within the system with the exception of one object.

This one object is lost at about −5 Myr as can be seen from the jump in the mean orbit of all the clones. The real object in this case remains within the same range as the clones for all the orbital elements, with the exception of a short divergence in inclination.

The evolutions of the orbits of the clones for object 6 are shown in Fig. 4. Again these objects do not stay close to their starting positions, but do remain well within the limits of the simulation. In this case the two objects which are lost from the system are not lost due to ejection. In fact they are lost when they collide with Jupiter, the first collision occurring at about −21 Myr and the second at about −6 Myr. Again the real object evolves in a way that is similar to the evolution of the clone objects. The fact that in these three cases the

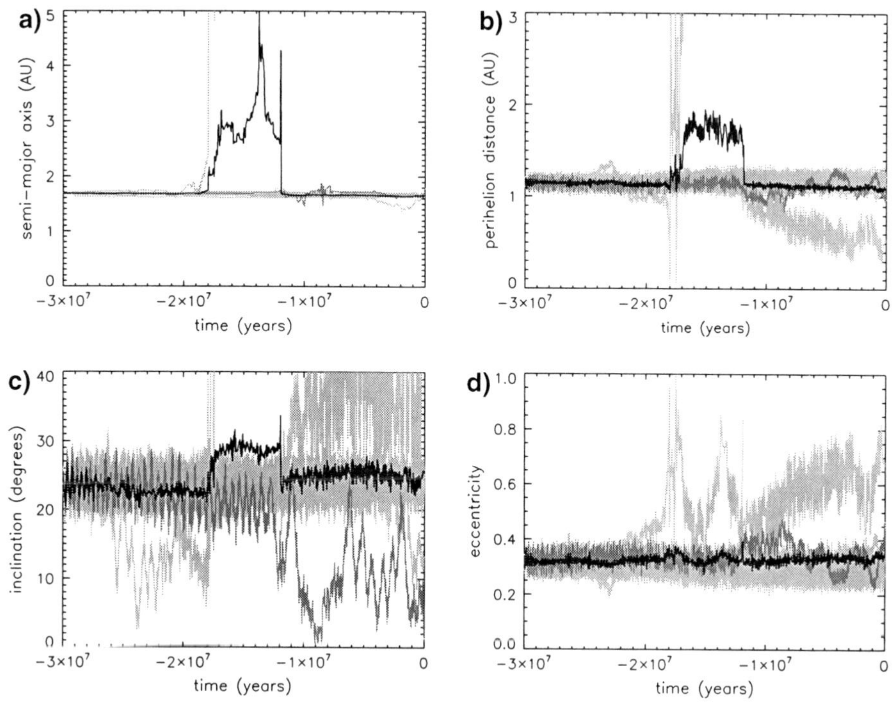

Fig. 2 Evolution of the orbital elements of object 1 (MORP288) from Table 2. The light grey lines show the evolution of the clone objects, the black line shows the evolution of the mean for all the objects remaining in the simulation at each time and the dark grey line shows the evolution of the associated object from Table 2

real object and most of the clone objects evolve in a somewhat similar way suggests that the region of space that they occupy is indeed a stable region.

That of these three objects, the orbit which appears to lie in the most stable region is a real object may be significant. At the present time, MORP 288 has orbital elements slightly different from the average as shown in Fig. 2. Specifically MORP 288 has a slightly smaller perihelion distance and hence a larger eccentricity, it also has a substantially lower inclination than the mean of the clones in Fig. 2. It is not inconceivable that MORP 288 could have originally been in an orbit matching the starting point of the clones from Fig. 2 and then been one of the few to have moved from this stable zone. At the starting epoch in all three cases the inclination of the objects is quite high, over 20°. Looking at the current lists of known Near Earth Objects shows us that such a high inclination is not extremely uncommon in the present population. The number of objects that have inclinations ≥20° is about 28% among the 4,500 known Apollo, Aten and Amor asteroids[3].

In Table 2 it can be seen that objects 1 and 3 share similar perihelion distances, eccentricities and inclinations. These are also the two sets where nine clones survive. If we exclude the one clone object from each case which does not survive for the whole integration and the two main objects, which do not behave like most of their clones, we can see that the evolutions of many of the remaining objects are similar. In group 1, eight of the nine surviving objects evolve in a similar way and four objects from group 3 also evolve in

[3] Minor Planet Center: http://cfa-www.harvard.edu/iau/lists/Unusual.html

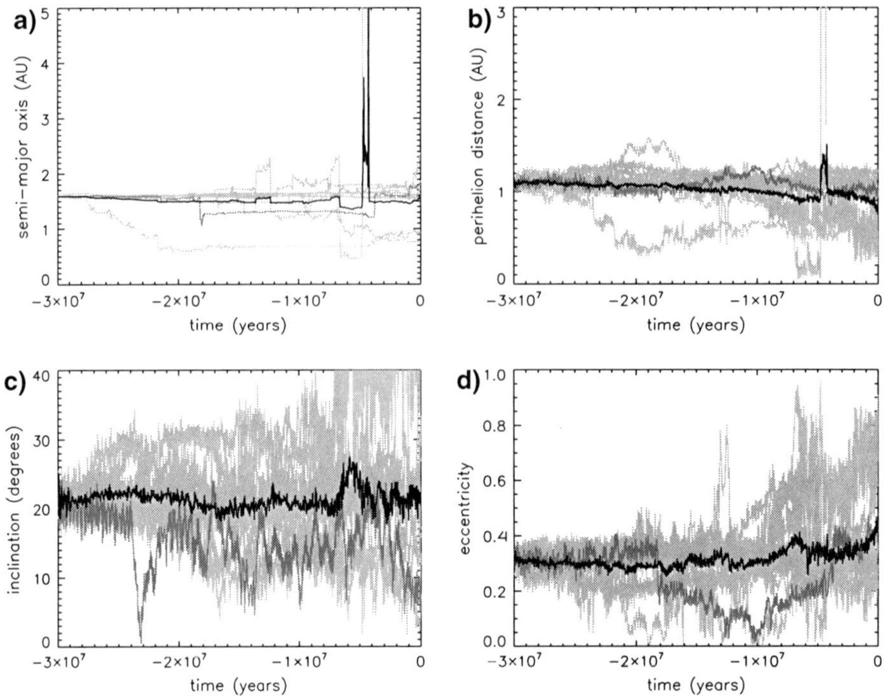

Fig. 3 Similar plots to Fig. 2, but for object 3 from Table 2

a similar way to these eight. These twelve objects are plotted in Fig. 5 along with the mean of their orbital elements calculated at each time interval. There are some slight variations in the evolutions, for example one object's eccentricity and inclination diverges from the rest near the end of the simulation, however, the mean remains fairly steady for all elements.

The semi-major axes of these twelve objects remain within the range $1.6 < a < 1.7$ AU, whilst the eccentricities remain in the range $0.25 < e < 0.40$ and inclinations remain roughly in the range $18° < I < 28°$. The final mean values for the orbital elements are $q = 1.14$ AU, $e = 0.31$ and $i = 22.7°$. These are all quite different from the values for the objects in Table 1.

4 Discussion and Conclusions

We conclude from this study that it is possible for large objects whose orbits are not strongly influenced by non-gravitational forces to survive on fairly unchanged orbits within the inner Solar System for many millions of years. Thus the existence of meteorite streams is indeed plausible. The objects we investigated tended to be more stable on high inclination orbits, this is to be expected as objects on these higher inclination orbits spend less time within the ecliptic and so are less likely to undergo close encounters with the planets.

We also note that the meteorite MORP 288 is on a long lived orbit and could have existed as part of a meteorite stream for up to 30 Myr. Many of the clones surrounding MORP 288 and our clone object gp1cl05, when integrated forward in time, evolved in a

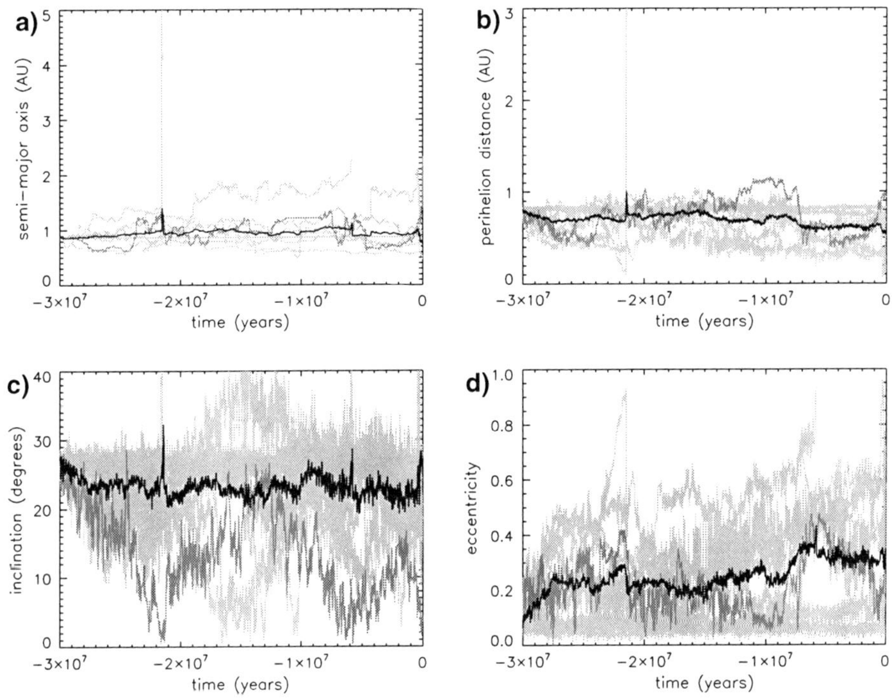

Fig. 4 Similar plots to Fig. 2, but for object 6 from Table 2

similar way and appear to all be in the same stable area. This indicates to us that the orbit of MORP 288 was stable for timescales up to a few 10^7 years and that in such an orbit a stream of meteorite sized objects could also remain on stable orbits for similar time periods.

The final orbits of the objects surrounding MORP 288 after they were integrated forward to the present epoch do not match those of the Innisfree-Ridgedale groups, in particular the eccentricities are slightly lower, the inclinations are somewhat larger and, crucially, the orbit is not Earth-crossing. The result, therefore, does not necessarily support the hypothesis that the groups in Table 1 are true meteorite streams. In fact, since these groups do not evolve coherently, the opposite is suggested, that the groups are likely to be the result of coincidence. What our result does suggest is simply that meteorite dropping streams can exist. In Figs. 5 and 2 it can be seen that objects can move away from the initial orbit and become Earth-crossing. Therefore, while the stream itself is not Earth-crossing, and hence has a chance to survive for a long time, differing perturbations on the individual stream members can create Earth-crossing objects which could be seen as meteoritic fireballs.

Pauls and Gladman (2005) conclude that three meteorite pairings, including the Innisfree-Ridgedale pairing, cannot survive on similar orbits for timescales of over a few 10^5 years. This would seem to contradict our findings of a potential stream lifetime ~ 100 times greater for objects similar to this paring. However, although we found this orbit from a starting point of investigating the Innisfree-Ridgedale group, the orbit of interest here is not the orbit of the Innisfree and Ridgedale fireballs. We suggest that the particular case of the orbit near MORP 288 may be an example of an island of orbital stability in an

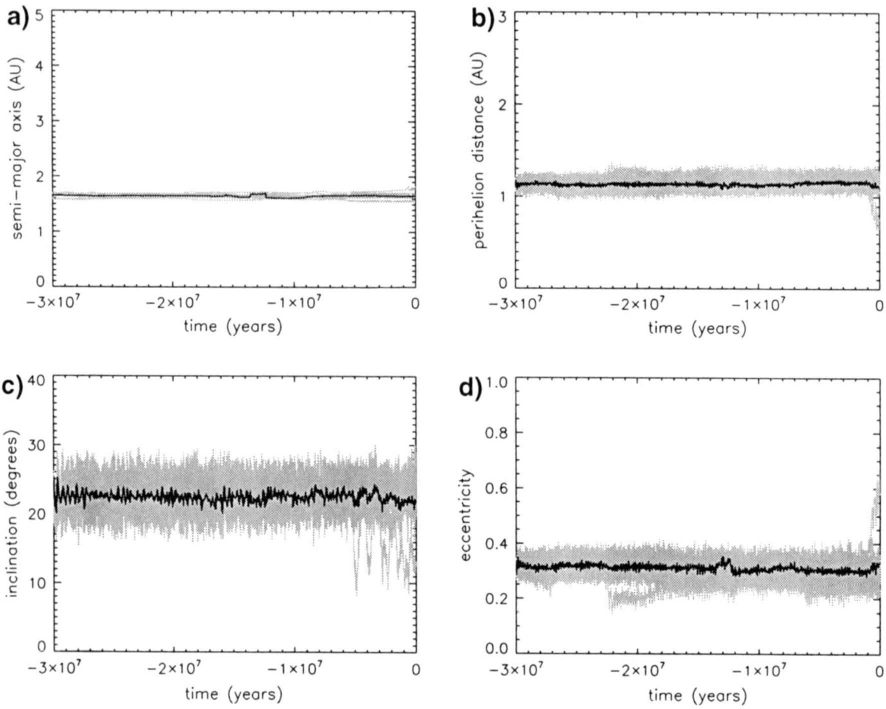

Fig. 5 The evolution of the eight clones of object 1 (MORP 288) and four clones of object 3 (gp1cl05) which evolve in a similar way. The grey lines are the clones themselves and the black line is the mean of the clone orbits

otherwise unstable region as described in the introduction. The inclination of MORP 288 at the ancient epoch is significantly higher than those of the three pairings studied by Pauls and Gladman (2005), $\approx 23°$ compared to $\approx 12°$, and the perihelion distance of the stream is greater than 1 AU, which will also contribute to its stability due to the factors mentioned above.

We accept the conclusion of Pauls and Gladman (2005) that the current fireball data set has not convincingly been shown to itself contain members of the same meteorite stream, and also note that the three pairings that they studied lose coherence over short time scales. However, we do not feel that their results rule out the possible existence of long lived meteorite streams. In fact, we feel that our results show that it is possible, and that the orbit of MORP 288 was an orbit where such a stream can persist. Of course, MORP 288 itself was not, and nor was any other meteorite, part of a known stream and Pauls and Gladman show that many meteorites with very similar orbits must be observed before a stream can be shown to exist above the level of a chance coincidence of orbits. Our idea that individual objects are pushed out of a orbit which is not Earth-crossing onto Earth-crossing orbits further complicates this. It may be that such a stream of meteorite sized particles will never be proven to exist by Earth-bound observers.

A caveat to the conclusions made here must be that the compositions of the objects in Table 1 are not know with certainty. Ceplecha and McCrosky (1976) discuss the problems of determining the initial fireball composition from analysis of the observations. They determined that is only possible to make a tentative classification of whether a fireball is produced by a chondritic object and is therefore strong enough to survive passage through the atmosphere. Therefore the assertion than each of the mentioned fireballs is indeed meteoritic must also be taken tentatively. Nonetheless this does not affect the main point of the conclusion; that streams of meteorite sized objects can exist.

Acknowledgements DCJ would like to acknowledge PPARC/STFC for the award of a studentship which allowed this work to be carried out. The authors also thank the handling editor, one anonymous reviewer and Jack Drummond for their constructive comments.

References

P.A. Bland, The impact rate on Earth. Philos. Trans. R. Soc. A. **363**, 2793–2810 (2005)

J. Borovička, P. Kalenda, The Morávka meteorite fall. 4: meteoroid dynamics and fragmentation in the atmosphere. Meteorit. Planet. Sci. **38**, 1023–1043 (2003)

R. Bulirsh, J. Stoer, Numerical treatment of ordinary differential equations by extrapolation methods. Numer. Math. **8**, 1–13 (1966)

Z. Ceplecha, R.E. McCrosky, Fireball end heights - a diagnostic for the structure of meteoric material. J. Geophys. Res. **81**, 6257–6275 (1976)

Z. Ceplecha, D.O. ReVelle, Fragmentation model of meteoroid motion, mass loss, and radiation in the atmosphere. Meteorit. Planet. Sci. **40**, 35–54 (2005)

J.E. Chambers, A hybrid symplectic integrator that permits close encounters between massive bodies. Mon. Not. R. Astron. Soc. **304**, 793–799 (1999)

J.D. Drummond, A test of comet and meteor shower associations. Icarus **45**, 545–553 (1981)

M.J. Duncan, H.F Levison, S.M. Budd, The dynamical structure of the Kuiper Belt. Astron. J. **110**, 3073–3081 (1995)

M.J. Duncan, H.F. Levison, M.H. Lee, A multiple time step symplectic algorithm for integrating close encounters. Astron. J. **116**, 2067–2077 (1998)

P. Farinella, R. Gonczi, Ch. Froeschlé, Cl. Froeschlé, The injection of asteroid fragments into resonances. Icarus **101**, 174–187 (1993)

I. Halliday, Detection of a meteorite 'stream' - observations of a second meteorite fall from the orbit of the Innisfree chondrite. Icarus **69**, 550–556 (1987)

I. Halliday, A.T. Blackwell, A.A. Griffin, The Innisfree meteorite and the Canadian camera network. J. R. Astron. Soc. Can **72**, 15–39 (1978)

I. Halliday, A.T. Blackwell, A.A. Griffin, Evidence for the existence of groups of meteorite producing asteroidal fragments. Meteoritics **25**, 93–99 (1990)

I. Halliday, A.A. Griffin, A.T. Blackwell, The Innisfree meteorite fall - a photographic analysis of fragmentation, dynamics and luminosity. Meteoritics **16**, 153–170 (1981)

I. Halliday, A.A. Griffin, A.T. Blackwell, Detailed data for 259 fireballs from the Canadian camera network and inferences concerning the influx of large meteoroids. Meteorit. Planet. Sci. **31**, 185–217 (1996)

G. Heusser, W. Hampel, T. Kirsten, O.A. Schaeffer, Cosmogenic isotopes in recently fallen meteorites. Meteoritics **13**, 492–493 (1978)

P. Jenniskens, 2003 EH_1 is the Quadrantid shower parent comet. Astron. J. **127**, 3018–3022 (2004)

D.C. Jones, I.P. Williams, M.D. Melita, The dynamics of objects in the inner Edgeworth-Kuiper Belt. Earth Moon Planets **97**, 435–458 (2005)

D.C. Jones, I.P. Williams, V. Porubčan, The Kappa Cygnid meteoroid complex. Mon. Not. R. Astron. Soc. **371**, 684–694 (2006)

L. Kornoš, J. Tóth, Orbital evolution of Příbram and Neuschwanstein. Earth Moon Planets. (This Volume) (2007)

A.K. Lavrukhina, A.V. Fisenko, E.M. Kolesnikov, Multiple fall of Příbram meteorites photographed: II. Preatmospheric size and radiation age of the Příbram chondrite. Bull. Astron. Inst. Czech. **25**, 122–126 (1974)

M.E. Lipschutz, S.F. Wolf, R.T. Dodd, Meteoroid streams as sources for meteorite falls: a status report. Planet. Space Sci. **45**, 517–523 (1997)

R.E. McCrosky, C.-Y. Shao, A. Posen, Prairie Network fireballs. I - general information and orbits. Meteoritika **37**, 44–59 (1978)

R.E. McCrosky, C.-Y. Shao, A. Posen, The Prairie network bolide data. II - trajectories and light curves. Meteoritika **38**, 106–156 (1979)

A. Pauls, B. Gladman, Decoherence time scales for "meteoroid streams". Meteorit. Planet. Sci. **40**, 1241–1256 (2005)

V. Porubčan, I.P. Williams, L. Kornoš, Associations between asteroids and meteoroid streams. Earth Moon Planets **95**, 697–711 (2004)

G.W. Wetherill, Multiple cosmic-ray exposure ages of meteorites. Meteoritics **15**, 386–387 (1980)

I.P. Williams, The dynamics of meteoroid streams, in *Chaos, resonance, and collective dynamical phenomena in the Solar System: Proceedings of the 152nd symposium of the IAU*, ed. by S. Ferraz-Mello (Kluwer, Dordrecht, Netherlands, 1992), pp. 299–313

I.P. Williams, G.O. Ryabova, A.P. Baturin, A.M. Chernitsov, The parent of the Quadrantid meteoroid stream and asteroid 2003 EH1. Mon. Not. R. Astron. Soc. **355**, 1171–1181 (2004)

S.F. Wolf, M.E. Lipschutz Meteoroid streams: Evidence for meteorites recovered on Earth. Earth Moon Planets **68**, 605–637 (1995)

Z. Wu, I.P. Williams, The Perseid meteor shower at the current time. Mon. Not. R. Astron. Soc. **264**, 980–990 (1993)

Motion of a Meteoroid Released from an Asteroid

Peter Vereš · Jozef Klačka · Ladislav Kómar · Juraj Tóth

Originally published in the journal Earth, Moon, and Planets, Volume 102, Nos 1–4.
DOI: 10.1007/s11038-007-9187-x © Springer Science+Business Media B.V. 2007

Abstract Evidence of asteroid surface features as regolith grains and larger boulders implies resurfacing possibility due to external forces such as gravitational tidal force during close planet encounters. Motion of a meteoroid released from an asteroid in the gravitational fields of the asteroid and the Earth is modeled. We are interested mainly in a distance between the meteoroid and the asteroid as a function of the time. Applications to Itokawa and some close approaching NEAs are presented.

Keywords Asteroid · Meteoroid release · Tidal force · Itokawa · Interplanetary dust particles

1 Introduction

There is evidence of many fine structures on the surface of small asteroids recently imaged by targeted space missions (Gaspra, Ida, Mathilde, Eros, Itokawa). The surface is covered by fine-grained regolith, rocks and boulders and has various features such as impact craters, grooves etc. The study of the physical properties of asteroid gives us information about porosity, internal structure, collisional history and origin. It has been suggested by the observation evidence of "spin rate barrier" at 2.2 h (Harris and Pravec 2006) that most of kilometer-size asteroids are rubble piles. Small monolithic fast rotating asteroids should not have regolith grains on their surface because of the higher centrifugal force relative to the asteroid's gravity. If other asteroids of comparable sizes and spin-rates as Itokawa were rubble piles they should have a regolith cover that could be easily released from the asteroid surface by external forces, such as gravitational tidal force from a planet, collision and YORP effect spin period speed-up. Sufficient tidal force during close approaches of such an asteroid with the planet could affect the asteroid to change shape or even disrupt it (Richardson et al. 2001).

P. Vereš · J. Klačka · L. Kómar · J. Tóth (✉)
Department of Astronomy, Physics of the Earth and Meteorology, Faculty of Mathematics, Physics and Informatics, Comenius University, 842 48 Bratislava, Slovakia
e-mail: toth@fmph.uniba.sk

As Olivier (1925) and Hoffmeister (1937) had already proposed such asteroids could be sources of meteoroids (from dust particles to boulders). Some of these asteroids could be dormant cometary nuclei with occasional comet-like activity, that is the formation of a transient coma (133P/Elst-Pizarro, 107P/Wilson-Harrington, 174P/Echeclus, 176P/Linear). Tidal disruption or collision among asteroids could also produce asteroidal meteoroids. Our goal is to investigate the mechanism of gravitational tidal force acting on an asteroid during its close encounter with the Earth. Asteroids close encounters are in some cases semi-periodic and could produce short and narrow streams and dust trails. At a certain distance the Earth's gravitational field could cause a change of the shape of the asteroid surface thereby relocating grains, rocks or boulders that could be released into interplanetary space as meteoroids. We searched for the minimum perpendicular distance of an asteroid from the Earth (impact parameter b) where surface particles would start to move and a distance where particles could escape the asteroid's gravitational field.

2 Model

We considered an asteroid of spherical shape with homogeneous density covered by regolith and one spin axis that is perpendicular to the plane of asteroid motion. An orthogonal coordinate system is set to the center of the Earth, which reduces the problem to two-dimensional space. We searched for a value of the impact parameter b, where a meteoroid of any size could escape the gravitational field of the asteroid. Solving the motion equation of the meteoroid due to the gravitational influences of the asteroid and the Earth, we get the time dependence of meteoroid distances from asteroidal mass center. Equations of motion are given by the following two expressions:

$$\frac{\mathrm{d}^2\mathbf{R}_A}{\mathrm{d}t^2} = -\frac{GM_p}{|\mathbf{R}_A|^3}\mathbf{R}_A, \tag{1}$$

$$\frac{\mathrm{d}^2\mathbf{R}}{\mathrm{d}t^2} = -\frac{GM_p}{|\mathbf{R}+\mathbf{R}_A|^3}(\mathbf{R}+\mathbf{R}_A) + \frac{GM_p}{|\mathbf{R}_A|^3}\mathbf{R}_A - \frac{GM_A}{|\mathbf{R}|^3}\mathbf{R}, \tag{2}$$

where $\mathbf{R}$ is the position vector of the meteoroid with respect to the asteroid center and $\mathbf{R}_A$ is the position vector of the asteroid with respect to the planet (Earth), M_p is the mass of the planet (Earth), M_A is the mass of the asteroid and G is the gravitational constant. Equation 2 is solved by numerical integration with an input of $\mathbf{R}_A$ obtained from the analytical solution of equation (1), which describes the motion of an asteroid on a hyperbolic orbit in the Earth's gravitational field.

A test meteoroid particle was placed on the equator of the rotating asteroid. We considered both prograde and retrograde rotation. At the large distance from the Earth $\mathbf{r}$, asteroid moves at velocity $\mathbf{v}$ along a line at a perpendicular distance b from the Earth's center of gravity. Also it is possible that lofted asteroid surface debris falls back onto the surface if it receives insufficient tidal pull to leave its gravitational field. In this case we consider the fall and the bounce of the debris as an inelastic collision. The fraction of the kinetic energy left to bounced meteoroid is small due to character of the regolith. The deflection of the meteoroid will be under the same angle to the vertical as it strikes the surface of the asteroid.

2.1 Specification of the Model

At first we studied a model body that is similar to Itokawa with following physical properties: mass (3.58 ± 0.18) × 10^{10} kg, effective diameter D = 326 m, spin period 12.132 h, heliocentric velocity 33.2 km/s and geocentric velocity 6 km/s (Abe et al. 2006; Fujiwara et al. 2006). We investigated the distance between the asteroid and the meteoroid as the function of time while the asteroid is approaching the Earth. We assumed that a meteoroid left the surface but very soon fell back at a different location on the equator. We assumed for the regolith surface that the fraction of the kinetic energy left to bounced meteoroid is 0.01. As the second step we have chosen several close approaching NEAs (Near Earth Asteroids) with a known spin period to investigate the possibilities of particles release and we proposed possible meteor radiants for such meteoroid bodies using software developed by Neslušan et al. (1998) and also Klačka (1999).

3 Results

The investigation using Itokawa's orbital and physical properties revealed that the impact parameter for particle escape from Itokawa is at 1,765,900 km, which is 4.5 lunar distances (mean Earth–Moon distance) and much more farther than Roche limit for rigid Itokawa (~ 1.8 Earth radii). A small increase or decrease of this value would change the particle release behavior (Fig. 1). In range of 1,000 km around 1,765,900 km there are cases ranging from small, single jumps to clear debris escape (steps A–F). Dynamics of particle lift-off could be very interesting. Since in 1905 Itokawa come close to 4.5 lunar distances from the Earth, potential meteoroid release should be considered possible. There is an observational evidence by Hayabusa space mission, that regolith moves in time and also boulders chance their position (Abe, pers. comm. 2006). We did not investigate total amount of matter escaped from the asteroid. The object spends limited time closer than impact parameter, limited amount of dust and particles will leave the surface. Although the asteroid would have undergone many close encounters with the Earth in the past (Yoshikawa and Michel 2006) there is still large amount of regolith on the surface. We investigate several close NEA encounters in order to determine if there might have been particle lift-off due to Earth's tidal force. We mainly studied NEAs with known spin periods that passes by the Earth within 10 lunar distances (source: http://www.neo.jpl.nasa.gov/neo/close.html). For NEAs without a known spin period we assumed spin period cut-off value of 2.2 h. Such bodies with diameters of several hundreds of meters have a centrifugal force lower than gravitational force and there is the possibility that the surface is covered by regolith. According to our simulations spin period affects particle lift-off behavior only slightly and that is why we did not take rotation spin-up/spin-down caused by tidal force into account. We found several candidates (Table 1), where particles could have left surface and even escaped the NEA gravitational field due to the Earth's tidal force. Such particles would spread due to different escape velocities, gravitational perturbations and non-gravitational forces, which could have created a short and narrow meteor stream. Its predicted radiant positions and time of maximum activity is shown in (Table 1). Meteors from such sources should have very low geocentric velocities (< 20 km/s), which can disadvantage the observation of meteors in the Earth atmosphere, but they might be present among the interplanetary dust particles that are collected in the Earth's atmosphere (Rietmeijer 2000; Rietmeijer and Jenniskens 2000; Rietmeijer and Warren 1994).

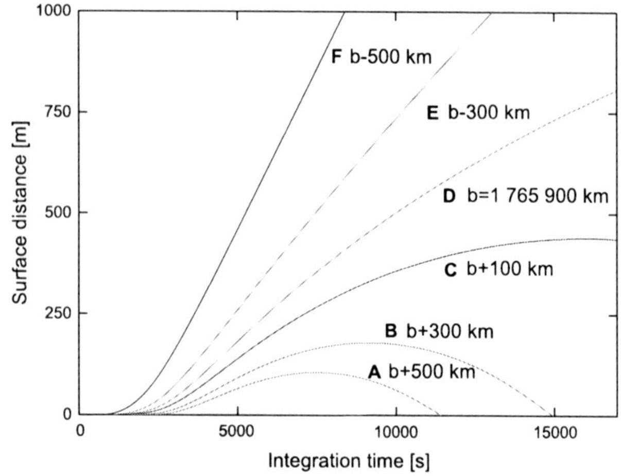

Fig. 1 Numerical simulation results: distances of released meteoroids from Itokawa as the function of time and impact parameter b; C curve corresponds to parabolic solution b = 1,765,900 km, range of A–F curves is within 1000 km around C curve solution

Table 1 Selected possible parent bodies of putative asteroidal meteor streams due to the Earth tidal force

Object [design.]	b[LD]	Date	H(1,0)	V_{geo} [km/s]	α [deg]	δ [deg]	$\lambda_{\odot}$ [deg]
2002 FD6	1.45	1911-Apr-06	22.2	11.36	202.0	21.6	17.0
2002 MN	0.3	2002-Jun-14	23.3	10.29	114.2	25.1	83.0
2005 XA8	0.6	2005-Dec-05	25.6	12.08	263.3	− 8.3	253.2
2006 DD1	0.3	2006-Feb-23	26.5	17.07	138.1	− 1.7	334.2
2007 JB21	1.6	2007-May-09	25.4	7.67	347.0	77.6	228.1
25143 Itokawa	4.5	1905-Jun-27	19.2	6.00	166.7	12.4	14.6

b-Impact parameter, date of close encounter, H(1,0)—Absolute magnitude of the asteroid, V_{geo}—Geocentric velocity, radiant coordinates (α, δ), $\lambda_{\odot}$—Solar longitude of maximum activity

4 Conclusions

We found that some asteroids during close encounters with the Earth could produce meteoroids when regolith particles, rocks or boulders were released from the surface by the gravitational influence of the Earth. We proposed possible radiants and activity time of some of these putative asteroidal meteor streams, despite there is a strong negative selection effect of observation because of their low geocentric velocities. Observation of meteors from such sources could confirm existence of asteroidal meteor streams and the way how they form.

Changing minimum distance during Earth encounter we found various behavior of meteoroid motion, not only meteoroid release, but also its position change on the surface of the parent NEA. Further investigation of particle lift-off behavior may bring interesting results. We suggest the tidal force of the Earth as an important mechanism of regolith movement on the asteroid and even responsible for asteroidal meteoroids and dust particles production.

Acknowledgments This work was supported by VEGA—the Slovak Grant Agency for Science (grant No. 1/3067/06 and 1/3074/06) and by Comenius University grant UK/379/2007. Authors are thankful to prof. Frans J. Rietmeijer for valuable suggestions and to reviewers for useful comments.

References

S. Abe, T. Mukai, N. Hirata, O.S. Barnouin-Jha, A.F. Cheng, H. Demura, R.W. Gaskell, T. Hashimoto, K. Hiraoka, T. Honda, T. Kubota, M. Matsuoka, T. Mizuno, R. Nakamura, D.J. Scheeres, M. Yoshikawa, Mass and local topography measurements of Itokawa by Hayabusa. Science **312**, 1344–1347 (2006)

A. Fujiwara, J. Kawaguchi, D.K. Yeomans, M. Abe, T. Mukai, T. Okada, J. Saito, H. Yano, M. Yoshikawa, D.J. Scheeres, O. Barnouin-Jha, A.F. Cheng, H. Demura, R.W. Gaskell, N. Hirata, H. Ikeda, T. Kominato, H. Miyamoto, A.M. Nakamura, R. Nakamura, S. Sasaki, K. Uesugi, The rubble-pile asteroid Itokawa as observed by Hayabusa. Science **312**, 1330–1334 (2006)

A.W. Harris, P. Pravec, Rotational properties of asteroids, comets and TNOs, in Proceedings of the 229th Symposium of the International Astronomical Union, ed. by D. Lazzaro, S. Ferraz-Mello,J.A. Fernández. Asteroids, Comets, Meteors, Búzios, Rio de Janeiro, Brasil, August 7–12, 2005 (Cambridge University Press, Cambridge, 2006), pp. 439–447

C. Hoffmeister, Die Meteore (in German) (Akademische Verlagsgesellschaft, Leipzig, 1937), p. 154

J. Klačka, Meteor Streams and Parent Bodies. astro-ph/9910044 (1999)

L. Neslušan, J. Svoreň, V. Porubčan, A computer program for calculation of a theoretical meteor-stream radiant. Astron. Astrophys. **331**, 411–413 (1998)

C.P. Olivier, Meteors. The Williams and Wilkins Company, Baltimore, 1925), p. 276

D.C. Richardson, Z.M. Leinhardt, H.J. Melosh, W.F. Bottke, E. Asphaug, gravitational aggregates: evidence and evolution, in Asteroids III, ed. by W.F. Bottke, A. Cellino, P. Paolicchi, R.P. Binzel (University of Arizona Press, Tucson, 2001), pp. 501–515

F.J.M. Rietmeijer, Interrelationships among meteoric metals, meteors, interplanetary dust, micrometeorites, and meteorites. Meteorit. Planet. Sci. **35**, 1025–1041 (2000)

F.J.M. Rietmeijer, P. Jenniskens, Recognizing Leonid meteoroids among the collected stratospheric dust, in Leonid storm research, ed. by P. Jenniskens, F.J.M. Rietmeijer, N. Brosch, M. Fonda (Kluwer Acad. Publ., Dordrecht, 2000), pp. 505–524

F.J.M. Rietmeijer, J.L. Warren, Windows of opportunity in the NASA Johnson Space Center Cosmic Dust Collection, in Analysis of interplanetary dust, ed. by M.E. Zolensky, T.L Wilson, F.J.M. Rietmeijer, G.J. Flynn. Amer. Inst. Physics Conf. Proc. 310 (Amer. Inst. Physics Press, Woodbury, 1994), pp. 255–275

M. Yoshikawa, P. Michel, Orbital evolution of asteroid (25143) Itokawa: it's past, present, and future, in 37th annual Lunar and Planetary Science Conference, League City, Texas, March 13–17, 2006, pp. 1545–1546

Searching for the Parent of the Tunguska Cosmic Body

Tadeusz Jan Jopek · Christiane Froeschlé · Robert Gonczi · Piotr A. Dybczyński

Originally published in the journal Earth, Moon, and Planets, Volume 102, Nos 1–4.
DOI: 10.1007/s11038-007-9156-4

Abstract We present the results of an extensive study of the Tunguska Cosmic Body (TCB) origin on dynamical grounds. To identify the TCB parent, or a plausible candidate, we applied the well-known concept of dynamical similarity whereby we have compared the geocentric and heliocentric dynamical parameters of a selected set of the Near Earth Objects (NEOs) and TCB particles. First, we made use the idea of Kresak by comparing geocentric coordinates of the TCB radiant with those of the NEOs. Second, we studied the long-term dynamical evolution of all NEOs and TCB particles searching for similarities between their heliocentric orbits. As a general result, we observed many more similar cases and a different pattern of the high orbital similarity among the TCB particles and the asteroid orbits than we did for comets.

Keywords Meteoroids · Minor planets · Comets · Methods · Numerical

1 Introduction

In papers by Trayner (1997), Vasilyev (1998) and Bronshten (2000), one can read that in 1926 and 1933 L.A. Kulik the scientific leader of the first scientific expedition to the Tunguska region (see e.g. Kulik 1927a, b, 1933, 1938) favoured 7P/Pons-Winnecke as the parent body of the TCB, in 1966 Fesenkov wrote about a similarity between the TCB and comet Mrkos. Zotkin (1969), and 9 years later Kresak (1978) (see also Asher and Steel 1998), hypothesised that the TCB was a fragment of comet 2P/Encke. The cometary nature of the TCB was strongly criticised by Sekanina (1983, 1998). Sekanina (1983) concluded that the body was not a comet but was most likely a small Apollo-type asteroid of 90–190 m across, with a probable mass of 10^{12}–10^{13} g, and a material density ~ 3 g/cm^3. This Sekanina hypothesis was criticised by Levin and Bronshten (1986) and Bronshten

T. J. Jopek (✉) · P. A. Dybczyński
Obserwatorium Astronomiczne UAM, ul Sloneczna 36, Poznan 60-286, Poland
e-mail: jopek@amu.edu.pl

C. Froeschlé · R. Gonczi
Observatoire de la Côte d'Azur, B.P. 4229, Nice 06304, France

(2000). Levin and Bronshten (1986) submitted that if TCB was a stony asteroid (density of ~3 g/cm^3), it would be subjected to progressive braking producing many solid fragments with a wide range of dimensions. When impacting the Earth the largest of those fragments would form impact craters or pits, but no traces of those were ever found. Chyba et al. (1993) suggested that the TCB was probably a fairly strong and dense asteroid-like object (60 m in diameter), but probably not as strong or dense as iron meteorites. Andreev (1990) using a small set of possible TCB orbits carried out an analysis that suggested the TCB was an Apollo asteroid. Bronshten (1999a) obtained a small set of possible TCB orbits consistent with a cometary origin, and a larger sample set consistent with Apollo like asteroids. Since neither macroscopic remnants nor craters were found, and considering mechanisms of disintegration and plausible estimates of physical parameters, such as the initial mass, initial velocity and energy of explosion, Bronshten (1999b, 2000) concluded that the TCB was of cometary nature. Farinella et al. (2001) who estimated the probabilities that the TCB came from a cometary source and an asteroidal source, found that only 17% of TCB orbits were of cometary origin and 83% of asteroidal sources. These results agreed with the findings of Andreev. It is clear that no single statistical approach will give a definite answer on the TCB nature.

Despite a low fraction of the cometary orbits among the possible TCB orbits, we believe there is still a fair chance that the TCB origin was associated with a comet disruption. In this paper we made an extensive study of the TCB origin based on the dynamical evolution of 3,311 possible TCB particles and 3,238 of NEOs (Near Earth Asteroids (NEAs) plus comets). Assuming a priori, that the TCB was a fragment resulting from partial disruption of a NEA or a comet, the aim of our paper was to find the possible parent bodies of the TCB.

2 The NEO and Possible TCB Orbital Data

The 2,656 NEA orbits were taken from NEO Dynamic Site (2004). We took the cometary data from Marsden and Williams (2003). The possible TCB orbits we calculated using the parameters given in Table 1. In order to make our study more extensive, we used larger intervals of these parameters than in the paper by Farinella et al. (2001). As in the present study the motion of particles was integrated, the TCB parameters were not corrected for the zenithal attraction of the Earth. All NEO and TCB orbits with $e \geq 1$ were excluded from our considerations. For all comets with the multi-apparitions in our sample we took only one apparition that was closest to the TCB event. For comet Biela we used all apparitions. In this manner we used the orbits of 582 comets, 2,656 NEAs and 3,311 TCB particles. The NEO orbits were pre-integrated to the epoch of the TCB event, 1908 06 30, 00:13:35 UT, (TDB JD = 2418122.509531).

3 Methods and Results

First we repeated the study of Kresak (1978), but instead of using the ecliptical radiant coordinates we applied two variables, that is, the geocentric speed U and its elongation θ from the Earth apex. Due to the invariant property of U and θ (Öpik 1976; Valsecchi et al. 1999), this approach allows us to find more candidates for the TCB parent. In Fig. 1 we show 1,375 NEOs and all TCB particles used in our study. Several points are obvious:

Table 1 The dynamic parameters of the Tunguska body adopted in this paper

Time (UT)	1908 06 30, $00^h\ 13^m\ 35^s$	
Location	60 deg 53′09″ N, 101 deg 53′40″ E	
Explosion altitude [km]	$E = 8.5$	
Azimuth [deg]	$A \in (97, 127)$	$\Delta A = 5.0$
Altitude [deg]	$a \in (3,28)$	$\Delta a = 1.0$
Velocity [km/s]	$V \in (14, 32)$	$\Delta V = 1.0$

The observed radiant horizontal coordinates A, a, V speed and E the explosion altitude above the Earth surface are given. The location coordinates refer to epicenter point of the explosion. The third column gives the grid steps in azimuth, altitude and velocity

- in the area occupied by the TCB particles we have many more NEAs than comets.
- with few exceptions, the comets and NEAs are quite well separated by the line corresponding to the orbits for which the semi-major axis $a = 3$ AU. And, as one can see, among the candidates for the TCB parent, we have also many asteroids with geocentric velocities close to 30 km/s.
- contrary to the results obtained by Kresak (1978), our analyses show comet Encke is seen in the diagram of Fig. 1.

The coincidence of the TCB and NEO points in the U–θ plane at the time of the Tunguska event, gives no certainty that at some time in the past these objects were a single body. The U–θ plane has a semi-invariant property only, and a pair of objects in AD 1908 that we are looking for may be placed at different locations in this plane.

The hypothesis that two objects once constituted a single body and that in some epoch they had separated with small relative velocities may be partly verified by the methods developed for searching the meteor streams and their parent bodies. That is, by checking, if their orbits are very similar and if the MOID (Minimum Orbital Intersection Distance) of these orbits has a small value. Checking may be carried out during the numerical

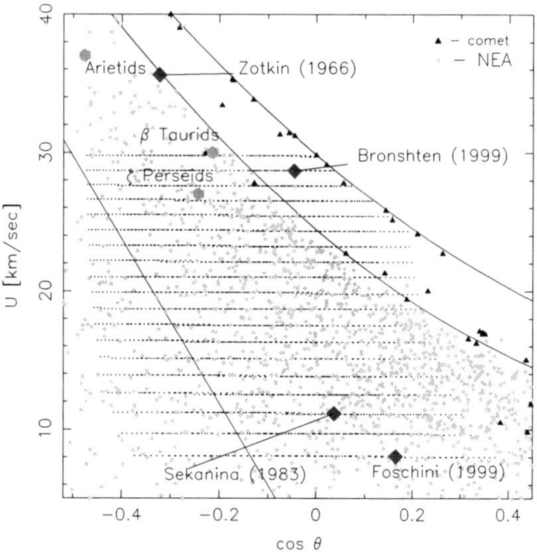

Fig. 1 Thirty-five comets, 1,340 asteroids and 3,311 TCB particles occupying similar region on the U–θ plane (see text). Comets are marked as solid triangles; the NEA's as small solid circles, the area of the TCB points is represented by parallel dotted lines. Also the positions of three major daytime meteor streams and four the possible TCB radiants are indicated. Comet Encke is shown very close to the left of the β Taurids meteor stream. The solid lines correspond to the parabolic limit and to two constant semi-major axes (3 AU; 1 AU). This figure is an extended modified version after Farinella et al. (2001)

integration of the motion of all the TCB and NEO objects. It is clear that by such approach we are testing only two necessary conditions of the hypothesis. Therefore all NEOs found in this way will have the status of being a candidate for the TCB parent. We made such a test in the second part of our study:

- starting from AD 1908, we integrated all NEO and TCB particles for 20 Kyrs in the past,
- for consecutive 20-year intervals, using D_{SH}-criterion by Southworth and Hawkins (1963) we determined the orbital similarity among all TCB–NEO pairs and the values their MOIDs. Also, every 20 years we searched for the closets similarity amongst all TCBs and NEOs. All pairs with the smallest D-values were analysed in details.

In general, we observed many more similar cases and a different pattern of high orbital similarity among TCB and NEA orbits than among TCB and comet orbits. Using the threshold $D_c = 0.1$ we found 125 of NEAs very similar to one of the TCB orbits and only four comets with such similarity, and this result is in agreement with the paper Farinella et al. (2001).

At each of 1,000 of the intermediate epochs the smallest D-value always occurred amongst the NEA and TCB orbits (mostly we observed $D < 0.05$). In the year 932 BC we found the minimum value $D = 0.0237$ among 2000 WK63 and TCB particle No 2207 ($A = 97°$, $a = 26°$, $V = 26$ km/s). In the year 632 BC the same object proved to be very similar ($D = 0.0306$) to TCB particle No 2208 ($A = 97°$, $a = 27°$, $V = 26$ km/s). In Fig. 2,

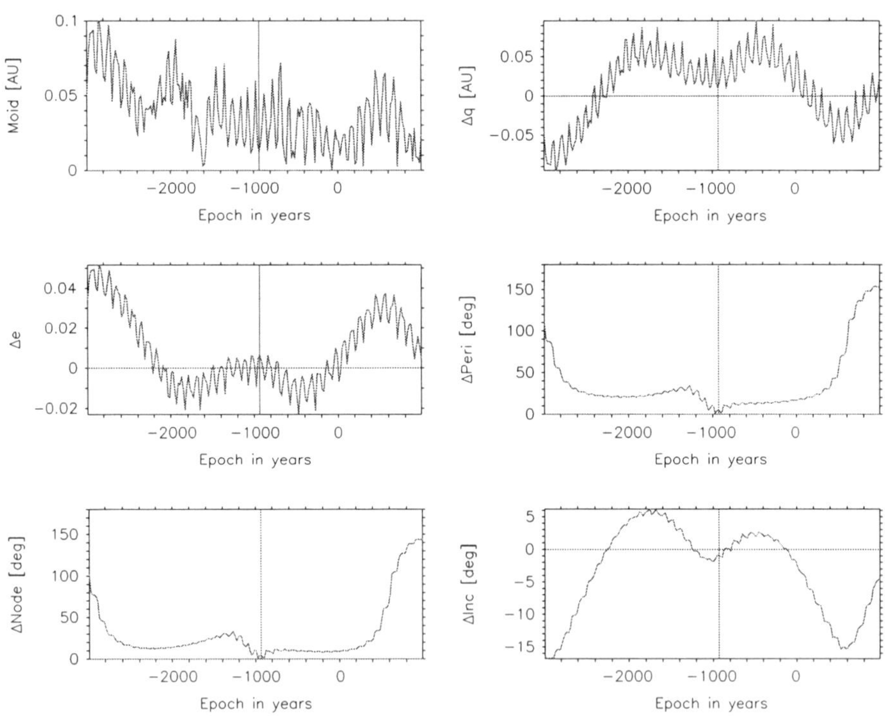

Fig. 2 The case of the highest orbital similarity: the TCB particle No 2207 and the NEA 2000 WK63. We plotted the values of their MOIDs and the orbital element differences during the past 4 Kyrs. A vertical line indicates the epoch of the highest orbital similarity

the MOIDs, and differences between the orbital elements are plotted for 2000 WK63 and TCB particle No 2207. In the vicinity of the epoch of greatest similarity Δq, Δe, $\Delta\omega$, $\Delta\Omega$, Δi are very small. The same type of the orbital coincidences we observed for more than 20 other NEAs and TCB particles. Also we have found that for some pairs of bodies, a very high orbital similarity was preserved for several hundred years.

4 Conclusions

In this study of the origin of the Tunguska body we have applied two methods based on the comparison of the geocentric or heliocentric parameters of 3,248 NEOs and 3,311 TCB particles. We did not find the TCB parent body. We found that from 20 years to 20 Kyrs in the past about 130 NEOs moved on the orbits that were highly similar (D-criterion) to the orbits of some of the TCB particles. Comet 2P/Encke cannot be linked to the TCB event but a dozen comets moved on orbits more similar to some of the TCB particles. In 16192 BC, TCB particle No 2205 and comet 97P/1906 V2 were on highly similar orbits ($D = 0.0701$). In 932 BC, 2000 WK63 and TCB particle No 2207 were on highly similar orbits ($D = 0.0237$). We showed that an NEA does not necessarily imply a low orbital velocity for the Tunguska Cosmic Body that could have been as high as 30 km/s.

Acknowledgements TJJ was partly supported by the KBN Project 2 PO3D 007 22, and his visit in Nice was funded by the Observatoire de la Côte d'Azur. The authors wish to thank Professor Frans J.M. Rietmeijer for his suggestions for the improvement of this paper. Also we acknowledge referees David Asher and the anonymous one for their comments and for meticulously checking the details of the manuscript.

References

G.V. Andreev, Was the Tunguska 1908 event caused by an Apollo asteriods? in *Asteriods, Comets, Meteors III*, ed. by C.I. Lagerkvist, H. Rickman, B.A. Lindblad. in *Proceedings of a meeting held at the Uppsala University*, 12–16 June 1989 (1990), pp. 489–492

D.J. Asher, D.I. Steel, On the possible relation between the Tinguska bolide and comet Encke. Planet. Space Sci. **46**, 205–211 (1998)

V.A. Bronshten, Trajectory and orbit of the Tunguska meteorite revisited. Meteorit. & Planet. Sci. **34**, A137–A143 (1999a)

V.A. Bronshten, The nature of the Tunguska meteorite. Meteorit. & Planet. Sci. **34**, 723–728 (1999b)

V.A. Bronshten, Nature and destruction of the Tunguska cosmical body. Planet. Space Sci **48**, 855–870 (2000)

Ch.F. Chyba, P.T. Thomas, K.J. Zahnle, The 1908 Tunguska explosion: atmospheric disruption of a stony asteroid. Nature **361**, 40–44 (1993)

P. Farinella, L. Foschini, Ch. Froeschlé, R. Gonczi, T.J. Jopek, G. Longo, P. Michel, Probable asteroidal origin of the Tunguska Cosmic Body. Astron. & Astrophys. **377**, 1081–1097 (2001)

L. Foschini, A solution for the Tunguska event. Astron. & Astrophys. **342**, L1–L4 (1999)

L. Kresak, The Tunguska object: a fragment of comet Encke? Bull. Astron. Inst. Czechoslov. **29**, 129–134 (1978)

L. Kulik, On the History of the Bolide of 1908 June 30. J. Russian Acad. Sci. **1927A**, 393–398 (1927a)

L. Kulik, On the fall of the podkamennaya tunguska meteorite in 1908. J. Russ. Acad. Sci. **1927A**, 399–402 (1927b)

L. Kulik, Preliminary results of the meteorite expeditions made in the decade 1921–1931. Works Lomonossoff Inst. Russ. Acad. Sci. **2**, 73–80 (1933)

L. Kulik, The meteorite of June 30, 1908 in Central Siberia. Astron. Soc. Pac. Leaf. **109**, 78–84 (1938)

B.Y. Levin, V.A. Bronshten, The Tunguska event and the meteors with terminal flares. Meteoritics **21**, 199–215 (1986)

B.G. Marsden, G.V. Williams, *Catalogue of Cometary Orbits 2003*, 15th edn. (Smithsonian Astrophysical Observatory, Cambridge, MA, 2003)
NEODYS site, (2004) http://newton.dm.unipi.it/cgi-bin/neodys/neoibo
E.J. Öpik, *Interplanetary Encounters* (Elsevier, New York, 1976), pp. 155
Z. Sekanina, The Tunguska event: no cometary signature in evidence. Astron. J. **88**, 1382–1414 (1983)
Z. Sekanina, Evidence for asteroidal origin of the Tunguska object. Planet. Space Sci. **46**, 191–204 (1998)
R.B. Southworth, G.S. Hawkins, Statistics of meteor streams, Smithson. Contr. Astrophys. **7**, 261–285 (1963)
Ch. Trayner, The Tunguska event. J. Br. Astron. Assoc. **107**, 117–130 (1997)
G.B. Valsecchi, T.J. Jopek, Cl. Froeschlé, Meteoroid stream identification: a new approach-I. Theory. Mont. Not. R. Astron. Soc. **304**, 743–750 (1999)
N.V. Vasilyev, The Tunguska problem today. Planet. Space Sci. **46**, 129–150 (1998)
I.T. Zotkin, Trajektorja i orbita Tungusskogo meteorita. Meteoritika **27**, 109–118 (1966) (in Russian)
I.T. Zotkin, Anomalnyje sumierki cvjazannyje s Tungusskim meteoritom. Meteoritika **29**, 170–176 (1969) (in Russian)

Orbital Evolution of Příbram and Neuschwanstein

Leonard Kornoš · Juraj Tóth · Peter Vereš

Originally published in the journal Earth, Moon, and Planets, Volume 102, Nos 1–4.
DOI: 10.1007/s11038-007-9213-z

Abstract The orbital evolution of the two meteorites Příbram and Neuschwanstein on almost identical orbits and also several thousand clones were studied in the framework of the N-body problem for 5,000 years into the past. The meteorites moved on very similar orbits during the whole investigated interval. We have also searched for photographic meteors and asteroids moving on similar orbits. There were five meteors found in the IAU MDC database and six NEAs with currently similar orbits to Příbram and Neuschwanstein. However, only one meteor 161E1 and one asteroid 2002 QG46 had a similar orbital evolution over the last 2,000 years.

Keywords Meteorite · Meteoroid · Asteroid · Příbram · Neuschwanstein

1 Introduction

It is almost 50 years since the fall (April 7, 1959) and recovery of the Příbram meteorite (Ceplecha 1961), the first meteorite with a precisely known heliocentric orbit (Table 1). Later, the fall of the Neuschwanstein meteorite was observed on April 6, 2002 and it was successfully recovered (Oberst et al. 2004). It was shown that both meteorites were moving on similar orbits (Spurný et al. 2003), but the question about their origin remains unanswered. Moreover, their different meteoritic types, Příbram being an H5 ordinary chondrite (Ceplecha 1961) with cosmic-ray exposure age 12 Myr (Stauffer and Urey 1962) and Neuschwanstein an EL6 enstatite chondrite with cosmic-ray exposure age 48 Myr (Bishoff and Zipfel 2003; Zipfel et al. 2003), makes their common origin very problematic. It is a challenge for the scientific community to explain the dynamical and physical evolution of these two meteorites. Earlier, the existence of asteroidal-meteoritic streams was suggested by Halliday et al. (1990). Recently, the observation of Neuschwanstein led Spurný et al. (2003) to suggest a heterogeneous meteoritic stream in the orbit of Příbram. On the other

L. Kornoš · J. Tóth (✉) · P. Vereš
Department of Astronomy, Physics of the Earth and Meteorology, Faculty of Mathematics, Physics and Informatics, Comenius University, Mlynská dolina, 842 48 Bratislava, Slovak Republic
e-mail: toth@fmph.uniba.sk

DOI: 10.1007/978-0-387-78419-9_9

Table 1 Orbital elements (eq. 2000.0) of Příbram and Neuschwanstein (Spurný et al. 2003)

	Příbram	Neuschwanstein
a	2.401 ± 0.002 AU	2.40 ± 0.02 AU
e	0.6711 ± 0.0003	0.670 ± 0.002
q	0.78951 ± 0.00006 AU	0.7929 ± 0.0004 AU
Q	4.012 ± 0.005 AU	4.01 ± 0.03 AU
ω	241.750° ± 0.013°	241.20°± 0.06°
Ω	17.79147°± 0.00001°	16.82664° ± 0.00001°
i	10.482° ± 0.004°	11.41° ± 0.03°

hand, a statistical analysis by Pauls and Gladman (2005) showed that the occurrence of pairs as close as Příbram and Neuschwanstein is at the 10% level, which is consistent with random chance. Recently, Jones and Williams (2007) studied the possible existence of meteoritic streams. Trigo-Rodríguez et al. (2007), performing orbital and spectral analyses, found three meteorite-dropping bolides, which may well be associated with the Near Earth Asteroid 2002 NY40. In the present paper, we analyze possible associations of meteors and NEAs with Příbram and Neuschwanstein and also their orbital evolutions on a time scale of 5,000 years. Also, we discuss the possible common origin of Příbram and Neuschwanstein.

2 Associations with Příbram and Neuschwanstein

The heliocentric orbits of Příbram and Neuschwanstein are almost identical (Table 1), but the errors in the orbital elements of Neuschwanstein are about 1 order of magnitude larger compared to Příbram. However, both orbits are relatively precise and the D-criterion of Southworth and Hawkins (1963), $D_{SH} = 0.03$, indicates a very close similarity.

We have searched for possible members of a meteoroid stream, to be associated with the meteorites, in the IAU Meteor Database of Photographic Orbits (Lindblad et al. 2003) based on $D_{SH} \leq 0.2$ (cf. Jones et al. 2006). There were five meteoroids found, which are listed in Table 2 (for details of the designations see Neslušan 2003) and compared to the orbit of Příbram. While the Příbram and Neuschwanstein entry masses were several hundred kilograms, the other meteoroids mentioned in Table 2 are very small. The photometric mass of the largest one, 161E1, is about 2,100 g.

Table 2 Orbital elements (eq. 2000.0), geocentric velocity V_g, geocentric radiant (RA and DC), magnitude and D-criterion of Příbram and Neuschwanstein meteorites (Spurný et al. 2003) as well as five meteoroids from the IAU Meteor Database (Lindblad et al. 2003)

Meteoroid	q (AU)	a (AU)	e	i (°)	ω (°)	Ω (°)	π (°)	V_g (km/s)	RA (°)	DC (°)	Mag	D_{SH}
Příbr	0.790	2.401	0.671	10.5	241.8	17.8	259.5	17.43	192.3	17.5	−19.2	–
Neusch	0.793	2.400	0.670	11.4	241.2	16.8	258.0	17.51	192.3	19.5	−17.2	0.03
012F1	0.776	2.217	0.650	0.7	244.6	16.6	261.3	16.41	183.3	0.2	−6.7	0.17
161E1	0.817	2.696	0.697	9.6	236.5	18.9	255.4	16.95	189.5	17.8	−10.8	0.06
079H1	0.863	2.757	0.687	8.9	228.7	19.8	248.4	15.43	185.4	20.6	2.4	0.15
130F1	0.774	2.867	0.730	16.1	242.5	20.2	262.7	19.93	200.3	22.6	−10.7	0.12
083H1	0.821	2.582	0.682	4.9	236.5	21.7	258.1	16.01	186.9	8.6	2.0	0.10

Table 3 Orbital elements (eq. 2000.0) of Příbram (Spurný et al. 2003) as well as six objects from the NEA database (Bowell 2007)

Name	q	a	e	i	ω	Ω	π	$H(1,0)$	D_{SH}
Příbram	0.790	2.401	0.671	10.5	241.8	17.8	259.5		
1998 SJ70	0.656	2.236	0.706	7.4	244.4	23.8	268.2	18.3	0.18
2002 EU11	0.746	2.397	0.689	2.9	274.5	346.3	260.8	20.9	0.15
2002 QG46	0.905	2.434	0.628	8.3	268.2	346.0	254.2	19.6	0.17
2003 RM10	0.755	1.847	0.591	13.7	287.0	341.6	268.6	20.2	0.20
2005 GK141	0.938	2.735	0.657	14.0	218.2	34.2	252.5	22.1	0.19
2005 RW3	0.754	2.107	0.642	2.7	218.9	49.4	268.3	22.8	0.18

$H(1,0)$ is the absolute magnitude of NEAs and D_{SH} is the D-criterion

Also we have searched for a possible parent body among Near Earth Asteroids. We have found six NEAs from the current (April 2007) Bowell (2007) database, within $D_{SH} \leq 0.2$. The osculating orbital elements compared to Příbram are listed in Table 3.

Similarity of osculating orbits is not enough to prove any association among the orbits mentioned above. Therefore we have looked for similarity in orbital evolution over the past 5,000 years. We have numerically integrated the motion of the Příbram and Neuschwanstein meteorites, the five meteoroids and six NEAs using the multi-step procedure of the Adams–Bashforth–Moulton 12th order method, with a variable step-length. The positions of the perturbing major planets were obtained from the JPL Ephemeris DE406.

Only the orbital evolution of the best associations are presented in Fig. 1. The D_{SH} between Příbram and Neuschwanstein is within 0.07 and also the difference in the longitude of perihelion is very small ($\Delta\pi \leq 3°$) during the integration time of 5,000 years. This indicates a very close orbital evolution between the two meteorites. Only one meteoroid 161E1 and one asteroid 2002 QG46 were found with reasonably similar evolution to the meteorites in the last 2,000 years or so. However, the orbital evolution of asteroid 2002 QG46 is not so close to Příbram. So we prefer only the meteoroid 161E1 as a possible association.

3 Clones of Příbram and Neuschwanstein

Pauls and Gladman (2005) integrated Příbram's orbit for several hundred thousand years and showed that the substantial decoherence of the modeled stream occurred in about 50,000 years. However, here we study the orbital evolution of clones covering the error intervals of Příbram's and Neuschwanstein's orbital elements in order to check the stability of their orbital regions.

We have distributed five values equidistantly within the error interval of each parameter (semimajor axis, eccentricity, inclination, argument of perihelion and mean anomaly). The sixth parameter, the longitude of node, remained fixed, being of two orders better precision. Using the combinations of five values in five orbital parameters, 3,125 clones were obtained for each meteorite.

We have numerically integrated the clones of Příbram and Neuschwanstein over the past 5,000 years. The orbital evolution of all clones is more or less similar and stable. The clones of Příbram are less spread at the end of integration due to the smaller initial dispersion. The largest dissimilarity in the orbital evolution is caused by different initial semimajor axes of

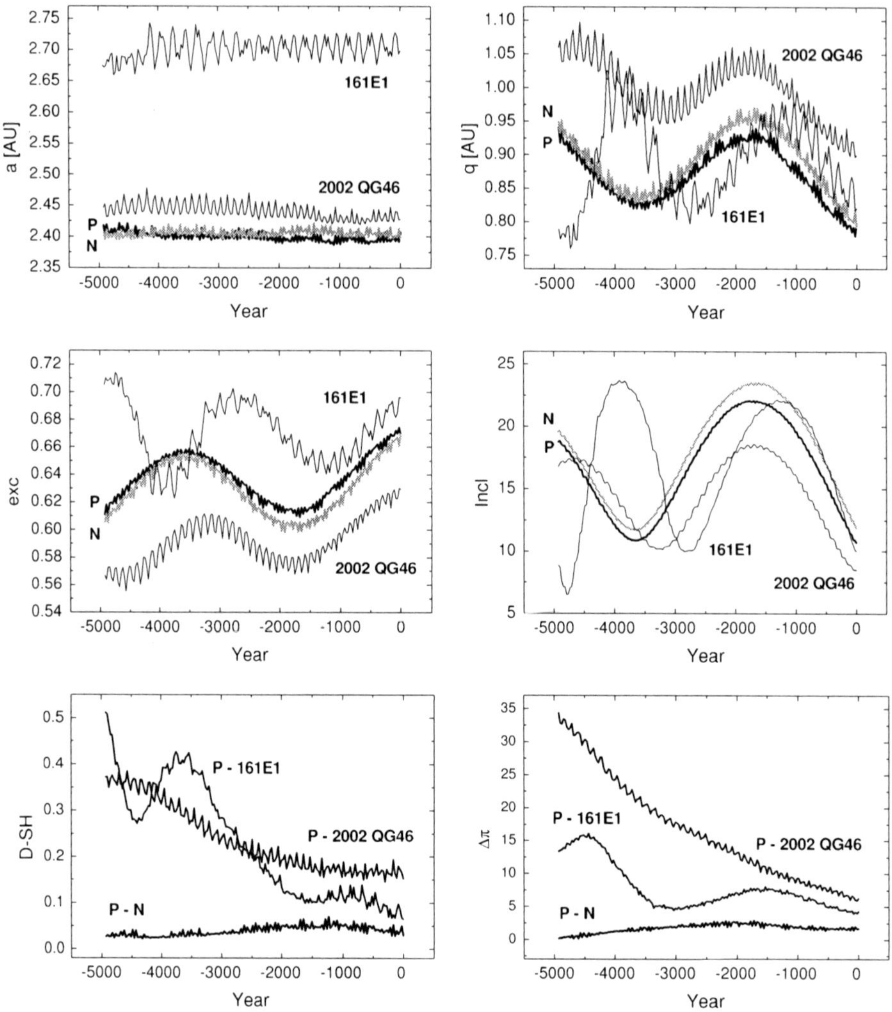

Fig. 1 The orbital evolution in semimajor axis a, perihelion distance q, eccentricity e, inclination i, D-criterion and difference in longitude of perihelion $\Delta\pi$ of Příbram (P), Neuschwanstein (N), meteor 161E1 and asteroid 2002 QG46

clones. A comparison of the orbital evolution of Neuschwanstein clones that have semimajor axes at the edges of the error interval ($a = 2.38$ AU and $a = 2.42$ AU) is presented in Fig. 2. As can be seen, the evolution of both sets of clones is very similar. Essentially the only difference is that the period of the variations in perihelion, eccentricity and inclination for the clones with $a = 2.42$ AU is shorter than for the clones with $a = 2.38$ AU. This is caused by the distance of the orbit from the orbit of Jupiter being smaller, as shown by Wu and Williams (1992). The descending nodes of almost all clones are stable and close to the Earth's orbit during the last 3,000 years. The longitude of the ascending node is dispersed by about 10° after 5,000 years of evolution. If we suppose that our clones represent a meteoroid stream, then it would have a similar dispersion of the orbital elements as that depicted in Fig. 2. The possible stream could be active for at least ±5 days around the date of the Příbram fall.

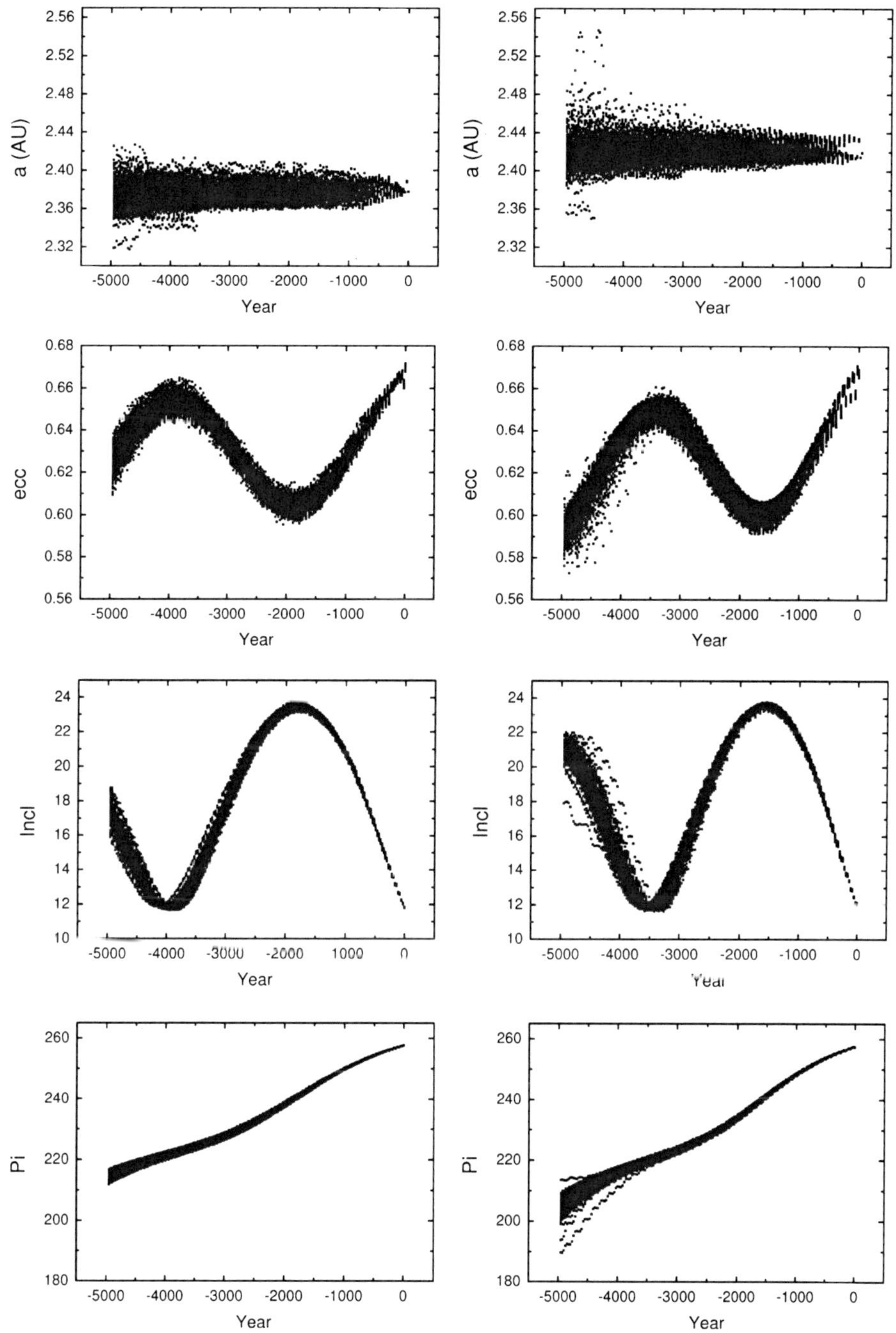

Fig. 2 The orbital evolution in semimajor axis, eccentricity, inclination and longitude of perihelion of clones of Neuschwanstein. The left set of graphs presents the clones for the initial semimajor axis $a = 2.38$ AU and the right set for $a = 2.42$ AU

Analysis of the orbital evolution has shown that 75% of the clones of Příbram and 84% of Neuschwanstein experienced close encounters with the Earth within 0.028 AU in the last 5,000 years. This distance is equivalent to a gravitational perturbation by Jupiter from a distance of 0.5 AU with respect to the perturbed body. Closer approaches caused a larger spread in the orbital elements at the end of the integration (Fig. 2). Some of the clones undergo more than one close approach to the Earth. Only a few clones encountered Mars also.

The results of the orbital integration of the clones of Příbram and Neuschwanstein show that the orbits are rather stable over several thousand years. A body with slightly different orbital elements from Příbram would then also have a similar evolution. Is it possible that Příbram and Neuschwanstein have such close orbits by chance?

We are interested in an occurrence of orbits of Příbram type in a five dimensional space of orbital elements. In our previous paper (Vereš et al. 2006), we generated and modeled 10^7 synthetic orbits of 10 m size bodies according to the NEA orbit distribution of Bottke et al. (2000) and population distribution of Stuart and Binzel (2004). A probability was found for the occurrence of each orbital element $(a, e, i, \omega, \Omega)$ within the error boundaries of Příbram and Neuschwanstein. Then the overall chance of this type of orbit occurring at random is the product of the probabilities in each element. The resultant probability is very small, only 2.75×10^{-11}.

When we extend this NEA synthetic population to smaller objects, of the initial radius of the Neuschwanstein meteoroid 0.3 m (ReVelle et al. 2004), we obtain a population with a cumulative number of 2.5×10^9 (Stuart and Binzel 2004) or 1.4×10^{11} (Brown et al. 2002) bodies. Then the expected occurrence of orbits within the error interval of Příbram and Neuschwanstein could be from 0.07 to 4 orbits depending on the real cumulative number in the NEA population.

4 Conclusions

If the real number of meteorite producing bodies of size $\sim$0.6 m in the NEA population is about 10^{11}, we would expect at least one very close pair in the Příbram region. This is in good agreement with conclusions of Pauls and Gladman (2005) that the occurrence of such close orbits is by chance. On the other hand, considering a more conservative assessment of 10^9 bodies in the NEA population, the probability of the existence of the Příbram and Neuschwanstein pair is very low. Moreover, this probability seems to be even smaller when we take into account the fact that both bodies entered the Earth's atmosphere within a time interval of 43 years, as was mentioned by Spurný et al. (2003).

Based on our dynamical investigation described above, we are in favour of the hypothesis of a common origin of the Příbram and Neuschwanstein meteorites from a heterogeneous parent asteroid. The close evolution of the two orbits over several thousand years is not a proof (e.g., Porubčan et al. 2004; Jones et al. 2006; Trigo-Rodríguez et al. 2007), but it does give significant support to suspicions about their common origin. The parent body of these meteorites could be a rubble pile asteroid which can possess heterogeneous material gravitationally aggregated after collisions. In another paper (Vereš et al. 2007) it has been proposed that relatively recent release of meteoroids from a parent asteroid by the Earth's tidal force is possible at substantially larger distances than the Roche limit. At such distances the differential gravitational influence would be insufficient to disperse the orbits of released meteoroids from the parent body. That is why we expect similar orbits of the parent body and Příbram and Neuschwanstein. We suppose that the different cosmic-ray ages of the meteorites are affected by having different cosmic radiation exposure times during which they were exposed on the surface of the "parent" body.

Acknowledgements This work was supported by VEGA—the Slovak Grant Agency for Science (grant No. 1/3067/06) and by Comenius University grant UK/401/2007. The authors are grateful to reviewers I. P. Williams and D. Asher for valuable suggestions.

References

A. Bishoff, J. Zipfel, Mineralogy of the Neuschwanstein (EL6) Chondrite – First Results (abstract 1212) 34th Lunar and Planetary Science Conference (2003)

W.F. Bottke, R. Jedicke, A. Morbidelli, B. Gladman, J.-M. Petit, Understanding the distribution of near-Earth objects. Science **288**, 2190–2194 (2000)

T. Bowell, Asteroid Orbital Element Database (2007), http://www.alumnus.caltech.edu/~nolan/astorb.html

P. Brown, R.E. Spalding, D.O. ReVelle, E. Tagliaferri, S.P. Worden, The flux of small near-Earth objects colliding with the Earth. Nature **420**, 294–296 (2002)

Z. Ceplecha, Multiple fall of Příbram meteorites photographed. I. Double-station photographs of the fireball and their relations to the found meteorites. Bull. Astron. Inst. Czech. **12**, 21–47 (1961)

I. Halliday, A.T. Blackwell, A.A. Griffin, Evidence for the existence of groups of meteorite-producing asteroidal fragments. Meteoritics **25**, 93–99 (1990)

D.C. Jones, I.P. Williams, High inclination meteorite streams can exist. Earth Moon Planet (2007). doi: 10.1007/s11038-007-9163-5

D.C. Jones, I.P. Williams, V. Porubčan, The Kappa Cygnid meteoroid complex. Mon. Not. R. Astron. Soc. **371**, 684–694 (2006)

B.A. Lindblad, L. Neslušan, J. Svoreň, V. Porubčan, IAU Meteor Database of photographic orbits version 2003. Earth Moon Planet **93**, 249–260 (2003)

L. Neslušan, IAU Meteor Database of Photographic Orbits (2003), http://www.astro.sk/~ne/IAUMDC/Ph2003/DATA2003/document.txt

J. Oberst, D. Heinlein, U. Köhler, P. Spurný, The multiple meteorite fall of Neuschwanstein: circumstances of the event and meteorite search campaigns. Meteorit. Planet. Sci. **39**, 1627–1641 (2004)

A. Pauls, B. Gladman, Decoherence time scales for "meteoroid streams". Meteorit. Planet. Sci. **40**(8), 1241–1256 (2005)

V. Porubčan, I.P. Williams, L. Kornoš, Associations between asteroids and meteoroid streams. Earth Moon Planet **95**, 697–712 (2004)

D.O. ReVelle, P.G. Brown, P. Spurný, Entry dynamics and acoustics/infrasonic/seismic analysis for the Neuschwanstein meteorite fall. Meteorit. Planet. Sci. **39**(10), 1605–1626 (2004)

R.B. Southworth, G.S. Hawkins, Statistics of meteor streams. Smithson Contr Astrophys. **7**, 261–285 (1963)

P. Spurný, J. Oberst, D. Heinlein, Photographic observations of Neuschwanstein, a second meteorite from the orbit of the Příbram chondrite. Nature **423**, 151–153 (2003)

H. Stauffer, H.C. Urey, Multiple fall of Příbram meteorites photographed. III. Rare gas isotopes in the Velká stone meteorite. Bull. Astron. Inst. Czech. **13**, 106–109 (1962)

J.S. Stuart, R.P. Binzel, Bias-corrected population, size distribution, and impact hazard for the near-Earth objects. Icarus **170**(2), 295–311 (2004)

J.M. Trigo-Rodríguez, E. Lyytinen, D.C. Jones, J.M. Madiedo, A.J. Castro-Tirado, I.P. Williams, J. Llorca, S. Vítek, M. Jelínek, B. Troughton, F. Gálvez, Asteroid 2002NY40 as a source of meteorite-dropping bolides. Mon. Not. R. Astron. Soc. (2007). doi:10.1111/j.1365-2966.2007.12503.x

P. Vereš, L. Kornoš, J. Tóth, Search for very close approaching NEAs. Contrib. Astron. Obs. Skalnaté Pleso **36**, 171–180 (2006)

P. Vereš, J. Klačka, L. Kómar, J. Tóth, Motion of a meteoroid released from an asteroid. Earth Moon Planet (2007). doi:10.1007/s11038-007-9187-x

Z. Wu, I.P. Williams, On the Quadrantid meteoroid stream complex. Mon. Not. R. Astron. Soc. **259**, 617–628 (1992)

J. Zipfel, B. Spettel, Schönbeck T, H. Palme, A. Bischoff, Bulk chemistry of the Neuschwanstein (EL6) chondrite – First results (abstract 1640) 34th Lunar and Planetary Science Conference (2003)

Meteors in the IAU Meteor Data Center on Hyperbolic Orbits

M. Hajduková Jr.

Originally published in the journal Earth, Moon, and Planets, Volume 102, Nos 1–4.
DOI: 10.1007/s11038-007-9171-5

Abstract The hyperbolic meteor orbits among the 4,581 photographic and 62,906 radar meteors of the IAU MDC have been analysed using statistical methods. It was shown that the vast majority of hyperbolic orbits has been caused by the dispersion of determined velocities. The large proportion of hyperbolic orbits among the known meteor showers strongly suggests the hyperbolicity of the meteors is not real. The number of apparent hyperbolic orbits increases inversely proportional to the difference between the mean heliocentric velocity of meteor shower and the parabolic velocity limit. The number of hyperbolic meteors in the investigated catalogues does not, in any case, represent the number of interstellar meteors in observational data. The apparent hyperbolicity of these orbits is caused by a high spread in velocity determination, shifting a part of the data through the parabolic limit.

Keywords Meteoroid · Meteor shower · Hyperbolic · Interstellar

1 Introduction

The present work is based on the meteor orbits data collected in the IAU Meteor Data Center (MDC). The database contains 4,581 photographic and 62,906 radar meteor orbits (Lindblad et al. 2005). Among the photographic orbits, there are 527 (11.5%) orbits with eccentricity $e > 1$. Radar orbits contain 1,875 (2.98%) hyperbolic orbits.

The proportion of hyperbolic orbits differs in different catalogues in the MDC and shows a dependence on the quality of observations (Štohl 1971). From a detailed analysis of the hyperbolic orbits in MDC photographic and radar data (Hajduková 1994; Hajduková and Paulech 2007), it was made clear that many hyperbolic orbits are probably a consequence of errors in determination of the meteor velocity or other parameters. Many conclusions based on the highly hyperbolic orbits derived from radar observations do not take into account the sensitivity of radar methods of velocity determination (Hajduk 2001)

M. Hajduková Jr. (✉)
Astronomical Institute of the Slovak Academy of Sciences, 84504 Bratislava, Slovak Republic
e-mail: astromia@savba.sk

and, therefore, they do not allow us to have much confidence to the derived results, especially those concerning interstellar sources of high velocity meteors. Our task here will be to estimate the limits of the possible errors.

2 Hyperbolic Orbits and Meteor Showers

One of the best ways of proving false hyperbolicity is to find the hyperbolic meteors among those meteors fulfilling the criteria of belonging to meteor showers.

There are 527 hyperbolic orbits in the photographic catalogues of the MDC and approximately 50% of them belong to known meteor showers. The proportion of hyperbolic orbits in the database is different in different showers. Among the total of 832 photographic Perseids, there are 224 hyperbolic orbits, which represent 27%, but among the 386 Geminids (v_{HGem} = 36.6 km s^{-1}), there is only 1 case of hyperbolic orbit. The problem of Perseids is that their heliocentric velocity (41.72 km s^{-1}) differs from the parabolic limit v_p = 42.14 km s^{-1} only by 0.4 km s^{-1} and, hence, a small error in the velocity determination may result in a designation of hyperbolicity of orbit. The present work examines five meteor showers having heliocentric velocities v_H close to the parabolic limit $\sqrt{2}v_0$, where v_0 is the Earth's velocity. The data are shown in Table 1. In our analysis the shower characteristics given by Ceplecha et al. (1998) were used in the same way as they were earlier when searching the MDC photographic data (Hajduková 2002).

Figure 1 shows an increasing number of hyperbolic orbits as inversely proportional to the difference ($\Delta v = v_H - \sqrt{2}v_0$) between the parabolic velocity and the mean velocity v_H of particular shower meteors. Figure 1 explains that the proportion of hyperbolic, or, as we can now say, of formal hyperbolic orbits, increases with the increasing heliocentric velocity ($N_{e>1}/N = f(v_H)$) of a particular shower, approaching the parabolic velocity limit.

Figure 2 shows eccentricities and geocentric velocities for photographic data of the 5 selected meteor showers. The plots show a considerable dispersion in both parameters. The largest spread of values is seen for the Perseids, reaching a value of 15 km s^{-1}, which correspond to errors in velocity of 17–20%. This is a strong finding, because the authors of individual photographic catalogues of orbits usually speak about a standard error in the geocentric velocity determination corresponding to the value of $\pm$ 0.5 km s^{-1}. The dependence of eccentricities on non-atmospheric velocities for radar data in Fig. 2 shows much larger scatter, following from a much smaller number of shower meteors in radar catalogues.

The observation of "hyperbolic meteors" among the shower meteors suggests that a similar effect of erroneously-determined "hyperbolic orbits" should also be ascribed to the

Table 1 Selected shower meteor data from the photographic MDC catalogues

Shower	No of meteors	No of hyp. meteors	Hyperbolic meteors (%)	Mean geoc. vel. $\bar{v}_G$ (km s^{-1})	Mean helioc. vel. $\bar{v}_H$ (km s^{-1})	$\Delta v = \bar{v}_H - \sqrt{2}v_0$
Lyrids	17	6	35	47	41.92	– 0.2
Perseids	835	224	27	59	41.70	– 0.4
Orionids	72	21	29	67	41.52	– 0.6
Leonids	36	5	14	71	41.43	– 0.67
Eta Aquarids	16	1	6	66	40.96	– 1.13

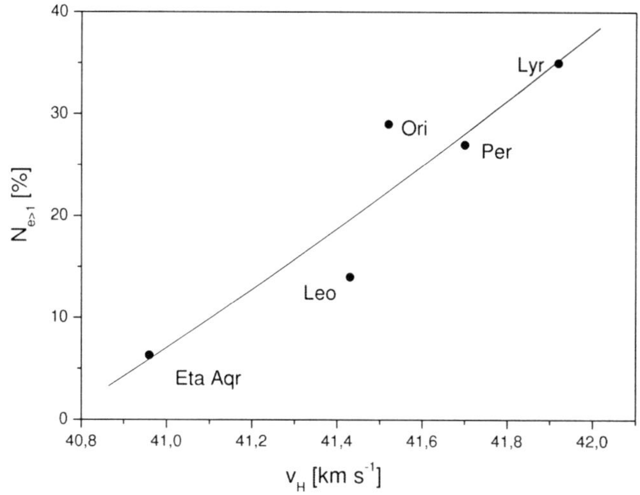

Fig. 1 The dependence of the contribution of hyperbolic meteor orbits in the selected meteor showers in the MDC photographic data on the mean heliocentric velocity of particular meteor showers. This dependence is clear proof that hyperbolic orbits among shower data are the consequence of the error distribution in the velocity determinations

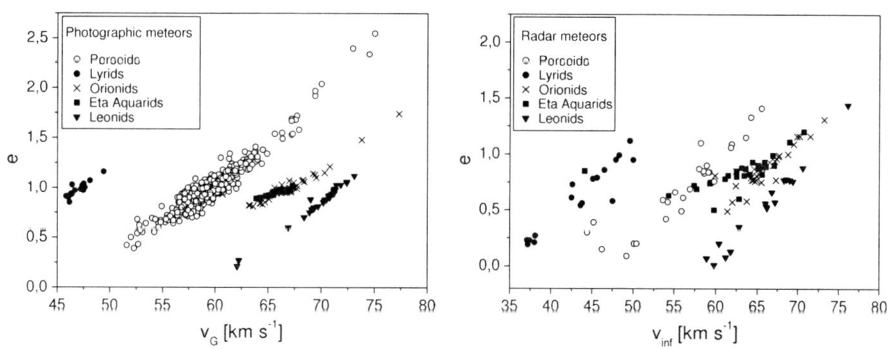

Fig. 2 Eccentricities and velocities of the 5 selected meteor showers in the photographic (left) and radar (right) data show that errors in velocity determination can reach the values ~10 km s^{-1}

sporadic meteors, at least for those the velocities of which are not too far from the parabolic limit.

3 The Velocity Distribution

The different precision of measurements, depending on the quality of observations, causes a natural spread in the velocity distribution. The shape of this spread gives a scattered gaussian distribution, which in the vicinity of the parabolic limit of the velocity, as it is in cases of investigated showers, exceeds the difference Δv between the mean heliocentric velocity of a particular meteor and the parabolic velocity, resulting in it being designated a

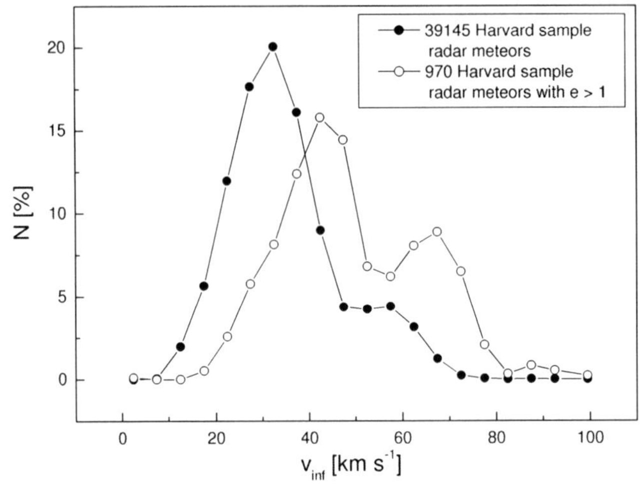

Fig. 3 The velocity distribution (normalized to 100%) of all the meteors and hyperbolic meteors for the Harvard radar sample of the MDC

"hyperbolic orbit". The velocity distribution of the Harvard 39 145 radar meteors and 970 hyperbolic meteors of the same sample, visualized here in the same proportion, is shown in Fig. 3. The velocity distribution of meteors with eccentricity $e > 1$ follows exactly the distribution of all meteors, but they are shifted by about 10 km s^{-1} towards the larger velocities along the whole scale. The only logical explanation for the observed shift between both sets of data is that the hyperbolicity of the set of meteors with $e > 1$ is caused by a high spread in velocity determinations, shifting a part of the data through the parabolic limit. The suggestion that the errors in the determination of v_{inf} from radar observations for high velocity meteors may be as large as 10 km s^{-1} does have some independent support. We have found it through the analysis of meteors belonging to the known meteor showers. A similar shift of the velocities of meteors with $e > 1$ from all the meteors was found by Kashcheyev and Kolomiyets (2001), who analysed 250 000 radar meteors from Kharkov.

4 Conclusions

Statistical analysis of the IAU Meteor Data Center photographic and radar meteor data shows that the vast majority of orbits in catalogues recorded as hyperbolic with $e > 1$ are only the consequence of measurement errors and their hyperbolicity is not real. The above identification of measurement errors as large as 10 km s^{-1} in the velocity of high velocity meteoroids has some consequences:

Firstly, the flux of interstellar meteors is much lower than it was declared by the authors of catalogues, or believed in some analyses of these observations. Secondly, the number of hyperbolic meteors (with $e > 1$) in the investigated catalogues of IAU MDC does not in any case represent the number of interstellar meteors in observational data. Hyperbolic meteors cannot be automatically counted as interstellar meteors without making detailed analysis of the data. It is clear that interstellar meteors may be present also within the error bars, however, they cannot be identified.

Acknowledgements This work was supported by the Scientific Grant Agency VEGA, grant No 3067.

References

Z. Ceplecha, W.G. Elford, D.O. Revelle, R.L. Hawkes, V. Porubčan, M. Šimek, Meteor phenomena and bodies. Space Sci. Rev. **84**, 327–471 (1998)

A. Hajduk, On the very high velocity meteors. in *Meteoroids 2001 Conference, ESA SP - 495*, ed. by B. Warmbein, (Kiruna, 2001), pp. 557–559

M. Hajduková Jr., On the frequency of interstellar meteoroids. Astron. Astrophys. **288**, 330–334 (1994)

M. Hajduková Jr., Shower meteor data in the IAU MDC. Acta Astron. et Geophys. Univ. Comenianae, **XXIV**, 33–39 (2002)

M. Hajduková Jr., T. Paulech, Hyperbolic and interstellar meteors in the IAU MDC radar data. Contrib. Astron. Obs. Skalnate Pleso **37**, 18–30 (2007)

B.L. Kashchejev, S.V. Kolomiyets, Interstellar particle detection and selection criteria of meteor streams. in *Meteoroids 2001 Conference, ESA SP - 495*, ed. by B.Warmbein, (Kiruna, 2001), pp. 643–650

B.A. Lindblad, L. Neslušan, V. Porubčan, J. Svoreň IAU Meteor Database of photographic orbits - Version 2003. Earth, Moon and Planets **93**, 249–260 (2005)

J. Štohl, On the problem of hyperbolic meteors. Bull. Astron. Inst. Czechosl. **21**, 10–27 (1971)

Meteoroid Stream Searching: The Use of the Vectorial Elements

Tadeusz J. Jopek · Regina Rudawska · Przemysław Bartczak

Originally published in the journal Earth, Moon, and Planets, Volume 102, Nos 1–4.
DOI: 10.1007/s11038-007-9197-8

Abstract In this initial study, we propose a new distance function D_V involving heliocentric vectorial orbital elements. The function measures differences between: the orbital energies, the angular momentums vectors and the Laplace vectors. In comparison with the widely used D_{SH} criterion of Southworth and Hawkins, D_D criterion of Drummond and their hybrid D_H by Jopek, the new function contains one invariant with respect to the principal secular perturbation: the orbital energy. The new function proved to be useful in the classification amongst the IAU2003 meteoroids which we searched for streams by D_V function and also using D_{SH} and D_N-function given by Valsecchi et al. For major streams, the results agree very well. For minor, and near-ecliptical streams the results sometimes differ markedly.

Keywords Meteoroids · Meteoroid streams · Methods: data analysis

1 Introduction

Southworth and Hawkins (1963) developed a distance function D_{SH}—the function of the orbital similarity named by them D-criterion—an important component of the computer meteoroid stream searching algorithm. Drummond (1979, 1981) introduced its modification D_D and Jopek (1993) proposed an alternative hybrid D_H. All D-functions are taken to be distances in a five-dimensional space, whose coordinates are heliocentric orbital elements e, q, ω, Ω, i. Other variations of the original D_{SH} function were given by (Steel et al. 1991; Asher et al. 1993), where instead five-dimensional space the authors use only three-dimensions q, e, i or a, e, i. In Valsecchi et al. (1999) the authors introduced D_N function involving four quantities U, $\cos\theta$, ϕ and λ; first three borrowed from Öpik's theory of close encounters (see Öpik 1976; Valsecchi et al. 1999): the geocentric velocity U, and the angles θ, ϕ defining the antiradiant direction in the geocentric ecliptic rotating reference frame, located at the longitude λ at the time of the meteor observation. Not long ago

T. J. Jopek (✉) · R. Rudawska · P. Bartczak
Obserwatorium Astronomiczne UAM, Sloneczna 36, 60-286 Poznan, Poland
e-mail: jopek@amu.edu.pl

Neslušan (2001) introduced C-criterion based on the difference between the orbital momentum vectors (per unit mass) of two orbits.

In the present paper we introduce yet another distance function D_V defined in the domain of the heliocentric orbital elements. Following direction pointed out by Neslušan we propose to use the full set of the vectorial elements. In the following sections we describe the new function, and present the results of its application to Lindblad et al. (2003) photographic meteor catalogue.

2 New Distance Function D_V

We take the vector $(\mathbf{h}^T, \mathbf{e}^T, E)^T$ which consists of vectorial elements: the angular momentum vector $\mathbf{h}$, the Laplace vector $\mathbf{e}$ and the energy constant E (see e.g. Breiter and Ratajczak (2005)). In the units AU, AU/day and the mass of the Sun, these quantities are defined by equations:

$$\mathbf{h} = (h_1, h_2, h_3)^T = \mathbf{r} \times \dot{\mathbf{r}} \tag{1}$$

$$\mathbf{e} = (e_1, e_2, e_3)^T = \frac{1}{\mu}\dot{\mathbf{r}} \times \mathbf{h} - \frac{\mathbf{r}}{|\mathbf{r}|} \tag{2}$$

$$E = \frac{1}{2}\dot{\mathbf{r}}^2 - \frac{\mu}{|\mathbf{r}|} \tag{3}$$

where $\mu = k^2$, k is the Gauss constant, whereas $\mathbf{r} = (x, y, z)$, $\dot{\mathbf{r}} = (\dot{x}, \dot{y}, \dot{z})$ are the heliocentric vectors of the position and velocity of the meteoroid. Describing the ith meteoroid by the set of vectorial elements:

$$\mathbf{O}_i = (\mathbf{h}_i^T, \mathbf{e}_i^T, E_i)^T = (h_{i1}, h_{i2}, h_{i3}, e_{i1}, e_{i2}, e_{i3}, E_i)^T$$

we define a new distance function, measuring dynamical similarity among two meteoroids, as:

$$\begin{aligned} D_V^2 =& w_{h1}(h_{i1} - h_{j1})^2 + w_{h2}(h_{i2} - h_{j2})^2 + 1.5\, w_{h3}(h_{i3} - h_{j3})^2 \\ &+ w_{e1}(e_{i1} - e_{j1})^2 + w_{e2}(e_{i2} - e_{j2})^2 + w_{e3}(e_{i3} - e_{j3})^2 + 2\, w_E(E_i - E_j)^2 \end{aligned} \tag{4}$$

where w_{hk}, w_{ek}, w_E are suitably defined weighting factors.

In comparison with D_{SH}, D_D, D_H, inclusion of the orbital energy, due to its invariant properties, brings important advantage into D_V. However inclusion of the difference between the eccentricities is a disadvantage when one of the orbits undergo significant Kozai perturbations. Also, D_V slightly overestimates the differences in the orientation of two orbits, similarly to the D_D criterion, as was pointed out in Jopek (1993). Despite of disadvantages mentioned above, which we meet also in D_{SH}, D_D, D_H-functions, we have found that D_V criterion is very useful in classification of the meteoroids. In Sect. 4 we describe the results of the cluster analysis applied with D_V, D_{SH}, D_N amongst the photographic meteors taken from the IAU Meteor Data Center.

To normalize contribution of each term in D_V, following Southworth and Hawkins (1963); Porubčan (1977), we propose the weights

$$w_E = (2\sigma_E)^{-2}, \quad w_{hk} = (2\sigma_{hk})^{-2}, \quad w_{ek} = (2\sigma_{ek})^{-2}, \quad k = 1, 2, 3 \tag{5}$$

where σ_E, σ_{hk}, σ_{ek} are expected standard deviations of the corresponding vectorial elements in a stream. However, due to invariant properties of E and semi-invariant character of h_3 by introducing multipliers 2 and 1.5 we have increased their influence on the resulting D-value given by definition (4).

To estimate the standard deviations (5), we simulated formation and the dynamical evolution for several meteoroid streams: the Perseids, Leonids, Orionids and Geminids. The particles ejection model, their evolution were slightly different to those used by Williams and Wu (1993, 1994). Next, having the corresponding distributions, the averaged standard deviation of each vectorial element has been found, and the resulting values, on several epochs are given in Table 1. As can be noticed, for a given epoch, dispersions of the vectorial element are not the same, they differ up to 4–5 orders. In searching of the meteoroid streams we applied the set of weight coefficients corresponding to the epoch 4,000 years after the stream formation.

3 The Meteor Data and the Stream Searching Method Used in this Study

We used 4,097 photographic meteors extracted from the computer files `geo2003.dat` and `orb2003.dat` downloaded from the IAU Meteor Data Center (Lindblad et al. 2003). Before being used for the classification, the available 4,581 meteor data were examined to check their internal consistency by the method slightly different to those described in Jopek et al. (2003). The test failed 306 times, and all this data as well as the orbits with $e > 1.1$ were rejected. The meteor data were analysed using the distance functions D_{SH}, D_N and D_V. First, the set of 4,097 meteors was pre-classified (single neighbour linking technique, D_{SH} function and a rough estimate of the threshold) obtaining the sporadic sample of 2,699 meteors. Using this sample, for each distance function the threshold of the dynamical similarity was found by the method similar to (Jopek and Froeschlé (1997); Jopek et al. (1999, 2003), see Table 2. Next, alike as in Jopek et al. (2003) we processed all 4,097 meteors accepting all streams of nine or more members detected with the reliability level 99%.

Table 1 The mean values of standard deviations of components of the vectorial elements of the typical meteoroid stream and the sporadic background

Epochs (years)	0	2,000	4,000	5,000	6,000	Sporadic
σ_{h_1}	2.5×10^{-5}	3.9×10^{-4}	9.8×10^{-4}	1.3×10^{-3}	1.5×10^{-3}	6.8×10^{-3}
σ_{h_2}	2.4×10^{-5}	3.4×10^{-4}	5.9×10^{-4}	6.8×10^{-4}	7.9×10^{-4}	7.8×10^{-3}
σ_{h_3}	2.8×10^{-5}	1.7×10^{-4}	4.1×10^{-4}	5.1×10^{-4}	6.2×10^{-4}	1.2×10^{-2}
σ_{e_1}	2.8×10^{-3}	6.6×10^{-3}	1.1×10^{-2}	1.4×10^{-2}	1.8×10^{-2}	5.1×10^{-1}
σ_{e_2}	2.8×10^{-3}	7.1×10^{-3}	1.6×10^{-2}	2.0×10^{-2}	2.3×10^{-2}	4.9×10^{-1}
σ_{e_3}	2.0×10^{-3}	1.3×10^{-2}	2.3×10^{-2}	2.7×10^{-2}	3.1×10^{-2}	3.4×10^{-1}
σ_E	7.1×10^{-7}	7.6×10^{-7}	9.8×10^{-7}	1.1×10^{-6}	1.2×10^{-6}	4.3×10^{-4}

The values for the typical stream were obtained as results of simulation. The values for sporadic meteoroids we calculated using IAU 2003 data. In columns 2–6, for several epochs, the averaged values were obtained using standard deviations for the Perseids, Orionids, Leonids and Geminids stream. The last column contains the standard deviations obtained for vectorial elements of the sporadic subset of the IAU 2003 meteor data

Table 2 The values of thresholds $D_{c,M}$ and their uncertainties applied in the meteoroid association tests amongst 4,097 meteoroids

M	D_{SH}	D_N	$D_V \times 10^{-1}$
8	0.0610 ± 0.0009	0.1146 ± 0.0018	0.2414 ± 0.0025
9	0.0644 ± 0.0007	0.1201 ± 0.0031	0.2575 ± 0.0032
10	0.0670 ± 0.0005	0.1237 ± 0.0046	0.2707 ± 0.0040
11	0.0689 ± 0.0004	0.1260 ± 0.0060	0.2815 ± 0.0048
12	0.0704 ± 0.0005	0.1275 ± 0.0070	0.2906 ± 0.0053
13	0.0715 ± 0.0007	0.1288 ± 0.0075	0.2985 ± 0.0056
14	0.0727 ± 0.0008	0.1303 ± 0.0074	0.3057 ± 0.0056
15	0.0741 ± 0.0008	0.1326 ± 0.0064	0.3128 ± 0.0052

They correspond to the reliability level $W_M = 99\%$ and are given for each stream population M, and for each distance function: D_{SH}, D_N, D_V. In the case of D_V the thresholds have been calculated using definition (4)

4 Results of the Classifications With D_{SH}, D_N, D_V

Using the function D_{SH} we detected 14 streams, combining 36% of the 4,097 orbits; in the search made using D_N, 17 streams were detected and the stream component included 46% of the sample; in case of the D_V we obtained 12 streams forming 36.2% of the sample.

The main results of all searches are summarized in Table 3. In general D_V seems to work more similarly to D_{SH} rather then to D_N. With D_N function (based on two 3-body secular semi-invariants), more streams and more members of the given stream have been identified.

In Table 3 the best agreement of the results we see for the Geminids, Leonids, Lyrids, December Monocerotids, Perseids and Quadrantids, which were detected in all searches. However with D_{SH} we found 24 Leonids, of which only 16 were found and D_V, plus an additional one only.

Using D_N function, the Orionids and η Aquariids were identified, practically, as two separate groups of 67 and 15 members. With D_V and D_{SH} the Orionids and η Aquariids form a single group of 56 and 72 members, respectively. The N-branch of α Capricornids was identified in all searches, the S-branch members of this stream significantly less numerous, were found only with D_N and D_V. Opposite results we have in case of δ Aquariids: with D_{SH} both N, S branches were found as two separated streams of 13 and 36 members. With D_N function, both branches formed one group of 84 members, with D_V we identified only 34 members of the Southern branch. Also, with D_{SH}, the κ Cygnids was identified as two groups of 15 and 20 members. With D_N and D_V functions only one stream was found consisting of 56 and 41 members.

The most complex result we observed in case of the Taurids. Using D_N we have found main group of 271 members and a second small one of 21 members. The main group contains many N and S Taurids as well as quite a lot of χ Orionids, as already flagged in the original IAU 2003 catalogue. Also with D_{SH} we identified two groups of Taurids, the main one of 152 members and small one of 14 members. The main group found with D_N included all Taurids detected by D_{SH} and all but four members identified with D_V. Using D_V, one group of Taurids was found, it included 138 members of the main group found with D_{SH} and 11 members of the second group detected with D_{SH}-function.

The last rows of Table 3 lists four streams detected with D_N only: the Virginids, σ Hydrids, ω Piscids and Cassiopeiids.

Table 3 Meteoroid streams detected in three searches

Name	Code	M_{SH}	M_N	M_V	N-SH	V-SH	V-N	$D_{c,SH}$	$D_{c,N}$	$D_{c,V} \times 10^{-1}$
Geminids	67	381	390	380	381	369	380	0.0689	0.1275	0.3128
Leonids	92S	24	32	17	24	16	17	0.0689	0.1201	0.3057
Lyrids	1	13	14	13	13	12	13	0.0564	0.0637	0.2906
Dec. Monocerotids	87S	13	22	12	13	11	12	0.0715	0.1201	0.2815
Perseids	14	630	705	650	630	619	650	0.0741	0.1146	0.3057
Quadrantids	45	51	57	49	51	51	49	0.0741	0.1260	0.2707
η Aquariids	115S	14	–	11	In 281N	11	In 281N	0.0727	–	0.2707
Orionids	115S	58	67	45	58	45	44		0.1288	
η Aquariids-2	281N	–	15	–	In 115S	–	–	–	0.1201	–
Orionids-2	281N	–	1	–	–	–	–	–		–
α Capricornids (N)	11VN	40	77	46	36	36	43	0.0689	0.1326	0.3057
α Capricornids (S)	11VN	–	9	3	0	–	3			
δ Aquariids (N)	388S	13	23	–	13	–	–	0.0704	0.1326	–
δ Aquariids (S)	373S	36	61	34	36	32	34	0.0689		0.3128
κ Cygnids	15	15	56	41	15	15	39	0.0741	0.1326	0.2985
κ Cygnids2	412S	20	–	–	In 15	–	–	0.0704	–	–
Taurids (N)	44N	46	106	54	46	42	51	0.0741	0.1303	0.3128
Taurids (S)	44N	106	165	113	106	96	112			
Taurids-2 (N)	139N	–	14	–	–	–	–	–	0.1275	–
Taurids-2 (S)	139N	–	7	–	–	–	–			–
Taurids-3 (S)	69V	14	–	–	In 44N		–	0.0689	–	
Virginids (N)	57N	–	15	–	–	–	–	–	0.1303	–
Virginids (S)	57N	–	3	–	–	–	–	–		–
σ Hydrids	103N	–	21	–	–	–	–	–	0.1260	–
ω Piscids (S)	463N	–	9	–	–		–	–	0.1201	–
Cassiopeiids	704N	–	16	–	–	–	–	–	0.1288	–

The first column gives the stream name, the second its code, the third, fourth and fifth ones give the number of members M_{SH}, M_N and M_V identified by, respectively, D_{SH}, D_N and D_V, for the threshold values $D_{c,M}$ given, respectively, in the ninth, tenth and eleventh columns; finally, the columns six, seven and eight, give the number of common members N-SH, V-SH, V-N. The flags S, N and V sometimes added to the stream codes denote that the latter refers to the search made with D_{SH}, D_N and D_V respectively. The absence of the flag means that the code is the same for all distance functions

5 Conclusions

The new D_V-function proved to be useful in the classification of the IAU2003 photographic meteoroids. In comparison with D_{SH} and D_N-function, for major streams the results agree very well. For minor, and near-ecliptical streams the results of the identification may differ markedly. As was expected, the distance functions D_V, D_N, D_{SH} are not mutually equivalent. However, the main results obtained with the D_V function are more similar to those obtained by D_{SH} than by D_N criterion. In this study we begin initial investigations of the new D-criterion and the authors are aware that several points need future study e.g. the final shape of the D_V-function and suitable choice of the weighting coefficients.

Acknowledgements In this study, majority of calculations were done at Poznań Supercomputing and Networking center. The authors wish to acknowledge referees Giovanni Valsecchi, Paul Wiegert and Vladimir Porubčan for their comments, suggestions which improved this paper.

References

D.J. Asher, S.V.M. Clube, D.I. Steel, Asteroids in the Turid complex. MNRAS **264**, 93 (1993)

S. Breiter, R. Ratajczak, Vectorial elements for the Galactic disc tide effects in cometary motion. MNRAS **364**, 1222 (2005)

J.D. Drummond, On the meteor/comet orbital discriminant D, in *Procceedings Southwest Regional Conference on Astronomy Astrophysics*, vol. 5, Little Rock Arkansas, ed. by P.F.Gott, P.S. Riherd, pp. 83–86 (1979)

J.D. Drummond, Earth-orbit-approaching comets and their theoretical meteor radiants. Icarus **47**, 500–517 (1981)

T.J. Jopek, Remarks on the meteor orbital similarity D-criterion. Icarus. **106**, 603–607 (1993)

T.J. Jopek, Cl. Froeschlé, A stream search among 502 TV meteor orbits. An objective approach. Astron. Astrophys. **320**, 631–641 (1997)

T.J. Jopek, G.B. Valsecchi, Cl. Froeschlé, Meteoroid stream identification: a new approach-II. Application to 865 photographic meteor orbits. MNRAS **304**, 751–758 (1999)

T.J. Jopek, G.B. Valsecchi, Cl. Froeschlé, Meteoroid stream identification: a new approach-III. The limitation of statistics. MNRAS **344**, 665–672 (2003)

B.A. Lindblad, L. Neslušan, V. Porubčan, J. Svoreň, IAU meteor data center, photographic database, version 2003. Earth Moon Planets **93**, 249–260 (2003)

L. Neslušan, A sketch of an orbital-momentum-based criterion of diversity of two keplerian orbits. in *Proceedings of the US/European Cellestial Mechanics Workshop*. Poznañ, Poland, July 3–7, 2000, pp. 365–366 (2001)

E.J. Öpik, Interplanetary Encounters (Elsevier, New York, 1976)

V. Porubčan, Dispersion of orbital elements within the Perseid meteor stream. Bull. Astron. Inst. Czech. **28**, 257–266 (1977)

R.B. Southworth, G.S. Hawkins, Statistics of meteor streams Smithson. Contr. Astrophys. **7**, 261–285 (1963)

D.I. Steel, D.J. Asher, S.V.M. Clube, The structure and evolution of the Taurid complex. MNRAS **251**, 632–648 (1991)

G.B. Valsecchi, T.J. Jopek, Cl. Froeschlé, Meteoroid stream identification: a new approach-I. Theory. MNRAS **304**, 743–750 (1999)

I.P. Williams, Z. Wu, The Geminid meteor stream and asteroid 3200 Phaethon. MNRAS **262**, 231–248 (1993)

I.P. Williams, Z. Wu, The current Perseid meteor shower. MNRAS **269**, 524–528 (1994)

Directional Variation of Sporadic Meteor Activity and Velocity

M. D. Campbell-Brown

Originally published in the journal Earth, Moon, and Planets, Volume 102, Nos 1–4.
DOI: 10.1007/s11038-007-9152-8

Abstract The majority of small (millimeter size) meteoroids striking the Earth every year belong to the sporadic sources: the helion/antihelion, apex and toroidal sources. Radar data from the CMOR facility near London, Ontario, Canada provides five years of sporadic activity information and velocity distributions at two degree resolution, allowing the fine structure of each source to be investigated. We have used five years of orbital data to investigate the directional dependence of the activity and the velocity distribution of the sporadic meteoroid population on a two degree scale. These data can be used to investigate the origin of the sporadic meteoroid sources.

Keywords Meteors · Radar · Sporadic meteors

1 Introduction

The sporadic meteoroid environment, consisting of those meteoroids which do not belong to showers, is mainly concentrated in six sources: the helion and antihelion sources, the north and south apex, and the north and south toroidal sources. Previous studies of sporadic activity have tended to focus on the activity of these six sources. Poole (1997) used radar data to statistically measure the activity of the helion and antihelion sources for each month of the year. Using more precise data from the CMOR radar, Campbell-Brown and Jones (2006) found very similar results for the helion and antihelion sources, showing that seasonal atmospheric effects are not important contributers to the measured rates, since the activity curves are the same in both hemispheres. The latter work also measured the activities of the north toroidal and north apex sources.

The orbital characteristics of sporadic meteoroids were studied in detail by Galligan and Baggaley (2004), using the AMOR (Advanced Meteor Orbit Radar) system. They used a total of approximately 5×10^5 meteor orbits, with a limiting magnitude of about +14.

M. D. Campbell-Brown (✉)
Department of Physics and Astronomy, University of Western Ontario, 1151 Richmond St, London, ON, Canada
e-mail: mcampb33@uwo.ca

DOI: 10.1007/978-0-387-78419-9_12

This study corrected observing biases of the AMOR system, and also took into account collisional probabilities. Radiant distributions of sporadic meteors have also been studied using high power, large aperture radars; Chau et al. (2007) collected 1.7×10^5 orbits from 14 days and observed all the sporadic sources.

The Canadian Meteor Orbit Radar (CMOR) has been running for five years with both calibrated single station data (14 million echo profiles collected) and orbits (2.3 million computed). With this quantity of data, the activity and orbital parameters of the sporadic sources can be examined at high resolution, allowing the sources to be studied in greater detail and the sporadics which occur outside the sources to be investigated. The speeds and orbital parameters of sporadic meteors as a function of heliocentric radiant can also be explored in detail.

2 Observations

The CMOR system has been in operation with orbital capabilities since May 2002. The 29 MHz system, with a peak power of 6 kW, has a limiting meteor magnitude of about +8. The pulse repetition factor is 532 pulses per second. Further details of the system can be found in Jones et al. (2005). For the current study, multistation orbital data from the 29 MHz system have been used.

The orbital data gives very accurate information about the trajectory and speed of each meteor, but cannot directly be used to look at the time variation of meteoroid flux. The orbital system relies on microwave links to two remote stations (6 and 8 km from the main site), which are affected by atmospheric conditions. The number of orbits calculated each day therefore depends not only on the number of meteoroids striking the atmosphere, but on the local, ground-level weather conditions. In a future study, single station data will be used to look at the small scale variations in flux.

3 Results

Figure 1 shows the uncorrected number of meteors observed whose radiants, in heliocentric coordinates, fell inside each two degree bin of latitude and longitude. In this and the following plots, the equator falls along the ecliptic plane, and the apex of the Earth's way is

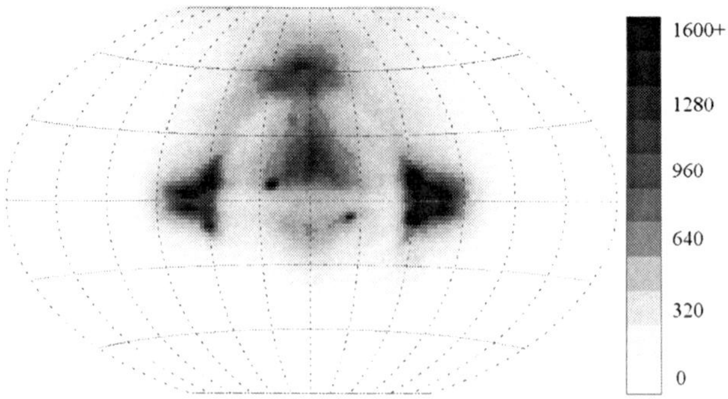

Fig. 1 Number of meteoroids per two degree radiant bin, in heliocentric coordinates. Average over five full years of CMOR orbital data (2002–2007)

at the centre, with the sun 90 degrees to the left, and the anti sun point 90 degrees to the right. The antapex point is at the edges of the plot.

The five sporadic sources visible from CMOR's latitude show up as expected. The helion and antihelion sources produce the majority of the meteors observed by CMOR. In addition to the sources, there is a concentration of sporadic meteor radiants on a ring joining the helion, antihelion and north toroidal sources, which appears to connect to the south toroidal source as well. Showers have not been removed, and show up clearly on the plots. Most major showers, such as the Geminids (in the antihelion source) and the Arietids (in the helion source), are embedded in the sporadic sources, but a few are visible on the edges. For example, the Eta Aquariids are at the bottom left of the north apex source, the Perseids at the top left, and the Orionids are to the right of the south apex source.

The position of the sources compare well with previous work (Chau et al. 2007; Galligan and Baggaley 2004; Jones and Brown 1993). The helion and antihelion sources are, when showers are included, distinctly elongated along the ring: ignoring these lobes (which correspond to the positions of significant showers and may therefore vanish when showers are removed), their centers are on the ecliptic, almost exactly 20 degrees in from the helion and antihelion points. The apex sources are rather triangular in our data, but show a vertical region of enhancement in the centre (visible mainly on the better observed north apex source) similar to that seen by Chau et al. (2007). The centres of the sources are approximately 15 degrees from the ecliptic, in good agreement with other studies. The north toroidal source is centered just below 60 degrees latitude, very close to previous studies. The ring feature has not, to our knowledge, been seen before.

In Fig. 2, the same plots are shown corrected for CMOR's observing biases. In particular, the collecting area of the radar for each radiant, the attenuation due to initial trail radius, finite velocity, Faraday rotation, and the pulse repetition factor are calculated for each meteor, and the echo is weighted accordingly. The collecting area correction tends to decrease the radiant rate in the northern part of the plot, since these radiants are always above the horizon, and to increase the rate in the southern part of the plot, where radiants are up only a short time each day. This is most obvious in the north toroidal and south apex sources in Figs. 1 and 2. The initial trail radius correction (which accounts for echoes which are missed because of destructive interference between the near and far parts of the ionized train, when the meteor occurs high in the atmosphere) gives a higher weighting to faster meteors, which are less likely to be observed; it increases the strength of the apex sources, which are composed mainly of high speed, retrograde meteoroids. Faraday

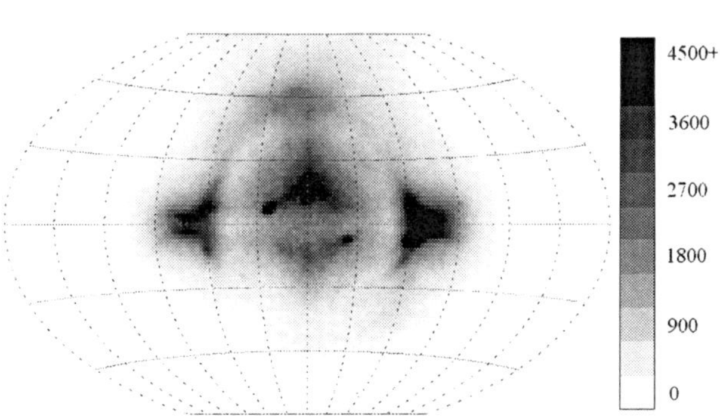

Fig. 2 Same data as in Fig. 1, corrected for radar collecting area, initial trail radius, finite velocity effects, and Faraday rotation

rotation affects the helion source more than the others, since the ionosphere is only significant at meteor heights during the day; even so, this effect is not strongly visible in the data correction. The finite velocity effect affects mostly slow moving meteors; it is negligible compared to the other corrections. The pulse repetition frequency bias affects short lived meteors which can only be sampled a few times and are therefore likely to escape detection; it is also a minor correction in our data. All of the atmospheric effects were computed according to the formalism in Ceplecha et al. (1998). Note that Fig. 2 is merely scaled to account for these biases: it is not a plot of true flux. Because of the scaling, the absolute values are higher than in the previous plot.

The average geocentric speeds of meteoroids in each two degree bin are shown in Fig. 3. These are speeds corrected for atmospheric deceleration (see Brown et al. (2004) for details), and with the gravitational effects and rotation of the Earth removed. As expected, the speeds are highest for the retrograde meteoroids in the apex direction and lowest near the antapex. The noise which is present at the southern edge of the plot and around the antapex is due to the relatively small number of meteors with radiants in these areas. The ring visible in the activity plots is also visible in the speed plot, as a slight decrease in the steady increase of speeds toward the apex. Note that the sporadic sources are not clearly defined in velocity space. The apex sources, as in previous studies (Galligan and Baggaley 2004; Jones and Brown 1993; Chau et al. 2007), consists of high speed (between 70 and 45 km/s) meteoroids. The north toroidal source has average speeds approximately 35 km/s, and the helion and antihelion sources have average speeds which vary from 20 to 35 km/s, also consistent with previous studies.

One explanation for the ring of sporadic radiants, inside which is a depleted zone with few radiants, can be found in the collision probabilities of meteoroid orbits with the Earth. The collision probability depends on the orbital parameters; a weighting proportional to the collisional probability of each meteor was calculated following the method of Galligan and Baggaley (2004) (note there is a typographical error in Eq. 43 of that paper), according to Eq. 1. The scaled probabilities for meteors in each bin were then averaged. Figure 4 shows the average log of the scaled collision probability.

$$n_c \sim \frac{v_\infty^2}{v_g^2 a^{1.5} \sin i} \sqrt{\frac{3 - a^{-1} - 2\sqrt{a(1-e^2)} \cos i}{2 - a^{-1} - a(1-e^2)}} \tag{1}$$

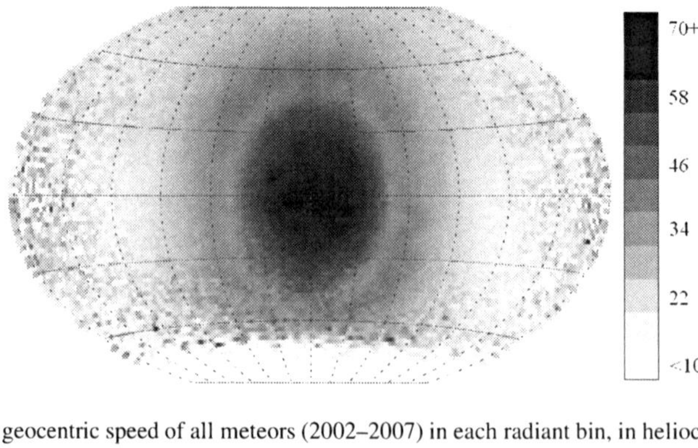

Fig. 3 Average geocentric speed of all meteors (2002–2007) in each radiant bin, in heliocentric coordinates (km/s)

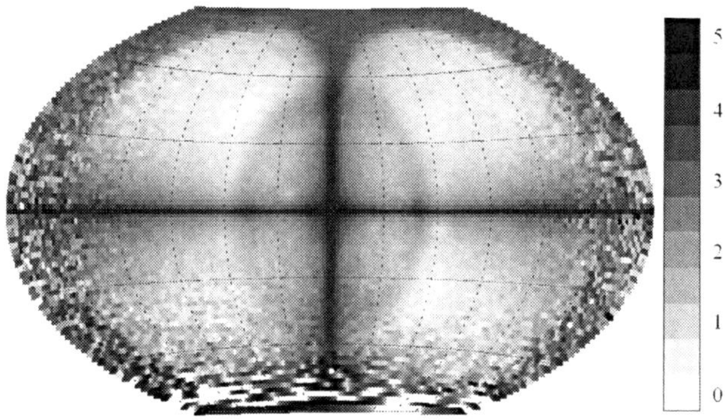

Fig. 4 Average log of collisional probability with the Earth of all meteors (2002–2007) in each radiant bin, in heliocentric coordinates

In addition to the expected high collision probabilities for meteoroids with radiants on the ecliptic and in the apex direction, there is a slight enhancement in the probability in a ring surrounding the apex, with a radius of about 50 degrees. Even a slight enhancement of the collisional probability could result in a significant depletion of meteoroids in these types of orbits, if it is associated with shorter collisional lifetimes with respect to the zodiacal cloud. One possible interpretation of the decrease in the apparent number of meteoroids at the ecliptic and in the ring is that the collisional lifetime (both with the Earth and each other) is short compared to the replenishment timescale for these particles. This is suggestive that the present sporadic meteoroid environment is not in a steady state, a conclusion also reached by Grün et al. (1985). More work needs to be done to verify such a conclusion, however.

The sporadic meteoroid complex, as observed from the Earth, is defined by a combination of collisional lifetimes and collisional probability with the initial distribution given by the parent bodies (likely comets, for most sporadic meteoroids of this size). A careful study of sporadic radiants and orbital parameters will allow the origins and dynamics of this population to be studied in detail. In the future, showers will be removed from our data set to make the sporadic contributions clearer; in addition, the relationship between showers and sporadic meteors from similar radiants will be investigated, to explore how stream meteoroids become part of the sporadic complex. Collisional lifetimes will be calculated for our sporadic meteors, which will give additional insight into the origins of the sporadic sources, as will studying the potential parent bodies. The radiant distributions will also be weighted to the same limiting mass, which will tend to reduce the number of radiants in high speed regions; high speed meteoroids produce more ionization and are therefore detectable at lower masses than slow meteoroids.

Seasonal variations in the orbital parameters and radiant distribution of sporadic meteors will also be examined in a future work, to see what non-uniformities in the population may reveal about the origins. We expect to see a continuum linking tightly clustered young meteoroid streams, old meteoroid streams which are more diffuse in radiant and node longitude, and features in the sporadic sources too diffuse to be considered streams.

Acknowledgments Thanks to R. Weryk for assistance with the plots. Thanks also to David Galligan and an anonymous reviewer for helpful comments at the review stage.

References

P. Brown, J. Jones, R.J. Weryk, M.D. Campbell-Brown, EM&P **95**, 617–626 (2004)
M.D. Campbell-Brown, J. Jones, MNRAS **367**, 709–716 (2006)
Z. Ceplecha, J. Borovička, W.G. Elford, D.O. Revelle, R.L. Hawkes, V. Porubčan, M. Šimek, SSR **84**, 327–471 (1998)
J.L. Chau, R.F. Woodman, F. Galindo, Icarus **188**, 162–174 (2007)
D.P. Galligan, W.J. Baggaley, MNRAS **353**, 422–446 (2004)
E. Grün, H.A. Zook, H. Fechtig, R.H. Giese, Icarus **62**, 244–272 (1985)
J. Jones, P. Brown, MNRAS **265**, 524–532 (1993)
J. Jones, P. Brown, K. Ellis, A. Webster, M. Campbell-Brown, Z. Krzemenski, R. Weryk, P&SS **53**, 413–421 (2005)
L.M.G. Poole, MNRAS **290**, 245–259 (1997)

Meteor Showers Originated from 73P/Schwassmann–Wachmann

Shun Horii · Jun-ichi Watanabe · Mikiya Sato

Originally published in the journal Earth, Moon, and Planets, Volume 102, Nos 1–4.
DOI: 10.1007/s11038-007-9224-9

Abstract The nucleus of the Comet 73P/Schwassmann–Wachmann had been split into many fragments at least past two returns. Since the related dense dust trail has been detected in the space infrared observation, the strong activity of the meteor shower is highly expected in the future. We applied the so-called dust-trail theory to this interesting object, and obtained several results on the future encounter with the dust trail. In this paper we introduce our results on the forecasts.

Keywords Comets: individual(73P/Schwassmann–Wachmann) · Meteoroids · Meteors · Solar system

1 Introduction

Meteor showers occur when the Earth passes through dense trail of meteoroids. When the comet 73P/Schwassmann–Wachmann returned in 1995, this comet had been split into some nuclei (Sekanina et al. 1996), and more than 50 fragments are discovered by now (Fuse et al. 2007). According to NASA Mission News "Spitzer Telescope Sees Trail of Comet Crumbs" which was announced on May 10, 2006, the related dense dust trail has been detected in the space infrared observation, so we can expect that these dense dust trails will cause the meteor showers in the future.

Actually, in the past, there were meteor storms originated from split comets like this. One of the famous cases is a historical storm of the Andromedids produced by the comet 3D/Biela which broke apart in 1842/1843 are known (Jenniskens and Vaubaillon 2007).

According to previous studies, first, it is predicted that the trail of dust ejected before the breakup of 73P/Schwassmann–Wachmann that happened in 1995 approaches the Earth

S. Horii (✉)
The Graduate University for Advanced Studies (SOKENDAI), 2-21-1 Osawa, Mitaka, Tokyo 181-8588, Japan
e-mail: shun.horii@nao.ac.jp

J. Watanabe · M. Sato
National Astronomical Observatory of Japan, 2-21-1 Osawa, Mitaka, Tokyo 181-8588, Japan

DOI: 10.1007/978-0-387-78419-9_13

very closely in 2022 and 2049 (Wiegert et al. 2005). Additionally, as for the trail of dust ejected after this breakup, it is predicted that it approaches the Earth very closely in 2022 (Lüthen et al. 2001), but the situation after 2022 is not known yet. So, we tried to analyze this interesting comet with the so-called dust-trail theory, and examined the possibility of future meteor showers originated from the dust generated after this breakup.

2 The Method of Our Calculation of the Dust Trails

We applied the most simple approach of the dust-trail theory (e.g., Asher 2000), which was described by Sato (2003) in detail. Each trail was assumed to be formed by meteoroids ejected during the perihelion passage of the parent comet. The trail was calculated by test particles ejected parallel to the comet motion, both ahead of and behind the comet. At first, the ejection velocity was set to be within the range between -120 and $+120$ m/s, where "+" is in the direction of the comet's motion and "$-$" in the opposite direction. Although we usually set this ejection velocity to be within the range between -30 and $+30$ m/s, since meteoroids that we considered in this study were ejected from the split nuclei of the comet, these meteoroids were likely to have higher ejection velocity than usual, so we considered the range of this ejection velocity wider. Integration was carried out by using the Runge–Kutta–Fehlberg method together with Encke's method. In calculating the perturbations, we included the three largest main-belt asteroids (Ceres, Pallas and Vesta) in addition to the eight planets, Pluto and the Moon, using the JPL ephemeris DE406. We did not take the effect of radiation pressure on the meteoroids into account in our calculation. We used the comet orbital elements calculated by Kinoshita (2007) and these are available on his Web page. In this study, we consider meteoroids ejected from only nucleus C. We studied the trails generated from 1995 to 2065 and approaching the Earth between 1995 and 2070.

3 Results

As the results of our calculation, we found that several trails will approach the Earth closely in the future. Table 1 shows the data of dust trails which will approach the Earth within 0.02 AU. V_g is the expected geocentric velocity before the gravitational focusing of the Earth. When the value of V_g is higher, smaller dust also comes to be observed as a meteor. r_D and r_E are the heliocentric distances of the dust trail's descending node and of the Earth at the same longitude, and $r_E - r_D$ gives the distance of the approach of dust trail to the Earth. As a rough and empirical criterion, in the case where its approach is within 0.01 AU, meteors originated from its trail are probably observed, and in the case where within 0.001 AU, meteor showers and even meteor storms possibly occur. The parameter f_M is called "the mean anomaly factor", and represents the degree of the stretch of the trail (McNaught and Asher 1999). The values of f_M in Table 1 were calculated by $f_M = \Delta M_0/\Delta M$, where ΔM is the difference between the mean anomaly values of both ends of a given part of the trail at the time when it approaches the Earth, and ΔM_0 is that of the same part of the trail, but at the first revolution without perturbations. Generally the trail lengthens with time. Since the density of dusts decreases as the trail lengthens, the lower the value of f_M becomes, the less dense the dust trail becomes. Therefore, the spatial density of a trail encountered by the Earth depends on V_g, ejection velocity, $r_D - r_E$, and f_M, and these four parameters are important in discussing the possibility of future meteor showers.

Table 1 The data of dust trails which will approach the Earth within 0.02 AU

Observation year	Ejection year	Expected peak time (UT)	Solar longitude (J2000.0)	$r_D - r_E$ (AU)	Ejection velocity (m/s)	f_M	Expected position of radiant (J2000.0)		V_g (km/s)
							α (deg.)	δ (deg.)	
2021	1995	May 14.72 17:23	53.841	−0.0037	−118.90	0.033	201.58	+8.79	12.84
2022	1995	May 31.21 04:59	69.448	−0.00038	−26.71	0.24	209.48	+28.13	12.10
	2001	May 31.26 06:15	69.499	−0.0058	−30.26	0.23	209.04	+28.27	12.20
	2006	May 31.30 07:04	69.532	−0.011	−33.43	0.32	208.70	+28.36	12.27
	2011	May 31.34 08:12	69.577	−0.017	−43.97	0.47	208.30	+28.48	12.35
	2017	May 31.37 08:55	69.607	−0.018	−88.83	0.94	208.35	+28.56	12.33
2032	1995	May 19.49 11:51	58.613	+0.0085	−105.95	0.044	204.38	+10.62	12.20
2037	1995	May 18.27 06:33	57.159	+0.0077	−109.46	0.062	203.56	+9.68	12.30
2042	1995	May 16.99 23:53	55.657	−0.00073	−112.76	0.021	202.42	+9.05	12.62
2043	1995	May 21.72 17:18	59.968	+0.0098	−94.19	0.10	206.98	+13.22	12.37
2047	1995	May 16.10 02:31	54.529	−0.0010	−115.41	0.019	201.98	+8.68	12.76
2048	1995	May 20.89 21:25	59.858	+0.014	−102.01	0.045	205.67	+11.80	12.05
	2001	May 27.31 07:24	66.023	+0.00023	−105.42	0.00058	209.38	+21.63	12.20
2064	2001	May 17.55 13:13	56.542	+0.0025	−37.21	0.029	205.78	+14.77	12.90
	2006	May 16.02 00:27	55.065	−0.0010	−33.56	0.028	204.48	+13.87	13.05
	2011	May 14.44 10:39	53.546	+0.0013	−31.76	0.087	203.80	+12.57	13.12
	2017	May 13.03 00:36	52.176	+0.0071	−38.99	0.076	203.29	+11.37	13.13
	2022	May 10.46 11:03	49.698	+0.019	−48.47	0.15	202.42	+9.16	13.16
	2027	May 11.66 15:44	50.853	+0.018	−51.58	0.0063	201.47	+9.92	12.92
	2033	May 11.62 14:49	50.816	+0.0073	−62.35	0.21	201.15	+9.82	13.00
	2038	May 11.48 11:37	50.687	+0.0018	−67.76	0.23	200.86	+9.81	13.12
	2043	May 11.54 13:04	50.746	−0.0025	−83.95	0.17	200.60	+9.99	13.17
	2049	May 11.63 15:00	50.824	−0.0076	−110.81	0.10	200.25	+10.24	13.21

Looking at the values in Table 1, at first, it is in 2022 that we notice the most possible chance of the meteor storm. Figure 1 shows the location of the dust trails in the ecliptic plane and the path of the Earth in 2022. The shape of each trail is assumed to be like a tube whose diameter is 0.001 AU. Especially, the dust trail ejected in 1995 will approach the Earth as closely as 0.00038 AU. This result correspond well to the results of the previous

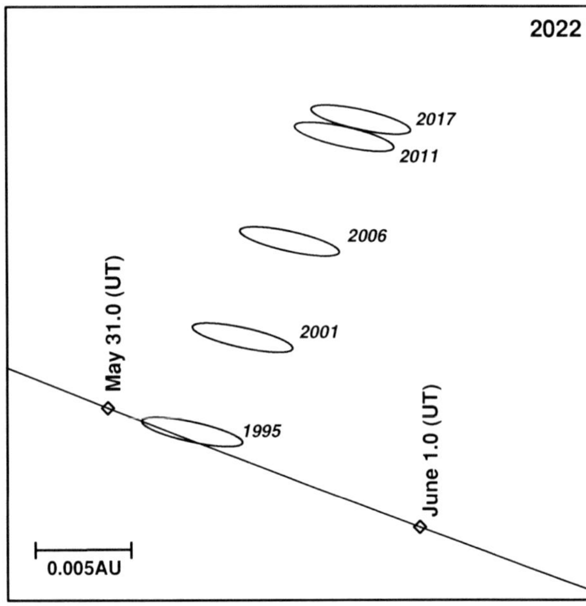

Fig. 1 The location of the intersection with the ecliptic plane of the dust trails is shown. The shape of each trail is assumed to be like a tube whose diameter is 0.001 AU. The continuous line represents the path of the Earth in 2022

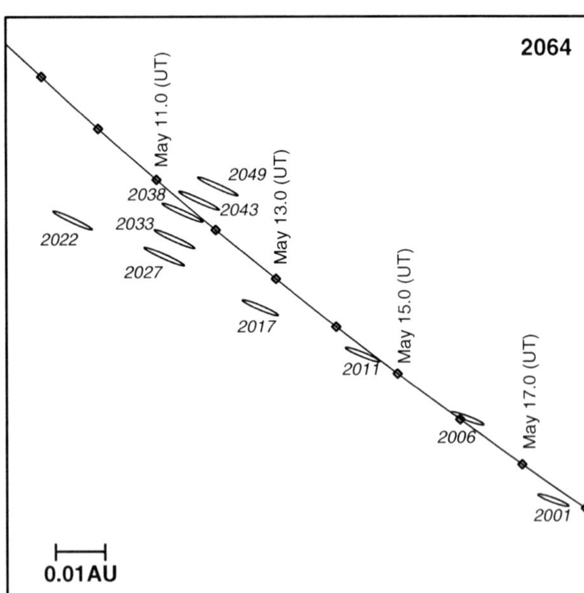

Fig. 2 Same as Fig. 1, but in 2064

study (Lüthen et al. 2001). And its ejection velocity is not so high, and its value of f_M is not so low. But it is a disappointing point that the value of V_g is lower than general meteor showers. Still, in 2022, the meteor due to this dust trail is highly expected. The other chance may be the encounter in 2064, which is shown in Fig. 2. No dust trail will approach the Earth as closely as the above-mentioned dust trail in 2022. Additionally, both ejection velocity and f_M are not in ideal condition compared with 2022. Still, it is interesting that the dust trails will approach the Earth one after another for a few days. In addition to the encounters in these 2 cases, looking at Table 1, we can expect that meteors will be observed in several other epochs.

4 Concluding Remarks

According to the results of our calculation, the several dust trails will approach the Earth closely in the future. So, we can expect that meteors originated from 73P/Schwassmann–Wachmann will be observed in several years, especially in 2022. At this stage, we considered only nucleus C in our calculation. This comet probably continues to be split into many fragments. We immediately have to examine whether there are meteors expected from the other nuclei of this comet.

References

D.J. Asher, in *Proc. Int. Meteor Conf.*, Frasso Sabino, Italy, 23–26 Sept, 1999, ed. by R. Arlt (IMO, Belgium, 2000), p. 5

T. Fuse, N. Yamamoto, D. Kinoshita, H. Furusawa, J. Watanabe, PASJ (Publ. Astron. Soc. Japan) **59**, 381–386 (2007)

P. Jenniskens, J. Vaubaillon, Astron. J. **134**, 1037–1045 (2007)

K. Kinoshita (Comet Orbit Home Page, 2007), http://www9.ocn.ne.jp/~comet/. Accessed 1 Feb 2007

H. Lüthen, R. Arlt, M. Jäger, WGN J. Int. Meteor Organ. (JIMO) **29**(1), 15–28 (2001)

R.H. McNaught, D.J. Asher, WGN J. Int. Meteor Organ. (JIMO) **27**, 85 (1999)

M. Sato, WGN J. Int. Meteor Organ. (JIMO) **31**, 59 (2003)

Z. Sekanina, H. Boehnhardt, H.U. Käufl, K. Birkle, JPL Cometary Sciences Group Preprint Series. 183 (1996)

P.A. Wiegert, P.G. Brown, J. Vaubaiillon, H. Schijns, MNRAS **361**, 638–644 (2005)

The Lyrid Meteor Stream: Orbit and Structure

Vladimir Porubčan · Leonard Kornoš

Originally published in the journal Earth, Moon, and Planets, Volume 102, Nos 1–4.
DOI: 10.1007/s11038-007-9188-9 © Springer Science+Business Media B.V. 2007

Abstract A filamentary structure in the Lyrid meteor stream based on photographic orbits available in the IAU Meteor database is identified and studied. About 17 Lyrids are found in the database and the stream mean orbit is derived. The shower radiant is compact, of a size 2° × 1.5°. Applying a stricter limiting value for the Southworth-Hawkins D-criterion, two distinct filaments in the stream, on a short and a long period orbit, are separated. To confirm their consistency as filaments, their orbital evolution over 5,000 years is investigated.

Keywords Lyrid meteor stream · Meteoroids · Photographic meteor orbits

1 Introduction

The April Lyrids are known as producing a weak meteor shower with a visual ZHR at maximum of about 5–10. However, occasionally they exhibit an enhancement of activity, exceeding 100 meteors per hour (Lindblad and Porubčan 1992). The maximum generally appears on April 21–22 with the radiant at a right ascension of 272° and declination of 34°. The stream is genetically associated with comet C/1861 G1 Thatcher. By looking in the literature for the Lyrid outbursts reported in the last two centuries, remarkable features are found. For example, it is evident that all the peaks were of a short duration (2 h maximum) and at almost the same solar longitudes, approximately a quarter of a day before the annual peak (Lindblad and Porubčan 1992). These occasional activity enhancements indicate a filamentary structure in the stream.

Older observations also seem to indicate a 12-year periodicity of the enhanced Lyrid maxima (Guth 1947). Arter and Williams (1995) looked for an explanation of the observations, and the results of a study of the stream's evolution in which model particles were released from comet Thatcher indicate an occurrence of higher density parts of the stream in a 12-year cycle (Arter and Williams 1997).

V. Porubčan (✉) · L. Kornoš
Comenius University Bratislava, Bratislava 84228, Slovak Republic
e-mail: porubcan@fmph.uniba.sk

2 Orbit and Filaments of the Lyrids

In the present catalogues of meteor orbits, the most accurate observations come from photography. However, there is only a small number of accurate orbits available for a detailed analysis of the Lyrid stream. Kresák and Porubčan (1970) for their analysis of the shower radiant and orbit had at their disposal only seven Lyrids. Lindblad and Porubčan (1991) derived the mean orbit of the stream and motion of the radiant on the basis of 14 Lyrids.

The current version of the Meteor Data Center catalogue of photographic orbits contains the orbital elements of 4,581 meteors (Lindblad et al. 2005). The catalogue has increased by about 1,000 new orbits. This provided a chance to derive a more precise mean orbit and radiant of the stream.

Lyrid stream members were identified by using a computerized stream search procedure utilizing the Southworth-Hawkins D-criterion (Southworth and Hawkins 1963). For a limiting value of $D = 0.20$, 17 meteors belonging to the Lyrids, from the period April 21 to 25, were found. The Lyrid radiant daily motion and radiant ephemeris derived from the 17 stream members is

$$\alpha = 272.3^\circ + 0.802^\circ(L_s - 32.5^\circ), \delta = 33.4^\circ - 0.155^\circ(L_s - 32.5^\circ) \quad (1)$$

The daily motion in right ascension and declination was found by a least squares solution; L_s is the solar longitude of the time of observation for equinox 2000.0, and 32.5° is the solar longitude of the maximum activity. The size of the radiant area corresponds to the dispersion of the orbits and thus to the stream structure: the Lyrid radiant, allowing for the daily motion and centered on the α and δ given by (1), reaches a size of about $2^\circ \times 1.5^\circ$.

We have also searched for possible filaments in the stream by lowering the limiting value of D. For $D = 0.025$ two separate groups in very different orbits (short and long period ones) were obtained. Each filament is formed by four meteors, but in filament 2 meteor 048C1 is on a hyperbolic orbit and was therefore excluded from the sample. The mean orbits of the filaments are listed in Table 1 together with the mean Lyrid orbit. Table lists also the orbit, theoretical meteor radiant (α, δ) and geocentric velocity (V_g) of comet Thatcher. As the semimajor axes of Lyrids have a range corresponding to orbits from very short period up to hyperbolic ones, the mean a in Table 1 was calculated from the mean e and q.

In the next step we have verified the reality of the filaments by making a backward integration of their orbits. The two filaments and Lyrid mean orbit were numerically

Table 1 Orbital elements and radiants of the Lyrid filaments, mean Lyrid orbit and comet C/1861 G1 Thatcher (eq. 2000.0), with corresponding standard deviations *s.d.*

Object	Q [AU]	a [AU]	q [AU]	e	i (°)	ω (°)	Ω (°)	π (°)	α (°)	δ (°)	V_g [km/s]	n
filament 1	21.4	11.15	.914	.918	79.2	216.0	31.6	247.6	271.4	33.0	46.26	4
s.d.	-	-	.002	.018	.4	.3	.7	.8	.4	.2	.14	
filament 2	141.4	71.15	.925	.987	80.4	213.0	32.8	245.8	273.0	33.3	47.35	3
s.d.	-	-	.002	.009	.1	.4	.7	.2	.5	.2	.11	
mean orbit	91.2	46.05	.921	.980	79.7	213.8	32.5	246.3	272.3	33.4	47.03	17
s.d.	-	-	.007	.072	1.0	1.6	1.0	1.5	1.3	.6	.91	
Thatcher	110.4	55.682	.9207	.9835	79.77	213.45	31.87	245.32	272.0	33.5	47.08	-

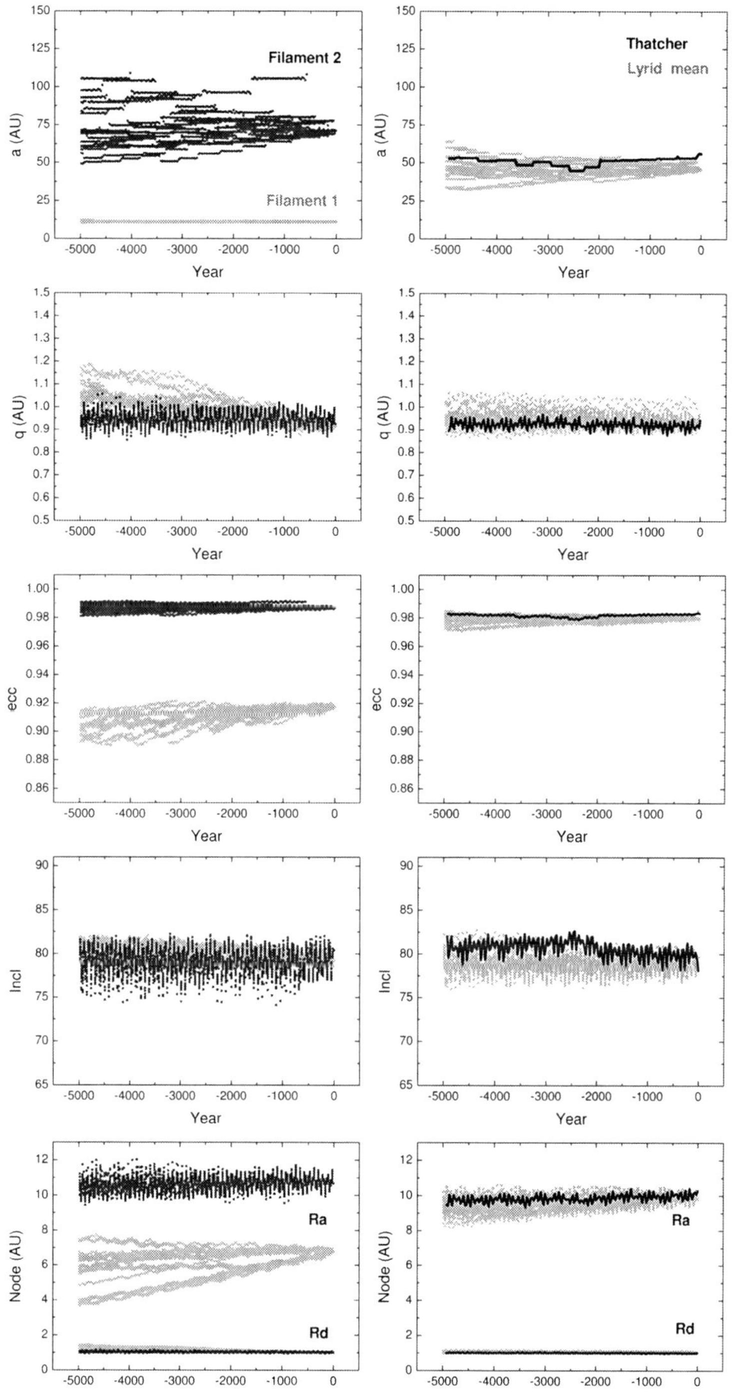

Fig. 1 Orbital evolution of the Lyrid filaments 1 and 2 (left plots), the Lyrid mean orbit and parent comet Thatcher (right plots) over the last 5,000 years

integrated back 5,000 years. In this integration, each orbit is represented by 18 modeled particles distributed equidistantly by 20° in mean anomaly.

The orbital evolution of filaments 1 and 2 and the mean Lyrid orbit, as well as that of comet Thatcher, is shown in Fig. 1, where the plots show the evolution in semimajor axis a, perihelion distance q, eccentricity e, inclination i and heliocentric distances Ra and Rd of the ascending and descending nodes over the last 5,000 years.

3 Discussion and Conclusions

In the present study the stream and comet are analysed from a short-term point of view (5,000 years). The orbit of the stream and comet are shown to be relatively stable. The Lyrids are influenced by three dominant bodies that are shaping the structure of the stream: Jupiter, Saturn and Earth.

Filament 1 (April 21–22, mean $L_s = 31.6°$) coincides quite well with the outburst peak preceding the annual maximum (Lindblad and Porubčan 1992) and is more influenced by perturbations. The ascending node of the filament is at a heliocentric distance of 6.8 AU and exhibits a relatively large dispersion caused by the strong influence of Jupiter. Several modeled particles have repeated close approaches to the Earth and Jupiter during the period of integration.

Filament 2 (April 22–23, mean $L_s = 32.8°$) is closer to the annual maximum. The evolution of the mean Lyrid orbit and filament 2 is rather stable. Both orbits undergo similar evolution, almost identical with the evolution of the parent comet Thatcher. The ascending nodes are close to the orbit of Saturn. A few modeled particles have close approaches to Saturn and Earth.

In conclusion, we have identified two filaments among photographic Lyrid meteor orbits. They are moving in very distinct orbits with mean periods of revolution of about 40 and 600 years. The filaments are result of gravitational modification of the stream structure by planets and are influenced mostly by Jupiter with a significant contribution from Saturn. The descending nodes of all the investigated orbits are stable and close to the Earth's orbit during the whole period of integration.

Acknowledgements Thanks to David Asher and Josep Trigo-Rodriguez for helpful comments at the refereeing stage. This work was supported by the Scientific Grant Agency VEGA, grant No. 3067.

References

T.R. Arter, I.P. Williams, The April Lyrids. Mon. Not. R Astron. Soc. **277**, 1087–1096 (1995)
T.R. Arter, I.P. Williams, Periodic behaviour of the April Lyrids. Mon. Not. R Astron. Soc. **286**, 163–172 (1997)
V. Guth, On the periodicity of Lyrids. Bull. Astron. Inst. Czechosl. **1**, 1–4 (1947)
L. Kresák, V. Porubčan, The dispersion of meteors in meteor streams. I. The size of the radiant areas. Bull. Astron. Inst. Czechosl. **21**, 153–170 (1970)
B.A. Lindblad, V. Porubčan, The orbit of the Lyrid meteor shower. Bull. Astron. Inst. Czechosl. **42**, 354–359 (1991)
B.A. Lindblad, V. Porubčan, Activity of the Lyrid meteor. in Asteroids stream., *Comets, Meteors 1991*, ed. by A. Harris, E. Bowell (Flagstaff, 1992) pp. 367–370
B.A. Lindblad, L. Neslušan, V. Porubčan, J. Svoreň, IAU Meteor Database of photographic orbits - version 2003. Earth Moon Planets **93**, 249–260 (2005)
R.B. Southworth, G.S. Hawkins, Statistics of meteor streams. Smithson Contr. Astrophys. **7**, 261–285 (1963)

Model Radiants of the Geminid Meteor Shower

Galina O. Ryabova

Originally published in the journal Earth, Moon, and Planets, Volume 102, Nos 1–4.
DOI: 10.1007/s11038-007-9180-4 © Springer Science+Business Media B.V. 2007

Abstract This paper describes the final stage of the study of the Geminid meteoroid stream formation and evolution using the nested polynomials method reported by Ryabova (in: Warmbein (ed.) Meteoroids 2001, Proc. of the Internat. Conf., Kiruna, Sweden, 6–10 August 2001; MNRAS 375:1371–1380, 2007). In the previous work we discussed possibility to calibrate the model using the shape of the model activity profiles and configuration of orbital parameters. Here we show that the radiant structure also could be utilized for this purpose, since the model radiant structure has a very specific pattern. Model area of radiation does not have a "classical" prolate linear shape, and the configuration of activity centers has a "V" shape. During one night of simulated observations several activity centers could be observed. The model produced maps of the velocity distribution in the radiant area

Keywords Meteoroids · Methods: numerical

1 Introduction

Ryabova (2007) presented a qualitative model of the Geminid meteoroid stream from the Geminid's parent body (asteroid (3200) Phaethon) in order to study the main features of its structure and explain the processes which are responsible for that structure. It was shown that the structure of a model stream of collisional (Ryabova 1989), or eruptive (Bel'kovich and Ryabova 1989) origins does not agree with the observations of this meteor shower. The activity profile for the Geminid shower has the specific bimodal shape, which is expected by a cometary model of the stream generation (Ryabova 2001). So there are strong grounds to believe that the stream has a cometary origin. Its formation probably occurred during a relatively short time of one or several cometary revolutions (Lebedinets 1985). The very stable shape of the shower activity profile during last 60 years (Rendtel 2004) could be

G. O. Ryabova (✉)
Research Institute of Applied Mathematics and Mechanics, Tomsk State University, Tomsk 634050, Russia
e-mail: ryabova@niipmm.tsu.ru

indirect evidence of that. The age of the stream was previously estimated at 2,000 ± 1,000 years (Ryabova 1999). The uncertainty of the stream's age gives rise to a peculiar problem, which is that the location of the model cross-section on the ecliptic could be shifted from the location of the real stream.

Ryabova (2007) described how the observed shapes of the flux density and mass index *s* profiles and the configurations of the shower orbital elements, can help in the calibration of the model. In this paper we concentrate on the model radiant structure.

2 Model

The methodology applied in this work is fairly simple and is regularly used (see for e.g. Fox et al. 1983; Brown and Jones 1998; Vaubaillon et al. 2005). For a given parent body orbit we choose points of ejection according to chosen scheme of the meteoroid stream generation (collisional, eruptive, cometary etc.). Then the ejection velocity vector is obtained for every model meteoroid, and the meteoroid orbit is calculated. Evolution of the orbit is calculated from the moment of ejection till the present.

For this work we used Whipple (1951) formula to calculate the ejection velocity value, while the directions of ejections were assumed to be distributed uniformly in the sunlit hemisphere. The ejection points were distributed uniformly around the parent body orbit, that fits reasonably well to dust production rate proportional to r^{-4}, where r is heliocentric distance. Assuming the age of the stream 2,000 years (Ryabova 1999), the orbit of asteroid (3200) Phaethon calculated for the epoch JD1721206.3 (0.407 AD) has been used as the reference orbit. Meteoroid ejections were modeled for two streams of spherical particles (density 1 g cm^{-3}) with masses of $m_3 = 2.14 \times 10^{-3}$ g and $m_4 = 2.14 \times 10^{-4}$ g. For short we will refer to these streams and their showers as "stream m_3" or "shower m_4". The evolution of the test particle orbits was calculated using nested polynomials. In detail the method and model used was described in Ryabova (2007).

To explain how we use a radiant structure to calibrate the model, we should consider the Geminid model cross-section in the ecliptic plane. Figure 1 illustrates that the shower activity profiles and the profiles of the mass index *s* will be strongly dependent on the location where the Earth crosses the stream (shown as lines *A*, *B*, *C*, *D* and *E* in Fig. 1, *A* being along the Earth orbit). It was shown earlier that the Geminid meteoroid stream consists two layers (Ryabova 2001, 2007). The origin of the layers is that the orbital characteristics of the particles ejected from the parent comet are different for when the comet approaches perihelion and when it moves away from perihelion. In the small panel designated "Stream m_3" (Fig. 1), which displays the cross-section of the corresponding model stream, the pre-perihelion layer is shown by gray color, and the post-perihelion layer by black color. The layers cross approximately along the mean orbit of the model stream. The Earth's orbit thus intersects two different dust layers resulting in two different shower activity maxima. If we consider only differential[1] showers, the first maximum of each consists mainly of pre-perihelion meteoroids, the second maximum will be mainly post-perihelion meteoroids. The distance between the first and the second maxima depends on the mass of meteoroids (see small panels in Fig. 1), because the ejection velocity is larger

[1] The differential shower/stream is defined as a shower/stream of particles with a definite meteoroid mass, for example m_3 or m_4. The cumulative shower/stream consists of particles having masses larger than some minimal mass.

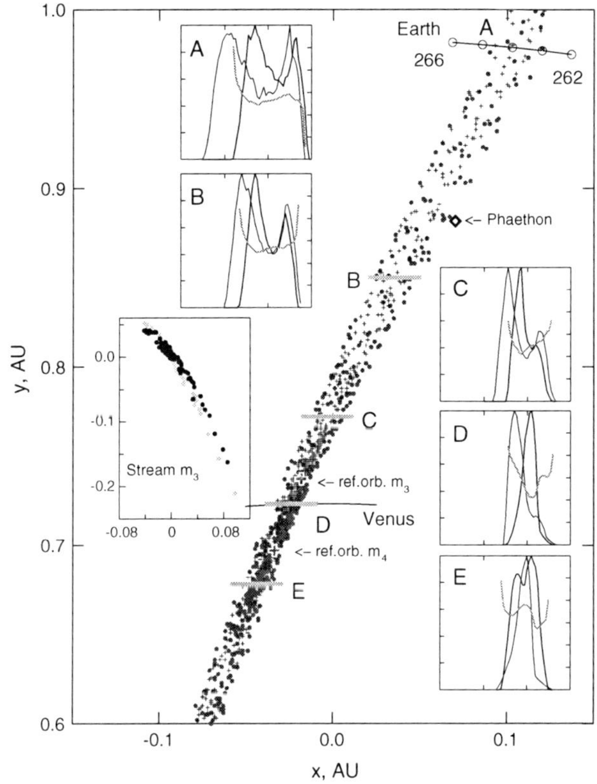

Fig. 1 Geminid's model cross-section in the ecliptic plane for orbits of particles with masses m_3 (+) and m_4 (•). *A* designates the Earth's orbit in the interval 262°–266° in solar longitudes. Other sections are designated by *B–E*. In the small panels, designated by *A–E*, activity profiles, i.e. flux density variations along the Earth's orbit, for particle masses m_3 (thick line) and m_4 (thin line), and a profile for mass index *s* (thickest line) are shown. The distance between the tick marks on the abscissa-axes of small panels is equal to 1°. The profiles are calculated along the corresponding sections. The small panel designated by "Stream m_3" demonstrates pre- and post-perihelion layers (see text) in the cross-section of the model stream m_3 at the descending node of its mean orbit, designated "ref. orb. m_3". The plane of the plot is normal to velocity vector of the orbit in the node. The abscissa-axis is directed away from the Sun, the scales on both axes are in AU. Figure 1 was modified after Ryabova (2007, Fig. 5) and Ryabova (2006, Fig. 1)

for smaller particles. In the cumulative shower the separation of pre- and post-perihelion meteoroids is not so distinct.

The location of stream in the model presented here is shifted from in the real stream because (1) the exact age of the stream is unknown, (2) we used polynomial approximations instead of a precise method of numerical integration (Ryabova 2007). However, the dependence of the shower activity profiles, the profiles of the mass index and radiant patterns on the location could help us to fit the model. For example, if we were to find that the observed radiant pattern fits the model pattern for section *C* and differs from all others, we could then suggest that the Earth should cross the stream near the section *C*. To put it differently, we should move the model in such a way that the section *C* coincided with the real shower activity area on the Earth's orbit. It is obvious that the task is unrealizable if the patterns are similar each other.

3 Discussion

Let us consider the radiant areas corresponding to sections *A* and *C* in Fig. 1. Figure 2 shows the geocentric equatorial coordinates (α, δ) of the radiants used in our model stream. In this figure and all other figures presenting model radiants, the axes scales are not shown because the model stream is not yet calibrated. The model streams m_3 and m_4 are quite similar in structure, apart from dispersion in the stream m_4 is larger, because ejection velocities for smaller particles are larger. So we may consider a radiant structure pattern for any of the model streams, m_3 or m_4.

The comparisons of the modeled results with observations qualitative only. Each of the sections in Fig. 2 show very specific and different patterns for the modeled Geminid radiant structure for the various possible ways of the Earth's orbit crossing through the stream. For an interval of observations, which is less than the full length of the shower, the pattern inevitably changes, because the stream structure (both for a model or a real stream) changes along the Earth's orbit. For example for only one night of observations or 0.5° in solar longitude the radiant structure is shown in Fig. 3; the pattern in Fig. 3 consists of the radiants from the left panel of Fig. 2 related to the selected interval. When selected observation time were to be around the first activity of the maximum, the model predicts that we would observe pre-perihelion meteoroids (Ryabova 2007, Fig. 13), and indeed overwhelming majority of radiants we see in Fig. 3 are from pre-perihelion orbits.

We found that the model radiant area has two centers of activity that correspond to the two maxima of the activity curves as shown in Fig. 2. The location of the activity centers for the model showers m_3 and m_4 (Fig. 4) has a V-shape. But we should take into consideration the following. Firstly, the right center is more intensive than the left one for both showers (Fig. 4). Secondly, the maps for streams m_3 and m_4 contain 5,000 radiants each, while for a real meteor shower flux density for particles with masses m_4 should be 10^s

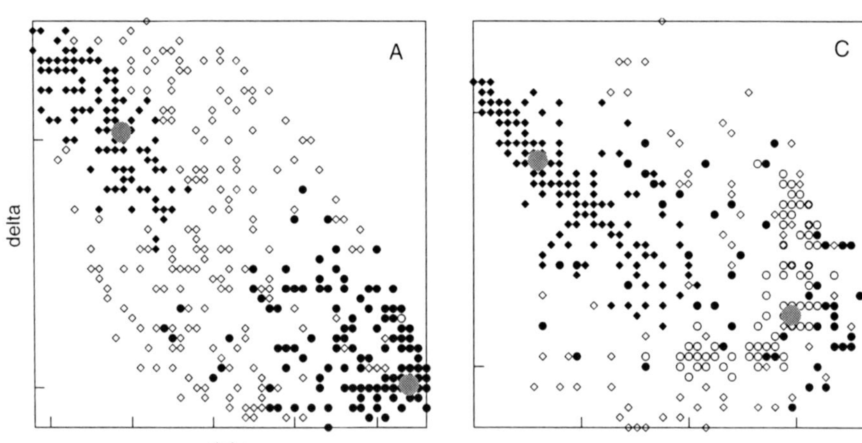

Fig. 2 Geocentric equatorial coordinates for the 500 random model particles forming the activity profiles of the stream m_4 in sections *A* (Ryabova 2007; Fig. 12, reproduced by courtesy of MNRAS) and *C*. Two large circles mark the first and the second maxima of activity. The distance between the tick marks on the axes is equal to 0.5°. Cells are designated depending on the true anomaly of ejection point υ_e on the cometary orbit: (*black diamonds*) $180° < \upsilon_e < 270°$, i.e. nucleus moving from aphelion; (*empty diamonds*) $270° < \upsilon_e < 360°$, i.e. approaching to perihelion; (*black circles*) $0° < \upsilon_e < 90°$, passed perihelion; (*empty circles*) $90° < \upsilon_e < 180°$, moving to aphelion

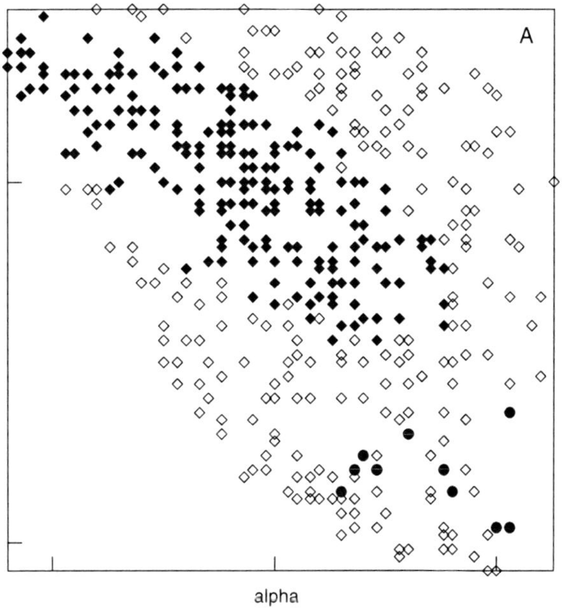

Fig. 3 The same, as in Fig. 2, but positions only for meteoroids observed during the first maximum of the shower m_4

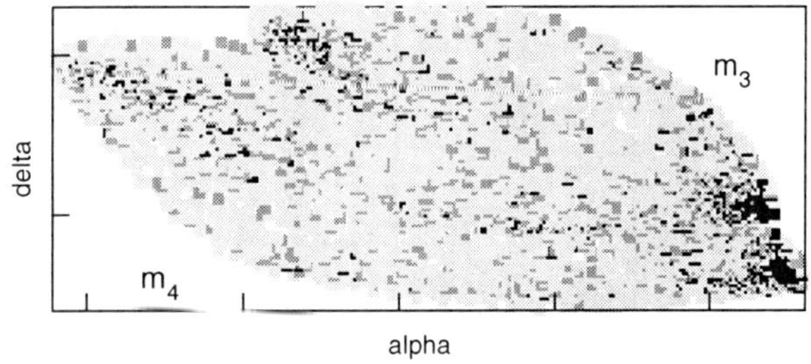

Fig. 4 The superimposed maps of radiant activity for showers m_3 and m_4. Section *A*. Highest activity is designated by black color. The distance between the tick marks on the axes is equal to 0.5°

times of flux density for particles with masses m_3. Finally, we usually do not observe 10 thousand radiants in the Geminid meteor shower. So the left side of "V" can stay unnoticed in the real shower.

Figure 5 shows how the model predicts that we may observe several activity centers during one night of Geminid shower observation. This resulting radiant activity is based on 10,934 radiants in Fig. 5a and 2,373 in Fig. 5b. The amount of test particles for model streams m_3 and m_4 is 10 million for each.

According to data of radar observations (Sidorov and Kalabanov 2001, 2002; Kalabanov et al. 2002) several activity spots were found in the area of Geminids radiation; the authors interpreted them as microshowers of unknown origin. But statistics of their data is rather insufficient to make any conclusions. The lack of observational data is a persistent problem. One of the best samples[2] of radiants (and orbits) we managed find so far are video

[2] Another good sample obtained by video observation of meteors at the Ondřejov observatory could be found in Koten et al. (2003). We did not use it here because too small number of meteors with estimated photometric masses $\sim 10^{-4}$ g ($N = 6$) does not allow to compare this subsample with the subsample for meteors with estimated photometric masses $\sim 10^{-3}$ g ($N = 41$).

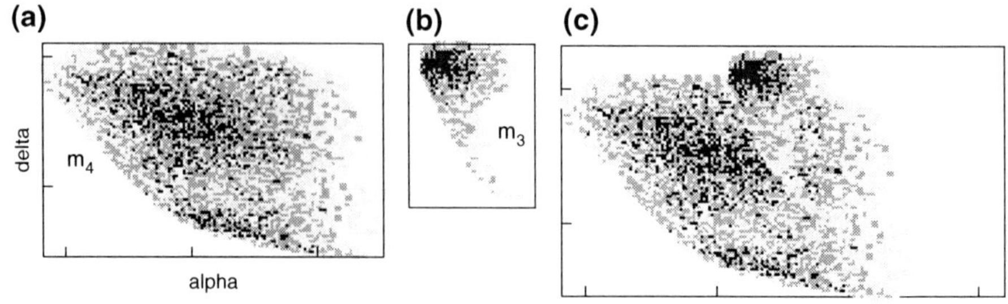

Fig. 5 The map of radiant activity for meteoroids with mass m_4 (**a**), m_3 (**b**), and combined (**c**). Period of "observation" is 0.5° in solar longitude around the first maximum of the activity curve for the shower m_4, section *A*. Gradations of gray shows the density of radiants (black is the highest). The distance between the tick marks on the axes is equal to 0.5° in (**a**) and (**c**). The sides of the panel in (**b**) embracing square are equal approximately 0.4°. Unmarked axes are for the same (α, δ) parameters

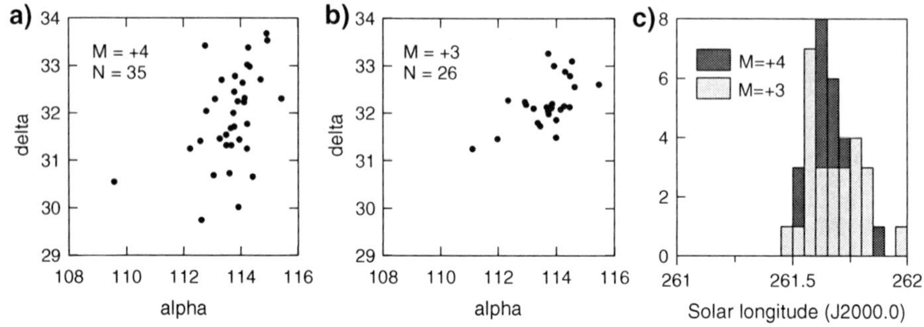

Fig. 6 Radiant positions for DMS video orbits (see text) are shown for Geminid meteors of estimated visual magnitude (**a**) $M = +4$ and (**b**) $M = +3$. *N* is number of radiants for each sample. Distribution of the same orbits in solar longitude is shown in (**c**)

observations completed by the Dutch Meteor Society[3] (Lignie and Betlem 1997; Lignie 1998), is shown in Fig. 6. Taking into account that in the model we considered the differential showers, i.e. with a definite particle mass, we have to make comparison within a narrow mass range. Thus in Fig. 6 the radiant positions (α, δ) are shown Geminid meteors of estimated visual magnitudes[4] $M = +3$ and $M = +4$. DMS observations are concentrated in a narrow range of solar longitudes (Fig. 6c) between the maxima. This case of observed meteor activity should be compared with the narrow-range patterns obtained by the model that were shown in Figs. 3 and 5. But we cannot make direct comparisons between these structures obtained by the model and from observations because the statistics in case of the observations are poor. Another possible issue could be observational bias although any such selection effects were not studied yet.

Figure 7 displays the geocentric velocity distributions for sections *A* and *C*. Section *A* shows distinct regions occupied by different velocity intervals, but in section *C* there are no specific intervals. The results in Fig. 7 highlight that when we will have statistically

[3] http://www.dmsweb.org

[4] We do not consider still unsolved problem of so called "mass scale", i.e. correspondence of the meteor mass and magnitude for different methods of observations. So the choise of $M = +3$ and $M = +4$ is rather arbitrary: just faint meteors.

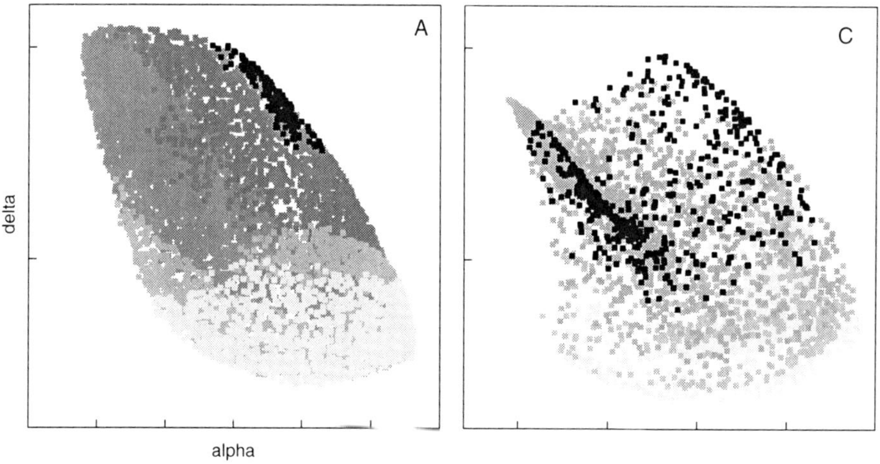

Fig. 7 The maps of geocentric velocity distribution in radiant areas for sections *A* and *C* of the shower m_4. The darker areas correspond to larger velocity. The distance between the tick marks on the axes is equal to 0.5°. Unmarked axes are for the same (α, δ) parameters

reliable meteor observations that are free of observational selection effects, such as the geometrical one, comparison with the model predictions as were described in this paper can be made successfully. Such systematic comparisons will serve further calibrations of the model to determine the scale of the observed structures found in the radiant distribution of simulated meteors in the model stream. For example, an issue that needs to be resolved concerns the scale of the model radiants in Figs. 2–7 and the precision at which individual meteor radiants can be determined observationally, e.g. Koten et al. (2003) and Campbell-Brown (2007).

Acknowledgements This work was supported by RFBR Grant N 05-02-17043. I wish to thank the Organizers of the Meteoroids 2007 conference for financial support, and also thank Dr. Tadeusz Jopek and the anonymous reviewer for helpful comments. I am extremely indebted to my Editor, Dr. Frans J.M. Rietmeijer, for his tireless efforts to improve the manuscript.

References

O.I. Bel'kovich, G.O. Ryabova, Formation of the Geminid meteor stream with the disintegration of a cometary nucleus. Sol. Syst. Res. **23**, 98–102 (1989)

P. Brown, J. Jones, Simulation of the formation and evolution of the Perseid meteoroid stream. Icarus **133**, 36–68 (1998)

M. Campbell-Brown, The meteoroid environment: shower and sporadic meteors, in *Dust in Planetary Systems*, ed. by H. Krueger, A. Graps. Proc. of Working Group, Kauai, Hawaii, USA, 26–30 September 2005 (2007), pp. 11–21 [ESA SP-643]

K. Fox, I.P. Williams, D.W. Hughes, The rate profile of the Geminid meteor shower. MNRAS **205**, 1155–1169 (1983)

S. Kalabanov, V. Sidorov, A. Stepanov, Structure of area of radiation of Geminids meteor shower and its vicinities on celestial sphere. One or many showers?, in *Asteroids, Comets, Meteors—ACM 2002*, ed. by B. Warmbein. Proc. of Internat. Conf., Berlin, Germany, 29 July–2 August 2002 (ESA Publications Division, Noordwijk, Netherlands, 2002), pp. 165–168 [ESA SP-500]

P. Koten, P. Spurný, J. Borovička, R. Štork, Catalogue of video meteor orbits. Part 1. Publ. Astron. Inst. Acad. Sci. Czech Rep. **91**, 1–32 (2003)

V.N. Lebedinets, Origin of meteor swarms of the Arietid and Geminid types. Sol. Syst. Res. **19**, 101–105 (1985)
M. de Lignie, Mass segregation in the Geminid meteoroid stream as seen from recent photographic and video observations. Radiant **20**, 58–60 (1998)
M. de Lignie, H. Betlem, Simultane videometeoren van de Geminidenactie 1996 (in Dutch). Radiant **19**, 111–114 (1997)
J. Rendtel, Evolution of the Geminids observed over 60 years. Earth Moon Planets **95**, 27–32 (2004)
G.O. Ryabova, On possibility of the Geminid meteoroid stream generation during crater-forming collision of asteroids (in Russian). Astron. Geod. Tomsk Gos. Univ. Tomsk **15**, 182–189 (1989)
G.O. Ryabova, Age of the Geminid meteoroid stream (review). Sol. Syst. Res. **33**, 224–238 (1999)
G.O. Ryabova, Mathematical model of the Geminid meteor stream formation, in *Meteoroids 2001*, ed. by B. Warmbein. Proc. of the Internat. Conf., Kiruna, Sweden, 6–10 August 2001 (ESA Publications Division, Noordwijk, Netherlands, 2001), pp. 77–82 [ESA SP-495]
G.O. Ryabova, Meteoroid streams: mathematical modelling and observations, in *Asteroids, Comets, Meteors*, ed. by D. Lazzaro, S. Ferraz Mello, J.A. Fernandez. Proc. of the 229th IAU Symp., Buzios, Rio de Janeiro, Brasil, 7–12 August 2005 (Cambridge University Press, Cambridge, 2006), pp. 229–247
G.O. Ryabova, Mathematical modelling of the Geminid meteoroid stream. MNRAS **375**, 1371–1380 (2007)
V. Sidorov, S. Kalabanov, The discret solution of a quasy-tomography problem for construction of radiant distribution of meteors by results of radar goniometer measurements, in *Meteoroids 2001*, ed. by B. Warmbein. Proc. of the Internat. Conf., Kiruna, Sweden, 6–10 August 2001 (ESA Publications Division, Noordwijk, Netherlands, 2001), pp. 21–26 [ESA SP-495]
V. Sidorov, S. Kalabanov, Heterogeneity of sporadic meteor complex as the rich data for possible prediction of comets, asteroids and other bodies, in *Asteroids, Comets, Meteors – ACM 2002*, ed. by B. Warmbein. Proc. of Internat. Conf., Berlin, Germany, 29 July–2 August 2002 (ESA Publications Division, Noordwijk, Netherlands, 2002), pp. 149–152 [ESA SP-500]
J. Vaubaillon, F. Colas, L. Jorda, A new method to predict meteor showers. II. Application to the Leonids. Astron. Astrophys. **439**, 761–770 (2005)
F.L. Whipple, A comet model II. Physical relations for comets and meteors. Astrophys. J. **113**, 464–474 (1951)

The Orionid Meteor Shower Observed Over 70 Years

Jürgen Rendtel

Originally published in the journal Earth, Moon, and Planets, Volume 102, Nos 1–4.
DOI: 10.1007/s11038-007-9192-0

Abstract Visual Orionid meteor data dating back to 1944 were transformed into the standard format of the Visual Meteor Data Base (VMDB) of the International Meteor Organization (IMO) for systematic analysis. The strong 2006 Orionid return with a very low population index ($r = 1.6$) and a peak ZHR of 60 (about 2.5 of the average peak strength) resembled meteor showers connected with the returns of resonant meteoroids. An investigation of data dating back to 1928 yielded similar rate enhancements in 1936, further supporting the assumption that meteoroids trapped in the 1:6 resonance with Jupiter caused the unusual 2006 Orionid return.

Keywords Meteors · Meteor showers · Orionids · Outburst · Resonant meteoroids

1 Introduction

The Orionid meteor shower is one of the two meteor showers associated with comet 1P/Halley. Its typical maximum ZHR is of the order of 20–25. The broad maximum lasts usually from October 20 to 24. The activity profile during this period is not smooth but shows several submaxima. The years surrounding the comet's latest perihelion passage yielded no rate enhancement (Porubčan et al. 1991) because of the large minimum distance between the orbits of the parent comet and the Earth. Other outbursts, such as in 1993 (Rendtel and Betlem 1993) were due to isolated particle concentrations far distant from the comet that may occur when the comet is far from its perihelion position (Jenniskens 2006). Most modelling attempts were made 1985/1986 during 1P/Halley's latest perihelion (e.g. McIntosh and Jones 1988) and later by Ryabova (2003) with a summary of modelling attempts. Speculations about particles in resonant orbits date back to Hajduk (1970).

J. Rendtel
International Meteor Organization, PF 600118, 14476 Potsdam, Germany

J. Rendtel (✉)
Astrophysical Institute Potsdam, An der Sternwarte 16, 14482 Potsdam, Germany
e-mail: jrendtel@aip.de

J.M. Trigo-Rodriguez et al. (eds.), *Advances in Meteoroid and Meteor Science*.
DOI: 10.1007/978-0-387-78419-9_16

2 Orionid Data

Since the 80s visual data of meteor showers are available on a global scale as a result of standardized observing techniques and reporting procedures. They are stored the Visual Meteor Data Base (VMDB) of the International Meteor Organization (IMO). The continuous data collection started in 1988, but for some showers older data was added. The Orionids with a well defined radiant and no other active meteor showers occurring at the same time and same area in the sky, fulfill the criteria to be included in the database.

Of course, the information about an individual visual meteor is limited: the shower association of a single meteor is based on its direction, angular velocity and apparent trail length. In the case of reasonable meteor rates (i.e. more than about 15 shower meteors per hour), the statistics is not sensitive against mis-aligned meteors. To keep the data set consistent and avoid effects from low activity periods, we concentrate our analysis to the near-maximum period between $\lambda_{\odot} = 206°$ and $212°$ (all values of the Solar longitude in the text refer to equinox 2000.0).

Data were added from the sources listed in Table 1. The Skalnaté Pleso data (Štohl and Porubčan 1981) include detailed lists for each individual observer that can be directly transformed into the VMDB format. Data lists published by Prentice prior to 1939 (Prentice 1936, 1939) contain only meteor totals for entire nights and are therefore not suitable for rate calculations. Lovell (1954) showed the processed rates from Prentice and Alcock on p. 289. Additional data may be reconstructed from the American Meteor Society publications by Olivier (1935) found in the process of writing this paper, but they are not yet included in the VMDB.

While we were able to calibrate the data back to 1944 according to the present VMDB standards, this was not possible for all older data. In most cases we have to restrict conclusions about the activity of the Orionids at earlier returns to relative values. The number of non-shower meteors reported in the individual count intervals is used to check the qualitative value of the Orionid rates.

3 Population Index Profiles

The Orionid showers typically produce faint meteors while the fraction of fireballs is small as compared with other meteor showers (Rendtel et al. 1995). The population index r was derived from the magnitude data using an adaptive interval length. Data is sampled until 100 meteors are in a bin for an individual value of r. Details are described by Arlt and

Table 1 Orionid data sources added to the VMDB

Source	Years	Remarks
Meteor News (USA)	1979–1987	Reports of raw data
West Australian Meteor Society	1979–1986	Raw data lists
ZHR Bulletin (Hungary)	1984–1986	Raw data lists
Arbeitskreis Meteore (Germany)	1979–1987	Original forms
Skalnaté Pleso (Slovakia)	1944–1950	Raw data lists (Štohl and Porubčan 1981)
Loreta (Italy)	1936	Count data (Millman 1936)
Prentice, Alcock (UK)	1928–1939	Processed data (Lovell 1954)

Barentsen (2006). The interval length was chosen as short as possible in order to detect short term variations which may indicate fine structures in the stream. The population index can be transformed into the mass index by

$$s = 2.5 \log r + 1 \tag{1}$$

if the shape of the light curve does not change over the magnitude range being considered.

Minimum values of the population index are given in Table 2. If possible, data were analysed per return to avoid averaging over structures or annual peculiarities. Values for the population index commonly listed in the literature are 2.9 (Rendtel et al. 1995) or around 2.5 (Dubietis 2003). Our analysis showed that despite all variations from one return to the next a minimum of the population index r can be found repeatedly at 207.9° ± 0.15° in the annual profiles. As an example, we show the profile of the Orionids 1995 in Fig. 1. Whether the smaller r-values in 1993 and 2006 (Table 2) can be regarded as an indication for a 12-year periodicity as was suggested by Hajduk (1970) must remain unanswered in the absence of sufficient data.

The 2006 Orionid return yielded a very distinct data set of 12,012 visual Orionids recorded by 58 observers within 389 h. This allowed us to analyse the shower characteristics with high temporal resolution of 0.1° over most of the period of maximum activity. Numerous bright fireballs were recorded by different techniques (see Spurný and Shrbený 2007, or Trigo-Rodriguez et al. 2007, for example) and both the minimum and average r-values were extremely low (see Table 2 and Rendtel 2007). This resembled very much the parameters of the Leonids 1998 (Tóth et al. 1998; Asher et al. 1999) and the June-Bootids 1998 (Rendtel et al. 1998; Arlt et al. 1999; Asher and Emel'yanenko 2002) and gave rise for the assumption that the Orionids 2006 were caused by resonant meteoroids (Rendtel 2007).

4 Activity Profiles

For the calculation of the Zenithal Hourly Rate (ZHR) we applied the population index profiles of the same year, or the nearest return that yielded sufficient data for the calculation of a profile. The ZHR is calculated

$$\mathrm{ZHR} = N_{\mathrm{ORI}} \frac{r^{6.5-\mathrm{lm}}\, F}{T_{\mathrm{eff}}\, \sin^{\gamma} h_R} \tag{2}$$

Table 2 Lowest value of the population index r of the Orionids near the rate maximum

Year(s)	Data	Minimum r	Remarks
1979–1989	133	2.6	Few magnitude data; average profile
1991, 1992, 1994	204	2.3	Annual data not sufficient
1993	122	2.0	
1995	263	2.2	
1997–2001	155	2.6	Leonid years—ORI less attractive
2002–2005	158	2.1	Too few data per year
2006	215	1.6	Entire period $r \approx 1.9$ and high ZHR

The second column "data" gives the number of magnitude distributions included in the analysis

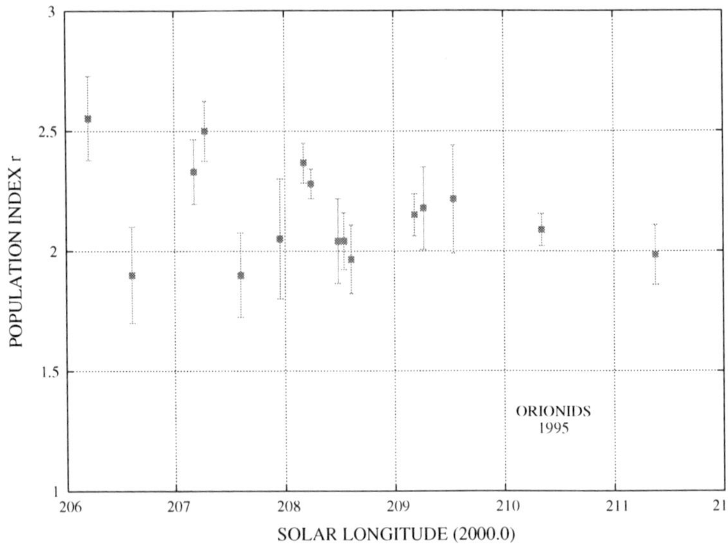

Fig. 1 Profile of the population index r for the near maximum interval between $\lambda_\odot = 206°$ and 212° of the 1995 Orionid return

with N_{ORI} being the number of Orionid meteors observed in the selected interval, the population index (r) at the time of the observation, lm the limiting magnitude, F a geometric correction for obstructions, T_{eff} the effective observing time and h_R the radiant elevation. Values of $\gamma > 1$ were proposed to account for different entry angles of meteors (Zvolánková 1983). For this study we selected only intervals with lm better than 5.8 and a radiant elevation of at least 20°. A detailed investigation was made to ensure that there are no spurious artifacts in the selected periods when the observing region changed, for example, from western European to North American locations with different radiant elevations (Rendtel 2007). For this analysis we used a zenith coefficient $\gamma = 1$; values of $\gamma > 1$ changed the smooth profile and introduced overcorrected ZHRs for radiant elevations in the range between 20° and 45°. Arlt and Barentsen (2006) described in detail the procedure of averaging the individual ZHR values to obtain the profile.

The long-term average maximum ZHR of the Orionids reached values of 20–25 in the period between $\lambda_\odot = 207°$ and 211°. Since we had no magnitude data available for the Orionids prior to 1979, we applied an average $r = 2.4$ to the 1944–1950 data. The maximum ZHR for each return, or group of returns in case of too few annual data, is shown in Fig. 2. The horizontal error bars indicate that data of several years are combined to calculate an average ZHR.

In 2006 we found not only a distinct particle size distribution as shown by the population index profile in Fig. 3 but also strongly enhanced ZHR values with peaks of ZHR$\approx$60, i.e. about 2.5 times the usual peak ZHR (Fig. 4). The highest ZHR coincided with the times of the lowest population index, i.e. the densest regions of the stream were characterized by large meteoroids. The only exception was the very late ZHR peak at $\lambda_\odot = 211.8°$ which coincided with a high r-value and thus represented a different particle population.

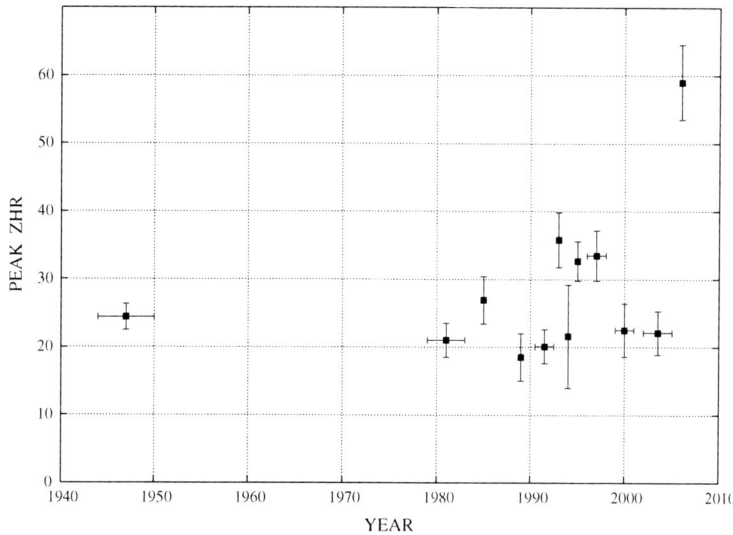

Fig. 2 Maximum ZHRs of the Orionid returns in the period 1944–2006. A horizontal error bar indicates that the value represents an average over several Orionid returns

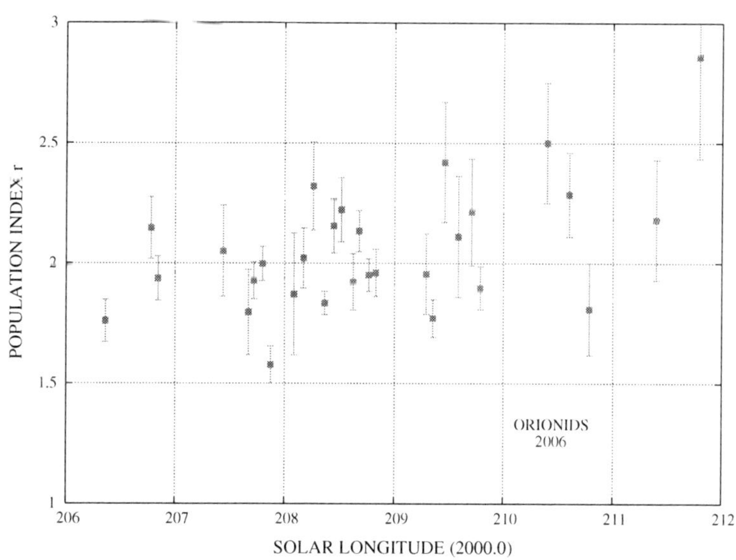

Fig. 3 Profile of the population index r for the near-maximum interval $\lambda_{\odot} = 206°–212°$ of the 2006 Orionids

5 Discussion

The unusual 2006 Orionid return with its particle size distributions and duration of enhanced activity similar to the Leonids 1998 and the June-Bootids 1998 led to the

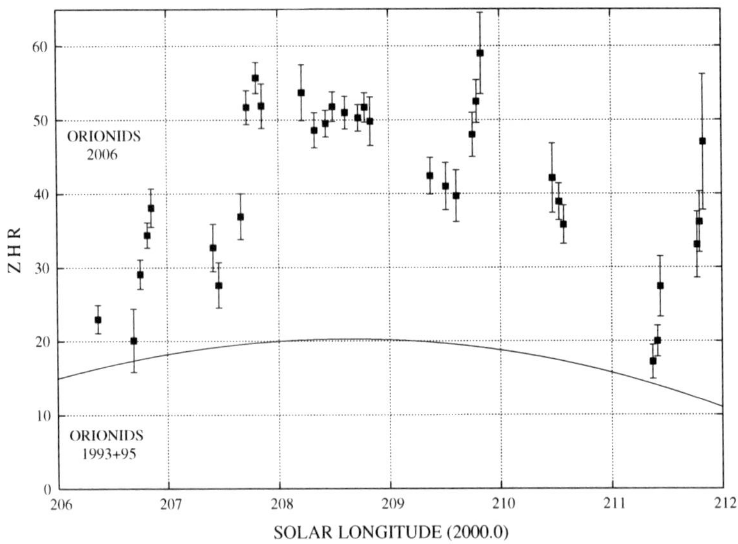

Fig. 4 ZHR-profile of the Orionids 2006 showing the same interval as in Fig. 3. For comparison the average ZHR of the 1993 and 1995 returns is shown as a line

assumption that it was caused by meteoroids in resonant zones. Since the parent comet 1P/Halley is not resonant with Jupiter, it may deliver meteoroids into all resonances over a long time. As the minimum distance between the comet and the Earth orbits is quite large, the question is, which resonance could be the most probable one? Model calculations by Emel'yanenko (2001) indicate the 1:6 resonance because it has the greatest width of all discussed resonance zones. Consequently, we could find Orionid rate enhancements that occurred six Jupiter revolutions earlier, i.e. about 72 years earlier than 2006. As described in the Sect. 2, the analysis of old data is not straightforward as in the case of those covering the last six decades. The strongest hint on Orionid rates exceeding the neighbouring years by a factor of about three is found in observations made by Loreta in Bologna in 1936 (Millman 1936). Unfortunately, no data of the same observer is available for other returns. Hence a calibration is difficult. Still, the numbers of non-Orionid meteors reported indicates that the Orionid rates indeed were significantly enhanced. Similar hints can be found in Lovell (1954), although no such calibration is possible and the original papers by Prentice (1936, 1939) do not allow further comparison. So we have to restrict our analyses to relative rates rather than ZHRs as shown in Fig. 5. The fact that the rich 1936 return is by 2 years off from the exact 1:6 ratio may indicate that it extends over a longer portion along the orbit and we observed perhaps different ends of the meteoroids being trapped in the resonant zone. If this is the case, there is a chance of enhanced rates also in 2007 as calculated by Sato and Watanabe (2007). According to their model, the meteoroids near the resonant zone have orbital periods between 70 and 72 years. Just before the manuscript was finished in October 2007, enhanced Orionid rates as expected from the width of the resonance zone and the modelling have been recorded during the 2007 return.

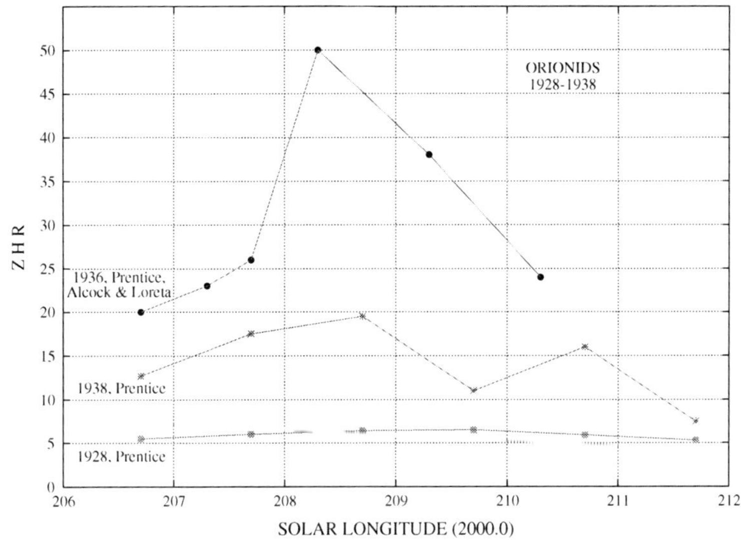

Fig. 5 Orionid rates of the 1928, 1936, and 1938 returns derived from visual observations. Data of Prentice and Alcock are summarized by Lovell (1954); data of Loreta are published by Millman (1936). Although these are no modern ZHRs, we find a distinct increase of the activity during the 1936 Orionids

6 Conclusions

Over most of the period from 1944 to 2006 which has been analysed from a comprehensive data set, the population index r and maximum ZHR of the Orionid meteors show rather small variations with average values of $r \approx 2.4$ and ZHR $\approx$ 20–25. The Orionid 2006 data described the passage of the Earth through a different particle population. The duration of the significantly enhanced activity lasted from $\lambda_{\odot} = 207.6°–210.7°$ and the magnitude data can be compared with encounters of resonant meteoroids. The unusual meteoroid size distribution was confirmed by photographic data (cf. Spurný and Shrbený 2007; Trigo-Rodriguez et al. 2007) and video data. Recent model calculations by Sato and Watanabe (2007) strongly suggest that the 2006 Orionids are due to meteoroids ejected from 1P/Halley more than 2,900 years ago. The rate increase found in the 1936 data (Fig. 5) emphasizes the 1:6 resonance with Jupiter as the responsible region. The long time lapse between the meteoroid ejection and the observation allowed the meteoroids to move close enough to the Earth's orbit.

Acknowledgements We thank all observers for sending their data to the IMO's VMDB. Most of the 2006 data input was performed by Rainer Arlt with the assistance of Javor Kac. Pierre Bader, Frank Enzlein, Ulrich Sperberg and Roland Winkler made substantial additions, especially of the older data. David Asher of Armagh gave very useful comments on the role of stream meteoroids in resonant orbits. Thanks also to Peter Jenniskens, Robert Hawkes and Frans Rietmeijer for useful comments during the referee process.

References

R. Arlt, G. Barentsen, Bulletin 21 of the International Leonid Watch: global analysis of visual observations of the 2006 Leonid meteor shower. IMO J. WGN **34**, 163–168 (2006)

R. Arlt, J. Rendtel, P. Brown, V. Velkov, W.K. Hocking, J. Jones, The 1998 outburst and history of the June Boötid meteor shower. Monthly Notices Roy. Astron. Soc. **308**, 887–896 (1999)

D.J. Asher, V.V. Emel'yanenko, The origin of the June Bootid outburst in 1998 and determination of cometary ejection velocities. Monthly Notices Roy. Astron. Soc. **331**, 126–132 (2002)

D.J. Asher, M.E. Bailey, V.V. Emel'yanenko, Resonant meteoroids from Comet Tempel-Tuttle in 1333: the cause of the unexpected Leonid outburst in 1998. Monthly Notices Roy. Astron. Soc. **304**, L53–L56 (1999)

A. Dubietis, Long-term activity of meteor showers from Comet 1P/Halley. IMO J. WGN **31**, 43–48 (2003)

V.V. Emel'yanenko, Resonance structure of meteoroid streams. in *Proceedings of the Meteoroids 2001 Conference*, ed. by B. Warmbein. ESA Publications Division (ESA SP-495) Noordwijk, Kiruna, 6–10 August 2001, pp. 43–45

A. Hajduk, Structure of the meteor stream associated with comet Halley. Bull. Astron. Inst. Czechosl. **21**, 37–45 (1970)

P. Jenniskens, *Meteor Showers and their Parent Comets* (Cambridge University Press, Cambridge, 2006), 802 pp

A.C.B. Lovell, *Meteor Astronomy* (Clarendon Press, Oxford, 1954), 463 pp

B.A. McIntosh, J. Jones, The Halley Comet meteor stream—numerical modelling of its dynamic evolution. Monthly Notices Roy. Astron. Soc. **235**, 673–693 (1988)

P.M. Millman, Observations of the Orionids in 1936. J. Roy. Astron. Soc. Can. **30**, 416–418 (1936)

C.P. Olivier, Report of the American Meteor Society for 1919–1925. Publ. Leander McCormick Obs. **5**, 1–49 (1935)

V. Porubčan, A. Hajduk, B.A. McIntosh, Visual meteor results from the International Halley Watch. Bull. Astron. Inst. Czechosl. **42**, 199–204 (1991)

J.P.M. Prentice, The radiants of the Orionid meteor shower. J. Br. Astron. Assoc. **46**, 329–336 (1936)

J.P.M. Prentice, The radiants of the Orionid meteor shower. J. Br. Astron. Assoc. **49**, 148–153 (1939)

J. Rendtel, Three days of enhanced Orionid activity in 2006—meteoroids from a resonance region? IMO J. WGN **35**, 41–45 (2007)

J. Rendtel, H. Betlem, Orionid meteor activity on Oct. 18, 1993. IMO J. WGN **21**, 264–268 (1993)

J. Rendtel, R. Arlt, A. McBeath, *Handbook for Visual Meteor Observers* IMO Monograph No. 2, International Meteor Organization, Potsdam, 1995

J. Rendtel, R. Arlt, V. Velkov, Surprising activity of the 1998 June Bootids. IMO J. WGN **26**, 165–172 (1998)

G. Ryabova, The Comet Halley meteoroid stream: just one more model. Monthly Notices Roy. Astron. Soc. **341**, 739–746 (2003)

M. Sato, J. Watanabe, Origin of the 2006 Orionid outburst. Publ. Astron. Soc. Japan **59**, L1–L4 (2007)

P. Spurný, L. Shrbený, Exceptional fireball activity of Orionids 2006. Earth Moon Planets (2007, this issue)

J. Štohl, V. Porubčan, Orionid meteor shower—activity and magnitude description. Contr. Astr. Obs. Skalnaté Pleso **10**, 39–51 (1981)

J. Tóth, L. Kornoš, V. Porubčan, Photographic Leonids 1998 observed at Modra observatory. Earth Moon Planets **82/83**, 285–294 (1998)

J.M. Trigo-Rodriguez, J.M. Madiedo, J. Llorca, P.S. Gural, P. Pujols, T. Tezel, The 2006 Orionid outburst imaged by all-sky CCD cameras from Spain: meteoroid spatial fluxes and orbital elements. Monthly Notices Roy. Astron. Soc. **380**, 126–132 (2007)

J. Zvolánková, Dependence of the observed rate of meteors on the Zenith distance of the radiant. Bull. Astron. Inst. Czechosl. **34**, 122–128 (1983)

Activities of Parent Comets and Related Meteor Showers

Jun-Ichi Watanabe · Mikiya Sato

Originally published in the journal Earth, Moon, and Planets, Volume 102, Nos 1–4.
DOI: 10.1007/s11038-007-9193-z

Abstract The activity of a meteor shower is thought to be proportional to the activities through time of the parent comet. Recent applications of the dust trail theory provide us not only with a new method to forecast the occurrences and intensities of shower activities, but it is also offers a new approach to explore the history of past activities of the parent comet by retro-tracking its associated meteor showers. We introduce the result of an effort for relating meteor shower activities to the parent comet activities for which we chose the October Draconids and comet 21P/Giacobini-Zinner in this paper.

Keywords Meteor showers · October Draconids · Phoenicids · Comets · 21P/Giacobini-Zinner · P/Blanpain · Activity relation

1 Introduction

The interrelation between comets and meteor streams is one of the important topics in meteor science. The question is not only the identification of the parent body for a particular meteor stream, but also to the understand the physical aspects of cometary and meteor shower activities. The recent advancement of the dust trail theory that was effectively applied by Kondrat'eva and Reznikov (1985) made it possible to predict meteor shower activities precisely, and to give us the opportunity to observe even minor showers, such as the June Bootids (Kasuga et al. 2004; Jenniskens 2004), when we know their parent comet (see Watanabe 2004 for a review). It seems also possible to apply the proven successful dust trail theory to study the history of past activities of a parent comet by mapping its meteor streams. This "inverse" approach may be a good tool for studying the physical evolution of the comets in the inner solar system.

J.-I. Watanabe (✉) · M. Sato
National Astonomical Observatory of Japan, 2-21-1, Osawa, Mitaka, Tokyo 181-8588, Japan
e-mail: jun.watanabe@nao.ac.jp

M. Sato
e-mail: Mikiya.Sato@nao.ac.jp

DOI: 10.1007/978-0-387-78419-9_17

The validity of this approach was already demonstrated for the Phoenicid meteor shower and the parent object 2003WY25 (Watanabe and Sato 2005), which is thought to be a dormant fragment from comet P/Blanpain(1819 W1). The "inverse" application of the dust trail theory to this case successfully reproduced the historical outburst in 1956 (Watanabe and Sato 2005; Jenniskens and Lyytinen 2005). At the epoch of the outburst it was revealed that a bundle of the trails that had formed from the late eigteenth through the early nineteenth centuries came close to the Earth's orbit on December 5 (Watanabe and Sato 2005). The bundle consisted of the trails formed mainly between 1743 and 1808, but especially the trails formed between 1760 and 1803 came close to the Earth's orbit within 0.00045 AU. This indicates that the parent object was definitely active enough to eject meteoroids until early in the nineteenth century, including 1819 when this parent body was observed as an active comet. Because the parent comet was only witnessed during 1819, the strong cometary activity during this apparition, such as an outburst or fragmentation event may have caused the strong display in 1956 (Jenniskens and Lyytinen 2005). The parent object 2003 WY25 is either a dormant comet or a comet entering a dormant phase. Jewitt (2006) found an extremely faint coma around this object. The history of this object's activity can be traced by inspecting the Phoenicid meteor shower activity, which then becomes a "fossil record" of its past activities.

2 Future Activities of Phoenicids

We carried out the calculations for future Phoenicid dust trails, and surveyed the situation of the dust trails that come close to the Earth's orbit within 0.003 AU. We found two cases of possible Phoenicid activities expected in 2008 and 2014. The latter has better condition because of the lower ejection velocity and the concentration of five closely spaced trails (Fig. 1).

It is clear that a bundle of the trails formed in early twentieth century will come close to the Earth's orbit, especially the five trails from 1909 through 1930 (Fig. 1). That is, assuming the parent object was active enough to eject meteoroids in this period, the 2008 activity would be due to the just one trail formed in 1866. If we see any Phenicid meteor shower activity in 2008, we will know that the parent object 2003 WY25 was still active in the middle of the nineteenth century. When the dust trail theory will have successfully

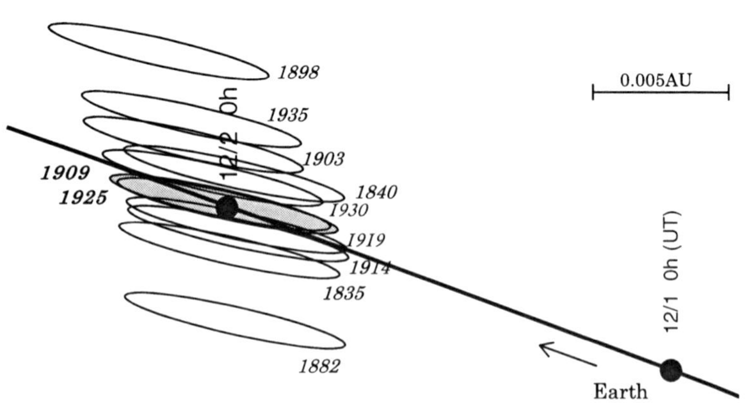

Fig. 1 The geometrical relationship between the dust trails of the Phoenicids and Earth's orbit in 2014

predicted both these shower activities, we will have gained confidence we could also apply the theory backwards using showers as "fossils" of the past parent body activity.

3 Relation between Cometary and Corresponding Meteor Shower Activities: October Draconids

In order to clarify the quantitative relation between cometary and the corresponding meteor shower activities we need an appropriate sample set. It has to satisfy two conditions: (1) strong meteor shower outbursts were observed and the attributed dust trails for the outbursts can be identified and (2) the parent comet was observed as an active comet at the epoch of the formation of these dust trails. It is generally difficult to find such example of meteor showers that satisfy both conditions. For example, in the case of the Leonids we can easily find out various records of the most of its meteor showers related to trails dating back to 802 AD (Asher et al. 1999) but not for the parent comet 55P/Tempel-Tuttle which was firts seen briefly in 1366, and well observed only in 1865 and had its next apparition in 1965.

Among the short periodic comets, the most suitable example appropriate for our purpose is comet 21P/Giacobini-Zinner and the associated October Draconids. The outburst observed in 1998 is explained simply by the single dust trail produced in 1926. This comet was discovered in 1900 and observed as an active comet in 1926, which means that we may exploit this situation as a basis to relate apparitions of the meteors in October Draconids to activity of the parent comet. We selected two other apparitions of October Draconid meteor storms observed in 1933 and 1946. The former was attributed to two dust trails produced in 1900 and 1907, while the latter was caused by six trails from 1900 through 1933. The *fM* value for each trail is calculated by the dust trail theory (Asher 1999). The parameter *fM* is the degree of the extension of the trail, and is derived by $fM = \Delta t_0/\Delta t$, where Δt time needed for the trail passing of the ecliptic plain, and Δt_0 is the same but at the first return, without considering the perturbation (Asher 1999). The *fM* value is basically proportional to n^{-1}, where n is the number of returns. Hence the *fM* is a measure of the meteoroid density within a trail, namely a large *fM* value indicates a strong shower; a low value corresponds to a weak meteor display. The strength of a meteor shower can be generally expected by the summation of the *fM* values of the attributing trails.

Figure 2 shows the relation between the observed ZHR and *fM* values. The observed ranges of the ZHR applied here are $10{,}000 \pm 2{,}000$ for 1933 storm, and $12{,}000 \pm 3{,}000$ for 1946 storm (Jenniskens 1995). Although we find the general trend that the observed ZHR increases with larger *fM*, the "estimated" line, normalized by the 1998s case, does not coincide with the 1933 and 1946 storm levels. The observed ZHR was much higher than was "estimated" by this method. The 1998 storm was caused by a single trail of 1926. Inspecting the average brightness of the parent comet during this apparition of 1926, we find that the parent comet was fainter than other decades (Vsekhsvyatskij 1964). The dust production rate should be larger if the comet was more active. In order to consider this effect, we surveyed the absolute magnitude derived from archived data, and corrected as $\Sigma\ (fM \times \Delta Q(H10))$, where Q(H10) is the dust production rate factor relative to the 1926s case. The result is shown in Fig. 3 that shows a generally better agreement between the observed and "estimated" ZHR for storms within the error of calculation, although the 1933s storm is still a little bit higher value than "estimated" line.

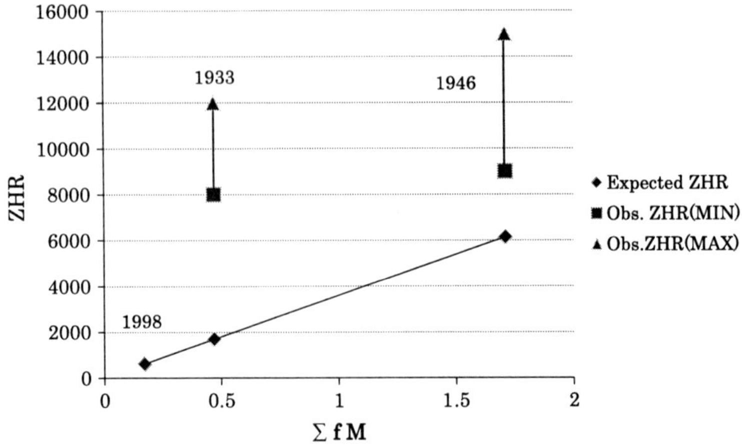

Fig. 2 The relation between the ZHR and *fM* value. The solid line is the expected relation when normalized to the 1998 outburst. The vertical lines are the range of the observed ZHR. Both the 1946 and 1933 storms are higher than expected

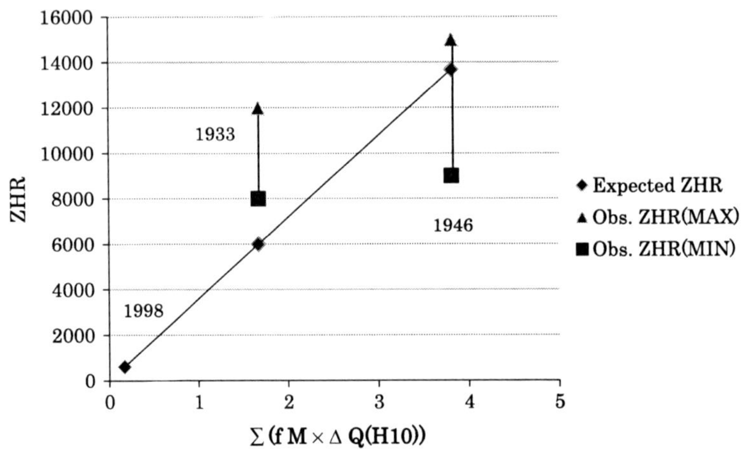

Fig. 3 Same as Fig. 2, but the "corrected" *fM* value by the absolute magnitude of the parent comet. The 1933 and 1946 storm activities are well within the expected level of the 1998 activity case

It should be noted that there are many factors that should be included in this correction. We assume the same size distribution for the dust particles, and here we neglected the effect of the dust ejection velocity. There are additional factors we should consider when trying to apply such relation to other meteor shower and parent comet relationships. The dust to gas ratio will be different among comets. Although 21P/Giacobini-Zinner has a typically average value for the dust to gas ratio, it is also famous for being a chemically peculiar type of comets known as carbon-depleted comets (A'Hearn et al. 1995). The coma of this comet is abundant in larger size grains (Lara et al. 2003). In order to establish the relation between the strength of meteor showers and cometary activities, we should obtain many samples appropriate for such purpose in the future.

4 Concluding Remarks

As shown for the Phoenicids, meteor activity may be used for studying past cometary activity of the parent comet. In order to do this, it is important to establish the quantitative relationship between cometary and meteor shower activities. We showed here a result of an effort to clarify the relationship for the October Draconids and their parent comet 21P/Giacobini-Zinner. More work needs to done on a larger set of related comet-meteor activities to improve the accuracy of applying the "inverse" dust trail theory. The predicted October Draconids activity in 2011 will be mainly due to the 1887 and 1900 trails. Unfortunately we do not have any data on the cometary brightness in 1887. Assuming the cometary activity in 1887 was similar to that in 1900, when the comet was much brighter than 1926, our method applied to the October Draconids 2011 activity predicts that the ZHR will be about 600. It will be around 200 when we neglect the cometary activity difference between 1887–1900 and 1926.

5 Historical Note

While travailing in the Indian Ocean as a member of a Japanese expedition on December 5, 1956, Prof. J. Nakamura observed the sudden appearance of a Phoenicid meteor shower. Watanabe and Sato (2005) noticed an apparent non-negligible discrepancy in the radiant point based on his observations. In 2006 we met with Prof. J. Nakamura, now retired in his home. He expressed his unfamiliarity at the time using the standard astronomical method of plotting meteors in a star chart. He had only a small-size star chart available that was inadequate to determine the precise radiant position.

Acknowledgements We would like to thank Dr. Frans J. M. Rietmeijer for his kind advices as a guest editor. We would also like to thank two referees, Dr. Juraj Toth and anonymous, for thier constructive comments on our paper. This research was partially supported by the Ministry of Education, Science, Sports and Culture, Grant-in-Aid for Scientific Research (C), 195404490002, 2007.

References

M.F. A'Hearn, R.L. Millis, D.G. Schleicher, D.J. Osip, P.V. Birch, The ensemble properties of comets: results from narrowband photometry of 85 comets, 1976–1992. Icarus **118**, 223–270 (1995)

D. Asher, Leonid dust trail theories, ed. by R. Arlt, Proc. International Meteor Conference, Frasso Sabino, Italy, International Meteor Organization, (1999), pp. 5–21

D.J. Asher, M.E. Bailey, V.V. Emel'Yaneko, Resonant meteoroids from Comet Tempel-Tuttle in 1333: the cause of the unexpected Leonid outburst in 1998. Mon. Not. R. Astron. Soc. **304**, L53–L56 (1999)

P. Jenniskens, Meteor stream activity. 2: meteor outbursts. Astron. Astrophys. **295**, 206–235 (1995)

P. Jenniskens, 2004 June Bootids: video images and low-resolution spectra of 7P/Pons-Winnecke debris. WGN **32**, 114–116 (2004)

P. Jenniskens, E. Lyytinen, Meteor showers from the debris of broken comets: D/1819 W1 (Blanpain), 2003 WY25, and the Phoenicids. Astron. J. **130**, 1286–1290 (2005)

D. Jewitt, Comet D/1819 W1 (Blanpain): not dead yet. Astron. J. **131**, 2327–2331 (2006)

T. Kasuga, J. Watanabe, N. Ebizuka, T. Sugaya, Y. Sato, First result of June Bootid meteor spectrum. Astron. Astrophys. **424**, L35–L38 (2004)

E.D. Kondrat'eva, E.A. Reznikov, Comet Tempel-Tuttle and the Leonid meteor swarm. Solar Syst. Res. **19**, 96–101 (1985)

L.-M. Lara, J. Licandro, A. Oscoz, V. Motta, Behaviour of comet 21P/Giacobini-Zinner during the 1998 perihelion. Astron. Astrophys. **399**, 763–772 (2003)

S.K. Vsekhsvyatskij, *Physical Characteristics of Comets*. (Israel Program for Scientific Translations, Jerusalem, 1964), pp.351–400
J. Watanabe, Meteor streams and comets. Earth Moon Planet. **95**, 49–61 (2004)
J. Watanabe, M. Sato, Phoenicids in 1956 revisited. Publ. Astron. Soc. Jpn **57**, L45–L49 (2005)

Search for Past Signs of October Ursae Majorids

Štefan Gajdoš

Originally published in the journal Earth, Moon, and Planets, Volume 102, Nos 1–4.
DOI: 10.1007/s11038-007-9196-9

Abstract A new meteroid stream—October Ursa Majorids—was announced by Japanese observers on Oct. 14–16, 2006 (Uehara et al. 2006). Its weak manifestation was detected among coincidental major meteor showers (N/S Taurids, Orionids), as its meteors radiated from a higher placed radiant on the northern sky. We have tried to find out previous displays of the stream throughout available meteor orbits databases, and among ancient celestial phenomena records. Although we got no obvious identification, there are some indications that it could be a meteor shower of cometary origin with weak/irregular activity, mostly overlayed by regular coincidental meteor showers. With a procedure based on *D*-criterion (Southworth and Hawkins 1963) we found a few records in IAU MDC database of meteor photographic orbits which fulfill common similarity limits, for October Ursae Majorids. However, their real association cannot be established, yet. With respect to the mean orbit of this stream, we suggest for its parent body a long-period comet.

Keywords Meteor streams · October Ursae Majorids · New meteor stream identification

1 Introduction

October Ursa Majorids (OUM) is a new meteor stream whose weak (but clear) manifestation was reported by Japanese observers within their video network observations on Oct.14–16, 2006 (Uehara et al. 2006). Meteors of the stream radiated from a position R.A. = 144.8° and Dec. = 64.5°, with the geocentric velocity $V_g = 54.1$ km s^{-1}. Simultaneously observed meteors yield the mean orbit (read: average of individual meteors elements) of the stream with parameters $a = 5.9$ AU, $q = 0.979$ AU, $e = 0.875$, $\omega = 163.7°$, $\Omega = 202.1°$, and $i = 99.7°$ (J2000.0). With respect to the renewed meteor

Š. Gajdoš (✉)
Department of Astronomy, Physics of the Earth, and Meteorology, Faculty of Mathematics, Physics and Informatics, Comenius University, Mlynská dolina 842 15, Bratislava
e-mail: gajdos@fmph.uniba.sk

DOI: 10.1007/978-0-387-78419-9_18

stream nomenclature rules (Task Group for Meteor Shower Nomenclature, IAU Commission 22), we will use a correct name for this stream: October Ursae Majorids (OUM).

In a searching for past displays of OUM, we examined sources of meteor orbits and meteor phenomena (second section). In third section, we are looking for an admissible type of a parent body which could either fit orbital characteristics of the new meteor shower or, in assumed specific circumstances resulted in formation of the meteoroid stream. All data we used are in Equinox (J2000.0).

2 Survey of Common Sources

2.1 Established Showers in a Vicinity of OUM

In a search for coincidental streams, we firstly had look at the summary tables published by Jenniskens (2006), which besides of his own results encompass a compilation of meteor streams from many authors. The tables list basic orbital data of cometary and suspected asteroidal streams, their major apparition peaks, as well as bibliographic sources of the data.

Working list of cometary meteor showers therein (Table 7), presents two streams exactly in the time of the OUM apparition. Daytime phi-Virginids (DFV, #240, ecliptic helion source) have its peak on Oct. 15 at the solar longitude $\lambda_{\odot} = 202.0°$ while the gamma-Puppids (GPU, #239) peak on Oct. 16 at the solar longitude $\lambda_{\odot} = 202.7°$. DFV is a broad daytime stream and GPU do not match Oct. Ursae Majorids radiant at all (Dec. = $-44.0°$). There are two other well-known showers in a week intervals before and after the OUM: Oct. Draconids (DRA, #9) on Oct. 8 ($\lambda_{\odot} = 195.1°$) and Oct. Ursae Minorids (OUI, #241) on Oct. 21 at $\lambda_{\odot} = 208.0°$. Having in mind some analogy with OUM detection, we shouldn't omit recent case of the October Camelopardalids (OCT, #281) on Oct. 5, 2005 (Jenniskens 2005).

In the Working list of possible asteroidal meteor showers therein (Table 9), we identified a single shower in the mid of October with a few members only—delta-Cygnids (DCY, #282). Again, radiant coordinates differ more than allowed, and both do not fulfill the coordinates of the OUM.

Besides of highlighted streams, there are many minor meteor showers with low or irregular activity in a period from mid of September to the end of October (corresponding solar longitudes $\lambda_{\odot} = \sim 170°–220°$). We point out that OUMa displayed on a background of approaching Orionids, as well as ongoing long-lasting Taurid meteor complex, in 2006. As mentioned by Uehara et al. (2006) already, if we should search for the OUMa pattern we have to do it over a miscellaneous background activity. Therein, a percentage of detected OUMa meteors was at the level up to 9%, the lowest among above mentioned streams in the period Oct. 10–20, considering other minor streams activity as sporadic. We ascertained in this survey that no hitherto known meteoroid stream could be associated with the OUMa, as none has suitable orbital and/or geophysical parameters.

2.2 A Search for Ancient OUMa Activity

Jenniskens (2006) presents historic meteor phenomena reports in his Table 1, also, originally collected by many authors. Looking at the solar longitudes, there are listed two timely lagged sightings around 202°: on Oct. 9.7, 1798 at $\lambda_{\odot} = 199.4°$, and second one identified with

Table 1 An appearance time and orbital parameters of listed meteors, and their *D*-criterion value

Meteor	Date	q (AU)	a (AU)	e	ω (°)	Ω (°)	i (°)	D
On012	Sep. 28, 2002	0.899	47.000	0.981	142.4	185.4	101.4	
331J1	Oct. 09, 1953	0.998	17.737	0.943	177.0	196.3	104.0	(0.137)
022K1	Oct. 14, 1958	0.994	9.204	0.892	173.5	200.8	98.5	0.080
OUMa	Oct. 15, 2006	0.979	5.920	0.875	163.7	202.1	99.7	0.079
100E1	Oct. 16, 1979	0.986	−2.158	1.457	168.9	202.3	98.6	[0.013]
037E2	Oct. 24, 1995	0.939	−8.758	1.107	153.5	210.3	112.8	

The sign minus of semimajor axis denotes a hyperbolic orbit. A case of the IAU MDC bolide 100E1 is analyzed in text, separately

Orionids in year 288 ($\lambda_\odot = 206.1°$ and $207.5°$), visible in two subsequent days (September 25.3 and 26.7). Next are records on September 25.7, 930, September 23.7, 585, and Sep. 27.0, 903, at the solar longitudes 201.9°, 202.4°, and 203.2°, all identified as Orionids. Description of sightings from Oct. 9, 1798, ($\lambda_\odot = 199.4°$)—*Stars flew all around, next few nights too*—evoke a hope. It should mean solar longitudes close to the OUMa.

2.3 An Inspection of the IMO Video Database

In examination of the IMO Video Meteor Database we refer to precise complete analysis by Molau (2006). Along with used analysis procedure description, he lists detected meteor showers in his Table 2. There, among known showers have been identified six sporadic sources (N/S Apex, N/S Toroidal, Helion, Antihelion). Based on radiant coordinates and velocity we tried to find some sign of regular stream, in the list. But, among a cluster of mainly sporadic and known sources in September and October data, no fitting candidate was found. As to video observations, noteworthy is that Trigo-Rodríguez et al. (2007) detected no OUMa activity either by the all-sky cameras or by the video cameras patrol.

2.4 A Search for Past and Recent Activity of OUMa in Radio Observations

In a search for past activity of the OUMa, radio observations are very proper. However, Orionids dominate in the second half of October, thus common radio campaigns start a little later as we need. From the same reason, the forward-scatter system with baseline Bologna-Modra operated on Oct. 13–28, 2002, and Oct. 14–28, 2003, with only accident overlapping of investigated OUMa activity period. Unfortunately, our moderate noisy data do not provided any clue. However, a dedicated analysis of observations from continually working radar systems around the world (where a radiant is above horizon, currently e.g., CMOR) could help reveal recent activity of OUMa, especially its presence in 2006. A more challenging is searching for 2006 display of the OUMa among the radio forward-scattering meteor data of the Global-MS-Net. Yrjölä and Jenniskens (1998) proved that in course of continuous observation by the same forward-scatter system, there is possible to detect an activity of zenithal hourly rate as low as 3–4 meteors, if the activity is insulated or on steady background. For example, recent video detection of Oct. Camelopardalids display in 2005 (Jenniskens 2005; only 10 days before OUMa and shifted ~20° in

Table 2 A geophysical and physical data on listed meteors

Meteor	R.A. (°)	Dec. (°)	V_g (km s^{-1})	H_b (km)	H_e (km)	Mag	Mass (g)	Type
On012	134.5	+63.70	55.8	110	79	−8.2	40	II/IIIa
331J1	124.3	+66.60	56.7	112	102	0.6	0.003	–
022K1	137.8	+67.70	54.2	100	88	−2.7	0.3	–
OUMa	144.8	+64.48	54.1	See	Graph	No	Data	–
100E1	150.6	+67.90	59.2	109	87	−10.3	100	–
037E2	153.9	+53.50	61.4	113	79	−7.9	953	–

R.A. and ~15° in Dec. off OUM radiant) was confirmed by proper use of this source. Our opinion is that including of the observability function (Hines 1955, 1958) into an identification procedure wouldn't be successful among coincidental minor meteor streams and rising Orionids rate.

2.5 A Search Among Precise Meteor Orbits

The most precise source for searching of possible OUMa related meteors is the IAU MDC database of photographic meteor orbits (Lindblad et al. 2005). Currently, the database contains homogenized orbital elements of 4,581 meteors along with geophysical data (radiant coordinates, geocentric and heliocentric velocities, beginning/end heights of meteors, photometric masses). In a course of the renewed search for the bolide meteor showers (Gajdoš and Porubčan 2005; Gajdoš 2005), we supplemented the MDC database with a few tens (state to March 31, 2006) of meteor orbits from unpublished database of the Interplanetary Section, Astronomical Institute AS CR in Ondřejov (Spurný 2007, Personal communication). The meteor orbits bear labels On*xxx*. The whole working data set is in equinox (J2000.0).

Due to initial knowledge on the OUMa activity, we handled broader interval of solar longitudes to catch material for analysis. We involved the iteration procedure used by Porubčan and Gavajdová (1994) which is based on *D*-criterion by Southworth and Hawkins (1963). We had look for meteors fitting the mean orbit of OUMa with the limit value of *D*-criterion $D_{SH} \leq 0.20$. The searching revealed a few meteor orbits similar to the OUMa, only. An explanation we see in high inclined retrograde orbit of the OUMa. In a triplet with the OUMa mean orbit appeared two meteor orbits labeled 022K1, 331J1, having appropriate values of *D*-criterion. Noteworthy, when mutual pairs of the three orbits are taken into consideration, latter two orbits fit better each other than 022K1 with the OUMa (Table 1 gives this value of *D*). In parenthesis, we add *D* value of 331J1 orbit from triplet. For the sake of completeness, other photographic orbits were subjected to an inspection. It resulted in three more high inclined orbits throughout investigated period (On012, 100E1, 037E2). Each of them is matter of our interest.

The tables above contain relevant meteors data taken into consideration: designation, apparition date and orbital parameters (perihelion distance, semimajor axis, eccentricity, and angular elements) are in the Table 1, as well as radiant coordinates, geocentric velocity, beginning/ending heights of meteors, their magnitudes and photometric masses, are in the Table 2, respectively. One bolide has an additional datum on type of the meteoroid. Tables list four meteor orbits from the IAU MDC database, the mean orbit of OUMa, and the orbit of the fireball "Velvary" (our working label On012).

If we omit a, e elements of 100E1, in first look there seems fairly compact group of three orbits in mid of October (022K1, OUMa, 100E1), and "wings" in late September (On012) and October (037E2). Bolide 100E1 apparently outstands with a hyperbolic orbit, whilst other parameters fit well. This orbit comes out from Ondřejov photographic program (Ceplecha and Rajchl 1965) and in the original version it had the parabolic eccentricity value. In consistent IAU MDC database (Lindblad et al. 2005) it has recalculated high hyperbolic eccentricity $e = 1.457$. This and original values indicate difficulties in meteor orbit processing. Due to less favorite circumstances (e.g., small number of photographic images for elaboration, unfavourable position of a bolide, small number of breaks on bolide trail, etc.) some parameters could embody more uncertainty. This feature of the IAU MDC database is pointed out by Hajduková (1994), in her searching for hyperbolic interplanetary particles, and generalized in Hajduková and Hajduk (2006). They demonstrated that the number of hyperbolic meteors found within meteor showers increases with a velocity approaching the hyperbolic limit of particular shower. As well known, in a process of some orbital parameters reduction, the meteor velocity determination is crucial. Thus, providing of an elliptical orbit we verified a case of 100E1. Putting its parameters a, e equal to those ones of the OUMa mean orbit, we examined theirs similarity. D-criterion value of such a hypothetic orbit in a triplet with OUMa and 022K1 was as low as 0.013 (see Table 1, in square brackets). Apparently, besides of elements a, e, orbit and geophysical parameters of bolide 100E1 show a high level of similarity with OUMa characteristics.

In the Fig. 1, beginning (H_b) and ending (H_e) heights of meteors are compared. Uehara et al. (2006) gave no maximum brightness height for listed meteors, thus we omitted it in both tables for IAU MDC meteors, too. Single plumb lines depict meteors used for OUMa mean orbit calculation. Lines with begin/end squares are from both IAU MDC/Ondřejov meteor databases. Reported OUMa meteors are of an unified display, with a slight tendency of their beginning heights increasing toward brighter meteors. Uehara et al. (2006) list visual magnitudes with errors $\pm 1^m$ as observed from more stations (we plot the arithmetic mean of individual magnitudes), while IAU MDC meteors bear photographic magnitudes. Both magnitudes differ, but considering generally accepted dependence we assume a larger photometrical mass for brighter meteoroids. Other meteors seems to have a comparable (331J1, 037E2, On012, 100E1) or lower (022K1) beginning heights. But, these data were taken by different cameras with different sensitivity what results in undervalued beginning (low) and overvalued (high) terminal heights. Both plotted groups of meteors (OUMa + IAU MDC) do not overlap a limit of ~ 130 km at which a thermal ablation of meteoroid start to dominate. Koten et al. (2004) pointed out that the beginning height increases with increasing photometric mass. This feature was reported for generally fainter meteors of several cometary meteor showers, and also for brighter meteors (Spurný et al. 2000). Due to similar behavior of plotted meteors, we suggest their common physical characteristics. Based on the two papers, and presented meteoroid type II/IIIa (Ceplecha 1988) of meteor On012 (Table 2), we may conclude that discussed meteoroids had a cometary origin.

3 Searching for Possible Parent Body

An other approach to the subject is plain search for a possible parent body based on simple similarity of orbits. A high inclination was a first indication. For the sake of completeness, we browsed groups of near-Earth bodies. In the recent NEA population (as to Nov. 12, 2007) in the JPL webpage (http://neo.jpl.nasa.gov) exist objects with inclination up to 72°, only, and single object (2007 VA_{85}) with retrograde orbit ($\sim 133°$), probably a dormand

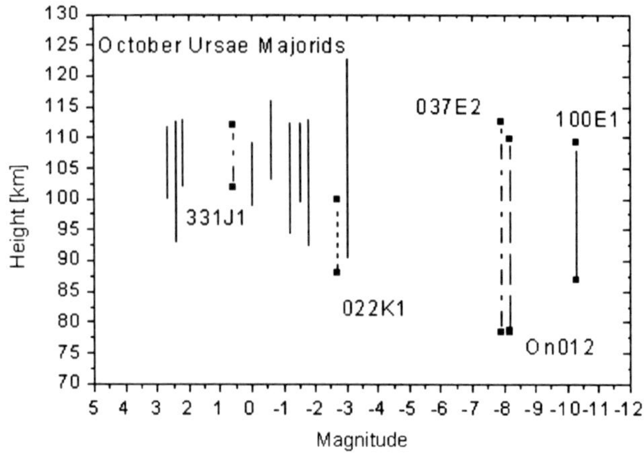

Fig. 1 Beginning and ending heights of investigated meteors

comet. Another search produced a cluster of about twenty cometary orbits fulfilling wider criteria: $i \sim 100° \pm 20°$, and $\omega \sim 165° \pm 30°$, and $\Omega \sim 200° \pm 20°$. We applied the code for calculation of theoretical meteoroid stream radiant by Neslušan et al. (1998) to finding out a meteor producing orbit. At the same time we got a minimal distance of the comet/Earth orbits. Although a number of examined orbits approach the Earth's one very close (e.g., C/1907 G1 less than 0.003 AU), no OUMa fitting radiant occurred, whereas more of them are of southern declination. More meaningful entry is distance of the comet orbit nodes from the orbit of the Earth. These data provided the *Catalogue of cometary orbits* (Marsden and Williams 2005). We checked the minimal distance of individual comets from the Earth at the time of the comet passage. Of investigated set, fairly close encounters experienced single-apparition comets C/1014 C1 (0.0407 AU) and C/1132 T1 (0.0447 AU). Considering orbital elements and geometrical conditions of cometary approach, we found, say, prototypes of possible parent comet of the OUMa meteoroid stream. Besides of above mentioned, they are C/1683 O1, C/1848 U1, C/1975 T2, C/1999 J3. The comet C/1975 T2 (Suzuki-Saigusa-Mori) has orbital period of about 446 years. This fact would be attractive in an assumption of meteor stream formation by regular activity of a long-period comet (e.g., April Lyrids and comet C/1861 G1 with period 415 years). In such a case, an unexpected 2006 display of OUMa we could interpret as a collision of the Earth with a filament of the stream.

4 Conclusions

We searched a display of reported OUMa meteor shower in a list of known meteor showers and among ancient records, as well as in the IMO video database. Besides of one indistinct record from Oct. 9, 1798, we found no obvious signs of OUMa activity in the past, in individual sources. We concluded that search for OUMa in the Global-MS-Net archive would be unsuccessful, even with using of the observability function. The IAU MDC database of photographic orbits provides us with several orbits similar to that of the OUMa mean orbit. We believe that meteoroids 331J1 and 022K1 are past members of assumed OUMa meteoroid stream, and also 100E1 is potential. The OUMa are of cometary origin. We suggest their parent body being a long-period, high inclined comet (nearly-isotropic, coming from the Oort cloud). The OUMa display like an irregular swarm, possibly with some filaments.

We think that along with independent confirmation, a dedicated inspection of CMOR data could help to reveal more details on the OUMa 2006 display. Anyway, aimed radar/video/photographic observations starting at begin of October would be suitable to monitoring of OUMa activity, in 2007 and after.

Acknowledgements This paper was worked out with a help of grant No.01/3074/06 of Slovak grant agency VEGA. We would like to thank to Pavel Spurný for provision of data from his bolide database (in preparation). We are very thankful to referee for valuable comments.

References

Z. Ceplecha, Earth's influx of different populations of sporadic meteoroids from photographic and television data. Bull. Astron. Inst. Czech. **39**, 221–236 (1988)

Z. Ceplecha, J. Rajchl, Programme of fireball photography in Czechoslovakia. Bull. Astron. Inst. Czech. **16**, 15–22 (1965)

Š. Gajdoš, Bolide meteor streams, PhD thesis, (in Slovak, 2005)

Š. Gajdoš, V. Porubčan, Bolide meteor streams, in *Dynamics of Populations of Planetary Systems*, IAU Coll.197, ed. by Z. Knežević, A. Milani (Cambridge University Press, Cambridge, 2005), pp. 393–398

M. Hajduková Jr., On the frequency of interstellar meteoroids. Astron.Astrophys. **288**, 330–334 (1994)

M. Hajduková Jr., A. Hajduk, Mass distribution of interstellar and interplanetary particles. Contrib. Astron. Obs. Skalnaté Pleso. **36**, 15–25 (2006)

C.O. Hines, Diurnal variations in the number of shower meteors detected by the forward-scattering or radio waves – Part I. Theory. Can. J. Phys. **33**, 493–503 (1955)

C.O. Hines, Diurnal variations in the number of shower meteors detected by the forward-scattering or radio waves – Part III. Ellipsoidal theory. Can. J. Phys. **36**, 117–126 (1958)

P. Jenniskens, October Camelopardalids, CBET 309 (2005)

P. Jenniskens, *Meteor Showers and Their Parent Comets*. (Cambridge University Press, Cambridge, UK, 2006)

P. Koten, J. Borovička, P. Spurný, H. Betlem, S. Evans, Atmospheric trajectories and light curves of shower meteors. Astron. Astrophys. **428**, 683–690 (2004)

B.A. Lindblad, L. Neslušan, V. Porubčan, J. Svoreň, IAU Meteor Database of photographic orbits. Earth Moon Planets **93**, 249–260 (2005)

B.G. Marsden, G.V. Williams, *Catalogue of Cometary Orbits*, 16th edn. Solar, Stellar & Planetary Sciences Division, Smithsonian Astrophysical Observatory (2005)

S. Molau, How good is the IMO working list of Meteor showers? A complete analysis of the IMO Video Meteor Database. *Proceedings of IMC, Roden*, September 14–17 (2006)

L. Neslušan, J. Svoreň, V. Porubčan, A computer program for calculation of a theoretical meteor-stream radiant. Astron. Astrophys. **331**, 411–413 (1998)

V. Porubčan, M. Gavajdová, A search for fireball streams among photographic meteors. Planet. Space Sci. **42**(2), 151–155 (1994)

R.B. Southworth, G.S. Hawkins, Statistics of Meteor Streams. Smithson. Contr. Astrophys. **7**, 261–285 (1963)

P. Spurný, H. Betlem, K. Jobse, P. Koten, J. Van't Leven, New type of radiation of bright Leonid meteors above 130 km. Meteorit. Planet. Sci. **35**, 1109–1115 (2000)

J.M. Trigo-Rodríguez, J.M. Madiedo, A.J. Castro-Tirado, J.L. Ortiz, J. Llorca, J. Fabregat, S. Vítek, P.S. Gural, B. Troughton, P. Pujols, F. Gálvez, Spanish Meteor Network: 2006 continuous monitoring results. WGN **35**(1), 13–22 (2007)

S. Uehara, SonotaCo, Y. Fujiwara, T. Furukawa, H. Inoue, K. Kageyama, K. Maeda, H. Muroishi, S. Okamoto, T. Masuzawa, T. Sekiguchi, M. Shimizu, H. Yamakawa, Detection of October Ursa Majorids in 2006. WGN **34**(6), 157–162 (2006)

I. Yrjölä, P. Jenniskens, Meteor stream activity VI. A survey of annual meteor activity by means of forward meteor scattering. Astron. Astrophys. **330**, 739–752 (1998)

The P/Halley Stream: Meteor Showers on Earth, Venus and Mars

Apostolos A. Christou · Jeremie Vaubaillon · Paul Withers

Originally published in the journal Earth, Moon, and Planets, Volume 102, Nos 1–4.
DOI: 10.1007/s11038-007-9201-3 © Springer Science+Business Media B.V. 2007

Abstract We have simulated the formation and evolution of comet 1P/Halley's meteoroid stream by ejecting particles from the nucleus 5000 years ago and propagating them forward to the present. Our aim is to determine the existence and characteristics of associated meteor showers at Mars and Venus and compare them with 1P/Halley's two known showers at the Earth. We find that one shower should be present at Venus and two at Mars. The number of meteors in those atmospheres would, in general, be less than that at the Earth. The descending node branch of the Halley stream at Mars exhibits a clumpy structure. We identified at least one of these clumps as particles trapped in the 7:1 mean motion resonance with Jupiter, potentially capable of producing meteor ourbursts of ZHR $\sim$ 1000 roughly once per century.

Keywords 1P/Halley · Mars · Venus · Meteors · Meteor outbursts · Meteor showers

1 Introduction

1P/Halley, the archetype for the Halley-type comet class, is one of the most extensively studied cometary bodies. It has been observed at 30 perihelion returns since 239 BC. This, and the relative regularity of its orbital evolution, have allowed the reconstruction of its

A. A. Christou (✉)
Armagh Observatory, College Hill, Armagh BT61 9DG, Northern Ireland, UK
e-mail: aac@arm.ac.uk

J. Vaubaillon
IMCCE, Observatoire de Paris, 77 Avenue Denfert-Rochereau, 75014 Paris, France
e-mail: vaubail@imcce.fr

J. Vaubaillon
CalTech/IPAC/SSC, 1200 E. California Blvd, Pasadena, CA 91125, USA

P. Withers
Center for Space Physics, Boston University, 725 Commonwealth Avenue, Boston, MA 02215, USA
e-mail: withers@bu.edu

orbital motion back to 1404 BC, over a millenium before the first observations (Yeomans and Kiang 1981). The comet's nucleus was recently investigated in situ by a flotilla of Russian and European spacecraft at its return to perihelion in 1986. Because the nodes of its orbit are near 1 AU, Halley meteoroids intercept the Earth both before and after perihelion, resulting in two distinct meteor showers, the October Orionids and the May η Aquariids. The comet's orbit also approaches the orbits of Venus and Mars, raising the possibility that meteors can occur at those planets when the latter pass through the Halley meteoroid stream. To investigate this point we have simulated the evolution of the Halley stream and characterised its planet-intercepting component at Earth, Mars and Venus.

2 Method

Our method is that of Vaubaillon et al. (2005a; b). An ensemble of test particles is ejected from the cometary nucleus and propagated forwards in time under planetary perturbations and size-dependent non-gravitational forces until a planet intercept occurs. Values of those cometary parameters required to translate our model particle fluxes into actual meteoroid fluxes were assumed as follows: $[Af\rho] = 17378$ cm (a measure of the dust production rate) at perihelion (Feldman et al. 1987; A'Hearn et al. 1995), nuclear radius $r_N = 7.5$ km, fraction of active area $f = 0.3$ (van Nes 1986), differential size distribution index s = 3.25. It is important to note that this s is related to the differential mass distribution index s_m (cf. Eq. C4 in Vaubaillon et al. (2005a)) but it is *not* s_m.

The simulation of the generation and evolution of the meteoroid stream was run on 5–50 parallel processors at CINES (France). We ejected 5×10^4 test particles distributed over five size bins (100 μm–10 cm) during a single perihelion passage. The starting state vector of the comet, from which the test particles are ejected, was derived by taking the reference orbit of comet Halley at the 239BC perihelion passage from JPL HORIZONS (small body code: 900001; Giorgini et al. 1996) and integrating it backwards to a perihelion state vector at 2924 BC. At this point we expect the location of the comet in its orbit to be significantly randomized with the true comet position differing by several decades in mean anomaly from our starting position. However, previous works have shown the structure of the stream's projection on the ecliptic to be insensitive to the position of the comet itself (McIntosh and Jones 1988; Ryabova 2003). It is rather dependent on the orbit evolution which is regular over this timescale and dominated by precession of the lines of apses and nodes (McIntosh and Hajduk 1983).

Several provisos should apply to interpreting our results. Firstly, our estimated ZHR is strictly applicable only to the case of the Earth (Koschack and Rendtel 1990a, b) but should be representative of observable meteor activity in the Martian atmosphere for such fast meteoroids (Adolfsson et al. 1996). At Venus, it should be treated as a lower limit due to the intrinsic capacity of that atmosphere to produce brighter meteors than the Earth's (Christou 2004; McAuliffe and Christou 2006). Our ZHR estimate also depends on the size of the sampling area, on the planetary orbital plane, over which the particles are counted to estimate the flux (the ΔT quantity in Vaubaillon et al. (2005a) multiplied by the planet's orbital velocity). Here we adopted $\Delta T = 20$ h as it yields the best agreement with the observed relative activity between the two Halley branches at the Earth. Results for all three planets are summarised in Table 1.

In order to estimate the meteoroid flux density required for the ZHR calculation we binned together planet-approaching meteoroids between two consecutive perihelion passages of the comet (e.g. between 1910 and 1986) and averaged the result over the

Table 1 Characteristics of the Halley meteor showers both at the ascending (HAN) and descending (HDN) nodes on Venus, Earth and Mars as derived from our simulations

Branch	Planet	Max (°)	Activity arc (°)	v (km sec^{-1})	ZHR
HDN (EAQ)	E	41	38–48	67	20
HAN (ORI)	E	211	208–213	67	13
HDN	M	26	26–29	55	62
HAN	M	220	216–224	55	2
HDN	V	65	61–73	80	4

Column 2 identifies the relevant planetary body as Earth (E), Venus (V) or Mars (M). Columns 3 and 4 give the activity maximum and duration in terms of the solar longitude (λ_s). Column 5 provides the velocity of the meteoroids at atmospheric entry. Column 6 gives the estimated ZHR

number of planetary years in that period. This method should be valid when the structure of the shower remains unaltered from year to year over this timescale but breaks down when outburst activity dominates the flux.

3 Results

3.1 Earth

The structure of the Halley stream at the Earth's orbit has been modelled extensively in previous works (McIntosh and Hajduk 1983; McIntosh and Jones 1988; Wu and Williams 1993; Ryabova 2003). Here we have used more particles and let them evolve longer than previously attempted but our purpose is different: a realistic end result at the Earth, as gauged by the level of agreement with those works, will bolster the validity of our findings at Venus and Mars. The structure we find is shown in Fig. 1. The "lopsided tadpole" form of the Halley descending node (HDN) branch, responsible for the η Aquariid shower, is

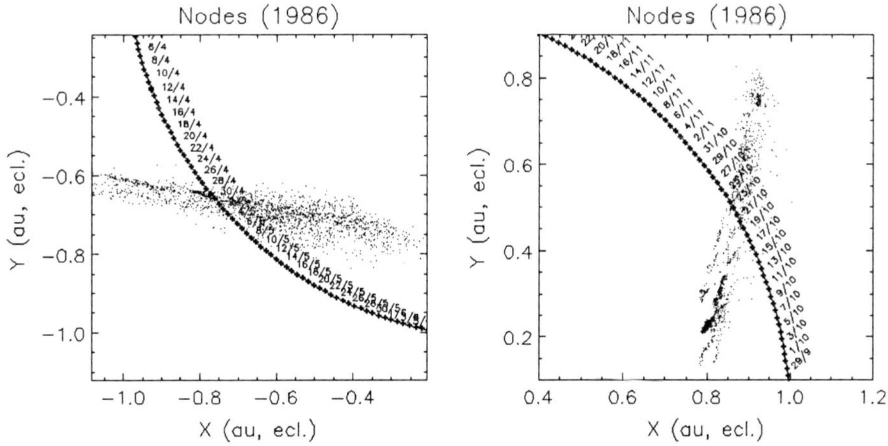

Fig. 1 Ecliptic nodes of particles ejected in 2924 BC that approached the Earth between 1910 and 1986. Left panel: Descending nodes related to the η Aquariids. Right panel: Ascending nodes related to the Orionids

reminiscent of Fig. 4 of McIntosh and Jones (1988) and Fig. 8 of Ryabova (2003). The head of the tadpole is thought to be rich in larger particles mimicking the dynamical evolution of the comet. The tail is composed of smaller particles lagging behind the main body of the stream, their orbits having suffered significant differential precession due to non-gravitational forces. We also see filamentary structure, also reported in those works. The width of the stream at the Earth's orbit as given in Table 1 is contained within the observed duration of the shower (37°–51°; Hajduk et al. 2002; Dubietis 2003). Our model maximum occurs ~4 days earlier than observed (Rendtel 1997; Hajduk et al. 2002), possibly due to the stream having suffered more differential precession than recently-ejected material. Taking this result at face value implies that this shower, as presently observed, contains meteoroids ejected 5000 years ago, probably a small fraction compared to the accumulated population from previous and subsequent perihelion passages. The model results for the Halley Ascending Node (HAN) branch, responsible for the October Orionids, are generally quite similar with those of McIntosh and Jones and Ryabova but there are also differences. The main body of our model HAN meteoroids clearly intersects the orbit of the Earth, in agreement with the conclusion by those authors that the Orionid meteoroids we currently observe were ejected from the nucleus of P/Halley before 1404 BC. As found for the η Aquariids, our model gives a shorter duration than the observations. The stream model maximum is at 211°, 3 days later than observed. Combined with the earlier than observed η Aquariid maximum, this indicates some precession in ω between the observed and the model shower. Several dense concentrations or clumps of Earth-intercepting material are also evident. These may be associated with the trapping of Halley particles in mean motion resonances with Jupiter as recently reported (Trigo-Rodriguez et al. 2007; Sato and Watanabe 2007; Rendtel 2007). A global analysis of the Halley resonant meteoroid complex will be the subject of a future paper; we do, however, mention one particular case in some detail in the section discussing our results at Mars.

3.2 Venus

At Venus we find that only the HDN branch is active, the HAN branch passing well outside the Venusian orbit (Fig. 2). The duration of the resulting meteor shower would be 12° in orbital longitude or 7 days (e.g. 24/06/2007–05/07/2007). Looking at the evolution of this part of the stream over time, we find that the meteoroids started to intercept the Venusian orbit only recently, around 500 AD, their nodes moving progressively inwards and counterclockwise on the Venusian orbital plane.

3.3 Mars

Both branches of the Halley stream appear to be active at the Martian orbit. The distribution of nodes of Mars-intercepting meteoroids (Fig. 3) indicates that the two Martian showers may have different activity profiles, even taking into account the overlap with meteoroid trails from other perihelion returns of the comet.

The particle flux profile of the HAN branch appears symmetric with a maximum at $\lambda_s = 220°$ and an overall duration of ~8° (15 days) in solar longitude. The equivalent period in the Martian Calendar adopted by the atmospheric science community (Clancy et al. 2000) is $L_s = 321°–329°$ (e.g. 29 Sep–5 Oct, 2007) and repeats every Martian year. Extrapolating from our earlier comparison between the predicted and

Fig. 2 As Fig. 1 but for Venus. Particle nodes are plotted for the Venusian orbital plane. At the heliocentric distance of Venus, Halley particles intercept the planet at the descending node only

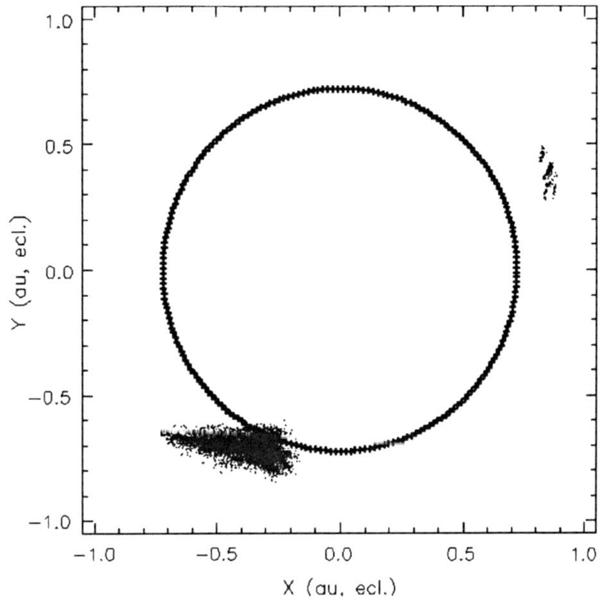

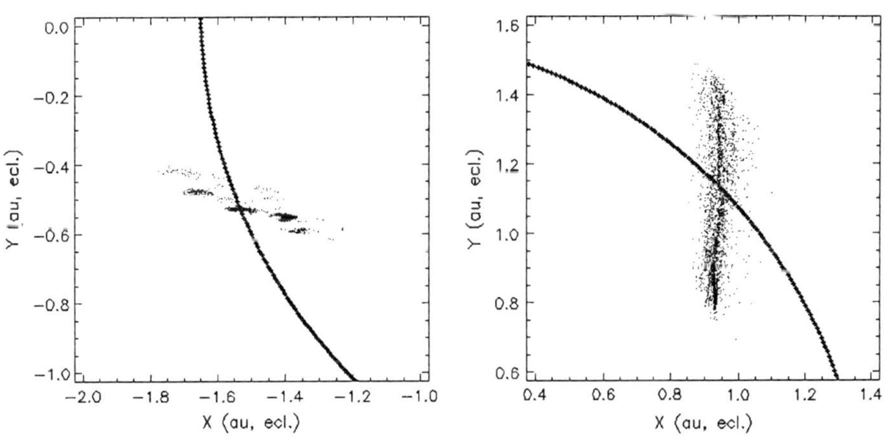

Fig. 3 As Fig. 1 but for Mars. Particle nodes are plotted for the Martian orbital plane. These intercept the planet both before and after perihelion, resulting in two distinct showers. Note the clumpy nature of the outbound branch compared to that of the inbound one

observed properties of the two branches at the Earth, we expect the actual shower to be longer in duration and its maximum several degrees of longitude in advance of our estimated value.

The HDN branch, on the other hand, exhibits a clumpy structure. Over the 40 Martian orbital periods that elapsed from 1910 to 1986, only 4 contained Mars-approaching material. Moreover, the ZHR of $\sim$60 reported in Table 1 comes from a single encounter between Mars and a dense clump of $<$1 cm meteoroids in 1972 when

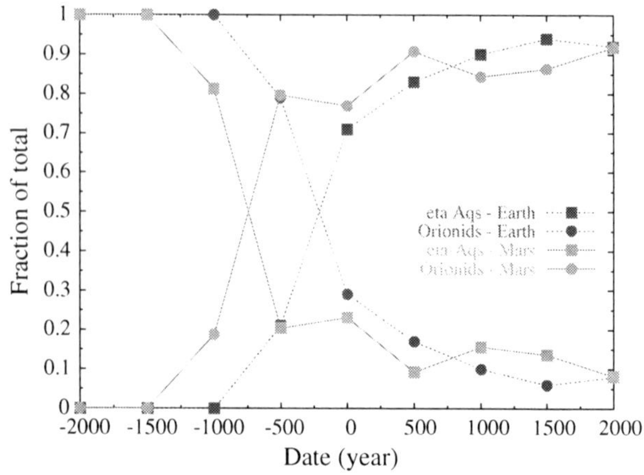

Fig. 4 Model variation in activity of the HAN and HDN branches at the Earth and Mars over the past four millenia. At the Earth, the two showers switch in activity during the latter half of the first millenium BC. After this time, HDN becomes the dominating branch. For Mars, a similar switch occurs during the first half of the first millenium BC but in the opposite sense so that the HAN branch becomes predominant

the ZHR reached 1200. A more careful inspection revealed that such encounters occur at $L_s = 113°–116°$ and follow a pattern, that is, they occur in 83-year intervals and only when Jupiter is in a particular segment of its orbit ($\lambda = 275°–323°$). This lead us to suspect that the origin of these outbursts were Halley meteoroids trapped in the 7:1 mean motion resonance with Jupiter. This was confirmed by verifying that the angle $7\lambda_{\text{Halley}} - \lambda_{\text{Jupiter}} - 6\varpi_{\text{Halley}}$ librates for many of these particles. It is important to emphasise that, since (a) we do not, in fact, know that the nucleus of comet Halley reached perihelion in 2924 BC and (b) the position-sensitive nature of the resonant trapping and confinement of cometary particle trails (Asher et al. 1999), we cannot issue any forecasts on Martian meteor storms. Our work does show, however, that the existence of dense resonant structures in the Halley stream is dynamically possible over long periods of time, in this case $\sim$5000 years. This is especially relevant given recent observations of Orionid outbursts at the Earth attributed to resonant structures (Trigo-Rodriguez et al. 2007; Sato and Watanabe 2007; Rendtel 2007). It suggests that Mars may be a prime observing location for studying the Halley resonant complex. Observations of such outbursts at Mars may also be used to infer the position of the comet several thousand years before its present observational arc.

Finally, we investigated how the level of activity from the two branches varies, according to our model, over the past several thousand years at Mars and the Earth. Counting the fraction of planetary years, between two successive perihelion returns of the comet, in which planet-intercepting meteoroids were detected, we found a duality between the two branches (Fig. 4). The HAN branch at the Earth and HDN branch at Mars appear to be the only active branches before 1500 BC. After that time, the other two branches begin to pick up in activity and a switch occurs, during the first and second halves of the first millenium BC for Mars and the Earth respectively. This leads to the present situation with HDN being the dominant branch at the Earth and HAN at Mars.

Acknowledgments The authors wish to thank the CINES team for the use of the super-computer. Part of this work was carried out during JV's visit to Armagh Observatory in March 2006 funded by PPARC Grant PPA/V/S/2003/00049. Astronomical research at the Armagh Observatory is funded by the Northern Ireland Department of Culture, Arts and Leisure (DCAL).

References

L.G. Adolfsson, B.A.S. Gustafson, C.D. Murray, The Martian atmosphere as a meteoroid detector. Icarus **119**, 144–152 (1996)

M.F. A'Hearn, R.L. Millis, D.G. Schleicher, D.J. Osip, P.V. Birch, The ensemble properties of comets: results from narrowband photometry of 85 comets, 1976–1992. Icarus **118**, 223–270 (1995)

D.J. Asher, M.E. Bailey, V.V. Emelyanenko, Resonant meteoroids from comet Tempel-Tuttle in 1333: the cause of the unexpected Leonid outburst in 1998. Mon. Not. R. Astron. Soc. **304**, L53–L56 (1999)

A.A. Christou, Prospects for meteor shower activity in the venusian atmosphere. Icarus **168**, 23–33 (2004)

R.T. Clancy, B.J. Sandor, M.J. Wolff, An intercomparison of ground-based millimeter, MGS TES, and Viking atmospheric temperature measurements: seasonal and interannual variability of temperatures and dust loading in the global Mars atmosphere. J. Geophys. Res. **105**, 9553–9572 (2000)

A. Dubietis, Long-term activity of meteor showers from comet 1P/Halley. WGN **31**(2), 43–48 (2003)

P.D. Feldman, M.C. Festou, M.F. A'Hearn, 13 co-authors, IUE observations of comet P/Halley—evolution of the ultraviolet spectrum between 1985 September and 1986 July. Astron. Astrophys. **187**, 325–331 (1987)

J.D. Giorgini, D.K. Yeomans, A.B. Chamberlin, P.W. Chodas, R.A. Jacobson, M.S. Keesey, J.H. Lieske, S.J. Ostro, E.M. Standish, R.N. Wimberly, JPL's on-line solar system data service. Bull. Am. Astron. Soc. **28**, 1158 (1996)

A. Hajduk, M. Hajdukova, V. Porubčan, G. Cevolani, One hundred years of observations of the comet Halley meteor stream. in *Proc. Asteroid Comets Meteors 2002 Conf.*, ed. by B. Warmbein, 29 July–2 August 2002, Berlin, Germany. ESA SP-500, (Noordwijk, Netherlands, 2002), pp. 113–116

R. Koschack, J. Rendtel, Determination of spatial number density and mass index from visual meteor observations I. WGN **18**, 45–58 (1990a)

R. Koschack, J. Rendtel, Determination of spatial number density and mass index from visual meteor observations II. WGN **18**, 119–140 (1990b)

J.P. McAuliffe, A.A. Christou, Modelling meteor ablation in the venusian atmosphere. Icarus **180**, 8–22 (2006)

B.A. McIntosh, A. Hajduk, Comet Halley meteor stream: a new model. Mon. Not. R. Astron. Soc. **205**, 931–943 (1983)

B.A. McIntosh, J. Jones, The Halley comet meteor stream: numerical modelling of its dynamic evolution. Mon. Not. R. Astron. Soc. **235**, 673–693 (1988)

J. Rendtel, The eta-Aquarid meteor shower in 1997. WGN **25**(4), 153–157 (1997)

J. Rendtel, Three days of enhanced Orionid activity in 2006—meteoroids from a resonance region? WGN **35**(2), 41–45 (2007)

G. Ryabova, The comet Halley meteoroid stream: just one more model. Mon. Not. R. Astron. Soc. **341**, 739–746 (2003)

M. Sato, J. Watanabe, Origin of the 2006 meteor outburst. Publ. Astron. Soc. Jpn. **59**, L21–L24 (2007)

J.M. Trigo-Rodriguez, J.M. Madiedo, J. Llorca, P.S. Gural, P. Pujols, T. Tezel, The 2006 Orionid outburst imaged by all-sky CCD cameras from Spain: meteoroid spatial fluxes and orbital elements. Mon. Not. R. Astron. Soc. **380**, 126–132 (2007)

P. van Nes, Giotto encounters comet Halley. Ruimtevaart **35**, 1–8 (1986)

J. Vaubaillon, F. Colas, L. Jorda, A new method to predict meteor showers I. Description of the model. Astron. Astrophys. **439**, 751–760 (2005a)

J. Vaubaillon, F. Colas, L. Jorda, A new method to predict meteor showers II. Application to the Leonids. Astron. Astrophys. **439**, 761–770 (2005b)

Z. Wu, I.P. Williams, Comet P/Halley and its associated meteoroid stream, in *Meteoroids and their Parent Bodies*, ed. by J. Stohl, I.P. Williams (Astronomical Inst., Slovak Acad. Sci., Bratislava, 1993) pp. 77–80

D.K. Yeomans, T. Kiang, The long-term motion of comet Halley. Mon. Not. R. Astron. Soc. **197**, 633–646 (1981)

Multi-station Video Orbits of Minor Meteor Showers

José M. Madiedo · Josep M. Trigo-Rodríguez

Originally published in the journal Earth, Moon, and Planets, Volume 102, Nos 1–4.
DOI: 10.1007/s11038-007-9215-x

Abstract During 2006 the SPanish Meteor Network (SPMN) set up three automated video stations in Andalusia for increasing the atmospheric coverage of the already existing low-scan-rate all-sky CCD systems. Despite their initially thought complementary nature, sensitive video cameras have been employed to setup an automatic meteor detection system that provides valuable real-time information on unusual meteor activity, and remarkable fireball events. In fact, during 2006 SPMN video stations participated in the detection of two unexpected meteor outbursts: Orionids and Comae Berenicids. The three new SPMN stations guarantee almost a continuous monitoring of meteor and fireball activity in Andalusia (Spain) and also increase the chance of future meteorite recoveries. A description of the main characteristics of these new observing video stations and some examples of the trajectory, radiant and orbital data obtained so far are presented here.

Keywords Meteors · Meteor showers

1 Introduction

High-sensitivity video devices have been commonly used for the study of the activity of meteor streams. These provide useful data for the determination, for instance, of radiant, orbital and photometric parameters (Koten 1999; Koten et al. 2003, 2007; Molau et al. 1997; Molau 2004; de Lignie and Jobse 1996). With this aim, multiple-station video observations of major and minor meteor showers have been systematically performed since

J. M. Madiedo (✉)
Facultad de Ciencias Experimentales, Universidad de Huelva, Huelva 21071, Spain
e-mail: madiedo@uhu.es

J. M. Trigo-Rodríguez
Institut de Ciències de l'Espai–CSIC, Campus UAB, Facultat de Ciències, Torre C5-parell-2[a],
08193 Bellaterra, Barcelona, Spain

J. M. Trigo-Rodríguez
Institut d'Estudis Espacials de Catalunya (IEEC), Edif. Nexus, c/Gran Capità, 2-4,
Barcelona 08034, Spain

June 2006 within the framework of the SPanish Meteor Network (SPMN). For this purpose, three new automated video stations supported by Universidad de Huelva have been set up in Andalusia. These are endowed with high sensitivity wide-field video cameras that achieve a meteor limiting magnitude of about +3. The new stations have increased the coverage performed by the low-scan all-sky CCD systems operated by the SPMN. Wide field video systems achieve a time accuracy of about 0.01 s for determining the appearance of meteor and fireball events. This work provides an overall description of the activity of several meteor showers observed during the first year of operation of the SPMN video stations. Particularly, our present efforts are specially dedicated to obtaining accurate heliocentric orbits to link these meteoroid streams with their progenitor bodies. Our research program is particularly focused in the coverage of scarcely known minor meteoroid streams, and the study of bright meteorite-dropping bolides. A new software package is being developed by the SPMN in order to help with the analysis of the huge amount of data recorded so far.

2 Description of the Observing Stations and Procedures

The three new SPMN video stations (Table 1) are endowed with different models of Watec and Mintron cameras. These are high-sensitivity devices that employ a black and white 1/2″ Sony interline transfer CCD image sensor. Fast aspherical lenses (f0.8–f1.2) are attached to the video cameras in order to maximize image quality and detect meteors as faint as magnitude +2/+3. Their focal length ranges from 3.8 to 12 mm. Dew removers are also attached to the optics when the systems must operate below dew point and, in order to increase the signal to noise ratio, thermoelectrical coolers (Peltier systems) are also attached to the cameras when operation temperature is high (above 25°C). The whole system was designed to be fully portable in order to setup mobile stations when necessary. In fact, the Cerro Negro video station (Table 1) is a mobile station.

The observing stations are automatically switched on and off at sunset and sunrise, respectively. The images taken by the cameras at 25 fps and with a resolution of 720 × 576 pixels are continuously sent either to a set of videocassette recorders (VCR) or to PC computers through a video capture card. Every VCR stores the whole observing session on VHS tapes that are processed later on in order to extract from them the video sequences containing meteor trails. However, the computers run a software (UFOCapture, by SonotaCo, Japan) that automatically registers meteor trails and stores the corresponding video frames on hard disk. In any event, before the signal from the cameras reaches the computers or the VCRs, a video time inserter that employs a GPS device (KIWI-OSD, by PFD Systems) inserts time information on every video frame. This allows us to measure time in a precise way (about 0.01 s) along the whole meteor path. In addition, some of the cameras have also a diffraction grating attached to the optics in order to record meteor

Table 1 Geographical location of the fixed and mobile video stations located in Andalusia and operated by the SPMN

SPMN station #	Station (Province)	Longitude	Latitude	Altitude (m)
1	Seville (Seville)	05°58′50″ W	37°20′46″ N	28
2	Cerro Negro (Seville)	06°19′35″ W	37°40′19″ N	470
3	El Arenosillo (Huelva)	07°00′00″ W	36°55′00″ N	30

spectra and perform the chemical characterisation of corresponding meteoroids. Astrometric reduction was made by hand, comparing the *X*, *Y* positions of stars and meteors in every frame of the sequences with meteors. Note that we don't use any software for automatic astrometry of the images. This crucial step was made via direct measurements of the pixel position, as in all our previous work (see e.g. Trigo-Rodríguez et al. 2002, 2004, 2006b). The *Network* software also allows us to predict the position of every meteor from each station by assuming the typical values of ablation height (Trigo-Rodríguez et al. 2002, 2004b). A search through the database of meteors that appeared during the same observing interval and in the proper position allowed the unequivocal identification of common double-station meteors. Once identified, the software estimates the atmospheric trajectory and radiant for each meteor. From the astrometric measurements of the shutter breaks and the trajectory length, the velocity of the meteoroid was derived. The average value of observed velocities for each shutter-break was obtained, and the preatmospheric velocity V_∞ was taken from the velocity measured in the earliest frames of the video sequences. In the last step, we determine the orbital elements from our trajectory and radiant data by using the *MORB* program Ceplecha et al. (2000). As a consequence of the observational data reduction effort, reliable trajectory and orbital data was obtained and is presented in Sect. 3.

3 Results

Because of extraordinary weather conditions in Spain during 2006, it was possible to obtain a huge amount of data related to the activity of minor and major streams. The number of nights studied by the video network was 150, corresponding to about 1,400 h of effective time, and about 3,000 meteors/month. Consequently, it is easy to understand that most of the results obtained so far still need to be reduced. In order to help with this, we are developing a software package called Amalthea which can perform in shorter time different tasks related to astrometry and photometry. We focus here on data obtained from several major streams, but especially on poorly studied meteoroid streams.

Since June 2006, the new SPMN video stations have registered an important number of bright fireballs belonging to different meteor showers (see e.g. Trigo-Rodríguez et al. 2007a) and also participated in the confirmation of the Orionid outburst in October 2006 (Trigo-Rodríguez et al. 2006, 2007b). SPMN video cameras also followed the activity of 2006 Leonids (Jenniskens 2006b).

As many of the meteor trails have been registered simultaneously by several of these observing stations, it has been possible to obtain the orbital elements of the corresponding meteoroids. Some of the results obtained so far are shown in Tables 2 and 3. Thus, for instance, in August 2006 we could observe several members of the π Eridanids (ERI). The data show a radiant position close to the estimated for the proposed parent body (see Table 7 of Jenniskens 2006a).

Jenniskens (2006a) proposes an activity period for September ε Perseids (SPE) in mid-September (Table 7, radiant#208), past observations suggest extended activity much earlier (Trigo-Rodríguez 1989). A clear example would be the video meteor SPMN070806 (Tables 2 and 3) that seems to be associated with this diffuse stream. Its radiant position has a small difference in declination that would be produced either by high dispersion or radiant drift. In any case, it is clear that the first week of September requires additional coverage of the activity of minor showers radiating from Auriga and Perseus.

Table 2 Trajectory and radiant data of 2006 SPMN video meteors. Equinox (2000.0)

SPMN code	Stream	M_v	H_b (km)	H_{max} (km)	H_e (km)	α_g (°)	δ_g (°)	V_∞ (km/s)	V_g (km/s)	V_h (km/s)
050806	ERI	−3	105.8	–	84.0	53.4 ± 0.3	−14.5 ± 0.3	59.9 ± 0.3	58.5	39.9
070806	SPE	−1	113.6	–	94.0	53.3 ± 0.3	30.4 ± 0.3	63.8 ± 0.3	62.6	36.0
101006	NTA	+1	106.5	102.7	86.5	31.0 ± 0.2	17.7 ± 0.2	23.5 ± 0.2	21.1	34.8
011106	LEO	+1	97.3	–	84.3	154.8 ± 0.3	25.3 ± 0.7	72.0 ± 0.2	70.8	42.1
061106	LEO	−3	99.0	96.8	87.4	155.0 ± 0.3	25.0 ± 0.7	72.2 ± 0.3	71.0	42.3
071106	LEO	−1	107.2	–	95.1	156.3 ± 0.3	21.5 ± 0.3	72.1 ± 0.2	70.9	42.2
031206	ALY	−1	98.2	–	88.6	135.33 ± 0.19	38.93 ± 0.10	56.3 ± 0.2	55.1	47.5
071206	COM	−1	99.8	–	92.7	180.2 ± 0.3	27.6 ± 0.3	63.8 ± 0.3	62.8	39.9
081206	HYD	−2	111.6	–	94.8	135.4 ± 0.3	−2.5 ± 0.3	58.9 ± 0.3	58.1	45.0

Stream association is given using IMO standard labels (Mv: visual magnitude; H_b, Hmax, He: height corresponding to the beginning point, maximum brightness point, and ending point, respectively; α_g, δ_g: right ascension and declination of the geocentric radiant; $V\infty$: pre-atmospheric meteor velocity; V_g: geocentric meteor velocity; V_h: heliocentric meteor velocity)

Table 3 Orbital elements of imaged meteors. Equinox (2000 00)

SPMN code	q (AU)	$1/a$ (AU^{-1})	e	i (°)	ω (°)	Ω (°)
050806	0.816 ± 0.007	0.19 ± 0.03	0.848 ± 0.022	117.2 ± 0.5	54.49 ± 1.23	336.46192 ± 0.00001
070806	0.813 ± 0.011	0.52 ± 0.03	0.580 ± 0.022	153.4 ± 0.8	242 ± 2	156.52746 ± 0.00002
101006	0.507 ± 0.007	0.636 ± 0.016	0.677 ± 0.007	4.26 ± 0.19	282.54 ± 1.12	201.46599 ± 0.00023
011106	0.9880 ± 0.0005	0.023 ± 0.022	0.977 ± 0.021	156.3 ± 0.9	177.8 ± 1.5	236.46875 ± 0.00008
061106	0.9878 ± 0.0007	0.008 ± 0.032	0.99 ± 0.03	156.8 ± 1.1	177.3 ± 1.9	236.46878 ± 0.00010
071106	0.9865 ± 0.0018	0.016 ± 0.022	0.985 ± 0.022	156.9 ± 0.8	185.1 ± 2.3	236.50565 ± 0.00007
031206	0.2052 ± 0.0020	−0.508 ± 0.019	1.104 ± 0.005	84.1 ± 0.5	300.5 ± 0.4	273.06619 ± 0.00001
071206	0.688 ± 0.014	0.24 ± 0.03	0.832 ± 0.019	134.3 ± 0.6	250.2 ± 1.9	273.18564 ± 0.00002
081206	0.228 ± 0.006	−0.25 ± 0.03	1.057 ± 0.007	110.2 ± 0.8	119.6 ± 1.0	93.18892 ± 0.00001

In October 2006 we detected several members of the Andromedids (AND), the Northern Taurids (NTA) and the ν Aurigids (NAU) minor meteor showers. The results calculated for the ν Aurigids (Trigo-Rodríguez et al. 2007b, MNRAS) show orbital similitude with the orbital elements obtained by Sekanina (1976). It was completely unexpected to find members of this meteoroid stream so early in the month because the previously reported activity period is October 20–22 (Jenniskens 2006a). However, other single station meteors recorded by the SPMN video cameras were also well associated by alignment and angular velocity with this radiant. The SPMN video stations were also able to record in November the activity of 2006 Leonids (Jenniskens 2006b). Three orbits obtained during the outburst are shown in Table 3.

On December 24–25, 2006 SPMN video cameras recorded a high activity of the Comae Berenicids (COM). Four video cameras operated from two stations in Seville province (Spain) recorded the event. In particular, two Watec video cameras operated under dark skies with fields of view $88° \times 56°$ and $57° \times 43°$ and limiting meteor magnitudes of +3 recorded 12 Comae Berenicids meteors between 3 h 30 m and 4 h 30 UTC. A simulation taking into account sensor sensitivity, geometric loss, radiant altitude and position, as well as particle distribution ($r = 2.0 \pm 0.4$, for $N = 25$) provided a maximum COM meteoroid flux of 4×10^{-3} ($m_{6.5}$/km^2/h) with corresponds to an equivalent (human) ZHR $= 60 \pm 25$, about 10 times the activity expected for this minor shower in such date. Accurate single-station astrometry reveals that this activity comes from an apparent radiant located in RA: $181 \pm 2°$ and DEC: $+26 \pm 2°$. SPMN 071206 exhibit a similar radiant to the derived from single-station data. This activity is in agreement with additional forward scatter meteor observations performed by the SPMN from Cerro Negro (Seville) using a computer-controlled ICOM IC-PCR1500 radio scanner attached to a 1/2 wave vertical antenna and a Hamtronic LNK-50 preamplifier. This system was tuned to 55.249 MHz, and the whole observing session was recorded on hard disk. However, these data could not be contrasted with other sources. Alastair McBeath (Society for Popular Astronomy, England) pointed out that a possible confirmation of this data is an anomalous peak observed in the 3–4 h UTC interval by Gaspard de Wilde from Belgium (Mc Beath, pers. comm.). Although a clear confirmation of this high activity has not been obtained we think it deserves to be mentioned. During the December monitoring we also recorded a detectable activity from the α Lyncids (ALY) and the σ Hydrids (HYD) radiants, and at least one orbit has been obtained for these (SPMN031206 and SPMN081206, respectively). About 80% of the December multiple-station data remains to be reduced.

4 Conclusions

The SPMN has increased its atmospheric coverage in the south of Spain by means of three new video stations. These allow achieving a spatial resolution of about 1 arc min, a time resolution of about 0.01 s and a meteor velocity accuracy that ranges between 0.2 and 0.4 km/s. Besides, the distance separating the two main cores of the network (Catalonia and Andalusia) guarantee almost a continuous monitoring of meteor and fireball activity, which has provided a huge amount of data during the last year (in total, eight observing stations are in operation). We have also given a few examples of how the SPMN multiple-station observations can provide valuable orbital information on minor meteoroid streams. Many of the data obtained so far still need to be reduced, although a new software package is being developed by our network to help with this task. On the other hand, the SPMN video systems are currently under expansion in order to improve the devices used to register meteor spectra and also to include daytime monitoring cameras.

Acknowledgements The authors thank Dr. Alberto J. Castro-Tirado (IAA-CSIC) for providing the required infrastructure for setting up the El Arenosillo SPMN video station. *Universidad de Huelva* (*UHU*) and *Consejo Superior de Investigaciones Científicas* (*CSIC*) provided financial support during this work. JMT-R also thanks MEC for a JdC research grant.

References

Z. Ceplecha, P. Spurný, J. Borovička, *Meteor Orbit (MORB) Software* (Ondrejov Observatory, Czech Republic, 2000)

M. de Lignie, K. Jobse, Double-station video observations of the 1995 quadrantids. WGN J. IMO **24**(1–2), 20–26 (1996)

P. Jenniskens, *Meteor Showers and Their Parent Comets*. (Cambridge Univ. Press, Cambridge, UK, 2006a)

P. Jenniskens, Leonids meteors 2006, in Central Bureau for Astronomical Telegrams–CBET #710 (2006b)

P. Koten, Photometry of TV meteors, in *Proceedings of the Meteoroids '98 Conference*, eds. by W.J. Baggaley, V. Porubcan (Astron. Inst., Slovak Acad. Sci., Bratislava, 1999), pp. 149–152

P. Koten, P. Spurný, J. Borovička, R. Štork, Catalogue of video meteor orbits 1. Publ. Astron. Inst. ASCR **91**, 1–32 (2003)

P. Koten, J. Borovička, P. Spurný, R. Štork, Optical observations of enhanced activity of the 2005 Draconid meteor shower. Astron. Astrophys. **466**(2), 729–735 (2007)

S. Molau, The AKM video meteor network, in *Proceedings of the Meteoroids 2001 Conference*, (2004) pp. 315–318

S. Molau, M. Nitschke, M. de Lignie, R.L. Hawkes, J. Rendtel, Video observations of meteors: history, current status, and future prospects. WGN J. IMO **25**(1), 15–20 (1997)

Z. Sekanina, Statistical model of meteor streams IV. A study of radio streams from the synoptic year. Icarus **27**, 265–321 (1976)

J.M. Trigo-Rodríguez, The epsilon-Perseids in 1987 and 1988. WGN J. IMO **17**(4), 156–158 (1989)

J.M. Trigo-Rodríguez, J. Llorca, J. Fabregat, On the origin of the 1999 Leonid storm as deduced from photographic observations. Earth Moon Planets **91**, 107 (2002)

J.M. Trigo-Rodríguez, A.J. Castro-Tirado, J. Llorca, J. Fabregat, V.J. Martínez, V. Reglero, M. Jelínek, P. Kubánek, T. Mateo, A. de Ugarte Postigo, The development of the Spanish Fireball Network using a new all-sky CCD system. Earth Moon Planets **95**, 553 (2004)

J.M. Trigo-Rodríguez, J. Llorca, E. Lyytinen, J.L. Ortiz, A. Sánchez Caso, C. Pineda y, S. Torrell, 2002 Leonid storm fluxes and related orbital elements. Icarus **171**, 219–228 (2004b)

J.M. Trigo-Rodríguez, et al., Orionid meteors, in Central Bureau Astronomical Telegram, CBAT #698, IAU (2006)

J.M. Trigo-Rodríguez, J. Llorca, A.J. Castro-Tirado, J.L. Ortiz, J.A. Docobo, J. Fabregat, The Spanish fireball network. Astron. Geophys. **47**(6), 26 (2006b)

J.M. Trigo-Rodríguez, J.M. Madiedo, A.J. Castro-Tirado, J.L. Ortiz, J. Llorca, J. Fabregat, S. Vítek, P.S. Gural, B. Troughton, P. Pujols, F. Gálvez, Spanish meteor network: 2006 continuous monitoring results. WGN J. IMO **35**, 13–22 (2007a)

J.M. Trigo-Rodríguez, J.M. Madiedo, J. Llorca, P.S. Gural, P. Pujols, T. Tezel, The 2006 Orionid outburst imaged by all-sky CCD cameras from Spain: meteoroid spatial fluxes and orbital elements. Mon. Not. R. Astron. Soc. **380**, 126–132 (2007b)

Exceptional Fireball Activity of Orionids in 2006

Pavel Spurný · Lukáš Shrbený

Originally published in the journal Earth, Moon, and Planets, Volume 102, Nos 1–4.
DOI: 10.1007/s11038-007-9210-2 © Springer Science+Business Media B.V. 2007

Abstract We report exceptional fireball activity of the Orionid meteor shower in 2006. During four nights in October 2006 the autonomous fireball observatories of the Czech part of the European Fireball Network (EN) recorded 48 fireballs belonging to the Orionids. This is significantly more than the total number of Orionids recorded during about five decades long continuous operation of the EN. Based on precise multi-station photographic and radiometric data we present accurate atmospheric trajectories, heliocentric orbits, light curves and basic physical properties of 10 Orionid fireballs with atmospheric trajectories that were long enough and, with one exception, were observed from at least three stations. Seven were recorded in within a 2-h interval in the night of 20/21 October. Their basic parameters such as radiant positions and heliocentric orbits are very similar. This high fireball activity originated from a very compact geocentric radiant defined by $\alpha = 95.10° \pm 0.10°$ and $\delta = 15.50° \pm 0.06°$. These fireballs most likely belonged to a distinct filament of larger meteoroids trapped in 1:5 resonance with Jupiter. From detailed light curves and basic fireball classification we found that these meteoroids appertain to the weakest component of interplanetary matter.

Keywords Orionids · Fireball · Trajectory · Radiant · Orbit · Light curve

1 Introduction

The Orionid meteor shower is a relatively strong and stable regular annual shower with a peak visual hourly rate of 15–30 meteors and broad maximum generally occurring on October 20–23. The parent is comet 1P/Halley, now in an orbit passing a far +0.151 AU from Earth (Jenniskens 2006). The last return of parent comet 1P/Halley caused a renewed interest in the study of both streams originating from this comet, i.e. Orionids and η-Aquariids. Most theoretical works and modeling attempts (Hajduk 1970; McIntosh and Hajduk 1983; Hughes 1987; McIntosh and Jones 1988; Wu and Williams 1993) as well as

P. Spurný (✉) · L. Shrbený
Astronomical Institute of the Academy of Sciences, Ondřejov Observatory, Ondrejov, Czech Republic
e-mail: spurny@asu.cas.cz

DOI: 10.1007/978-0-387-78419-9_21

increasing number of meteor observations mostly by visual amateur observers date around this return. However no enhanced Orionid activity was observed and it was concluded that enhanced rates are not connected with the parent comet returns (Porubčan et al. 1991) but are due to isolated particle concentrations not necessarily in the comets vicinity. The latest outburst was observed in 1993 (Rendtel and Betlem 1993). Emel'Yanenko (2001) in his theoretical work explains enhanced and short-term activity of a shower with a libration of a meteoroid stream, and for Orionids he presents three sets of parameters that describe three possible resonant zones for this shower (Table 1). These librating particles create a resonance substream. The 2006 Orionid activity was caused according to Sato and Watanabe (2007) by the dust trails formed by meteoroids ejected from 1P/Halley in years 1265, 1197 and 910 BC and trapped in the 1:6 mean motion resonance with Jupiter. The exceptional activity of 2006 Orionids Rendtel (2007) is also ascribed to the meteoroids from a resonant zone with the most favorable resonance the 1:6 with Jupiter. Rendtel (2007) also mentions the decrease of the population index, r. During the maximum of the activity the value of r was 1.6 but the long-term average is 2.3–2.9. This fact confirms that the meteors observed during the Orionid 2006 maximum deviated significantly from the average Orionid meteors and on average are significantly brighter. A similar effect was observed during the 1993 Orionid outburst (Jenniskens 1995).

2 Instrumentation, Observations and Data Processing

The multi-station photographic observation of fireballs in fireball networks represents a very efficient and precise method how to record the atmospheric interactions of larger meteoroids. During the short moment of meteoroids ablation we can determine their atmospheric trajectories, orbits, light curves and basic physical properties. One of the most advanced operational fireball network is the Czech part of the European Fireball Network, where each station is equipped with the newest generation camera, modern and sophisticated completely Autonomous Fireball Observatory(AFO) (Spurný et al. 2007). The AFO imaging system consists of a Zeiss Distagon fish-eye objective (f/3.5, f = 30 mm) and a large-format sheet film. All AFOs are equipped with a rotating shutter close to the focal plane to determine fireball velocity. At present we operate 10 stations almost uniformly deployed across the territory of the Czech Republic. The typical precision of measurement of any individual point on the luminous atmospheric trajectory for fireballs up to approximately 200 km distance from the stations is about 10–15 m. This precision is proportionally decreasing with the distance of fireball from stations. In some ideal cases we can determine reliably fireballs at a distance of about 500 km from our territory. It enables us to observe fireballs over large parts of Central Europe. Except direct fireball imaging each AFO includes also an all-sky brightness sensor (radiometer) with sampling rate of 500

Table 1 Parameters of principal resonance zone near orbit of the Orionid meteor shower (Emel'Yanenko 2001)

$j':j$	a (AU)	Δa (AU)
1:6	17.19	1.0
1:5	15.22	0.9
1:4	13.12	0.8

a is the semimajor axis at the center of the $j':j$ resonance, Δa is the width of the resonance zone measurements per second. Therefore, along with the accurate time of fireball passage and its duration, we also obtain a very detailed light curve. These sensors reliably work even under a cloudy sky, so we have basic information about fireball luminosity and its very approximate location even without photographic records.

This system was in full operation during the 2006 Orionid enhanced activity. During four nights from October 20 we recorded a total of 48 bright Orionids although during this time our observation was strongly affected by unstable weather conditions. The entire data set includes multi-station photographic fireballs and those recorded photographically only from one station, some that were too short (only the brightest flare) to make it possible to compute the trajectory or to determine the velocity, and others that were recorded only by brightness sensors due to a cloudy sky.

Only 10 recorded fireballs were long and bright enough to be recorded photographically from more than one station so that we can precisely determine all important parameters describing their atmospheric trajectories, heliocentric orbits and basic physical properties. In this paper we deal only these multi-station fireballs. All presented Orionid fireballs were measured and processed using our standard procedures (Borovička et al. 1995; Ceplecha 1987).

3 Atmospheric Trajectories, Light Curves and Physical Properties

Atmospheric trajectories for the Orionid fireballs presented here were determined from all available images (Table 2). Because Orionid meteors are very fast, their atmospheric trajectories are often very short (last column in Table 2). This can decrease the precision in determination of other critical parameters. However, in all cases the fireballs listed in Table 2 were recorded from more than two stations, which highly increases the reliability of the data here presented. The only exception being MET06 that was recorded from only two stations. Following the format in the tables, fireballs MET01, MET08 and MET09 were recorded from three stations, MET02, MET07 and MET10 from four stations, MET04 from five stations, MET05 from six stations, and finally MET03 even from eight stations of the Czech Fireball Network.

Table 2 Atmospheric trajectories of Orionids 2006 fireballs

Meteor no.	Date	Time (UT)	H_B (km)	λ_B (°)	φ_B (°)	H_E (km)	λ_E (°)	φ_E (°)	L_{obs} (km)
MET01*	21.10.	0:36:13	105.6	14.544	49.020	85.3	14.283	49.116	30.0
MET02*	21.10.	0:50:44	104.3	12.027	48.362	90.3	11.855	48.428	20.7
MET03*	21.10.	0:57:58	108.9	14.979	50.343	85.3	14.712	50.469	34.1
MET04*	21.10.	1:46:13	100.7	19.520	49.816	82.0	19.414	49.924	23.7
MET05*	21.10.	1:50:42	108.2	15.037	49.285	78.1	14.838	49.453	39.0
MET06*	21.10.	2:01:45	101.8	17.015	49.406	89.4	16.937	49.492	15.0
MET07*	21.10.	2:35:48	110.8	16.290	49.303	86.4	16.218	49.446	30.2
MET08	21.10.	23:52:10	106.9	17.032	50.117	88.2	16.720	50.194	30.3
MET09	22.10.	2:01:29	114.3	18.341	48.984	77.8	18.176	49.191	40.5
MET10	22.10.	22:55:46	108.0	19.496	49.825	87.5	19.026	49.885	40.5

H is the height above sea level, λ and φ are the geographical coordinates, L_{obs} is the length of observed trajectory. The subscript "B" denotes values at the beginning point of the atmospheric trajectory, the subscript "E" at the end point. The fireballs from the filament are denoted by asterisk

The results on atmospheric trajectories are collected in Tables 2–4. The time of meteor beginning, geographical position, beginning and end heights and length of observed atmospheric trajectory are presented in Table 2. The beginning heights range from 100 to 114 km and the terminal heights from 78 to 90 km corresponding to a range of the observed trajectory lengths from 15 to 40 km. Physical data of these fireballs are presented in Tables 3 and 4. Zenith distances for the end point, initial velocities (mean measured velocity without deceleration), maximum absolute photographic magnitudes, initial photometric masses, PE coefficients that describe the empirical end heigh criterion and fireball types according to the classification of (Ceplecha and McCrosky 1976) are shown in Table 3.

Table 3 Physical data on Orionids 2006 fireballs

Meteor no.	ZD_E (°)	V_∞ (km/s)	M_{max} (mag)	m_∞ (g)	PE	Type
MET01*	47.7	67.6	−5.1	0.5	−5.10	II/IIIA
MET02*	46.8	67.7	−8.4	6	−5.93	IIIB
MET03*	45.6	67.7	−6.0	3	−5.44	IIIA
MET04*	37.7	67.8	−10.0	40	−5.73	IIIB
MET05*	38.7	67.8	−8.2	7	−5.13	II/IIIA
MET06*	37.0	67.6	−3.0	0.1	−5.20	II/IIIA
MET07*	34.9	67.6	−7.9	3	−5.61	IIIA/IIIB
MET08	52.6	67.8	−6.0	2	−5.48	IIIA
MET09	36.1	67.4	−8.3	10	−5.19	II/IIIA
MET10	60.0	67.5	−6.9	4	−5.47	IIIA

ZD_E is the zenit distance of the radiant at the end point of the atmospheric trajectory, V_∞ is the initial velocity, M_{max} is the maximum absolute magnitude, m_∞ is the initial photometric mass, PE is the coefficient that describes the empirical end height criterion and designates the type of fireball (Ceplecha and McCrosky 1976). The fireballs from the filament are denoted by asterisk

Table 4 Heights and durations of the flares and dynamic pressures

Meteor no.	H_{MF} (km)	H_{TF} (km)	p_{MF} (MPa)	p_{TF} (MPa)	ΔT_{mf} (ms)	ΔT_{tf} (ms)
MET01*	94	85	0.007	0.033	45	6
MET02*	94	–	0.007	–	65	–
MET03*	91	85	0.012	0.033	80	10
MET04*	89	82	0.017	0.055	60	13
MET05*	85	78	0.033	0.110	65	5
MET06*	–	89	–	0.016	–	7
MET07*	87	–	0.023	–	19	–
MET08	–	88	–	0.020	–	8
MET09	81	78	0.065	0.105	15	9
MET10	90	87	0.013	0.022	40	12

H_{MF} is the height at maximum of the brightness and H_{TF} is the height at the terminal flare. p_{MF} is the dynamic pressure at maximum of the brightness and p_{TF} is the dynamic pressure at the terminal flare. ΔT_{mf} and ΔT_{tf} are durations of the maximum and the terminal flares. Duration of the flare presented here is the full width at half maximum. The fireballs from the filament are denoted by asterisk

The rotating shutter hides one half of the meteor trail so in most of cases the short-term flares are not visible in the photographs at all. These flares are obvious in the light curves from the AFO brightness sensors. This is well documented in Fig. 1 showing the photographic images and radiometric light curves for all 10 Orionid fireballs listed in tables. From known durations of the fireballs and their approximate light curve profile, both derived from the photographic records and from durations of the AFOs light curves, we were able to derive instantaneous heights of flares for each fireball in the atmosphere. Values of these quantities (if visible in the light curve) are presented in Table 4. Since the overlap of the light curves is approximate, the heights listed are rounded-off to kilometers. The duration of the flares varies from several tens of milliseconds (near the mid part of the trajectory) to only several milliseconds in terminal flares. A typical Orionid light curve has a broader maximum and one much shorter very pronounced terminal flare. From such type of light curve we can infer that the material of the Orionid meteoroids easily disintegrates first into bigger particles which gradually ablate and create a longer middle peak and near the end of its trajectory the remaining part of the initial meteoroid completely disintegrates into a large amount of very small particles which ablate and evaporate very quickly. From the values listed in Tables 3 and 4 it is evident that all ten presented Orionid meteoroids consist of very weak and fragile material that is usually assumed to be of cometary origin which corresponds with known parent body—comet 1P/Halley.

4 Radiants and Orbital Elements

Geocentric radiant positions and orbital elements for all 10 Orionid fireballs are tabulated in Table 5. All values are given in the J2000.0 equinox. The fireballs are arranged according to date and time of occurrence, which is given by increasing values of the ascending node. It is evident that the first seven fireballs which were recorded in within a 2-h interval on October 21st all have very similar values. Therefore we conclude that these meteoroids belonged to one very compact filament which slightly differs from a regular background Orionids (Lindblad and Porubčan 1999; de Lignie and Betlem 1999).

The compactness of geocentric radiants of fireballs belonging to this new filament is shown in Fig. 2 along with the mean geocentric radiant value, the three Orionids recorded during two following nights (MET08–MET10), and the mean radiant positions determined by Lindblad and Porubčan (1999) from IAU MDC photographic data and by de Lignie and Betlem (1999) from DMS video data. Mean orbital elements for filament and both published data sets are also listed in Table 5. Although the differences are not too significant in the statistical sense we can still find some distinctions. The radiant position of the 2006 filament is systematically shifted by about 0.1° or 0.6° (depending on source of data) to higher right ascensions and 0.2° to lower declinations. Also some orbital elements are slightly different: the filament meteoroids have about 0.01–0.03 AU larger perihelion distances and about 0.3°–0.6° smaller inclinations. As shown in Table 5 and Fig. 2 some characteristics of the three Orionids recorded during two following nights differ from the filament. However, this difference is not so obvious to completely exclude the possibility that these meteoroids could belong to the filament (it would need statistically larger set of data).

As shown in Table 5 our values significantly differ from those published by Trigo-Rodríguez et al. (2007). They reported Orionid fireballs recorded in the same time interval in the night of October 20/21, 2006, which means that the orbital characteristics of the meteors in this narrow time slot should be very similar. We found that none of the three Orionid meteors detected by the Spanish Meteor Network (SPMN) match the parameters of

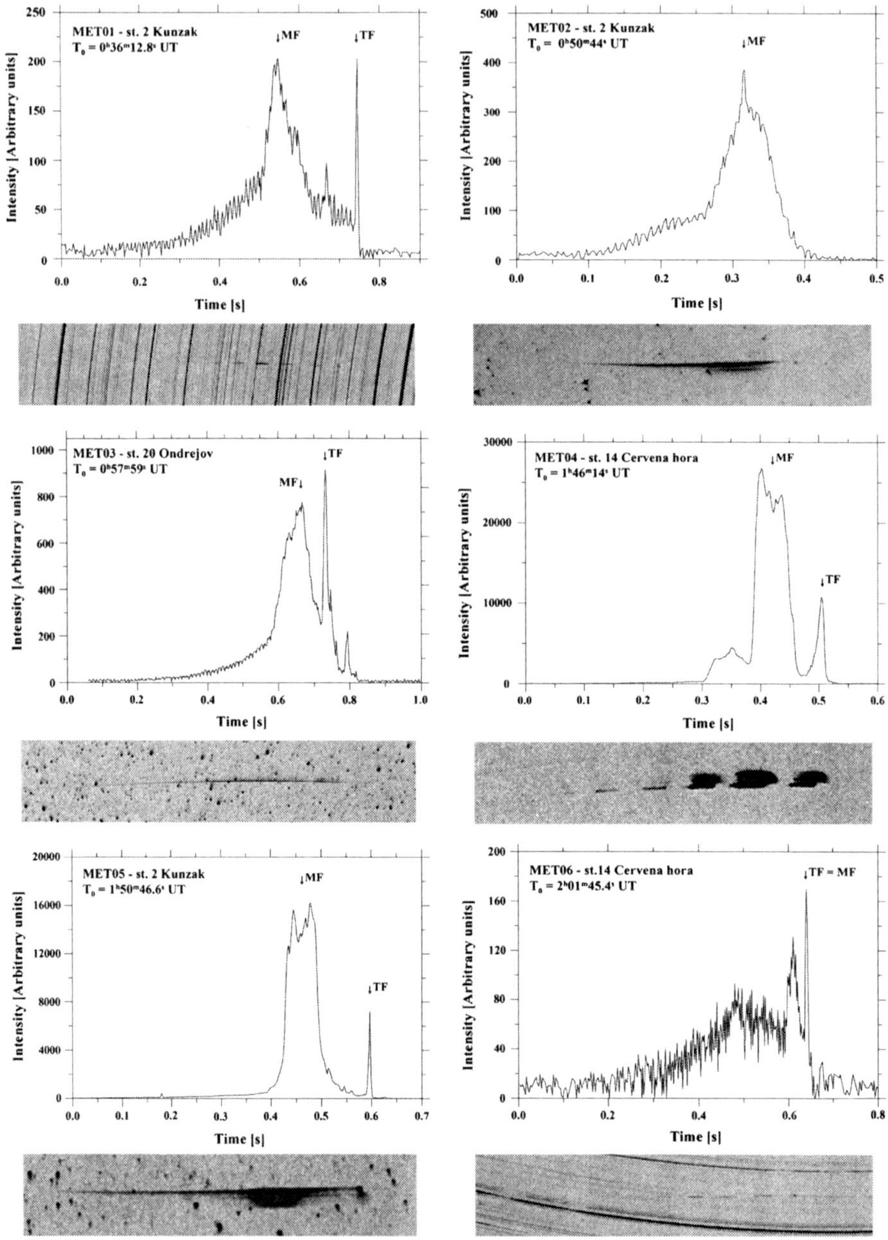

Fig. 1 Light curves from AFO's brightness sensors and images of the Orionid fireballs from all-sky cameras. MF means the position of the maximum brightness, TF the position of the terminal flare (see Table 4). The images from fixed cameras display star trails and interruptions of the meteors caused by rotating shutter (15 breaks/second). The guided images were taken by guided all-sky camera at Ondřejov Observatory and show the entire fireball trails. All fireballs flew from left to right in the images

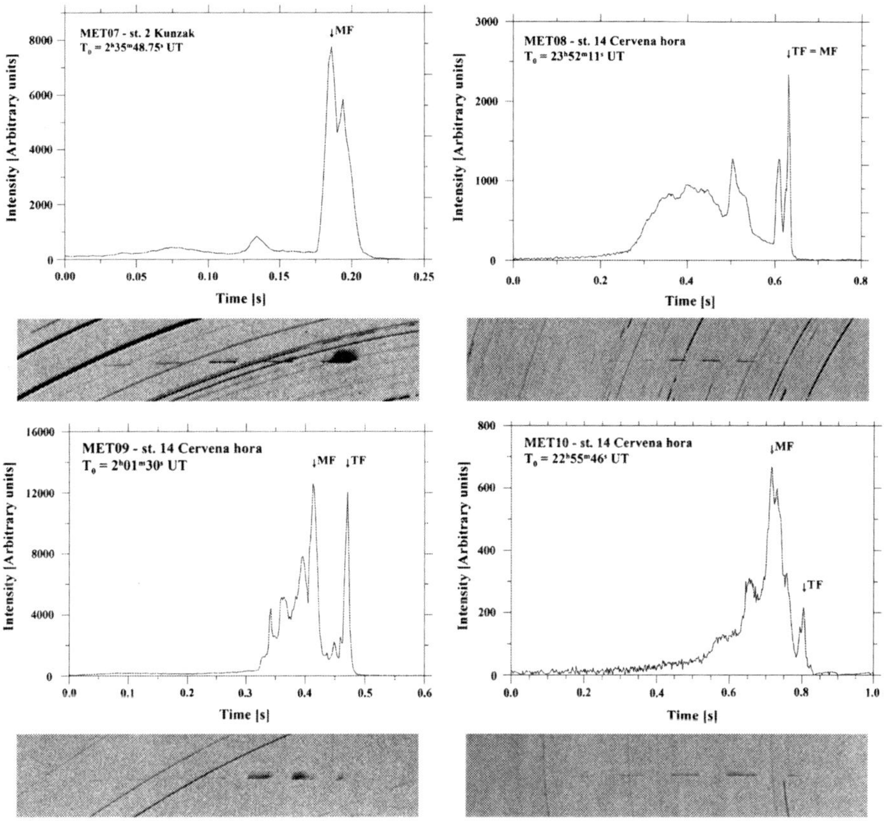

Fig. 1 continued

our filament (Table 5) and that also the radiants and orbital elements of these three Orionids significantly differ among each other. The SPMN results are plainly contradicting the consistent results from our observations of Orionid fireballs within the same short period of time. The discrepancy is unexpected in the face of the very high precision that was reported by Trigo-Rodríguez et al. (2007) that was supported by very low standard deviations of each individual entry. We suggest that the discrepant SPMN results might be either due to an overestimation of the precision in the measurements or a systematic error in computations.

The mean heliocentric orbit of the Orionid 2006 filament has semimajor axis $a = 14.8$ AU, eccentricity $e = 0.959$, inclination $i = 163.71°$, perihelion distance $q = 0.603$ AU and argument of perihelion $\omega = 78.7°$. According to Emel'Yanenko (2001) (Table 1) it is probable that particles from this filament were in the 1:5 resonance with Jupiter. From detailed analysis of visual observations of the 2006 outburst as well as another Orionid outbursts observed in twentieth century, Rendtel (2007) suggested that these outbursts could be caused by particles from the 1:6 resonance. Similarly Sato and Watanabe (2007) ascribe the 2006 Orionid activity to particles from the 1:6 mean motion resonance with Jupiter. However we have no arguments from our study presented here to decide which value is unambiguously correct. Although all our presented values are determined with high precision and reliability we know that the least-precise value is the

Table 5 Radiants and orbital elements (J2000.0) of 2006 Orionids fireballs

Meteor no.	α_G (°)	δ_G (°)	V_G (km/s)	a (AU)	e	q (AU)	ω (°)	Ω (°)	i (°)
MET01*	94.98	15.58	66.7	13.8	0.957	0.600	79.2	27.38605	163.8
MET02*	95.19	15.56	66.8	14.4	0.958	0.606	78.4	27.39710	163.9
MET03*	95.17	15.46	66.8	13.8	0.956	0.605	78.6	27.40109	163.6
MET04*	95.24	15.54	66.9	16.2	0.962	0.608	78.1	27.43440	163.9
MET05*	95.04	15.48	66.9	17.3	0.965	0.604	78.5	27.43752	163.7
MET06*	94.98	15.45	66.7	14.2	0.958	0.599	79.3	27.44516	163.5
MET07*	95.10	15.44	66.7	13.7	0.956	0.601	79.0	27.46867	163.6
MET08	95.84	16.25	66.9	15.7	0.962	0.595	79.6	28.35028	165.3
MET09	95.41	15.53	66.5	13.9	0.958	0.582	81.3	28.43983	163.6
MET10	96.43	15.90	66.6	14.3	0.959	0.584	81.0	29.30975	164.5
Filament	95.10	15.50	66.79	14.8	0.959	0.603	78.7	27.42	163.71
σ	0.10	0.06	0.09	1.4	0.003	0.003	0.4	0.03	0.14
IAU MDC	95.2	15.8	66.52	14.4	0.961	0.576	81.9	28.4	164.0
σ	2.6	0.7	1.16	8.6	0.057	0.039	5.2	3.4	1.3
Video DMS	94.45	15.74	66.6	14.6	0.961	0.590	80.5	27.4	164.3
SPMN11006	90.92	17.00	65.8	25.0	0.980	0.491	91.4	27.34436	165.55
SPMN21006	92.75	15.79	66.2	21.7	0.975	0.540	85.8	27.50407	163.48
SPMN51006	93.81	16.00	66.6	18.5	0.969	0.571	82.3	27.36133	164.2

(α_G, δ_G) is the geocentric radiant, V_G is geocentric mean velocity without atmospheric drag (not mesurable on our records). Data for the filament, IAU MDC (Lindblad and Porubčan 1999), DMS video meteors (de Lignie and Betlem 1999) and SPMN meteors (Trigo-Rodríguez et al. 2007) are mean values, σ is standard deviation for each entry. The fireballs from the filament are denoted by asterisk

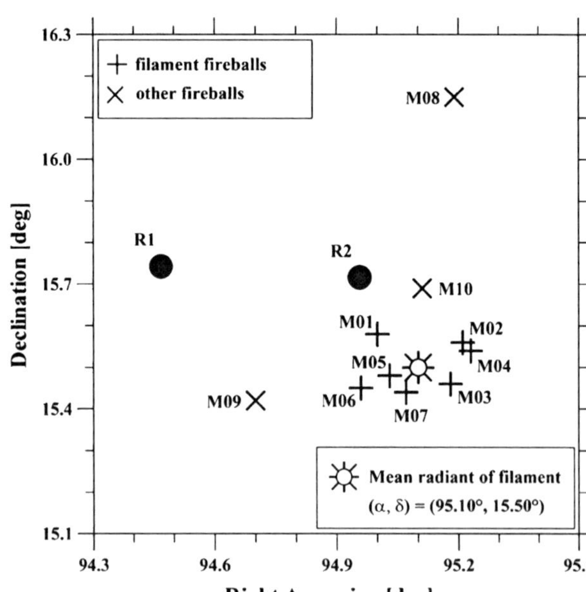

Fig. 2 Geocentric radiants of 2006 Orionid fireballs (J2000.0). The radiants are normalised to the node 27.4°, with a radiant drift of 0.70 dRA/Dsol and 0.11 dDec/Dsol. Mean radiant of filament is computed from meteors MET01 to MET07. R1 is the radiant position according to (de Lignie and Betlem 1999) and R2 according to (Lindblad and Porubčan 1999), both normalised to the node 27.4°

semimajor axis (i.e. also period), which is strongly affected by the uncertainty of entry velocity that is objectively difficult to determine with sufficient precision. It is caused by the fact that Orionids are very fast meteors and their atmospheric trajectories are relatively short. Therefore it will certainly need further study to decide this discrepancy.

5 Conclusions

We present results on atmospheric trajectories, orbits, light curves and physical properties of 10 Orionid fireballs recorded by cameras of the Czech Fireball Network during high Orionid activity in three nights of October 2006. The main conclusions are as follows.

(a) We determined the precise mean radiant position and orbital elements of the very distinct filament that produced the observed outburst of Orionid activity in morning hours of 21st October 2006. We found that this filament only slightly differs from mean shower characteristics determined from IAU MDC photographic data by Lindblad and Porubčan (1999) or from DMS video data (de Lignie and Betlem 1999). Our values significantly differ from the values for the 2006 Orionid outburst published by Trigo-Rodríguez et al. (2007).

(b) From single station photographic data and radiometric records we observed unusual high activity of bright Orionids over relatively long period during four consecutive nights from October 20–24. The lack of multi-station photographic data from second, third and fourth observing nights was partly caused by bad weather conditions over the Czech Republic and partly also by a decreasing number of brighter meteors. In second and third night we recorded only three fireballs from more than one station and their orbital characteristics slightly differ from the filament (Table 5). They rather better correspond to the background values (Lindblad and Porubčan 1999).

(c) From very consistent mean values of orbital elements of fireballs belonging to the conspicuous filament of the 2006 Orionid outburst we found that this high activity could be caused by meteoroids trapped in 1:5 resonance.

(d) According to analysis of light curves and atmospheric penetration ability defined by PE coefficient we found that all recorded Orionid meteors do not significantly differ among each other and belong to the weak and fragile component of interplanetary matter, as expected since the Orionids are associated with comet 1P/Halley.

Acknowledgements We greadfully acknowledge thorough and helpful comments from both reviewers Dr. P. Jenniskens and J. Rendtel as well as editor Dr. F. Rietmeijer. This work was partly supported by institution research plan AV0Z10030501 of the Astronomical Institute of the Academy of Sciences of the Czech Republic and partly by the European project ORIGINS MRTN-CT-2006-035519.

References

J. Borovička, P. Spurný, J. Keclíková, A new positional astrometric method for all-sky cameras. A&A **112**, 173–178 (1995)

Z. Ceplecha, R.E. McCrosky, Fireball end heights—a diagnostic for the structure of meteoric material. J. Geophys. Res. **81**, 6257–6275 (1976)

Z. Ceplecha, Geometric, dynamic, orbital and photometric data on meteoroids from photographic fireball networks. Bull. Astron. Inst. Czech. **38**, 222–234 (1987)

M. de Lignie, H. Betlem, A double-station video look on the October meteor showers. WGN J. IMO **27**, 195–201 (1999)

V.V. Emel'Yanenko, Resonance structure of meteoroid streams, ed. by B. Warmbein, in *Proceedings of the Meteoroids 2001 Conference, Kiruna, Sweden*, August 6–10, 43–45 (2001)
A. Hajduk, Structure of the meteor stream associated with comet Halley. Bull. Astron. Inst. Czech. **21**, 37 (1970)
D.W. Hughes, P/Halley dust characteristics—a comparison between Orionid and Eta Aquarid meteor observations and those from the flyby spacecraft. A&A **187**, 879–888 (1987)
P. Jenniskens, Meteor stream activity. 2: Meteor outbursts. A&A **295**, 206–235 (1995)
P. Jenniskens, *Meteor Showers and their Parent Comets* (Cambridge University Press, Cambridge, 2006), p. 790
B.A. Lindblad, V. Porubčan, Orionid meteor stream. Contrib. Astron. Observatory Skalnaté Pleso **29**, 77 (1999)
B.A. McIntosh, A. Hajduk, Comet Halley meteor stream—a new model. MNRAS **205**, 931–943 (1983)
B.A. McIntosh, J. Jones, The Halley comet meteor stream—numerical modelling of its dynamic evolution. MNRAS **235**, 673–693 (1988)
V. Porubčan, A. Hajduk, B.A. McIntosh, Visual meteor results from the International Halley Watch. Bull. Astron. Inst. Czech. **42**, 199–204 (1991)
J. Rendtel, H. Betlem, Orionid meteor activity on October 18, 1993. WGN J. IMO **21**, 264–268 (1993)
J. Rendtel, Three days of enhanced Orionid activity in 2006—meteoroids from a resonance region? WGN J. IMO **35**, 41–45 (2007)
M. Sato, J. Watanabe, Origin of the 2006 Orionid outburst. Publ. Astron. Soc. Jpn **59**, L21–L24 (2007)
P. Spurný, J. Borovička, L. Shrbený, Automation of the Czech part of the European fireball network: equipment, methods and first results, in *Proceedings of the IAU Symposium*, vol 236, ed. by G.B. Valsecchi, D. Vokrouhlický, pp. 121–130 (2007)
J.M. Trigo-Rodríguez, J.M. Madiedo, J. Llorca, P.S. Gural, P. Pujols, T. Tezel, The 2006 Orionid outburst imaged by all—sky CCD cameras from Spain: meteoroid spatial flux and orbital elements. MNRAS **380**, 126–132 (2007)
Z. Wu, I.P. Williams, Comet P/Halley and its associated meteoroid streams. in *Proceedings of the International Astronomical Symposium*, ed. by J. Stohl, I.P. Williams, Meteoroids and their parent bodies Smolenice, Slovakia, July 6–12, 1992, 77 (1993)

Video Observations of the 2006 Leonid Outburst

Pavel Koten · Jiří Borovička · Pavel Spurný · Stephen Evans · Rostislav Štork · Andrew Elliott

Originally published in the journal Earth, Moon, and Planets, Volume 102, Nos 1–4.
DOI: 10.1007/s11038-007-9157-3

Abstract We carried out double station observations of the Leonid meteor shower outburst, which occurred in the morning hours of November 19, 2006. Using image-intensified cameras we recorded approximately 100 Leonid meteors. As predicted, the outburst was rich especially in fainter meteors. The activity profile shows that the peak of the outburst occurred at 4:40 ± 0:05 UT. The maximum reached flux was 0.03 meteoroids km^{-2} hod^{-1} for meteors brighter than +6.5 magnitude.

Keywords Meteors · Meteor showers · Leonids

1 Introduction

The most recent perihelion passage of the Leonid meteor shower parent comet–comet 55P/ Tempel-Tuttle—occurred in 1998. Significant meteor storms were observed during several subsequent years (McNaught et al. 1999). Astronomers obtained a huge volume of the meteor data and used it not only for determination of the meteoroids properties, but also for more detailed and precise models of the dust trail encounters with the Earth.

Although the season of intense storms associated with the 1998 perihelion passage of the parent comet is several years past, a small outburst was predicted for 2006. McNaught et al. (1999) predicted that the encounter with the trail created in 1932 would occur on November 19, 4:45 UT. According to Lyytinen et al. (2000) the maximum activity was predicted to occur at 4:48 UT. A new method of Vaubaillon et al. (2005), Vaubaillon and Colas (2006) also expected significant activity of the Leonid meteor shower in the morning hours of November 19. All models predicted that the stream should be abundant in smaller particles resulting in fainter meteors.

P. Koten (✉) · J. Borovička · P. Spurný · R. Štork
Astronomical Institute of the Academy of Sciences, Fričova 298, Ondřejov 251 65, Czech Republic
e-mail: koten@asu.cas.cz

S. Evans · A. Elliott
British Astronomical Association, Burlington House, Piccadilly, London W1J 0DU, UK

DOI: 10.1007/978-0-387-78419-9_22

2 Data Acqusition

The 2006 Leonid meteor shower was observed within the standard double station program of meteors observation, using video techniques, which has been carried out in the Czech Republic since 1998. Another double station experiment, which also contributed to this report, was made in the United Kingdom.

2.1 Observations and Instrumentation

The Czech double station experiment was carried out on the base Ondřejov–Kunžak. The distance between stations it 92.5 km and azimuth of the southern station is 340°. Instrumentation at each station consisted of 2nd generation image intensifiers (Mullard XX1332), commercial S-VHS camcorders and a lens. At the Kunžak station an Arsat 50 mm/F1.4 objective lens was used which provided a circular field-of-view (FOV) with a diameter of about 44°. The meteor limiting magnitude (MLM) of this configuration was +5.0^m. The camera was aimed at an elevation of 48°. Projection of the FOV at a height of 108 km yields an effective collection area of 16,700 km^2. The height was chosen as the mean height of the maximum brightness of the observed double station Leonid meteors. A different lens (Jupiter 85 mm/F2.0) was used at the Ondřejov station. This configuration provides also a circular FOV with diameter 30° and MLM + 6.0^m. The collection area of this camera, aimed at an elevation of 50°, was 10,600 km^2.

The second double station experiment was performed in the United Kingdom. The stations were located at Moreton in Marsh and Warton. The distance between them is 211.5 km and the azimuth of the southern station is 337°. Both stations were equipped with Watec 902H CCD video cameras and 12 mm f/0.8 lenses providing a rectangular FOV 30° × 23°. MLM of such configuration is +4.0^m. Again, S-VHS tapes were used for recording.

2.2 Data Processing

All the recorded data were processed using our standard procedures. Firstly the records were searched using the MetRec software (Molau 1999). Then all likely Leonids were digitalized and measured by the self-automatic software MetPhoto (Koten 2002). Atmospheric trajectories and heliocentric orbits were calculated for the double station meteors (Borovička 1990). The single station meteors were processed too. Shower association was determined in consideration of angular velocity and the distance of the prolonged path from the Leonid radiant Δ. Meteor was usually accepted as a Leonid when $\Delta < 3°$. The precision of the atmospheric trajectory of such single station meteor is significantly lower in comparison with the double station ones. Whereas the error of the height determination for the double station meteor is only a few tenths of a kilometre, in the case of the single station ones this could be up to 5 km. Thus such data are useful especially for the activity profile determination but not for atmospheric trajectory analyses.

In total we recorded 40 double station Leonid meteors—27 from the Czech Republic, 13 within the UK experiment. Moreover 62 single station meteors recorded in Czech Republic were accepted as Leonid shower members.

2.3 Weather Conditions

It is necessary to note that the observations in the Czech Republic were hampered by the unfavourable weather. At the beginning of the observation period (after 0 UT) the sky at Kunžak station was overcast. The situation got better at this station and the sky became clear before 4 UT, i.e. before the activity started to increase. The Ondřejov station had very good weather from the beginning but the most critical hours were obscured by fog. Thus only brighter meteors were detected at this station during the maximum.

3 Activity and Flux of Leonids

3.1 Activity Profile

Figure 1 shows the curve of the Leonid activity as recorded from both the Kunžak and Moreton stations. All single and double station meteors were included. Corrected hourly rate (cHR) was calculated as the number of Leonid meteors in the field of view divided by the cosine of the actual zenith distance of the radiant. No other correction was applied.

Activity curve at the Kunžak station shows a well defined peak with a maximum between 4:30 and 5:00 UT. The activity started just before 4:00 UT and steadily increased up to the maximum. The descending branch seems to be less steep but the observation stopped after 5:15 UT because the sky became too bright due to dawn. Moreton camera recorded brighter meteors in comparison with Kunžak. The peak of the activity occurred a little bit later than in Czech Republic but due to the low number of meteors it is impossible to draw any conclusion from this fact. Nevertheless the more western position of this

Fig. 1 Activity curve of Leonids in 15 minutes intervals as recorded from Kunžak (upper plot) and Moreton (lower plot) stations

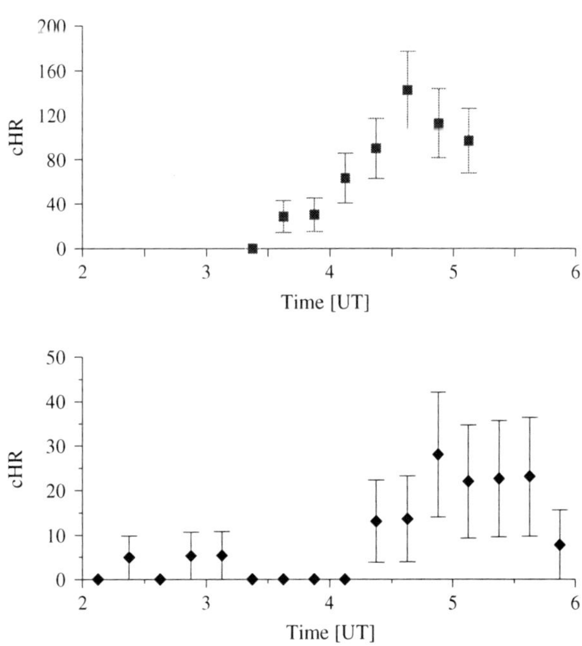

station enabled a longer observation period. We can see that the enhanced activity continued at least one hour after the maximum.

To be able to determine the time of maximum activity more precisely we constructed an activity profile with meteors binned into 5 minute intervals (Fig. 2). From this plot we can determine that the maximum of activity occurred at 4:40 ± 0:05 UT. The corresponding solar longitude is λ_o = 236.609±0.002° (J2000).

3.2 Mass and Population Index

The cHR values calculated in Sect. 3.1 are specific for each used camera, because they depend on the FOV and the MLM of the camera. To convert cHR into flux of meteoroids we need to know the effective collection area of the camera (given in Sect. 2.1) and the population index.

Figure 3 shows the distribution of the masses for all single and double station meteors recorded by the Kunžak camera. The masses were computed from meteor light curves (Koten et al. 2004). From the slope of the linear fit we can determine the mass distribution index s = 1.9. Note that there is a bias against the detection of the fainter meteors (i.e. smaller masses), which resulted in a nonlinear behaviour of the plot for smaller masses. The obtained mass distribution index results in a population index r = 2.3.

3.3 Flux of Meteoroids

Finally, we can calculate the flux of meteoroids per km^{-2} and h^{-1}. The value Φ_{MLM} is necessary to convert to the standard limiting magnitude of +6.5 to enable comparison with other observations and with visual data too. Calculation shows that the flux of meteoroids up to limiting magnitude +6.5 (which corresponds to meteoroid mass of ~ 10^{-5} g) reaches maximum value of $\Phi_{+6.5}$ = 0.030 ± 0.007 km^{-2} h^{-1}.

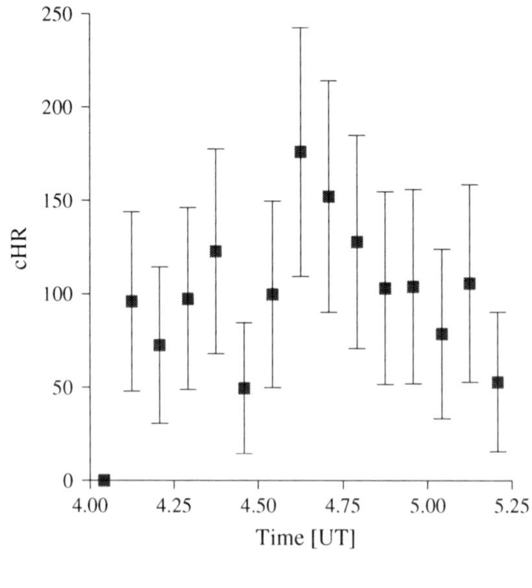

Fig. 2 Detail of the activity profile around the maximum as observed from Kunžak station. Meteors are binned into 5 minutes intervals

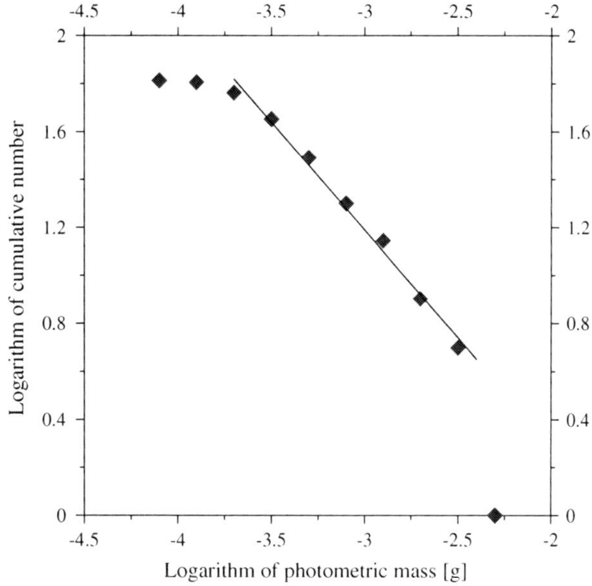

Fig. 3 Distribution of the masses of Leonid meteoroids for camera at Kunžak station

4 Discussion

We can confirm from our video observation that high activity of the Leonid meteor shower occurred in the morning hours of November 19, 2006 as was predicted by several papers, e.g. (McNaught et al. 1999), (Lyytinen et al. 2000), (Vaubaillon et al. 2005). We can only offer congratulations to the authors of these models for very precise timing of the maximum activity. The difference between predicted and observed peak of activity was only few minutes! Not only the time of the maximum but also the estimated meteor rates and mass (or size) of the meteoroids agreed. For example (Lyytinen et al. 2000) expected particles about 1/4 mm in diameter and (Vaubaillon and Colas 2006) particles in range between 0.1 mm and 0.5 mm. These sizes are comparable with the meteors we really detect by image intensifiers. Although the activity did not reach the high levels of recent years, it still significantly exceeded the annual number of Leonid meteors.

Our planet met with the particles ejected from the parent comet in 1932 twice in recent years (McNaught et al. 1999). The first encounter occurred on November 18, 2000 around 8 UT. Data on the resulting Leonid storm were reported by Brown et al. (2002). Therefore we have unique opportunity to compare the properties of both 2000 and 2006 encounters.

The geometry of the 2000 and 2006 encounters was different as is seen from the models of McNaught et al. (1999) and Lyytinen et al. (2000). Earth met a more dispersed part of the 1932 stream in 2006. Thus the activity should be lower. If we compare absolute numbers of both encounters, we can confirm this fact. Flux of the meteoroids brighter than +6.5 magnitude reached a value of 0.15 meteoroids km^{-2} h^{-1} in 2000. This flux was five times higher than in 2006.

The derived mass distribution index was not as high as was expected. The value of 1.9 is higher than the same quantity reported by Brown et al. (2002) for the 2000 encounter (s = 1.7), but the difference is not large. Nevertheless smaller particles were observed in 2006.

Acknowledgements This work was supported by the grant KJB300030502 from the Grant Agency of Academy of Sciences of Czech Republic and by the Academy of Sciences scientific project AV0Z10030501.

References

J. Borovička, The comparison of two methods of determining meteor trajectories from photographs. Bull. Astr. Inst. Czechlosl. **41**, 391–396 (1990)

P. Brown, M. Campbell, R. Suggs, W. Cooke, C. Theijsmeijer, R.L. Hawkes, J. Jones, K.J. Ellis, Video and radar observations of the 2000 Leonids: evidence for a strong flux peak associated with 1932 ejecta?. Mon. Not. R. Astron. Soc. **335**, 473–479 (2002)

P. Koten, Software for processing of meteor video records, In *Proceedings of the Asteroids, Comets, Meteors 2002 conference*, ed. by B. Warmbein, (ESA SP-500, Berlin, 2002), pp. 197–200

P. Koten, J. Borovička, P. Spurný, S. Evans, H. Betlem, Atmospheric trajectories and light curves of shower meteors, Astron. Astrophys. **428**, 683–690 (2004)

E.J. Lyytinen, T. van Flandern, Predicting the strenght of Leonid outbursts. Earth, Moon Planets **82**, 149–166 (2000)

R.H. McNaught, D.J. Asher, Leonid dust trails and meteor storms, WGN. J. Int. Meteor. Org. **27**, 85–102 (1999)

S. Molau, The meteor detection software MetRec, in *Proceedings of the Meteoroids 1998 conference*, ed. by W.J. Baggaley, V. Porubčan (Bratislava, 1999), pp. 131–134

J. Vaubaillon, F. Colas, http://www.imcce.fr/en/ephemerides/phenomenes/meteor (2006)

J. Vaubaillon, F. Colas, L. Jorda, A new method to predict meteor showers: II. Application to the Leonids, Astron. Astrophys. **439**, 761–770 (2005)

Predictions for the Aurigid Outburst of 2007 September 1

Peter Jenniskens · Jérémie Vaubaillon

Originally published in the journal Earth, Moon, and Planets, Volume 102, Nos 1–4.
DOI: 10.1007/s11038-007-9174-2 © Springer Science+Business Media B.V. 2007

Abstract The September 2007 encounter of Earth with the 1-revolution dust trail of comet C/1911 N1 (Kiess) is the most highly anticipated dust trail crossing of a known long period comet in the next 50 years. The encounter was modeled to predict the expected peak time, duration, and peak rate of the resulting outburst of Aurigid shower meteors. The Aurigids will radiate with a speed of 67 km/s from a radiant at R.A. = 92°, Decl. = +39° (J2000) in the constellation Auriga. The expected peak time is 11:36 ± 20 min UT, 2007 September 1, and the shower is expected to peak at Zenith Hourly Rate = 200/h during a 10-min interval, being above half this value during 25 min. The meteor outburst will be visible by the naked eye from locations in Mexico, the Western provinces of Canada, and the Western United States, including Hawaii and Alaska. A concerted observing campaign is being organized. Added in proof: first impression of the shower.

Keywords Meteor shower · Meteoroid stream · Long period comet · Comet C/1911 N1 (Kiess) · Aurigids

1 Introduction

Long-period comets, such as C/1995 O1 (Hale-Bopp), have long been thought to be responsible for some of the largest impact craters on Earth (e.g., Zimbelman 1984; Weissman 1990, 2007). They have a mean impact probability of 2.2×10^{-9} per perihelion passage if the perihelion distribution is uniform and the inclination distribution is random, accounting for a small but important fraction of the potential impacts on Earth. They can be

Prepared as a contribution to the conference proceedings of "Meteoroids 2007", to be published in the journal "Earth, Moon, and Planets".

P. Jenniskens (✉)
Carl Sagan Center, SETI Institute, 515 N. Whisman Road, Mountain View, CA 94043, USA
e-mail: pjenniskens@mail.arc.nasa.gov

J. Vaubaillon
Spitzer Science Center, Caltech, 1200 East California Boulevard, Pasadena, CA 91125, USA

DOI: 10.1007/978-0-387-78419-9_23

big and tend to approach the Earth from all possible directions at random times, typically at a high velocity (56.4 km/s most probable value according to Weissman (2007)). They offer little advanced warning, except for a trail of dust particles in their orbit released during the previous return to the Sun. The Earth encounters these dust trails on occasion, causing brief 1–2 h meteor showers (Jenniskens et al. 1997).

Now, the imminent encounter with the dust trail of comet C/1911 N1 (Kiess) may teach us how long-period comets lose large dust grains, how to translate the observed dust trail crossings into physical data of the parent comet, and even to find more evidence for the hypothized "pristine crust" of a comet.

Since the confirmed detection of the predicted (Jenniskens 1995) outburst of 1995 alpha-Monocerotids from an unknown long-period comet and the subsequent Leonid storms, the basic physical principles behind these transient showers are understood (Kondrat'eva and Reznikov 1985; Jenniskens 1997; Jenniskens et al. 1997, McNaught and Asher 1999; Lyytinen 1999). Dust ejected from the parent comet is dispersed due to small differences in orbital period from ejection speed and radiation pressure, causing some particles to return earlier than others. Upon return, the thin stream of dust wanders in and out of Earth's path due to planetary perturbations by the major planets, which work slightly differently on particles at different positions along the dust trail. A meteor shower outburst is observed only when the trail is in the Earth's path at the very moment when Earth passes the node of the particles (for a review see Jenniskens 2006).

In the case of so called *intermediate long-period comets* such as Kiess, with orbital periods of 200–10,000 years, the trail is so much perturbed that the second revolution dust trail is dispersed beyond recognition. As a result, the outburst meteors observed in prior encounters of the dust trail in 1935, 1986, and 1994 all date from the last time (approximately 2,000 years ago) that the comet was near the Sun (Lyytinen and Jenniskens 2003). In 2007, that same dust trail will shower Earth again.

Until now, the alpha-Monocerotid shower is the only encounter with the dust trail of a long-period comet observed by modern instrumental techniques (Spurný et al. 1995; Rendtel 1995; Znojil and Hornoch 1995; Borovička and Spurný 1995; Rendtel et al. 1996; Langbroek 1996; Šimek 1996; Jenniskens and Docters van Leeuwen 1997; Jenniskens et al. 1997). Interestingly, these meteoroids were very unusual. They were found to be almost completely lacking in sodium (Borovička et al. 2002, 2005) and penetrated relatively deep in Earth's atmosphere (Jenniskens et al. 1997). Presumably because material was sampled that came from a "pristine crust" caused by exposure to cosmic rays at the time of cold storage in the Oort cloud. Short period comets such as the parent of the Leonid shower have long lost this pristine crust.

We do not know if comet Kiess still had its pristine crust and lost some of it 2,000 years ago, perhaps now causing similar unusual meteors in 2007. It is interesting, however, that George Zay and Bob Lunsford, the only two visual meteor observers to witness the 1994 Aurigid outburst, described the outburst Aurigids as having a greenish or bluish look to them, while being more white outside this interval (Zay and Lunsford 1994). That suggests that the meteoroids produced unusually strong iron and magnesium atom line emissions from ablating metal atoms, relative to the air plasma emissions in the orange and red. This could point towards a different particle morphology of outburst Aurigids than those of other Aurigids seen outside the outburst, because the ratio of metal atom to air plasma emissions is a function of how the meteoroid matter is ablated.

2 Predictions for the 2007 Aurigid Shower

We investigated the distribution of dust from comet C/1911 N1 (Kiess) using a comet ejection model developed by Crifo and Rodionov (1997), based on the model by Whipple (1951), and calculated rigorously the planetary perturbations on the particles from the point of ejection until intersection with Earth's orbit (for a full review of the method see Vaubaillon et al., 2005a, b). Earlier results were published in Jenniskens and Vaubaillon (2006, 2007a, b), each publication aimed at a different audience.

One million meteoroids in five bins of mass were ejected from the comet orbit in 83 BC, which is the perihelion time of the nominal comet orbit (Minor Planet Center comet orbit database; Marsden et al. 1978) when integrated backward in time. Forward integration, from 83 BC until the current perihelion return, confirms that planetary perturbations occur only on the inward leg. As a result, the overall motion of the dust trail is not sensitive to the adopted perihelion time of the comet in that previous return, although the precise position is.

All particles that are at the descending node ± 2 months before Earth encountered this point are included in Fig. 1,which shows the point where each particle crossed the ecliptic plane. Also shown is the position of Earth in 1-h intervals (on the dates listed in Table 1 below). Relatively bright meteors responsible for visual meteors are shown as small dots. Large dots show the distribution of faint +6 magnitude meteors detected by video cameras and some radar systems.

Much of the shape of the distribution in Fig. 1 is the result of motion of the dust trail over a 4-month period. In 1935 and 1994, planetary perturbations caused the trail to move rapidly from outside to inside Earth orbit over the months around the outburst. In 1986, the trail moved from inside to outside Earth orbit. In contrast, the trail will be nearly stationary in 2007. The daily motion of the dust trail relative to Earth orbit can be removed by fitting a first or second order polynomial to the X and Y positions as a function of perihelion time and then interpolate each position back to the perihelion time corresponding to the particles encountered by Earth. This is justified, because the dust trail shows no gaps or strong density variations in the periods $+2$ months from the encounter times.

The result is shown in Fig. 2. Table 1 summarizes the statistical data of the calculated encounters in their usual meaning (Jenniskens 2006), including the expected time of the peak, the width of the shower, the mean miss distance $\Delta(E - D)$, the initial difference in semi-major axis of the meteoroid orbit relative to that of the comet (Δa), and the dilution factor f_M. Observed parameters of past Aurigid showers are summarized in Table 2.

After correcting for the trail motion (Fig. 2), we find that the model predicts a trail position that is, within a fraction of the width of the trail, at the same distance from Earth orbit $\Delta(E - D)$ in all years. Because of that, we are confident that the shower will return based on it having been seen in 1935, 1986, and 1994.

One objective of future work is to understand why the model puts the trails always just inside Earth orbit, as noticed before in the early results by Lyytinen and Jenniskens (2003). In the absence of previous sightings of a shower, these effects make it difficult to predict meteor showers from other long-period comets. Some of that discrepancy could come on account of an uncertainty in the orbital period of the comet orbit, or it could be due to the specifics of comet dust ejection. The dust trails of Leonid parent comet 55P/Tempel-Tuttle, for example, were calculated with the same ejection model +0.00077 AU too far outward than observed (Jenniskens 2006, Fig. 15.33). Tempel-Tuttle's orbit is well known. Hence, that discrepancy is thought to be due to ejection conditions being slightly different than those in the Crifo model. No effort has yet been made to improve the dust ejection model accordingly.

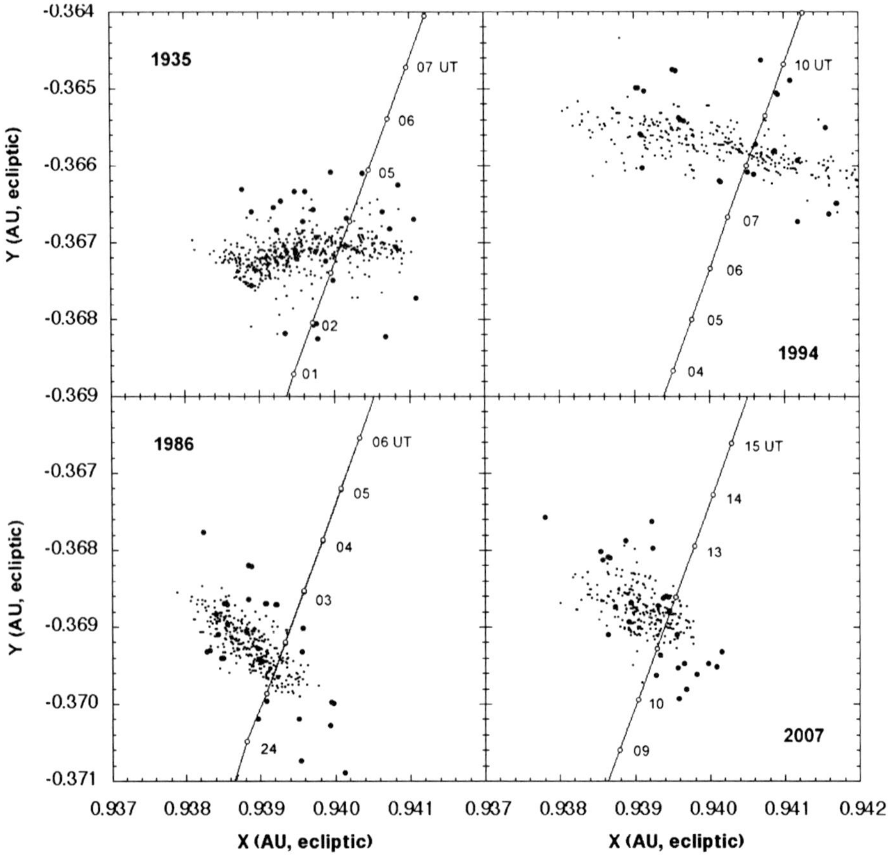

Fig. 1 Position of the node of the model 1-revolution Aurigid stream particles that are within 2 months from passing the ecliptic plane at the time of past Aurigid outbursts (grain diameter: • = 0.1–0.2 cm (faint ∼ +6 magnitude meteors); . = 0.2–2 cm (bright +3 to −3 magnitude meteors)

Table 1 Calculated circumstances for the encounter with the 1-revolution (83 BC) trail of C/1911 N1 (Kiess) at the time of Aurigid outbursts

Year (AU)	$\Delta(E-D)$ (AU)	Δa *(AU)*	f_M	Sol. Long. (°, J2000)	Date	Time (UT)	FWHM (min)	Moon phase
2007	−0.0003863	6.9726	0.005810	158.561	Sep. 01	11:36	25	0.8
1994	−0.0008137	6.0279	0.004612	158.738	Sep. 01	08:01	33	0.1
1986	−0.0003673	5.4497	0.016433	158.530	Sep. 01	01:38	27	0.6
1935	−0.0005241	1.7459	0.031045	158.656	Sep. 01	03:05	35	0.6

The observed peak time in past Aurigid encounters was off by −1 min in 1935, −16 min in 1986, and −7 min in 1994. Therefore, our best estimate for the peak time, 11:36 UT, has an uncertainty of about ±20 min. The predicted encounter time makes the shower favorable for viewing from the western states and provinces of the US and Canada, and

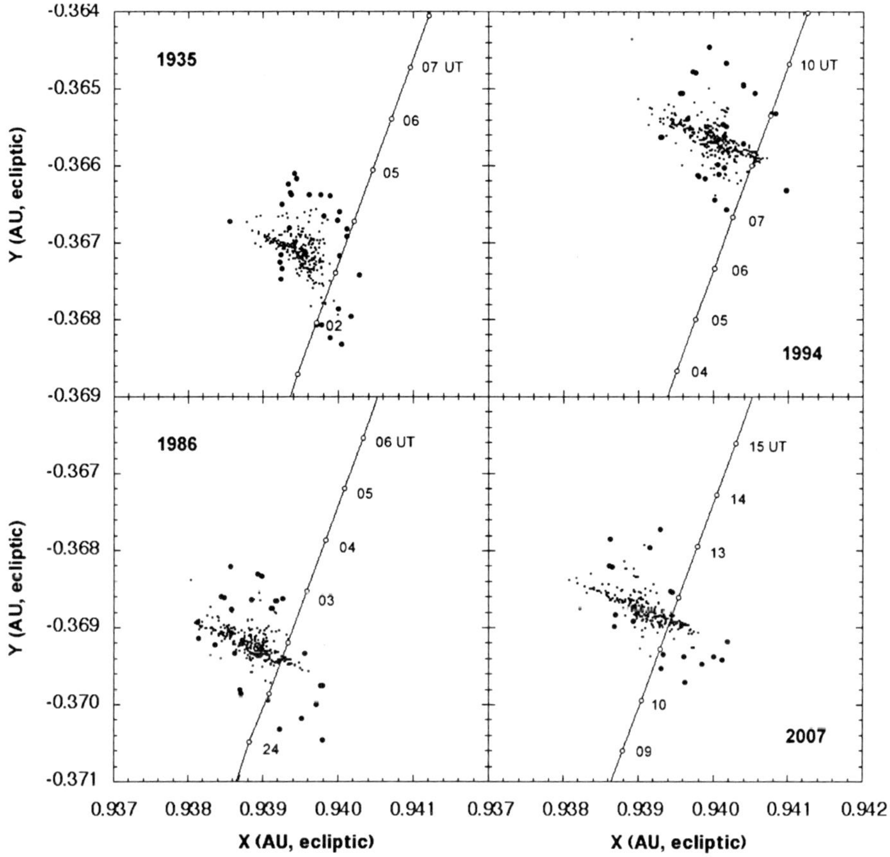

Fig. 2 Position of the node of the model 1-revolution Aurigid stream particles after correcting for motion of the trail

Table 2 Observed parameters of past Aurigid outbursts

Year	Sol. Long. (°, J2000)	Date	Time (UT)	ZHR (/h)	FWHM (min)	R.A. (J2000)	Decl. (J2000)	<*m*> (average magnitude)
1994	158.733	Sep. 01	07:54	200 ± 25	~30	–	–	+1.13
1986	158.519	Sep. 01	01:22	200 ± 25	28 ± 7	90.5	+34.6	+0.54
1935	158.656	Sep. 01	03:04	≥100	31 ± 13	86.3	+40.5	+2.62

western Mexico, where the radiant will be high in the sky just before dawn in the early morning of September 1 (Fig. 3).

The density of particles in the stream in 2007 will be the nearly the same as in the 1986 and 1994 returns. Unfortunately, data of past outbursts were hampered by bad observing circumstances. Rates continued to rise when twilight interfered in Germany and the Czech Republic during the 1935 outburst (Teichgraeber 1935; Guth 1936), from which we have ZHR > 100/h. The single eyewitness of the 1986 outburst, Istvan Tepliczky of Hungary, derived an average ZHR = 39.6 ± 8.1 from the period between the first and last Aurigid

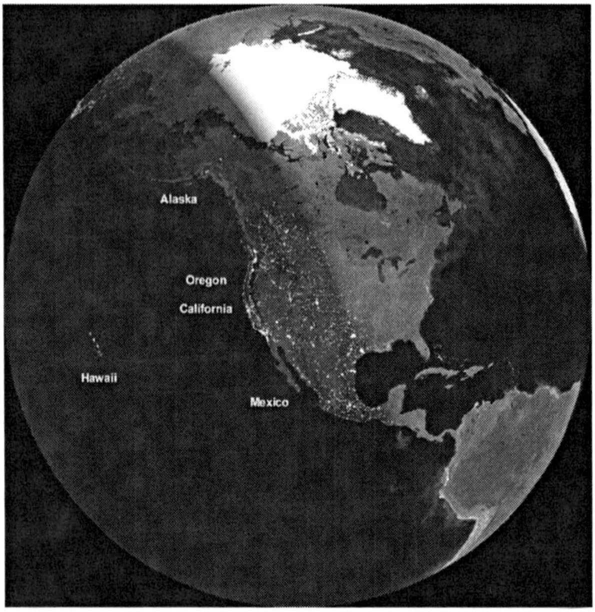

Fig. 3 Earth as seen from the perspective of the approaching dust grains at the peak of the predicted meteor outburst on September 1, 2007

(00:47–02:12 UT), during which 24 Aurigids were seen (Tepliczky 1987), which corresponds to a peak ZHR = 200 ± 25/h based on 10-min intervals. The rate measurement in 1994 was hampered by a low radiant elevation. For the hour between 7:22 and 8:22 UT, with the radiant being at 13° elevation at 7:49 UT, Zay and Lunsford (1994) calculated a ZHR = 55/h (Zay) and 37/h (Lunsford), respectively. Again, the rate varied strongly during that interval. In small 10-min intervals, we calculate a peak ZHR of again about 200 ± 25 per hour. Hence, meteor rates in 2007 are expected to increase to ZHR = 200/h in a short time interval at the center of the outburst depending on the exact position of the trail crossing.

Based on past Leonid storm observations, the width of the trail is expected to be wider if we pass further from the trail center (Jenniskens 2006). Given that the trail will be at the same location as in past returns, we can use the sparse data from past observations to predict what to expect from the 2007 encounter. Each of the past observed showers lasted about 1.5 h, with a Full Width at Half Maximum ~28 min. For the center of the trail, our model predicts FWHM = 27.3 min for 1986 and 32.9 min for 1994, in good agreement. If Earth will pass through the trail center, then the 2007 return would have a FWHM duration of about 25 min.

In 2007, we will be only 15% further from the comet than in 1994. Our model (Fig. 4) shows that large particles can make it out to this position in the trail, and the model predicts that meteoroids as big as 20 cm in diameter (−10.6 magnitude, according to Eq. C.12 in Jenniskens 2006) may be observed and meteors down to magnitude +4 should be abundant. The model also predicts that a lack of relatively faint meteors (1 mm, or +6.6 magnitude meteors) be present in the stream, mainly because the smaller meteoroids are ejected at higher speed. Past magnitude estimates of the Aurigids are summarized in Table 3 and show that most were in the range −3 to +3 magnitude. We expect that to be so again in 2007. A Moon 4 days past full will not dampen the display much.

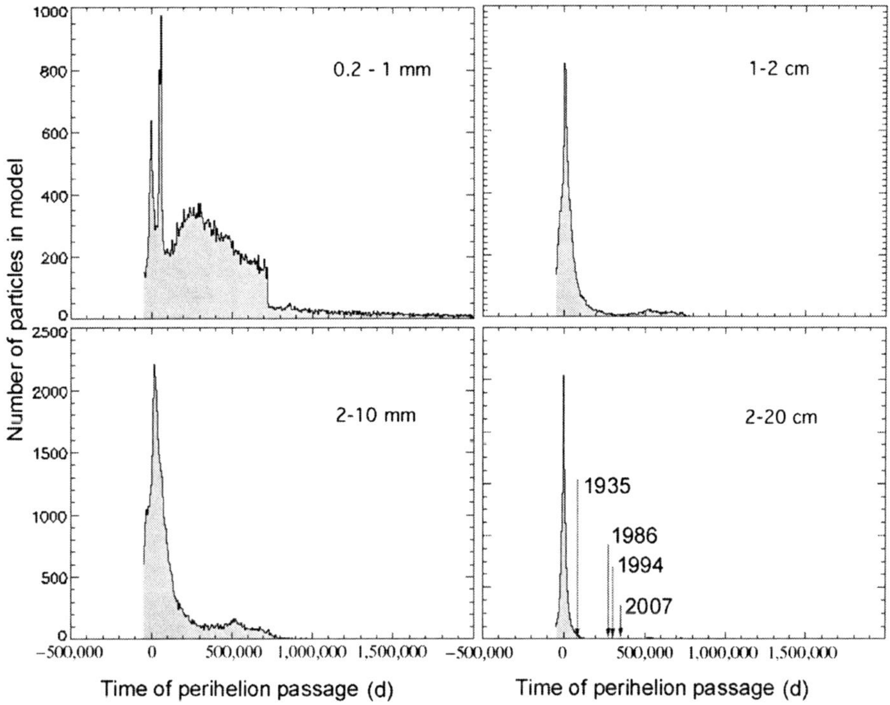

Fig. 4 Particle distribution along the comet dust trail for different particle diameters. The location where Earth crossed the trail in 1935, 1986, 1994, and 2007 is indicated in the last panel. The vertical axis gives the number of particles in the model, which is a relative measure of the particle density along the dust trail

Table 3 Observed magnitude distributions (Teichgraeber 1935; Tepliczky 1987; Zay and Lunsford 1994)

	−4	−3	−2	−1	0	1	2	3	4	5	Observer[a]
Aurigids											
1994	0	0	1	0	1	8	6	2	0	0	LUNRO
1994	0	0	1	0	1	9	3	3	2	1	ZAYGE
1986	1	0	0	5	7	3	6	1	1	0	TEPIS
1935	–	–	–	2	2	1	4	9	4	1	VR
Sporadics											
1994	0	1	1	0	3	3	18	24	10	2	LUNRO
1994	0	1	0	2	6	5	16	21	8	6	ZAYGE

[a] IMO Observer Codes: LUNRO = Bob Lunsford (California), ZAYGE = George Zay (California), and TEPIS = Istvan Tepliczky (Hungary). Also: VR = Vrátník at the Stefanik Sternwarte in Prague

3 Future Work

The short duration and the abundance of zero and +1 magnitude meteors will make for a very impressive shower. Accurate measurements of the trail width as a function of particle mass will help validate predictions in the model. Note how faint meteors are expected to have a wider stream cross section. The particle size distribution will calibrate the dispersion of meteoroids along the comet dust trail. The mass of the comet nucleus is a free parameter in the model and may be derived from these measurements. This mass can then be compared to mass estimates based on the brightness of comet Kiess in 1911.

Of interest too is the annual shower activity associated with comet Kiess, because the intensity of the shower and duration hold clues about how many revolutions the comet has completed since being captured. This Aurigid shower (IAU #206) is known from only three meteoroid orbits (Jenniskens 2006). Dubietis and Arlt (2002a, b) calculated a Zenith Hourly Rate curve with a peak of about 7 meteors/hour, but it is not clear whether the visual observers were sufficiently capable of discriminating the annual Aurigids from other apex source meteors at that time.

It will also be interesting to study the light curves, the penetration depth, and the spectra of the Aurigid meteors for clues about the presence of pristine comet crust material. It is not certain that the dust of Kiess contains such unusual meteoroids. If such meteors lacking in sodium are found, the material properties of this dust should be investigated to derive the density of the crust material and its main element composition. Unusual meteors should be looked for in both the outburst and the annual Aurigid component.

The fast meteors are an impact hazard to satellites in orbit (Beech and Brown 1993). At the peak of the Aurigid shower, the influx rate of fast meteoroids of several mm size will increase briefly by a factor of ~ 100, more so for large particles and for spacecraft surfaces oriented towards the Earth's apex, but the flux will not rise so high that an impact is certain, even considering the whole surface area of active satellites.

4 Added in Proof

These predictions were presented at the Meteoroids 2007 conference in June. Results were published in a paper in EOS, Transactions of the AGU on August 7 (Jenniskens and Vaubaillon 2007b). Subsequently, JPL issued a new orbit for comet Kiess derived from a more restricted dataset of observations (JPL-3), calculated by Jon Giorgini of NASA/JPL, which resulted in a perihelion time of 4 A.D. $\pm$ 40 years. We repeated the calculations and found that a 4 A.D. ejection date put the dust trails in Earth's path in 2007, implying that this was a better solution for the comet's perihelion time (Jenniskens et al. 2007). The peak time now was 11:33 $\pm$ 20 min UT.

The Meteoroids 2007 conference helped coordinate the airborne and ground-based observing campaign. NASA Ames facilitated the deployment of two privately owned Gulfstream V aircraft, which provided a team of 24 researchers an opportunity to observe the Aurigid shower from 47,000 ft altitude, where extinction near the horizon is low and a large surface area can be monitored (Fig. 5). Participating researchers in this *Aurigid Multi-Instrument Aircraft Campaign* (Aurigid MAC) were from the USA, UK, France, Germany, and The Netherlands, many with past experience in Leonid MAC and spacecraft reentry missions. The main purpose of the mission was to measure the duration, peak time, and particle size distribution in the dust trail accurately by imaging as many meteors as possible with cameras sensitive to meteors of different brightness. And to measure meteor light curves, penetration depth, and optical spectra, in search of evidence that some of these meteors may be pieces of the original cosmic-ray-produced crust of the comet.

The shower peaked at 04:15 $\pm$ 5 min PDT, earlier than our predicted 04:33 PDT $\pm$ 20 min (Jenniskens and Vaubaillon 2007b), but in line with the shower in 1935, 1986 and

Fig. 5 Composite image of Aurigid shower with 15 Aurigids observed from one of the two aircraft by Jason Hatton (ESA/ESTEC). The meteors span the period 11:04:44–11:50:54 UT 1st September 2007 (a 46-min period covering the peak)

1994, which also appeared to be slightly earlier than predicted. The shower may have been wider than expected, about FWHM = 0.68 h instead of the predicted 0.42 h, which could imply that we passed slightly further from the trail center than expected.

At the peak, meteors were detected at a Zenith Hourly Rate of about 130/h, within a factor of two from the anticipated rate. This number is a small improvement on the rates reported in near-real time, which peaked at 100/h. In an effort to inform satellite operators about the shower's activity in near-real time, a team of four amateur meteor observers kept a tally of Aurigid meteors by means of a video headset display hooked to intensified cameras positioned at the windows and by using an automated counting tool. The Zenith Hourly Rates were phoned in every 10–15 min and immediately posted on our mission website (http://aurigid.seti.org). Figure 6 shows the rates calculated from those reports after a first re-evaluation of the calibration. The rates were simply scaled to the response from the Perseid shower 2 weeks earlier. We flew a test flight at that time, involving one aircraft, during which the Perseid shower was observed under no Moon conditions. The Aurigid shower rates are not expected to scale precisely in the same way, due to the different magnitude size distribution index, and so the rates are expected to slightly change again in the final result.

At the time of writing, in late September, the tally of Aurigid optical spectra (400–800 nm) was 44 individual Aurigids, and that number is still rising. Many Aurigids have a strong forbidden line of oxygen at 577 nm, which may account for some meteors appearing greenish, rather than being due to the production of metal atom emissions as we thought before. The Aurigid meteoroids appear to contain more sodium than the alpha Monocerotids. It is too early to tell if the sodium content is anomalous compared to that of Leonids and Perseids, as potential evidence of a comet's pristine crust. There is some indication, perhaps, that the morphology of the grains was more sintered, because, at first glance, we do not see the early release of sodium relative to magnesium and iron that was common among 1–4 revolution dust trail Leonids, and thought to be a sign of sodium-containing minerals being efficiently exposed to heat by a finer fragmentation process.

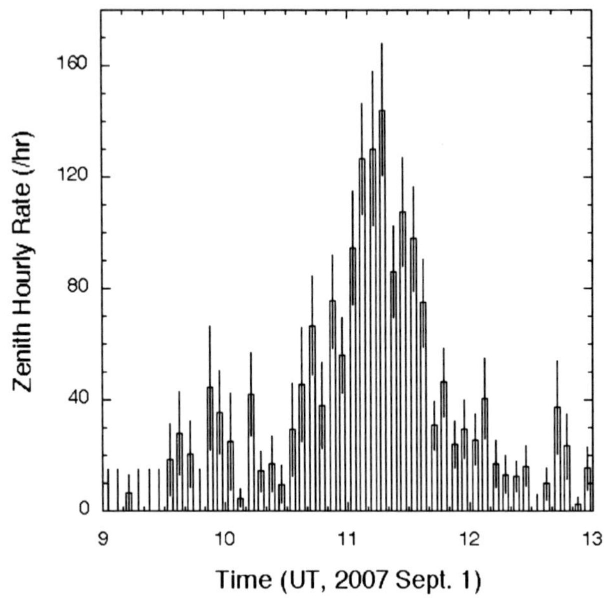

Fig. 6 Zenith Hourly Rates of the Aurigid shower in 5-min intervals as derived from near-real time counts by the flux team of Aurigid MAC

Few meteors were observed from both planes in a favorable geometry for calculating penetration depth accurately, due to a misalignment of the aircraft trajectories at the time of the peak. The stereoscopic alignment was restored only during the declining tail of the shower. Fortunately, many Aurigids were imaged by ground teams at Lick Observatory and Fremont Peak, in an effort led by Tolis Apostolos (Armagh Observatory). These observations were supported by scattered ground-based observers in the wider Bay Area, where amateur astronomers photographed many Aurigids using digital cameras and low-light-level cameras. At the time of writing, at least 24 Aurigids are known to have been recorded from two or more sites simultaneously, sufficient for a detailed analysis of penetration depths.

Acknowledgements This paper was improved by comments from editor Frans Rietmeijer and reviewers Peter Brown and David Asher. We thank operators at CINES (France) for their help with the super-computer used to do the simulations. The Aurigid Multi-Instrument Aircraft Campaign (Aurigid MAC) was sponsored by NASA Ames Research Center and the NASA Planetary Astronomy Program. This was the first deployment of the two Gulfstream V aircraft in a research mission. We received tremendous support from the aircraft operators of *H211 LLC* and organizers at *NASA Ames Research Center*, and thank all that made this mission possible.

References

M. Beech, P. Brown, Impact probabilities on artificial satellites for the 1993 Perseid meteoroid stream. MNRAS **262**, L35–L36 (1993)

J. Borovička, P. Spurný, The visual observations of the outburst of the 1995 alpha-Monocerotids in Ondrejov. JIMO **23**, 203–205 (1995)

J. Borovička, P. Spurný, P. Koten, Evidences for the existence of non-chondritic compact material on cometary orbits. Warmbein B. (Ed.) ESA Publications Division. ESA SP **500**, 265–268 (2002)

J. Borovička, P. Koten, P. Spurný, J. Bocek, R. Stork, A survey of meteor spectra and orbits: evidence for three populations of Na-free meteoroids. Icarus **174**, 15–30 (2005)

J.F. Crifo, A.V. Rodionov, The dependence of the circumnuclear coma structure on the properties of the nucleus. Icarus **129**, 72–93 (1997)

A. Dubietis, R. Arlt, Annual activity of the alpha Aurigid meteor shower as observed in 1988–2000. JIMO **30**, 22–31 (2002a)

A. Dubietis, R. Arlt, The current delta Aurigid meteor shower JIMO **30**, 168–174 (2002b)

V. Guth, Über den meteorstrom des kometen 1911 II (Kiess) (in German). Astron. Nachr. **256**, 27–28 (1936)

P. Jenniskens, Good prospects for alpha-Monocerotid outburst in 1995. JIMO **23**, 84–86 (1995)

P. Jenniskens, Meteor stream activity. IV. Meteor outbursts and the reflex motion of the Sun. Astron. Astrophys. **317**, 953–961 (1997)

P. Jenniskens, Meteor Showers and Their Parent Comets (Cambridge University Press, Cambridge, UK, 2006), 790 pp

P. Jenniskens, G. Docters van Leeuwen, The alpha-Monocerotids meteor outburst: the cross section of a comet dust trail. Planet. Space Sci. **45**, 1649–1652 (1997)

P. Jenniskens, J. Vaubaillon, The 2007 September 1 Aurigid meteor storm. Dissertatio Cum Nuncio Sidereo III. No. 6. IAU General Assembly, Prague, p. 1 (2006)

P. Jenniskens, J. Vaubaillon, Aurigid predictions for 2007 September 1. JIMO **35**, 30–34 (2007a)

P. Jenniskens, J. Vaubaillon, An unusual meteor shower on September 1, 2007. EOS, Trans. AGU **88** (August 7 issue), 1–2 (2007b)

P. Jenniskens, H. Betlem, M.C. de Lignie, M. Langbroek, The detection of a dust trail in the orbit of an Earth threatening long-period comet. Astrophys. J. **479**, 441–447 (1997)

P. Jenniskens, J. Vaubaillon, S. Nakano, 2007 Aurigid meteors, in *CBET 1045*, ed. by D.W.E. Green (Central Bureau for Astronomical Telegrams, Smithsonian Astrophysical Observatory, Cambridge, MA, 2007)

E.D. Kondrat'eva, E.A. Reznikov, Comet Tempel-Tuttle and the Leonid meteor swarm. Solar Syst. Res. **19**, 96–101 (1985)

M. Langbroek, De alpha Monocerotiden uitbarsting van 21/22 november 1995 (in Dutch) Radiant J. Dutch Meteor Soc. **18**, 122 124 (1996)

E. Lyytinen, Leonid predictions for the years 1999–2007 with the satellite model of comets. Meta Res. Bull. **8**, 33–40 (1999)

E. Lyytinen, P. Jenniskens, Meteor outbursts from long-period comet dust trails. Icarus **162**, 443–452 (2003)

B.G. Marsden, Z. Sekanina, E. Everhart, New osculating orbits for 110 comets and analysis of original orbits for 200 comets. Astron. J. **83**, 64–71 (1978)

R.H. McNaught, D.J. Asher, Leonid dust trails and meteor storms. JIMO **27**, 85–102 (1999)

J. Rendtel, Activity burst of alpha Monocerotids on November 22, (1995). JIMO **23**, 200–230 (1995)

J. Rendtel, P. Brown, S. Molau, The 1995 outburst and possible origin of the alpha-Monocerotid meteoroid stream. MNRAS **279**, L31–L36 (1996)

M. Šimek, The 1995 alpha-Monocerotids from radar observation at Ondrejov. JIMO **24**, 88–89 (1996)

P. Spurný, J. Borovička, A. Gomez, L. Bellot Rubio, A. Roman, F. Reyes, J. Rendtel, S. Molau, G. Forti, R. Haver, R. Gorelli, Z.A. Nagy, K. Srenczky, I. Tepliczky, J. Gerbos, P. Rapavy, V. Hrusovsky, C. Steyaert, M. de Meyere, Alpha Monocerotid meteors 1995, in *IAU Circular* 6265, ed. by B.G. Marsden (Minor Planet Center, 1995)

A. Teichgraeber, Unerwarteter meteorstrom (in German). Sterne **15**, 277 (1935)

I. Tepliczky, The maximum of the Aurigids in 1986. JIMO **15**, 28–29 (1987)

J. Vaubaillon, F. Colas, L. Jorda, A new method to predict meteor showers. I. Description of the model. Astron. Astrophys. **439**, 751–760 (2005a)

J. Vaubaillon, F. Colas, L. Jorda, A new method to predict meteor showers. II. Application to the Leonids. Astron. Astrophys. **439**, 761–770 (2005b)

P.R. Weissman, The Oort cloud. Nature **344**, 825–830 (1990)

P.R. Weissman, The cometary impactor flux at the Earth. in *Near Earth Objects, our Celestial Neighbors: Opportunity and Risk.* Proceedings of IAU Symposium 236, ed. by G.B. Valsecchi, D. Vokrouhlický (Cambridge University Press, Cambridge, UK, 2007), pp. 441–450

F.L. Whipple, A comet model. II. Physical relations for comets and meteors. Astrophys. J. **113**, 464–474 (1951)

G. Zay, R. Lunsford, On a possible outburst of the 1994 α-Aurigids. JIMO **22**, 224–226 (1994)

J.R. Zimbelman, Planetary impact probabilities for long-period comets. Icarus **57**, 48–54 (1984)

V. Znojil, K. Hornoch, Observing the alpha-Monocerotids from Lelekovice. JIMO **23**, 205–206 (1995)

Characterization of the Meteoroid Spatial Flux Density during the 1999 Leonid Storm

Peter S. Gural · Peter Jenniskens

Originally published in the journal Earth, Moon, and Planets, Volume 102, Nos 1–4.
DOI: 10.1007/s11038-007-9176-0 © Springer Science+Business Media B.V. 2007

Abstract The November 18, 1999 Leonid storm was rich in meteors and well observed by airborne intensified video cameras aimed low in the sky which enabled enhanced meteor counts over ground-based observations. The two- and three-dimensional distribution of meteoroids was investigated for signs of clustering that could provide evidence of meteoroid fragmentation shortly after lift-off from the parent comet 55P/Tempel-Tuttle, or much later due to space weathering. Analysis of the video tapes yields a refined estimation of the mass ratio during the peak of s = 1.65 and spatial flux density of 0.5 particles/km^2 greater than those causing visual magnitude +6.5 during the 5 min centered around the peak of the storm. Furthermore, the projection of the individual trails into three-dimensional Heliocentric coordinates, shows non-homogeneity of the stream on spatial scales from hundreds to thousands of kilometers.

Keywords Meteor shower · Meteoroid stream · Dust trail · Comet · Comet dust ejection · Dust fragmentation · Space weathering

1 Introduction

The degree of particle clustering in freshly ejected meteoroid streams, such as the 1899 dust trail of comet 55P/Tempel-Tuttle that caused the 1999 Leonid storm, is important because of its promise to provide evidence for fragmentation of meteoroids shortly after lift-off from a comet's surface. Such indications first became evident when the Stardust spacecraft flew near comet 81P/Wild 2. There the impact dust collector of the mission has encountered regions of high meteoroid density with particles in the range 10^{-11}–10^{-4} g, mixed with regions of sparse activity where the density was a factor of 100 million times lower (Tuzzolino et al. 2004; Clark et al. 2004; Green et al. 2007). That fragmentation

P. S. Gural (✉)
SAIC, 14668 Lee Road, Chantilly, VA 20151, USA
e-mail: peter.s.gural@saic.com

P. Jenniskens
Carl Sagan Center, SETI Institute, 515 N. Whisman Road, Mountain View, CA 94043, USA

DOI: 10.1007/978-0-387-78419-9_24

could have been due to a later sublimation of ices that were initially contained in the meteoroid at the time of ejection.

In addition, the known age of the stream makes it possible to study effects of space weathering in the meteoroid's orbital revolution around the sun (Whipple 1963; Mukai et al. 2001; Nikolova and Jones 2001; Trigo-Rodriquez et al. 2005; Jenniskens 2006). Most notably, the extremes of heating near perihelion and cooling in the outer extent of the orbit, combined with the fragile nature of the particles, could potentially cause fragmentation of the meteoroids in space. If such fragmentation can be measured, the rate of decay of the stream can be measured.

The first direct evidence for space weathering was found by Kinoshita et al. (1999), who observed a very short duration, dense cluster of meteors during the 1997 Leonid shower resulting from a fragmentation that happened shortly before the cloud of particles hit the Earth. Calculations based on the extent of the spatial distribution of the fragments implied that the breakup occurred around perihelion passage about 6 days prior to entry (Watanabe et al. 2002). In addition, Singer et al. (2000) has argued that there were periodic variations in the Leonid flux rates during the 1999 Leonid storm on a 7-min time scale. All other studies of cluster analysis have so far come up empty.

In a previous study of the frequency of detected Leonids during the 1999 Leonid storm, we had found no evidence for clustering on a 1–30 s time scale (Gural and Jenniskens 2000). However, we only considered a 15-min time interval and attempted a one-dimensional correlation of meteoroids in time only. In this paper, we analyze a more extended portion of the storm profile and perform a higher order correlation analysis in two and three dimensions of space and time.

2 Dataset and Processing

The November 18, 1999 Leonid storm was observed from two aircraft, the ARIA and FISTA (Jenniskens et al. 2000). During the time of the Leonid storm peak near 2:00 UT, both were flying west-northwest near 37°E longitude, 21°N latitude, and at 11 km altitude. On each aircraft were deployed intensified video cameras with 30° × 40° rectangular fields of view (FOV), a circular field cutoff diameter of 38°, limiting magnitude of +6.5, with six cameras pointed at low elevation and two at high elevation, covering the air cap both north and south of the aircraft.

The ARIA video sequences AL50R and AL50F were studied in depth for the analysis herein. These represent two adjacent fields of view looking south spanning 160–240° in azimuth and 2–33° in elevation angle. The results expand on our initial analysis of the camera AL50R (Gural and Jenniskens 2000). Using a high altitude aircraft platform and low look elevation is advantageous due to the enhanced surface area observed without atmospheric extinction losses. The left-looking camera (AL50R) saw meteors from 150 km to 900 km distance since its FOV orientation dipped very low in elevation angle. The right-looking camera (AL50F) was aimed slightly higher and thus only covered 100–500 km in distance. Both cameras combined in azimuth covered a 600-km wide air cap. Therefore, clustering on the scales up to several hundred km was deemed to be observable within this data subset.

The software MeteorScan (Gural 1997) was used to detect meteors in the video. To determine the software's detection performance, a manual method of detecting meteors was applied to a 40 min segment of AL50F. This latter technique (Gural and Jenniskens 2000) involved playing a video loop of 1 s duration with an analyst mouse clicking on both the beginning and end points of each meteor, advancing one-half second, and repeating the process until completion of the processing period. We concluded that MeteorScan was able to detect only 75% of the manually discerned meteors, having missed mostly low elevation (short and slow movers) and multiple meteors occurring simultaneously. Since the total counts versus time correlated very well between the two methods, the automated scans were used in comparing cameras for large-scale spatial analysis where nearest neighbor spacing was not an issue. For small-scale spatial analysis, however, the manual detection of meteors was used with its associated higher accuracy end point positions (limited to the two south-looking cameras).

3 Mass Index Estimation

For the clustering analysis, 10-min intervals were chosen when the flux was relatively constant and the aircraft did not undergo any turns or banks (that is, the FOV pointing stayed within 2° in azimuth and elevation). The period of time chosen fell within one-half hour of the peak of the storm near 2:00 UT, from 1:54:53 to 2:04:53 UT. We also analyzed an "off-peak" period, at half the observed flux, from 2:23:07 to 2:33:07 UT.

A unique by-product of the analysis steps described above is that the azimuth and elevation map of the meteor endpoints shows an increasing density of meteors at decreasing elevation angles. When run through a meteor simulation (Gural and Jenniskens 2000; Gural 2002; Gural 2004), it was found that below 25°, the curves of counts versus elevation angle change significantly with population or mass index (Fig. 1). If the mass index is low, then there will be a higher percentage of bright meteors and more will be seen at lower elevation angles relative to higher elevations and the curves will rise more steeply. Thus by fitting the measurements to the simulated elevation counts, we can estimate the population index independently of the magnitude estimation for each meteor.

What was found is that the measurements fell near a population index $r = 1.75 \pm 0.07$ (mass index $s = 1.61 \pm 0.04$) for the period around the meteor storm peak. Repeating the

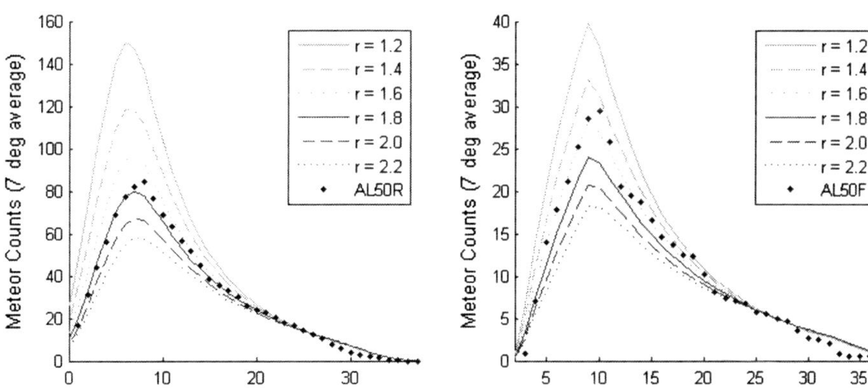

Fig. 1 Mass index estimation from elevation counts around the peak of the Leonid storm (left) and one-half hour later (right). Solid lines are simulated counts and points are measurements with both averaged over a 3° sliding window

analysis at a later time in the collection that was one-half hour after the storm peak, the population index was slightly lower at r = 1.55 ± 0.07 (mass index s = 1.48 ± 0.04). The change is very small, and perhaps not significant, indicating very little change in mass index across the stream.

The implication is that the various size particles ejected from the comet possess the same velocity distribution and do not spatially separate across the orbit.

4 Along-orbit Analysis: Temporal Correlations Only

As a first step, we applied the one-dimensional temporal correlation analysis described by Gural and Jenniskens (2000) to the newly expanded dataset. The human observer often sees meteors arrive in pairs and flurries and we addressed the question of whether those apparent bursts of activity are significant. This analysis essentially focused on a single spatial dimension that corresponds to the spread of meteoroids along the orbit. In Fig. 2 is plotted a histogram of meteor spacing in time for both a simulated and measured dataset associated with camera AL50F. The temporal resolution for the binning was set to one-tenth of a second or equivalently 4 km spatial scales in heliocentric coordinates. Based on Fig. 2, no difference can be seen relative to a purely random distribution of particles. The same result was reported before (Gural and Jenniskens 2000).

It was proposed that the large number of particles incident at the storm peak could perhaps be washing out any small-scale clustering. Thus the analysis was repeated for a later time period at only half the peak flux rate. Again, no difference was seen relative to a random distribution.

We also examined the larger scale temporal distribution of meteoroids in the Leonid stream. This was done by comparing the automatically detected flux counts across several of ARIA's cameras. As seen in Fig. 3, a combination of the four ARIA cameras shows

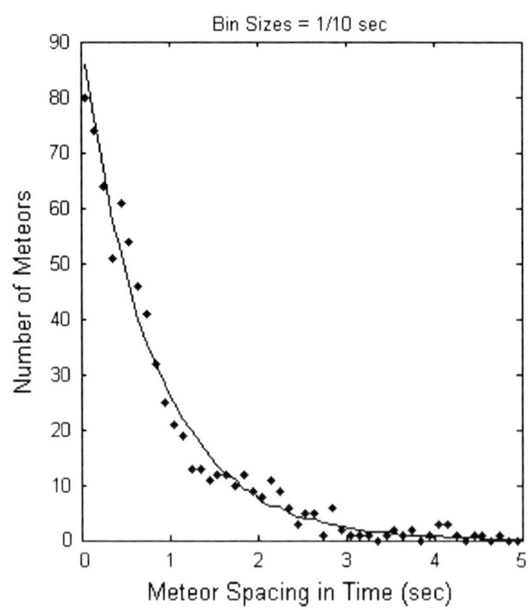

Fig. 2 Temporal cluster analysis of camera AL50F showing a histogram of meteor spacing for the storm peak near 2:00 UT. Solid line is a random simulated meteor distribution whereas points are measurements spanning a 10-minute period

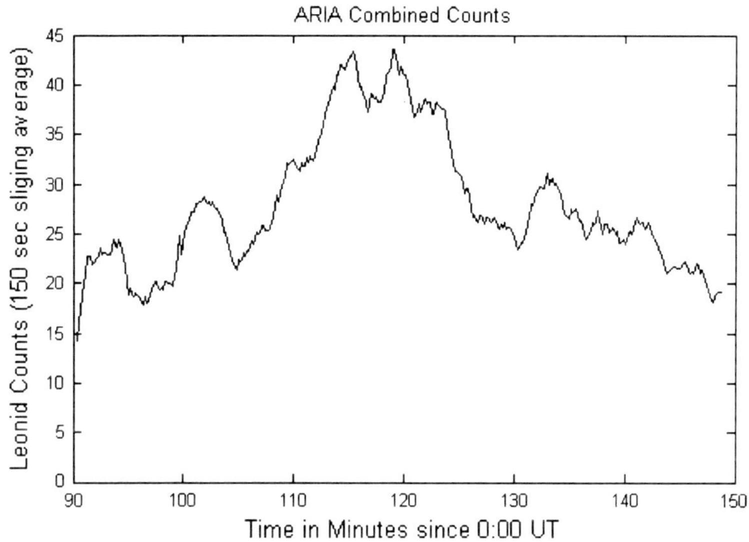

Fig. 3 Combined and averaged (2.5-minute sliding mean) counts from four ARIA low elevation intensified cameras. Subtract 1 min to convert to topocentric time

20% flux variations with surges of meteors evident during 4-minute long windows of heightened activity.

Each of the cameras showed correlations in the density fluctuations on longer time scales of minutes, as previously reported by Singer et al. (2000), who had temporal separations of individual maxima in the 6–9 min range. Singer et al. had attributed periodicity to the enhancements, but that is not evident in our data.

5 Cross-section Analysis: Spatial Correlations Only

The end point positions of each meteor measured on the focal plane can be converted into a three-dimensional spatial distribution of particle locations through a series of coordinate transformations. Astrometry of several frames provided the plate coefficients to convert from focal plane row and column to standard coordinates. Furthermore, obtaining the celestial coordinates of the image center, Julian date, latitude, longitude, and local sidereal time, we could compute the azimuth and elevation from the observer's geocentric inertial coordinate system. Setting the mean end point height of all observed Leonids to 95 km, yielded range and thus x, y, z, and time. Next, the product of the radiant velocity and the time difference (relative to the interval start time) was subtracted from each meteor's geocentric position to produce a three-dimensional mapping. That mapping is in a heliocentric-based coordinate system with origin at the observer and all measurements synchronized to the same time instant. Several analyses were attempted to search for clustering.

The plane normal to the Leonid stream orbit is a two-dimensional spatial surface through which the stream meteoroids pass. When examined in a heliocentric coordinate system that plane represents a cross section of the stream density. If there was a boulder break-up, then a mini-dust trail should form, and when projected down onto that plane would appear as a tight clumping of meteors that should exceed the densities expected from a simple random distribution of particles in space. Given the motion vector of the Earth's way (apex motion), one can determine the heliocentric Leonid orbit velocity vector in local observer's coordinates and thus project the three-dimensional map described earlier onto the plane normal to the heliocentric motion of the stream. This was computed and is shown in Fig. 4 for cameras AL50R and AL50F. Note that when this figure was compared to an equivalent figure derived using a random sample distribution (for the same camera geometry and flux count) as in Fig. 5, the visual degree of clumping is approximately the same.

To put a quantitative test to the spatial distribution, two evaluations were performed. We examined the spacing between nearest neighbors (Fig. 4, left) and we looked for density enhancements of meteors on a spatial scale of 100 km (Fig. 4, right). As seen in Fig. 4, both the spacing between nearest neighbors in the cross-sectional plane and the density of meteors on the scale of 100 km are no different than a random distribution.

6 Three-dimensional Spatial Analysis: Temporal + Spatial

When reviewing the video record via a second-by-second playback mode, it appears that meteors are indeed clumped on small spatial scales. This may be an effect of the arrival of random meteors giving the illusion of clustering and how easily the human response can be fooled into wanting to associate these in clusters. However, when playing the tapes it is quite evident that meteors very often appear together in one portion of the FOV, followed by a pause and then a flurry of meteors closely spaced in another region of the FOV, thus giving the impression of clustering.

An analysis of the three-dimensional spacing between meteors was performed in heliocentric coordinates. Taking adjacent-in-time meteors and placing their relative spacing into histogram bins of 20 km and 60 km sizes, yields the plots in Fig. 6 left and right respectively. When compared to a random distribution, there is evidently an enhancement of spacing in the 100–200 km range amounting to 10% of the total particle

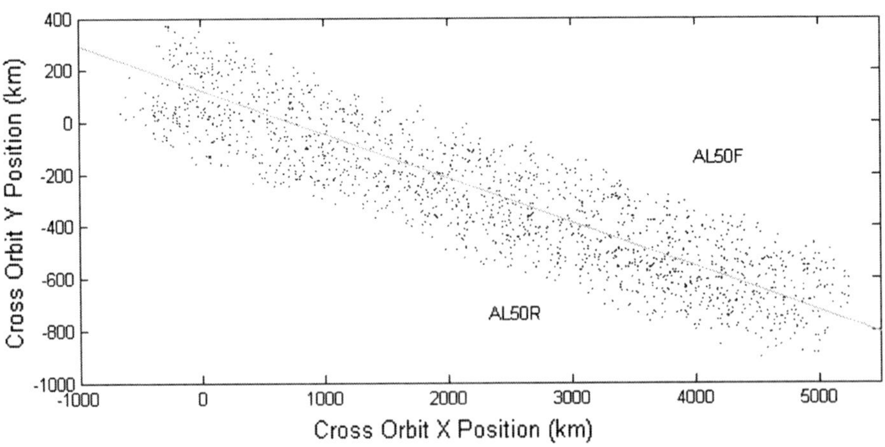

Fig. 4 The cross section of the Leonid stream covering a span of 10 min around the peak

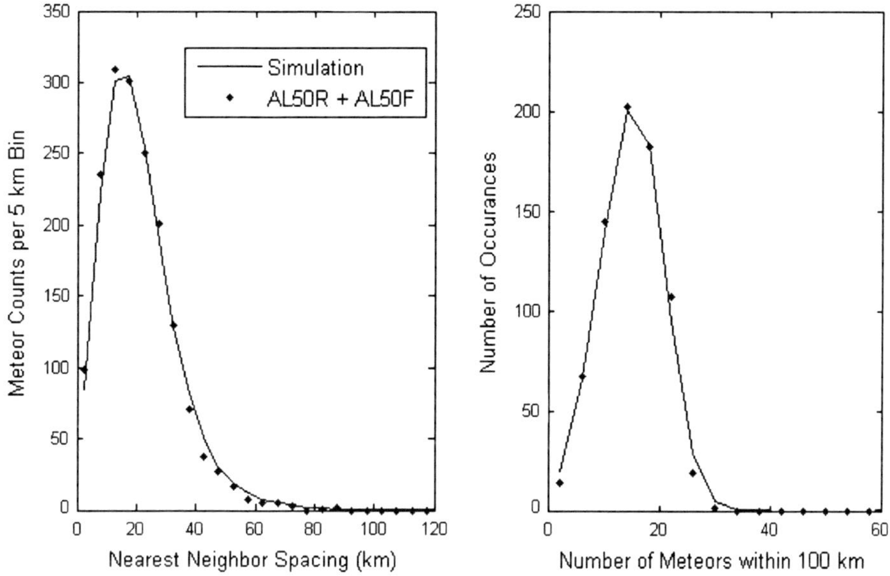

Fig. 5 Left: Nearest neighbor spacing in stream cross-section. Right: Meteor density in 100 km extent around each meteor. Solid lines are from a random spatial distribution of meteors, points are from the heliocentric positions of the measurements

count. To address concern once again that the flux level was too high and may be causing an artificial enhancement, the analysis was redone during the "off-peak" period with the same level of enhancement and spatial scales just described.

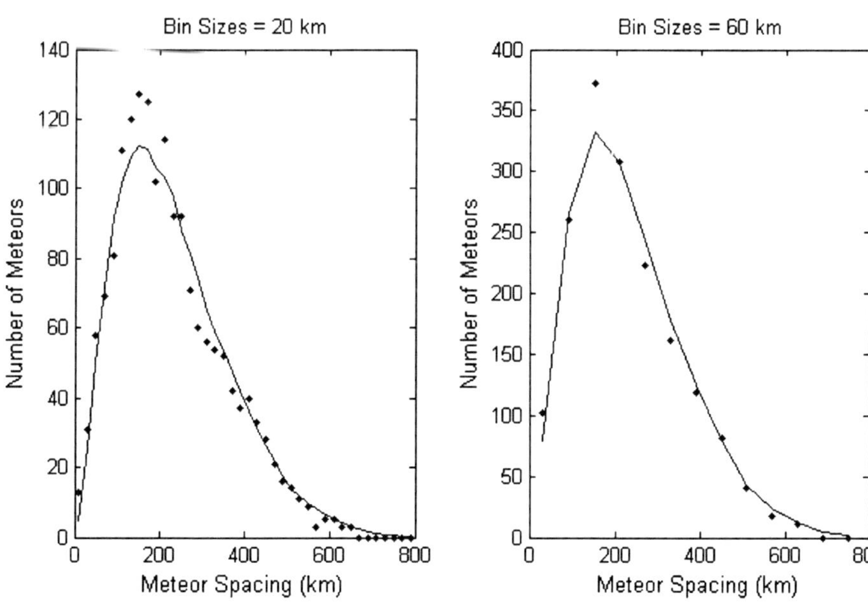

Fig. 6 Histograms of three-dimensional spacing between adjacent-in-time and space meteors during the storm peak at two different bin sizes

7 Discussion

The flux density fluctuations on timescales of several minutes along the comet orbit, and our newly discovered enhancement of 100–200 km spacing in three dimensions, point to some degree of fragmentation of meteoroids in the Leonid stream.

If the 6–9 min correlation can be attributed to boulder fragmentation after lift-off from the comet, then a typical disruption speed (assumed to be isotropic) of 32 cm/s is required if the break-up occurred shortly after ejection from the comet. If the disruption had occurred long after ejection into the interplanetary medium, then the relative speed would need to be higher to still have a 6–9 min correlation scale.

The disruption speed was calculated as follows. When a boulder fragments, the clusters of particles will evolve into an elongated dust trail after a single orbit due to the differential orbital period caused by the ejection speed and relatively more powerful solar radiation pressure on the smaller particles. The spreading will increase with each subsequent orbit. For ejection at perihelion, the induced change in semi-major axis (a) by an ejection speed (v) at an angle ϕ away from the direction of boulder motion (V), with $v_p = v \cos(\phi)$, is $\Delta a/a = 2(1+e)/(1-e) * v_p/V$ (Williams 2001), which corresponds to $\Delta a = 0.0020$ AU for an adopted value of v = 20 cm/s (using V = 41.34 km/s, a = 10.338 AU, e = 0.90555 for comet 55P/Tempel-Tuttle). This corresponds to a temporal change of 85 h after one orbit and 256 h after three orbits. Thus the entire debris trail would pass by the Earth's orbit over approximately 500 h, but the Earth actually passes through the dust trail for a much shorter period of time.

The transverse spreading across the stream's orbit turns out to be only a function of the initial ejection speed. Each particle moves along a unique orbit and will, in the absence of planetary perturbations, return to the point of fragmentation. As a result, the width of the debris trail will depend on where along the orbit the dust was ejected (true anomaly ν), but will not change over time. The dispersion in node will result from the transverse component of the ejection speed $v_t = v \sin \phi$, with ϕ the ejection angle out of the plane of the comet orbit. The longitude of the node will be changed by $\Delta\Omega \sim \sin(\omega + \nu) / \sin(i) * v_t/V$ rad (Williams 2002). Thus a 20 cm/s transverse speed would result in a nodal change of only $\Delta\Omega \leq 0.0009°$ (<80 s in Earth's path), corresponding to a width of less than 2,400 km.

Hence, it is the transverse spreading of the trail that limits the temporal duration of a flux anomaly. Due to the shallow 18° angle of encounter, it takes Earth $\leq$ 128 s/sin(18°) = 7 min to travel through a dust trail from a boulder disrupted at 32 cm/s, but that boulder's debris is spread over an elliptical area of size 3,800 × 12,500 km, the size of Earth. This explains the lack of clumping on a few hundred kilometer spatial scale when the dust trails are projected onto the plane of the ecliptic (Fig. 3).

On the other hand, the excess of correlated particle distances in the range 100–200 km when viewing the data in three dimensions (Fig. 6) implies that fragmentation is an ongoing process. The correlation detected is that of pairs and triplets of meteors moving in the same path, but now somewhat dispersed along the comet orbit. For two grains to separate by 100–200 km at 32 cm/s relative speed takes only 0.5 years implying very recent fragmentation some time after ejection from the comet. Radiation pressure alone can induce differences in orbital period that result in significant dispersion after just a single orbit. If this is the main dispersive mechanism, then pairs of correlated meteors should show the smaller fragment typically trailing the larger one. This conjecture could be verified through a more detailed analysis of the video collection.

8 Further Work

Future work with this unique dataset would involve estimation of the meteor magnitudes directly from the video record to determine the mass index independently and test the cause of fragmented particle dispersion. Errors inherent in the current estimation of each meteor's position in three-dimensional space can be reduced to resolve the question of spatial correlations on small scales. One could also extend the analysis to the north-facing Aria and FISTA video collections which have the advantage of volumetric overlap and thus can provide more definitive meteor end heights.

Acknowledgments The 1999 Leonid MAC mission was supported by NASA Planetary Astronomy and Astrobiology programs and executed by the 452nd Flight Test Squadron of the U.S. Air Force at Edwards Air Force Base, with support of NASA Ames Research Center and the SETI Institute.

References

B.C. Clark, S.F. Green, T.E. Economou, S.A. Sandford, M.E. Zolensky, N. McBride, D.E. Brownlee, Release and fragmentation of aggregates to produce heterogeneous, lumpy coma streams. J. Geophys. Res. **109**, E12 (2004).CiteID E12S03

P.S. Gural, An operational autonomous meteor detector: development issues and early results. JIMO **25**, 136–140 (1997)

P.S. Gural, in *Meteor Observation Simulation Tool. Proceedings of the International Meteor Conference* ed. by M. Triglav, A. Knöfel, C. Trayner (International Meteor Organization, 2002), pp. 29–35

P.S. Gural, A human visual perception model and its impact on Population Index estimation, ZHR, and best look direction. JIMO **32**, 97–108 (2004)

P.S. Gural, P. Jenniskens, Leonid storm flux analysis from one Leonid MAC video AL50R. Earth Moon Planets **82/83**, 221–247 (2000)

S.F. Green, N. McBride, M.T.S.H. Colwell, J.A.M. McDonnell, A.J. Tuzzolino, T.E. Economou, B.C. Clark, Z. Sekanina, P. Tsou, D.E. Brownlee, Stardust Wild 2 dust measurements. ESA SP **643**, 35–44 (2007)

P. Jenniskens, *Meteor Showers and their Parent Comets*. (Cambridge University Press, Cambridge, U.K., 2006), Chap. 31

P. Jenniskens, S.J. Butow, M. Fonda, The 1999 Leonid multi-instrument aircraft campaign—an early review. Earth Moon Planets **82–83**, 1–26 (2000)

M. Kinoshita, M. Marayuma, T. Sagayama, Preliminary activity of Leonid meteor storm observed with a video camera in 1997. Geophys. Res. Lett. **26**, 41–44 (1999)

T. Mukai J. Blum A.M. Nakamura R.E. Johnson O. Havnes, Physical processes on interplanetary dust. in *Interplanetary Dusts*, ed. by E. Grün B.Å.S. Gustafson S.F. Dermott H. Fechtig (Springer-Verlag, Berlin, 2001), pp. 445–507

S. Nikolova, J. Jones, Lifetimes of meteoroids in interplanetary space: the effect of erosive collisions and planetary perturbations. ESA SP **495**, 581–585 (2001)

W. Singer, S. Molau, J. Rendtel, D.J. Asher, N.J. Mitchell, U. von Zahn, The 1999 Leonid meteor storm: verification of rapid activity variations by observations at three sites. MNRAS **318**, L25–L29 (2000)

J.M. Trigo-Rodriguez, H. Betlem, E. Lyytinen, Leonid meteoroid orbits perturbed by collisions with interplanetary dust. Astrophys. J. **621**, 1146–1162 (2005)

A.J. Tuzzolino, T.E. Economou, B.C. Clark, P. Tsou, D.E. Brownlee, S.F. Green, J.A.M. McDonnell, N. McBride, M.T.S.H. Colwell, Dust measurements in the coma of comet 81P/Wild 2 by the Dust Flux Monitor Instrument. Science **304**, 1776–1780 (2004)

J.I. Watanabe, I. Tabe, H. Hasegawa, T. Hashimoto, T. Fuse, M. Yoshikawa, S. Abe, B. Suzuki, Meteoroid clusters—evidence of fragmentation in space. ESA SP **500**, 277–279 (2002)

F.L. Whipple, Meteoric erosion in space. Smiths. Contr. Astrophys. **7**, 239–248 (1963)

I.P. Williams, The determination of the ejection velocity of meteoroids from cometary nuclei. ESA SP **495**, 33–42 (2001)

I.P. Williams, The evolution of meteoroid streams. in *Meteors in the Earth's Atmosphere*, ed. by E. Murad, I.P. Williams (Cambridge University Press, Cambrdige, 2002) pp. 13–32

On the Substantial Spatial Spread of the Quadrantid Meteoroid Stream

K. Ohtsuka · M. Yoshikawa · J. Watanabe · E. Hidaka · H. Murayama · T. Kasuga

Originally published in the journal Earth, Moon, and Planets, Volume 102, Nos 1–4.
DOI: 10.1007/s11038-007-9217-8 © Springer Science+Business Media B.V. 2007

Abstract We explored the substantial spatial spread of the Quadrantid stream, based on the backward integration of orbital motions of the Quadrantids, impulsively perturbed by Jupiter. We found that the Jovian impulses can widely spread out them in the early twentieth century, especially their perihelia extended by a factor of $\sim$90 than those at the observed epoch. We regarded the spread as the intrinsic one of the Quadrantid stream itself.

Keywords Meteors · Individual (Quadrantids) · Orbital evolution

1 Introduction

The giant planets, such as Jupiter and Saturn, strongly perturb a motion of closely approaching small solar system body. For example, Jupiter generates the gravitational impulses in the meteoroid streams having highly-inclined orbit, like the Quadrantids (McIntosh 1991; Ohtsuka et al. 1995; Jenniskens 2006) that here we investigate and the Perseids (Trigo-Rodriguez et al. 2005).

The Jovian impulses, acting on the Quadrantids in around their aphelia, are important mechanics for the orbital motions of the Quadrantids as well as the secular and resonant perturbations, since the Jovian impulses drive the orbital evolution very rapid. The orbits of

K. Ohtsuka (✉) · E. Hidaka · H. Murayama
Tokyo Meteor Network, 1–27–5 Daisawa, Setagaya-ku, Tokyo 155–0032, Japan
e-mail: ohtsuka@jb3.so-net.ne.jp

M. Yoshikawa
ISAS/JAXA, 3–1–1 Yoshinodai, Sagamihara, Kanagawa 229–8510, Japan

J. Watanabe
National Astronomical Observatory, Osawa, Mitaka, Tokyo 181–8588, Japan

T. Kasuga
Institute for Astronomy, University of Hawaii, 2680 Woodlawn Drive, Honolulu, Hawaii 96822–1897, USA

such Quadrantids should be further scattered subsequently. Ohtsuka et al. (1995) concluded that the 1987 Quadrantid swarm, photographically observed by the Tokyo Meteor Network (TMN), was indeed impulsively perturbed by Jupiter in 1984, half an orbital period before the observations (hereafter, we call such an "Impulsively Perturbed Quadrantid" as "IPQ"). Two other IPQ swarms were also photographically observed in 1963 by the Tokyo Astronomical Observatory (TAO, the present National Astronomical Observatory) team (Ohtsuka et al. 1995) and in 1999 by the TMN team again. Therefore, now we have three sets of the IPQ orbit data recorded at every 12-year intervals (i.e., = heliocentric orbital period of Jupiter), except for no record in 1975.

Integrating their orbital motions backward, we investigated their evolutional behavior characteristics analytically and numerically. As a result, we achieved some interesting findings, among which here we deal with the spatial spread of the IPQ orbits caused mainly by the Jovian impulses, obviously wider in the early twentieth century than those at the observed epoch. It is very important for studying the formation process of the Quadrantid stream complex to evaluate the spatial spread of the Quadrantids.

2 Data and Numerical Analysis

First we selected out 13 IPQs as a data sample from among our IPQ orbit database. Since all of them were long-trail meteors, we could determine their no-atmospheric velocities (V_{∞}) well on the basis of the exponential curve fitting for the atmospheric deceleration. For that reason, their orbital parameters were precisely reduced by running our original software "METEOR J2000", Ver. 2.0. These IPQ orbital data are listed in Table 1, where the column heads from left to right are: meteor code number; Ep = osculation (observed) epoch in JDT (add 2400000 to this); M = mean anomaly in degree; a = semimajor axis in AU; e = eccentricity; ω = argument of the perihelion in degree; Ω = ascending node in degree; i = inclination in degree. The 1963 and 1987 IPQ orbital data have already been reduced by Ohtsuka et al. (1995), however, they were re-reduced for this study. All of these IPQ meteoroids are in the mass of gram-order or more.

Next we integrated their orbital motions backward for $\sim$2 centuries, using the "SOLEX", Ver. 9.1 package, developed by Vitagliano (1997), which based on a 16th-order polynomial extrapolation method, the Bulirsh-Stoer integrator. The initial data of the IPQs were taken from Table 1. The integrator can process very close encounters by a routine of the time step of automatic adjustments precisely, in which its truncation and round off errors are entirely negligible for our very short-term integration. Therefore, the SOLEX integrator should be sufficiently reliable to perform our study.

3 Results and Concluding Remarks

In our knowledge, a spatial spread of the meteoroid stream has traditionally been considered as a flat ring of Earth-focusing at near its perihelion, as against enlarging around its aphelion. Its shape has long been accepted as that of a typical meteor stream by many investigators. However, they investigated the stream structure in the Earth-crossing part only, which satisfies the following condition of the relation among a, e, and ω, of meteor:

$$R = r = \frac{a(1 - e^2)}{1 \pm e \cos \omega} \sim 1, \tag{1}$$

Table 1 Heliocentric orbital data of the IPQs at each initial (observed) epoch

Code no.	Ep 2400000+	M	a	e	ω	Ω 2000.0	i
1963 TAO							
M6301	38033.27860	1.49	3.025	0.677	169.54	283.2046	72.08
M6302	38033.29980	1.20	3.226	0.697	170.67	283.2266	69.76
M6303	38033.34233	0.83	2.846	0.655	174.72	283.2700	71.19
1987 TMN							
TN20	46799.26362	0.59	3.144	0.688	175.62	283.0284	73.22
TN21	46799.26609	0.88	2.962	0.669	174.06	283.0309	73.50
TN22	46799.29343	0.47	2.987	0.671	176.81	283.0591	71.46
TN24	46799.32814	0.82	3.096	0.683	174.05	283.0946	72.38
TN25	46799.34030	0.57	2.838	0.654	176.40	283.1070	72.60
1999 TMN							
T9901-01	51182.25141	0.53	3.368	0.708	175.60	282.9325	70.76
T9901-02	51182.25769	0.44	2.986	0.671	177.02	282.9389	72.69
T9901-03	51182.29133	1.26	2.983	0.672	171.37	282.9734	72.12
T9901-05	51182.32375	0.29	3.229	0.696	177.71	283.0066	73.08
T9901-06	51182.35801	1.64	2.971	0.672	168.79	283.0417	72.09

where R and r are respectively Earth's and meteor's heliocentric (nodal) distance ~ 1 AU. Therefore, the selection effect, defined by Eq. 1, does not provide us any substantial spatial spread information for every meteoroid stream (Babadzhanov and Obrubov 1987).

Meanwhile, some investigators have ever attempted to find clues about the substantial spatial spread of meteoroid streams, modelling the stream structure. As for the Quadrantids: e.g., the long-term perturbation cycle model by Babadzhanov and Obrubov (1987), structured by the evolutional passageway of a parent candidate, Comet 96P/Machholz; the dust trail model by Vaubaillon et al. (2006), from another stronger parent candidate, Amor asteroid 2003 EH1 (for the association with the Quadrantids, see also Jenniskens 2004; Williams et al. 2004). These modeled structures seem very likely.

Here we explored the substantial spatial spread of the Quadrantids, based on the backward integration of orbital motions of these actually observed IPQs. The IPQs had sometimes encountered Jupiter in the integration time-span. Consequently, the Jovian impulses can rapidly spread out the IPQs with mass > 1 g. In the early twentieth century, the perihelia extended their width up to at least ~ 0.6 AU in the range between 0.8 and 1.4 AU. Thus, they were widely distributed by a factor of ~ 90 than those of ~ 0.007 AU (range between 0.976 and 0.983 AU) at the observed epoch. Such a spreading tendency is recognized in not only near the perihelia but also around the whole orbit area, except for near the aphelia where both spreads are comparable with each other, ranging over ~ 1 AU across the Jovian orbit. It should be also noted that since the scatter of Ω is very small then, $283°.4 < \Omega < 284°.2$, thus the spread is predominant horizontally rather than vertically in the orbital plane. The IPQs' spread at epoch JDT 2420000.5 (1913 August 13.0 TT) and the traditional spread at the initial epoch, where all the IPQs multiply plotted, are respectively illustrated at the *left* and the *right* in Fig. 1. Hence, the stream shape seems a thin-layered "donut (or torus)" in the evolutional passageway structure of the Quadrantid stream complex, which is quite similar to the theoretical one of Vaubaillon et al. (2006).

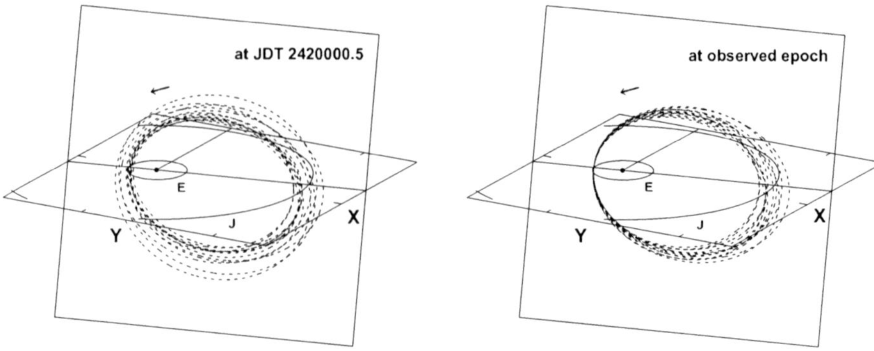

Fig. 1 Distant view of the IPQs' orbital spread at JDT 2420000.5 (*left*) and at the initial (observed) epoch (*right*). The orbits E and J mean those of Earth and Jupiter, respectively

Although the secular perturbation also advances the orbital evolution of the Quadrantids, the IPQs in the early twentieth century would still remain in near-Quadrantid evolutional phase in the orbital evolution of the Quadrantid stream complex. Therefore, if the Quadrantid complex meteoroids are evenly distributed over the whole space of their evolutional passageway, we may regard the IPQs' spread in those days as the intrinsic spatial spread of the Quadrantid meteor stream itself. If so, we'd better re-examine the total flux, mass, and volumes of the Quadrantid stream, considering the substantial orbital spread.

References

P.B. Babadzhanov, Yu.V. Obrubov, Publ. Astron. Inst. Czechosl. **67**, 141–150 (1987)
P. Jenniskens, AJ **130**, 3018–3022 (2004)
P. Jenniskens, Meteor Showers and Their Parent Comets, Chap 20 (2006)
B.A. McIntosh, in Comets in the Post-Halley Era, vol. 1, eds. by R.L. Newburn, Jr. et al. (Kluwer, Dordrecht, 1991), pp. 557–591
K. Ohtsuka, M. Yoshikawa, J. Watanabe, PASJ, **47**, 477–486 (1995)
J.M. Trigo-Rodriguez, et al. Earth Moon Planets, **97**, 269–278 (2005)
J. Vaubaillon, P. Lamy, L. Jorda, MNRAS **370**, 1841–1848 (2006)
A. Vitagliano, Cel. Mech. Dyn. Astron. **66**, 293–308 (1997)
I.P. Williams, et al. MNRAS **356**, 1171–1181 (2004)

Lunar Gravitational Focusing of Meteoroid Streams and Sporadic Sources

Peter S. Gural

Originally published in the journal Earth, Moon, and Planets, Volume 102, Nos 1–4.
DOI: 10.1007/s11038-007-9195-x

Abstract Recent work on the gravitational focusing of meteoroid streams and their threat to satellites and astronauts in the near-Earth environment has concentrated on Earth acting as the gravitational attractor, totally ignoring the Moon. Though the Moon is twelve-thousandths the mass of the Earth, it too can focus meteors, albeit at a much greater distance downstream from its orbital position in space. At the Earth–Moon distance during particular phases of the Moon, slower speed meteoroid streams with very compact radiant diameters can show meteoroid flux enhancements in Earth's immediate neighborhood. When the right geometric alignment occurs, this arises as a narrowed beam of particles of approximately 1,000 km width. For a narrow radiant of one-tenth degree diameter there is a 10-fold increase in the level of flux passing through the near-Earth environment. Meteoroid streams with more typical radiant sizes of 1° show at most two times enhancement. For sporadic sources, the enhancement is found to be insignificant due to the wide angular spread of the diffuse radiant and thus may be considered of little importance.

Keywords Meteor flux · Gravitational focusing

1 Introduction

Focusing of meteoroids behind planetary bodies is a well known phenomenon that has been published by several authors (Divine 1992; Divine et al. 1993; Staubach et al. 1997; Peterson 1999; Jones and Poole 2006). Common to all is the understanding that a gravitating body can locally enhance a meteoroid's stream flux by thousands and in the case of Earth focusing, is a safety consideration when geosynchronous Earth orbiting (GEO) satellites or lunar based astronauts are in alignment with a stream's radiant and the Earth. In such situations, the enhanced flux rate can pose a serious threat to equipment and life. The converse geometry however is typically ignored. That is, when the Moon comes between the meteoroid stream radiant and the near-Earth region extending out to the GEO

P. S. Gural (✉)
SAIC, 14668 Lee Road, Chantilly, VA 20151, USA
e-mail: peter.s.gural@saic.com

belt. The reason quoted is that the Moon is only 0.012 times the Earth's mass and thus is less influential on bending meteoroid streams into focused beams of particles. This is indeed true but not altogether negligible since the weaker gravitational force simply focuses the meteoroids further downstream. In the lunar focusing geometry, the region of concern is actually 384,000 km away from the gravitating body rather than merely 43,000 km as in the case of Earth focusing of meteoroids into the GEO belt.

2 Geometry and Simulation

The fundamental geometry of focusing places the gravitating body along the line of the meteoroid stream radiant and examines points downstream of the oncoming flux. The region directly behind the body is shielded so there is no flux. Further down stream a cross-over point occurs where meteoroids that just skim the body on opposite sides meet. On either side of that point, the off-axis flux can be enhanced but only minimally. However, further downstream of that closest cross-over point is where significant enhancement can occur and is typically characterized by a very narrow flux region of high density.

A Monte Carlo simulation was developed to explore the flux distribution for streams and sporadic sources given the spread in the radiant direction, spread in velocities associated with the source, and random starting position sufficiently upstream of a gravitating body's influence. The step-by-step procedure is detailed in Appendix A. It is essentially characterized as a random draw (from user defined probability density functions) of a particle's state vector which is then propagated through the classical hyperbolic orbital equations to a measurement point downstream. This is repeated for billions of particles and the number count in a fixed downrange shell with spherical segments of four kilometer off-axis spacing is accumulated. Flux is computed per unit area and normalized by the incident flux. Note that the result is axially symmetric with respect to the radiant direction when the statistics are high enough thus only requiring the count accumulation be done in a single off-axis arc length vector.

The simulation was first tested on a simple Earth based focusing example with no spread in radiant or velocity (collimated case) as shown in Fig. 1 and favorably compared to results previously reported (Peterson 1999). A more recently published paper (Jones and Poole 2006) provided comparison plots from two other authors. For a perfectly collimated beam of meteoroids the Monte Carlo simulation was found to meet that paper's "Opik test" of Earth impacting flux enhancement. It also matched closely the curves labeled Devine and J&P (Fig. 4, Jones and Poole 2006) for collimated off-axis results at moderate to large values of F, which is the situation encountered in this paper. An analytic solution was also derived for the perfectly collimated case and the current simulation matched both the on and off axis flux enhancement expected downstream. Furthermore, for the case of a diffuse radiant or non-collimated beam of particles, the simulation matched the published on-axis results (Jones and Poole 2006) and also showed an off-axis half-power angular spread in the downstream flux that corresponds to the upstream radiant diameter modeled.

3 Lunar Focusing of Meteoroid Streams

The focusing of meteoroid streams by the Moon, at the Earth's position downstream, occurs for stream velocities of 34 km/s or less. Higher than that and the convergence point behind the Moon is located at a range greater than the Earth–Moon distance and is of little

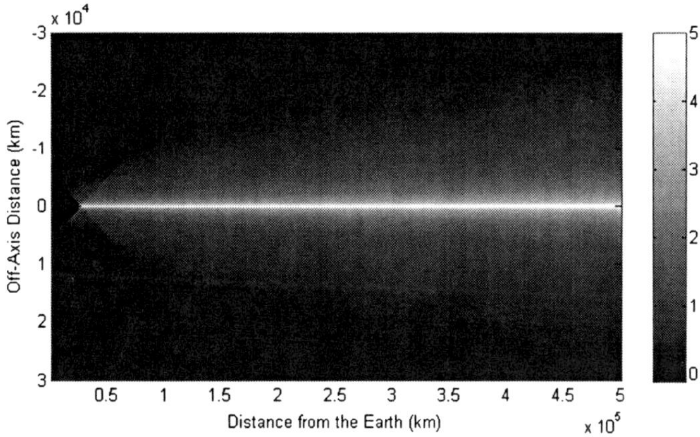

Fig. 1 Two-dimension plot of meteoroid flux distribution for Earth focusing of a 25 km/s stream with no radiant spread in angle and velocity. The color bar is the log of the ratio of observed flux to incident flux

consequence, even if there is some angular spread in the radiant. Shown in Fig. 2 is a cross sectional plot of flux enhancements for a meteoroid stream of 20 km/s given various radiant diameters. The cross section is taken at a range of 384,000 km from the Moon and represents the stream's behavior near the Earth ignoring any gravitational effects of the Earth. The top curve is the flux enhancement for the perfectly collimated case and represents the upper limit one could expect to observe. The middle curve of Fig. 2 is for a very narrow radiant of one-tenth degree diameter and is representative of a young stream that has taken just a few revolutions around the Sun. An example is the observed limited radiant spread of the 1999 Leonids (Betlem et al. 2000). The bottom curve is for an older more typical stream such as the Orionids or the Geminids where the radiant diameter is closer to 1° (De Lignie and Betlem 1999). Although the Lunar flux enhancement is orders of magnitude lower than the equivalent focusing by the Earth, there is still an enhancement of 10 times the incoming stream's spatial density for very young streams and nearly two times for older streams. The region of highest flux density occurs in a fairly narrow region less than 1000 km wide.

In addition, the flux enhancement is expected to increase with decreasing velocity of the meteoroids as the slower stream particles are more easily bent by the Moon's gravity. This is shown in Fig. 3 for collimated, narrow, and more typical sized radiant diameters. The results indicate the greatest concern would be for a Lunar alignment with a very young meteoroid stream of low to moderate speed. The narrow radiant curve falls off beyond 35 km/s due to the shielding of the higher speed streams in the shadow region downstream of the Moon. Note that these simulations were all run for a meteoroid stream containing a Gaussian spread of the velocities, whose standard deviation was 5% of the nominal entry velocity.

It may be argued for the scenario described in this paper that a three-body solution needs to be considered. To test the efficacy of the current two-body approach for Lunar focusing, where effectively the Earth's gravitation influence had been ignored, the simulation was modified to a more general propagation model. Following Appendix A, a hyperbolic orbit was still computed for meteoroids passing near the Moon but they were now only propagated 66,000 km downrange generating new position and velocity components for an intermediate state vector strictly under the Moon's influence. At this point

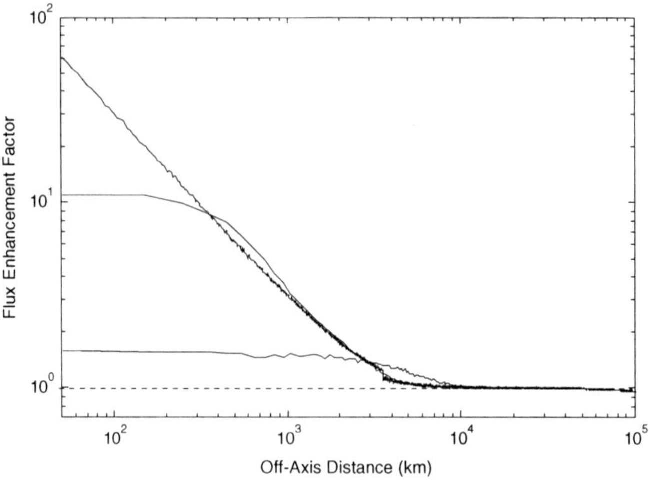

Fig. 2 Flux enhancement cross-section of a 20 km/s meteoroid stream downrange 384,000 km from the Moon. Top curve is a perfectly collimated stream's flux enhancement, middle curve the flux contribution for a one-tenth degree radiant diameter, and the bottom curve for 1° radiant spread

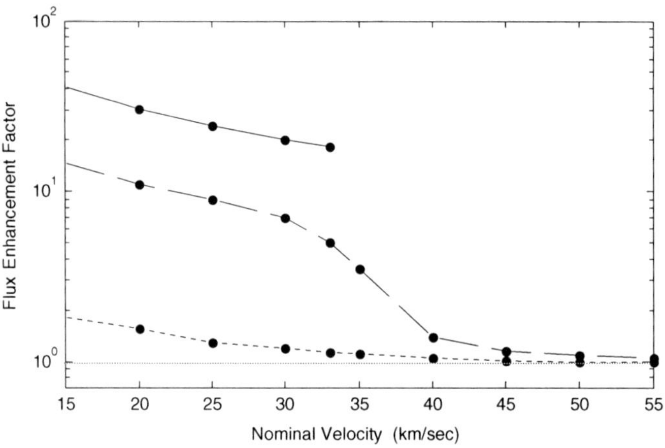

Fig. 3 On-axis flux enhancement at the Earth–Moon distance as a function of stream velocity for collimated (solid), 0.1° radiant diameter (dashed), and 1° diameter meteoroid streams (dotted). Collimated was taken 100 km off-axis

the Earth's gravitational influence was assumed to take over and a second hyperbolic orbit around the Earth was solved using the new state vector as input. The flux was then accumulated in surface planes slicing through various regions of the Earth's immediate neighborhood out to the GEO belt. What was found was that the modified simulation produced the same angular spreading and flux values as the two-body solution with the flux enhancement tube merely shifted closer towards the Earth. Thus the focused stream is acted on uniformly from a central force and behaves like the phenomena of zenith attraction on a meteoroid stream's radiant. The only disparity arises when the focused streamlet passes within three Earth radii and is viewed on the leeward side of our planet. In that limited area behind the Earth, the focus tube's cross section gets elongated or stretched in the Earth-streamlet direction, but with little reduction in the flux enhancement.

Thus, a meteoroid stream can be weakly focused to a greater extent than several authors have previously stated and can present a short term threat when the Moon aligns within a few degrees of a young stream's radiant. The alignment angle is determined by the diameter of the region of concern. For the GEO belt of 43,000 km radius, the geometry requires the Moon to be within 6° of the radiant. Searching through the Moon's position for the next 20 years, there are only three meteor showers that fit that criterion within 3 days of their peak flux. They all have ZHRs less than 20 and include: the Mu Sagittariids in 2008, 2012, 2016, 2019, 2022; the Capricornids in 2012, 2015; and the Chi Orionids in 2009, 2012, 2017, 2020.

For those meteor streams that have velocities greater than 34 km/s, the near-Earth environment sits in a shielded zone behind the Moon when perfect alignment occurs and no meteors would be seen. However, when the Moon is slightly off the radiant, no flux enhancement would occur but the trajectories of the meteors seen will have non-radiant paths due to the perturbing effect by the Moon. This would be an interesting sight to behold. The years and streams when this is possible include: the Eta Aquariids in 2010, 2013; the Southern Delta Aquariids in 2018, 2021; the Northern Delta Aquariids in 2009, 2014, 2017; and the Orionids in 2013, 2016.

Sporadic sources were not expected to have any significant enhancement due to their large radiant diameters and far greater spread in their velocity distribution. For the simulation analysis, the sporadic radiant diameter was chosen to be 5° in diameter and a mean velocity of 35 km/s with standard deviation of 10 km/s. The results indicate less than 1% cross-sectional flux enhancement and thus no different than the normal background sporadic flux.

4 Summary

In geometric alignments with young meteor streams whose radiant position lies near the ecliptic, the Moon can enhance flux by a factor of up to 10 times over a 1,000 km wide beam for showers possessing a velocity of less than 34 km/s and up to two times for older streams with wider radiant diameters. Though not dramatic, this enhancement is greater than that alluded to by other authors. The enhancement for the sporadic meteoroid sources is found to be of insignificant consequence to the near-Earth environment.

Appendix A—Monte Carlo Simulation

Each randomly drawn particle is propagated through a hyperbolic orbit with a single gravitational center of mass. A starting plane is defined normal to the nominal radiant direction and 20 body radii upstream from the center of mass—sufficiently upstream so the gravitational influence on the starting trajectory is negligible. For each particle there is computed:

State Vector

A starting position drawn uniformly from a 100,000 × 100,000 km area. The starting velocity magnitude is drawn from a Gaussian distribution whose mean is the nominal entry velocity of the meteoroid source and its associated standard deviation. The angular spread in the radiant is obtained from two independent Gaussian distributions of velocity components mutually orthogonal to the nominal radiant direction. For a specified radiant diameter, the distribution generates orientation vectors that fall within that diameter for 39% of the particles.

Plane of Propagation

The velocity unit vector and starting-position-to-gravitating-body-center unit vector defines the plane of propagation. The cross product yields the normal to that plane which in turn defines the in-plane unit vector normal to velocity.

Hyperbolic Trajectory Parameters

The state vector can now be solved in the propagation plane for impact parameter Rc, semi-major axis a, eccentricity e, deflection angle d, and tested to ensure that the closest point of approach exceeds the body radius plus atmospheric height.

Downstream Counting

For each downstream radii of observation, an analytic solution is computed for the hyperbolic true anomaly theta (Kaplan 1976) and the resultant distance off the radiant axis. The bin counter is incremented for the associated arc length which is done along spherical shells centered on the gravitating body to simplify the hyperbolic propagation solution for position. This could also be done analytically for a planar surface but involves slightly more computation. Note that the flux is defined through a spherical shell centered on the gravitating body but at long ranges like the Earth–Moon distance this is equivalent to the flux through a surface normal to the nominal radiant direction.

Flux Computation

Each spherical segment bin (assumes axial symmetry at high counting statistics) is normalized by the segment's area to obtain local flux and is in turn normalized by the incident flux (total simulated meteoroids divided by allowable starting area). These steps are executed in MATLAB for several billion simulated trajectories.

References

H. Betlem, P. Jenniskens, P. Spurny, G. Docters Van Leeuwen, K. Miskotte, C. R. Ter Kuile, P. Zarubin, C. Angelos, Precise trajectories and orbits of meteoroids from the 1999 Leonid meteor storm. Earth Moon Planets **82–83**, 277–284 (2000)

M. De Lignie, H. Betlem, A double-station video look on the October meteor showers. WGN J. IMO **27**(3/4), 195–201 (1999)

N. Divine, Meteoroid focusing at a planet. JPL Interoffice Memorandum 5217-92-86 (1992)

N. Divine, E. Grun, P. Staubach, Modeling the meteoroid distribution in interplanetary space and near-Earth. Proceedings of the first European conference on space debris, ESA SD-01 (1993), pp. 245–250

J. Jones, L. M. G. Poole, Gravitational focusing and shielding of meteoroid streams. MNRAS **375**(3), 925–930 (2006)

M. H. Kaplan, *Modern Spacecraft Dynamics and Control* (John Wiley & Sons, New York, 1976), pp 91–95

G. E. Peterson, *Dynamics of Meteor Outbursts and Satellite Mitigation Strategies* (The Aerospace Press, El Segundo, 1999), pp 75–100

P. Staubach, E. Grun, R. Jehn, The meteoroid environment near Earth. Adv. Space Res. **19**, 301–308 (1997)

Comparison of Meteoroid Flux Models for Near Earth Space

Gerhard Drolshagen · Valeri Dikarev · Markus Landgraf · Holger Krag · Wim Kuiper

Originally published in the journal Earth, Moon, and Planets, Volume 102, Nos 1–4.
DOI: 10.1007/s11038-007-9199-6

Abstract Over the last decade several new models for the sporadic interplanetary meteoroid flux have been developed. These include the Divine-Staubach and the Dikarev model. They typically cover mass ranges from 10^{-18} g to 1 g and are applicable for model specific Sun distance ranges between 0.1 AU and 20 AU Near 1 AU averaged fluxes (over direction and velocities) for all these models are tuned to the well established interplanetary model by Grün et al. However, in many respects these models differ considerably. Examples are the velocity and directional distributions and the assumed meteoroid sources. In this paper flux predictions by the various models to Earth orbiting spacecraft are compared. Main differences are presented and analysed. The persisting differences even for near Earth space can be seen as surprising in view of the numerous ground based (optical and radar) and in situ (captured Inter Stellar Dust Particles, in situ detectors and analysis of retrieved hardware) measurements and simulations.

Keywords Meteoroids · Meteoroid Flux models · Near Earth space · Interplanetary meteoroid models · Space dust · IMEM · Divine model · Divine-Staubach model · Impact flux

G. Drolshagen (✉) · W. Kuiper
TEC-EES, ESA/ESTEC, Keplerlaan 1, 2201 AZ Noordwijk, The Netherlands
e-mail: gerhard.drolshagen@esa.int

V. Dikarev
MPI For Solar System Research, Katlenburg-Lindau, Germany

M. Landgraf · H. Krag
ESA/ESOC, Darmstadt, Germany

W. Kuiper
RheaTech Ltd., London, UK

1 Introduction

Any assessment of particle impact risks to spacecraft in orbit requires reliable meteoroid population models. Over the last decade, new models for the sporadic interplanetary meteoroid flux have been developed. These models cover the full velocity range and particle diameters from sub-microns to cm.

In this paper, flux predictions by the various models to Earth orbiting spacecraft are compared. The main focus is on the velocity and directional distributions and the implemented meteoroid sources. Main differences are presented and discussed.

1.1 Model Description

Several models, each having their own population source characteristics, are used for the comparison. An overview is given in Table 1.

The Grün interplanetary flux model (Grün et al. 1985) assumes an isotropic meteoroid distribution which is based on lunar crater, zodiacal light and in situ measurement data. For the conversion of crater sizes to particle masses, a constant velocity of 20 km/s was used. The Grün model is frequently used with added velocity distributions, such as from SSP 30425 (Kessler et al. 1994) or Taylor (Taylor 1995), to include directional effects. SSP 30425 is a velocity distribution, developed for the International Space Station. Therefore, it is valid for Low Earth Orbits (LEO) only. Taylor used data from the Harvard Radio Meteor Project (HMRP) to develop a velocity distribution, which is valid for near Earth orbits and the interplanetary space near 1 AU.

The Divine interplanetary model (Divine 1993) was one of the first models with non-isotropic distributions. The model is based on five different populations each having separable distributions in particle mass, inclination, eccentricity, and perihelion distance.

Staubach (Staubach et al. 1996; Grün et al. 1997) upgraded Divine's model using new data from GALILEO and ULYSSES dust detectors. Solar radiation pressure was added as a second perturbation force and an additional population, Inter Stellar Dust (ISD), was implemented.

With the IMEM/Dikarev model (Dikarev et al. 2005a, b, c), an attempt was made to construct a meteoroid model, based on the physical effects that influence meteoroid orbit and sources, in addition to fitting model predictions to observations. The ISD population was adopted from the Divine-Staubach model, with a re-normalisation to take additional ULYSSES dust detector data into account.

1.2 Test Cases

The different meteoroid models are compared for LEO (400 km circular or bit and 51.6 inclination) and Geostationary orbit (GEO). First the flux to a Randomly Tumbling Plate

Table 1 Meteoroid models used for comparison

Meteoroid model	Year of release	Applicable mass domain	Applicable regime
Grün et al.	1985	10^{-18}–100 g	Around 1 AU from sun
Divine	1993	10^{-18}–1 g	0.1–20 AU from sun
Divine-Staubach	1996	10^{-18}–1 g	0.1–20 AU from sun
IMEM/Dikarev	2003	10^{-18}–1 g	0.1–10 AU from sun

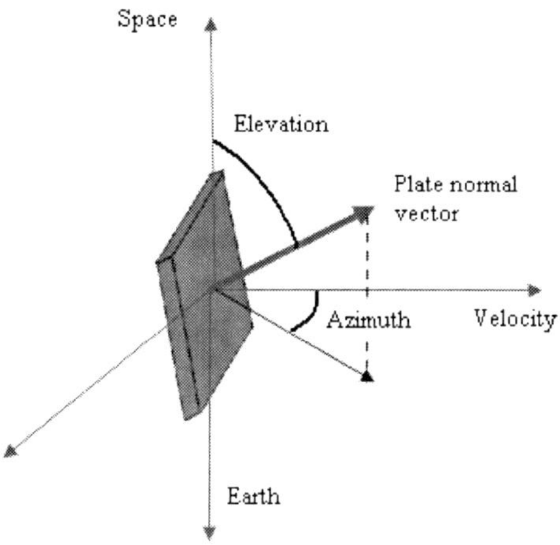

	Azimuth [°]	Elevation [°]
Ram	0	90
Starboard	90	90
Wake	180	90
Space	0	0

Fig. 1 Definition of reference frame and orientations

(RTP) is predicted for a mass range of 10^{-15}–1 g. More detailed information is obtained by comparing the directional dependence of the models. For two mass thresholds, 10^{-12} g and 10^{-3} g, the flux from all models is predicted for oriented plates facing towards ram, starboard, wake and space, respectively (see Fig. 1 for definition of orientations).

Finally, normalised velocity distributions are compared for both orbits and different mass thresholds.

2 Results

The results for the RTP analysis are shown in Fig. 2. The calculated fluxes include the effects from Earth shielding and gravitational attraction. The model by Grün et al. has been combined with the Taylor/HRMP velocity distribution. All fluxes are for an orbiting spacecraft in a 400 km LEO.

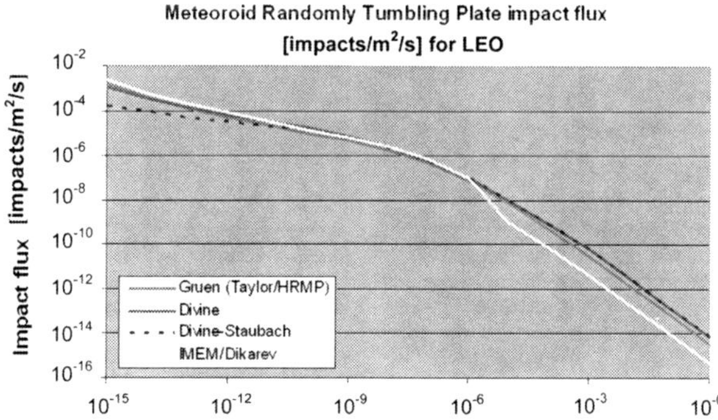

Fig. 2 Predicted meteoroid fluxes to one side of a randomly tumbling plate in LEO

The flux predictions to a RTP agree quite well for all models. This is not really surprising as all models analysed have fitted the random plate flux near 1 AU to the interplanetary model by Grün et al. In the low mass regime ($<10^{-12}$ g), the Divine-Staubach model is predicting lower fluxes compared to the other models. The Dikarev model predicts lower fluxes for meteoroid masses larger than 10^{-5} g. In the Dikarev model, flux results are based on crater volume, which is proportional to the kinetic energy of impacting particles. The IMEM/Dikarev model assumes higher impact velocities for the larger masses than the 20 km/s which were assumed by Grün et al. To be consistent with the crater data used by Grün et al. this leads to lower fluxes for a given fixed mass compared with the Grün model.

Tables 2–5 give predicted fluxes for orbiting surfaces with four different fixed orientations relative to the spacecraft velocity vector.

Directional effects result from model characteristics and from the orbital motion of the spacecraft. All models predict the highest flux for the ram facing surface and the lowest for the wake direction. Similar to the RTP analysis, the fluxes from the Divine-Staubach and IMEM/Dikarev models differ from those predicted by the other models at certain mass regimes.

Table 2 Directional dependence for m $\geq 10^{-12}$ g in LEO

Flux [impacts/m^2/s] for m $\geq 10^{-12}$ g and LEO				
Model	Ram	Starboard	Zenith	Wake
Grün (Taylor/HRMP)	1.09E−04	5.13E−05	7.48E−05	1.53E−05
Divine	1.41E−04	8.31E−05	9.55E−05	1.67E−05
Divine-Staubach	6.66E−05	5.18E−05	5.04E−05	1.38E−05
IMEM/Dikarev	1.79E−04	1.01E−04	1.28E−04	2.03E−05

Table 3 Directional dependence for m $\geq 10^{-3}$ g in LEO

Flux [impacts/m^2/s] for m $\geq 10^{-3}$ g and LEO				
Model	Ram	Starboard	Zenith	Wake
Grün (Taylor/HRMP)	5.98E−11	2.82E−11	4.10E−11	8.40E−12
Divine	1.27E−10	6.50E−11	8.19E−11	8.60E−12
Divine-Staubach	1.27E−10	6.50E−11	8.19E−11	8.59E−12
IMEM/Dikarev	8.55E−12	9.98E−13	6.86E−12	2.02E−12

Table 4 Directional dependence for m $\geq 10^{-12}$ g in GEO

Flux [impacts/m^2/s] for m $\geq 10^{-12}$ g and GEO				
Model	RAM	Starboard	Zenith	Wake
Grün (Taylor/HRMP)	5.78E−05	4.10E−05	4.05E−05	2.72E−05
Divine	7.77E−05	4.95E−05	4.92E−05	2.69E−05
Divine-Staubach	4.62E−05	1.82E−05	3.34E−05	2.31E−05
IMEM/Dikarev	9.28E−05	4.86E−05	5.99E−05	3.28E−05

Table 5 Directional dependence for m ≥ 10^{-3} g in GEO

Flux [impacts/m^2/s] for m ≥ 10^{-3} g and GEO				
Model	RAM	Starboard	Zenith	Wake
Grün (Taylor/HRMP)	3.17E−11	2.25E−11	2.22E−11	1.49E−11
Divine	5.70E−11	4.74E−11	3.29E−11	1.33E−11
Divine-Staubach	5.70E−11	4.74E−11	3.29E−11	1.33E−11
IMEM/Dikarev	6.95E−12	4.25E−12	4.94E−12	3.45E−12

The Divine and Divine-Staubach models predict equal fluxes for m >10^{-3} g. The upgrade of the Divine model by Staubach only influences the lower meteoroid mass regime.

Figures 3 and 4 show the normalised velocity distributions for LEO and mass thresholds m > 10^{-12} g and m > 10^{-3} g. The IMEM/Dikarev, Divine and Divine Staubach models have a build-in velocity distribution resulting from the source terms. The distributions denoted by SSP 30425 and Taylor/HRMP can be used with the isotropic distribution of the Grün model. The rather artificial SSP 30425 distribution was developed for engineering purposes of the Space Station Programme. It was one of the earliest developments of a velocity distribution and never aimed at scientific accuracy.

The velocity distributions from the different models differ considerably. In the high (>10^{-3} g) mass regime, the normalised velocity distribution graph for the IMEM/Dikarev model indeed peaks at higher impact velocities compared to the other models.

For the lower mass threshold, the velocity distributions of the Divine-Staubach and IMEM/Dikarev models have local maxima between 50 km/s and 65 km/s. These result from the ISD population which makes a noticeable contribution for smaller masses. The magnitude of this ISD contribution and the impact velocity depends on the yearly season. The models for the ISD populations assume a fixed velocity of 26 km/s relative to the sun and a fixed arrival direction in a sun centered ecliptic reference system (77° longitude and −3° latitude for IMEM/Dikarev). The impact fluxes and velocities of ISD particles are then determined by the motion of the ISD particles and the Earth relative to the sun. The additional spacecraft motion introduces the double peaks between 48 km/s and 65 km/s in Figs. 3 and 5. The results in Figs. 3–5 are for 21 March when the ISD contribution and

Fig. 3 Velocity distributions for LEO and m > 10^{-12} g

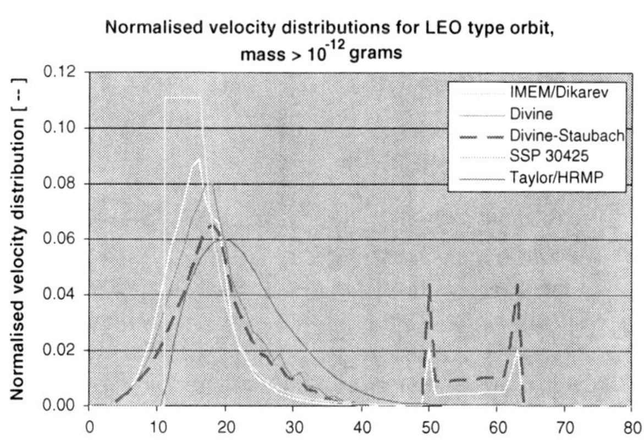

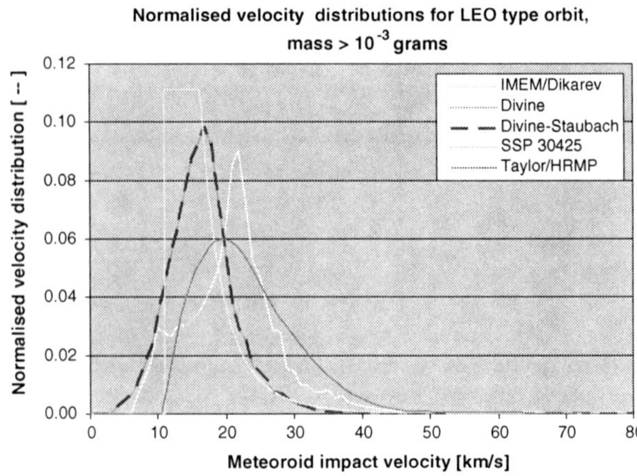

Fig. 4 Velocity distributions for LEO and m > 10^{-3} g

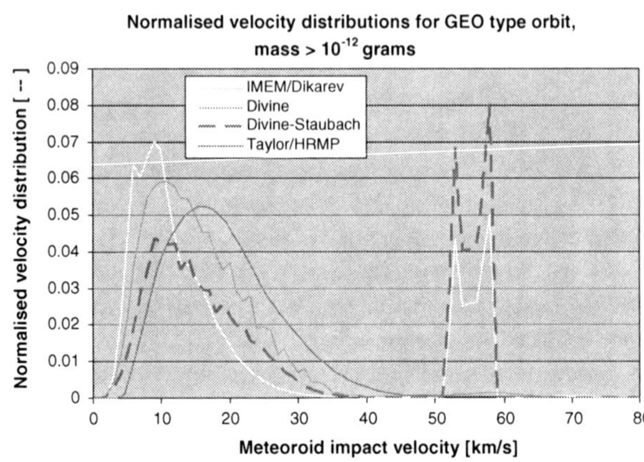

Fig. 5 Velocity distributions for GEO and m > 10^{-12} g

relative velocities are near maximum. The models also predict vanishing ISD fluxes for larger masses as is evident from the absence of this population for m > 10^{-3} g (Fig. 4).

The normalised velocity distributions for GEO and m > 10^{-12} g are presented in Fig. 5. Compared to LEO, the GEO distributions are shifted towards lower impact velocities. This is a direct result of the reduced gravitational attraction from Earth and lower spacecraft velocity in GEO. Velocity distributions for other mass thresholds show a similar behavior when compared for LEO and GEO.

3 Conclusions

The meteoroid fluxes predicted for randomly oriented plates in near Earth orbits agree well for all models. For these models, the measurement data from the vicinity of the Earth has been refitted—since this, to a large degree, overlaps with the data already used by Grün et al. the flux levels correspond well for the near Earth space. Some differences were found in the lower and higher meteoroid mass regimes. The IMEM/Dikarev model predicts lower meteoroid fluxes for the higher mass regime compared to the other models, which is a direct consequence of the higher velocities, assumed by this model.

Directional and velocity distributions of the various models are quite different indicating persistent uncertainties. Differences for Sun distances away from 1 AU will be larger still. Near Earth meteoroid flux predictions are validated by data sets from ground observations and in-flight measurements. At other interplanetary distances, these data sets become scarce and the discrepancies will become larger.

This paper did not perform an exhaustive comparison of all existing meteoroid models. The new MEM model (Jones 2004) is based on data from the Canadian CMOR radar. It is mainly based on cometary sources and applicable for the mass range 10^{-6}–10 g and for Sun distances between 0.2 AU and 2 AU.

Even near Earth increased efforts should be made to measure the full meteoroid population, including the complete range of velocities. Present optical and radar measurements of meteors are strongly dominated by the high velocity tail of the meteoroid population.

The present comparison of flux predictions near Earth from existing models shows a clear need for additional measurements and simulations in order to derive a reliable model for the population of interplanetary and interstellar meteoroids.

References

V. Dikarev, E. Grün, M. Landgraf, R. Jehn, Update of the ESA meteoroid model, in *Proceedings of the 4th European Conference on Space Debris*, (ESA SP-587, 2005a), pp. 271–277

V. Dikarev, E. Grün, J. Baggaley, D. Galligan, M. Landgraf, R. Jehn, The new ESA meteoroid model. Adv. Space Res. **35**(issue 7), 1282–1289 (2005b)

V. Dikarev, E. Grün, J. Baggaley, D. Galligan, M. Landgraf, R. Jehn, Modeling the sporadic meteoroid background cloud. Earth Moon Planets **95**, 109–122 (2005c)

N. Divine, Five populations of interplanetary meteoroids. J. Geophys. Res. **98**, 17029–17048 (1993)

E. Grün, H.A. Zook, H. Fechtig, R.H. Giese, Collisional balance of the meteoritic complex. Icarus **62**, 244–272 (1985)

E. Grün, P. Staubach, M. Baguhl, D.P. Hamilton, H.A Zook, S. Dermott, B.A. Gustafson, H. Fechtig, J. Kissel, D. Linkert, G. Linkert, R. Srama, M.S. Hanner, C. Polanskey, M. Horanyi, B.A. Lindblad, I. Mann, J.A.M McDonnell, G.E. Morfill, G Schwehm, South–North and radial traverses through the interplanetary dust cloud. Icarus **129**(issue 2), 270–288 (1997)

J. Jones, Meteoroid engineering model—final report. NASA/MSFC internal report SEE/CR-2004, 400 (2004)

D.J. Kessler, R.C. Reynolds, P.D. Anz-Meador, Space station program natural environment definition for design, international space station alpha. NASA SSP 30425, Revision B, national aeronautics and space administration space station program office, Houston, TX, USA (1994)

P. Staubach, E. Grün, R Jehn, The meteoroid environment near Earth. Adv. Space Res. **19**(issue 2), 301–308 (1996)

A.D. Taylor, The harvard radio meteor project meteor velocity distribution reappraised. Icarus **116**, 154–158 (1995)

Dynamical Effects of Mars on Asteroidal Dust Particles

Ashley J. Espy · Stanley F. Dermott · Thomas J. J. Kehoe

Originally published in the journal Earth, Moon, and Planets, Volume 102, Nos 1–4.
DOI: 10.1007/s11038-007-9200-4

Abstract Asteroidal dust particles resulting from family-forming events migrate from their source locations in the asteroid belt inwards towards the Sun under the effect of Poynting-Robertson (PR) drag. Understanding the distribution of these dust particle orbits in the inner solar system is of great importance to determining the asteroidal contribution to the zodiacal cloud, the accretion rate by the Earth, and the threat that these particles pose to spacecraft and satellites in near-Earth space. In order to correctly describe this distribution of orbits in the inner solar system, we must track the dynamical perturbations that the dust particle orbits experience as they migrate inwards. In a seminal paper Öpik (1951) determines that very few of the μm-cm sized dust particles suffer a collision with the planet face as they decay inwards past Mars. Here we re-analyze this problem, considering additionally the likelihood that the dust particle orbits pass through the Hill sphere of Mars (to various depths) and experience potentially significant perturbations to their orbits. We find that a considerable fraction of dust particle orbits will enter the Hill sphere of Mars. Furthermore, we find that there is a bias with inclination, particle size, and eccentricity of the particle orbits that enter the Martian Hill sphere. In particular the bias with inclination may create a bias towards higher-inclination sources in the proportions of asteroid family particles that reach near-Earth space.

Keywords Zodiacal cloud · Zodiacal dust · Dust dynamics

1 Introduction

The dominant source of the zodiacal cloud has been debated for many years. We know from observations that it must have asteroidal and cometary components, but in what relative proportions? In order to fully understand the contributions of these sources to the zodiacal cloud, we must be able to model the cloud and match the observed structure.

A. J. Espy (✉) · S. F. Dermott · T. J. J. Kehoe
Department of Astronomy, University of Florida, Gainesville, FL 32611, USA
e-mail: ashley@astro.ufl.edu

We will focus on the asteroidal contribution here. The dust bands, which are a fine structure component superimposed on the broad background cloud, are known to be asteroidal in origin because they result from the breakup of the parent body of asteroid families. Modeling of these dust bands provides information on the orbits, sizes, and amount of dust producing the bands. However, the dust bands are constrained outside 2 AU, due to the action of the secular and mean-motion resonances that disperse the dust band particles into the background cloud. In order to recreate the structure of the zodiacal cloud observed from Earth orbit and constrain the dust particle orbits as they reach near-Earth space and pose threats to spacecraft and satellites, we must define the distribution of dust orbits from their source regions in the main belt inwards to 1 AU. The orbits of the asteroidal dust particles change as they evolve inwards under PR drag due to three main effects: the dispersion from secular and Jovian mean-motion resonances inside 2 AU, trapping in mean-motion resonances with Mars and the Earth, and the scattering of particles by Mars. These effects are each dependent on particle size and are more pronounced for larger particles that migrate inward more slowly. The trapping in resonances with the Earth is well studied (e.g., Dermott et al. 1994) as is the dispersion from the secular and Jovian mean-motion resonances (e.g., Kehoe et al. 2007). We focus here on the scattering effects of Mars on the dust particle orbits through calculations based on the theoretical argument presented by Öpik (1951).

2 Method

Following the method of Öpik (1951) on determining collision probabilities of dust particles with planets on eccentric orbits, we have generalized the results to allow for eccentric orbits of the dust particles as well. The survival fraction of particles that do not undergo a close planetary encounter, χ, is calculated relative to the PR drag timescale, Δt, of the eccentric particle orbit for a specific particle diameter and its lifetime from encounters, T, and is given by

$$\chi = e^{-\Delta t/T}. \tag{1}$$

This collisional lifetime is found from the probability of collision per orbit of the particle, P, for a specific relative velocity, inclination, and sphere of action, τ, which we vary from the face of the planet (for collisions) up to the Hill sphere of the planet (for potentially significant orbital effects). The probability of collision is given by,

$$P = \frac{\tau^2 U}{\pi \sin i \cdot |U_x|}. \tag{2}$$

U and U_x represent the relative velocity for the specific geometry of interaction of the orbits and are given by,

$$U_x^2 = 2 - \frac{1}{A} - A(1 - e^2) + \frac{4}{9}e_0^2, \tag{3}$$

$$U^2 = 3 - \frac{1}{A} - 2\sqrt{A(1 - e^2)}\cos i + \frac{4}{9}e_0^2. \tag{4}$$

where e is the particle eccentricity, e_0 is the planet eccentricity, and A is a/a_0, the ratio of the particle to planet heliocentric distance which we assume to be unity to allow for interaction.

3 Results

Using the theoretical method described in the previous section, we investigated the fraction of asteroidal dust particles of various diameters, orbital inclinations, and orbital eccentricities that would enter Mars Hill sphere and thus experience potentially significant orbital perturbations. The Hill sphere approximates the gravitational sphere of influence of an astronomical body (e.g., Murray and Dermott 1999), thus the Martian Hill sphere is the region in which Mars can cause gravitational perturbations. The deeper into the Hill sphere that a particle penetrates (smaller percentage thereof), the greater the potential orbital effects on that particle. We chose specifically to look at inclinations of $\sim 2°$ and $\sim 10°$, since these represent the locations of the recent asteroid family formations of Karin and Veritas, respectively. These families are known to be the sources of the "near-ecliptic" and "ten-degree" zodiacal dust bands discovered by the Infrared Astronomical Satellite (IRAS), and thus are known to be the sources of dust that migrates into the inner solar system. We confirm Öpik's result of a survival fraction of ~ 1 for direct interactions of the dust particles of all sizes of interest with the face of Mars. However, we also find that when the region of interest is increased from the face of Mars to its Hill sphere, the results are quite different, with a very large fraction of all particle sizes at the inclinations of interest passing through this Hill sphere and thus suffering potentially significant perturbations to their orbits.

3.1 Inclination

Figure 1 shows the survival fraction as a function of inclination for a range of particle sizes for different interaction regions of interest. Panel A shows the level of interaction of the particles with the Hill sphere of Mars. All of the 1,000 μm particles enter the Hill sphere regardless of the inclination of their orbit, but the fraction of smaller particles that enter the Hill sphere will be a function of inclination and thus the percentage of particles affected will be different for the different asteroid families/dust bands. Since the low inclination particles are more strongly affected than the higher inclination particles, this may introduce a bias in the asteroidal family dust particles reaching the inner solar system unperturbed, since likely many more of the Karin particles will suffer significant orbital perturbations than the higher inclination Veritas particles. Panels B and C show the same information for closer approaches to Mars, within 50% and 10% of the Hill sphere respectively. Panel D shows the likelihood of a particle striking the face of the planet, the problem studied by Öpik.

3.2 Size

In addition to the bias with inclination, we also see a bias with particle size for the fraction of orbits which travel through the sphere of influence of Mars. Most particles $\geq$100μm travel through the Hill sphere of Mars, but as the the penetration level increases, a bias towards the largest particles entering the region begins to show (Fig. 1). The larger the particle, the more likely the probability to pass close to the planet, due to their longer PR drag lifetimes. Since the size range of 100–200 μm provides most of the cross-sectional area and mass at 1 AU (Grün et al. 1985), the strong dependence of this particular particle

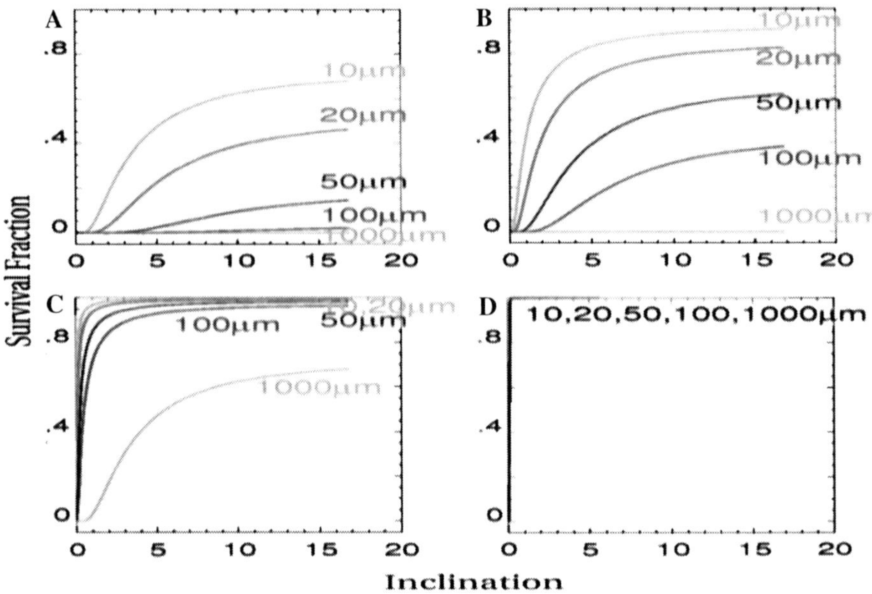

Fig. 1 Survival versus inclination for Hill Sphere (**a**), 50% (**b**), 10% (**c**), planet face (**d**)

size range on inclination could create a bias in the asteroid family particles contributing to the infrared emission of the zodiacal cloud.

3.3 Eccentricity

We also investigated the dependency of the interaction with Mars on dust particle orbits with different eccentricities. We find that the percentage of interaction is highly dependent on size with the largest particles and highest eccentricities most likely to enter the Hill sphere. At large eccentricities, though, the likelihood of interaction for all particle sizes converges, meaning that dust on highly eccentric orbits would not show the size bias in survival rates that lower eccentricity orbits have. This analysis may not extend to very low eccentricities near circular.

4 Summary

We conclude here that Mars may have a significant effect on dust particle orbits. Because the particles don't penetrate very deeply into the Martian Hill sphere, the result of the gravitational interactions will likely be that of modest orbital changes rather than removal through ejection from the system or accretion (survival fraction ~ 1 for the planet face). The effects of the perturbations will increase will passing distance; thus those particles penetrating deepest into the Hill sphere will be expected to have the largest perturbations to their orbits. Furthermore, the dependence on inclination will cause a bias in the effects on dust particles from different asteroid families and the dependence on size will effect how the size-frequency distribution of asteroidal particles changes with heliocentric distance. In the near future we will test these theoretical results against direct numerical simulations. In order to describe the complicated geometry of orbital interactions with solvable equations, we make some assumptions about the interaction (for example, as to what level certain orbital crossings can be considered linear) that need to be tested. The theoretical results presented here inform us of the likelihood of an interaction but not the outcome. Numerical simulations will reveal the extent of the effects of the gravitational interactions with Mars on the dust particle orbits. In addition, they will allow us to test what percentage of these particles get trapped into mean-motion resonances with Mars and how that might affect the overall picture of gravitational scattering by Mars. Inclusion of these effects into our dynamical modeling will be another step towards a complete dynamical model of the zodiacal cloud.

References

S.F. Dermott, S. Jayaraman, Y.L. Xu, B.A.S. Gustafson, J.C. Liou, A circumsolar ring of asteroidal dust in resonant lock with the Earth. Nature **369**, 719–723 (1994)

E. Grün, H.S. Zook, H. Fechtig, R.H. Giese, Collisional balance of the meteoritic complex. Icarus **62**, 224–272 (1985)

T.J.J. Kehoe, S.F. Dermott, L.M. Mahoney-Hopping, The effect of inter-particle collisions on the dynamical evolution of asteroidal dust and the structure of the zodiacal cloud.,Dust and Planetary Systems, ESA SP-243, 81–85. ESA Publications Division, Noordwijk (2007)

C.D. Murray, S.F. Dermott, Solar System Dynamics, 116–117. Cambridge University Press, UK (1999)

E.J. Öpik, Collision probabilities with the planets and the distribution of interplanetary matter. Proc. Royal Irish Acad. **54**, 165–199 (1951)

Determination of the Velocity of Meteors Based on Sinodial Modulation and Frequency Analysis

Felix Bettonvil

Originally published in the journal Earth, Moon, and Planets, Volume 102, Nos 1–4.
DOI: 10.1007/s11038-007-9166-2

Abstract In meteor photography the velocity of meteors is generally obtained from a chopper which blocks periodically the incident light beam in front of the camera lens. In this paper I examine modulation of the meteor trail instead with a sinodial function and use frequency analysis to compute accurately the mean atmospheric velocity.

Keywords Photography · Velocity · Orbital elements · High-precision · All-sky · Fireball patrol · Camera · Deceleration · Shutter · Fisheye lens

1 Introduction

In 2006 I described a digital All-sky camera for fireball patrol work based on a Nikon Coolpix 4500 camera with a FC-E8 fisheye converter lens (Bettonvil 2006a). The idea of using a digital camera for fireball work was based on the computation of a fireball trajectory from scanned photographic film with only medium resolution (2722 × 2338 pixels, approx. 5′/pixel) which gave acceptable orbits (Bettonvil 2006b). The medium resolution affected in particular the accuracy of the velocity.

Commonly rotating shutters are used to measure the velocity, as already demonstrated by Elkin more then a century ago (Millman 1936). They consist of a wheel, rotating in front of the lens with a constant speed and having a number of segments which block the incident light beam periodically, imaging the meteor trail on the camera as a dashed line. The edges of the breaks form the measurement points. From the distances between them and the known number of revolutions per second of the shutter, the semi-instantaneous and mean angular velocity can be calculated as well as the atmospheric velocity expressed in km/s in the case the meteor trail has been captured from two different observing locations.

F. Bettonvil (✉)
Astronomical Institute, Utrecht University, Princetonplein 5, 3584 CC Utrecht, The Netherlands
e-mail: F.C.M.Bettonvil@astro.uu.nl

F. Bettonvil
Netherlands Foundation for Research in Astronomy (ASTRON), Dwingeloo, The Netherlands

DOI: 10.1007/978-0-387-78419-9_29

The 4-Megapixel camera has an image scale of 6′/pixel. If we assume a typical angular velocity for the meteor of 10°/s and a shutter speed of 10 breaks/s, one period measures only 10 pixels; 5 for the blocked and 5 for the unblocked part, which makes accurate measurement of the edges rather difficult. The uncertainty which is typically obtained for the mean atmospheric velocity is in the order of a few percent.

2 Sinodial Modulation

In Bettonvil (2007) I described the idea of modulation of the meteor trail with a triangular or sinodial function for measurement of the mean velocity. Fourier analysis techniques can reveal the dominant modulation frequency and thus the velocity. The advantage is that all pixels contribute in the measurement and not only the pixels locating the shutter breaks. Persistent trains, slow variations in brightness and/or flares give only minor problems because of their different frequency regime. The method leads as well to better results in the case of fore-shortened trails because the merging of breaks does not harm much.

3 Simulations

Simulations have been done to study the method in more detail. Artificial meteors with trail lengths of 100 pixels (which are typical for the described camera) were created including variations in brightness, background signal, saturation and noise. They were modulated with a sinodial function and then 8-bit digitized. With a Fast Fourier transform the frequency spectrum was computed with wavelengths from 2 to 100 pixels. The strongest peak, representing the modulation, was easily detected. A value for this frequency was obtained by fitting a Gaussian function through the peak (Fig. 1) which has some width due to the fact that the data sample is not infinite and sample and shutter frequency generally not exactly match, but which has no major influence on the frequency. The obtained values differed 0.1–0.4% from the real modulation frequency. An estimation of the uncertainty was computed from the differences of the Gauss fit with the data points in the frequency spectrum. It gave values from 0.2 until 0.5%, hence it was concluded that it is a good approximation for the error.

A further test was done with a real video meteor which showed breaks with a sinodial-modulated appearance and which had similar path length (Fig. 2) and which was approx.

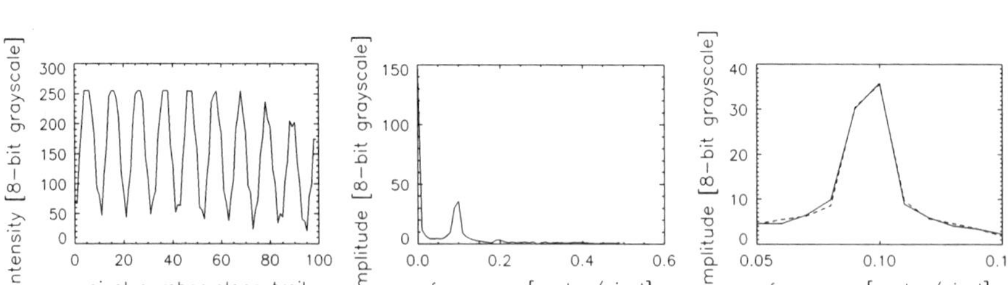

Fig. 1 A simulation of velocity determination of a meteor with sinodial modulation and FFT analysis. An artificial meteor trail is modulated with a sinodial function (left). Then the frequency spectrum is computed (center) which reveals clearly a peak around the modulation frequency. A Gaussian fit through the peak (right graph, dashed contour) is used to locate the peak frequency, which is in this case 0.096 cycles/pxl

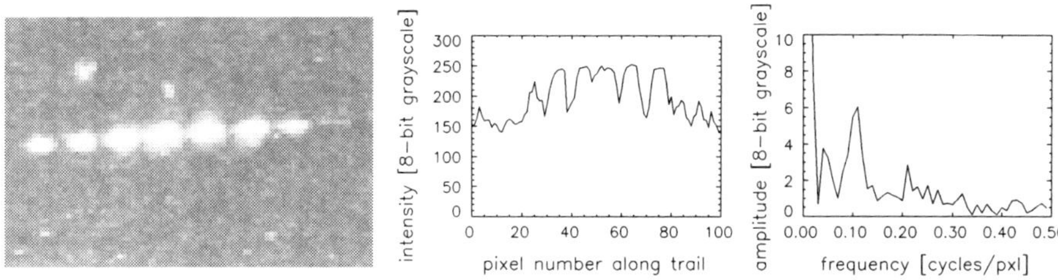

Fig. 2 Example of velocity determination with FFT analysis using real data: (left) image of a meteor trail taken with a video camera; (center) variation of the intensity along the trail as a function of pixel number; (right) frequency spectrum. The dominant frequency is 0.103 cycles/pxl, the computed uncertainty 0.5%

perpendicular to the observer, hence with negligible foreshortening. The obtained uncertainty was of the same order.

As a final test, the trail was processed again but with the modulation information extracted in the conventional way by measuring the position of the begin point of every break and the frequency computed from their relative distances (including a weight factor which gave two neighboring points a lower weight than two points further apart). The obtained accuracy was 1.7%; the difference with the FFT method 2%.

4 Test Setup

Two prototypes of sinodial choppers have been made based on the idea described in Bettonvil (2007). The first is a conventional rotating shutter but consisting of many small blades with different size, sampling a sinodial function (Fig. 3, left). The second prototype modulates the light with two linear polarizers. One of them is fixed; the other rotating with a constant speed (Fig. 3, right). When the axes of linear polarization of the two polarizers are parallel, light is transmitted; when they are perpendicular, the light is blocked. The advantage of polarizers is that they give a continuous modulation, disadvantage that the total transmission is lower because of the use of only one linear polarization state. Moreover, meteors should not show polarization effects themselves.

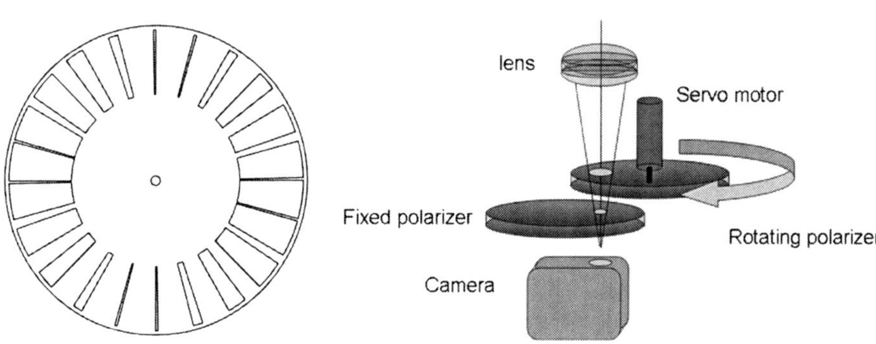

Fig. 3 Two alternatives for sinodial choppers: (left) mechanical shutter which is split up in 24 segments sampling a sinodial function; (right) use of two linear polarizers. One is fixed, the other rotates with a constant speed

5 Discussion and Conclusions

It has been shown that sinodial modulation and frequency analysis are a good alternative for the common method of velocity determination and that it leads to a higher precision. It was found that fast modulations (30 cycles/s, the highest used in the simulations) gave the best accuracies.

The presented FFT analysis does, strictly spoken, not account for deceleration. Deceleration broadens the modulation peak in the frequency spectrum from which the mean velocity is computed. For fast meteors this is acceptable when calculating orbital elements (Betlem et al. 1999); for others and especially fireballs not. There we have to restrict ourselves to the first half of the trajectory which shows minimal deceleration in order to find more realistic pre-atmospheric velocities (Bettonvil 2006b). The FFT analysis however can be extended with Windowed Fourier Transformations and Wavelet analysis techniques which are capable of disentangling the deceleration. A quantitative study in terms of accuracy is foreseen.

Although invented for all-sky work, the presented method is naturally of advantage for cameras with (much larger) image scales too. With 10× longer focal lengths it would be realistic to obtain errors below 0.1% which is comparable with the best of other high precision orbit determination work (Betlem et al. 1999; Kohoutek 1959).

References

F. Bettonvil, Digital all-sky cameras II: a new method for velocity determination, in *Proceedings of the International Meteor Conference 2006*, ed. by F. Bettonvil, J. Kac (International Meteor Organization, Hove, 2007) in print

F. Bettonvil, A digital all-sky camera, in *Proceedings of the International Meteor Conference 2005*, ed. by L. Bastiaens, J. Verbert, J. Wislez, C. Verbeeck (International Meteor Organization, Hove, 2006a) pp. 90–98

F. Bettonvil, Orbit calculation of the August 15, 2002 fireball over the Netherlands, in *Proceedings of the International Meteor Conference 2005*, ed. by L. Bastiaens, J. Verbert, J. Wislez, C. Verbeeck (International Meteor Organization, Hove, 2006b) pp. 171–178

H. Betlem, J. Jenniskens, J. Van 't Leven, C. Ter Kuile, C. Johannink, H. Zhao, C. Lei, G. Li, J. Zhu, S. Evans, P. Spurný, Very precise orbits of 1998 Leonid meteors. Meteorit. Planet. Sci. **34**, 979–986 (1999)

L. Kohoutek, On the precision of the photographical determination of the geocentric meteor velocity. Bull. Astron. Inst. Czech. **10**, 120–134 (1959)

P. Millman, The importance of meteor photographs taken with a rotating shutter. J. R. Astron. Soc. Can. **30**, 101–103 (1936)

Chapter 2. Observation Techniques and Programs

The Canadian Meteor Orbit Radar Meteor Stream Catalogue

Peter Brown · Robert J. Weryk · Daniel K. Wong · James Jones

Originally published in the journal Earth, Moon, and Planets, Volume 102, Nos 1–4.
DOI: 10.1007/s11038-007-9162-6 © Springer Science+Business Media B.V. 2007

Abstract The Canadian Meteor Orbit Radar is a multi-frequency backscatter radar which has been in routine operation since 1999, with an orbit measurement capability since 2002. In total, CMOR has measured over 2 million orbits of meteoroids with masses greater than 10 μg, while recording more than 18 million meteor echoes in total. We have applied a two stage comparative technique for identifying meteor streams in this dataset by making use of clustering in radiants and velocities without employing orbital element comparisons directly. From the large dataset of single station echoes, combined radiant activity maps have been constructed by binning and then stacking each years data per degree of solar longitude. Using the single-station mapping technique described in Jones and Jones (Mon Not R Astron Soc 367:1050–1056, 2006) we have identified probable streams from these single station observations. Additionally, using individual radiant and velocity data from the multi-station velocity determination routines, we have utilized a wavelet search algorithm in radiant and velocity space to construct a list of probable streams. These two lists were then compared and only streams detected by both techniques, on multiple frequencies and in multiple years were assigned stream status. From this analysis we have identified 45 annual minor and major streams with high reliability.

Keywords Meteor streams · Radar · Meteoroids

1 Introduction

The recognition of meteor streams as debris from cometary activity stands as one of the seminal discoveries in meteor astronomy over the last two centuries (Burke 1986). Meteor streams trace the present and past activity of their parent bodies and therefore form an important link to understanding the dynamical and physical evolution of comets. Additionally, recognizing associations between streams and asteroids (the Geminids and 3200

P. Brown (✉) · R. J. Weryk · D. K. Wong · J. Jones
Department of Physics and Astronomy, University of Western Ontario, London, ON, Canada N6A 3K7
e-mail: pbrown@uwo.ca

Phaethon, the Quadrantids and 2003 EH1 for example) may provide insight into transition objects which straddle the comet—asteroid boundary.

The starting point for all stream studies is separating stream meteoroids, (which usually have a single common parentage), from the sporadic background; this implies associating individual meteoroid orbits with other meteoroid orbits. Historically this has been accomplished through use of a dissimilarity criterion employing orbital elements and/or radiant/velocity measurements of individual meteoroids (cf. Valsecchi et al. 1999; Jopek et al. (1999) and references therein). These various criteria (and associated critical limit) have been applied to a few distinct datasets where individual meteoroid orbits have been measured (such as the Harvard Super-Schmidt photographic data) in an effort to identify probable streams.

It is clear that such approaches work well in identifying the major streams; almost all the various dissimilarity criteria "extract" the dozen or so most active streams, though the particulars of period of activity, radiant location/drift may vary somewhat. Where significant differences arise, however, is in minor stream identification. Here earlier datasets might have only a few orbits and errors in these orbits make separation from the sporadic background problematic.

The solution to this problem is to perform searches on meteoroid orbit datasets which are large enough that the many minor streams are readily identifiable purely on statistical grounds. A recent attempt along these lines by Galligan and Baggaley (2002) identified half a dozen streams in the Advanced Meteor Orbit Radar (AMOR) collection of 0.5 million orbits. The small number of positive stream detections in the AMOR dataset is undoubtedly a function of the small particle sizes observed by AMOR ($\sim$40 μm), the sporadic background vastly outnumbering stream meteoroids at such small meteoroid sizes (cf. Jenniskens 2006). The technique and methodology used by Galligan and Baggaley (2002) forms the basis of the present work which applies a related technique to identification of meteor streams in the radar data collected by the Canadian Meteor Orbit Radar (CMOR).

2 Equipment and Data Collection Techniques

The CMOR (43.264 N, 80.772 W) is a triple frequency (17.45, 29.85 and 38.15 MHz) backscatter system with all three systems operating simultaneously at a pulse repetition frequency of 532 Hz and 6 kW peak transmit power. Each radar system has a receiving interferometry array which permits measurement of echo direction. The vertically directed transmit and receive antennae have a combined beam pattern which is nearly all-sky; the effective beamwidth to the 3 dB points is located 45° from the zenith. The 29.85 MHz unit also has two outlying stations connected to the main site via UHF datalinks. These outlying stations record reflections from points along some meteor trails distant from the main site specular reflection point. Combining the relative timing of the detections from these two remote sites relative to the main site and the interferometric information permits a complete reconstruction of the meteor velocity vector. Additional details of the system and the data collection architecture are described in more detail in Jones et al. (2005) and Webster et al. (2004).

The radar is effectively sensitive to meteors with apparent radio magnitudes (cf. McKinley 1961) $\sim$+8. Figure 1 shows the distribution of apparent magnitudes and masses (as determined from the mass-magnitude-velocity relation of Verniani (1973)) for all meteoroids with determined orbits. The mean magnitude for echoes with measured

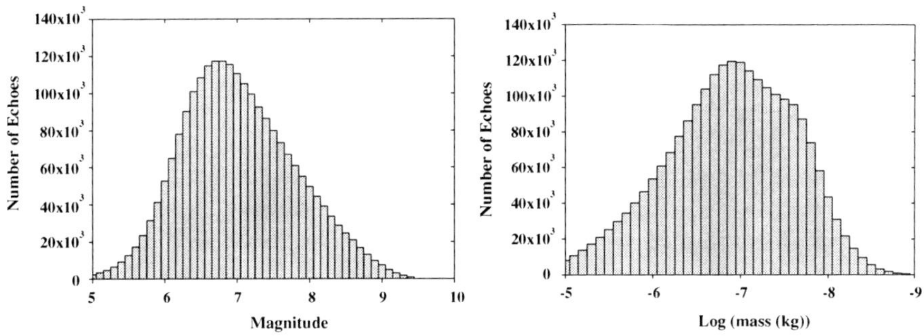

Fig. 1 Magnitude distribution (left) and mass distribution (right) for all CMOR echoes having determined orbits

orbits is +7 and the mean mass is 1.3×10^{-7} kg; this is a lower limit as in the radar analysis it is assumed that the radar reflection specular point for any particular meteor ionization train coincides with the point of maximum ionization.

3 Errors and Biases

The detection of stream structure in CMOR data requires measurement of an interferometric location (altitude and azimuth of reflected RF pulse) for a given echo and an echo time. This information, together with the station location is sufficient to perform single station radiant mapping in equatorial coordinates (cf. Jones and Jones 2006). In addition to these data, two time-of-flight measurements from the outlying station on 29.85 MHz are needed to uniquely define a velocity vector for individual meteor echoes.

Errors in time of echo occurrence are essentially negligible, having precisions of order ms and accuracies <1 s respectively. Interferometric errors are less than 1° for echoes with elevations above 20° (Jones et al. 1998). Since the interferometric error increases rapidly below this elevation, only echoes with elevations greater than 20° are used. Specifically, a mean error value in the interferometry <0.3° was determined by Jones et al (1998) through simulations using signal-to-noise ratios above 20 dB. From an analysis of single station radiants, Jones and Jones (2006) concluded that the effective spread in radiant positions due to interferometry errors was ~1.2°. Finally, a direct comparison between simultaneously detected electro-optical meteors and radar echoes determined a mean error <0.2° for echoes with elevations above 50° (Weryk and Brown 2007). The random errors in directionality are the main source of uncertainty; comparison between directions determined between 29 and 38 MHz yield a mean and median systematic difference in echo directions of <0.1°. This is consistent with the observation that the receiver phases on all three systems drift by no more than 0.7° in a given day based on twice hourly automatic measurements over many years, with the mean drift <0.3° throughout much of the year. This is a consequence of the temperature compensation employed for the radar systems, which ensures the hardware experience diurnal temperature variations <2°C throughout most of the year.

For individual orbit measurements, the primary source of error in velocity and radiant measurement is in the estimation of the difference in time of occurrence between the remote sites and the main site. Typically this amounts to 1–2 pulses (1.9–3.8 ms).

However, slower meteoroids have much shallower rise times often resulting in larger absolute time errors (cf. Jones et al. 2005 for details of the time pick algorithm). This is partially compensated by the lower velocity (and hence longer time delays between the stations)—the opposite situation exists for faster meteoroids such that the relative error remains approximately constant as a function of velocity. The resulting errors amount to 5–10% in velocity and 1–2° in radiant direction. These error margins have been independently validated through comparison between simultaneously observed optical and radar events (Weryk and Brown 2007) which show even smaller differences, but apply on average to higher SNR echoes. The dependence of the actual error with SNR is still to be investigated with these simultaneous measurements.

Finally, we note that showers with high velocity meteoroids or with steep mass distribution indices (i.e. a preponderance of small meteoroids) will be more difficult to detect due to the effects of the echo height ceiling. Similarly, very slow streams will be selected against due to the high velocity dependence of ionization (cf. Ceplecha et al. 1998). Our all-sky coverage tends to limit the biases introduced by the beam pattern to purely declination effects. Radiants at high declinations have larger daily integrated collecting areas than those at low declinations (cf. Campbell-Brown and Jones 2006). For all the stream detections which follow we have not corrected for this declination sensitivity—in a future paper concerned solely with stream activity this collecting area bias will be examined in detail.

The single largest remaining systematic uncertainty in our measurements is V_∞, the velocity at the top of the atmosphere. Our measurement of in-atmosphere velocity is subject to the effects of deceleration. For a particular echo this will be a function of the height at which our velocity measurement is made. Brown et al. (2005) have presented an empirical method of correcting CMOR data for height-deceleration effects as a function of velocity, but this relies on use of previous video/photographic estimates of stream V_∞ as a means of calibration. Thus our velocity estimates for the 11 streams used in the calibration are not strictly independent measurements. The streams used for calibration are listed in the summary stream table (Table 1).

4 Stream Detection Methodology and Analysis

Detection of streams in CMOR data was performed as a two stage process. Each quasi-independent stage produced a list of possible radiants and a final master list was constructed from only streams which were detectable with both techniques, generally having comparable radiant locations and radiant drifts as determined by the two methods providing sufficient numbers of echoes were available for each technique.

In the first stage, all single station echoes on all three frequencies were combined into individual solar longitude bins from all years of observations. In total, 26.3, 18.3 and 11.0 million echoes were recorded on 17, 29 and 38 MHz systems respectively. This produces an equivalent single solar year, the assumption being that the streams of interest are active from the same radiant positions each year. Each solar longitude bin is then processed using the convolution technique described in Jones and Jones (2006). The result for each degree of solar longitude is a map of the relative radiant activity in equatorial coordinates. The relative radiant activity (or relative strength) quantifies local enhancement in the number density of echo radiants over size scales of 1–2° following the procedure outlined in Jones and Jones (2006). By examining several strong, known streams near the tails of their activity period, we determined a cutoff value in the scaled radiant activity

Table 1 Stream list. Single station values and WC (wavelet coefficient analysis) are shown, together with geocentric radiant coordinates at the time of WC maximum. Streams highlighted in bold have been used as calibrations to compute a mean deceleration correction. Only showers having W_c in at least a single degree bin in solar longitude more than 10σ above background are included in this compilation

Stream name	Single station analysis								Wavelet analysis											
	λ_{max}	λ_{start}	λ_{end}	Relative activity	$\Delta\alpha$	$\pm\Delta\alpha$	$\Delta\delta$	$\pm\Delta\delta$	λ_{max}	λ_{start}	λ_{end}	$W_{c\ max}$	$\Delta\alpha$	$\pm\Delta\alpha$	$\Delta\delta$	$\pm\Delta\delta$	αmax (J2000)	δmax (J2000)	Vg (km/s)	σVg (km/s)
April Lyrids	**32.5**	**30**	**34**	**36**	**0.78**	**0.16**	**−0.42**	**0.21**	**32.5**	**31**	**33**	**26.8**	**1.50**	**0.06**	**−0.30**	**0.06**	**272.3**	**32.6**	**47.3**	**4.1**
Daytime April Piscids	32.5	30	36	26	0.90	0.08	0.37	0.05	24.5	16	33	18.1	0.90	0.04	0.39	0.03	3.8	5.5	28.9	3.4
Eta Aquariids	**44.5**	**33**	**64**	**277**	**0.69**	**0.01**	**0.33**	**0.01**	**45.5**	**35**	**59**	**285.2**	**0.69**	**0.01**	**0.33**	**0.00**	**338.0**	**−0.7**	**64.6**	**6.2**
Southern Daytime May Arietids	47.5	30	61	30	0.97	0.03	0.30	0.02	46.5	23	63	23.6	0.92	0.01	0.32	0.01	28.4	7.7	28.3	3.3
Northern Daytime omega Cetids	47.5	30	59	29	0.97	0.03	0.31	0.01	45.5	16	58	38.4	0.95	0.01	0.36	0.01	9.0	17.3	36.8	4.1
Southern Daytime omega Cetids	49.5	34	59	32	0.90	0.03	0.43	0.02	45.5	18	62	42.5	0.93	0.01	0.44	0.01	20.5	−6.1	36.9	3.9
Daytime Arietids	**76.5**	**65**	**93**	**255**	**0.63**	**0.02**	**0.19**	**0.01**	**74.5**	**64**	**88**	**169.8**	**0.60**	**0.02**	**0.19**	**0.01**	**41.7**	**23.6**	**39.1**	**4.2**
Daytime Zeta Perseids	83.5	47	95	41	1.01	0.02	0.25	0.01	74.5	58	88	32.0	1.00	0.02	0.20	0.01	57.4	23.4	26.4	3.9
Southern June Aquilids	80.5	78	81	28	−0.25	0.34	0.46	0.16	80.5	78	82	31.1	0.02	0.23	0.30	0.12	304.7	−32.8	38.6	3.4
Daytime Gamma Taurids	81.5	77	93	32	0.82	0.11	0.34	0.04	85.5	70	98	19.2	0.82	0.02	0.27	0.01	56.7	11.5	36.4	3.7
Daytime Epsilon Perseids	95.5	92	107	24	0.77	0.08	0.42	0.06	95.5	92	107	13.0	0.78	0.05	0.15	0.04	58.2	37.9	44.8	4.4
Daytime Beta Taurids	96.5	82	103	45	0.88	0.07	0.05	0.01	93.5	90	100	25.8	0.89	0.08	0.04	0.05	82.0	20.0	27.4	3.1
Vulpeculids	106.5	104	109	22	0.60	0.21	−0.36	0.26	106	104	106	17.2	1.50	0.40	0.45	0.38	326.3	14.7	29.9	3.2
Northern June Aquilids	108.5	85	114	29	0.81	0.02	0.20	0.02	102	77	117	44.8	0.83	0.01	0.16	0.01	310.4	−4.9	38.4	3.9

Table 1 continued

Stream name	Single station analysis								Wavelet analysis											
	λ_{max}	λ_{start}	λ_{end}	Relative activity	$\Delta\alpha$	$\pm\Delta\alpha$	$\Delta\delta$	$\pm\Delta\delta$	λ_{max}	λ_{start}	λ_{end}	$W_{c\ max}$	$\Delta\alpha$	$\pm\Delta\alpha$	$\Delta\delta$	$\pm\Delta\delta$	αmax (J2000)	δmax (J2000)	Vg (km/s)	σVg (km/s)
Beta Equuleids	108.5	104	113	20	0.84	0.18	–0.45	0.19	107	104	112	16.4	0.69	0.09	–0.28	0.14	321.5	8.7	31.6	3.1
July Sigma Cassiopeiids	110.5	100	113	25	0.95	0.24	0.37	0.11	105.5	102	110	13.9	0.70	0.19	0.37	0.07	343	49.6	38.9	3.3
Psi Cassiopeiids	116.5	112	120	27	0.62	0.26	0.68	0.18	118	110	124	22.1	1.22	0.15	0.43	0.06	11.9	65.4	44	4.6
Alpha Capricornids	**124.5**	**102**	**130**	**36**	**0.69**	**0.02**	**0.24**	**0.01**	**124**	**116**	**128**	**19.9**	**0.66**	**0.02**	**0.28**	**0.02**	**302.9**	**–9.9**	**22.2**	**2.3**
Southern Delta Aquariids	**125.5**	**115**	**155**	**356**	**0.77**	**0.01**	**0.28**	**0.01**	**127**	**115**	**145**	**342.4**	**0.78**	**0.01**	**0.25**	**0.01**	**341.0**	**–16.1**	**41.1**	**3.8**
Piscis Austrinids	125.5	121	138	26	0.53	0.10	0.17	0.06	126.5	125	131	11.0	0.89	0.09	0.16	0.09	347.9	–23.7	44.1	3.7
Southern Iota Aquariids	129.5	125	149	22	0.93	0.05	0.34	0.02	129.5	128	133	11.5	0.36	0.12	–0.14	0.07	332.9	–14.7	30.5	3.1
Daytime Xi Orionids	131.5	116	138	19	0.81	0.06	–0.09	0.03	131.5	131	133	6.5	0.90	NA	–0.40	NA	102.9	16.6	45.4	4.2
Northern Delta Aquariids	137.5	131	159	22	0.70	0.03	0.33	0.01	139	128	155	19.7	0.75	0.03	0.28	0.02	344.9	2.2	37.7	4.3
Perseids	139.5	123	142	103	1.35	0.07	0.23	0.02	140	134	142	74.5	1.23	0.09	0.27	0.07	46.9	56.9	62.1	7.2
Northern Iota Aquariids	167.5	113	182	33	0.84	0.01	0.33	0.01	160	145	176	18.8	0.80	0.02	0.33	0.02	356.0	3.0	28.6	3.6
Daytime Kappa Leonids	178.5	164	189	40	0.63	0.03	–0.31	0.02	183	171	193	21.7	0.55	0.02	–0.26	0.02	161.5	15.4	43.3	4.5
Daytime Sextantids	**187.5**	**180**	**193**	**83**	**0.69**	**0.02**	**–0.58**	**0.02**	**187**	**174**	**194**	**61.9**	**0.70**	**0.03**	**–0.51**	**0.01**	**154.6**	**–1.4**	**31.84**	**3.3**
Southern Taurids	**191.5**	**166**	**236**	**56**	**0.82**	**0.01**	**0.25**	**0.01**	**197**	**172**	**218**	**49.9**	**0.82**	**0.01**	**0.29**	**0.01**	**31.0**	**8.0**	**27.92**	**3.7**
October Draconids	195.5	195	195	20	0.00	0.00	0.00	0.00	196	195	195	24.4	0.00	0.00	0.00	0.00	261.7	54.8	19.7	2.3

Table 1 continued

Stream name	Single station analysis								Wavelet analysis											
	λ_{max}	λ_{start}	λ_{end}	Relative activity	$\Delta\alpha$	$\pm\Delta\alpha$	$\Delta\delta$	$\pm\Delta\delta$	λ_{max}	λ_{start}	λ_{end}	$W_{c\ max}$	$\Delta\alpha$	$\pm\Delta\alpha$	$\Delta\delta$	$\pm\Delta\delta$	αmax (J2000)	δmax (J2000)	Vg (km/s)	σVg (km/s)
Orionids	**209.5**	**196**	**222**	**132**	**0.84**	**0.02**	**0.03**	**0.01**	**208**	**198**	**221**	**96.3**	**0.78**	**0.01**	**0.03**	**0.01**	**94.7**	**15.5**	**66.4**	**6.3**
Northern Taurids	223.5	201	236	31	0.90	0.02	0.24	0.01	225	207	235	21.9	0.88	0.01	0.19	0.02	53.3	21.0	28.1	2.9
Leonids	**236.5**	**230**	**238**	**82**	**0.51**	**0.15**	**–0.48**	**0.14**	**237**	**228**	**238**	**29.2**	**0.63**	**0.08**	**–0.27**	**0.06**	**155.0**	**21.6**	**69**	**6.8**
November omega Orionids	247.5	226	256	78	0.74	0.01	–0.02	0.01	246	230	253	63.9	0.74	0.02	–0.06	0.01	90.2	15.5	43.5	3.9
Geminids	**260.5**	**243**	**269**	**817**	**1.12**	**0.01**	**–0.16**	**0.01**	**262**	**244**	**267**	**568.0**	**1.10**	**0.02**	**0.17**	**0.02**	**112.8**	**32.1**	**35**	**3.8**
December Monocerotids	262.5	253	266	21	0.53	0.05	–0.05	0.05	262	252	264	21.3	0.63	0.03	–0.11	0.05	102.6	8.1	41.5	3.7
Ursids	270.5	270	270	29	0.00	0.00	0.00	0.00	271	270	270	39.5	0.00	0.00	0.00	0.00	222.0	74.6	37.6	5.1
Sigma Serpentids	275.5	271	282	18	0.74	0.08	–0.16	0.11	276	261	279	26.0	0.75	0.02	–0.14	0.03	242.8	–0.1	42.67	4
January Leonids	280.5	278	284	42	0.99	0.10	–0.36	0.03	283	280	284	37.6	0.66	0.11	–0.14	0.05	148.3	23.9	52.7	4.4
Omega Serpentids	280.5	270	281	25	0.60	0.06	0.08	0.10	275.5	271	279	16.5	0.76	0.05	0.11	0.18	242.7	0.5	38.9	3.4
Quadrantids	**283.5**	**281**	**286**	**238**	**–0.01**	**0.32**	**0.38**	**0.32**	**284**	**279**	**285**	**237.5**	**0.72**	**0.05**	**–0.55**	**0.23**	**231.7**	**48.5**	**42**	**4**
Alpha Cetids	290.5	280	294	17	0.69	0.05	–0.19	0.03	285.5	281	289	15.6	0.65	0.07	–0.17	0.06	127.6	–7.9	43.6	3.9
Theta Coronae Borealids	296.5	294	300	24	1.57	0.17	–0.91	0.11	297	293	303	63.2	0.70	0.16	–0.06	0.09	232.3	35.8	38.66	4.5
Lambda Bootids	295.5	283	299	27	1.36	0.13	–0.62	0.05	296	285	297	34.4	0.88	0.08	–0.69	0.03	219.6	43.2	41.75	4.2
Zeta Coronae Borealids	295.5	291	303	18	0.69	0.15	0.05	0.11	295	291	303	23.0	0.69	0.09	–0.11	0.08	244.8	31.1	44.25	4.3
Alpha Antiliids	313.5	308	321	24	0.91	0.05	–0.38	0.06	316	299	320	30.7	0.84	0.03	–0.36	0.03	162.7	–12.6	44.75	4.3

corresponding to the average sporadic background which could be reliably used to separate noise from stream signals. This cutoff (15) was then used throughout the single station analysis. Figure 2 shows the maximum scaled radiant activity for each solar longitude degree with streams identified.

For each degree of solar longitude, all radiants above the threshold value are recorded. Finally, a potential list of streams is computed taking local maxima found from this single station mapping and comparing it to solar longitude bins before and after each interval. If additional local radiant maxima are found within +2° in right ascension and declination per degree solar longitude from the original solar longitude bin the individual maxima are linked and recorded as a possible stream. Linkages lasting more than three consecutive degrees and/or having maxima above 20 lasting at least two bins were then automatically identified. From this list, potential streams showing consistent radiant drifts in α,δ were saved and then manually examined to further reduce false detections. Cross-comparisons were made among the three frequencies with the requirement that 29 and 38 MHz show consistent radiant positions and radiant drifts for a stream to be recognized. Data from 17 MHz suffers from terrestrial broadcast interference during the day and hence the radiant maps from 17 MHz were used only as a backup means of confirmation not a primary selection filter. This procedure rapidly identified most major streams previously recorded in other lists as well as some new streams/minor streams.

In the second detection stage, all individual geocentric radiants measured using the time-of-flight technique on 29.85 MHz were examined separately for potential clustering. In total 2.5 million orbits were available for this stage of the analysis. The geocentric radiant locations and geocentric velocity of the orbit of each meteor together with the time of occurrence were used to detect enhancements relative to the sporadic background.

The identification of streams in this second stage was done by performing a wavelet transform on the geocentric radiant data. These data were further partitioned by geocentric velocity in bins corresponding approximately to the upper limit of the average error values in velocity (10%). Our searches were conducted across these velocity bins with varying probe sizes, the probe sizes representing the angular scale size of the wavelet probe (the

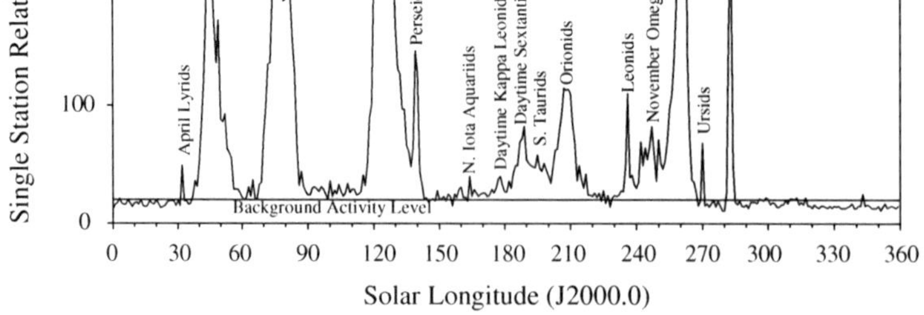

Fig. 2 The maximum in single station radiant activity for each degree of solar longitude throughout the year. Prominent streams (activity above background as defined in the text) are marked

clustering size scale the particular probe is most sensitive). The wavelet transform and its application to radar meteor data is described in Galligan and Baggaley (2002).

We proceed by computing wavelet coefficients in a grid made up of 0.5° increments in sun-centred ecliptic longitude (λ–λ_o) and latitude (β). We use discrete wavelet probe sizes of 1,2,3,4 and 8° and with geocentric velocity partitions from 14–75 km/s, spaced with bin intervals of 10% in speed. These combinations of wavelet coefficients are computed for each degree of solar longitude, with all meteoroid orbits from all years combined. Each degree of solar longitude has on average 12,000 individual radiant measurements.

For each wavelet grid location (λ-λ_o, β) within a given probe size and velocity combination, the median and standard deviation of that particular grid point is computed over all degrees of solar longitude. Points more than 3σ above the median are discarded and the process iteratively repeated until no more points forming the median at that grid location lie outside the 3σ bounds. Local maxima in wavelet coefficients are then identified as those points more than 3σ above the final median estimate and where the density of radiants is above a minimum threshold, which was empirically determined to be 20 radiant points per square degree.

These local maxima are then linked across individual solar longitude bins, provided the local maxima in adjacent solar longitude bins are separated by less than 2.5° in grid space or less than 3.5° if separated by two degrees of solar longitude. These potential streams are then further refined by requiring that at least one of the local maxima is $>10\sigma$ above the median background. Single radiant points visible for only one solar longitude bin were considered potential streams if the local maxima was 15σ above the median background.

With all these chains identified in solar longitude, velocity, probe size partitions, many duplicate potential streams appeared. Having identified the location in grid space where the potential radiants are located, the solar longitude bin of the stream maximum was then identified and wavelet coefficients computed in 1 km/s steps centred at the grid location maximum. It was found that the velocity bin where the maximum relative wavelet coefficient was determined was generally insensitive to medium-scale probe sizes (2,3,4°). The process of selecting a best-fit velocity and separation of the stream from the background in this manner is shown in Fig. 3a and b for the S. Delta Aquariid stream.

Once the best fit velocity bin was identified, the same procedure was applied in the solar longitude bin of the maximum (partitioned at the wavelet bin closest to the peak velocity) and different probe sizes from 0.1 to 20° at 0.1° intervals applied at the grid point maximum. The resulting curve identified the best fit probe size as the probe-size where the maximum wavelet coefficient was computed. This process is shown in Fig. 4 for the S. Delta Aquariid stream.

Once all potential stream chains were identified by this procedure, they were checked additionally for positive daily mean radiant drift in α and consistent drifts for all points in both α,δ.

5 Results and Conclusions

A total of 45 streams were identified using the above described search criteria. Table 1 lists the stream names, period of detectable activity based on our criteria (for both single station and wavelet transformed data), radiant positions and radiant drifts and best-fit geocentric velocity. The errors are 1 σ bounds for the velocity fits and for the linear regression in α,δ.

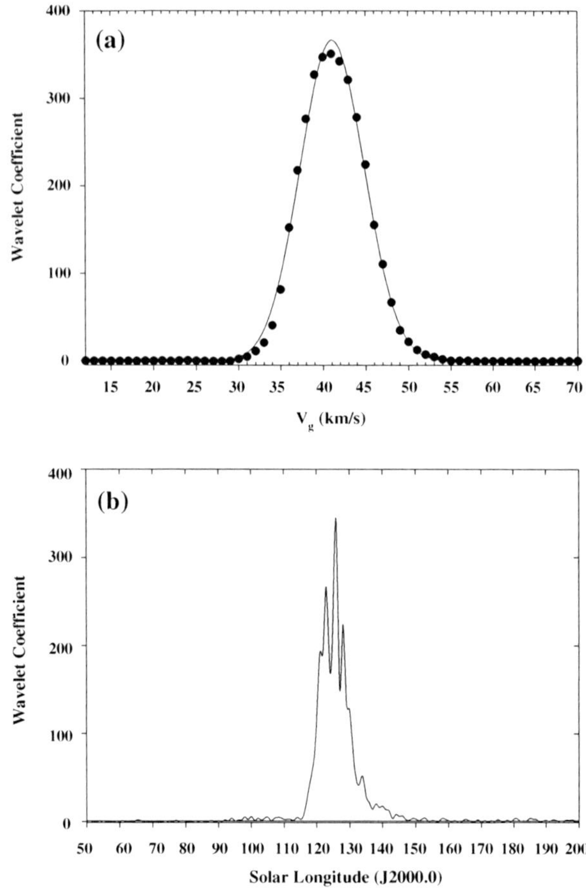

Fig. 3 (**a**) (top): Wavelet coefficients (WC) in 1 km/s velocity steps partitioned for a probe size of 3° centred at the radiant at the time of maximum for the S. δ Aquariid stream. The fit is Gaussian with a peak at 41.1 km/s and standard deviation of 3.8 km/s. (**b**) (below): Wavelet coefficients at a probe size of 3° for velocities between 39 and 45 km/s at the sun-centred grid coordinates of the maximum of the S. δ Aquariid stream computed for each degree of solar longitude. Here the 3σ detection threshold is at a WC of 5.3

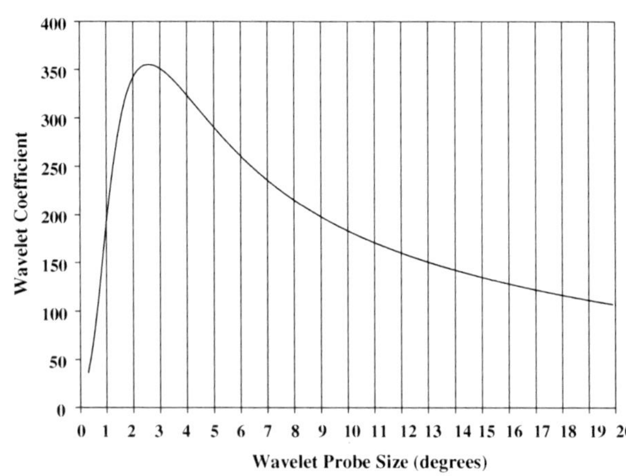

Fig. 4 Wavelet coefficients as a function of probe size for velocities between 39 and 45 km/s at the sun-centred grid coordinates of the maximum of the S. δ Aquariid stream on the day of maximum (λ_o = 126°)

Note that for several streams a higher order fit is clearly needed for the radiant drift in α,δ (such as the Taurids and S. Delta Aquariids). However, we apply linear fits for ease of comparison with other literature sources.

A good portion (73%) of our detected streams have been previously described in the literature. A total of 12 of our 45 detected streams are unreported in the literature or have weak associations to previously poorly characterized streams making associations difficult.

In comparing the present list to previous primary data source stream lists, it is apparent that many individual streams have previously been detected or classified as multiple streams, when in fact the present work suggests they have a single stream origin. This is the result of the intermittent periods of coverage for many earlier surveys (particularly radar surveys), which often required manual operation. We also note that the relatively small number of stream detections as compared to the large number recognized by the IAU working list suggests either many streams have shallow mass distribution indices and are not highly populated at the smaller masses detectable by the radar or that the streams do not exist.

A more complete description of the CMOR meteor stream working list, together with detailed orbital data (and orbital element variations) as well as a more detailed analysis of the various streams from our data will appear separately. We also intend to examine the flux and mass distribution indices for all streams reported here. Relaxation of the relatively strict stream selection criteria applied here, (particularly for wavelet analysis), results in significantly more (almost double) the number of identified streams. Together with the increasing number of CMOR orbits (now more than 3 million) we expect to significantly increase our signal-to-noise detection thresholds for minor streams in the future.

Acknowledgments PGB thanks the Natural Sciences and Engineering Research Council, the Canada Research Chairs program and the Meteoroid Environment Office of NASA for funding support. J. Baggaley and an anonymous reviewer provided helpful comments to the first version of this manuscript.

References

J.G. Burke, *Cosmic Debris: Meteorites in History*. (University of California Press, Berkeley, 1986)

P.G. Brown, J. Jones, R.J. Weryk, M.D. Campbell-Brown, Earth Moon Planets **95**, 617–626 (2005)

M.D. Campbell-Brown, J. Jones, Mon. Not. R. Astron. Soc. **367**, 709–716 (2006)

Z. Ceplecha, et al., Space Sci. Rev. **84**, 327–471 (1998)

D.P. Galligan, J. Baggaley, in *Dust in the Solar System and Other Planetary Systems, Proc. IAu Colloq. 181.*, eds. by S.F. Green, I.P. Williams, J.A.M. McDonnell, N. McBride (Pergamon, Oxford, 2002), pp. 42–47

P.J. Jenniskens, *Meteor Streams and Their Parent Comets*. (Cambridge University Press, 2006)

J. Jones, P. Brown, K.J. Ellis, A.R. Webster, M.D. Campbell-Brown, Z. Krzemenski, R.J. Weryk, Planet Space Sci **53**, 413–421 (2005)

J. Jones, W. Jones, Mon. Not. R. Astron. Soc. **367**, 1050–1056 (2006)

J. Jones, A.R. Webster, W.K. Hocking, Rad. Sci. **33**, 55–65 (1998)

T.J. Jopek, G.B. Valsecchi, Cl. Froeschle, Mon. Not. R. Astron. Soc. **304**, 751–758 (1999)

D.W.R. McKinley, *Meteor Science and Engineering*. (McGraw Hill, 1961)

G.B. Valsecchi, T.J. Jopek, Cl. Froeschle, Mon. Not. R. Astron. Soc. **304**, 743–750 (1999)

F. Verniani, J. Geophys. Res. **78**, 8429–8462 (1973)

A.R. Webster, P.G. Brown, J. Jones, K.J. Ellis, M. Campbell-Brown, Atmos. Chem. Phys. Discuss **4**, 1181–1201 (2004)

R.J. Weryk, P.G. Brown, *Comparisons of Simultaneously Detected Electro-optical and Radar Meteors, Meteoroids2007*. (Barcelona, 2007) this issue

Infrasonic Observations of Meteoroids: Preliminary Results from a Coordinated Optical-radar-infrasound Observing Campaign

Wayne N. Edwards · Peter G. Brown · Robert J. Weryk · Douglas O. ReVelle

Originally published in the journal Earth, Moon, and Planets, Volume 102, Nos 1–4.
DOI: 10.1007/s11038-007-9154-6 © Springer Science+Business Media B.V. 2007

Abstract Recent observations using the newly installed Elginfield infrasound array in coordination with the Southern Ontario all-sky meteor camera network and Canadian Meteor Orbit Radar (CMOR) has shown that the number of meteors producing infrasound at the Earth's surface is more frequent than previously thought. These data show the flux of meteoroids capable of producing infrasound at the ground is at least 1/month and is limited to meteors with peak visual brightness above –2. Comparisons to current meteor infrasound theory show excellent agreement with amplitude and period predictions for weakly non-linear shock waves using a realistic vertically inhomogeneous atmosphere. Similar predictions show isothermal assumptions underestimate the amplitude by orders of magnitude.

Keywords Atmosphere · Infrasound · Meteor · Meteoroid · Shock waves

1 Introduction

Observations of infrasonic sound from large >1 m diameter meteoroids are well documented and are becoming more common with the inception of the International Monitoring System's (IMS) global network of infrasound stations. This coincides with a general revival of infrasound as a monitoring tool (e.g. ReVelle 1997; Brown et al. 2002;

W. N. Edwards (✉)
Department of Earth Sciences, University of Western Ontario, 1151 Richmond Street, London, ON, Canada N6A 5B7
e-mail: wedwards@uwo.ca

P. G. Brown · R. J. Weryk
Department of Physics and Astronomy, University of Western Ontario, 1151 Richmond Street, London, ON, Canada N6A 3K7

D. O. ReVelle
Atmospheric, Climate and Environmental Dynamics, Meteorological Modeling Team, Los Alamos National Laboratory, MS D401, P.O. Box 1663, Los Alamos, NM 87545, USA

Klekociuk et al. 2005; Edwards et al. 2006). Yet similar infrasonic observations of smaller (<10 cm diameter) meteoroids remain rare: only a handful of these are fully documented (Kraemer 1977; Brown et al. 2007). Other observations of infrasound from small meteoroids tend to be coincident signals that are detected after a bright meteor event but lack trajectory information—this limits confidence in these associations (McIntosh et al. 1976; ReVelle and Whitaker 1999; Le Pichon 2002). With the sparseness of this dataset the result is that fundamental meteor infrasound theory (ReVelle 1974, 1976) and its predictions have remained generally untested and unconstrained by observation for more than 30 years.

To address this need, an ongoing coordinated campaign is underway using the Southern Ontario Meteor Network's (SOMN) all-sky camera system, the Canadian Meteor Orbit Radar (CMOR) and the newly installed Elginfield infrasonic array (ELFO) to monitor meteor generated infrasound from common regional meteor events. Such events refer to modestly bright fireballs (–2 and brighter) and have the advantage that they are at close range to the observing station such that the acoustic signal is not significantly attenuated or dispersed. Typical ground-projected distances are within ~200 km (depending on meteor altitude) or before the acoustic/infrasonic sound refracts back up into the atmosphere due to the tropospheric temperature/sound speed gradient. The goal of this long-term campaign is to determine or constrain: (1) the flux of infrasound producing meteoroids at the Earth's surface, (2) the altitudes at which these infrasonic waves are being generated, (3) the fundamental physics of shock production during hypersonic flight of meteoroids and (4) the relationship between a meteoroid's kinetic energy and surface observations of period and amplitude. The following sections will describe the preliminary results of the first 1½ years of this campaign.

2 Equipment and Methodology

The equipment used in this campaign are all component parts of the SOMN. This includes six all-sky video meteor cameras (Weryk et al. 2007), CMOR (Jones et al. 2005) and the four element infrasound station ELFO located in Elginfield, Ontario (43.1907°N, 81.3152°W, 322 m, Fig. 1). These first two systems provide triggers for potential infrasound producing meteors and direct when to search for signals in the continuous pressure data logged by ELFO. In the case of the SOMN all-sky cameras, these triggers are bright visual meteors ($M_P < -2$), while for CMOR, they are obvious cases of meteor head-echo detection (i.e. radar reflections from the plasma surrounding the meteoroid which are automatically detected at a rate of ~3/week).

In the case of potential meteor detection the analysis methodology to confirm the observation and detection is as follows:

(1) All-sky camera (multi-station): Event time, trajectory, radiant, light curve, velocity and mass are determined using standard reduction methods (Ceplecha and McCrosky 1976; Ceplecha 1987; Ceplecha et al. 1998; Borovička 1990).
(2) All-sky camera (single-station): Event time and direction are determined.
(3) Radar detection: Event time, range rate, duration and signal interferometry are recorded and used to determine 3D meteor trajectory and velocity.
(4) Comparison: Observed infrasonic back azimuth and propagation time are compared to the range and direction to the observed meteor trajectory. If these are found to be consistent with the meteor source, the event is logged as a probable detection.

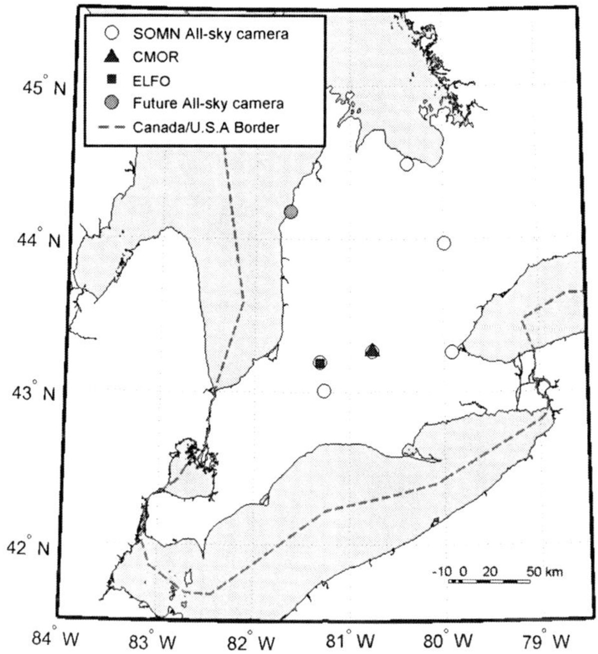

Fig. 1 Map of the Southern Ontario Meteor Network of all-sky cameras and relative locations of CMOR and the Elginfield observatory infrasound array, ELFO

(5) Reconstruction: Atmospheric data collected includes: (a) UKMO stratospheric assimilated data (Swinbank and O'Neill 1994) profile for the day of observation, (b) mesospheric winds measured from 82–98 km by CMOR (Hocking 1997), (c) MSISE-00 (Hedin 1991) and HWM93 (Hedin et al. 1996) atmospheric temperature, pressure and horizontal wind models. These measurements/models are fused together and used to reconstruct the best fit atmospheric propagation conditions at the time of the infrasonic detection.

(6) Ray tracing: Using the observed locations on the trajectory, rays are propagated towards ELFO, through the reconstructed atmosphere and the resulting travel times, azimuths and incidence angles are compared to those observed (Edwards and Hildebrand 2004). If a solution exists which reproduces the observations well (travel time delay, arrival azimuth, arrival angle), the observation is then confirmed and the source altitude delimited.

3 Observations

Between January 23, 2006 and June 6, 2007 (nearly 17 months of operation at ELFO) the total number of probable infrasound detection from regional meteors is 18. Of these 18 events, 13 have been coincident with a multi-station meteor detection affording complete trajectory and mass/energy estimation. Details of the detections are given in Table 1. Even with slightly more than a dozen confirmed multi-camera detections one significant conclusion can be drawn; infrasound from moderately sized (<10 cm diameter) is significantly

Table 1 Current (as of June 2007) coordinated observations and detections of regional meteor generated infrasound using the ELFO infrasound array in conjunction with the Southern Ontario Meteor Network all-sky cameras and the Canadian Meteor Orbit Radar

Meteor (YYYYMMDD)	Time (UTC)	MP_{max}	M_P (kg)	V_∞ (km/s)	Arrival time (UTC)	Duration (s)	Δp (Pa)	τ_{peak} (s)	H_S (km)	Type	M_I (kg)
20060213	08:49:25	R	–	12.70 ± 0.09	08:53:33.8	3–4	0.212 ± 0.079	0.300 ± 0.033	69.0 ± 0.2	Q	
20060302	06:28:14	SS	–	–	06:41:54.5	3–4	0.105 ± 0.005	0.595 ± 0.019	–	–	–
20060305	05:15:37	−9.7 ± 0.1	11.6 ± 1.3	18.65 ± 0.04	05:21:27.8	8–9	0.156 ± 0.028	0.151 ± 0.017	60.0 ± 4.5	B	0.477
20060405	03:03:27	−6.9	6.80	18.71 ± 0.11	03:10:01.6	4–5	0.166 ± 0.024	0.211 ± 0.076	67.9 ± 2.0	B	4.78
20060419	04:21:28	−5.9 ± 0.2	0.217 ± 0.024	19.02 ± 0.03	04:27:17.9	~1	0.061 ± 0.032	0.297 ± 0.050	79.3 ± 1.3	B	2.68
20060419b	07:05:57	−4.2 ± 0.6	0.135 ± 0.060	14.21 ± 0.07	07:10:34.8	~0.5	0.137 ± 0.048	0.113 ± 0.031	55.6 ± 4.1	B	0.103
20060805	08:38:50	−14.8 ± 1.1	2.6 (7.6/0.92)	70.08 ± 0.27	08:46:00.0	45	0.65 ± 0.16	1.530 ± 0.136	91.7 ± 2.6	Q	1.49
20060901	06:44:49	SS	–	–	06:48:19.7	15	0.054 ± 0.010	0.383 ± 0.032	–	–	–
20061021	03:42:07	R	–	–	03:56:05.0	30	0.0270 ± 0.0008	0.820 ± 0.170	–	–	–
20061101	06:46:12	−9.5 ± 0.4	0.058 ± 0.029	56.8 ± 3.4	06:55:00.7	10	0.037 ± 0.007	1.048 ± 0.073	88.7 ± 0.1	Q	0.262
20061104	03:29:30	−7.4 ± 0.6	0.038 ± 0.026	29.93 ± 0.12	03:35:25.0	~0.5	0.084 ± 0.020	0.177 ± 0.018	66.3 ± 0.2	B	0.083
20061121	10:45:46	−13.5 ± 1.0	0.33 (0.78/0.14)	76.5 ± 5.8	10:54:22.5	20	0.028 ± 0.005	1.110 ± 0.182	103.3 ± 0.1	N	0.211
20061223	06:27:26	<−13.7	>146	23.36 ± 0.04	06:37:33.5	32	0.058 ± 0.027	0.580 ± 0.032	82.3 ± 1.5	B	12.2
20070102	10:42:03	−6.24 ± 0.01	0.0151 ± 0.0011	41.03 ± 0.19	10:51:42.7	~3	0.041 ± 0.004	0.921 ± 0.358	88.4 ± 0.9	N	3.48
20070125	10:02:05	−6.7 ± 0.7	0.0126 ± 0.0081	68.63 ± 0.09	10:08:42.2	5	0.036 ± 0.004	1.213 ± 0.107	98.6 ± 3.0	B	0.438
20070129	00:49:51	R + SS	–	–	00:55:27.0	1.5	0.197 ± 0.062	0.466 ± 0.051	–	–	–
20070421	09:21:01	−8.5 ± 0.7	0.19 ± 0.11	35.77 ± 0.35	09:31:38.6	1.5	0.015 ± 0.001	0.650 ± 0.057	95.8 ± 2.1	N	0.989
20070511	07:41:14	−5.7 ± 0.4	0.0013 ± 0.0002	64.72 ± 0.31	07:48:34.7	3.5–4	0.012 ± 0.002	0.716 ± 0.097	102.1 ± 9.0	B	0.035

Column definitions: Meteor/Time—meteor event date and occurrence time, MP_{max}—maximum panchromatic meteor magnitude, M_P—panchromatic mass, V_∞—initial velocity, Arrival time/duration—infrasonic signal arrival at ELFO and signal duration, Δp—maximum signal overpressure, τ_{peak}—measured period, H_S—estimated infrasound source height, M_I—infrasonic mass. Categories of infrasonic observations (type) are: B—Ballistic, Q—Quasi-ballistic, N—Non-ballistic. Peak panchromatic magnitudes for each event are given where known, while single station and radar observations are marked with SS or R, respectively more common than has been previously observed. Previous attempts at monitoring regional meteor infrasound at the Springhill Meteor Observatory (McIntosh et al. 1976), and by the Prairie Network (Kraemer 1977) resulted in only one confirmed sporadic meteor and one probable Geminid meteor infrasonic signal after ~5 years of observation. Our current observations show that the flux is at least two orders of magnitude higher than these early observations suggest. We suspect that significant advances in computer technology and digital signal processing since the late 1970s and early 1980s, has likely made identification of these sometimes very weak signals (Table 1) far easier.

The bulk of the observed detections in Table 1 tend to be from sporadic meteors. Some shower meteors, however, have been identified and include a suspected Orionid (20061021) and a confirmed Leonid (20061121) and Quadrantid (20070102). Additionally upon inspection of ray tracing geometry (i.e. deviation of the ray vector from the meteor trajectory), the observations may be placed into one of three categories:

(1) Ballistic—rays which propagate approximately 90° ± 20° (Brown et al. 2007) from the meteor trajectory, consistent with cylindrical blast wave theory (ReVelle 1976).
(2) Quasi-ballistic—rays which border, but do not fall within, the ballistic regime (110°–125° deviation). Although often appearing to have ballistic shock features (i.e. N-type waves), these deviations are sufficiently large that they cannot be categorized as ballistic waves within uncertainties in the trajectory or model atmosphere.
(3) Non-ballistic—rays which appear to emit from an omni-directional or point-like source (e.g. due to fragmentation or the blunt end of a meteoroid's hypersonic shock).

Having identified potential source mechanisms for the observed meteor infrasound, we attempt to reproduce the ballistic and quasi-ballistic observations using current theory constrained by the known trajectory & source regions (ReVelle 1974, 1976).

4 Comparison with Theory

The theoretical developments of meteor generated infrasound were developed initially by ReVelle (1974) based upon research in weak shock propagation and cylindrical blast waves and applying these to high altitude, hypersonic meteor sources. While initial treatments (ReVelle 1976) were based upon simplifying isothermal atmosphere assumptions, a method was provided as to how the theory could be extended to more realistic vertically inhomogeneous atmospheres (ReVelle 1974). Using a top down methodology, we use the ray geometry and the observed meteor trajectory, photometric mass and velocity to calculate the predicted theoretical signal overpressure (amplitude), Δp, and period using both an isothermal and vertically inhomogeneous atmosphere.

Theoretical results for the nominal geometries for ballistic and quasi-ballistic observations are shown in Fig. 2. Theoretical uncertainties shown are the variation in predicted Δp and period accounting for uncertainties in geometry, mass and velocity. While predicted periods show similar scatter about the observed periods on the order of factors of 2 or 3 for either model atmosphere (Fig. 2b/d), due to general insensitivity of the theoretical period on model conditions (see ReVelle 1976 Eqs. 15 and 34), predicted weak shock Δp show significant differences (Fig. 2a/c). Using an isothermal atmosphere, Δp is consistently underestimated by at least two orders of magnitude if treated as a propagating weak shock wave. A vertically inhomogeneous, however, (thermally stratified, range independent) atmosphere produces similar agreement in Δp as seen for period. Treating the propagating wave as a weak shock that transitions to a linear wave (not shown) produces

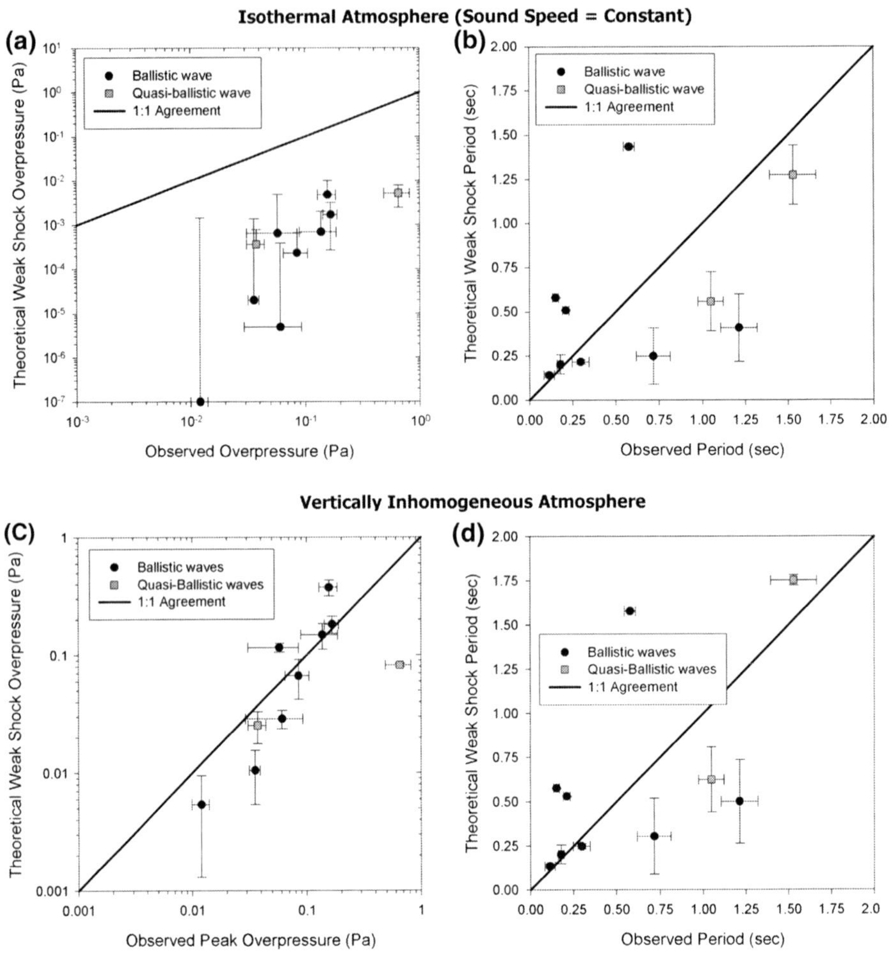

Fig. 2 Comparison of cylindrical weak shock theory of meteor generated infrasound (ReVelle 1974, 1976) with the observations of ballistic and quasi-ballistic infrasound at ELFO for those meteors with known trajectories. (**a/b**) For an isothermal atmosphere and (**c/d**) vertically inhomogeneous (layered) atmosphere

Δp*'s* for isothermal and inhomogeneous models that consistently underestimate and overestimate, respectively, the observed Δp by at least an order of magnitude, due to the slower decay of linear waves over weak shocks ($x^{-1/2}$ vs. $x^{-3/4}$ respectively where x is scaled distance from the source, R/R_0).

We conclude, therefore, that observations of regional meteor infrasound so far recorded by ELFO are consistent with theory assuming predominantly weakly non-linear shock waves propagate to the ground. Knowing this we attempt a bottom-up treatment, determining the mass of each source meteoroid (independent from photometric methods) by least squares fitting the sum of the residuals between theoretical Δp and period and those observed, using vertically inhomogeneous cylindrical blast wave theory with the meteoroid mass as the sole variable (the remaining parameters being constrained by observation). The resulting infrasonic mass for ballistic and quasi-ballistic observations (Table 1) is found to relate to the photometric mass roughly as:

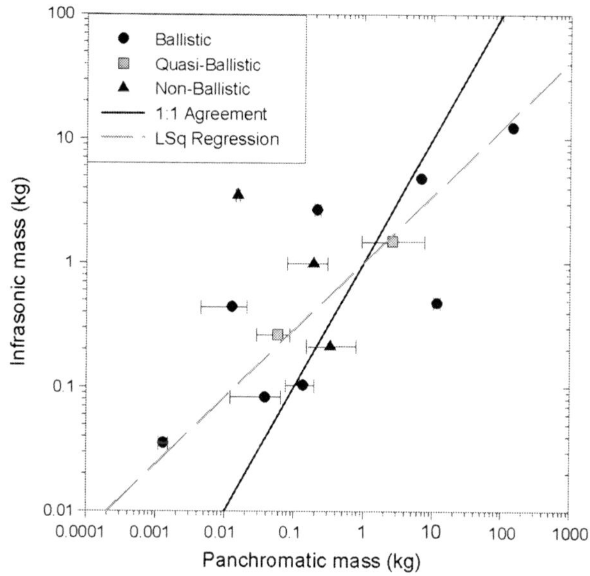

Fig. 3 Comparison of infrasonic mass to panchromatic mass. Infrasonic mass determined by fitting theoretical signal amplitude and period to observed signal. Panchromatic mass determined through standard light curve analysis (Ceplecha et al. 1998)

$$M_I = 1.00 \pm 0.60 M_P^{0.54 \pm 0.12} \tag{1}$$

where M_I is the infrasonic mass and M_P is the photometric mass in kilograms (Fig. 3).

5 Discussion and Conclusions

Recent observations of regional meteor infrasound using the newly installed ELFO infrasound array in tandem with the SOMN all-sky cameras and CMOR have discovered that the flux of meteor generated infrasound from small meteoroids <10 cm in diameter is far higher than previously observed. Although final calibration of the flux is in progress (awaiting corrections for observing biases and sky coverage), the flux appears to be at least 1 event/month for meteor panchromatic magnitudes –2 and brighter. Interestingly, ReVelle (1974) predicted ~20 events in a winter season might be detectable. The current results are a lower limit as no daytime events or meteors occurring during bad weather are detected. This suggests that ReVelle's (1974) estimate may be quite close to the actual effective detection limit. Work continues on correlating radar observed meteor head-echoes with the infrasound, preliminary results suggesting a handful of positive correlations/detections.

Recent observations using the European Fireball Network (Oberst et al. 1998) and the IMS station I26DE (Brown et al. 2007) showed that high altitude infrasound from shower meteors was possible. This more complete and general survey suggests that not only is infrasound production possible at these altitudes, it appears that it is quite common for meteors <–4 magnitude with at least 8 of 18 signals originating at altitudes >80 km. Similar to the results of Brown et al. (2007), ballistic shocks make up the majority of observations with non-ballistic observations being far more rare and often associated with meteors undergoing gross fragmentation.

Using the current observations of ballistic and quasi-ballistic meteor generated infrasound we are at last able to begin testing and constraining the theoretical predictions developed by ReVelle (1974) more than 30 years ago. We find that most of our observations fit with ELFO recording the arrival of weakly non-linear shock waves. Yet only when using a realistic vertically inhomogeneous atmosphere can the theory reasonably fit the overpressure observations. Isothermal atmosphere assumptions underestimate this amplitude by at least two orders of magnitude. Such a significant finding implies that the kinetic energy (E)–overpressure relationship (Ceplecha et al. 1998),

$$E = 11.5\pi\rho_m R'^3 \left(\Delta p / \sqrt{p_Z p_g}\right)^4 \frac{C_S^3}{V} \tag{2}$$

first introduced by (Brown et al. 1996) and based upon these isothermal assumptions, will significantly over estimate meteoroid kinetic energy by orders of magnitude when observed Δp are input. Note that in (2) and later (3), ρ_m, is the meteoroid density; V, meteoroid velocity; R' is the range from observation point to the meteor and p_Z and p_g are the ambient pressures at the source height and ground, respectively. Such over-the-top and/or highly variable estimates of kinetic energy have been commonly reported when applying this formula to real data (e.g. Brown et al. 1996; ReVelle and Whitaker 1999; Brown et al. 2007). In contrast, the equivalent kinetic energy–period (τ) relationship (Ceplecha et al. 1998),

$$E = \left(\frac{\pi}{12}\right)\left(\frac{\tau}{1.579}\right)^4 \left(\frac{\rho_m C_S^7}{VR'}\right) \tag{3}$$

also based upon isothermal assumptions, should produce more robust energy estimates than overpressure despite the severe dependence on the sound speed, C_S. This is due in part to the relative insensitivity of the period (in the current theory) to the atmospheric model chosen (ReVelle 1974, 1976) and because the speed of sound varies relatively little (305 m/s ± 15%) between $\sim$120 and 50 km altitude, where most of the meteor infrasound observed so far is generated (Table 1).

As the current multi-instrumental campaign to observe infrasound from these smaller, more common, regional meteors continues, it will finally be possible to explore the infrasound and hyper-velocity meteoroid relationship both quantitatively and statistically, placing limits on current theory and potentially revising our current understanding of the shock mechanism at the source. Additionally, further work on ballistic observations should also provide another method to constrain other poorly known quantities in hyper-velocity meteoroid-atmosphere interactions such as luminous efficiency and bulk density through use of infrasonic energy estimates and infrasonic mass in tandem with electro-optical observations. This should be possible for ballistic observations since, due to cylindrical propagation, ballistic waves originate along a finite section of the meteor trail and thus be related to the physical properties of the meteoroid (mass/size) along that section. Infrasound in meteor science is now steadily progressing from its early theoretical confines to both observational and practical applications.

References

J. Borovička, The comparison of two methods of determining meteor trajectories from photographs. Bull. Astron. Inst. Czech. **41**, 391–396 (1990)

P. Brown, A.R. Hildebrand, D.W.E. Green, D. Pagé, C. Jacobs, D. ReVelle, E. Tagliaferri, J. Wacker, B. Wetmiller, The fall of the St-Robert meteorite. Meteorit. Planet. Sci. **31**, 502–517 (1996)

P.G. Brown, R.W. Whitaker, D.O. ReVelle, E. Tagliaferri, Multi-station infrasonic observations of two bolides: signal interpretation and implications for monitoring of atmospheric explosions. Geophys. Res. Lett. **29** (2002). doi: 10.1029/2001GL013778

P.G. Brown, W.N. Edwards, D.O. ReVelle, P. Spurny, Acoustic analysis of shock production by very high-altitude meteors—I: infrasonic observations, dynamics and luminosity. J. Atmos. Solar-Terres. Phys. **69**, 600–620 (2007)

Z. Ceplecha, Geometric, dynamic, orbital and photometric data on meteoroids from photographic fireball networks. Bull. Astron. Inst. Czech. **38**, 222–234 (1987)

Z. Ceplecha, R.E. McCrosky, Fireball end heights: a diagnostic for the structure of meteoric material. J. Geophys. Res. **81**, 6257–6275 (1976)

Z. Ceplecha, J. Borovička, W.G. Elford, D.O. ReVelle, R.L. Hawkes, V. Porubčan, M. Šimek, Meteor phenomena and bodies. Space Sci. Rev. **84**, 327–471 (1998)

W.N. Edwards, A.R. Hildebrand, SUPRACENTER: Locating fireball terminal bursts in the atmosphere using seismic arrivals. Meteorit. Planet. Sci. **39**, 1449–1460 (2004)

W.N. Edwards, P.G. Brown, D.O. ReVelle, Estimates of meteoroid kinetic energies from observations of infrasonic airwaves. J. Atmos. Solar-Terres. Phys. **68**, 1136–1160 (2006)

A.E. Hedin, Extension of the MSIS thermospheric model into the middle and lower atmosphere. J. Geophys. Res. **96**, 1159–1172 (1991)

A.E. Hedin, E.L. Fleming, A.H. Manson, F.J. Schmidlin, S.K. Avery, R.R. Clark, S.J. Franke, G.J. Fraser, T. Tsuda, F. Vial, R.A. Vincent, Empirical wind model for the upper, middle and lower atmosphere. J. Atmos. Terres. Phys. **58**, 1421–1447 (1996)

W.K. Hocking, Strengths and limitations of MST radar measurements of middle-atmosphere winds. Ann. Geophys. **15**, 1111–1122 (1997)

J. Jones, P. Brown, K.J. Ellis, A.R. Webster, M. Campbell-Brown, Z. Krzemenski, R.J. Weryk, The Canadian Meteor Orbit Radar: system overview and preliminary results. Planet. Space Sci. **53**, 413–421 (2005)

A.R. Klekociuk, P.G. Brown, D.W. Pack, D.O. ReVelle, W.N. Edwards, R.E. Spalding, E. Tagliaferri, Y.B. Bernard, J. Zagari, Lidar, satellite and acoustic measurements of an asteroidal airburst in Earth's atmosphere. Nature **436**, 1132–1135 (2005)

D.R. Kraemer, Infrasound from accurately measured meteor trails, Ph.D. Dissertation, Ann Arbor, University of Michigan, MI (1977)

A. Le Pichon, J.M. Guerin, E. Blanc, D. Raymond, Trail in the atmosphere of the 29 December 2000 meteor as recorded in Tahiti: characteristics and Trajectory reconstitution. J. Geophys. Res. 107 (2002). doi: 10.1029/2001JD001283

B.A. McIntosh, M.D. Watson, D.O. ReVelle, Infrasound from a radar-observed meteor. Can. J. Phys. **54**, 655–662 (1976)

J. Oberst, S. Molau, D. Heinlein, C. Gritzner, M. Schindler, P. Spurny, Z. Ceplecha, J. Rendtel, H. Betlem, The "European Fireball Network": current status and future prospects. Meteorit. Planet. Sci. **33**, 49–56 (1998)

D.O. ReVelle, Acoustics of meteors—effects of the atmospheric temperature and wind structure on the sounds produced by meteors. Ph.D. Dissertation, Ann Arbor, University of Michigan, MI (1974)

D.O. ReVelle, On meteor-generated infrasound. J. Geophys. Res. **81**, 1217–1230 (1976)

D.O. ReVelle, in Annals of the New York Academy of Sciences, Near-Earth Objects—The United Nations International Conference, ed. by J.L. Remo. Historical detection of atmospheric impacts by large bolides using acoustic-gravity waves, vol 822 (New York Academy of Sciences, New York, NY, USA, 1997), pp. 284–302

D.O. ReVelle, R.W. Whitaker, Infrasonic detection of a Leonid bolide: 1998 November 17. Meteorit. Planet. Sci. **34**, 995–1005 (1999)

R. Swinbank, A.A. O'Neill, Stratosphere–Troposphere Data Assimilation System. Mon. Weather Rev **122**, 686–702 (1994)

R.J. Weryk, P.G. Brown, W.N. Edwards, Z. Krzeminski, S.H. Nudds, A. Domokos, D.L. Welch, The Southern Ontario meteor camera network. Earth Moon Planets (2007) this issue

Determination of Meteoroid Orbits and Spatial Fluxes by Using High-Resolution All-Sky CCD Cameras

Josep M. Trigo-Rodriguez · José M. Madiedo · Peter S. Gural · Alberto J. Castro-Tirado · Jordi Llorca · Juan Fabregat · Standa Vítek · Pep Pujols

Originally published in the journal Earth, Moon, and Planets, Volume 102, Nos 1–4.
DOI: 10.1007/s11038-007-9207-x

Abstract By using high-resolution, low-scan-rate, all-sky CCD cameras, the SPanish Meteor Network (SPMN) is currently monitoring meteor and fireball activity on a year round basis. Here are presented just a sampling of the accurate trajectory, radiant and orbital data obtained for meteors imaged simultaneously from two SPMN stations during the continuous 2006–2007 coverage of meteor and fireball monitoring. Typical astrometric uncertainty is 1–2 arc min, while velocity determination errors are of the order of 0.1–0.5 km/s, which is dependent on the distance of each event to the station and its

J. M. Trigo-Rodriguez (✉)
Institute of Space Sciences (CSIC), Campus UAB, Facultat de Ciències, Torre C5-parell-2ª, 08193 Bellaterra, Barcelona, Spain
e-mail: trigo@ieec.uab.es

J. M. Trigo-Rodriguez · J. Llorca
Institut d'Estudis Espacials de Catalunya (IEEC), Edif. Nexus, c/Gran Capità, 2-4, 08034 Barcelona, Spain

J. M. Madiedo
Facultad de Ciencias, Universidad de Huelva, Huelva, Spain

P. S. Gural
Science Applications International Corp, 14668 Lee Road, Chantilly, VA 20151, USA

A. J. Castro-Tirado · S. Vítek
Instituto de Astrofísica de Andalucía (IAA-CSIC), Camino Bajo de Huétor 50, 18008 Granada, Spain

J. Llorca
Institut de Tècniques Energètiques, Universitat Politècnica de Catalunya, Diagonal 647, ed. ETSEIB, 08028 Barcelona, Spain

J. Fabregat
Observatori Astronòmic, Universitat de València, Paterna, València, Spain

P. Pujols
Grup d'Estudis Astronòmics (GEA) and Agrupació Astronòmica d'Osona, Barcelona, Spain

particular viewing geometry. The cameras have demonstrated excellent performance for detecting meteor outbursts. The recent development of automatic detection software is also providing real-time information on the global meteor activity. Finally, some examples of the all-sky CCD cameras applications for detecting unexpected meteor activity are given.

Keywords Meteors · Meteoroids · Meteoroid stream · All-sky camera · CCD camera · Astrometry · Heliocentric orbits

1 Introduction

Trigo-Rodríguez et al. (2004) had reported on the first steps in the deployment of low-scan-rate, all-sky CCD cameras that operate up to + 2 to + 3 limiting magnitude for meteors. A general overview of observations recorded in 2006, including the main meteor activity highlights and a list of bright fireballs, has been published in previous papers (Trigo-Rodríguez et al. 2006a, b, 2007). The establishment of the SPanish Meteor Network (SPMN) was keyed off the development and deployment of innovative low-scan-rate, all-sky CCD cameras (Trigo-Rodríguez et al. 2004a). Presented herein are the trajectory and orbital results obtained with those systems during the years 2006 through 2007, that had undergone an equipment design upgrade that incorporated internal devices for accurately measuring meteor velocities. Results to be shown were obtained by the newer all-sky CCD stations set up in Catalonia during 2006. In particular, that year was excellent for our network due to favourable weather (generally dry with extraordinary clear weather conditions during autumn and winter) that guaranteed almost a continuous record of meteor activity.

We provide examples of trajectory and orbital results of meteors in the [−10, + 3] magnitude range, with limiting magnitudes substantially fainter than conventional all-sky photographic cameras (see e.g. review by Spurný et al. 2007). These data were obtained by using the low-scan rate, all-sky CCD cameras operated in Catalonia (Table 1). The reason for this selection is because the Catalan stations were built with internal rotating shutters to get accurate measurements of meteor velocities (Trigo-Rodríguez et al. 2007). We will focus here on some trajectory and orbital examples obtained by using our all-sky CCD cameras. In a parallel paper we also present video data obtained from three complementary video stations developed in Andalusia (Madiedo and Trigo-Rodríguez 2007).

Table 1 Stations of the SPMN involved in this work

Station #	Station (Province)	Longitude (E)	Latitude (N)	Alt. (m)	Imaging system
1	Montsec, OAdM (Lleida)	00°43′46″	42°03′05″	1570	AS
2	Montseny (Girona)	02°31′14″	41°43′17″	300	AS
3	Folgueroles (Barcelona)	02°19′33″	41°56′31″	580	WF

Acronyms for the different imaging systems are: AS (low-scan-rate CCD all-sky camera), and WF (low-scan-rate CCD wide-field camera)

2 Instrumentation, Data Reduction and Observation Sites

A description of the low-scan-rate, CCD cameras was published previously in Trigo-Rodríguez et al. (2004). The cameras are operated in a sequential imaging mode, each making 90 second exposures followed by a typical readout time of 30 or 15 s, depending on whether the readout is controlled by a parallel or USB port respectively. Currently, the CCD imagery is analyzed using a set of image processing modules that have recently been developed. The initial stage consists of obtaining a frame difference among two sequential images, with the result that stars appear as continuous or trailed tracks (non-guided images) characterized by a positive streak followed by an equal length negative streak (Fig. 1a). The background, in the periphery of the all-sky view where obstacles such as buildings and trees are visible, is effectively removed by the frame differencing. Since our subtraction occurs on the raw images without any image transformations, each differenced element from the two images has the same pixel response and the background mean is effectively eliminated without the need for flat fielding the frames. The residual variance will depend on the response of each pixel and ideally should be flat fielded after the difference is taken, but since we do not have a flat field available, this step is ignored and can be justified because of the spatial culling in the next step. A threshold is estimated (mean $\pm$ 2 $\times$ standard deviation) for a local region around each pixel exceedance with high and low outliers removed. Thus any pixel that possesses a large variance due to either hot or cold pixel response, is eliminated from the threshold estimation. For the algorithm to operate successfully without flat fielding, the majority of pixels must have nearly equivalent responses across the focal plane with only a few hot or cold pixels in existence. This is typically a good assumption for most CCD sensors.

Using the above mentioned procedure, meteor detection consists of peak finding after a spatial re-mapping of pixel exceedances above or below the local thresholds through the use of the Hough Transform (HT) (see e.g., Hough 1962; Duda and Hart 1972). Using a

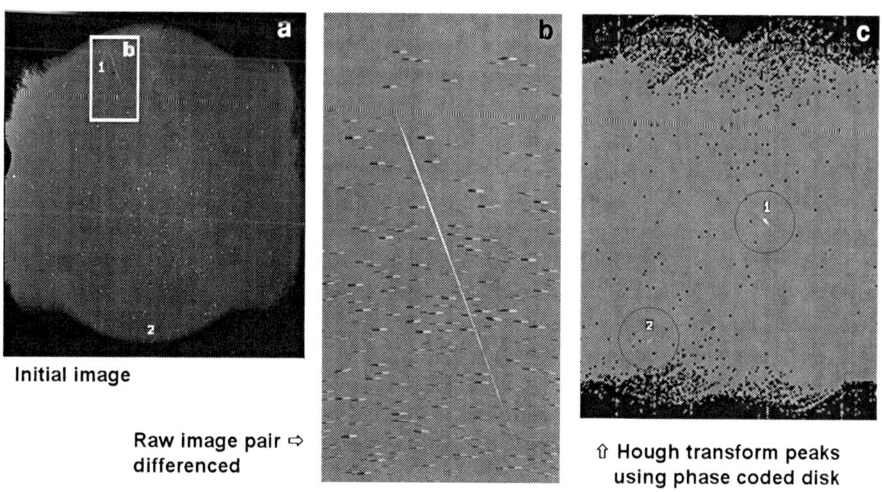

Fig. 1 Meteor detection steps. Plot (**a**) shows a representative all-sky image with stars, background, and two meteors (marked as 1 and 2). Plot (**b**) shows the difference frame where star trails have equal parts positive (white) and negative (black) components whereas meteors are predominantly only one colour. Plot (**c**) is the Hough space output showing two peaks at the angle and offset of the lines aligned with the meteor (for more details see Gural 2007)

modification to the traditional HT, we apply a phase-coded disk to each pixel exceedance to estimate the local orientation of the track. A phase coded disk (see Clode et al. 2004) is a convolution kernel that when applied to a binary image, determines the orientation angle of a line within a local region encompassed by the spatial extent of the kernel. Given the position of an exceedance pixel and angle of the line through that pixel, only a single cell in the Hough transform space is incremented rather than a hypothesized set of all angles through that pixel in traditional HT processing. This helps avoid the generation of Hough self noise (the "butterfly" pattern) and its associated reduced probability of detection.

When applying the HT to the SPMN imagery, star trails of both unguided exposures are mutually cancelling in the Hough space accumulator (Fig. 1b) since they have equal positive and negative contributions after the frame difference and nearly the same position and orientation angle. Additionally, a mechanism of false alarm mitigation is implemented that excludes regions in Hough space for every line detection found (Fig. 1c) thus helping to avoid multiple detections of the same meteor. This assumes that two meteors are unlikely in the same location and direction, which does not affect the final goal, that is, saving the image for future visual inspection/confirmation. Images with positive identification are saved and can be later re-examined in order to confirm the identification of the suspected streak as a meteor, and also to look for additional tracks by visual inspection. More details on the software are given in Gural (2007).

Meteor photometry of the CCD images is a field that we currently have under study. CCD cameras allow an analyst to estimate stellar and meteor magnitudes very accurately since they provide photon counts for every pixel. In all-sky CCD imaging we adopt an approach where meteor magnitudes are derived by comparing the intensity level of the pixels close to the maximum luminosity of the meteor trail and nearby stars. Taken into account is the different angular velocity of the meteors as a function of the distance to the radiant, and the typical duration of flares. An example of a photometric curve is given in Fig. 2. Additional extinction corrections are also considered to get the absolute magnitude of meteors that appear below 30° of altitude.

Positional measurements of the meteor track from the CCD images are currently performed by using the Maxim DL software package (http://www.cyanogen.com/). Accurate astrometry of medium sized field-of-view images is obtained for both stars and meteors by using the polynomic astrometric method published by Steyaert (1990). However, all-sky images have a more complex astrometric reduction to account for the image distortion of wide-eye lenses (Ceplecha 1987; Borovička et al. 1995). Both methods are currently implemented in our Amalthea software (Madiedo and Trigo-Rodríguez 2007) that has been written in C and follows the general numerical procedures described by Press et al. (1992). The new Amalthea software has wider applicability that our previous *Network* software (Trigo-Rodríguez et al. 2002, 2004b). With the present angular resolution (~ 1 arc minute) of the camera systems, the equatorial coordinates of the meteors are computed with an astrometric accuracy of approximately 0.01°, which also determines the apparent and geocentric radiant of any meteor. Reconstruction of the atmospheric trajectory and radiant is performed by using the method of intersecting planes developed by Ceplecha (1987). From the astrometric measurements of the shutter breaks along the trajectory and chopping rate, the velocity of the meteoroid is derived. The velocity measured for each shutter-break was obtained as well as the preatmospheric velocity V_{∞} from the velocity measured in the earliest portion of each meteor trajectory (usually in the first 3 or 5 breaks, when deceleration is weak). At present we do not apply deceleration corrections to the measured velocity, but we plan to apply a dynamic model of the meteor flight in the future by using Amalthea software. Finally, in order to determine orbital elements from our trajectory data

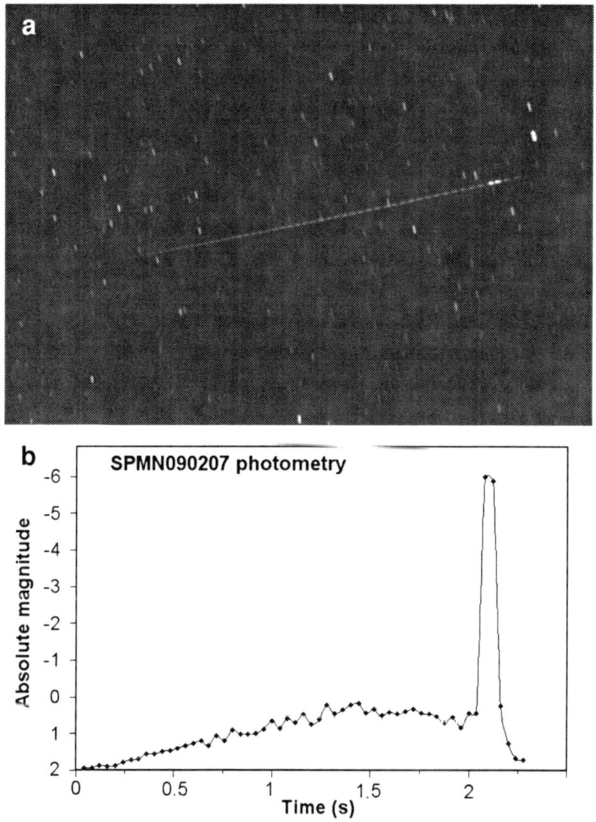

Fig. 2 Photometry of the SPMN090207 "Tordera" fireball. Each point represents the integrated signal computed for each different shutter break, once the background has is removed for considering only the light emitted by the fireball. (**a**) Image taken from Montseny all-sky camera, and (**b**) Absolute visual magnitude plotted as function of time. For star identification, Castor is the brightest star in the field, nearby the fireball's main flare

we used the MORB program provided by Ceplecha et al. (2000) from the Ondrejov Observatory in the Czech Republic.

3 Observations: Spatial Fluxes, Trajectory, Radiant and Orbital Data

Continuous monitoring of meteor activity using our all-sky CCD cameras allows for the determination of meteoroid spatial fluxes for minor showers. An example of unexpected activity identified from all-sky CCD imaging is the 2006 Orionid outburst that we quickly reported to the community (Trigo-Rodríguez et al. 2006, 2007b). By estimating the meteor magnitudes from the images, we obtained a magnitude distribution for the night of Oct. 20–21, 2006 (Trigo-Rodríguez et al. 2007a). From the derived population index ($r = 1.4 \pm 0.4$, $N = 33$) we determined an incident flux of 415 meteoroids brighter than magnitude + 6.5 per km^2 using a high fidelity meteor simulation (Gural and Jenniskens 2000; Molau et al. 2002). The meteor simulation determines the incoming spatial flux by modelling the camera's sensitivity characteristics, the geometric loss terms, the change in radiant position, the meteoroid stream's particle distribution, and associated entry parameters, to determine the meteoroid incident density that is required to match the actual counts seen by the sensor's field of view and look orientation. Using this flux value and the same simulation tool, but with human visual perception parameters and loss models substituted (Gural 2004), we estimated a visual (human) ZHR $= 50 \pm 15$ (Trigo-Rodríguez et al. 2007). In order to compute the spatial flux of meteoroids producing meteors brighter than +6.5 per km^2, as well as an equivalent visual Zenithal Hourly Rate (ZHR), the count rates must be obtained from the all-sky systems for a consistent limiting magnitude of detection.

Multiple-station detections of meteors, as opposed to single station observations, allows the computation of accurate trajectory and orbital data when imaged by all-sky CCD cameras. Table 2 shows the absolute magnitude (M_v) in the visual range, the height for beginning, maximum and terminal light (H_b, H_{max}, H_{end}), the geocentric radiant (α_g, δ_g to Eq. 2000.00), the convergence angle (Q) of the meteor seen from both stations and the infinity, geocentric, and heliocentric velocities (V_∞, V_g, V_h). Note that only those meteors with $Q > 20°$ were considered. From the radiant position, apparition time, and velocities estimated for the meteors listed in Table 2, the orbital elements were derived and shown in Table 3.

4 Discussion

Determination of spatial number densities for meteor showers in near real time from all-sky CCD imagery is considered a significant advance in the meteor community. We have already applied the regional coverage of the SPMN cameras to get quick reports of unusual meteor activity, as e.g. those reported for the 2006 Orionid outburst (Trigo-Rodríguez et al. 2006a) or the Kappa Cygnid fireball display (Trigo-Rodríguez et al. 2007c). We also reported that there was no evidence of background meteor activity from the dust trail of comet C/1911 N1 (Kiess) a few hours before the peak reported from United States (Jenniskens et al. 2007).

Direct astrometry on CCD images can provide accurate astrometry of meteors. When they are observed from several stations, trajectory and orbital data can be derived in just a few hours. The observational uncertainties in trajectory data and orbital elements reported in Tables 2 and 3 are similar to those reported by small photographic camera networks but they are still less accurate than those obtained by European Network (EN) photographic cameras (see e.g. Spurný et al. 2004). The main reason for this is the order of magnitude better spatial resolution of the EN all-sky photographic cameras, as was already pointed out in our previous work (Trigo-Rodríguez et al. 2004). In spite of this, some of the presented orbital data are perfectly comparable because the geometry of the meteor's appearance also plays an important role. On the other hand, developing techniques in higher resolution pixel interpolation (e.g. Quine et al. 2007) can be useful for reducing astrometric uncertainties.

The data shown in Tables 2 and 3 demonstrate that all-sky CCD cameras can collect valuable trajectory and orbital information on meteors. The quick reduction of data has been demonstrated to be important for alerting our community of unusual meteor activity on short time-scales (Trigo-Rodríguez et al. 2006, 2007b, c). Although the number of trajectories and orbital results presented herein has been limited for brevity, we think that the sample is well representative of meteors spanning different velocities, magnitudes, and

Table 2 Trajectory and radiant data, Equinox (2000.0)

SPMN Code	Stream	M_v	H_b (km)	H_{max} (km)	H_e (km)	α_g (°)	δ_g (°)	Q (°)	V_∞ (km/S)	V_g (km/S)	V_h (km/S)
101006	Taurid	+3	105.4	82.8	80.6	31.12 ± 0.10	16.83 ± 0.07	21	29.5 ± 0.3	27.5	35.6
021006	Orionid	−5	119.5	94.8	89.2	92.75 ± 0.16	15.79 ± 0.05	37	67.2 ± 0.2	66.2	41.7
011206	Geminid	−10	112.6	–	44.4	106.45 ± 0.18	30.82 ± 0.13	20	36.7 ± 0.3	35.2	36.9
301206	Sporadic	−2	104.2	–	78.6	147.51 ± 0.12	−21.15 ± 0.11	21	39.9 ± 0.2	38.4	26.2
010107	Jan. Drac.	+1	98.5	85.1	79.1	274.75 ± 0.15	50.06 ± 0.24	30	24.4 ± 0.2	21.4	35.4
090207	Sporadic	−6	87.1	67.3	65.8	69.95 ± 0.04	88.47 ± 0.04	53	15.20 ± 0.05	10.3	34.6
010707	Sporadic	−3	87.6	83.0	71.7	251.20 ± 0.44	−2.3 ± 0.3	56	12.20 ± 0.05	4.8	32.7
020707	β Androm?	+2	113.6	–	102.8	22.4 ± 0.3	37.74 ± 0.09	49	44.00 ± 0.2	42.4	20.6

Table 3 Orbital elements of imaged meteors, Equinox (2000.00)

SPMN Code	q (AU)	1/a (AU^{-1})	e	i (°)	ω (°)	Ω (°)
101006	0.333 ± 0.004	0.575 ± 0.015	0.809 ± 0.007	4.68 ± 0.12	299.31 ± 0.22	200.51705 ± 0.00013
021006	0.540 ± 0.005	0.046 ± 0.019	0.975 ± 0.010	163.48 ± 0.13	85.8 ± 0.8	27.50407 ± 0.00017
011206	0.181 ± 0.003	0.494 ± 0.017	0.911 ± 0.004	16.4 ± 0.4	315.5 ± 0.3	261.94012 ± 0.00007
301206	0.277 ± 0.003	1.259 ± 0.008	0.652 ± 0.002	84.1 ± 0.4	145.2 ± 0.5	98.2297 ± 0.00014
010107	0.9364 ± 0.0017	0.623 ± 0.009	0.417 ± 0.007	36.1 ± 0.3	146.6 ± 0.7	300.64133 ± 0.00007
090207	0.98622 ± 0.00003	0.674 ± 0.004	0.335 ± 0.004	16.48 ± 0.09	183.47 ± 0.10	320.6410 ± 0.0002
010707	0.9894 ± 0.0007	0.764 ± 0.007	0.244 ± 0.007	2.85 ± 0.11	210.1 ± 0.4	114.8138 ± 0.0007
020707	0.328 ± 0.007	0.488 ± 0.009	0.511 ± 0.009	115.0 ± 0.5	0.9 ± 0.3	124.59432 ± 0.00009

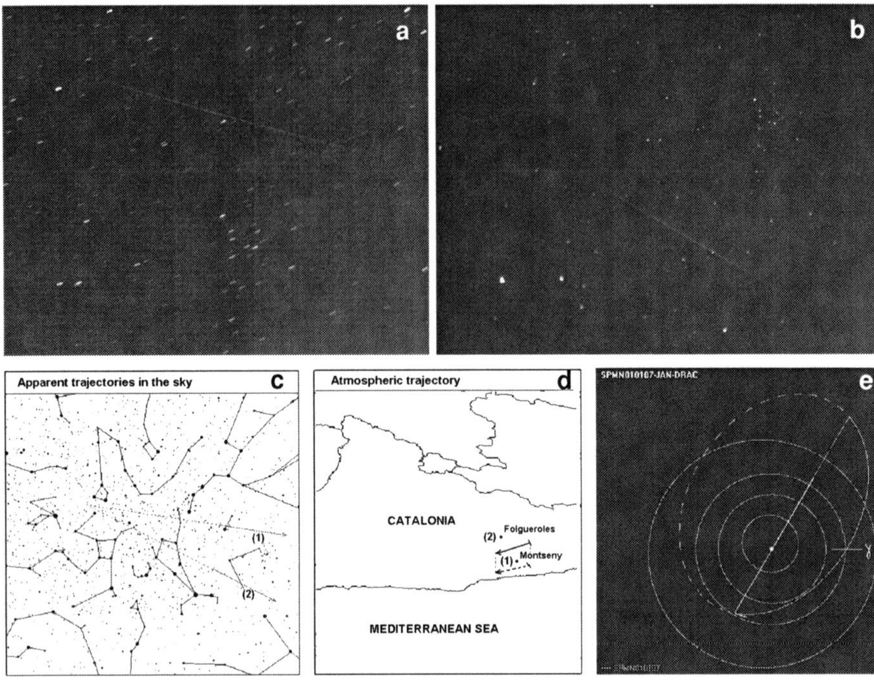

Fig. 3 Part of the all-sky CCD image centred in the +1 magnitude January Draconid (SPMN010107) recorded from (**a**) Station (1) Montseny with a shutter providing 25 breaks/s, and (**b**) Station (2) Folgueroles (**c**) Apparent trajectory of the meteor from both stations. (**d**) Atmospheric trajectory and its projection on the ground. (**e**) The heliocentric orbit of the meteoroid projected on the ecliptic plane, where the orbits of the terrestrial planets are shown for comparison

properties from major showers, minor showers, and sporadics. Continuous monitoring also permits searching for and studying poorly observed minor showers. A good example is a + 1 magnitude meteor presented in Fig. 3 (labelled as SPMN010107 in Table 2). Initially identified as a sporadic when compared to orbital elements obtained by Sekanina (1976), we later found similarity with a stream identified by Kronk (1988) as the January Draconids. Many of the meteors contained in Tables 2 and 3 exhibited magnitudes well below the typical limit of conventional (photographic) all-sky cameras. At the bright bolide extreme, astrometry has been found to be feasible since saturation of the imaging chip usually occurs over just a small portion of the fireball path. By way of example we included the orbit of a very bright Geminid bolide (SPMN011206) recorded during the extraordinary 2006 Geminid display (Fig. 9 of Trigo-Rodríguez et al. 2007), and the orbit of a −6 sporadic event (SPMN 090207).

5 Conclusions

Several examples were presented of the capability of the recently developed all-sky CCD camera systems described in Trigo-Rodríguez et al. (2004) for both trajectory and orbital determination of meteors. The deployment of complementary detection systems (video, forward scatter, infrasound, etc...) in a wide area network, are our future goals for achieving multi-instrument studies like those previously reported (Spurný et al. 2004; Weryk et al. 2007). At the current stage of development for the SPMN, the main conclusions of this paper are:

(a) Low-scan-rate, all-sky CCD cameras applied to meteor monitoring (Trigo-Rodríguez et al. 2004) can provide very valuable information on spatial fluxes of meteor showers. Detection of unexpected meteor activity is feasible due to the better sensitivity of the current CCD cameras over previous all-sky systems reaching meteor limiting magnitudes of + 2 to + 3.
(b) Reliable trajectory, radiant and orbital data of the imaged meteors can be computed on shorter time-scales (a few hours) than other existing processing systems with exception of real-time video systems.
(c) Subsequent advances in automatic meteor detection, stream association, data analysis, and other computational reduction tasks will provide important progress in improving our real-time alarm capabilities. All this will contribute to an increase in the amount of data available on poorly studied meteor showers that would be very valuable to the meteor community.

Acknowledgements These results were achieved by using wide-field automatic digital cameras described in the Spanish patent application number 200501127, filed May 2005, and later continued in the PCT document number PCT/ES06/070057. The 2006 development of the internal mechanism for obtaining meteor velocities by J.M.T-R is under consideration for a patent. The authors would like to thank *Instituto Nacional de Técnica Aeroespacial (INTA)* and *Consejo Superior de Investigaciones Científicas (CSIC)* for the development of the all-sky CCD camera prototype in 2002. Finally, J.M.T.-R thanks *Ministerio de Educación y Ciencia (MEC)* for a JdC grant.

References

J. Borovička, P. Spurný, J. Keclikova, A new positional astrometric method for all-sky cameras. Astron. Astrophys. Suppl. Ser. **112**, 173–178 (1995)
Z. Ceplecha, Geometric, dynamic, orbital and photometric data on meteoroids from photographic fireball networks. Bull. Astron. Inst. Czechosl. **38**, 222–234 (1987)
Z. Ceplecha, P. Spurny, J. Borovička, *MORB software to determine meteoroid orbits*. (Ondrejov Observatory, Czech Republic 2000)
S. Clode, E. Zelniker, P. Kootsookos, V. Clarkson, A phase coded disk approach to thick curvilinear line detection. In 2004 XII. European Signal Processing Conference, Vienna, Austria, September 2004 (2004) pp. 1147–1150
R.O. Duda, P.E. Hart, Use of the Hough transformation to detect lines and curves in pictures. Comm. ACM **15**(1), 11–15 (1972)
P. Gural, P. Jenniskens, Leonid storm flux analysis from one Leonid MAC video AL50R. Earth Moon Planet. **82/83**, 221–247 (2000)
P. Gural, A human visual perception model and its impact on population index estimation, ZHR, and Best Look Direction. WGN, J. IMO **32**(4), 97–108 (2004)
P. Gural, Algorithms and software for meteor detection. Earth Moon Planet. (2007) doi:10.1007/s11038-007-9161-7
P.V.C. Hough, Methods and means for recognizing complex patterns. U.S. Patent 3,069,654 (1962)
Jenniskens et al., 2007 Aurigid Meteors. Central Bureau Astronomical Telegram, CBAT #1045, IAU (2007)
G.W. Kronk, *Meteor showers: A descriptive catalogue*. (Enslow Publ., Hillside, USA, 1988)
A.B.C. Lovell, *Meteor Astronomy* (OUP, Oxford, 1954)
J.M. Madiedo, J.M. Trigo-Rodríguez, Multi-station video orbits of minor meteor showers. Earth Moon Planet. (2007). doi:10.1007/s11038-007-9215-x
S. Molau, P. Gural, O. Okamura, Comparison of the American and the Asian 2001 Leonid meteor store. WGN J. Int. Meteor Org. **30-1**, 3–21 (2002)
W.H. Press, S.A. Teukolsky, W.T. Vetterling, B.P. Flannery, *Numerical Recipes in C* (Cambridge University Press, NY, 1992)

B.M. Quine, V. Tarasyuk, H. Mebrahtu, R. Hornsey, Determining star-image location: a new sub-pixel interpolation technique to process image centroids. Computer Phys. Comm. **177**, 700–706 (2007)

Z. Sekanina, Statistical model of meteor streams IV. A study of radio streams from the synoptic year. Icarus **27**, 265–321 (1976)

P. Spurný, J. Borovička, P. Koten, Multi-instrument observations of bright meteors in the Czech Republic. Earth Moon Planet. **95**, 569–578 (2004)

P. Spurný, J. Borovička, L. Shrbeny, Automation of the Czech part of the European fireball network: equipment, methods and first results. in *Near Earth Objects, our Celestial Neighbors: Opportunity and Risk*, ed. by G.B. Valsecchi and D. Vokrouhlicky, Cambridge Univ. Press. Proceedings IAU Symposium No. 236, 121–130 (2007)

C. Steyaert, *Photographic Astrometry* (Edited by the International Meteor Organization, Belgium, 1990)

J.M. Trigo-Rodríguez, J. Llorca, J. Fabregat, On the origin of the 1999 Leonid storm as deduced from photographic observations. Earth Moon Planet. **91**, 107 (2002)

J.M. Trigo-Rodríguez, A.J. Castro-Tirado, J. Llorca, J. Fabregat, V.J. Martínez, V. Reglero, M. Jelínek, P. Kubánek, T. Mateo, A. de Ugarte Postigo, The development of the Spanish Fireball Network using a new all-sky CCD system. Earth Moon Planet. **95**, 553 (2004a)

J.M. Trigo-Rodríguez, J. Llorca, E. Lyytinen, J.L. Ortiz, A. Sánchez Caso, C. Pineda, S. Torrell, 2002 Leonid storm fluxes and related orbital elements. Icarus 171, 219 (2004b)

J.M. Trigo-Rodríguez et al., Orionid Meteors. Central Bureau Astronomical Telegram, CBAT #698, IAU (2006a)

J.M. Trigo-Rodríguez, J. Llorca, A.J. Castro-Tirado, J.L. Ortiz, J.A. Docobo, J. Fabregat, The Spanish Fireball Network. Astron. Geophys. **47**, 6–26 (2006b)

J.M. Trigo-Rodríguez, J.M. Madiedo, A.J. Castro-Tirado, J.L. Ortiz, J. Llorca, J. Fabregat, S. Vítek, P.S. Gural, B. Troughton, P. Pujols, F. Gálvez, Spanish Meteor Network: 2006 continuous monitoring results. WGN J. IMO **35**(1), 13–22 (2007a)

J.M. Trigo-Rodríguez, J.M. Madiedo, J. Llorca, P.S. Gural, P. Pujols, T. Tezel, The 2006 Orionid outburst imaged by all-sky CCD cameras from Spain: meteoroid spatial fluxes and orbital elements. Monthly Notices Roy. Astron. Soc. **380**, 126–132 (2007b)

Trigo-Rodríguez J.M. et al., Kappa Cignids 2007. Central Bureau Astronomical Telegram, CBAT #1055, IAU (2007c)

R.T. Weryk, P.G. Brown, A. Domokos, W. Edwards, Z. Krzeminski, S.H. Nudds, D.L. Welch, The Southern Ontario all-sky meteor camera network. Earth Moon Planet. (2007). doi:10.1007/s11038-007-9183-1

The Southern Ontario All-sky Meteor Camera Network

R. J. Weryk · P. G. Brown · A. Domokos · W. N. Edwards · Z. Krzeminski · S. H. Nudds · D. L. Welch

Originally published in the journal Earth, Moon, and Planets, Volume 102, Nos 1–4.
DOI: 10.1007/s11038-007-9183-1

Abstract We have developed an automated network of all-sky CCD video systems to detect medium–large meteoroids ablating over Southern Ontario, Canada. The system currently consists of five stations with the largest baseline being 180 km. Each site runs a video rate recorder with sufficient resolution to determine meteoroid trajectories with a typical precision of about 300 m but no worse than 1 km. The sensitivity of the camera is close to a stellar visual magnitude of +1 which allows for astrometric calibrations using field stars. Photometric procedures have also been developed and tested. The system has a limiting magnitude for meteors of about −2 with the current detection algorithm.

Keywords Meteors · All-sky · Detection · Real-time

1 Introduction

The value of regional networks of all-sky meteor cameras has been demonstrated by several groups (cf. Spurný 1994; Trigo-Rodríguez et al. 2006). These networks enable physical and dynamical processes of meteor ablation to be studied in detail, orbits computed, and in a few cases have provided astronomical context and flight behaviour of meteorite falls (cf. Spurný et al. 2003). The system described in this paper is a fully automatic network (which presently has five cameras with two more planned) that is capable of detecting centimetre-sized meteoroids ablating over Southern Ontario, Canada. The largest baseline for the five stations is 180 km, providing atmospheric overlap for most events by multiple cameras.

R. J. Weryk (✉) · P. G. Brown · A. Domokos · Z. Krzeminski · S. H. Nudds
Department of Physics and Astronomy, University of Western Ontario, London, ON, Canada N6A 3K7
e-mail: rjweryk@uwo.ca

W. N. Edwards
Department of Earth Science, University of Western Ontario, London, ON, Canada N6A 5B7

D. L. Welch
Department of Physics and Astronomy, McMaster University, Hamilton, ON, Canada L8S 4M1

DOI: 10.1007/978-0-387-78419-9_33

The primary purpose of the all-sky network is to act as the main trigger for the multi-instrumental (radar, infrasound, optical) Southern Ontario Meteor Network. Metric and photometric data from the all-sky cameras provide trajectory and energy estimates for each meteor event to be compared with similar estimates from other sensors. These optical data when combined with radar, infrasound, and high-speed photometers can provide strong constraints for numerical entry models (such as ReVelle 2005).

The secondary goal of the all-sky network is to provide orbits and atmospheric data for any meteoroids, to characterise enhancements in large meteoroids from streams (eg. Taurids in 2005) and to survey the overall orbital distribution and physical characteristics of centimetre-sized meteoroids.

2 Hardware Design

The cameras used are HiCam HB-710E with Hole Accumulation Diode (HAD) CCDs. Each camera is equipped with a Rainbow L163VDC4 1.6–3.4 mm f/1.4 lens. The unit is enclosed in a weather proof enclosure with 30 cm diameter acrylic domes. Each camera is set up such that the sky coverage is slightly less than all-sky (complete down to about 30° elevation), improving the pixels-per-degree scale for higher elevation meteors (to about 0.25°/pixel). The camera system operates only at night and connects to PC computers running the GNU/Linux operating system. Video is digitised (using off-the-shelf frame-grabber cards) with 640 $\times$ 480 resolution at 30 frames per second (RS-170 video standard) by Brooktree 878 video capture cards. This has sufficient resolution to determine meteor trajectories with typical precision of 300 m and worse-case fits of 1 km. The high density of stations in the network often results in bright meteors being detected at three or more stations, aiding in improved accuracy of trajectory fits. The stellar sensitivity of the camera is close to +1 magnitude for a single video frame, with effective meteor sensitivities near −2 magnitude.

3 Software Design

The detection algorithm works in real time by comparing the current video frame against an earlier frame (currently set to ten frames). The images are compared on a pixel-by-pixel basis and the number of pixels whose intensity has increased by at least a set threshold value (determined by the user) are counted. For typical sky brightness levels in Southern Ontario, we find that a threshold setting of 70 digital units works well, and prevents the system from triggering on random noise. When the pixel sum has reached a set limit for a predefined number of frames (currently 12 pixels for two consecutive frames), an event is triggered. The event ends when the pixel sum drops below a separate limit (currently set to five pixels for three consecutive frames). The system pads the detected event with extra video frames to compensate for those instances for when the meteor may still be present, but below the trigger sensitivity of the system.

The pixel location of a meteor is found through a centre-of-mass algorithm weighted by the increase in pixel intensity. Equation (1) shows the form of the algorithm for the x-coordinate. Here, v_x is the pixel intensity difference between frames for a particular pixel, and all pixels corresponding to the meteor are summed over. This method is used on other all-sky camera networks. It has proven to work well when compared to manual centroiding of a meteor in each video frame. We have compared this technique with several other centroiding algorithms and manual positional picks and find agreement in almost all cases. Exceptions are typically very bright events where flaring or blooming/optical artifacts can distort the centroiding algorithm and manual intervention often produces better results. Additional work on finding an automated method that works for all meteor cases is ongoing.

$$c_x = \frac{\Sigma v_x x}{\Sigma v_x} \quad (1)$$

4 Calibration and Analysis

Astrometric calibration of each camera is accomplished by first stacking video frames (currently we use 1200 frames) and then correlating and fitting the pixel and celestial coordinates for each star using the all-sky fitting routine of Borovička et al. (1995). This has proven to accurately map the all-sky image with typical stellar residuals in the sub-pixel range (for stars used in the fit) of 0.1/deg above 20° elevation. In general, astrometric fits rapidly degrade at lower elevations. Where possible, we avoid low elevation solutions.

As a meteor is detected, the astrometric position in each video frame is computed in real time using one of the previously computed calibrations and recorded.

The meteor appearance time is calibrated using Network Time Protocol (NTP) which can read from a local or remote GPS receiver. Using a local GPS receiver, time accuracies on the order of 50 μs can be obtained, much smaller than the interframe time of 33 ms.

Instrumental magnitudes are calculated using standard aperture photometry (Mighell 1999). A circular aperture of a variable size is set to surround the position of the meteor in each video frame. Pixel values are summed within the aperture and the background level (determined from median combining 25 frames where the meteor is not present) is subtracted. The resulting pixel sum is then transformed into an instrumental magnitude scale, with an arbitrary zero level. To obtain calibrations from the instrumental scale to the various Johnsons-Cousins filter bands, m_x (where x = UBVRI), field stars brighter than +2.5 are compared to their bright star catalog values (Hoffleit and Jaschek 1982). To extend the calibration to very bright magnitudes (brighter than −2) the planets Jupiter, Venus, and the Moon are used along with colour indices of the Sun. Using this method, the instrumental magnitude is found to relate linearly to the UBVRI magnitudes, with the best correlation found for the R band where the bandpass coverage of the HAD CCD response is greatest. Regressions are taken for instrumental and red-band magnitudes for each individual camera and applied to the raw data to move from an instrumental magnitude to an apparent magnitude. Lastly, we attempt to determine the magnitude offset for the HAD-based camera system to that of the Panchromatic in order to directly compare our observations with those of previous authors. To this end, we use published meteor spectra of several sporadic and shower meteors (e.g. Borovička and Betlem 1997; Borovička and Zamorano 1995; Carbary et al. 2004; Spurný et al. 2004) and apply the standard methods of synthetic photometry (Straizys 1996) to deduce the color indices for meteors between the HAD and the Panchromatic bandpasses. We find that for most meteors HAD − Pan = 0.5. One notable exception is Perseid meteors that have strong Ca^+ and Fe/Mg lines in the B band where the panchromatic is more sensitive than the HAD CCD.

In addition to this colour term correction to produce equivalent panchromatic magnitudes, we correct for air mass, using a standard correction of 0.34 mag/AM. This has been found to be a good average value from empirical measurements at our sites.

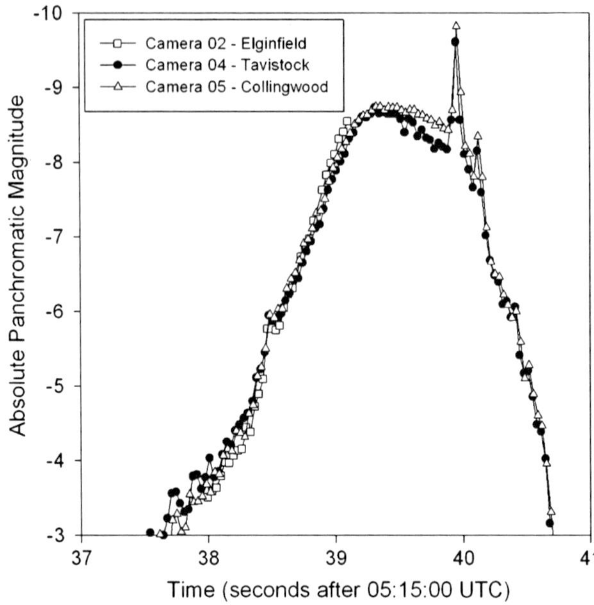

Fig. 1 Lightcurve for 20060305 event

Table 1 Velocity and orbital elements for the 20060305 event

Quantity	Value	Error	Units
beg lat	43.915	0.001	deg
beg lon	−81.740	0.001	deg
beg ht	77.2	0.1	km
end lat	44.114	0.002	deg
end lon	−81.701	0.002	deg
end ht	36.8	0.1	km
v_i	18.74	0.30	km/s
entry Z	41.7	0.5	deg
duration	3.2	0.1	s
RA	155.71	0.28	deg
Dec	2.58	0.61	deg
a	1.61	0.05	AU
e	0.54	0.02	none
incl	3.28	0.22	deg
ω	73.39	0.59	deg
Ω	164.41	0.00	deg
q per	0.74	0.01	deg
q aph	2.48	0.11	deg
mass	11.6	1.3	kg

5 System Operation

Once an event has been recorded to disk and has been astrometrically measured, it is also encoded into a video format for later verification. Each hour, all relevant data files are copied to a central server at the University of Western Ontario (UWO). Each morning, multistation matchups are automatically found, and the original uncompressed video frames for each multistation event are copied to the central server for additional analysis that is not yet automated (for example, photometric determination of mass and orbital parameters). Meteor events must be manually verified to filter out false triggers such as birds, thunderstorms, and aircraft. The system has been running since 2004 and has an average yearly event count exceeding 400 meteors (roughly half of which are multistation). False detections are low given the current triggering thresholds.

6 Results and Conclusions

Figure 1 shows an example lightcurve for an event detected at three stations on 5 March 2006. The variation in the measured lightcurve provides a means to gauge the inherent robustness of our reduction techniques when applied individually to each camera for a given event. The trajectory solution for this fireball, together with individual station velocity estimates and orbital information, is given in Table 1. At present we do not apply deceleration corrections to the observed average velocity to produce out-of atmosphere velocities; in the future we hope to arrive at such deceleration corrections by comparing modelled and observed lightcurves and metric data. In the example event, the initial speed was used to compute the orbit. These corrections are important as they allow for more accurate orbital determinations which allows for better matchups with potential parent objects. In certain cases, neglecting the deceleration can have a significant effect on the resulting orbit. The end height (36.8 km) and the speed at the end height (8 km/s) suggests that this fireball may have dropped a small meteorite in Lake Huron.

Acknowledgements The authors thank the two anonymous referees for providing constructive commentary on this paper.

References

J. Borovička, H. Betlem, Spectral analysis of two Perseid meteors. Planet. Space Sci. **45**, 563–575 (1997)

J. Borovička, P. Spurný, J. Keclikova, A new positional astrometric method for all-sky cameras. Astron. Astrophys. **112**, 173–178 (1995)

J. Borovička, J. Zamorano, The spectrum of a fireball taken with a 2-m telescope. Earth Moon Planets **68**, 217–222 (1995)

J.F. Carbary, D. Morrison, G.L. Romick, J.H. Yee, Spectrum of a Leonid meteor from 110 to 860 nm. Adv. Space Sci. **33**, 1455–1458 (2004)

D. Hoffleit, C. Jaschek, *The bright star catalogue*, 4th edn. (Yale University Observatory, New Haven, 1982), pp 472

K.J. Mighell, in *Algorithms for CCD Stellar Photometry, Astronomical Data Analysis Software and Systems VIII, ASP Conference Series*, ed. by D.M. Mehringer, R.L. Plante, D.A. Roberts, vol 172 (1999), pp. 317–328

D.O. ReVelle, Recent advances in bolide entry modelling: a bolide potpourri. Earth Moon Planets **97**, 1–35 (2005)

P. Spurný, Recent fireballs photographed in central Europe. Planet. Space Sci. **42**, 157–162 (1994)

P. Spurný, J. Borovička, P. Koten, Multi-instrument observations of bright meteors in the Czech Republic. Earth Moon Planets **95**, 569–578 (2004)
P. Spurný, J. Oberst, D. Heinlein, Photographic observations of Neuschwanstein, a second meteorite from the orbit of the Příbram chondrite. Nature **423**, 151–153 (2003)
V. Straizys, The method of synthetic photometry. Baltic Astron. **5**, 459–476 (1996)
J.M. Trigo-Rodríguez, J. Llorca, A.J. Castro-Tirado, J.L. Ortiz, J.A. Docobo, J. Fabregat, The Spanish fireball network. Astron. Astrophys. **47**, 6.26–6.28 (2006)

The IMO Virtual Meteor Observatory (VMO): Architectural Design

Detlef Koschny · Jonathan Mc Auliffe · Geert Barentsen

Originally published in the journal Earth, Moon, and Planets, Volume 102, Nos 1–4.
DOI: 10.1007/s11038-007-9216-9

Abstract This paper describes the progress on the Virtual Meteor Observatory (VMO), a database which is being developed at ESA/RSSD to store video meteor observations and their derived orbits. The VMO was triggered by a discussion which took place at the first Meteor Orbit Determination (MOD) workshop in Roden, The Netherlands, in September 2006. Representatives of 15 groups working on the determination of meteor orbits and working with the resulting orbits discussed the design and implementation of a database which would combine different meteor orbit datasets. From this the concept of the VMO was born, which will, in the long run, allow accessing meteor observations via the internet. In the beginning, it will focus on meteor orbit data obtained with video systems. This paper presents the architectural design of the database as it has been defined in the meantime.

Keywords Meteor · Orbit · Database · Archiving · Virtual Meteor Observatory

1 Introduction

Meteoroid orbits derived from meteor observations are relevant to a number of different scientific investigations, e.g., to compare modeling efforts of meteoroid streams with the actual meteoroid population in the solar system, or to put constraints on the dynamical behaviour of meteoroids and even their ejection mechanisms from their parent bodies. Meteor orbits are essential in studying whether meteoroids come from asteroidal or comet parent bodies (see e.g. Starczewksi and Jopek 2004).

Typically, there are groups focusing on observing meteors and computing orbits, while other groups concentrate on the interpretation of the orbit data. Thus, storing meteor orbits in such a way that makes the data easily accessible and traceable is important. A number of orbit databases exist already. For example, the IAU orbit database (Kornos 2007), the

D. Koschny (✉) · J. Mc Auliffe · G. Barentsen
Research and Scientific Support Department (RSSD), ESA/ESTEC, SCI-SM, Keplerlaan 1,
2200 Noordwijk, The Netherlands
e-mail: Detlef.Koschny@esa.int

database of the Japanese Meteor Science Seminar working group MSSWG,[1]the database of the Dutch Meteor Society DMS[2] and several others. When talking to the data users, one finds that several major drawbacks exist: the data is stored in a distributed manner, it is difficult to query, and there is often information missing, e.g., on the accuracy of the data.

An effort to improve this situation was started at the MOD Workshop in Roden, The Netherlands, in September 2006. There, Ryabova (2007) and Jenniskens (2007) presented examples of how meteor orbit data is used and outlined some of basic requirements of a meteor database. This paper derives from the previous discussions the 'user point of view', i.e., how will users want to store and access the data. It also presents the current architecture of the database, which will serve as a basis for the implementation.

2 Initial Concept Development and History

During the workshop, top-level user requirements for a meteor orbit database were defined by Ryabova (2007). To summarize, she emphasises that the data base shall allow to retrieve data for a well-defined mass range, the data shall have high accuracy and be traceable, and there should be a significant number of debiased data available. The detailed requirements were discussed in the workshop and are summarized in Koschny et al. (2007), a first definition of the database on which this paper continues is given in Barentsen 2007.

It was concluded that a centralised data storage and handling infrastructure for single-, double-, and multi-station meteor observations will be set up. This system will be in the form of an online database with inbuilt (and mostly autonomous) data validation and handling capabilities, as well as server-side routines for 'simultaneous event' identification and orbit calculation. The database will initially be implemented by the ESA/RSSD Meteor Research Group in collaboration with the International Meteor Organization and the members of the MOD Working Group.

3 User Perspective

An overview of the usage of the database is given in Fig. 1. The actual data files are indicated by dotted rectangles. There are several options for the user to interact with the database. On the simplest level, indicated on the top, the user sends his/her single station data to the VMO. The data is checked (as far as possible by automated routines) for good quality and ingested into the database.

Data quality checking will be implemented in several steps. First, simple consistency checks can be done, e.g. was the sun below the horizon during the observation (checking for errors in the time stamp); is the meteor position within the camera field of view, etc. In a second step, more advanced checks can be done, e.g., the computed errors can be checked on whether they are within acceptable limits, e.g. is the positional accuracy better than the pixel scale of the camera.

Software routines running as part of the system check the incoming data for potential simultaneous observations and identify such events. A built-in routine computes the orbits from these meteor events and puts them into the orbit data repository.

[1] http://www.imo.net/files/data/msswg/msswg.txt

[2] ftp://ftp.strw.leidenuniv.nl/pub/betlem/orbits/

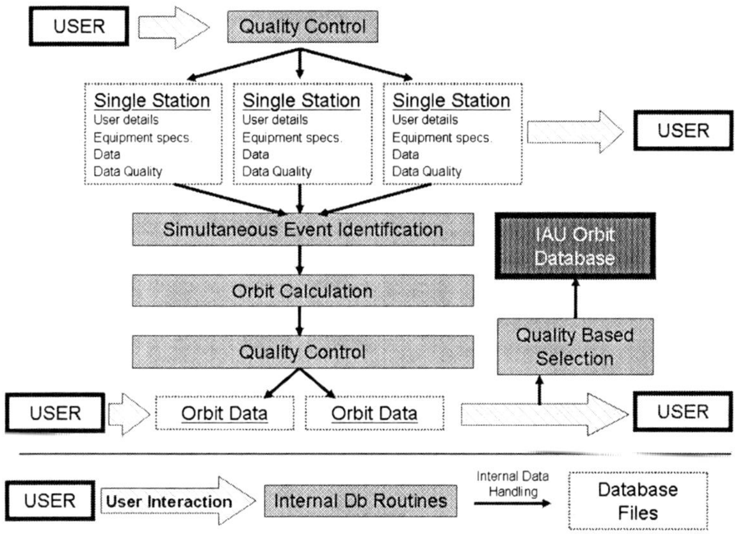

Fig. 1 The VMO functionality as seen from a user perspective

Alternatively, orbit data—e.g., output from UFOcapture[3]—can be sent to the database from users directly. Note that it would still be required to submit the corresponding single-station data to ensure the traceability of the data.

Internal data handling and processing routines will be made available to the user for incorporation into the user's software.

4 Work Flow

The workflow of the implementation happens in the following sequence.

1. Define the database architecture. So-called 'use cases' help in the definition. This point is addressed in the next section.
2. Define the database format for the single station meteor data and the orbit data. The definition is based on previous discussions as documented e.g., in Koschny et al. (2007).
3. Implement the single station database. While the current implementation of the database focuses on video meteors, the database structure is open for other meteor observation types (e.g., the observer description can be used for any observation, whether visual, video, telescopic, or radio).
4. Set up the user interface for the single station data, keeping in mind additional user interfaces e.g., for directly ingesting orbit data, and add quality control checking routines.
5. The database format for the orbit data will be implemented.
6. Software routines to perform simultaneous event identification and orbit calculations will be implemented.

[3] http://sonotaco.com/soft/e-index.html

We here present the results of item 1, the database architecture. Item 2 is already covered by Koschny et al. (2007). Items 3–6 will be subject of future papers.

5 Architectural Design

All the functionality requirements described in the previous sections have been converted to a database architecture, which is shown in Fig. 2. This architecture is currently being finalized and will form the starting point for the implementation.

The database is based on the Structured Query Language (SQL) standard which is a computer language designed for the retrieval and management of data in relational databases. It is implemented using the open source software package PostSQL. The actual data formats are defined via files in the Extended Markup Language (XML) format. The figure shows the different layers of the system. The central layer is called the 'developer layer'. In it, the VMO gives direct access to the database elements. This requires that user software base their data files on the XML definitions of the VMO. Alternatively, in the 'user layer', the software outputs their own data formats (as is the case for some of the existing meteor detection software such as MetRec (Molau 1999), MeteorScan (Gural 1997), or UFOCapture). A converter will then convert the output data to the VMO format.

All internal routines to access data and certain queries and computations will be provided to external users via web services based on the http protocol. This will allow users to interface to the VMO without having to know all the internals of the database. The transport of data will be performed via standardized XML files, similar to the data

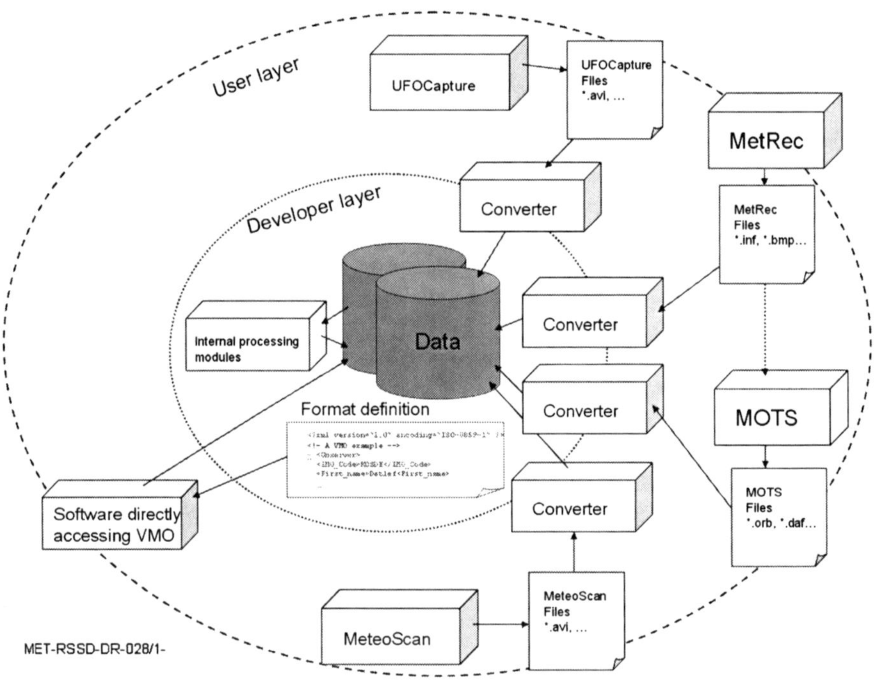

Fig. 2 Architecture of the Virtual Meteor Observatory (VMO)

definition files. Image files will be transmitted as separate files in one of the standard image file formats. To ensure that files do not get lost, all files belonging to one delivery (e.g., one observing night) have to be zipped together.

All single-station meteor events will be identified with a unique identifier. Internally, this will be an arbitrary hexadecimal number. This unique identifier will also be converted to a more human-readable identifier following the definition XXX_YYYY – MM – DD_CODE where XXX denotes the type of the meteor observation ('VID' for video, 'VIS' for visual, 'RDO' for radio, 'TEL' for telescopic), 'YYYY' is the year, 'MM' the month, 'DD' the day of the observation. 'CODE' is a code (hexadecimal or similar) for the meteor during that day. This will allow users to easier identify meteors in publications.

Similarily, the meteor orbits will also be assigned unique internal identifiers which will be converted to a more human-readable version in addition. These codes will allow the traceability of orbits to individual meteor events.

The user interface will be similar to the current interfaces of the International Meteor Organization. In addition, automated ingestion methods will be supported, e.g., direct database interface from software tools such as MeteorScan or UFOCapture would be possible. The implementation of the web pages will be done using the content management system *drupal*.[4]

6 Conclusion

This paper describes the architectural design of a so-called Virtual Meteor Observatory (VMO), and gives some background on the requirements for the database. The VMO is currently being implemented by the Meteor Research Group of ESA/RSSD in collaboration with the International Meteor Organization. The design phase is nearing its end, the descriptions given in this paper only need minor updates and will lead to the implementation phase of the VMO. Future work will encompass the implementation of the database and the user interface to ingest and retrieve data.

Acknowledgements The authors would like to thank all participants in the Roden workshop as well as all those who have joined and continue to contribute to the online discussion within the MODWG discussion forum. Thanks also to Maxim Khodachenko and Helmut Rucker from the EuroPlaNet N3 consortium, who support the ongoing activities not only conceptually but also by financially through the EuroPlaNet co-ordinated observations initiative. Thanks to J. Rendtel and an anonymous reviewer for their useful comments.

References

G. Barentsen, The unified meteor database: A generic archiving project for meteor data. in *Proceedings of the First Europlanet Workshop on Meteor Orbit Determination*, ed. by J. Mc Auliffe, D. Koschny. Roden, 11–13 September 2006, International Meteor Organisation, ISBN 978-2-87355-019-6, 101-104 (2007)

P. Gural, An operational autonomous meteor detector: Development issues and early results. WGN J. IMO **25**(3), 136–139 (1997)

P. Jenniskens, The IAU meteor shower nomenclature rules, in *Proceedings of the First Europlanet Workshop on Meteor Orbit Determination*, ed. by J. Mc Auliffe, D. Koschny. Roden, 11–13 September 2006, International Meteor Organisation, ISBN 978-2-87355-019-6 (2007), pp. 16–19

[4] http://drupal.org/

L. Kornos, The IAU meteor database. in *Proceedings of the First Europlanet Workshop on Meteor Orbit Determination*, ed. by J. Mc Auliffe, D. Koschny. Roden, 11–13 September 2006, International Meteor Organisation, ISBN 978-2-87355-019-6 (2007), pp. 52–54

D. Koschny, J. Mc Auliffe, G. Barentsen, The IMO Virtual Meteor Observatory (VMO)—A first definition, in *Proceedings of the First Europlanet Workshop on Meteor Orbit Determination*, ed. by J. Mc Auliffe, D. Koschny. Roden, 11–13 September 2006, International Meteor Organisation, ISBN 978-2-87355-019-6 (2007), pp. 105–123

S. Molau, The meteor detection software MetRec, in *Proceedings of the International Meteor Conference*, ed. by R. Arlt, A. Knoefel. Stara Lesna 20–23 August 1998, International Meteor Organisation, ISBN 2-87355-010-4 (1999), pp. 9–16

G. Ryabova, Possible use of meteor orbital data. In *Proceedings of the First Europlanet Workshop on Meteor Orbit Determination*, ed. by J. Mc Auliffe, D. Koschny. Roden, 11–13 September 2006, International Meteor Organisation, ISBN 978-2-87355-019-6 (2007), pp. 7–15

S. Starczewski, T.J. Jopek, Dynamical relation of meteorids to comets and asteroids. Earth Moon Planets **95** (1–4):41–44 (2004)

A New Bolide Station at the High Tatra Mountains

Jan Svoren · Pavel Spurny · Vladimir Porubcan · Zuzana Kanuchova

Originally published in the journal Earth, Moon, and Planets, Volume 102, Nos 1–4.
DOI: 10.1007/s11038-007-9179-x

Abstract The European Fireball Network (EN) is operating since 1963 and one of its stable stations, from the very beginning, is the station at the Skalnate Pleso Observatory in the High Tatras. The station is sited at a height of 1788 m. More than 2900 expositions has been made at the Skalnate Pleso station since 1964 and among them one significant and spectacular event was recorded—bolide Turji-Remety in 2001 followed by a fall of about 450 kg meteorite (Spurny and Porubcan [in: Warmbein (ed.) Asteroids Comets Meteors, 2002]). A systematic search for the meteorite was unsuccessful. The new station having an ideal horizon will be operating since July 2007 on the top of Lomnicky Stit (2636 m above the sea level). This station will be equipped with an Autonomous Fireball Observatory of the Astronomical Institute of the Czech Academy of Sciences, which are already utilized in the Czech part of the EN for several years.

Keywords Photographic observations of meteors · All-sky bolide camera

1 All-sky Camera at the Skalnate Pleso Observatory

Multi-station photographic observations rank among fundamental in meteor astronomy as provide the most detailed and precise information on the physical and orbital parameters of meteoroids. The successful photography and find of the Pribram meteorite in 1959 (Ceplecha 1961) was an impetus for the establishment of the all-sky photographic cameras network for fireball monitoring and possible recovery of meteorites in Czechoslovakia. The network operating since 1963 rapidly expanded to the European Fireball Network (EN) and at the end of the 1960s it consisted of 46 stations. Parallel with the EN there were running

J. Svoren (✉) · V. Porubcan · Z. Kanuchova
Astronomical Institute of the Slovak Academy of Sciences, 059 60 Tatranska Lomnica, The Slovak Republic
e-mail: astrsven@ta3.sk

P. Spurny
Astronomical Institute, Academy of Sciences of the Czech Republic, 251 65 Ondrejov, The Czech Republic

two other fireball networks—since 1964 the Prairie Network in the USA (McCrosky and Ceplecha 1969) and since 1971 the Meteorite Observation and Recovery Project (MORP) in western Canada (Halliday 1973). While both the networks in the USA and Canada have already stopped their operation several decades ago, the EN proceeds in the operation until present.

Since 2003, the Czech part of the EN underwent a significant improvement, where two new stations were built by the end of 2006 and at six stations older manual cameras were replaced by new full automatic cameras (Spurny et al. 2006).

One of the EN stable stations, from the very beginning, is the station at the Skalnate Pleso Observatory in the High Tatras (EN Station No. 22). The station is sited on the roof of the Skalnate Pleso Observatory at an altitude of 1788 m. It has gained many valuable data as it has been the most eastern running station of the EN for a long time.

During the period of operation over 40 years, the station as other stations of the EN, underwent several reconstructions. In the first years of operation each station was equipped with a commercial camera (effective focal length of 5.7 mm) and the sky reflected in a spherical mirror was photographed on a 35-mm film. A disadvantage of the device was a rather low positional accuracy and necessity to renovate the cover of the mirror more times in a year. In the 1990s, the commercial camera was replaced by a fish-eye camera (F-Distagon, f = 30 mm) enabling a direct photography of the whole sky on a 9 × 12 cm planfilm, removing thus the problems with the mirror and substantially increasing the accuracy. The best accuracy of the orbits obtained by the EN stations at present reaches 0.03° in the argument of perihelion, 0.01° in the inclination, less than 0.0002 AU in the perihelion distance and less than 7 m/s in the geocentric velocity (Spurny et al. 2006).

Nowadays, this is one of only two stations operating in Slovakia. The second one is located at the Comenius University Observatory at Modra. Up to now more than 2900 expositions have been taken at the Skalnate Pleso station since 1964, approximately 200 orbits were recorded and from them for about 20 orbits were calculated. Among them one significant and spectacular event was recorded on November 17, 2001—flight of the Turji-Remety bolide followed by an expected fall of a meteorite (Spurny and Porubcan 2002). The bolide flown over western Ukraine in direction to Slovakia. The fireball entered the atmosphere at a velocity of 18.5 km/s and began its luminous path of 106 km at a height of 81 km. The maximum brightness of the absolute photographic magnitude −18 reached at a height of 25 km (Fig. 1), terminated at 13.5 km near the Ukrainian village Turji-Remety,

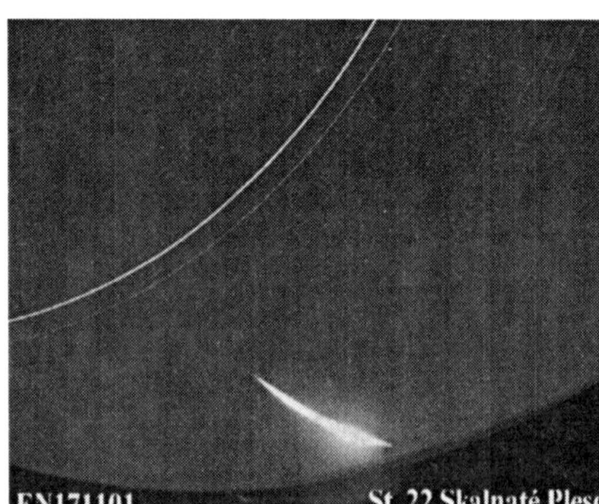

Fig. 1 The Turji-Remety event photographed at the Skalnate Pleso Observatory

Fig. 2 Autonomous photographic camera

which means that it is the deepest photographically recorded fireball in history. The calculated initial dynamical mass of the meteoroid was about 4300 kg, some part of it survived the flight through the atmosphere and dropped on the ground. The computed terminal mass was of the order of hundred of kilograms. From many visual observations, three main pieces were observed until the end of the luminous trajectory. According to the calculations the terminal masses of the pieces could reach over 100 kg each. Several expeditions were organized to recover the meteorites, but so far without a success (Toth et al. 2005).

2 Autonomous Fireball Observatory at the Lomnicky Stit

A disadvantage of the Skalnate Pleso position is a close mountain range shielding western horizon (the preferred direction where other stations of the EN are located) up to an elevation of 22°. The uncovered direction to west greatly increases chances and numbers of detection of common bolides even low at horizon for which reliable orbits can be calculated. Therefore, it was decided to move the station to a near only 1 km distant Lomnicky Stit (2636 m above the sea level), the site of the coronal station of the Astronomical Institute of the Slovak Academy of Sciences (Sakurai et al. 2004).

The new station having thus an ideal horizon will start its operation in July 2007 and will be equipped with an Autonomous Fireball Observatory of the Astronomical Institute of the Academy of Sciences of the Czech Republic (Fig. 2). The device has been already utilized in the Czech part of the EN for several years.

From a long-term study of the climate at the Lomnicky Stit it is evident that majority of clear nights and exposures can be expected in winter, when longer periods of inversion are more frequent. This advantage is superimposed by the fact that the adjacent station in the Czech Republic, Lysa Hora, which is in a favorable distance from the Lomnicky Stit of 80 km to the west, is also above the inversion because it is located in the mountains on the highest top of the Beskydy Mountains (1324 m above the sea level).

The camera has an internal heating important at almost Antarctic winter period conditions at the peak. The station is under a steady check by the staff of the close coronal station.

Acknowledgements This research was supported by Grant MRTN-CT-2006-035519 of the EU Program Marie Curie Actions—Research Training Networks and VEGA—the Slovak Grant Agency for Science (grants Nos. 7009 and 3067).

References

Z. Ceplecha, Multiple fall of Pribram meteorites photographed. Bull. Astron. Inst. Czechosl. **12**, 21–47 (1961)

I. Halliday, Photographic fireball networks. in *Evolutionary and Physical Properties of Meteoroids*, ed. by C.L. Hemenway, P.M. Millman, A.F. Cook (NASA SP-319, Washington, 1973), pp. 1–8

R.E. McCrosky, Z. Ceplecha, Photographic networks for fireballs. in *Meteorite Research*. ed. by P.M. Millman (Reidel Publ., Dordrecht, 1969), pp. 600–612

T. Sakurai, V. Rusin, M. Minarovjech, Solar-cycle variation of near-sun sky brightness observed with coronagraphs. Adv. Space Res. **34**, 297–301 (2004)

P. Spurny, V. Porubcan, The EN171101 bolide—the deepest ever photographed fireball. in *Asteroids Comets Meteors 2002*. ed. by B. Warmbein (ESA SP-500, 2002), pp. 269–272

P. Spurny, J. Borovicka, L. Shrbeny, Automation of the Czech part of the European fireball network: equipment, methods and first results. In: *Near Earth Objects, Our Celestial Neighbors: Opportunity and Risk, Proceedings of the IAU Symposium 236*, Vol. 2 (2006), pp. 121–130

J. Toth, J. Catlos, S. Gajdos, J. Vilagi, E. Demencik, D. Lorenc, Slovak expedition Turji-Remety. Meteor. Rep. SAS **26**, 56–64 (2005) (in Slovak)

TV Meteor Observations from Modra

J. Tóth · L. Kornoš · Š. Gajdoš · D. Kalmančok · P. Zigo · J. Világi · M. Hajduková Jr.

Originally published in the journal Earth, Moon, and Planets, Volume 102, Nos 1–4.
DOI: 10.1007/s11038-007-9160-8 © Springer Science+Business Media B.V. 2007

Abstract We present our experience and initial results of single-station observation using the new fish-eye TV system, as well as double station TV observation of the Geminids 2006 shower. The fixed fish-eye TV system was developed for monitoring meteor activity throughout the year. We discuss the astrometric precision of our observations using the UFOAnalyser software.

Keywords Fish-eye lenses · Meteor shower · TV meteor observations

1 Introduction

For more than 15 years, astrometric and photometric observations of asteroids and comets have been made at the Astronomical and Geophysical Observatory (AGO) of the Comenius University in Modra, as well as all-sky photographic meteor observations using Zeiss Distagon photographic cameras fitted with 3.5/30 mm fish-eye lenses (Zigo et al. 2006; Gajdoš et al. 2006). Two photographic meteor cameras operate at the site, one in fixed and another in a guided mode. Our station No. 21 Modra (Fig. 1a) is a part of the European Network (EN) for fireball detection coordinated by Ondřejov Observatory, Czech Republic. In the past, we had used TV cameras for major meteor shower observations (Tóth and Kornoš 2007). Recently we have developed a new fish-eye TV system to be used mainly for minor meteor shower observations.

J. Tóth (✉) · L. Kornoš · Š. Gajdoš · D. Kalmančok · P. Zigo · J. Világi
Department of Astronomy, Physics of the Earth and Meteorology Faculty of Mathematics, Physics and Informatics, Comenius University, Bratislava, Slovakia
e-mail: toth@fmph.uniba.sk

M. Hajduková Jr.
Astronomical Institute of Slovak Academy of Sciences, Dúbravská cesta 9, 845 04 Bratislava, Slovakia

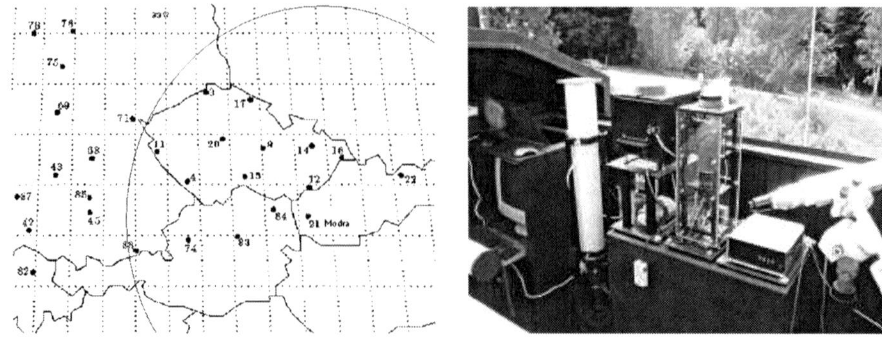

Fig. 1 (a) Part of European Fireball Network (Oberst et al. 1998), Modra station is marked by No. 21. The assumed observed radius is depicted by the circle. (b) The new fish-eye TV meteor system (in the middle) has started regular observations on April 1, 2007

2 The Fish-eye TV System

The new fish-eye TV meteor system (Fig. 1b) has started regular observations on April 1, 2007. The system consists of a fish-eye Canon 2.4/15 mm objective, 2″ Mullard image intensifier, Meopta 1.9/16 mm lens and a Watec 120N camera. The analog video signal is digitized in real time and analysed by "UFOCapture" software (author SonotaCo, http://www.sonotaco.com/e_index.html), which is able to detect any moving object including meteors. The resolution of the system is 720×540 (15 arcmin/px), corresponding to a field of view of $170° \times 140°$ (Fig. 2). The limiting stellar magnitude is $+5.5^m$ and meteors up to magnitude $+3^m$ are detected. The system operates autonomously.

The astrometric precision of this fish-eye TV system is quite good. The standard deviation for more than 50 stars reduced by the "UFOAnalyserV2" is less than 0.05° for zenithal distances up to 60°. Also we have tested the position accuracy for several stars and

Fig. 2 An example of the all-sky composite negative image obtained from an avi file of TV fish-eye system. The bright meteor, stars and Milky way are clearly visible. The limiting stellar magnitude is $+5.5^m$. South is up and west is on the left

planets near the horizon, where the astrometric precision of the measurement decreases to 0.5°. The fourth order polynomial expansion used by UFOAnalyserV2 is insufficient to correct fish-eye projection, especially near horizon.

3 Results

3.1 April 2007 Results

Our TV system, operating as a single station worked 27 nights (199 h) during April 2007, including nights with a bright Moon. In total 300 meteors were detected: 74 Lyrids, 24 meteors from antihelion source, 10 meteors from helion source, 20 meteors from other sources and 172 other sporadics. The identification of Lyrids were based in their radiant position and angular velocity. The sporadic meteor frequency was in the range of 0.5–3 meteors per hour, depending on the Moon phase, which reduced meteor rates due to the bright background of the sky. During this month, TV and photographic fish-eye cameras observed 7 fireballs. The brightest fireball was about –10th magnitude. The orbit of the "Kozmice" fireball from April 14 was also observed by our TV system and the EN photographic cameras for which Spurný (pers. comm. 2007) calculated its precise orbit as $a = 1.2780 \pm 0.0007$ AU, $e = 0.5073 \pm 0.0003$, $q = 0.62961 \pm 0.00014$ AU, $Q = 1.9263 \pm 0.0014$ AU, $\omega = 276.09° \pm 0.03°$, $\Omega = 24.36223° \pm 0.00001°$, $i = 7.934° \pm 0.009°$. The meteoroid, which had photometric mass of 3 kg, has a typical NEO orbit of Apollo type. The observed astrometric position of the fireball by TV and photographic method from Modra station correspond to each other within 3 arcmin.

3.2 Single Station Observations of the 2007 Lyrids

We observed the Lyrid activity for the period April 10–26. The activity profile during the night of maximum activity (April 22/23) was derived from fish-eye TV single station observations that were corrected to the radiant position (Fig. 3a). The discrepancy between our video and the IMO visual data for the Lyrid ZHRs is mainly caused by the Moon light, which reduced meteor rates during the first part of the night. We also obtained a single station radiant position at the time of maximum activity ($\alpha = 272.5°$, $\delta = 33.2°$, $\lambda_{\odot} = 33.2°$) for these 74 Lyrids (Fig. 3b). The diameter of the radiant area from the entire activity interval was 10°. The faintest observed Lyrid meteor was about magnitude $+3^{m}$ and the brightest one was about –6th magnitude.

3.3 Two Station Orbits from the 2006 Geminids

The activity of the Geminid meteor shower was monitored from two stations at Modra and Stupava that were both equipped with test non-intensified cameras (details in Tóth and Kornoš 2007) with 20° field of view.

Tóth and Kornoš (2007) presented the activity profile that was derived for this TV system and found that maximum activity peaked at 1:30 ± 0:30 UT December 14, 2006 and the single station radiant position was $\alpha = 114.0°$, $\delta = 33.3°$.

The radiant area derived from 31 Geminid meteor observations is very compact (less than 1°). This is indirect confirmation of the astrometric precision of "UFOAnalyser"

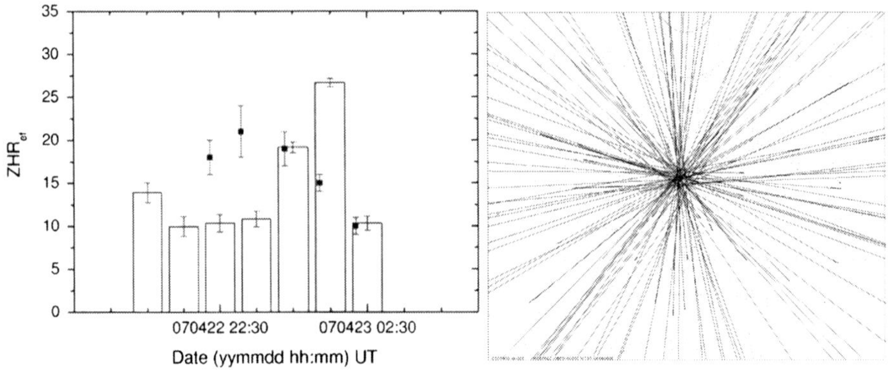

Fig. 3 (a) The activity profile (histogram) compared with the visual observations (black square) from IMO (Barentsen 2007). (b) Single station radiant of 74 Lyrids has diameter 10°. The scale of the image is about 100° × 100°

Table 1 The individual radiant positions and orbit parameters of two Geminids observed at 1:34:23 and 5:07:01 UT on December 14, 2006. The mean radiant and mean orbit of Geminid meteor shower are mentioned for comparison with our data

Name	α (°)	δ (°)	a (AU)	q (°)	e	i (°)	ω (°)	Ω (°)
20061214_013423	113.0	32.8	1.0	0.21	0.78	15.7	321.4	261.8
20061214_050701	113.7	31.9	1.1	0.17	0.85	18.5	323.9	262.0
Mean Geminid orbit	112.3	32.5	1.357	0.14	0.897	24.27	324.63	261.43

software, because Geminids are very well known for their compact radiant. Six Geminids were simultaneously observed from both stations. The heliocentric orbits for the two brightest meteors were computed by the "UFOOrbit" software (Table 1).

4 Conclusions

We described our first experience with the new fish-eye TV system as well as the "UFOCapture", "UFOAnalyser" and "UFOOrbit" software. We are able to capture meteor activity under suitable conditions and provide reliable data, although we do not have permanent and identical second station yet. We hope we will be able to provide TV meteor orbital data on the regular bases in the near future.

Acknowledgments This work was supported by the Scientific Grant Agency VEGA, grant No. 1/3067/06 and by Comenius University grant UK/401/2007. Authors are thankful to A. Galád, P. Kolény and M. Šebeň for valuable help with observations and construction of the TV system and also to Prof. Frans J. Rietmeijer and Dr. Margaret Campbell-Brown.

References

G. Barentsen, Lyrids 2007: first results http://www.imo.net/live/ lyrids2007 (2007)

Š. Gajdoš, D. Kalmančok, P. Zigo, P. Kolény, L. Kornoš, J. Tóth, A. Galád, M. Šebeň, J. Világi, EN station no. 21 – operation and results after 15 years of activity. Meteor Report **27**, 83–90 (English abstract) (2006)

J. Oberst, S. Molau, D. Heinlein, C. Gritzner, M. Schindler, P. Spurný, Z. Ceplecha, J. Rendtel, H. Betlem, The "European Fireball Network": current status and future prospects. Meteoritics and Planetary Science **33**, 49–54 (1998)

J. Tóth, L. Kornoš, in *Single and Double Station Meteor Observations Setups in Modra, Slovakia*, eds. J. Mc Auliffe, D. Koschny. Proceedings of the First Europlanet Workshop on Meteor Orbit Determination 2006, Roden, The Netherlands, 11–13 September 2006, (International Meteor Organization, 2007), pp. 46–52

P. Zigo, Š. Gajdoš, J. Tóth, in *Meteor observations at Modra Observatory*, eds. L. Bastiaens, J. Verbert, J.-M. Wislez, C. Verbeeck. Proceedings of the International Meteor Conference, Oostmalle, Belgium, 15-18 September, 2005, IMO, Belgium, (2006), pp. 37–41

The Armagh Observatory Meteor Camera Cluster: Overview and Status

Prakash Atreya · Apostolos Christou

Originally published in the journal Earth, Moon, and Planets, Volume 102, Nos 1–4.
DOI: 10.1007/s11038-007-9170-6

Abstract Armagh Observatory installed a sky monitoring system consisting of two wide angle (90° × 52°) and one medium angle (52° × 35°) cameras in July 2005. The medium angle camera is part of a double station setup with a similar camera in Bangor, ~73 km ENE of Armagh. All cameras use UFOCapture to record meteors automatically; software for off-line photometry, astrometry and double station calculations is currently being developed. The specifications of the cameras and cluster configuration are described in detail. 2425 single station meteors (1167, 861 and 806 by the medium-angle and the wide-angle cameras respectively) and 547 double station meteors were recorded during the months July 2005 to Dec 2006. About 212 double station meteors were recorded by more than one camera in the cluster. The effects of weather conditions on camera productivity are discussed. The distribution of single and double station meteor counts observed for the years 2005 and 2006 and calibrated for weather conditions are presented.

Keywords Meteor · Double station

1 Introduction

Meteor astronomy has implemented video techniques for all-night observations in addition to visual and photographic during the past decade. The IMO video meteor society (Molau 2001), Polish Fireball Network (Olech et al. 2006), Spanish Meteor Network (Trigo-Rodriguez 2007), Dutch Meteor Society (Miskotte and Johannink 2006) and Czech Meteor Network (Koten et al. 2006) are few of the meteor networks actively operating in Europe. These networks have thrived on the advancement of light sensitive video and CCD cameras with affordable prices. Detection and analysis software such as Metrec, Meteorscan and UFOcapture eases the tedious setup and encourages professional and amateur astronomers alike to set up meteor stations.

P. Atreya (✉) · A. Christou
Armagh Observatory, College Hill, Armagh BT61 9DG, Northern Ireland, UK
e-mail: atr@arm.ac.uk

Double station observations of meteors have rapidly increased during the last 5 years, with most of the networks mentioned above operating double or multiple stations. Multiple station observation are used to calculate atmospheric trajectories and heliocentric orbits of meteoroids accurately. The need for more double station observation was outlined by Trigo-Rodriguez et al. (2006).

2 Set-Up

With the intention to investigate meteor activity, improve understanding of poorly studied showers and investigate fireballs, Armagh Observatory installed a sky monitoring system in July 2005. It consists of two wide angle (FOV of 90° × 55°, 3.8 mm F0.8 focus lens) and one medium angle (FOV of 52° × 35°, 6 mm F0.8 focus lens) cameras. All three cameras are pointed at 60° altitude, and at azimuths of 60°, 150° and 330° for the medium angle (cam-1) and two wide angle (cam-2 and cam-3) respectively. The medium angle camera (cam-1) makes up a double station with similar camera from Bangor (Northern Ireland), approximately 73 km away, run by amateur astronomer Robert Cobain. Automated UFOCapture software is used by both stations to capture meteors.

Each video captured, whether containing a meteor or not, is visually inspected. The start\end time of the observation, the number of single\double station meteors detected and the hourly weather conditions are recorded in a log file everyday. Hourly weather are recorded as "clear" (if 90+% of the field of view is clear), "partly cloudy" (if 10+% of the field of view is clear and at least one star is visible) and "cloudy" (if less than 10% of the field of view is clear). The percentage of the clouds are estimated from the videos of meteors and the false detections manually. The total number of clear hours per day is calculated as the sum of number of clear hours and half the number of partly cloudy hours. Further detail on the location, cameras and the cluster specifications, and system of weather recordings can be found at Christou and Atreya (2007). Here we present the results from the first 18 months of operation (July 2005–December 2007).

3 Statistics for year July 2005–December 2006

3.1 Weather Record

Table 1 shows the weather record from Jul ′05 till Dec ′06, with column-2 showing the total number of hours of operation and column-3 showing the total number of hours of clear sky. The cameras operated 131 h/month (∼4.36 h/day) during Jun & Jul ′06, due to the short duration of the summer night. The operational hours of cameras increase gradually, reaching a maximum of 458 h/month (∼14.7 h/day) during Dec ′05. The total operational hour is 5168 h, which is ∼9.44 h/day for 1.5 years.

However, the total number of clear hours does not follow a similar distribution compared to that of total hours of operation. During the summer month of Jun ′06, the total number of clear hours is 69; about half of total hours of operation are clear. During Dec ′05, where the total number of hours of operation is three times compared to Jun ′06, the total number of clear hours is 115, only twice compared to Jun ′06. November has the most number of clear hours.

Table 1 The meteors record from July '05–Dec '06

Date	Total hours of operation	Clear hours of operation	Cam-1	Cam-2	Cam-3	Total number of meteors	Double station meteors
Jul 05	128	28	12	13	12	34	8
Aug 05	189	70	65	61	54	156	44
Sep 05	263	86	61	40	37	120	27
Oct 05	336	94	83	52	43	156	30
Nov 05	374	138	114	88	80	230	63
Dec 05	458	115	111	105	69	237	42
Jan 06	428	123	54	47	36	117	27
Feb 06	347	119	31	25	23	66	13
Mar 06	339	104	25	13	16	45	13
Apr 06	248	118	44	45	30	98	17
May 06	187	75	23	11	18	44	13
Jun 06	131	69	13	8	18	33	8
Jul 06	131	92	55	44	37	119	29
Aug 06	196	57	65	61	57	156	32
Sep 06	268	115	74	49	48	152	47
Oct 06	340	159	98	78	92	226	47
Nov 06	389	193	145	57	88	249	53
Dec 06	416	128	94	64	48	187	34
Total	5168	1883	1167	861	806	2425	547

Monthly readings of total hours of operation, total clear hours of operation, meteors recorded by cam-1, cam-2 and cam-3, total number of meteors and total number of double station meteors are shown in column 2–8

3.2 Single and Double Station Meteors

Table 1 shows the distribution of meteor recorded by 3 individual cameras during the first 18 months. Cam-1 (column 4) is medium angle, while cam-2 (column 5) and cam-3 (column 6) are both wide angle cameras. The total number of meteors captured by the three cameras are 1167, 861 and 806 respectively. The meteor count recorded by cam-1 is slightly higher than the other two wide angle cameras. This is mainly due to the fact that the wide angle units have a lower limiting magnitude, such that cam-1 can detect fainter meteors compared to cam-2 and cam-3. The three cameras are directed at different part of the sky, each separated by 90° with each other in azimuth. This could also show which part of the sky is more active during which time of the year.

Table 1 shows the total number of meteors (column 7) and double station meteors (column 8). During the months Feb '06 to Jun '06 only ~60 meteors/month were observed, with the exception of April '06 (due to the contribution from the Lyrids). Meteor activity was high during winter months with the highest achieved of 249 meteors during Nov '06. Double station meteors show a similar trend. The two maxima for double station meteors, 63 and 53, were observed during Nov '05 and Nov '06, respectively.

The trend of the number of meteors captured increasing in winter months and decreasing rapidly during summer months is predominant. This could be due to the fact that the number of clear hours of operation in summer is about one third compared to the summer months, and most of the dominant showers occur during winter months.

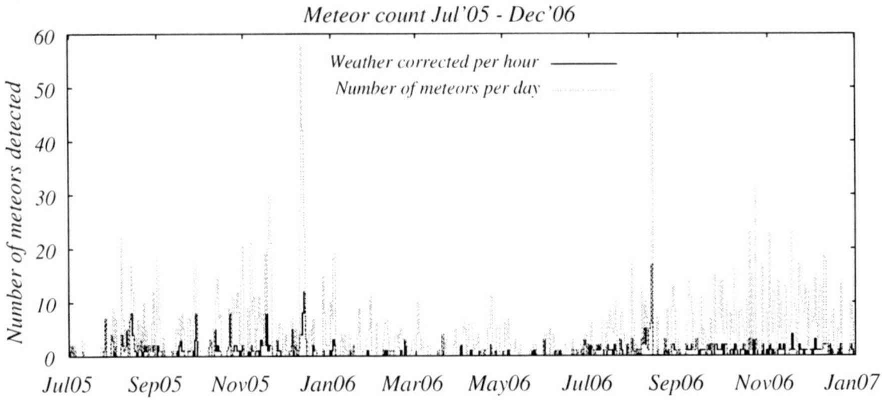

Fig. 1 Meteors observed per day (dashed line) and corrected for clear hours per hour (bold line) for Jul ′05–Dec ′06

3.3 Weather Correction

Figure 1 shows the distribution of meteors observed per day for the time period July ′05 to Dec ′06. The bold line shows meteor counts per hour corrected for the total number of clear hours of observation, while the dashed line shows the raw data, that is, meteors recorded per day. The dominant peak of Geminid (Dec ′05) and Perseid (Aug ′06) are evident in the data. Other peaks such as Perseids (Aug ′05), Leonids (Nov ′05), Quadrantids (Jan ′06), Orionids (Oct ′06) and Leonids (Nov ′06) can also be identified. However, the exact number of major and minor showers can only be identified after a detailed analysis of their radiants and velocities.

4 Discussion and Future Work

During the first 18 months, each camera was operated for $\sim$5100 h of observation of which 1883 h was clear sky, ie. a third of the total observational period was clear. 2425 single meteors were recorded in total, among them, 547 being double station. 212 of double station meteors were captured by two or more cameras from the cluster. In the year 2006, there were 1492 single station and 333 double station meteors observed. The number of meteors recorded by the Irish double station network is very positive result, taking into account that this was the first year of operation and the vagaries of the notorious Irish weather.

The software for the analysis of these meteors, aimed to calculate the trajectory and orbital parameters of the double station meteors, is in progress Atreya and Christou (2007). The result from these data would help to understand minor showers and the sporadic background, and their variation. The orbital data from the double station can contribute to the identification of parent bodies for some streams, and other related fields such as interplanetary dust processes, orbital dynamics and meteor showers in other planets can be studied (Christou 2004).

Acknowledgements The authors would like to thank John McFarland, Anthony Moraghan and everyone else who helped to keep the log for the weather and meteors updated daily. Astronomical research at Armagh Observatory is funded by the DCAL.

References

P. Atreya, A. Christou, Software for the photometric and astrometric analysis of video meteors. in *Proceedings of the IMC 2006*, in press (2007)

A. Christou, Prospects for meteor shower activity in the venusian atmosphere. ICARUS, **168**(1), 23–33 (2004)

A. Christou, P. Atreya, The Armagh Observatory meteor camera system. in *Proceedings of the IMC 2006*, in press (2007)

P. Koten, J. Borovicka, P. Spurn, S. Evans, R. Stork, A. Elliott, Double station and spectroscopic observations of the Quadrantid meteor shower and the implications for its parent body. MNRAS, **366**(4), 1367–1372 (2006)

K. Miskotte, C. Johannink, Taurids 2005: results of the Dutch Meteor Society. WGN **34**(1), 11–14 (2006)

S. Molau, The AKM video meteor network. in *Proceedings of the Meteoroids 2001 Conference*, pp. 315–318 (2003)

A. Olech, P. Zoladek, M. Wisniewski, M. Krasnowski, M. Kwinta, T. Fajfer, K. Fietkiewicz, D. Dorosz, L. Kowalski, J. Olejnik, K. Mularczyk, K. Zloczewski, Polish Fireball Network. in *Proceedings of the IMC 2005*, pp. 53–62 (2006)

J.M. Trigo-Rodrguez, J. Vaubaillon, E. Lyytinen, M. Nissinen, Multiple station meteor observations: an international program for studying minor showers exploring IMO potentiality. WGN **34**(2), 40–44 (2006)

J.M. Trigo-Rodriguez, J.M. Madiedo, A.J. Castro-Tirado, J.L. Ortiz, J. Llorca, J. Fabregat, S. Vitek, P.S. Gural, B. Troughton, P. Pujols, F. Galvez, Spanish Meteor Network: 2006 continuous monitoring results. WGN **35**(1), 13–22 (2007)

Algorithms and Software for Meteor Detection

Peter S. Gural

Originally published in the journal Earth, Moon, and Planets, Volume 102, Nos 1–4.
DOI: 10.1007/s11038-007-9161-7

Abstract An ever increasing variety of electronic instrumentation is being brought to bear in meteor studies and analysis, with unique meteor detection challenges arising from the attempt to do automated and near real-time processing of the imagery. Recent algorithm developments in the literature have been applied and implemented in software to provide reliable meteor detection in all-sky imagers, wide-field intensified video, and narrow field-of-view telescopic systems. The algorithms that have been employed for meteor streak detection include Hough transforms with phase coded disk, localized Hough transforms with matched filtering, and fast moving cluster detection. They have found application in identifying meteor tracks in the Spanish Fireball Network all-sky images, detailed analysis of video recordings during the recent Leonid meteor storms, and development of a detection/cueing technology system for rapid slew and tracking of meteors.

Keywords Meteor detection · Transient detection · Meteor tracking

1 Introduction

As meteor astronomy advances, it is relying more heavily on electro-optical (EO) instrumentation in place of human visual observations. Development of intensified low light video (Hawkes and Jones 1986) has extended the limiting magnitude of EO meteor detection beyond the normal human visual range and can be used for mass index estimation, flux, orbit calculation, and light curve analysis. Spectroscopy and high frame rate imaging of meteors is now possible with the demonstration by (Gural et al. 2004) of meteor tracking hardware providing composition and fragmentation information. With the advent of more sensitive focal planes, non-intensified applications are arising as well. Large format CCDs are used in all-sky fireball tracking with the potential for meteorite recovery as in the SPanish Fireball Network or SPFN (Trigo-Rodriguez et al. 2004) and Marshall Space Flight Center has begun a program in telescopic lunar meteoroid flash

P. S. Gural (✉)
SAIC, 14668 Lee Road, Chantilly, VA 20151, USA
e-mail: peter.s.gural@saic.com

monitoring (Suggs 2007), to examine the number density of boulder class meteoroids. All these represent transient detection problems in multi-frame image processing with a range of response times, image characteristics, noise statistics, and signal features.

2 Design Criteria and Software

In these examples, signals or the meteor's illuminated track, can last from milliseconds to seconds producing a spatial flash, a moving streak in video, or a line in a single image frame. The noise can be comprised of speckle features from an intensifier, cosmic ray artifacts, thermal, and dark current. Detection of signals in a noisy environment always involves a trade of probability of detection (Pd) for probability of false alarm (Pfa) and thus use of a priori knowledge can help decide the best algorithmic choice. However, processing throughput and the timeliness of detection can result in a less than optimal algorithm selection. Thus several questions must be answered in formulating any particular detection solution. Does the detection need to be done in real-time at the camera frame rate, or in near-real-time with a small latency, or at a later time offline in post-collection at the user's convenience? Must the detection be fast which usually requires a high SNR or robust to very low SNR events? What are the tolerance for and mitigation approach to false alarms? What are the processing capacity, storage needs, and interface requirements of the proposed computing system? What are the goals of the collection and the needs for calibration, post-analysis, and science exploitation that can drive the detection requirements?

To address these unique image processing needs, a variety of software packages have been developed such as MeteorScan, MetRec, UFOCapture (Molau and Gural 2005) for real-time video meteor detection, the SPFN MeteorDetector for all-sky fireball surveying, MeteorCue for meteor tracking and instrument pointing, LunarScan and LunaCon for lunar meteoroid impact flash detection, MachoScan for star occultation detection, and MeteorSim for meteor flux and human/video ZHR calculation. This does not represent a complete list of software (see Cheselka 1999 for an IRAS software package add-on for linear feature detection in astronomical imagery), but inherent in all is a requirement that transient detection algorithms be very fast, process large volumes of imagery, robust, and tuned towards the specific characteristics of the data and collection system. To maintain brevity in this paper only the meteor streak detection algorithms will be discussed and the reader is advised to check the references (Gural 2001, 2007; Parker et al. 2004) for discussions on flux simulation, lunar flash detection, and occultation detection respectively. In addition, a simple web search on the software names will turn up sites for downloading the aforementioned packages or alternatively contact this paper's author.

3 Detection Algorithms

The most common form of detection problem faced in the last decade has been to discover meteor tracks in video streams either live or pre-recorded. A typical meteor track is comprised of a streak lasting up to several video frames propagating in a linear fashion across space and time. There exists a variety of line detection algorithms published in the literature. The one with the best Pd, Pfa ratio is the matched filter (MF), where an object's motion is hypothesized for a particular starting point, speed, and direction, and has been applied to moving satellite and asteroid detection in a cluttered star field (Mohanty 1981;

Gural et al. 2005). Each frame set is shifted and stacked according to the motion hypothesis and the resultant summed frame tested against a threshold. However, the large number of potential motion hypotheses for meteors limits this technique except in those circumstances where the entire set of candidate motions is small. An alternative pseudo matched filter approach applied to meteors was published that included an angle hypothesis, median filter and sum technique (Torii et al. 2003) that again requires large hypothesis sets and is suitable for single frame processing but is too computationally burdensome for video frame rate processing.

Thus it is more common to see the application of the Hough Transform (HT) to the meteor line detection problem due to its relative speed improvement over the MF with only a slight loss in Pd. Unfortunately, there are a wide variety of HT algorithms to choose from and only those that the author has found to be relatively successful in meteor detection will be discussed herein. The basic concept behind the HT is to threshold an image, transform the pixel point exceedances from image space to Hough space (line orientation angle, origin offset), and finally locate peaks which have a correspondence to lines in the original image. Each HT can also be summed across several images to obtain the signal integration gain from a meteor with multi-frame duration. Depending on the HT algorithm chosen, the execution time can scale either linearly or quadratically by the number of pixel exceedances, trading speed, robustness, and sensitivity.

For even greater throughput in situations requiring real-time response, one can dispense with searching for linear features and instead simply locate clusters of pixels. Again one thresholds pixels, applies a fast cluster search algorithm, and searches for motion consistency between clusters found across several frames. The detection requires high SNR to mitigate false alarms and can be extended to situations where the transient leaves no temporal response as in meteoroid flash detection where one tries to match only a spatial signature.

Inherent in the above discussion is the underlying need to threshold the image for pixels containing a signal exceeding a noisy background level. The first stage of processing usually removes stationary features such as stars and background. This can be done through (1) removal of a mean or median which has lower noise characteristics but is harder to track in shifting scenes, (2) differencing adjacent in time frames resulting in a zero mean and finite variance estimate for each pixel with faster processing but higher noise levels, or (3) computationally loaded clutter suppression where the noise statistics are estimated and the image whitened through covariance estimation and inversion. With the stationary features removed and variance estimated, a threshold can be defined to flag the subset of pixels that then feed the detection algorithm. The next few sections will take these general concepts and apply them uniquely to each meteor detection problem.

4 All-sky Fireball Detection

The SPanish Fireball Network (SPFN) has begun operating a series of large format focal plane cameras (4 K × 4 K pixels) to detect fireballs and obtain atmospheric trajectories (Trigo-Rodriguez et al. 2004). The large number of pixels results in slow readout times so video rates are not possible in this setup. Instead the all-sky cameras integrate the star field for 3 min with a single exposure and rotating shutter, repeating this through the entire night. The all-sky nature of the scene contains a stationary horizon, star field rotation and trailing during the exposure, and has a changing background arising from the long time delay between frames. Initially failed approaches at detection included co-alignment

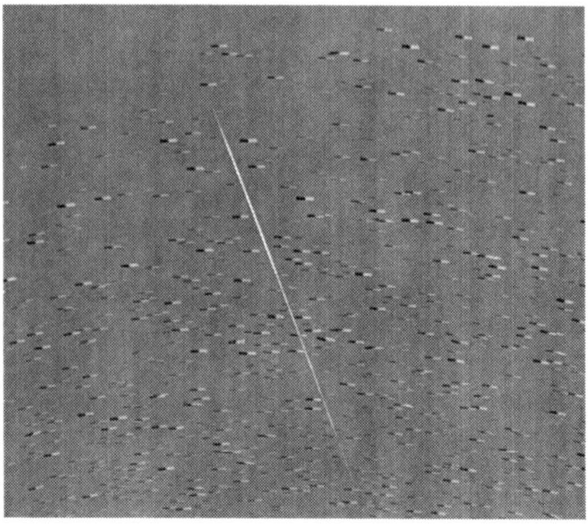

Fig. 1 Difference image of SPFN showing positive and negative star trails

(rotational registration) of frames to remove stars and the addition of a mask to screen out the background. However, bright star column bleed, lack of flat fielding, and residual registration errors caused a high level of false alarms.

The current algorithmic approach avoids the background subtraction and flat fielding issues by differencing two adjacent-in-time images directly. The trailed star tracks and meteors remain but the stars have both positive and negative trail components that over short time scales and equal exposure lengths appear as straight lines as in Fig. 1. The HT applied to the difference image results in each star effectively summing to near zero, whereas a meteor which appears in only one image has a net positive or negative Hough peak. To determine which subset of pixels to transform, a two-sigma clipped local mean and standard deviation is computed iteratively using a sliding window to obtain the estimate of the residual background noise statistics. Since the meteor detection can be processed off-line and reported on later, the algorithm uses a slower but more robust version of the Hough transform on each exceedance pixel where the orientation is obtained through the kernel convolution of a phase coded disk or PCD (Clode et al. 2004). Furthermore, false alarm mitigation and the desire to ensure that only a single detection of the same meteor is achieved, the peak finding algorithm employs Hough space region exclusion after each meteor is detected in the image.

5 Medium to Narrow Field Video Meteor Detection

The workhorse for EO meteor detection and analysis has been the medium field of view (FOV) camera system with fast low f-number lens, image intensifier, and frame-rate CCD macro-focused to the output face of the intensifier. Using standard off-the-shelf video components results in 25–30 frames per second (fps) with various image sizes depending on format (e.g. 480 × 640 pixels for NTSC). The limiting magnitude for these systems typically reaches +7 to +8 for a 40 degree FOV. One can trade FOV for sensitivity to fainter meteors, accuracy of track, and flux statistics. Software has been developed for real-time detection of meteors, but with the large storage capacity in modern day personal computers and/or the use of camcorders, the option for later off-line reduction of the imagery is possible and now done regularly.

The imagery is characterized by little to no change in the background between frames due to the high frame rate and slow movement of the star field in a wide FOV. For intensified systems there is a high noise speckle component which in earlier generation II systems was predominant near the center. A meteor typically appears in multiple frames propagating as a linear streak across the FOV. To remove stationary components, mean estimation and removal is preferable due to its lower noise variance, but frame differencing is usually done due to its legacy back to fast runtime in real-time operations. In either case, a variance is tracked for every pixel through an updating first order response filter that can be used to "whiten" the current image and threshold the frame for pixel exceedances. For linear time Hough transforms, the classic HT should not be used due to its "self-noise" issue and the PCD method is computationally too heavy to provide an analysis in a reasonable amount of time for the data volume of video imagery. A quadratic-in-time HT can be used that limits the transform to only local neighbor pixel pairs since a meteor's exceedance pixels are typically adjacent to each other (two points provide orientation and center offset and avoids Hough self noise). Summing multiple HTs in a short image sequence provides signal gain but requires that the absolute value of the difference be computed to avoid the meteor from canceling itself. To reduce false alarms, a subset of HT detected tracks are passed through a matched filter to enhance the robustness of the detection. The HT algorithms have difficulty with short and slow meteors since they do not produce spatial signal integration gain and lowering the detection threshold simply raises the Pfa to undesirable levels. As PC capabilities improve and the use of graphical processing units for image processing becomes more prevalent, the migration will be back to real-time processing and more robust algorithms like the PCD.

For ablation studies, narrow FOV video imaging has been tried recently which presents a slightly different signal environment for detection. The meteor may last at most two frames, can be electronically chopped (gated), and will have a longer streak in each frame. The same HT processing can be applied but the multi-frame integration must be limited to avoid adding too much noise when there is no signal present in non-meteor video frames.

6 Telescopic Meteor Tracking

To enhance the collection probability of capturing a meteor in narrow FOV instruments like spectrometers or high spatial resolution cameras, a meteor tracking system needs to be incorporated into a collection system's design. An approach using computer controlled mirrors to redirect the meteor's light to the instrument, requires very rapid response motors and a fast meteor detection/tracking system (Gural et al. 2004). The software must function on a medium FOV sensor matched to the steering system motion limits and operate at standard frame rates. The response time to capture a meteor before it fades out must include detection, mirror movement, and settle time to all occur within 100 ms of a meteor's first light. Such rapid response rules out the HT and thus a very fast clustering approach is called for. Although a clustering algorithm suffers from requiring high SNR for reliable detection, the nature of narrow FOV meteor work also requires bright meteors, so is compatible with the algorithm's poor light sensitivity. Frame differencing is chosen for its speed of stationary background removal coupled with noise variance tracking to provide thresholding for pixel exceedances. A fast region based clustering algorithm as demonstrated in Fig. 2, locates clumps of pixels in the odd and even fields separately of the interlaced video to provide higher time resolution for a meteor tracker. Cluster centroids are passed to an alpha-beta tracker which converts these positions to mirror angles.

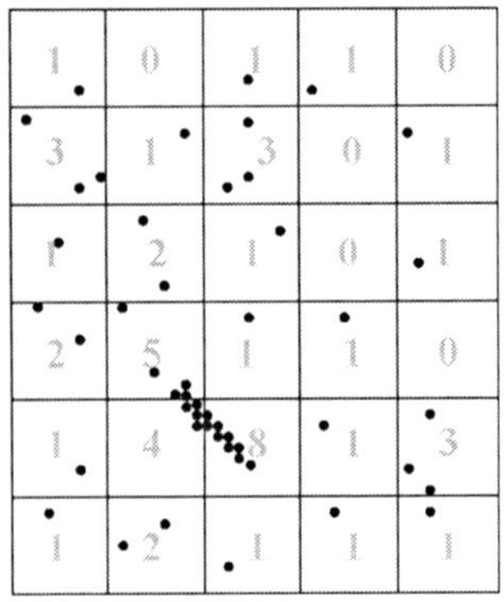

Fig. 2 Fast cluster counting in 32 × 32 blocks whose size corresponds to the maximum meteor motion for the typical intensified video image scale

Detection of a true track is declared early (after 2 out of 3 fields detected) to minimize the time to move the mirrors. This is feasible since one is willing to accept a higher Pfa to achieve a high Pd for the narrow FOV instrument.

7 Conclusion

In summary, this paper briefly discussed the algorithms used in many meteor detection systems in operation today accounting for trades in detection Pd, false alarm Pfa, sensor, signal, and noise characteristics, throughput requirements, and reporting timeliness. Details on the performance comparisons and quantitative analysis of these algorithms is beyond the scope of this paper with the goal having been to highlight the various algorithmic approaches possible given the wide operating characteristics of meteor imaging systems. Furthermore, detection is only the first phase of analysis in all meteor work, and calibration for mass index, flux estimation, and orbital elements requires application of additional image processing techniques and the refinement of high fidelity meteor simulation and analysis software.

References

M. Cheselka, in *Automatic detection of linear features in astronomical images. ASP Conf. Ser. Astronomical Data Analysis Software and Systems VIII*, vol. 172, eds. by D.M. Mehringer, R.L. Plante, D.A. Roberts (ASP: San Francisco, 1999) p. 349

S.P. Clode, E.E. Zelniker, P.J. Kootsookos, V.L. Clarkson, A phase coded disk approach to thick curvilinear line detection. Proceedings of EUSIPCO, Vienna, Austria, 2004, 1147–1150

P.S. Gural, Meteor observation simulation tool. Proceedings of the International Meteor Conference, Cerkno, Slovenia, Sept 20–23, 2001, pp. 29–35

P.S. Gural, Automated detection of lunar impact flashes. 2007 Meteoroid Environment Workshop, Huntsville, Alabama (http://see.msfc.nasa.gov/workshop07.htm). Accessed 4 Oct 2007

P.S. Gural, P.M. Jenniskens, G. Varros, Results from the AIM-IT meteor tracking system. Earth Moon Planets **95**, 541–552 (2004)

P.S. Gural, J.A. Larsen, A.E. Gleason, Matched filter processing for asteroid detection. Astron. J. **130**(4), 1951–1960 (2005)

R.L. Hawkes J. Jones, Electro-optical meteor observation techniques and results. Quart. J. Roy. Astron. Soc. **27**, 569–589 (1986)

N.C. Mohanty, Computer tracking of moving point targets. IEEE. Trans. Pattern Anal. Mach. Intell. **3**, 606–611 (1981)

L.C. Parker, R.L. Hawkes, P.S. Gural, Short Exposure Astronomical Techniques for Occultation Detection. J. Roy. Astron. Soc. Can. **98**, 120–127 (2004)

R. Suggs, The NASA lunar impact monitoring program. 2007 Meteoroid Environment Workshop, Huntsville, Alabama (http://see.msfc.nasa.gov/workshop07.htm). Accessed 4 Oct 2007

S. Molau P.S. Gural, A review of meteor detection and analysis software software. WGN, J. Int. Meteor. Organ. **33**:1, 15–20 (2005)

K. Torii M. Kohama T. Yanagisawa K. Ohnishi, The radiant structure of the Leonid meteor storm in 2001: observations with a telephoto lens system. Publication of the Astronomical Society of Japan **55**, L27–L30 (2003)

J.M. Trigo-Rodriguez, A.J. Castro-Tirado, J. Llorca, J. Fabregat, V.J. Martinez, V. Reglero, M. Jelinek, P. Kubanek, T. Mateo, A.D.U. Postigo, The development of the Spanish fireball network using a new all-sky CCD system. Earth Moon Planets **95**, 553–567 (2004)

"Falling Star": Software for Processing of Double-Station TV Meteor Observations

Pavlo Kozak

Originally published in the journal Earth, Moon, and Planets, Volume 102, Nos 1–4.
DOI: 10.1007/s11038-007-9223-x © Springer Science+Business Media B.V. 2008

Abstract Software named "Falling Star" has been developed for digital processing of double-station TV meteor observations. It was designed for measurement and calculation of both kinematic and photometric parameters of faint meteors observed with any video system. Data from video recordings are first digitized as standard AVI files, and then converted into the software's TVS (TV sequence) format. Additional astronomical information like date, time of observations, geographic position of point of the observation and horizontal coordinates of TV camera optical axis orientation are added to the files. These parameters allow the right ascension and declination of the optical center of camera for the moment of meteor flight to be calculated. "Falling Star" includes a range of automated procedures for the identification of frame stars with star catalogues, search of movable meteor-like objects inside frame, calculation of equatorial coordinates and photometry. Finally, meteor trajectory parameters, orbital elements and brightness curves are calculated. Errors of calculations are determined using Monte-Carlo method.

Keywords Meteors · TV observations · Method of digital processing · Software

1 Introduction

Meteor analysis software can be separated onto two categories: software for auto-detection of meteors in real time, and post-detection software for determination of meteor parameters from double-station observations. One of the most widely used software of the first type is MetRec (Molau 1998). Algorithms for both meteor trajectory and orbital elements determination and photometry have been developed by practically each meteor group carrying out double-station meteor observations, for instance (Hawkes 2002; Hawkes et al. 1993) and (Koten and Borovicka 2001; Koten 2002). This paper provides a brief overview of new

P. Kozak (✉)
Astronomical Observatory of Kyiv National Taras Shevchenko University, Kyiv, Ukraine
e-mail: kozak@observ.univ.kiev.ua
URL: www.observ.univ.kiev.ua

software named "Falling Star" which has been developed in our department over the past ten years. The main idea of the software development was to create a complex program allowing complete astrometric, kinematic and photometric processing of double-station meteor observations. A goal of the software was to create a program that would be independent of the observational system used.

2 Software Goals and Features

In order to achieve universality of the program we had to solve two fundamental problems: input of any digital format representing meteor images sequence, and using methods for astrometry and photometry applicable to wide range of modern video systems.

In order to avoid the limitation of a fixed file format we separate the software onto two applications: the program itself and the preliminary converter from file formats being used (in most cases AVI-files from video conversion, or bitmap data written directly from the detector to the hard drive) into TVS-file. Transformation of input data into TVS assumes the following: transfer of main file parameters (size, bits per pixel, number of frames, frame rate for standard PAL/SECAM or NTSC video or any other value if necessary, progressive or interlaced scanning type etc.) to the header of new file; extracting and transfer of intensity component for each pixel of the image if the frame-grabber doesn't provide for generation of purely monochrome picture, significantly decreasing in such a way the file size; separation of full frames onto odd and even half-fields in the case of interlaced scanning; providing TVS file with necessary comments; creation of additional subheader including date and time (UT) of observations; geographical coordinates of the observational point (latitude, longitude, altitude); orientation of the camera in horizontal coordinate system (azimuth and zenith angles); approximate angular size of FOV for the lens being used. We plan to provide the final version of the program with a description of the inner format of TVS file, which will allow the user to design a converter for other video formats.

Universality in methods for measurements and data processing of meteor images implies their correctness in results independently on the observational system being progressive or interlaced scanning, linear or non-linear signal response, presence or absence of afterimage and blooming, etc. For instance, we use super-isocon TV systems being of interlaced type, significantly non-linear and having long afterimage for dynamical objects, but most of modern video systems are assumed to be linear and relatively free of afterimages. How this approach is realized we will describe below in corresponding sections.

3 Processing of Meteor Image Sequences

The program consists of a series of procedures, which are consecutively called at meteor processing. Each subroutine saves the results of its calculations in separate files, which can be analyzed later. Processing of a meteor can be carried out manually, when each subroutine is called by a user, or automatically, when the application calls all subroutines consecutively. Execution of each procedure is accompanied by visualization. The time increase is not significant due to visualization, and the user can check visually the correctness of automated processing.

All operations are carried out separately over two TVS-files representing odd end even half-fields, supposing we use a system with interlaced scanning. The results of calculations for each file can be averaged after, or one higher resolution file created similar to the case of progressive scan. We will use the term "frames" for both half-fields for interlaced systems and full frames for progressive ones.

3.1 Mathematical Operations with Frames

Averaging is an operation over a sequence of $N \leq 100 \div 120$ frames (N given is for our TV system, and will vary with integration time, resolution and field of view). The goal of this averaging is to reduce the background noise $\sim \sqrt{N}$ in order to improve the precision of astrometry and photometry of star images. Number of frames for such an operation depends on system noise, but should be limited to prevent star shape deformation due to stellar drift. We recommend to average frames centered around the meteor (neither earlier nor later) to avoid problems with drift in measured coordinates of reference stars. The resultant frames are used only for star images processing.

Subtraction of frames is used to create frames where only the meteor image is present (star images disappear) in a similar manner to Hawkes et al. (2001). In the simplest case one can subtract a single frame taken just before the meteor appearance from each frame with a meteor. In this case the resultant fluctuations increase $\sim \sqrt{2}$, so we subtract the frame averaged on 40–50 frames before the meteor appearance.

The last main operation is the *summing* of frames where the meteor or its afterimage still exists. This frame is used for additional meteor measurements for radiant precise calculation and also can be used for photometry.

In addition to these basic operations there are many other mathematical and statistical procedures realized in the program: calculation of statistical distributions of intensities inside selected rectangular zone or entire frame; work with "photometer slit"; detaching and measuring of any photometric profiles etc.

3.2 Measurements of Stars and Meteor Image in the Frame

For measurements of rectangular coordinates of reference stars and their photometrical volumes the reductive method (Kozak 2001) or two-dimensional Gaussian fit can be used. We also recommend using the Gaussian fitting for the profiles on the meteor image. The position of a point corresponding to the meteor head is selected by a user (in current version of the program) using a collection of empirical rules. Supposing the meteor is a point object, and its PSF being of Gaussian type with known half-width determined from star images for the given lens, and considering the equations of charge accumulation and scanning we can generate the motion of PSF over the light detector (over the frame) with different meteor velocities and brightness. Additionally, we can change the brightness during the "meteor flight" to make the simulation more realistic. Using such empirical rules for processing of real meteors we can reach a precision of ± 0.5 pixels, which corresponds to the precision of star image measurements using the reductive method. In order to raise the precision we should try to find an exact solution of the resultant integral equation being an improperly posed problem. More detailed description of such an approach will be given elsewhere.

In order for star identification we have chosen two star catalogues: Tycho ACT RC 1997 (I/246) and AS CC 2 (I/280A) containing stars up to 12^m, which corresponds to our TV system sensitivity. The second one is preferable since it includes the spectral classes for some bright stars which are important for photometry. Taking data from the astronomical subheader of the TVS-file the program selects stars from the catalogue around the optical center, completely including the entire frame. The star map is plotting automatically using first the ideal projection of the sphere onto the plane and then adding necessary distortions. Identification of stars is realized automatically by means of comparison of their coordinates in observed and star map frames (Kozak 2001; Kozak et al. 2001).

3.3 Astrometry

The polynomial reduction models are used for the astrometric processing: linear polynomial, Deutsch 8 constants method (Deutsch 1965), full square polynomial and truncated cubic polynomial (12 constants), and can be found in (Kozak et al. 2001). The quality of each model was checked using stars of known position but assumed to be unknown objects. We evaluated the precision of astrometry in this way for variation of such parameters as reduction model, number of reference stars in some zone around the object, dimension of the zone, asymmetry of reference stars sample relatively the object, method for measurements of rectangular coordinates of the object and reference stars. The best reduction model for our system has been found to be linear one, and the reference stars zone being local. Precision of equatorial coordinates determination at pixel size of 4 arc min is 2–2.5 arc min, and decreases down to 1.5 arc min when Gaussian fitting of star shapes is used. The method of astrometric processing is presented in detail in (Kozak 2002).

3.4 Kinematics

Kinematics processing of a meteor is realized with the vector method developed by the author and described in (Kozak 2003). First the classic triangulation scheme is used, then, with the help of vector-matrix operations, we calculate all meteor trajectory parameters in the atmosphere proceed to a heliocentric coordinate system and compute the orbital elements of a meteor.

The input parameters for the kinematic processing are the equatorial coordinates of points on the meteor image calculated from both stations. In the express method we use a range of points ($N \approx 3 \div 20$) corresponding to the meteor head position on the image and respective relative time moments $(\alpha;\ \delta;\ t)_i$, $i = \overline{1, N}$. Another approach which allows an increase in the precision of radiant determination, and therefore all other parameters, makes additional use of an array of points along the meteor trajectory obtained from the summed frame $(\alpha;\ \delta)_j, j = \overline{1, M}$ (the image is similar to photographic one) where number of points is much higher ($M \approx 50 \div 350$). The Fig. 1 demonstrates the results of application of the two approaches.

3.5 Determination of Errors for Kinematical Parameters

Calculation of errors for all kinematical parameters of each individual meteor, using classic methods, is a very difficult problem due to a long chain of common non-linear transformations. Instead we use Monte-Carlo methods to statistically estimate probable errors in quantities. Application of regression analysis to the astrometric processing (Kozak 2002) provides us with mean values and standard deviations of equatorial coordinates for each

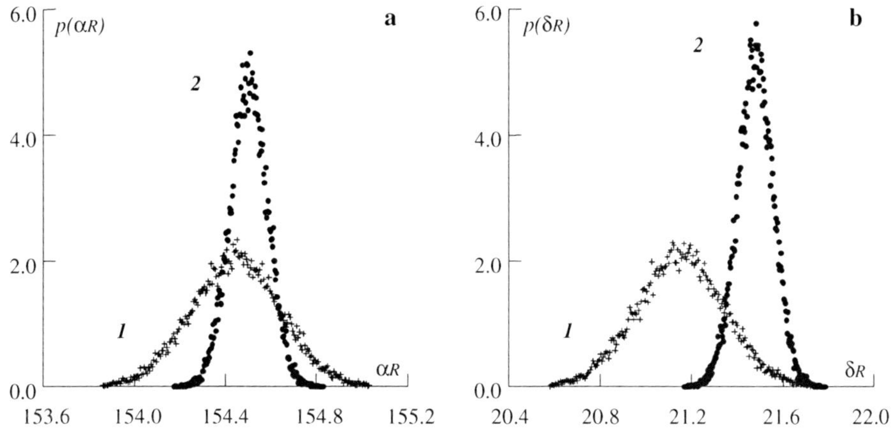

Fig. 1 Probability distribution p of errors for equatorial coordinates of geocentric radiant calculation (a Leonid 2002 meteor): (**a**) right ascension α_R (°); (**b**) declination δ_R (°); (1) express method using array of meteor head image points, (2) calculation using array of all points along the meteor trail image from the summed frame

kth measured point on the meteor image: $(\bar{\alpha}; \sigma_\alpha)_k, (\bar{\delta}; \sigma_\delta)_k$. The use of the astrometric test described above has shown near Gaussian error distribution, which can be completely described by these two parameters. Thus, we have a possibility to use as input parameters for kinematical processing not arrays of $(\alpha; \delta)_k$ as it was declared in section on kinematics but arrays of their statistical distributions with known parameters. Generating random values for each point in accordance with its distribution and sending them to kinematical processing procedure we will obtain respective random values for each investigated parameter (trajectory parameters, orbital elements, etc.). About 10–20 thousand steps are required to fully implement this and to obtain statistical distributions of errors for all calculated parameters. Then, having the statistical distribution for each parameter we can use not only its standard deviation as an error of the parameter, but also its average (or modal, median, if the distribution is asymmetrical) value as the physical value of the parameter itself. Detailed description of the method will be given elsewhere.

3.6 Photometry

The method for photometry has been developed using both theoretical and empirical techniques. The empirical approach consists in the use of results from well known experiment "artificial meteor", which is realized by means of the camera rotation with different angular velocities. Images of stars of different magnitudes moving on the frame (photocathode or CCD surface) with different linear velocities draw meteor-like trails being of different intensities. In the simplest approach we can construct a table of measured trail parameters as a function of stellar magnitude and image motion velocity. These results can be used to determine meteor magnitudes from stellar reference. One can do the measurements in single frames with meteor/star image motion or in summed frame, similarly to (Hawkes et al. 2001). This method is partly described in (Kozak et al. 2001). The method of pixel integration (Hawkes et al. 2001; Koten 2002), being applicable to linear response detectors, can also be realized in the program in manual mode of processing.

The theoretical approach, which is currently in testing mode and not included in the release of the software, includes the creation of the model for TV system functioning at the phases of charge accumulation and scanning, modeling meteor PSF motion across photocathode or CCD, etc. Possible non-linear response onto input signal and afterimage existence are input as model parameters.

4 Software Application Examples and Prospects

The quality of processing results depends on parameters of the meteor (angular length, velocity, distance of the radiant from optical center etc.), and on observational system parameters (focal length being responsible for spatial resolution, noise level), and ultimately on methods used. For demonstration of influence of these factors we will consider the trajectory parameters for 28 Leonids from 2002 calculated by means of this software (Kozak et al. 2007). During the first hour of the observations the Leonids radiant had been drifting consecutively through view fields of both TV systems, which adversely effected the triangulation processing precision: standard deviations were near 1° (the highest value being 2°.58) for radiant coordinates and 0.6–1 km (maximal value being 4.35 km) for the beginning heights. The situation radically changed when the radiant had left the fields of view: radiant determination errors decreased down to 0°.1–0°.4 (the lowest value being 0°.06), altitude errors down to 0.1–0.3 km (the lowest one being 0.07 km).

We expect to complete the theoretical approach to photometry and to complete automation of all procedures in the near future. Since the software is still developing, the author will be thankful for any recommendations and notes concerning its improvement.

Acknowledgements The author expresses his gratitude to Dr. Trigo-Rodriguez and Meteoroids-2007 organizing committee for kindly presented travel grant for visiting the conference. The author is indebted to Prof. Robert Hawkes for his valuable help in manuscript preparation, and English improvement.

References

A.N. Deutsch, On the reduction of photographic positions at an arbitrary optical center (in Russian). Astronomicheskiy J. XLII **5**, 1114–1116 (1965)

R.L. Hawkes, Detection and analysis procedures for visual, photographic and image intensified CCD meteor observations, in *Monograph on Meteors*, ed. by E. Murrad, I.P. Williams (Cambridge University Press, 2002), pp. 97–122

R.L. Hawkes, J.E. Bussey, S.L. MacPhee, C.S. Pollock, L.W. Taggart, Techniques for high resolution meteor light curve investigations, in *Proc Int Conf Meteoroids-2001*, ed. by Barbara Warmbein (ESA Publications Division, Noordwijk, 2001), pp. 281–286

R.L. Hawkes, K.I. Mason, D.E.B. Fleming, C.T. Stultz, Analysis procedures for two station television meteors, in *Proceedings of International Meteor Conference 1992*, ed. by D. Ocenas, P. Zimnikoval (Smolenice, Czechoslovakia, 1993), pp. 28–43

P. Koten, Software for processing of meteor video records, in *Abs Int Conf Asteroids, Coment, Meteors* (Berlin, Germany, 2002), p. 27

P. Koten, J. Borovicka, Light curves of fain meteors, in *Proc Int Conf Meteoroids-2001*, ed. by Barbara Warmbein (ESA Publications Division, Noordwijk, 2001), pp. 259–264

P.M. Kozak, Problem of identification of stars in the frame at digital processing of TV meteor observations (in Ukrainian). Visnyk Astron Shkoly **2**(1), 21–24 (2001)

P.M. Kozak, Analysis of the methods and precision of determination of the equatorial coordinates in digital reducing of TV observations of meteors (in Russian, English transl at www.allertonpress.com). Kinematika I Fizika Nebesnykh Tel **18**, 471–480 (2002)

P.M. Kozak, A vector method for the determination of trajectory parameters and heliocentric orbit elements of a meteor in TV observations (in Russian, English transl at www.allertonpress.com). Kinematika I Fisika Nebesnykh Tel **19**, 62–76 (2003)

P.M. Kozak, A.A. Rozhilo, Y.G. Taranukha, Some features of digital kinematic and photometrical processing of faint TV meteors, in *Proc Int Conf Meteoroids-2001*, ed. by Barbara Warmbein (ESA Publications Division, Noordwijk, 2001), pp. 337–342

P.M. Kozak, O.O. Rozhilo, V.G. Kruchynenko, A.M. Kazantsev, Y.G. Taranukha, Results from 2002 Leonid meteor storm TV observations in Kyiv. Adv Space Res **39**, 619–623 (2007)

S. Molau, The meteor detection software MetRec, in *Proc Int Conf. Meteoroids-1998*, ed. by W.G. Baggaley, V. Porubcan (Tatranska Lomnica, Slovakia, 1998), pp. 131–134

Updates to the MSFC Meteoroid Stream Model

Danielle E. Moser · William J. Cooke

Originally published in the journal Earth, Moon, and Planets, Volume 102, Nos 1–4.
DOI: 10.1007/s11038-007-9159-1

Abstract The Marshall Space Flight Center (MSFC) Meteoroid Stream Model simulates particle ejection and subsequent evolution from comets in order to provide meteor shower forecasts to spacecraft operators for hazard mitigation and planning purposes. The model, previously detailed in Moser and Cooke (Earth Moon Planets **95**, 141 (2004)), has recently been updated; the changes include the implementation of the RADAU integrator, an improved planetary treatment, and the inclusion of general relativistic effects in the force function. The results of these updates are investigated with respect to various meteoroid streams and the outcome presented.

Keywords Aurigids · Comet ejection · Draconids · Leonids · Meteor shower · Meteoroids · Model predictions · Numerical integration · Perseids · Stream model

1 Introduction

The NASA Meteoroid Environment Office has developed the Marshall Space Flight Center (MSFC) Meteoroid Stream Model to forecast meteor showers for Earth and Earth-orbiting spacecraft to provide information to spacecraft operators for hazard mitigation and mission planning. Changes to the model, previously presented in Moser and Cooke (2004), have recently been implemented. The updates include the use of a new numerical integrator, the inclusion of more planetary effects, and improvements in the calculation of planetary positions. General relativistic effects are also now taken into account. The immediate aim of this paper is to investigate the effect these updates had on various Leonid and Perseid

D. E. Moser (✉)
Meteoroid Environment Office, UNITeS Stanley Associates, NASA, Building 4487/EV13, Marshall Space Flight Center, Huntsville, AL 35812, USA
e-mail: danielle.e.moser@nasa.gov

W. J. Cooke
Meteoroid Environment Office, NASA, Building 4487/EV13, Marshall Space Flight Center, Huntsville, AL 35812, USA

streams in regards to peak time and duration, and to show the results of modeling the October Draconids and α-Aurigids with the model for the first time.

2 Model

2.1 Overview

In modeling particle ejection and subsequent evolution from comets, the workload is broken into three parts. First the test particles are created for each cometary perihelion passage, then their positions and velocities are integrated forward in time, and finally the particles are examined at specific times of interest. The first and third steps are detailed in Moser and Cooke (2004). It is the second step that will be discussed here as it is the one that has been affected by the recent update.

2.2 Updates

In the previous version of the MSFC Meteoroid Stream Model, a 4th order variable step-size Runge-Kutta (RK4) integrator was used to integrate meteoroid position and velocity forward in time. In this update, a 15th order RADAU integrator (Everhart 1985) has replaced the RK4. It is more accurate than the RK4, especially when close planetary approaches must be considered. Also of note is that it has been used successfully to determine the orbits of over 200 comets (Marsden et al. 1978; Everhart and Marsden 1983) and it is used by other stream modelers with good results (i.e. McNaught and Asher 2001; Vaubaillon 2002).

In the original model, the effects of radiation pressure, Poynting-Robertson drag, and the gravitational influences of 7 planets, Venus through Neptune, were taken into account. Mercury's mass was included in the mass of the Sun, and perturbations from the Earth–Moon barycenter were included, instead of treating the Earth and Moon separately. Jet Propulsion Laboratory's (JPL) DE406 (Standish 1998) was used to compute the positions of the planets: planetary positions were interpolated with a cubic spline subroutine from a look-up table of positions given every day from 1000 CE to 2150 CE.

For more accurate orbits, radiation pressure, Poynting-Robertson drag, and the gravitational influences of 8 planets, Mercury through Neptune, and 2 minor bodies, the Moon (treated separately from the Earth) and Pluto, are considered in the update; a general relativistic correction has also been added (Brumberg 1991). Resolving Mercury as a separate body made the largest improvement to the asteroid orbits used as test cases. An additional change to the model concerns the calculation of planet positions. JPL DE406 binary files along with publicly available subroutines making use of Chebychev polynomial interpolation valid from 3000 BCE to 3000 CE are now used. This interpolation scheme is more accurate and allows for the modeling of older streams, as is the case with the Perseids and Aurigids.

2.3 Inputs

Inputs to the model are shown in Table 1. For the Leonids, Perseids, Draconids, and Aurigids, the table lists the parent comet parameters, ejection power law, cap angle of

Table 1 Inputs to the MSFC Meteoroid Stream Model for the Leonid, Perseid, Draconid, and Aurigid meteor showers

Shower	Parent comet	Parent radius (km)	Ejection power law	Cap angle	No. of returns	No. of particles/return
Leonids	Tempel-Tuttle[a]	1.8	$r_h^{-5.0}$	30°	30	300,000
Perseids	Swift-Tuttle[b]	11.0	$r_h^{-6.0}$	60°	9	600,000
Draconids	Giacobini-Zinner[c]	1.7	$r_h^{-0.6}$	30°	28	300,000
Aurigids	Kiess[d]	3.0	$r_h^{-3.0}$	60°	1	550,000

The columns list the parent comet parameters (name and radius), ejection power law as a function of heliocentric distance, cap angle of ejection (an input in the ejection velocity function), the number of returns of the comet considered, and the number of particles ejected per return

[a] Radius from Hainaut et al. (1998). Ejection power law determined by fit from previous simulations

[b] Radius from Boehnhardt et al. (1996). Ejection power law from Fomenkova et al. (1995)

[c] Radius averaged from data given in Newburn and Spinrad (1989), Landaberry et al. (1991), and Churyumov and Rosenbush (1991). Ejection power law deduced from Hanner et al. (1992)

[d] No data is available for this comet. Kiess was taken to be an average long period comet and given properties as such—data on long period comets was taken from several sources and averaged

ejection (an input in the ejection velocity function), the number of returns of the comet considered, and the number of particles ejected per return. The comet parameters and ejection power law are determined from the literature. The physical properties of the modeled particles were determined from a uniform, random draw on log β, where β is the ratio of radiation pressure forces to the Sun's gravitational force. In the case of each shower, β ranged from $\sim 10^{-5}$ to 10^{-2}, resulting in a mass range between approximately 1 µg and 1 kg, assuming a density of 1000 kg m^{-3}. This particle size range is the range considered a threat to spacecraft.

3 Results and Discussion

An impact parameter (IP) is calculated for each particle approaching Earth within 1 week of the expected shower peak (Moser and Cooke 2004). The particle IPs, in effect the scaled probability that the particle will hit Earth, are summed in 0.005° or 0.01° solar longitude bins, depending on the shower. A Lorentzian is fit to the binned IP versus solar longitude—essentially the flux profile—in order to determine the time of the shower peak.

Figure 1 illustrates how the model update has improved the peak prediction time for the (a) 2001 and (b) 1999 Leonids. It has also improved the predicted duration of the 1999 Leonid storm. The model update does not improve peak prediction time for every stream, however. In the example in Fig. 2, the 1993 Perseids are better constrained by the previous version of the model, both in peak time and duration. It must be noted, however, that in the previous version of the model, 900,000 Perseid particles were simulated for each perihelion passage of the comet, as opposed to the updated model's 600,000 particles per return. As 9 cometary returns were considered, a total of 2.7 million *more* particles were integrated in the previous version. In general, simulating the ejection of more particles from the comet near perihelion yields a greater number of particles intercepted at Earth—and this, in turn, contributes to the overall shape and time of the shower peak. This difference in particle numbers could account for the lack of improvement to the Perseids after the model update, but more work is necessary to determine the culprit.

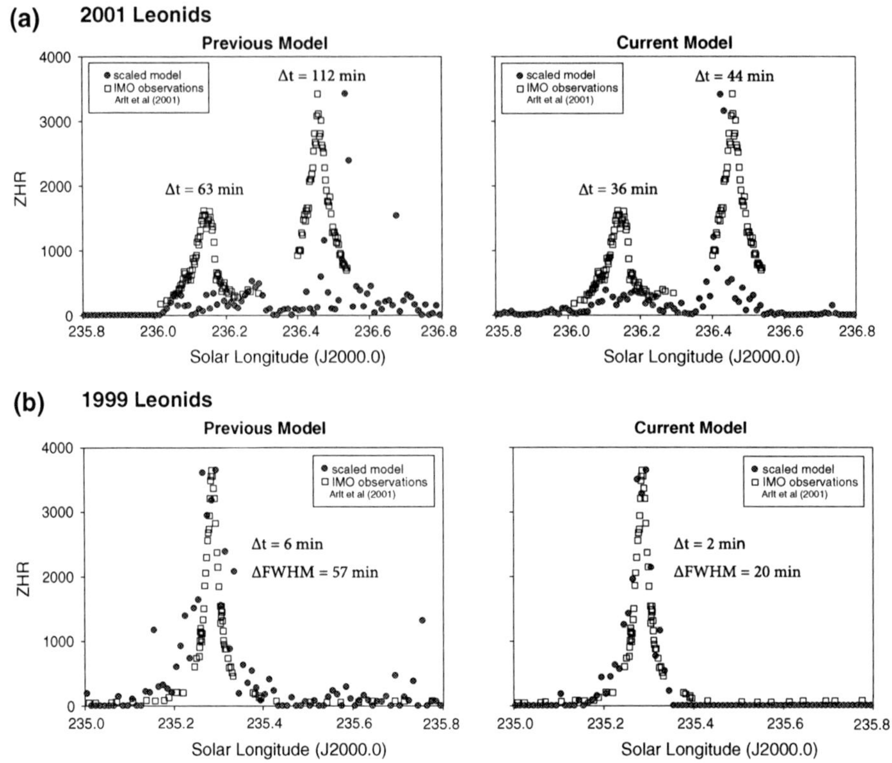

Fig. 1 Comparison of peak prediction times for the (**a**) 2001 Leonids and (**b**) 1999 Leonids. Each graph shows the scaled model IPs versus time alongside actual observations. The left panel gives results for the previous version of the model; on the right is the current version of the model. The Δts listed indicate the difference in time between the model's peak prediction and the observed peak time. The ΔFWHMs in part (**b**) indicate the difference in full width half maximum between predicted and observed. The current, updated model is an improvement over the previous model

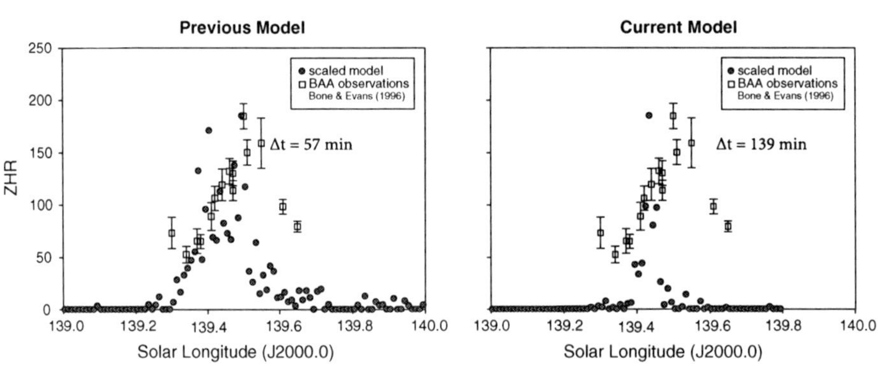

Fig. 2 Comparison of peak prediction times for the 1993 Perseids. See Fig. 1 for an explanation of the graphs. The previous version of the model is better than the current version, although the fact that 2.7 million more particles were studied in the previous version could account for this difference

Figures 3 and 4 show the results of modeling past Draconid storms/outbursts and Aurigid outbursts, respectively. This first attempt at modeling these streams was successful; the 1933, 1946, 1985, and 1998 Draconid peak times were predicted within 1 h of the observed time and the 1935, 1986, and 1994 Aurigid peaks were predicted within 15 min. The Draconid peak predictions can be further refined; the fact that the error is within 1 h is surprising, as the IP approach for the low inclination parent comet 21P/ Giacobini-Zinner was not thought to be valid (Moser and Cooke 2004). The upcoming 2007 Aurigid shower appears similar to the modeled 1935, 1986, and 1994 showers in the number of particles in the vicinity of Earth. It is therefore thought that the 2007 shower will be similar to the past showers in ZHR also: 40–50. It must be noted that the parent body C/ 1911 N1 (Kiess) is a long period comet and computations of its position are rough estimates at best.

4 Summary

Updates to the MSFC Meteoroid Stream Model better constrain the peak time and duration of the Leonid meteor showers. Improvements to the recent Perseid outbursts were not seen, though this may be accounted for by a failure to run the same number of particles as was done in the previous model version. The MSFC model was put to the task of modeling both

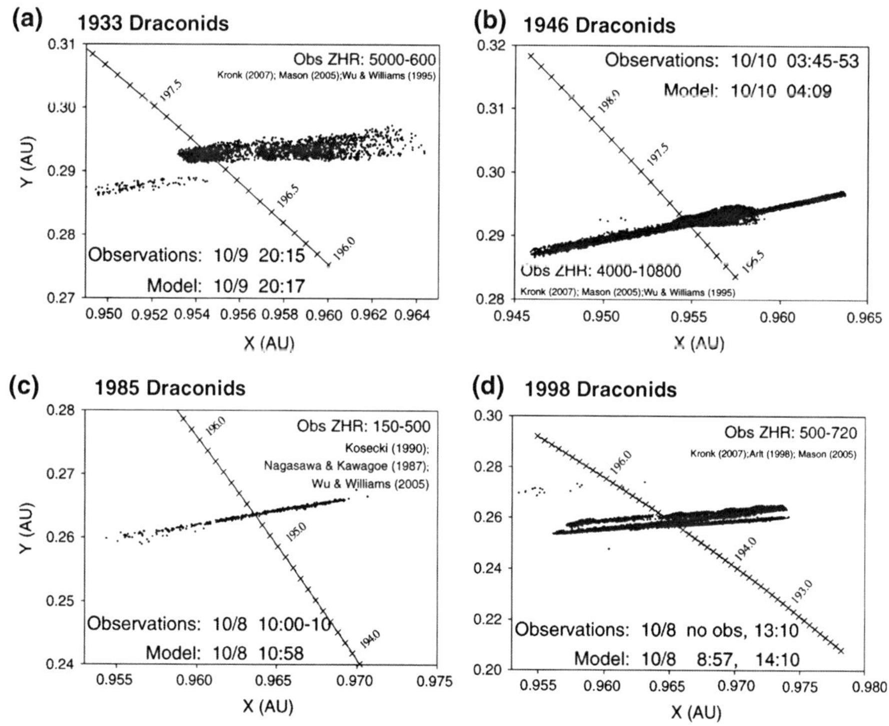

Fig. 3 Recent Draconid outbursts/storms. Each graph is a cross section plot in x-y ecliptic coordinates. The points indicate the nodal crossings of the modeled particles near Earth, represented by the solid line, during the various Draconid outbursts/storms. Peak observed times are listed along with the time the model predicts. Draconid peaks were predicted within 1 hour of the observed peak

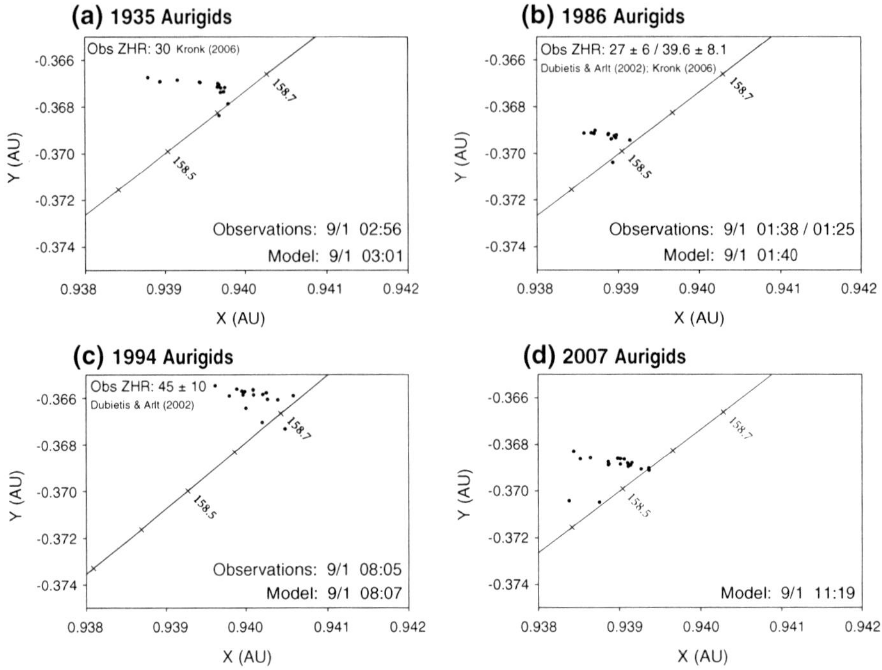

Fig. 4 Recent Aurigid outbursts. See Figure 3 for an explanation of the graphs. Aurigid peaks were predicted within 15 min of the observed peak

the Draconids and Aurigids for the first time this year. The Draconid outburst/storm and Aurigid peak predictions were surprisingly good. There was some concern about the Aurigids this year, but according to the model, the 2007 Aurigids will be on par with showers seen in 1935, 1986, and 1994: ZHR in the 40–50s (no storm).

Acknowledgements This work was supported by NASA contract NNM04AA02C. The authors wish to acknowledge the IMO; a number of their compiled observations were used a bases of comparison. Thanks also should go to Wade Batts, whose help reducing the new integrator's run-time was invaluable, and to Jeremie Vaubaillon, whose help and advice throughout the update was greatly appreciated.

References

R. Arlt, WGN J. IMO **26**(6), 256–259 (1998)
R. Arlt, J. Kac, V. Krumov, A. Buchmann, J. Verbert, WGN J. IMO **29**(6), 187–194 (2001)
R. Arlt, L.B. Rubio, P. Brown, M. Gyssens, WGN J. IMO **27**(6), 286–295 (1999)
H. Boehnhardt, K. Birkle, M. Osterloh, Earth Moon Planets **73**, 51–70 (1996)
N.M. Bone, S.J. Evans, J. Br. Astron. Assoc. **106**(1), 33–39 (1996)
V.A. Brumberg, *Essential Relativistic Celestial Mechanics* (Bristol, England: IOP Publishing Ltd., 1991), pp. 5–178
K.I. Churyumov, V.K. Rosenbush, Astron. Nachr. **312**(6), 385–391 (1991)
A. Dubietis, R. Arlt, WGN J. IMO **30**(1), 22–31 (2002)
E. Everhart, in *Dynamics of Comets: Their Origin and Evolutions*, eds. by A. Carusi, G.B. Valsecchi (1985), pp. 185–202
E. Everhart, B.G. Marsden, Astron. J. **88**, 135–137 (1983)

M.N. Fomenkova, B. Jones, R. Pina, R. Puetter, J. Sarmecanic, R. Gehrz, T. Jones, Astron. J. **110**(4), 1866–1874 (1995)
O.R. Hainaut, K.J. Meech, H. Boehnhardt, R.M. West, Astron. Astrophys. **333**, 746–752 (1998)
M.S. Hanner, G.J. Veeder, A.T. Tokunaga, Astron J. **104**(1), 386–393 (1992)
International Meteor Organization. 'IMO Meteor Shower Calendar 2007.' http://www.imo.net/calendar/2007. Cited 09 Jan 2007 (2007)
M. Kosecki, Icarus **88**, 122–128 (1990)
G.W. Kronk, 'The Alpha Aurigids.' http://comets.amsmeteors.org/meteors/showers/alpha_aurigids.html. Cited 06 Nov 2006 (2006)
G.W. Kronk, 'Draconid History.' http://comets.amsmeteors.org/meteors/showers/draconidhistory.html. Cited 08 Jan 2007 (2007)
S.J.C. Landaberry, P.D. Singh, J.A. de Freitas Pacheco, Astron. Astrophys. **246**(2), 597–602 (1991)
B.G. Marsden, Z. Sekanina, E. Everhart, Astron. J. **83**, 64–71 (1978)
J. Mason, J. Br. Astron. Assoc. **115**(5), 241 (2005)
R.H. McNaught, D.J. Asher, WGN J. IMO **29**(5), 156–164 (2001)
D.E. Moser, W.J. Cooke, Earth Moon Planets **95**, 141–153 (2004)
K. Nagasawa, A. Kawagoe, Icarus **70**, 138–145 (1987)
R.L. Newburn, H. Spinrad, Astron. J. **97**(2), 552–569 (1989)
E.M. Standish, JPL IOM 312.F-98-048 (1998)
J. Vaubaillon, WGN J. IMO **30**(5), 144–148 (2002)
Z. Wu, I.P. Williams, Planet Space Sci. **43**(6), 723–731 (1995)

The NASA Lunar Impact Monitoring Program

Robert M. Suggs · William J. Cooke · Ronnie J. Suggs · Wesley R. Swift · Nicholas Hollon

Originally published in the journal Earth, Moon, and Planets, Volume 102, Nos 1–4.
DOI: 10.1007/s11038-007-9184-0

Abstract NASA's Meteoroid Environment Office has implemented a program to monitor the Moon for meteoroid impacts from the Marshall Space Flight Center. Using off-the-shelf telescopes and video equipment, the Moon is monitored for as many as 10 nights per month, depending on weather. Custom software automatically detects flashes which are confirmed by a second telescope, photometrically calibrated using background stars, and published on a website for correlation with other observations. Hypervelocity impact tests at the Ames Vertical Gun Range facility have begun to determine the luminous efficiency and ejecta characteristics. The purpose of this research is to define the impact ejecta environment for use by lunar spacecraft designers of the Constellation manned lunar program. The observational techniques and preliminary results will be discussed.

Keywords Meteoroids · Lunar impacts · Space environments · Hypervelocity impact testing

---

R. M. Suggs (✉)
NASA, Space Environments Team, EV13, Marshall Space Flight Center, Huntsville, AL 35812, USA
e-mail: rob.suggs@nasa.gov

W. J. Cooke · R. J. Suggs
NASA, Space Environments Team and Meteoroid Environment Office, EV13, Marshall Space Flight Center, Huntsville, AL 35812, USA

W. R. Swift
Raytheon/MSFC Group, Space Environments Team, EV13, Marshall Space Flight Center, Huntsville, AL 35812, USA

N. Hollon
Jacobs Technology/MSFC Group, Space Environments Team, EV13, Marshall Space Flight Center, Huntsville, AL 35812, USA

N. Hollon
Villanova University, Villanova, PA 19085, USA

1 Introduction

Video observations of the Moon during the Leonid storm in 1999 (Dunham et al. 2000; Ortiz et al. 2000) confirmed that lunar meteoroid impacts are observable from the Earth. One probable Geminid impact was observed from lunar orbit by Apollo 17 astronaut Dr. Harrison Schmitt (NASA 1972). Since NASA's Constellation Program, which will place crews on the lunar surface for up to 6 months at a time, is currently in the preliminary design stages a new lunar impact ejecta environment model is needed. This exposure time is vastly increased over the Apollo Program and the risk from meteoroid impact ejecta must be better understood so that shielding on lunar spacecraft, spacesuits, and surface systems can be optimally designed. The existing model, NASA SP-8013 (NASA 1969), shows ejecta at a given particle size to be 10,000 times as abundant as primary meteoroids. This violates conservation of energy and is probably overly conservative which will result in lunar spacecraft designs with too much meteoroid shielding and hence too much weight.

Since our organization, NASA Marshall Space Flight Center's Natural Environments Branch, houses the Meteoroid Environment Office and the Constellation Program Environments and Constraints co-lead, we have the responsibility for defining the ejecta environment and have undertaken a program of observations, testing, and modeling to do so. Our first results were reported by Cooke et al. (2006, 2007).

2 Observational Technique

The observations are carried out at the Automated Lunar and Meteor Observatory located on-site at the Marshall Space Flight Center (latitude 34.66 N, longitude 86.66 W). The instruments consist of two Meade RCX400 14 inch (355 mm) diameter telescopes with Optec 0.33× focal reducers and StellaCam EX monochrome video cameras. The effective focal length is approximately 923 mm giving a horizontal field of view of 20 arc minutes covering approximately $4.5 \times 10^6\,km^2$ or 12% of the lunar surface (see Fig. 1). The limiting stellar magnitude at the 1/30 s frame rate is approximately 12. The video from the StellaCam EX is digitized using a Sony GV-D800 digital tape deck and sent by Firewire to a personal computer where it is recorded on the hard drive for subsequent analysis.

Fig. 1 Camera field of view and orientation

The observations are made of the earthshine portion of the moon when the sunlit portion is between 10% and 50% illuminated. This occurs on five nights and five mornings per month. We do not observe during phases less than 10% since the time between twilight and moon rise or set is too short. We do not observe during phases greater than 50% because the scattered light from the sunlit portion of the moon is too great and masks the fainter flashes. Large lunar features are easily visible in the earthshine and are used to determine the location of the impacts on the lunar surface.

The recorded video is analyzed using two custom programs. LunarScan (available at http://www.gvarros.com) was developed by Gural (2007) and modified to read the video files. The threshold for pixel exceedance is set to 3.5 times the standard deviation over the mean image. The mean and standard deviation are tracked on a frame by frame basis using a first order response filter for each pixel channel independently. The threshold exceedances are then examined using a spatial correlation filter that looks for a row containing an adjacent triplet of exceedances bordered two rows above or below by a pair of exceedances. The software finds flashes in the video which meet these criteria and presents them to a user who determines if they are cosmic ray impacts in the detector, sun glints from satellites between the Earth and the Moon, or actual meteoroid impacts. By requiring that a flash be simultaneously detected in both telescopes, cosmic rays and electronic noise can be ruled out. Some of the detected impacts were observed with only one telescope early in the program but only flashes which spanned more than two video frames and showed a proper light curve (abrupt brightness increase followed by gradual decay) were counted. There have also been two impacts independently observed by amateur astronomers using 8 inch (200 mm) telescopes (G. Varros, D. Clark private communication). For short flashes where satellite motion might not have been detectable, custom software was used to check for conjunctions with Earth orbiting satellites whose orbital elements are available in the unclassified satellite catalog (http://www.space-track.org). Since there is some probability that orbital debris or a classified satellite not listed in this catalog could cause such a short flash, another observing station has been constructed in northern Georgia about 100 km from MSFC. This will allow parallax discrimination between impact flashes and sun glints from manmade objects, even at geosynchronous altitude.

After detection and confirmation, another computer program, LunaCon, is used to perform photometric analysis (Swift et al. 2007). Background stars are used as photometric standards to determine the observed luminous energy of the flashes. Modifications to LunaCon to improve photometric calibration, determine observed lunar surface area (collecting area), and detection threshold are described in Swift et al. (2007).

3 Observational Results

A total of 54 impact flashes were observed between November 2005 and May 2007 (Fig. 2). These were observed in a total of approximately 190 h of observation. We assumed that impacts detected during the 3 days around the peaks of major meteor showers which were located on the portion of the Moon visible from the shower radiant (determined using LunarScan) were due to shower meteoroids. It is possible that sporadics caused some of these impacts but the rates increased so dramatically for the showers that it is likely they were actually shower meteoroids. Since the velocities and impact angles of shower meteoroids are well determined, we are currently following the technique of Rubio et al. (2000) to determine luminous efficiency using our Lyrid and Geminid impacts (11 impacts each). There were approximately 16 h of observation time during these shower periods and 27 possible shower impacts were seen giving approximately 1.6 flashes per hour.

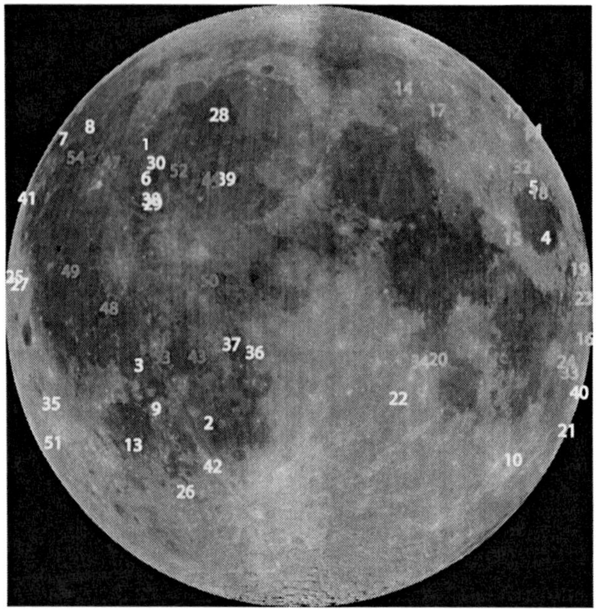

Fig. 2 Impact flashes observed between November 2005 and May 2007. Continuous monitoring was from April 2006 to May 2007. The yellow numbers are probably sporadics, the white is likely a Taurid, blue are Leonids, green are Geminids, and red are Lyrids. A complete list of candidate impacts is given at http://www.nasa.gov/centers/marshall/news/lunar/index.html

There were coincidentally a total of 27 likely sporadics; 21 were observed on the western hemisphere of the Moon (waxing phase) and six on the eastern (waning phase). Figure 3 shows the observation and impact geometry. The observed impact rate during waxing phases is approximately 0.19/h and during waning phases is 0.07/h. During waxing phases the observed portion of the Moon is exposed to the antihelion, north and south toroidal, and apex sporadic sources while during the waning phases the observed portion is exposed only to the antihelion and toroidal sources. The apex meteoroids are impacting the lunar far side which we cannot observe. Clearly, the higher speed apex meteoroids at 55 km/s deposit much more kinetic energy than a similar sized antihelion or toroidal meteoroid at 25 km/s making their observed rate much higher than their flux would indicate. Thus, a clear signature of the apex source is present even in this relatively small data sample.

The field of view of the camera encompasses approximately 10–12% of the total surface of the Moon. Assuming that helion meteoroids have the same flux as antihelion, this observed rate means that somewhere on the Moon, there are approximately 2–3 sporadic impacts per hour of sufficient energy to be observed from the Earth. These impacting meteoroids have masses of order 1 kg with a kinetic energy roughly equivalent to 200 kg of TNT. During meteor showers the rate increases dramatically, partially due to the flatter population index and hence larger percentage of larger particles.

4 Hypervelocity Impact Testing

In order to experimentally determine the luminous efficiency, a series of hypervelocity impact tests have been undertaken at NASA's Ames Vertical Gun Range. Pyrex spheres of

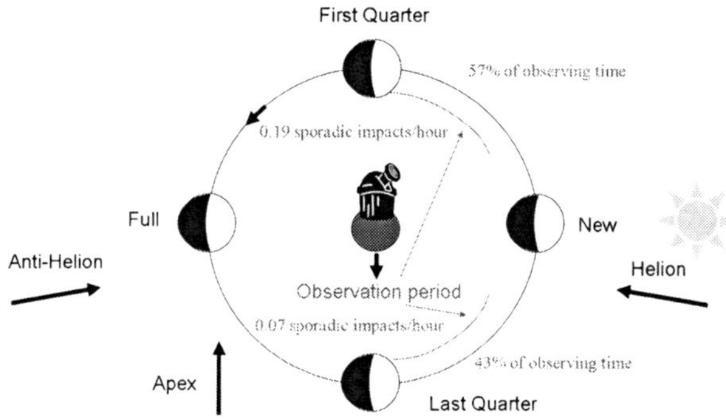

Fig. 3 Observation and major sporadic source geometry. The observed impact rate is higher near first quarter because the earthshine portion of the moon (dark part) is exposed to the apex, toroidal and antihelion sources. At last quarter the rate is lower since only the antihelion and toroidal source meteoroids impact the observed portion of the Moon. Note that the north and south toroidal sources are out of the plane of the page

¼ inch (6.3 mm) diameter were fired into ground pumice in a vacuum at speeds from 2.5 to 5.5 km/s and the impact flashes were recorded with the same StellaCam EX video cameras used for our lunar observations. Figure 4 shows the luminous efficiencies determined by the first series of shots in September 2006. The point in the upper right is the luminous efficiency $\eta = 2 \times 10^{-3}$ for Leonids determined by Rubio et al. (2000). All of the determinations of η have been plotted even though the impact angle was varied from 90° to 30° and the camera viewing angles varied between approximately 90° and 0°. Subsequent to these shots it was determined that the neutral density filters used to reduce the intensity of the impact flashes were not really neutral and had a factor of 10 higher transmissivity in the near infrared (where our cameras are sensitive and much of the thermal radiation from the impact is emitted) than in the visible. Thus the results in Fig. 4 are biased toward cooler impacts and should be treated as very preliminary. A second series of shots using truly neutral filters and lunar stimulant as the target material was

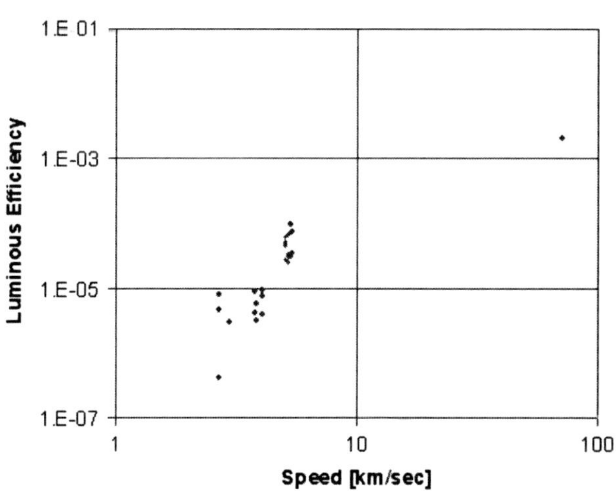

Fig. 4 Preliminary luminous efficiencies determined from Ames Vertical Gun Range tests (left hand side of figure) and from Leonids observations by Rubio et al. (2000) (point at upper right of figure)

completed in August 2007 and the analysis is underway. Future impact testing will be used to determine the mass flux, particle sizes, and particle velocities so that new cratering models can be validated and calibrated.

5 Conclusions

NASA Marshall Space Flight Center has begun a campaign to observe sporadic and shower meteoroid impacts on the Moon. The fluxes of large impactors will be determined using luminous efficiencies from hypervelocity impact testing and shower meteoroid impact statistics. Further impact testing coupled with cratering models to be developed during this research will be used to calculate ejecta characteristics from impacts of various energies. The observed impact flux, sporadic source directionality from the Meteoroid Engineering Model (McNamara et al. 2004), and a Monte Carlo and orbit generation model will be used to propagate the ejecta around the Moon. This engineering model of the ejecta environment will be used by space hardware designers to build the systems needed to explore and establish permanent bases on the Moon.

Acknowledgments The authors wish to acknowledge the meticulous and dedicated support of the following observers who recorded much of our video: Danielle Moser, Heather McNamara, Leigh Smith, Victoria Coffey, and Richard Altstatt. We also wish to thank Peter Schultz and Carolyn Ernst of Brown University, the staff of the Ames Vertical Gun Range, and Danielle Moser for their assistance during the hypervelocity impact testing.

References

L.R. Bellot Rubio, J.L. Ortiz, P.V. Sada, Luminous efficiency in hypervelocity impacts from the 1999 lunar Leonids. Astrophys. J. **542**, L65–L68 (2000)

W.J. Cooke, R.M. Suggs, R.J. Suggs, W.R. Swift, N.P. Hollon, Rate and distribution of kilogram lunar impactors Lunar and planetary science XXXVIII, Houston, Texas, LPI, Paper 1986 (2007)

W.J. Cooke, R.M. Suggs, W.R. Swift, A probable taurid impact on the moon. Lunar and planetary science XXXVII, Houston, Texas, LPI, paper 1731 (2006)

D. W. Dunham, B. Cudnik, D.M. Palmer, P.V. Sada, J. Melosh, M. Frankenberger, R. Beech, L. Pelerin, R. Venable, D. Asher, R. Sterner, B. Gotwols, B. Wun, D. Stockbauer, The first confirmed videorecordings of lunar meteor impacts. Lunar and planetary science conference XXXI, Houston, Texas, LPI, Paper 1547 (2000)

P. Gural, Automated detection of lunar impact flashes. Meteoroid environments workshop, NASA MSFC, Huntsville, Alabama (2007)

H. McNamara, R. Suggs, B. Kauffman, J. Jones, W. Cooke, S. Smith, Meteoroid Engineering Model (MEM): A meteoroid model for the inner solar system. Earth, Moon, Planets **95**, 123–139 (2004)

NASA, December 1972. "Apollo 17 air-to-ground communications transcript", http://www.jsc.nasa.gov/history/mission_trans/AS17_TEC.PDF p. 455

NASA SP-8013: 1969, Meteoroid environment—1969, near earth to lunar surface

J.L. Ortiz, P.V. Sada, L.R. Bellot Rubio, F.V. Aceituno, J. Aceituno, P.J. Gutierrez, U. Thiele, Optical detection of meteoroidal impacts on the Moon. Nature **405**, 921–923 (2000)

W.R. Swift, R.M. Suggs, W.J. Cooke, Algorithms for lunar flash video search, measurement, and archiving, this issue (2007)

Algorithms for Lunar Flash Video Search, Measurement, and Archiving

Wesley Swift · Robert Suggs · Bill Cooke

Originally published in the journal Earth, Moon, and Planets, Volume 102, Nos 1–4.
DOI: 10.1007/s11038-007-9226-7

Abstract Lunar meteoroid impact flashes provide a method to estimate the flux of the large meteoroid flux and thus their hazard to spacecraft. Although meteoroid impacts on the Moon have been detected using video methods for over a decade, the difficulty of manually searching hours of video for the rare, extremely brief impact flashes has discouraged the technique's systematic implementation. A prototype has been developed for the purpose of automatically searching lunar video records for impact flashes, eliminating false detections, editing the returned possible flashes, and archiving and documenting the results. Several utilities for measurement, analysis, and location of the flashes on the moon included in the program are demonstrated. Application of the program to a year's worth of lunar observations is discussed along with examples of impact flashes.

Keywords Moon · Lunar flash · TLP, Transient Lunar Phenomenon · Video analysis · Flash measurement

1 Introduction

NASA's Meteoroid Environment Office (MEO) has monitored the Moon for meteoroid impacts on a systematic basis since an initial detection (Cooke et al. 2006) in November 2005. These observations of the lunar earthshine for many as ten nights per month have yielded an immense quantity of video data and information on 74 impacts as of December 2007. Similar efforts in Europe have been persued by Ortiz et al. (2006, 2007).

W. Swift (✉)
MSFC ED44, Raytheon NASA/MSFC, Bldg 4487 C-151, Huntsville, AL 35812, USA
e-mail: Wesley.r.swift@nasa.gov

R. Suggs · B. Cooke
NASA/MSFC, EV13, Huntsville, AL 35812, USA

DOI: 10.1007/978-0-387-78419-9_42

2 Lunar Flash Video Search Method

The pioneering work in lunar flash search by Ortiz et al. (1999), Dunham et al. (2000), and Bellot Rubio et al. (2000a, b) was superseded by Gural whose LunarScan program (Cudnick et al. 2003) applied terrestrial video meteor techniques to the detection of Leonid lunar impacts. Further development was done by Swift to produce the program LunaCon, based on the atmospheric meteor analysis program Meteor44 (Swift et al. 2004, 2007). LunaCon was used for the detection of over 40 lunar impacts at the NASA Marshall Space Flight Center from November 2005 until the spring of 2007 (Suggs et al., this issue) and is currently used for the analysis and qualification of detected flashes (Cooke et al. 2007). Recent improvements in speed, sensitivity, and operability (Gural 2007) have resulted in LunarScan being the software of choice for lunar flash detection. Photometric analysis of detected flashes is performed as described below with LunaCon or with the aid of other astronomical photometric packages.

3 Lunar Flash Video Measurement

The goal of lunar flash measurement is to evaluate video impact flash surveys to estimate the mass flux of the impacting meteoroids. The flashes are optically unresolved so only intensity and duration information is available. The sparse, random nature of the events makes spectroscopy and more elaborate analysis techniques difficult. Thus one is left with flash photometry of each video image, the time between the flashes, and the area surveyed as information sources.

3.1 Surveyed Lunar Area

The Moon is a very large target and as such provides an excellent sensor for large mass, extremely small flux meteoroids which cause observable flashes. Since it can be many hours between flashes of a given magnitude, the observed area-time product (km^2-h) of null observations is a very significant part of the flux:

$$\text{Flux}_{\text{magnitude}} = \frac{N_{\text{magnitude}}}{\sum_{\text{obs}} (\text{Area} \times \text{Time})_{\text{magnitude}}}. \tag{1}$$

Methods have been devised for detecting and evaluating the lunar area visible in video images. The lunar limb is located in the image and solved for lunar disc center, (x_0, y_0), and radius, R, in image pixels. The spherical-moon weighted area in a pixel at radius r compared to the center pixel, A_c is given in Eq. 2:

$$\text{PixelAreaWeight}, W = \frac{1.0}{\text{Cos}\left(A\sin\left({}^{r}\!/\!_{R}\right)\right)} \tag{2}$$

The total lunar area in each image is the sum over all lunar pixels of the WxA_c product.

3.2 Lunar Flash Video background

The lunar mean brightness of the earthshine is significant because it forms the background from which the impact flashes must be detected and thus the limiting magnitude.

Significant factors in this limiting magnitude are the lunar mean intensity, *Lmi*, the maximum and minimum intensities, [Max, Min], the sky mean intensity in instrument units (IU), *Smi* and the seeing limited point spread function (*PSF*). The fitted area of the *PSF*, *Apsf* allows one to use the mean lunar surface intensity as a calibration transfer standard. The product of *Apsf* and *Lmi-Smi* yields the apparent lunar *PSF* intensity in IU, *Lpsfi*, a useful measure of the lunar background. The instrument sensitivity, *sen0*, defined for a zero magnitude star, is used to obtain the effective lunar *PSF* magnitude, *Lpsfm*, as follows:

$$Lpsfm_{\text{image}} = -2.5\text{Log}_{10}\left(\frac{Lpsfi}{sen0}\right) - 5\text{Log}_{10}\left(\frac{Apsf_{\text{image}}}{Apsf_{\text{star}}}\right) \quad (3)$$

A similar method is used to define the impact flash magnitude. It is also useful to evaluate the Contrast, Eq. 4, and the lunar *PSF* magnitude range, Range_{Lpsfm}, Eq. 5, since they define the limiting magnitude of our observations.

$$\text{Contrast}, C \equiv \frac{\text{Max} - \text{Min}}{\text{Max} + \text{Min}} \quad (4)$$

$$\text{Range}_{\text{Lpsfm}} = -2.5\text{Log}_{10}\left(\frac{Lpsfi}{sen0}\right) + [-2.5\text{Log}_{10}(1 + C^*), -2.5\text{Log}_{10}(1 - C^*)] \quad (5)$$

Range_{Lpsfm} approximates the mare and highland intensities covering a spot the same size as a lunar impact flash.

4 Lunar Flash Characterization from Video

Examination of the brighter observed lunar flashes shows the light curve to be well represented by an exponential decay curve. Unlike similar impacts onto solid targets, most of the initial plasma event is obscured and quenched by the regolith dust (Gault et al. 1964; Yanagisawa and Kisaichi 2002). This implies that video cameras observe the thermal emissions from the hot dust cloud and perhaps the evolving crater. Hypervelocity impact tests into simulated regolith at the NASA Ames Vertical Gun Facility (Ernst et al. 2004; Edwards et al., this issue) with these same cameras produce similar light curves from the extremely bright images of hot ejecta dust (Suggs et al., this issue).

4.1 Flash Characterization Method

A simple thermal decay can be represented by a time constant, α, and an initial peak value, I_0. The most reliable, measurable quantities observed in a lunar flash consist of the peak intensity, I_a, and the total intensity, I_T, of the flash. The peak intensity, I_a, depends on the camera exposure time and, to a lesser extent, on when the flash began in the exposure. For the case where the flash peak is the first exposure of period a,

$$I_a = \int_0^a i_0 e^{-\left(\frac{t}{\alpha}\right)} \text{dt} = i_0\alpha\left[1 - e^{-\left(\frac{a}{\alpha}\right)}\right] \Rightarrow \frac{I_a}{I_T} = 1 - e^{-\left(\frac{a}{\alpha}\right)} \quad (6)$$

Which yields an estimate of I_0 and α given the peak intensity, exposure time and the total intensity:

$$\alpha = \frac{a}{\ln\left(\frac{I_T}{I_T - I_a}\right)} \Rightarrow \text{Initial\_Intensity}, I_0 = {}^{I_T}/_{\alpha} \tag{7}$$

It is useful to let $\alpha \equiv 1$ when $I_a = I_T$ so that $I_0 = I_T$ as a special case. When the light curve of the flash is plotted as magnitude by the above methods, the decay is linear and the time constant, α, and initial flash magnitude, Im_f, are readily determined by point slope methods. From Im_f and α the total intensity and expected peak intensity for any given exposure time can be found independent of the properties of the camera that recorded the data.

4.2 Flash Characterization Example

The flash observed on May 1, 2006 at 2:34:40.05UT is presented as an example of the characterization of an exceptionally intense lunar impact. The intensity data in Fig. 1 (left plot) shows a reasonable fit to the log decay curve for an initial intensity of 6,500 IU with a time constant of 0.040 s. When the intensities are compared to a known star and plotted by magnitude, Fig. 1, (right plot) the linear fit to the magnitude decay is quite evident and the time constant and initial flash magnitude readily determined from the plot.

When the flash is of sufficient intensity and duration for fitting techniques to be used, as in Fig. 1, further analysis is possible. For example, the dim, second slope portion is consistent with the slower cooling rate of the surface of a crater. Only the examination of more of these rare, very bright flashes will tell.

5 Summary

Ground based lunar flash monitoring has evolved as a result of the regular observation program undertaken by the Meteoroid Environment Office of the NASA Marshall Space Flight Center. Methods have been developed for semi-automated lunar impact flash

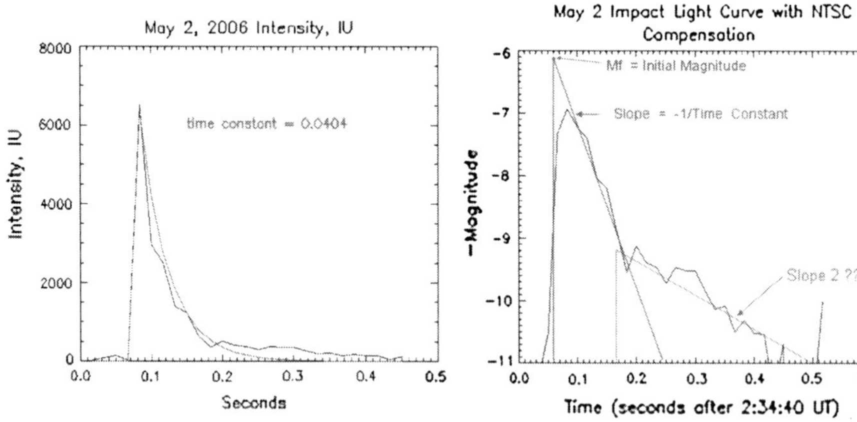

Fig. 1 May 1, 2006 Impact Intensity plot, left, and (-) Magnitude Plot, right. Time constant, $\alpha = 0.04$ s, initial intensity, $I_0 = 6{,}500$ IU, Initial Magnitude, $m_f = 6.1$. Note the second, dim portion of the curve with a slower decay possibly of crater surface origin

detection and software for this purpose is being made available for amateur observation. In order to evaluate the meteoroid flux from the observed impact rate, methods have been developed for finding lunar survey area which accounts for the spherical surface. The problems of intensity calibration of the observed flashes are partially resolved by a technique using the lunar intensity over the observed point spread function as a stellar calibration transfer standard which includes atmospheric effects within the image. A system for the characterization of impact flashes developed independent of instrumentation is described which works with the sparse data available in most observed flashes. 74 flashes have been archived as of December 2007.

References

L.R. Bellot Rubio, J.L. Ortiz, and P.V. Sada, Observation and interpretation of meteoroid impact flashes on the moon. Earth Moon Planet. vol. 82/83, p. 575–598 (2000a)

L.R. Bellot Rubio, J.L. Ortiz, P.V. Sada, Luminous efficiency in hypervelocity impacts from the 1999 lunar leonids. Astro. J. **542**, L65 (2000b)

W.J. Cooke, R.M. Suggs, W.R. Swift, A probable taurid impact on the moon. Lun. Planet. Sci. XXXVII (2006), Houston, Texas, LPI, paper 1731

W.J. Cooke, R.M. Suggs, R.J. Suggs, W.R. Swift, N.P. Hollon, Rate and distribution of kilogram lunar impactors. Lun. Planet. Sci. XXXVIII (2007), Houston, Texas, LPI, Paper 1986

B.M. Cudnick, D.W. Dunham, D.M. Palmer, A.C. Cook, R.J. Venable, P.S. Gural, The observation and characterization of lunar meteroid impact phenomena. Earth Moon Planet. **93**, 97–106 (2003)

D.W. Dunham et al., The first confirmed video recordings of lunar meteor impacts. Lunar and Planetary Science Conference XXXI, 2000. (Houston, Texas, Paper 1547, 2000)

D. Edwards, W. Cooke, D. Moser, W. Swift, Measurement of primary Ejecta from normal incident hypervelocity impact on lunar regolith simulant. Earth Moon Planet., this issue (2007). doi:10.1007/s11038-007-9198-7

C.M. Ernst, P.H. Schultz, Early-time temperature evolution of the impact flash and beyond. Lun. Planet. Sci. XXXV (2004), Houston, Texas, Paper 1986

D.E. Gault, E.D. Heitowit, H.J. Moore, Some observations of hypervelocity impacts with porous media, in *Proceedings of the Lunar Surface Materials Conference*, Boston, Massachusetts, May 21, 1963, Academic Press, (1964)

P. Gural, "Automated detection of lunar impact flashes," 2007 Meteoroid environments workshop, MSFC, Huntsville, Alabama, (2007)

NASA Meteoroid Environments Office, impact listings and LunarScan software, http://www.nasa.gov/centers/marshall/news/lunar/index.html

J.L. Ortiz, F.J. Aceituno, J. Aceituno, A search for meteoritic flashes on the Moon. A&A **343**, L57 (1999)

J.L. Ortiz, et al. Detection of sporadic impact flashes on the Moon: Implications for the luminous efficiency of hypervelocity impacts and derived terrestrial impact rates. Icarus **184**, 319–326 (2006)

J.L. Ortiz et al. Impact rates on earth from the study of sporadic impact flashes on the Moon, Poster 58, Meteoroids 2007 Conference, Barcelona, Spain, (2007)

R. Suggs et al. NASA's lunar meteoroid impact monitoring program. Earth Moon Planet., this issue (2007). doi:10.1007/s11038-007-9184-0

W.R. Swift, R.M. Suggs, W.J. Cooke, Meteor44 video meteor photometry. Earth Moon Planet. **95**, 533–540 (2004)

W.R. Swift, "LunaCon-software to detect lunar impacts," 2007 meteoroid environments workshop, MSFC, Huntsville, Alabama, (2007)

M. Yanagisawa, N. Kisaichi, Lightcurves of 1999 leonid impact flashes on the moon. Icarus. **159**, 31–38 (2002)

The Meteors, Meteoroids and Interplanetary Dust Program of the International Heliophysical Year 2007/9

Svitlana V. Kolomiyets · Mykola I. Slipchenko

Originally published in the journal Earth, Moon, and Planets, Volume 102, Nos 1–4.
DOI: 10.1007/s11038-007-9209-8

Abstract Under the title 'Meteors, Meteoroids and Interplanetary Dust', meteor research is included in the program of the International Heliophysical Year 2007/9.We list issues for coordinated meteor research within the framework of this global international program.

Keywords International Heliophysical Year (IHY) · Meteors · Meteoroids · Interplanetary dust medium

1 Introduction

The year 2007 marks the 50th Anniversary of the International Geophysical Year (IGY) and 50 years of space exploration. It also is the start of the International Heliophysical Year (IHY) 2007/9 that embraces atmospheric and solar-terrestrial physics, studies of other planets, the outer reaches of the heliosphere and interactions with the interstellar medium (Davila et al. 2001). The IHY activities described on the official IHY web site (http://ihy2007.org) include four key elements (1) coordinated research programs, (2) observatory/instrument development, (3) public outreach and (4) history/IGY Gold Program. IGY 1957/9 included meteor astronomy (Lovell 1954) in a direct response to the potential hazard to man-made satellites (Dubin 1960; Whipple 1958). Here we discuss science goals and programs.

2 Modern Meteor Science and Related Meteor Programs During IHY

Thanks to the effort of the first author and Discipline Coordinator, meteor research is officially included as an IHY program under the title 'Meteors, Meteoroids and Interplanetary

S. V. Kolomiyets (✉) · M. I. Slipchenko
Kharkiv National University of Radioelectronics, Lenin Avenue, 14, Kharkiv 61166, Ukraine
e-mail: s.kolomiyets@gmail.com

M. I. Slipchenko
e-mail: n.slip@kture.kharkov.ua

DOI: 10.1007/978-0-387-78419-9_43

Dust'. This title emphasizes a change of emphasis away from radar and optical techniques and observations from ground-based installations for ionospheric research (Davies 1957; Kolomiyets and Sidorov 2007). The opportunities of meteor astronomy have expanded today, and it is possible to study meteoroids at new levels (among others, Baggaley 2005; Green et al. 2002; Hawkes et al. 2005; Jenniskens 2005; Jenniskens et al. 2000; Murad and Williams 2002) and chapters in this volume (Trigo-Rodriguez et al. 2008).

2.1 Preservation of the Meteor Research Heritage

Modern meteor researchers have generally no access to non-English, e.g. Russian language, peer-reviewed publications and a situation exists that already existing knowledge and experience in aspects of meteor astronomy are not shared optimally. Of concern will be the preservation of the achievements of meteor science conducted since IGY (section V 'Ionosphere and Meteors') by publishing a book or a broad-based review papers with online access. Translation into English of key research monographs and papers spanning 50 years of meteor science in the Ukraine, Russia, and the other republics of the former Soviet Union, in particular meteor radar studies and meteor astronomy. Creating a new meteor database including meteor theories and hypotheses.

2.2 Development of Meteor Science

Programs to achieve this international development will include standardization of modern meteor research data with regard to the structure of databases, the meteor radar response function to convert to orbital distributions, and information on the comparability of different methods and different observational techniques. It will include revisions of existing models using the integrated databases, the establishment of an international course and/or manual of meteor astronomy and the organization of extended, international, collaborative observational programs during IHY 2007/9. The recently held international 'Meteoroids–2007' meeting in Barcelona (Spain) (Trigo-Rodriguez et al. 2008), and other conferences, will be platforms to promote modern meteor science internationally. With regard to meteor astronomy for the Developing World IHY 2007/9 encourages the creation of international meteor centers for outreach and promotion of meteor research, if possible, with a pilot center located in Kharkiv (Ukraine).

2.3 Meteors in the Terrestrial Atmosphere and Meteoroids in the Solar System (CIP 65)

The solution to the above-mentioned goals is possible only by joint efforts of research groups from many countries through the international meteor program or through a series of such programs. Since IHY's scientific activities will be organized via Coordinated Investigation Programs (CIPs), the first step in this direction will be CIP65 that was proposed by the first author. Details on these CIP65 are posted at http://ihy2007.org.uk/CIP_list.shtm. The entire worldwide meteor science community is invited in establishing and maintaining this program as part of IHY 2007/9.

Acknowledgements Authors would like to express their gratitude to the IHY Conveners for continuous support and to Drs Rietmeijer and M. Safonova for considerable help with English editing.

References

W.J. Baggaley, Interstellar dust in the solar system, in In Modern Meteor Science. An interdisciplinary view, ed. by R. Hawkes, I. Mann, P. Brown (Springer, Dordrecht, 2005), pp. 197–209

J.G. Davies, Radio observations of meteors. Adv. Electronics Electron Phys. **9**, 95–128 (1957)

J.M. Davila, A.I. Poland, R.A. Harrison, International Heliophysical Year. A program of global research continuing the tradition of previous international years, (IHY publication, 2001), 8 p. http://ihy2007.org/img/ihy.pdf

M. Dubin, Meteoric dust measured from Explorer 1. Planet. Space Sci., **2**, 121–129 (1960)

S.F. Green, I.P. Williams, J.A.M. McDonnell, N. McBride (eds.), Dust in the solar system and other planetary systems, COSPAR Colloquia Series, 15, (Pergamon Elsevier Science, 2002), 414 p.

R. Hawkes, I. Mann, P. Brown (eds.), *Modern Meteor Science. An interdisciplinary view*, (Springer, Dordrecht, the Netherlands, 2005), 732 p.

P. Jenniskens, On the future prospects of meteor detections (invited review), in *Modern Meteor Science. An interdisciplinary view*, ed. by R. Hawkes, I. Mann, P. Brown (Springer, Dordrecht, 2005), pp. 723–732

P. Jenniskens, F.J.M. Rietmeijer, N. Brosch, M. Fonda (eds.), *Leonid Storm Research* (Kluwer Academic Publishers, Dordrecht, 2000), 606 p.

S.V. Kolomiyets, V.V. Sidorov, IHY: Meteor astronomy and the new independent states (NIS) of the former Soviet Union, in *Proc. IAU Special Session 5*, ed. by J.B. Hearnshaw, P. Martinez (Cambridge University Press, Cambridge 2007), pp. 189–198

A.C.B. Lovell, Meteor Astronomy (Clarendon Press, Oxford 1954), 317 p.

E. Murad, I.P. Williams (eds.), *Meteors in the Earth's Atmosphere* (Cambridge University Press, Cambridge, 2002), 322 p.

J.M. Trigo-Rodriguez, F.J.M. Rietmeijer, J. Llorca and D. Janches (eds.), Advances in Meteoroid and Meteor Science. (2008)

F.L. Whipple, The meteoric risk to space vehicles. in *Proc. International Astronaut. Congress, Barcelona, 1957*, ed. by F. Hecht (Springer-Verlag, Vienna, 1958), pp. 418–428

Meteor Orbit Determinations with Multistatic Receivers Using the MU Radar

Yasunori Fujiwara · Yoshiyuki Hamaguchi · Takuji Nakamura · Masaki Tsutsumi · Makoto Abo

Originally published in the journal Earth, Moon, and Planets, Volume 102, Nos 1–4.
DOI: 10.1007/s11038-007-9150-x © Springer Science+Business Media B.V. 2007

Abstract The MU radar of RISH (Research Institute for Sustainable Humanosphere, Kyoto University), which is a MST radar (46.5 MHz, 1 MW peak power), has been successfully applied to meteor studies by using its very high versatility. The system has recently renewed with 25 channel digital receivers which significantly improved the sensitivity and precision of interferometer used in meteor observation. The transmission is now synchronized to GPS signals, and two external receiving sites with a ranging capability has additionally been operated in order to determine the trajectories and speeds of meteoroids.

Keywords Meteor · Radar observation

1 Introduction

Meteor orbits by radar observation have been mainly obtained with the method of Gill and Davies (1956), which was based on an idea of Kaiser (Hawkins 1964). Orbits of meteors are determined with the time difference of meteor echoes between the receivers at the main site (the same location of transmitter) and more than two additional sites within 10 km from the main site, together with the meteor velocities deduced from Fresnel diffraction patterns. Because ideal Fresnel diffraction patterns comprise <10% of available meteor observations (Davies and Gill 1960), this method can determine only a small portion of

Y. Fujiwara (✉) · Y. Hamaguchi
Nippon Meteor Society, 2-16-8 Mikunihonmachi Yodogawa-ku, Osaka 532-0005, Japan
e-mail: DHB15312@biglobe.ne.jp

T. Nakamura
RISH, Kyoto University, Uji, Kyoto, Japan

M. Tsutsumi
National Institute of Polar Research, Tokyo, Japan

M. Abo
Tokyo Metropolitan University, Hachioji, Tokyo, Japan

DOI: 10.1007/978-0-387-78419-9_44

orbits for the total number of meteors. Baggaley et al. (1994) improved on this method, and by using an extremely narrow beam antenna to determine the horizontal direction of the meteor's reflection point and an interferometer to determine its vertical direction, developed a method for determining the meteor's orbit without necessitating velocity determination via Fresnel diffraction pattern. The number of meteor orbits with a greater accuracy has increased significantly with this method. In this method, the main and external observation sites are wirelessly linked (microwave link) in order to measure the time difference in the reception of echoes. Instead of this microwave link, which entails extensive facilities, we have utilized a portable GPS receiving system for synchronizing the sites. Outline of the system and initial results are reported in this paper.

2 The Equipment

The principal radar system is the MU Radar of Kyoto University located at Shigaraki, Shiga-pref.(34.85 N, 136.11 E). The details and the layout of receiving antenna system of the MU Radar are shown in Fig. 1. The transmitting antenna pattern is doughnut-shaped instead of a sharp pencil-shaped and has the maximum gain at the zenith angle of 45° in order to detect meteor trails at low elevation angles efficiently (Nakamura et al. 1991). The MU Radar's 4 channel analog receiver was recently upgraded to a 25 ch digital receiver system. It can determine echo direction within 0.5° accuracy with its 25 ch interferometer. Clock time in the MU Radar system is controlled by GPS. The equipment of the outlying sites was developed and fabricated specially for this research. This equipment is composed of a GPS receiver, a device that emits a 2 ms time-coded pulse synchronized to GPS, a 45.6 MHz receiver, and a computer for making measurements/logs (Fig. 2). When the receiver output exceeds a threshold level, the received signal and 2 ms time synchronization signal undergo AD conversion and are stored to the computer. This enables the precise measurement of the echo time. In addition, because MU Radar transmission is

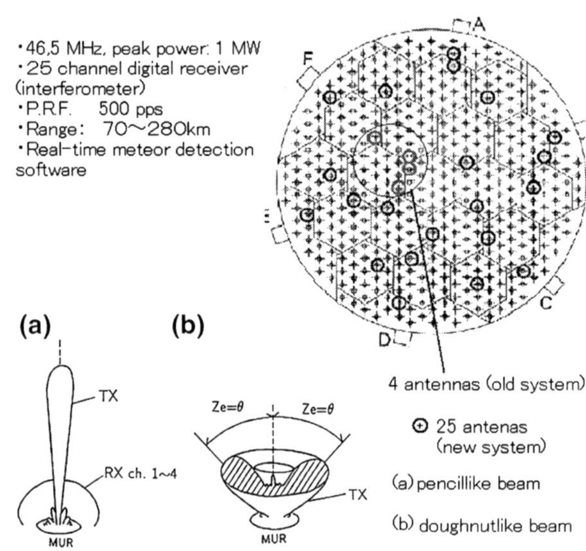

Fig. 1 The MU radar (main site)

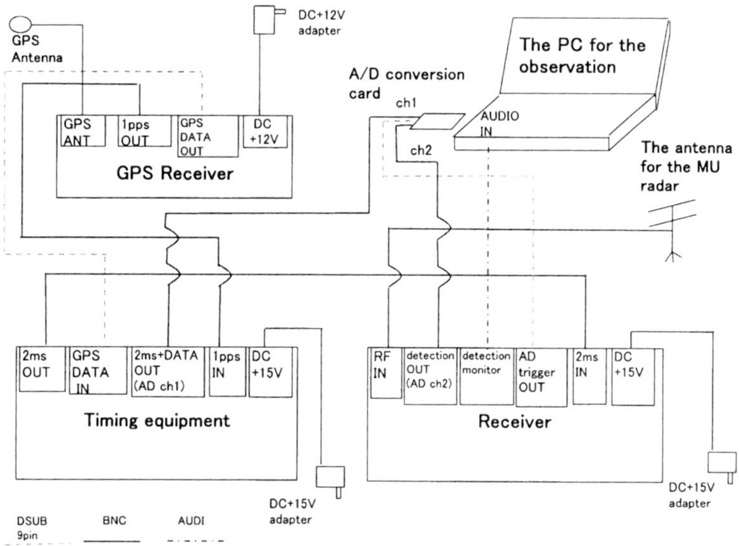

Fig. 2 The equipment of the outlying site

synchronized with the GPS clock time, it is also possible to measure the distances between the MU Radar, reflection point and the outlying sites.

3 Observation

Observations were carried out on December 14–16, 2006. The positional relations of the MU Radar and the outlying sites are shown in Fig. 3. The outlying sites used a two element Yagi antenna (directed to zenith). The number of meteors observed is shown in Table 1. In terms of the number of the MU Radar meteors, the improved new 25 ch system retrieved five times more echoes than the old 4 ch observations. At outlying site 2, the level of external noise was high so the number of retrieved echoes was low. Echo clock times were measured in 1 ms increments.

4 Results

In contrast to the MU Radar, which detects more than 50,000 meteor echoes per day, the outlying site detected only 1,000 meteor echoes per day. It is thought that this is due to the fact that the equipment's sensitivity is low compared to that of the MU Radar, and to the difficulty of setting the trigger level for echo detection. The method of determining trajectories and speeds from a combining interferometric and the time-delay techniques was the same formulas used by the Advanced Meteor Orbit Radar (AMOR) and the Canadian Meteor Orbit Radar (CMOR) (Baggaley et al. 1994; Webster et al. 2004; Jones et al. 2005). Figure 4 shows an example of a meteor echo observed at all three sites. Figure 5 show a plot of the radiant point distribution of 180 meteors observed on December 14. It can be seen clearly that radiant points are concentrated around expected value for the Geminids (right ascension (RA) 114°, declination (DC) +32°). Selecting

Fig. 3 The geographical relation of the MU radar and the outlying sites

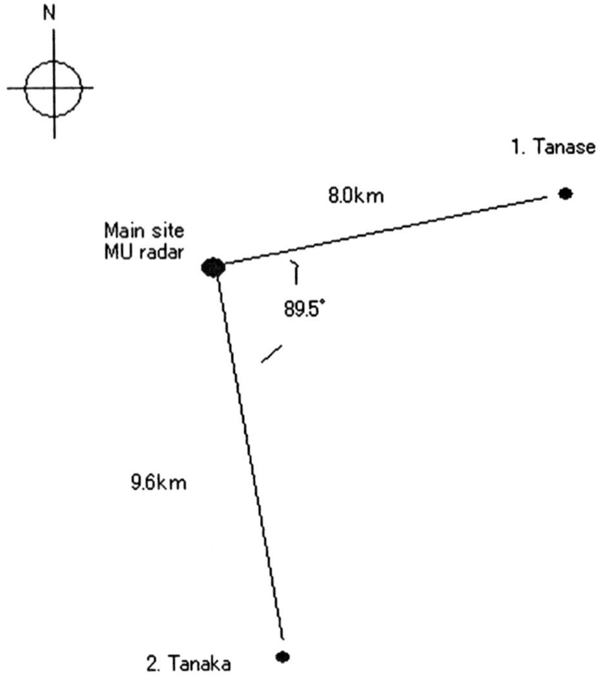

Table 1 Number of observed echoes

Data	Main site (MU Radar)	Site 1 (Tanase)	Site 2 (Tanaka)	*N*
14 (8 h)	>10,000	503	304	189
15 (24 h)	>50,000	1,899	1,011	688
16 (13 h)	>10,000	1,059	779	461
Total		3,461	2,124	1,338

Data: December 2005, ( ): observation time (hour)
N: Number of echoes observed at all three sites

meteors that are within 5° of this expected Geminid radiant, the average radiant point of 33 echoes is right ascension of 114.1°, declination of 32.9°, and the average velocity of 34.0 km/s, which agrees well with data from traditional photographic observations.

5 Summary and Future Plan

Meteor orbit determination was performed by developing portable equipment using GPS in place of the wireless links for the external observation points, and using it in conjunction with the MU Radar. A more effective meteor detection program is currently being developed. New external observation equipment with higher sensitivity is also under

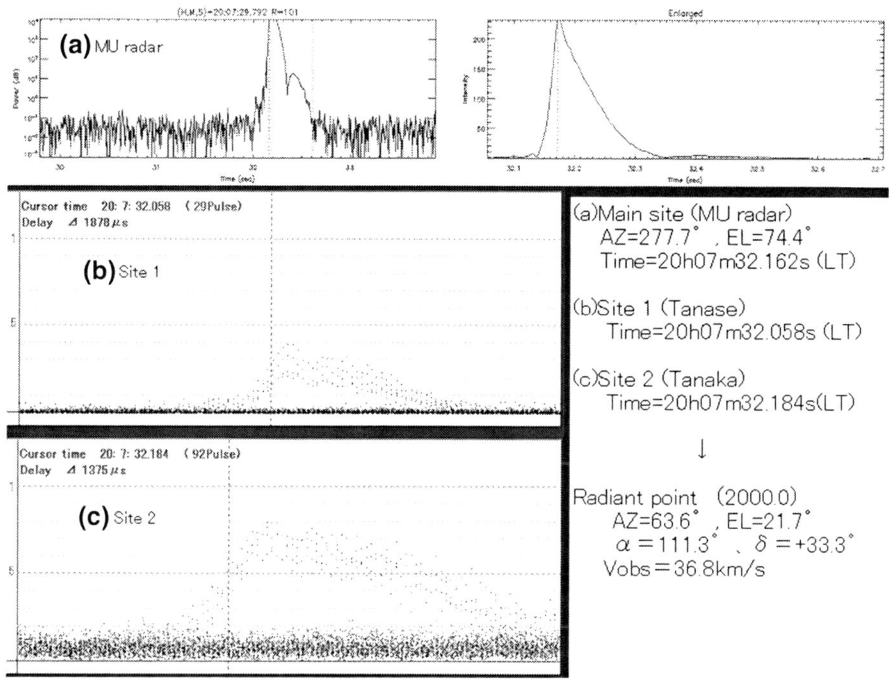

Fig. 4 Example of the meteor echo observed on all three sites showing the amplitude

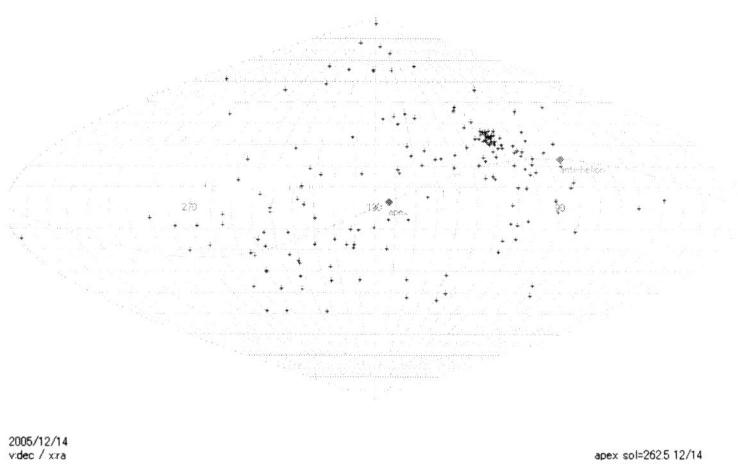

Fig. 5 The distribution of radiants of the 180 meteors observed 14 December 2005

development. There are plans to perform comparisons of results obtained from simultaneous observation with a TV system to verify precision.

Acknowledgment The authors wish to thank Mr. Tanaka and Mr. Tanase for offering us observation site. We also thank Mr. Ueda, Mr. Sagayama and Mr. Shiba for invaluable assistance.

References

W.J. Baggaley, R.G.T. Bennett, D.I. Steel, A.D. Taylor, Q. J. Roy. Astron. Soc. **35**, 293 (1994)
J.G. Davies, J.C. Gill, Mon. Not. R. Astron. Soc. **121**, 437 (1960)
J.C. Gill, J.G. Davies, Mon. Not. R. Astron. Soc. **116**, 105 (1956)
G.S. Hawkins, *Meteors, Comets, and Meteorites* (McGraw-Hill, New York, 1964)
J. Jones, P. Brown, K.J. Ellis, A.R. Webster, M. Campbell-Brown, Z. Krzemenski, R.J. Weryk, Planet. Space Sci. **53**, 413 (2005)
T. Nakamura, T. Tsuda, M. Tsutsumi, K. Kita, T. Uehara, S. Kato, S. Fukao, Radio Sci. **26**, 857 (1991)
A.R. Webster, P. Brown, J. Jones, K.J. Ellis, M. Campbell-Brown, Atmos. Chem. Phys. **4**, 679 (2004)

Physical Characteristics of Kazan Minor Showers as Determined by Correlations with the Arecibo UHF Radar

David D. Meisel · Johan Kero · Csilla Szasz · Vladimir Sidorov · Stan Briczinski

Originally published in the journal Earth, Moon, and Planets, Volume 102, Nos 1–4.
DOI: 10.1007/s11038-007-9203-1

Abstract In the northern hemisphere, the month of February is characterized by a lack of major meteor shower activity yet a number of weak minor showers are present as seen by the Kazan radar. Using the Feller transformation to obtain the distribution of true meteor velocities from the distribution of radial velocities enables the angle of incidence to be obtained for the single beam AO (Arecibo Observatory) data. Thus the loci of AO radiants become beam-centered circles on the sky and one can, with simple search routines, find where these circles intersect on radiants determined by other means. Including geocentric velocity as an additional search criterion, we have examined a set of February radiants obtained at Kazan for coincidence in position and velocity. Although some may be chance associations, only those events with probabilities of association > 0.5 have been kept. Roughly 90 of the Kazan showers have been verified in this way with mass, radius and density histograms derived from the AO results. By comparing these histograms with those of the "background" in which the minor showers are found, a qualitative scale of dynamical minor shower age can be formulated. Most of the showers are found outside the usual "apex" sporadic source areas where it is easiest to detect discrete showers with less confusion from the background.

Keywords Meteor shower · Sporadic source · AO · Kazan · HPLA · Radar

---

D. D. Meisel
SUNY Geneseo, Geneseo, NY, USA

J. Kero (✉) · C. Szasz
Swedish Institute of Space Physics, Kiruna, Sweden
e-mail: kero@irf.se

V. Sidorov
Kazan State University, Kazan, Russia

S. Briczinski
Penn State University, State College, PA, US

DOI: 10.1007/978-0-387-78419-9_45

1 Introduction

In a series of papers Sidorov et al. (see Sidorov et al. 2004a, b for example) and his coworkers at Kazan have resolved structure in the radar meteor sporadic background they called "microshowers" (=very weak minor showers). The configuration of these sources was determined using interferometric observations obtained with a classical trail scattering low-power, wide-beam, VHF radar operated by the Kazan State University. The Arecibo Observatory (AO) UHF radar on the other hand is a High Power Large Aperture (HPLA) facility and does not yet have full interferometric capability. In spite of this, the AO UHF radar has been used to study several important meteor properties using head echo scattering to determine radial velocities and in some cases decelerations (Janches et al. 2001; Mathews et al. 2003; Mathews 2004). These observations established that the majority of AO returns were from very small objects with radii on the order of microns, hence the name micrometeors was applied in publications. It was further determined that approximately 3% of the AO meteors were on hyperbolic heliocentric paths and hence candidates as interstellar particles (Meisel et al. 2002a, b). But because at AO, the angle that the meteor path makes with the beam axis is not directly observed, such observations, while reaching to extremely small masses, are always suspected of harboring unknown biases. In an effort to better understand the AO single beam data, a statistical reevaluation of the observing and data reduction methodologies was begun in 2004. Rather than being a guide to the methodologies themselves, this short paper illustrates one of the immediate consequences of the revised data reduction techniques including the ability to discern a beam inclination and its effects on derived meteoroid physical parameters. More detailed discussions of the methodologies are in preparation.

2 Sketch of the New Methodologies

There were three main breakthroughs in this data reevaluation. First a highly efficient and reliable automated meteor signature search algorithm was devised and perfected (Briczinski et al. 2006, 2007a,b). Next a pronounced correlation among decelerating meteors was discovered (Briczinski et al. 2007b) using the new search algorithm that tied together velocity (V), radius derived from the momentum equation utilizing the measured deceleration (r), and height ($\ln \frac{V^2}{r}$ versus height in the atmosphere). Residuals from this correlation have been used to estimate micrometeor mass densities at the time of radar visibility. The final breakthrough was the use of Feller's random vector theorem (Feller 1966) to transform observed radial velocities (and hence also decelerations) into actual meteor trajectory velocities. It also meant that derived quantities such as particle radius and mass density were estimates of the true quantities, not just upper limits.

Feller was able to reduce 3D random walks to 1D walks using a scheme of multidimensional projection. His random vector theorem uses the same technique to project 3D random vectors onto a fixed 1D axis. Thus if you know the random distribution of vector lengths in 3D you can by projection find the corresponding distribution in 1D, i.e. a radial direction. Feller showed that his transformation is invertible. Given an observed distribution of 1D vector components along a fixed axis, the theorem allows the reconstruction of the distribution of the vector lengths in 3D. Assigning a true length to each component is done through the standard statistical comparison of point position on the corresponding cumulative distributions.

Comparison of these transformed velocities with the original radial velocities gives the angles that the trajectory makes with the radar beam. In practice, subdividing the AO data into 2 h long segments seems to give the best balance between the needed Feller transform histogram resolution and the within-bin number of events for good Poisson statistics as needed for the proper operation of the transform technique. This 2 h partitioning of the data obtained Feb 24–26, 2006 08–15 h UTC was adopted for the results described here. The approximate mean error on each derived inclination angle as determined using the Feller procedure for these dates is $\pm$ 2.5°. The range of inclinations in the 2 h intervals are usually 0–40° to 0–60°. With inclinations available, all the derived quantities including mass, density, deceleration (and hence also radius), and velocity are estimates of the true quantities corrected rigorously for angle-to-the-beam effects. This also means that discovery of actual radiant directions is much enhanced compared with the down-the-beam assumptions that at best could tell inclinations in 15–25° wide bins and which had no corrections for angle made in the results as was done for the data used here.

3 Searching for Shower Associations

Without the Feller transformation, the work described here would not have been possible. Of course, the Feller transform applied to single, vertical, fixed beam observations does not result in a full 3D reconstruction, but only a 2D one. Thus while the beam inclination angle and the total meteor velocity can now be determined, the trajectory azimuth around the beam cannot. If the AO radar had a second beam of equal power sensitivity to that of the main beam but offset to it and pointed to a common volume with the main antenna, then a 3D reconstruction would be possible. But under present circumstances, instead of there being a point radiant for each trajectory, there is a circle of equal beam inclination angle (hereinafter called CEBIA for short) or in this case of zenith pointing, equal zenith distances as displayed in Fig. 1. In the absence of further information, we must calculate circle averaged orbital elements of a specified number of sample points whose means correspond to the previously studied cases obtained by assuming purely down-the-beam trajectories. But such a procedure will have large errors. Here we present an alternative that utilizes previously determined radiants as constraints.

If there are known radiants including velocities, as obtained by other independent optical or radar interferometric observations for example, we can straightforwardly determine the probability of source membership for each event. First the spatial intersection point of the observationally determined CEBIA and the great circle joining the antenna beam center and the radiant center gives the most likely place where the individual meteoroid comes from, if truly associated with the radiant. Comparison of the velocity values gives an independent filter from which a subset of meteors can be selected. The Kazan radiants having high probability of AO point associations are plotted as large open circles in Fig. 2 using an Aitoff projection centered on the apex. The positions and areas of the sporadic sources (open ellipses) are also indicated (Chau et al. 2007) with the ellipses representing standard deviation contours of meteor density around each center (star).

4 Determining Shower Membership

The postulated minor showers as mentioned above are specified in a geocentric, moving, ecliptic system of reference. The minor showers themselves are modeled as concentrations

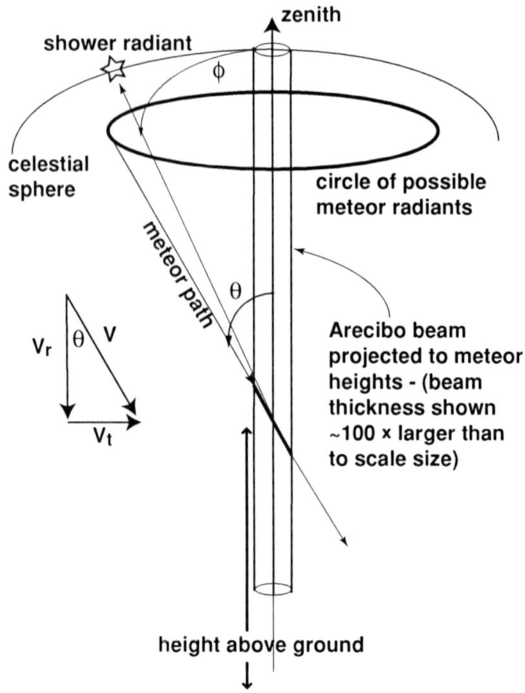

Fig. 1 Diagram showing the spatial configuration of the Arecibo beam (magnified here by about a factor of 100 for clarity), the meteor trajectory, the circle of possible radiants (denoted in the text as the CEBIA), and a shower radiant (shown as a five-point star) on the celestial sphere. Also shown are the velocity components of the meteor: V_r = the observed radial velocity, V_t = the inferred transverse velocity, and V = the meteor true or total velocity as inferred from the Feller transform (see the text). The angle θ = the inclination angle of the meteor path with respect to the beam axis (obtained by comparing V_r and V), the angle ϕ = the inclination of the shower radiant with respect to the beam axis. The probability of shower association is calculated from the absolute value of the minimum ϕ, θ difference. The actual meteor path as illuminated by the UHF radar is shown with a thick line segment. In actual practice, the meteor path can be offset from the beam axis (not shown). The offset angle can be estimated by a comparison of the observed duration with the duration calculated from the beam diameter and V_t

with the number densities of individual radiants specified by exponential radial functions. The algorithm we have adopted goes through the entire list of showers (in this case a special February list compiled from Kazan data by Sidorov to match the dates of the AO observations), whether deceleration was observed directly or not, and finds the source center that is closest to each of the resulting meteor intersection points (as described above). The resulting distance is then converted into a probability of membership resulting from an exponential probability assumption involving distance from the source center. In a parallel analysis, we have also investigated the associations of AO data with (a) the sporadic sources postulated by Jones and Brown (1993) and more recently detected with the HPLA radar at Jicamarca (see Chau et al. 2007) and (b)"major" showers as tabulated by Cook (1973) with the results mentioned here for comparison purposes. The sporadic source dimensions given by Chau et al. (2007) are used to estimate the variances of the exponentials for those identifications. For showers, the spatial probabilities are based on a 1/e radiant radius of 6°. This may seem a bit generous, but since small particles such as those that predominate the AO sample are subject to large perturbations it does not seem unreasonable.

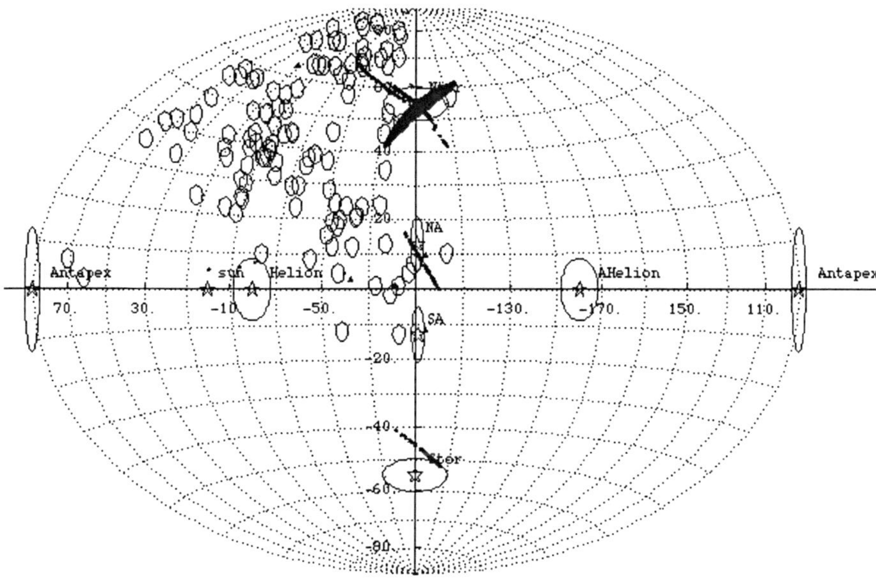

Fig. 2 Aitoff projection showing the positions of the AO CEBIA (circle of equal beam inclination angle) intersections relative to the sun-centered positions of the Kazan sources (open circles), the Jones and Brown (same as Chau et al.) sporadic sources (open ellipses with five pointed star symbols at the center) and the Cook radiants (small filled triangles). The AO point loci (strings of filled dots) attributed to sporadic sources are for Feb 24–26, 2006 during 08h–14h UTC on those days. Note that the Kazan sources for February do not fill the northern sky, but seem concentrated mostly in the north helion part of the sky and few overlap with the sporadic source positions. This makes it easier to distinguish them from the usual sporadic source distribution

Shower identifications involve not only spatial association but also velocity association. For the probability of velocity association, a Gaussian distribution centered on the velocity given for each shower is assumed with a 1/e width of $(2\ \text{km/s})^2$. This arises because the velocity errors at Kazan and Arecibo are both about 1 km/s or less. The total shower probability is the product of the spatial probability and the velocity probability. Sporadic source probabilities have only a spatial part because their velocity distributions are too wide to be useful as a criterion of association. In Fig. 2, the Kazan sources that had probabilities of association with AO intersections greater than 0.5 are plotted as open circles. Notice that individual intersection points for the Kazan radiants are NOT shown because it interferes with the visibilities of the radiant symbols (circles).

Since the use of the Feller theorem is a bit exotic, there may be some doubt concerning its validity. Thus the CEBIA intersection points (with spatial probability $P > 0.1$) for the Chau et al. (2007) sporadic source positions are also plotted as strings of small filled dots in Fig. 2. Note how the loci of the AO points cross the source areas with excellent agreement in the two northernmost areas. The same type of loci for the Southern Toroidal source seems to be a bit off but data is incomplete there because of a limitation by the southern horizon as seen in February from AO. The South Apex source seems to have no associated points. While there may be a true lack of micrometeors from that direction, it is more likely that the search method has problems when the positions of two closely spaced sources are aligned along nearly the same CEBIA relative to the beam center. In such cases, the source closest to the beam center will always have the highest probability of association. While micrometeors associated with the South Apex source were indicated by the search techniques, the probabilities obtained were below the chosen threshold. Because of this ambiguity, neither results for the South Apex or the South Torodial sporadic sources have been included in the analysis described below. It might seem that confirmation of the Feller method could be obtained by analyzing data obtained with a HPLA interferometer such as at Jicamarca or the tristatic EISCAT UHF system. But while such an experiment might be interesting to try, it must be pointed out that the Feller method needs higher data rates in a 1 or 2 h period than can be obtained at either of those facilities. Instead we intend to reanalyze AO data obtained during major showers as calibrations of the method as is standard practice with specular radars.

5 Data Analysis and Results

The Arecibo UHF radar is able to detect returns from meteoroid masses far smaller than those reported for most HPLA facilities and classical meteor interferometers. That means we can more reliably derive at each observing epoch, the individual physical properties (velocity, radius, density, and mass) of a much larger sample of events for each source than can usually be obtained directly from the Kazan radar interferometer observations themselves.

Once the AO particles that have the highest probability of association have been identified, histograms of the physical states of the associated particles can be constructed. The three quantities obtained from the revised AO analysis were mass, radius, and density. Although a formal error analysis has not yet been carried out, we estimate errors between 10% and 20% for these three quantities after corrections for inclination effects have been applied. Since there resulted associations for some 87 of 114 Kazan radiants, to display all these histograms (nearly 270, all will be available on the internet) would have been tiresome and not very informative. Our more meaningful comparison involves what are considered "core" objects ($P > 0.5$) versus "halo" objects ($0.5 > P > 0.1$). While the eye may spot subtle differences in the various histograms had we been able to present them, to quantitatively compare the resulting histogram plots, we used the Student t-test (Burington and May 1953) between two sample means since it incorporates the variances of each sample and is relatively insensitive to differing and possibly small sample numbers. The final results are expressible in terms of probabilities.

In actual practice, we examined the distributions of $\log_{10}$mass (kg), $\log_{10}$radius (μm) and $\log_{10}$density (g/cc) to obtain the probability of difference for each property. A single "total" probability of difference (P_{tot}) was obtained by taking the products of the three individual probabilities of the properties for each shower. Also we found the ratio of the number of "halo" events (N_{halo}) to the number in the "core" (N_{core}).

6 Interpretation and Discussion

In Fig. 3 we give only the two final quantities, $\log_{10}P_{\text{tot}}$ and $\log_{10}\frac{N_{\text{halo}}}{N_{\text{core}}}$. $\text{Log}_{10}P_{\text{tot}}$ is an indication of how different the "core" particle properties are from the "halo" ones while $\log_{10}\frac{N_{\text{halo}}}{N_{\text{core}}}$ is a crude measurement of how closely packed the "halo" is compared with the "core". We show data for the observed Kazan microshowers (black dots) and for the δ Cancri shower, the North Apex sporadic source (NA) and the North Toroidal sporadic source (Ntor) (open circles). It can be seen that the open circles are all to the right of the

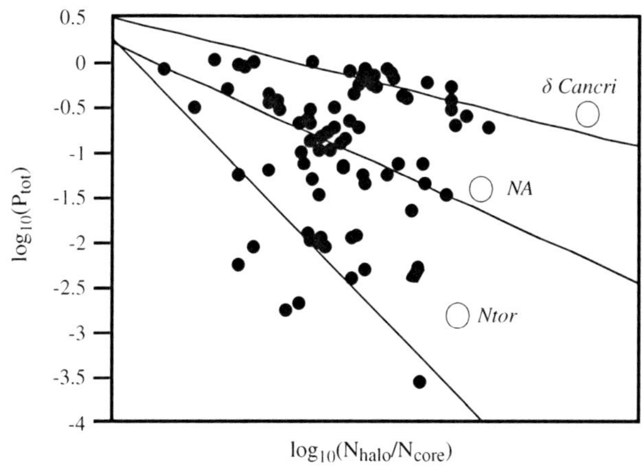

Fig. 3 Probability of different properties (P_{tot}) for halo and core particle as a function of shower/source concentration showing three possible sequences. The Kazan sources are shown as dots while the δ Cancri shower (Cook 1973), and the North Apex and the North Toroidal sources (Jones and Brown 1993) are shown as circles (in order from top of graph). Assuming a diffusion of meteors outward from the core center implies a time increase from left to right along each sequence and a time difference between sequences. See text for further details

other dots. Since N_{halo}/N_{core} is a measure of central concentration of each radiant, we conclude that the δ Cancri and sporadic sources are much less concentrated than the Kazan sources. The pattern of points is not random with what appears visually to be at least three sequences running from upper left to lower right. The reality of these sequences was verified using cluster analysis. To aid the discernment of these sequences, the points in each sequence were isolated and analyzed separately. Each sequence contains only one of the sporadic/δ Cancri points, but because the lines shown were determined by least squares, these extreme points played a critical role in the determination of the solutions. The linear correlation coefficients of the lines shown exceeded 0.9.

Given that the lines "diverge" from the upper left where the maximum contrast between the "core" regions and the "halo" regions occurs, a suggested explanation is that each sequence is a time-line along which radiant behavior "flows" from compact and contrasting to diffuse and more homogeneous. Apparently the three sequences depend on the rate of dissolution with the upper one "slow" and the lower one "rapid". The suggested temporal relationship along a sequence is that objects toward the right are older than objects to the left. Of course in the absence of a quantitative theory of meteor stream diffusion, age assignment cannot be done unambiguously. In ordinary diffusion, the rate of diffusion declines with time so that would mean the lower sequence is youngest with the top sequence being the oldest. But that would make the δ Cancri shower older than the sporadic sources which is not likely. That is why a quantitative dynamical study is needed. So why are there three sequences in the first place? Since the mean densities are all very close to a common value of about 0.5 g/cc, a characteristic of cometary material, it is speculated that these sequences are perhaps (in order) long period comets, short period comets, and sun-grazers, but more work is needed to establish such a relationship. Given the richness of this data shown by these preliminary t-test results, future ANOVA and multidimensional cluster analysis (Drummond 2000) seem warranted.

Acknowledgement Two of the authors (Johan Kero and Csilla Szasz) are financed by the Swedish National Graduate School of Space Technology.

References

S.J. Briczinski, C.-H. Wen, J.D. Mathews, J.F. Doherty, Q.-N. Zhou, Robust voltage fitting techniques for meteor Doppler determination. IEEE Trans. Geos. Remote Sens. **44** 3490–3496 (2006)

S.J. Briczinski, J.D. Mathews, D.D. Meisel, Applications of an automated micrometeor event searching routine. JASTP (2007a), in review

S.J. Briczinski, J.D. Mathews, D.D. Meisel, C.J. Heinselman, A comparison of automated-search meteor results from radar observations at AMISR Poker Flat, Søndrestrøm and Arecibo'. GRL (2007b), submitted

R.S. Burington, D.C. May, Handbook of Probability and Statistics. (Handbook Publishers, Sandusky, 1953)

J.L. Chau, R.F. Woodman, F. Galindo, Sporadic meteor sources as observed by the Jicamarca high-power large-aperture VHF radar. Icarus **188**, 162–174 (2007)

A.F. Cook, A working list of meteor streams. NASA Spec Publ **319**, 183–191 (1973)

J.D. Drummond, The D discriminant and near-earth asteroid streams. Icarus **146**, 453–475 (2000)

W. Feller, An introduction to probability theory and its applications, vol. 2 (Wiley, New York 1966), pp. 31–32

D. Janches, D.D. Meisel, J.D. Mathews, Orbital properties of the Arecibo micrometeoroids at earth interception. Icarus. **150**, 206–218 (2001)

J. Jones, P. Brown, Sporadic meteor radiant distribution: orbital survey results. MNRAS **265**, 524–532 (1993)

J.D. Mathews, Radio science issues surrounding HF/VHF/UHF radar meteor studies. JASTP **66**, 285–299 (2004)

J.D. Mathews, J. Doherty, C.-H.Wen, S.J. Briczinski, D.Janches, D.D. Meisel, An update on UHF radar meteor observations and associated signal processing techniques at Arecibo Observatory. JASTP **65**, 1139–1149 (2003)

D.D. Meisel, D. Janches, J.D. Mathews, Extrasolar micrometeors radiating from the vicinity of the local interstellar bubble. ApJ **567**, 323–341 (2002a)

D.D. Meisel, D. Janches, J.D. Mathews, The size distribution of Arecibo interstellar particles and its implications. ApJ **579**, 895–904 (2002b)

V. Sidorov, S. Kalabanov, S. Sidorova, I. Filin, Microshower Structure of the Meteor Complex. EM&P **95**, 155–164 (2004a)

V. Sidorov, S. Kalabanov, S. Sidorova, I. Filin, T. Filimonova, Associations of meteor microshowers or as the Kazan radar "SEES" radiants on northern celestial hemisphere. EM&P **95**, 165–179 (2004b)

Development of an Automatic Echo-counting Program for HROFFT Spectrograms

Kazuya Noguchi · Masa-yuki Yamamoto

Originally published in the journal Earth, Moon, and Planets, Volume 102, Nos 1–4.
DOI: 10.1007/s11038-007-9212-0

Abstract Radio meteor observations by Ham-band beacon or FM radio broadcasts using "Ham-band Radio meteor Observation Fast Fourier Transform" (HROFFT) an automatic operating software have been performed widely in recent days. Previously, counting of meteor echoes on the spectrograms of radio meteor observation was performed manually by observers. In the present paper, we introduce an automatic meteor echo counting software application. Although output images of the HROFFT contain both the features of meteor echoes and those of various types of noises, a newly developed image processing technique has been applied, resulting in software that enables a useful auto-counting tool. There exists a slight error in the processing on spectrograms when the observation site is affected by many disturbing noises. Nevertheless, comparison between software and manual counting revealed an agreement of almost 90%. Therefore, we can easily obtain a dataset of detection time, duration time, signal strength, and Doppler shift of each meteor echo from the HROFFT spectrograms. Using this software, statistical analyses of meteor activities is based on the results obtained at many Ham-band Radio meteor Observation (HRO) sites throughout the world, resulting in a very useful "standard" for monitoring meteor stream activities in real time.

Keywords Meteor · Radio meteor observation · Image processing · Software · Meteor echo · HRO · Forward-scattering radar

1 Introduction

Ham-band Radio meteor Observation (HRO) has been developed as VHF-band (30–300 MHz) forward-scattering radar since 1996 (Maegawa 1999). Recently, HRO has become one of the "standard" radio meteor observations and is widely performed by amateur meteor observers, as well as amateur radio communicators, all over the world. The observation of HRO is usually performed by a two-element antenna, a receiver, and a

K. Noguchi · M.-Y. Yamamoto (✉)
Kochi University of Technology, 185, Miyanokuchi, Tosayamada, Kami, Kochi 782-8502, Japan
e-mail: yamamoto.masa-yuki@kochi-tech.ac.jp

PC with a sound card. A beacon wave of 53.75 MHz is commonly used for HRO in Japan. The observation software "Ham-band Radio meteor Observation Fast Fourier Transform" (HROFFT) (developed by Kazuhiko Ohkawa) performs FFT processing each second to create a dynamic spectrum image every 10 min, as shown in Fig. 1. Powerful and useful software such as HROFFT enables amateur observers to build a simple automatic radio observatory for monitoring meteor activities. Recently, as an application of HRO, the HRO interferometer was developed by three teams in Japan (Ohkawa 2006; Maegawa et al. 2006; Yamamoto et al. 2007), and the HRO interferometer has operating since 2005 at Kochi University of Technology.

Since the HROFFT observation software creates a PNG image for every 10-min period, it produces 144 images per day and 4,320 images in 1 month. In the present case, a six-channel HRO system has been continuously operated since 2003 at Kochi University of Technology using a two-channel version of HROFFT. This version produces $4{,}320 \times 3 = 12{,}960$ sheets per month, and several meteor echoes are usually found on each HROFFT spectrogram, so that the observers have to analyze the enormous number of images in order to obtain the meteor activities.

Meteor echoes are easily found on the HROFFT spectrograms and are usually counted manually by HRO observers. Therefore, in order to analyze meteor activities from HRO data sets, the energy and time demanded of observers are significant. As a result, many HROFFT image archives are simply stored on a PC without being analyzed. In addition, manual echo-counting by individual observers with different counting criteria causes another problem in obtaining "standards" with respect to meteor activities. Ogawa et al., (2003) reported approaches for obtaining a "standard" based on global HRO observation data archives by being averaged for local dependences. More efforts in developing

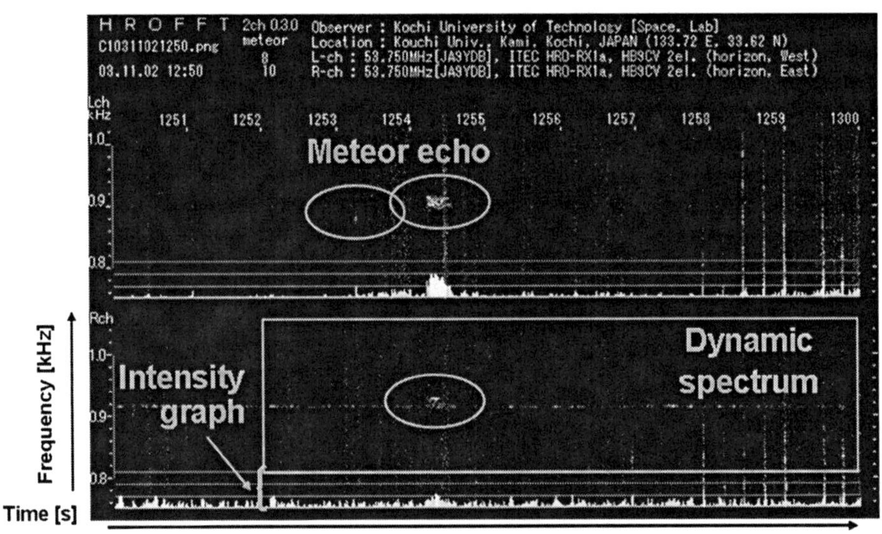

Fig. 1 Example image of a two-channel HROFFT spectrogram. The horizontal axis indicates the local time (s). The vertical axis of the spectrogram indicates the frequency (kHz), and vertical axis of the intensity graph indicates the relative power (dB). Ten-minute observations are recorded in the spectrogram image as 14-step colored dynamic spectra, and intensity graphs for each channel are added below. The beacon wave signal reflected by meteors is detected by receivers and down-converted into the audible frequency of approximately 900 Hz in the case of the usual HRO. Signals stronger than 10 dB in intensity are shown in yellow in the intensity graph, where 0 dB is set near the lowest noise level at each site

hardware and software were called in order to calibrate the absolute sensitivity of receiver and/or the parameter of software at each site because the sensitivity and noise level are different at each site and affected by local environment.

Therefore, the present paper proposes an automatic counting program in which an image processing technique is applied. The proposed program can provide a "standard" method of automatic counting for HRO spectrograms in order to obtain more useful outputs from global HRO data.

2 Development of "Meteor Echo Counter" Software

Meteor observation by forward-scattering radar is deeply affected by the geometrical configuration because the reflection region by each meteor trail should be a "mirror" for electromagnetic waves in three-dimensional space between transmitter and receiver. The configuration changes along the motion of radiant point for each meteor swarm. Especially in the case of zenith passage of the radiant point, the meteor echoes might be vanished because the reflection planes (mirrors) are created almost vertically with respect to the horizon. However, statistical analyses of meteor activities based on the results from several HRO observation sites around the world provides a very useful "standard" for monitoring meteor stream activities in real time. The software "Meteor Echo Counter" could provide an evolutional technique to speed up statistical processing. The development of "Meteor Echo Counter" software is described below.

The HROFFT spectrograms contain not only real meteor echoes but also noises of various types. These noises are categorized into, for example, vertical line noises by lightning discharges, horizontal line noises by artificial sources (such as interference by electronic power supplies or home electric appliances), airplane echoes, and ionospheric noises, as shown in Fig. 2. Therefore, applying the image-processing technique, the software is designed not only to count meteor echoes accurately and automatically but also to eliminate these types of noises. For example, when heavy ionospheric noises are received, as is shown in Fig. 2, the software will automatically skip counting meteor echoes on the exact image. Although vertical and horizontal line noises are easy to eliminate, airplane echoes are difficult to eliminate because of their complicated structures.

As a basic method by which to distinguish the above-mentioned typical meteor echoes, the program first performs the image binarization process to generate black and white images. Thereafter, the program begins to search each meteor echo in the following searching algorithm. The binarized echo images searched from left to right or up and down from any white points, as shown in Fig. 3, where the search range is represented by gray. If the search process can identify another white point, these two white points will be recognized as an independent echo combination, followed by a second search process starting

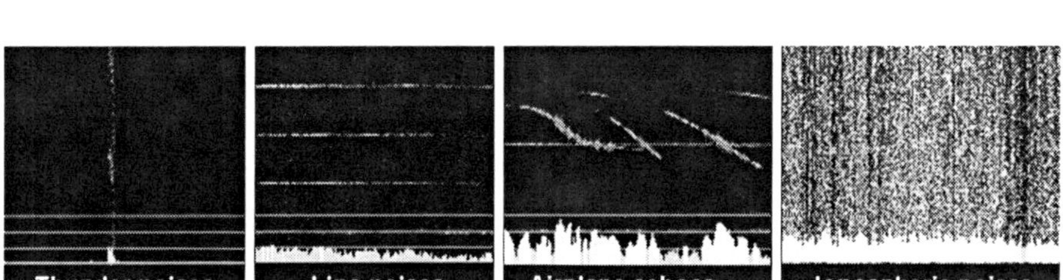

Fig. 2 Examples of various types of noises

from the adjacent white point. Through this procedure, the search range was designed as a vertically elongated diamond based on the characteristics of meteor echoes with a Doppler shift that varies with content, depending on the traveling speed of the meteor trail (scattering region) toward the observation site. Clear and precise counting of meteor echoes was established by applying this well-considered and tested method.

3 Results

The software automatically produces "processed result" images, as shown in Fig. 4. In the figure, distinguished from the various noises, each area surrounded by gray is treated as one echo. The red line indicates the portion of line type noises after their discriminations. When a long meteor echo (usually defined as 10 s or longer) is detected, the duration time of the echo is automatically calculated and displayed on the image at the upper left of each long echo, where the unit of time is seconds. The long echo on Fig. 4 was observed for 18 s, for example. The "echo-counting information result" and "meteor information result" are generated automatically (see Tables 1 and 2). The number of meteor echoes and the number of long echoes per channel are written into the text files of the "echo-counting information result". The software also outputs the detection time, the center frequency, the duration time, and the maximum signal strength on the dynamic spectrum for each meteor echo in the text files of the "meteor information result". In addition, the software can automatically generate activity graphs of meteor echoes every hour. Users can confirm the progress of the auto-counting process by watching these graphs in quasi real time. The number of meteor echoes usually depends on the observation environment, i.e., the noise environment of the receiving station (Rx), the intensity of the transmitting beacon waves at the transmitting station (Tx), and the distance between the Tx and Rx stations. In the echo-counting process, there are some threshold parameters to be used according to the environments of their receivers. In this software, users can change these parameters from GUI windows on demand.

4 Performance Assessment

In order to obtain a performance assessment of the software, we compared 10-day counting data between manual counting and automatic software counting, in detail. The results for the cases of observation sites with few noises and several noises are shown in Figs. 5 and 6, respectively. In both figures, the red line indicates the manual counting results and the blue line indicates the automatic counting results. The green and orange bars represent the number of average echoes longer than 10 s. Note that the observation sites of both figures are geographically different from each other. The former is 340 km distant from the Tx

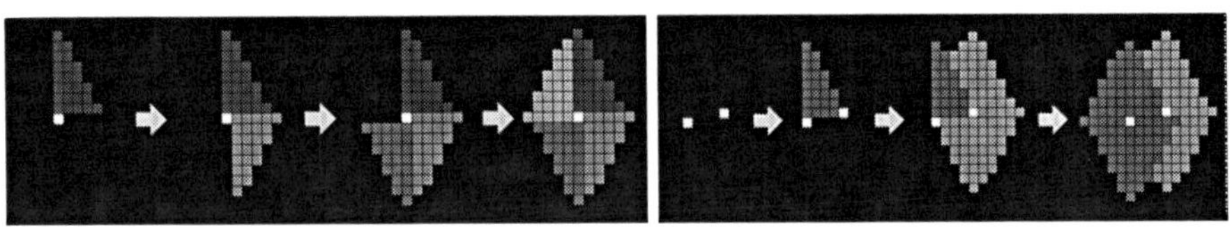

Fig. 3 Schematic diagram of the echo search procedure and its search range

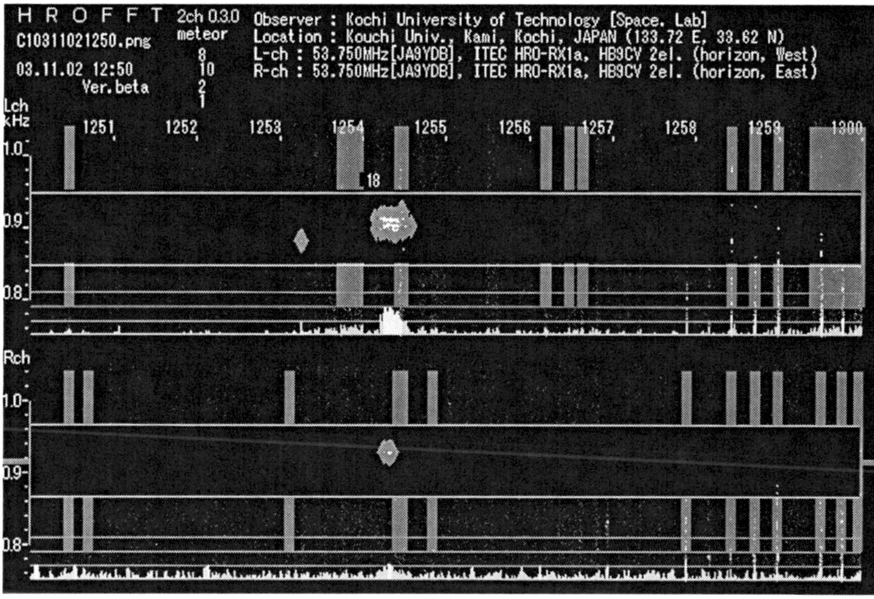

Fig. 4 Example of a processed result image. This image corresponds to the spectrogram of Fig. 1

Table 1 Example output of "echo-counting information result"

Filename	echo1	echo2	long1	long2	hikou1	hikou2	all1	all2	noip1	noip2	Yhei1	Yhei2
C10311021250	1	2	1	0	0	0	16016	16574	0	0	0	0
C10311022250	4	2	0	0	0	0	19917	12392	0	0	0	0
C10311030220	3	2	1	0	0	0	2023	2423	0	0	0	0
C10311030480	2	2	0	0	0	0	1384	1656	0	0	248	0
C10311030820	1	2	0	0	0	0	930	801	0	0	249	0
C10311030930	0	3	0	0	0	0	1013	1109	0	0	249	82
C10311050300	3	0	1	0	0	0	1198	1155	0	0	256	82

For channels 1 and 2, "echo" indicates the number of meteor echoes, "long" indicates the number of long-lasting meteor echoes, "hikou" indicates the number of airplane echoes, "all" indicates the total white pixels after the binarization process, "noip" indicates the flag of an extremely noisy spectrogram (the value will be 1 if the software skips counting because the threshold has been exceeded), and "Yhei" indicates the averaged position of observing frequency

site, whereas the latter is 200 km distant from the Tx site. The observation site of Fig. 5 is located in a rural area, whereas the observation site of Fig. 6 is located in an urban area near an international airport with heavy traffic. Therefore, the latter is much noisier, as noise from approximately five airplanes is identified on each HROFFT spectrogram. As a result, assuming an error range of less than two echoes between the automatic counting and the manual counting for a single image, the rate of agreement becomes 99% for the rural observation site and 81% for the urban one, respectively, as is shown in Figs. 5 and 6. A clear meteor activity profile, which is thought to be caused by the Geminids, was successfully identified by the automatic echo-counting software.

Table 2 Example output of "meteor information result". Luminance varies from 0 to 13 (14 steps)

Time (JS T)	Duration (s)	Doppler width (Hz)	Luminance (steps)
12:53:15	1	2	6
12:54:13	19	18	12
22:50:49	6	24	9
22:52:49	1	2	9
22:55:02	2	10	12

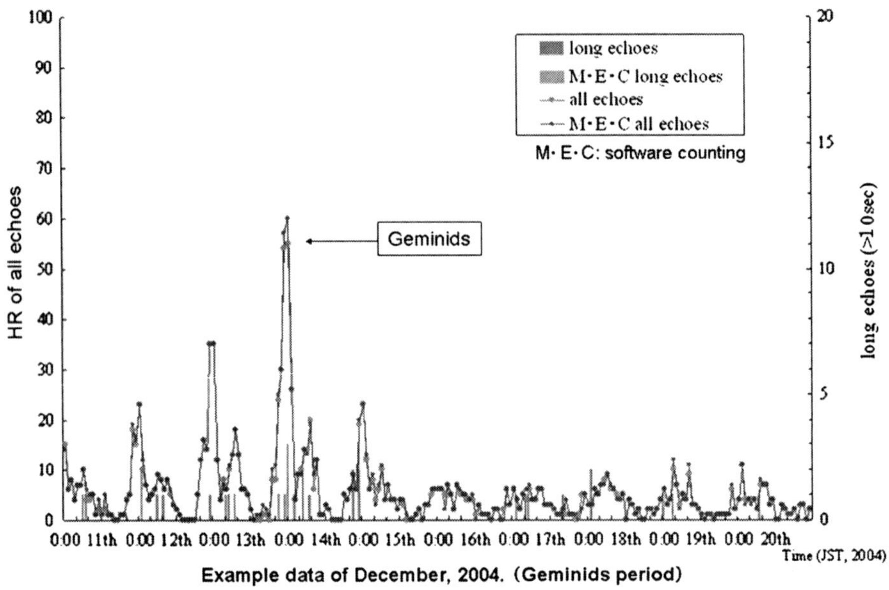

Fig. 5 Case in which the observation site is affected by few noises. "HR" denotes Hourly Rate (meteor echoes observed in 1 h), and "MEC" denotes Meteor Echo Counter (name of this software)

5 Discussion

Although some errors remain in the processes of auto-echo-counting on HROFFT spectrograms when the observation site is affected by numerous noises, an average coincidence of almost 90% was realized. The developed software "Meteor Echo Counter ver.1.0" has the capability of monitoring meteor activities even in observing an environment with several artificial noises and/or airplane echoes. Using a PC having a 2.4-GHz Pentium4 CPU with 496 MB of RAM (approximately 450 Mflops per second), the processing time of the auto-echo-counting process on the successive HROFFT data for 1 month was approximately 5 h. By applying the software to the HROFFT spectrogram data of Kochi University of Technology that has been archived for more than 2 years, meteor activity graphs were automatically produced with clear peaks near the timings of encounters of annual meteor storms of Quadrantids, η-Aquarids, Perseids, Orionids, and Geminids. In the above-mentioned PC environment, the processing time of the analyses for 2-year data was approximately 120 h. The software is able to perform automatic echo-counting, creating detailed meteor echo information without a placing a significant workload on observers.

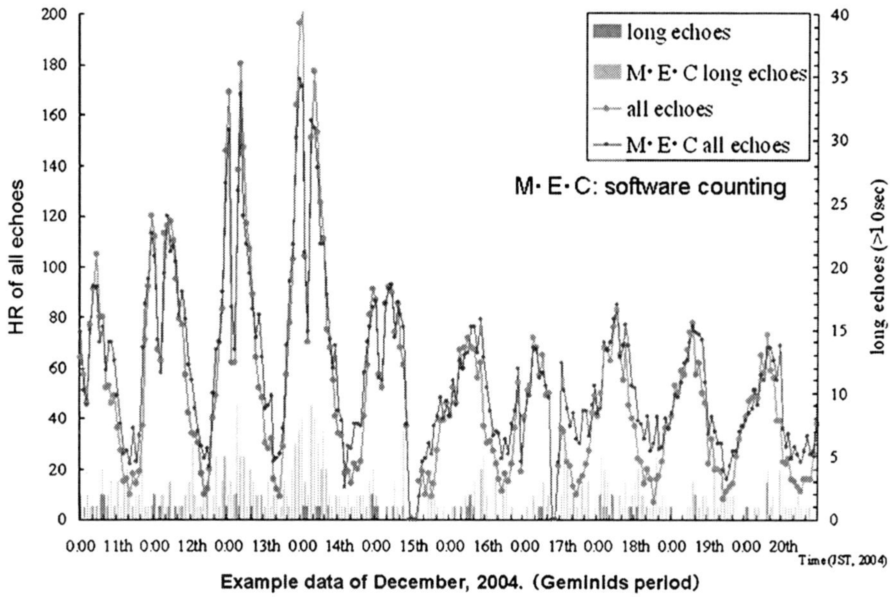

Fig. 6 Case in which the observation site is affected by several artificial noises

6 Conclusion

The software "Meteor Echo Counter ver.1.0" was developed as an automatic echo-counting program specified for the HROFFT spectrograms, providing the first "standard" software for automatic HRO echo counting for amateur observers. This software will help HRO observers to obtain scientific outputs from their HRO observations. The software is currently available on the Web (Noguchi 2007). Japanese and English versions of this software have already been released. Based on feedback from numerous HRO observers, this software will be distributed and used in global. Moreover, in combination with network software, Meteor Echo Counter ver.1.0 could produce a quasi-real-time meteor alert system for occasional meteor swarm activities in the near future.

Acknowledgments The authors would like to thank Mr. Masayoshi Ueda (The Nippon Meteor Society) for kindly providing his observation data and manual counting results for comparison. All users' feedbacks for improving the software algorithm are gratefully acknowledged.

References

K. Maegawa, HRO: a new forward-scatter observation method using Ham-band beacon. WGN **27**, 64–72 (1999)

K. Maegawa, S. Uno, H. Horiuchi, G. Okamoto, M.-Y. Yamamoto, Development of Radio Interferometer System for Meteor Observation, *Research reports of Fukui National College of Technology, Natural science and engineering*, vol. 39, (2006), pp. 31–36, in Japanese with English abstract

K. Noguchi, Meteor Echo Counter on web, (2007) http://www.gs.kochi-tech.ac.jp/115073w/

H. Ogawa, S. Toyomasu, K. Ohnishi, S. Amikura, T. Asahina, K. Miyao, K. Maegawa, Leonids 2001 by radio meteor observation all over the world. ISAS Rep. SP. **15**, 81–88 (2003)

K. Ohkawa, Meteor observation by interferometer, *Radio Meteor Observation Meeting 2006* (Hachioji, Tokyo, 2006) in Japanese

M.-Y. Yamamoto, H. Horiuchi, G. Okamoto, H. Hamaguchi, K. Noguchi, Development of HRO interferometer at Kochi University of Technology, *Proc. of Intl. Meteor Conf. 2006* (2007)

Chapter 3. Meteor-Atmosphere Interactions

What can We Learn about Atmospheric Meteor Ablation and Light Production from Laser Ablation?

R. L. Hawkes · E. P. Milley · J. M. Ehrman · R. M. Woods · J. D. Hoyland · C. L. Pettipas · D. W. Tokaryk

Originally published in the journal Earth, Moon, and Planets, Volume 102, Nos 1–4.
DOI: 10.1007/s11038-007-9186-y

Abstract Laboratory based laser ablation techniques can be used to study the size of the luminous region, predict spectral features, estimate the luminous efficiency factor, and assess the role of chemically differentiated thermal ablation. A pulsed Nd:YAG laser was used to ablate regions from ordinary and carbonaceous chondrite meteorites. CCD cameras and a digital spectroscope were used to measure the size and spectrum from the cloud of vaporised material. Scanning electron microscope (SEM) based energy dispersive x-ray spectroscopy (EDS) provided elemental abundance values in ablated and unablated regions. These results indicated some degree of differential ablation, with the most significant effect being significant loss of carbon from carbonaceous chondrites. This work suggests that a carbon matrix may play the role of the *glue* in the two component dustball model.

Keywords Meteors · Meteoroids · Meteorite · Methods: laboratory

1 Introduction

A number of aspects of meteor ablation and light production remain uncertain. For example, the size of the light production region is not well determined. The luminous efficiency factor and its velocity dependence for faint, fast meteors is debated, and as a consequence mass fluxes and associated space operation risks have considerable uncertainty. The chemical composition and physical structure of cometary origin meteoroids is not clear. While to some degree meteors must ablate in a chemically differentiated manner,

R. L. Hawkes (✉) · E. P. Milley · R. M. Woods · J. D. Hoyland · C. L. Pettipas
Physics Department, Mount Allison University, Sackville, NB, Canada
e-mail: rhawkes@mta.ca

J. M. Ehrman
Digital Microscopy Facility, Mount Allison University, Sackville, NB, Canada

D. W. Tokaryk
Department of Physics and Centre for Laser, Atomic and Molecular Sciences, University of New Brunswick, Fredericton, NB, Canada

the importance of this differential ablation is not clear. Laboratory based techniques, such as laser ablation of meteorites, can help to inform these topics. This paper builds on early meteorite ablation results (Milley et al. 2007) where we reported on ablation of ordinary chondrites.

2 Equipment

A pulsed YQD-101 Laser Photonics Nd:YAG laser was used at its first harmonic of 532 nm to ablate meteorites at 10 pps with an energy of 15 mJ per pulse and a pulse duration of 10 ns. A plano-convex F/4 lens with focal length of 100 mm was used to focus the laser light onto the sample. Laser light could be focussed to make holes in cardboard about 0.5 mm in diameter. Freshly cut meteorite samples were mounted perpendicular to the laser beam. Images were obtained using a Nikon D100 6.1 mega pixel digital SLR camera equipped with a Tamron 28–300 mm focal length F/3.5-6.3 lens. Images were taken in the RAW format under manual focus, exposure and aperture settings and an OD2 neutral density filter was used to reduce saturation. Spectroscopy was done using a DiVA (Digital Visible light Analyzer) Series 2 HWL digital spectrometer. Light was focussed onto the fibre optic cable using a F/1.9 16–100 mm Canon lens. A sharp 532 nm optical notch filter, Edmund Optics NT46-565, was used to block out the primary laser light. These initial experiments were performed at atmospheric pressure. SEM and EDS were used to measure size and elemental composition of the ablated and unablated regions. The samples were coated with approximately 10 nm of gold in a Hummer 6.2 sputtering unit and examined using a JEOL JSM-5600 SEM equipped with an Oxford Inca 200 EDS system. Images were acquired using an accelerating voltage of 10 kV and 48 mm working distance. The gold coating was needed to produce quality images on these nonconductive samples, although a consequence was that its presence masked some portions of the EDS elemental abundance graphs.

3 LIBS and Light Production

Laser Induced Breakdown Spectroscopy (LIBS) is a technique that employs a laser to heat a surface to vapourization thus creating an excited plasma with spectral features indicative of the composition of the material. Clegg et al. (2006) and Thompson et al. (2006) investigate the use of LIBS for the purpose of remote rock identification during planetary exploration. Their ChemCam, which uses LIBS, has recently been accepted for use on the 2009 Mars Science Laboratory.

Figure 1 shows a typical spectrum obtained in our experiment using the DiVA spectrometer while ablating the CV3 carbonaceous chondrite Allende. Spectra were obtained in a dark environment by averaging 400 samples. Several spectral features, which are commonly observed in meteor spectra (Ceplecha et al. 1998), have been identified. LIBS techniques, when done under suitable conditions, offer promise to help define the type of meteorite composition which best match meteor spectra. One of the primary reasons for undertaking this work was to study the size of the light production region in meteor ablation. Recently we have developed gated image intensified telescopic optics equipment which allows one to study features of the order of 1 m at meteor heights (Hawkes et al. 2005; Kaiser et al. 2005). Fragmentation seems to be a common process, for at least some meteoroid sizes, but it is important to know the size of the luminous region from single

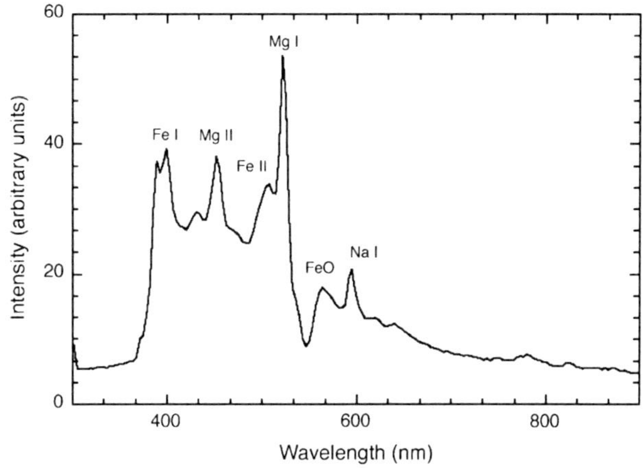

Fig. 1 Spectrum after atmospheric pressure LIBS of CV3 Allende meteorite. Optics blocks region below about 350 nm, and sensitivity and calibration are poor in the infrared region. A sharp notch filter at 532 nm distorts spectrum in that region

object ablation to interpret wake-based fragmentation studies (Hawkes et al.2005; Fisher et al. 2000; Shadbolt and Hawkes 1995).

We present in this paper high resolution images of the luminous region during laser meteorite ablation at atmospheric pressure. In general there is an intense central roughly circular bright region, which is surrounded by a less regular more strongly blue region (see Fig. 2a). The region of light for the pictures was generally on the order of 2 mm and consisted of a blue-purple glow (probably strong in the near ultraviolet), getting fainter towards the edges of the light region. When we integrated the light output for the luminous region there were not significant differences between trials. When camera settings are changed to prevent saturation the dimmer outer edge is lost and the central region is seen to not be perfectly circular (see Fig. 2b). In a few trials, more than one light region was present. We postulated that some of the material being ablated gets ejected from the surface resulting in multiple light production regions. We show one such multiple image in Fig. 2c.

Ultimately these techniques will be helpful in predicting the expected size of the meteor light production region in the absence of fragmentation. Using telescopic optics it has been shown that in many cases the width of the luminous column is no more than about a meter (Kaiser et al. 2005). When laser ablation is done under partial vacuum resulting in free

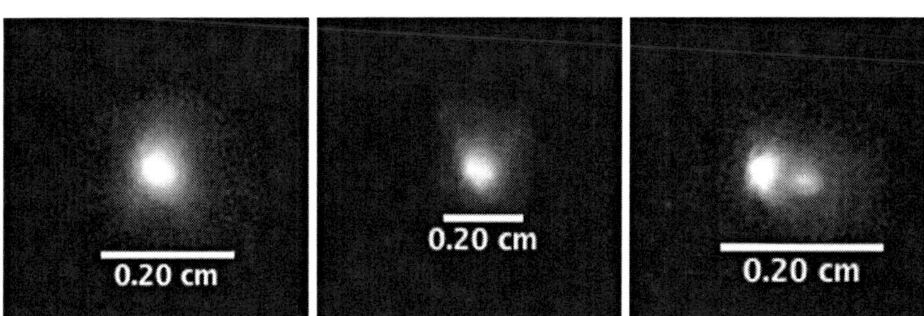

Fig. 2 The left image (**a**) shows a typical output with a saturated circular central region surrounded by a deep blue less regular outer region. In the central image (**b**) reduced sensitivity to prevent saturation, shows that the central region is deep blue and not completely circular. In the right image (**c**) is an example of a multiple light production region

molecular flow, it will be possible to scale (using the mean free path values) the observed luminous region size to that expected at meteor altitudes. The ablation conditions here are not free molecular flow, but may be directly applicable to conditions of ablation of meteorites or re-entry vehicles at much lower heights. Laser ablation can also be used to validate the results obtained by hydrodynamic simulations of atmospheric ablation of meteors (Popova et al. 2000; Boyd 2000).

4 Differential Ablation

From both meteor spectra (Borovicka et al. 1999) and from lidar measurements (von Zahn et al. 1999) there is an indication that meteoroids do not ablate as one homogeneous mixture, but more volatile materials ablate at greater heights. The lidar measurements studied the distributions of K, Ca and Fe, and found strongly differentiated ablation layers. The meteor spectra work suggested that Na ablated preferentially early, and the authors suggested that it might play the role of a binding agent in a two component dustball model. A number of studies have used equilibrium vapourization models to simulate this (McNeil et al. 2001; Schaefer and Fegley 2005). We studied the degree of differential ablation with our equipment by comparing, using the EDS feature of the SEM, the elemental abundances in the centre of the ablation pit to an unablated region. The EDS samples several μm below the surface, which will partially mask surface differential ablation. Figure 3 shows an EDS plot for the center of the ablation pit and the unablated region. The strong preferential ablation of carbon is obvious. The increase in Fe, Mg and Si is due to ablation of C leaving more of these materials available in the region which interacts with the electron beam.

5 Discussion

One purpose of this paper is to alert the community to the potential value of laser ablation techniques in characterization of meteor ablation and luminosity processes. Even under atmospheric pressure, LIBS may be helpful as an analog to meteor spectral studies and chemical composition determination. With reductions in gas pressure to free molecular flow conditions, one should be able to scale the size of the light production region using mean free path considerations.

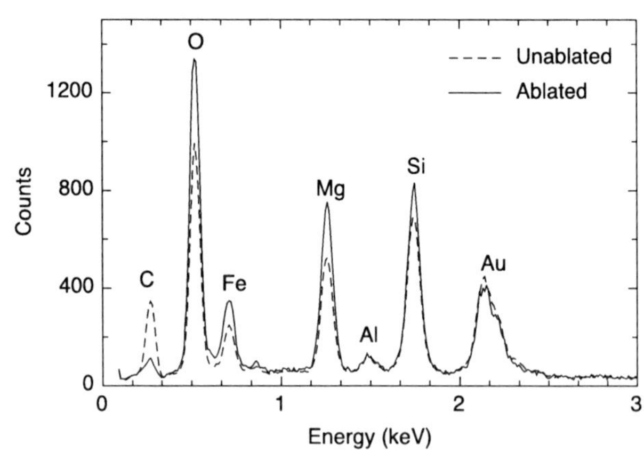

Fig. 3 EDS results for the region at the centre of the ablation pit and an average of regions in the same sample which were not laser ablated. These results are for Allende CV3 carbonaceous chondrite

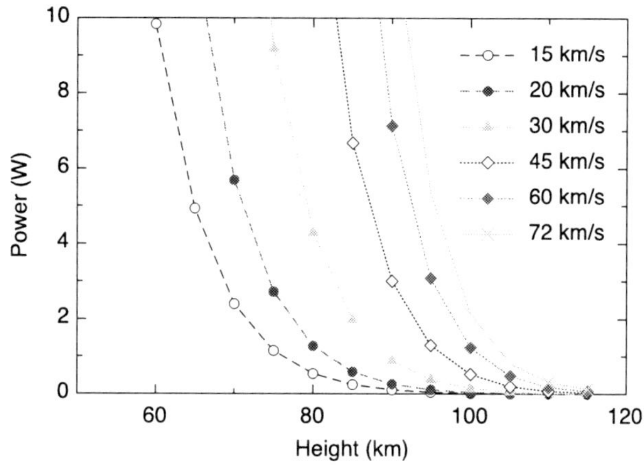

Fig. 4 Power rate to an 80 μm spherical meteoroid under different height and velocity conditions

While these initial results have provided some interesting trends, it is important to make the ablation rate more gentle and near continuous to simulate what happens to meteors in the Earth's atmosphere. To show the laser powers required, we have taken the dimensions of the focussed laser spot, and studied the power input from atmospheric collisions of a similar (80 μm radius) spherical meteoroid under various height and velocity conditions. The results are shown in Fig. 4. This suggests that the average power rate is not particularly high. While short pulse high energy lasers are suitable for LIBS, for other aspects a more continuous output laser is more appropriate. A near infrared CW laser would provide controllable heating with a minimum of other effects such as fluorescence. Diode laser arrays (DLA) or diode pumped solid state lasers (DPSSL) provide convenient devices for this application, as do CW CO or CO_2 lasers if longer wavelengths (near 5 or 10 μm respectively) were desired. The quantitative dustball model (Hawkes and Jones 1975) predicts that cometary meteoroids are collections of fundamental grains held together by a more volatile component which ablates first, releasing the grains during atmospheric flight. A variety of pieces of evidence favour this two component dustball model (Fisher et al. 2000) The differential ablation of the carbon component suggests that carbon itself may be the *glue* which holds dustball meteoroids together.

Acknowledgements We acknowledge support from the Natural Sciences and Engineering Research Council of Canada (NSERC) and the New Brunswick Innovation Fund. We thank Ralf Bruning and Marc Vallee for assistance with meteorite sample preparation.

References

I.D. Boyd, Earth Moon Planets **82–83**, 93 (2000)
J. Borovicka, R. Stork, J. Bocek, Meteorit. Planet. Sci. **34**, 987 (1999)
Z. Ceplecha, J. Borovicka, W.G. Elford et al., Space Sci. Rev. **84**, 327 (1998)
S.M. Clegg, R.C. Wiens, M.D. Dyar et al., LPS XXXVIII, **1338**, 1960 (2006)
A.A. Fisher, R.L. Hawkes, I.S. Murray et al., Planet. Space Sci. **48**, 911 (2000)
R.L. Hawkes, P.G. Brown, N.R. Kaiser et al., Earth Moon Planets **95**, 587 (2005)
R.L. Hawkes, J. Jones, Mon. Not. R. Astron. Soc. **173**, 339 (1975)
N. Kaiser, P.G. Brown, R.L. Hawkes, Earth Moon Planets **95**, 579 (2005)
W.J. McNeil, R.A. Dressler, E.J. Murad, Geophys. Res. **106**(A6), 10447 (2001)
E.P. Milley, R.L. Hawkes, J.M. Ehrman, Mon. Not. R. Astron. Soc. **382**, L67 (2007)
O.P. Popova, S.N. Sidneva, V.V. Shuvalov, A.S. Strelkov, Earth Moon Planets **82–83**, 1098 (2000)

L. Schaefer, B. Fegley, Earth Moon Planets **95**, 413 (2005)
L. Shadbolt, R.L. Hawkes, Earth Moon Planets **68**, 493 (1995)
J.R. Thompson, R.C. Wiens, J.E. Barefield et al., JGR **111**, E05006 (2006)
U. von Zahn, M. Gerding, J. Hffner et al., Meteorit. Planet. Sci. **34**, 1017 (1999)

Reanalysis of the Historic AFTAC Bolide Infrasound Database

Douglas O. ReVelle · Elizabeth A. Sukara · Wayne N. Edwards · Peter G. Brown

Originally published in the journal Earth, Moon, and Planets, Volume 102, Nos 1–4.
DOI: 10.1007/s11038-007-9173-3

Abstract We have recently digitized and partially reanalyzed the historic bolide infrasonic database. These 10 events were originally detected by the U.S. Air Force Technical Applications Center (AFTAC) from ~1960 to 1974. In this paper we present the first preliminary reanalysis results for two of the 10 bolide events, namely the Revelstoke bolide of 3/31/1965 as well as the Prince Edward Islands (P.E.I). S. African bolide of 8/03/1963, which were among the largest bolides detected during this time period. These bolides have been investigated initially since they are most likely to have had a significant effect on the computed global influx rate of ReVelle (Global Infrasonic Monitoring of Large Bolides, pp 483–490, 2001) as indicated in Brown et al. (Nature, 420:314–316, 2002). We are in the process of recomputing all relevant infrasonic propagation quantities such as plane wave back azimuth, signal velocities, power spectra, spectrograms, as well as energy estimates using multiple techniques. In a future paper we will present a complete digital reanalysis of the AFTAC bolide infrasonic data and its final resulting global bolide influx implications.

Keywords Bolides · Infrasound · AFTAC · Atmospheric acoustic-gravity waves · Global influx rate of largemeteor-fireballs

1 Introduction and Overview

1.1 Bolide Infrasound

ReVelle (1976, 2001, 2004, 2007a, b) has interpreted and analyzed bolide infrasound data from numerous sources. We have also used this data in combination with a number of other

D. O. ReVelle (✉)
Atmospheric, Climate and Environmental Dynamics Group, Earth and Environmental Sciences Division, Los Alamos National Laboratory (LANL), Los Alamos, NM 87545, USA
e-mail: revelle@lanl.gov

E. A. Sukara · W. N. Edwards · P. G. Brown
Department of Physics and Astronomy, University of Western Ontario (UWO), London, ON Canada N6A 3K7

DOI: 10.1007/978-0-387-78419-9_48

detection techniques as well (photographic, satellite, seismic, etc.) in order to compute the bolide mass and source energy, (cf. ReVelle et al. 2004). The primary source of bolide infrasound is from the propagation of the line source cylindrical blast wave modified by fragmentation processes. This source type can only occur for sufficiently large masses in hypersonic near-continuum flow where the line source blast radius, R_o, is directly proportional to the dominant wave period. Secondarily, wake turbulence as well as close range supersonic flight can also produce audible sounds, but acoustic amplitudes are generally insufficient to be observed at ranges >100 km, unlike the blast waves, which have been detected reliably out to 14,000 km for very large events.

1.2 The Historic AFTAC Database

As summarized in ReVelle (1997), between 1960 and 1974, U.S. Air Force Technical Applications Center (AFTAC) recorded acoustic-gravity waves (AGW) at very great ranges from numerous explosive sources including those from "airwave objects" as the bolide signals were first designated (ReVelle 1997). Infrasonic signals constitute only the higher frequency part of this atmospheric AGW spectrum. ReVelle and Wetherill (1978) were able to obtain this data from the US Air Force and subsequently work began on its implications for the global influx rate. Almost all events had their detections and energy levels confirmed by other available methods. One uncorroborated events was the S. African bolide (near the S. African Prince Edward Islands (P.E.I) in the far S. Atlantic region-hereafter just the S. African event) currently being reanalyzed. This very large bolide and the famous Canadian Revelstoke meteorite fall were among the very largest events (ReVelle 1997) in this database with source energies of the S. African and Revelstoke bolides of 1,100 kt and 26 kt respectively (where1 kt, TNT equivalent = 4.185×10^{12} J). These large events largely control the slope of the resultant global influx curve determined using infrasonic detection techniques. For this reason alone, they are now being extensively reexamined.

Each array consisted of at least four microbarometer pressure sensors that were placed 6–12 km apart at a minimum of 16 locations worldwide. The sensor response ranged from many minutes down to 8.2 Hz which was subsequently filtered in two specific wave bands, namely: (i) Internal gravity wave band: 440–44 s and (ii) Infrasonic wave band: 25 s to 8.2 Hz. Signal processing was done using completely analog methods including cross-correlation until 1972–1974 when digital techniques were introduced. Numerous sources were detected such as earthquakes, volcanic eruptions, aurora, bolides, Microbaroms, etc. In addition, the 10 bolide events that were detected were independently verified by additional detection techniques (seismic, VLF, i.e., very low frequency radio wave emission from a bolide, etc.). This dataset included a very small $\sim$0.20 kt event over Western India, a single very large and very remote 1,100 kt event, the fall of the Revelstoke meteorite ($\sim$26 kt) and two 10–15 kt events that fell in the middle east almost exactly one day apart. Only the Revelstoke and S. African bolides will be discussed further below.

2 Previous Analyses

2.1 Previous Analyses of AFTAC and Other Bolide Infrasonic Records

Wetherill and ReVelle (1978) and ReVelle (1980, 1997, 2001) analyzed the numerous records from these events. Through 1997 only the original AFTAC data were analyzed, but by 2001 a number of additional detections were also available that had been recorded at the Los Alamos National Laboratory regional infrasound arrays. These newer regional scale events were much smaller in source energy than those recorded previously on global scales. The new bolide detections allowed a set of analysis relations predicted by ReVelle (1976, 2001) and by ReVelle et al. (2004) for weak shock waves propagating from idealized line sources at relatively close ranges to be examined. These relations are now in routine use for regional scale infrasonic propagation and bolide detection studies within the Physics and Astronomy Department at the University of Western Ontario (Brown et al. 2007; Edwards et al. 2007; ReVelle 2007b).

2.2 Global Influx Results

Earlier Wetherill and ReVelle (1978) as well as ReVelle (1980, 1997, 2001) analyzed the data to predict the global meteoroid influx rate. They determined an influx rate that exceeded, but was close to that predicted in Brown et al. (2002), i.e., a constant power law type solution. These previously computed relations are summarized below:

(i) Infrasound (ReVelle 2001): Final compromise result

$$N(\geq E_s) = 5.66 \cdot E_s^{-0.724};\ r^2 = 0.954 \tag{1}$$

(ii) Satellite data ground-truthed with infrasound/meteorite falls (Brown et al. 2002):

$$N(\geq E_s) = 3.70 \cdot E_s^{-0.90};\ r^2 - 0.99724 \tag{2}$$

We also note finally that Bland (2005) has recently provided an overview summary of the global terrestrial meteoroid influx rate. Also, Ortiz et al. (2006) have presented lunar impact flash data that seem to be in better agreement with the infrasonic influx rate deduced by ReVelle, but is subject to luminous efficiency parameter uncertainties and needs additional calibrations to be fully accepted in our opinion.

3 New Research Efforts

3.1 Signal Digitization

During 2006, LANL and UWO decided to digitize the analog chart records. This arduous task was almost completely undertaken by one of us (Sukara). Please note that all source-observer ranges quoted below are the computed great circle paths on a spherical earth.

3.2 Reanalysis of the AFTAC Bolide Events

3.2.1 Analysis: March 31, 1965, Revelstoke Bolide: Arrays MF and PD Reconstruction of the Event

In Fig. 1a and b, we plot the filtered time series using the high frequency (HF) pass-band: From 25s to 8.2 Hz.

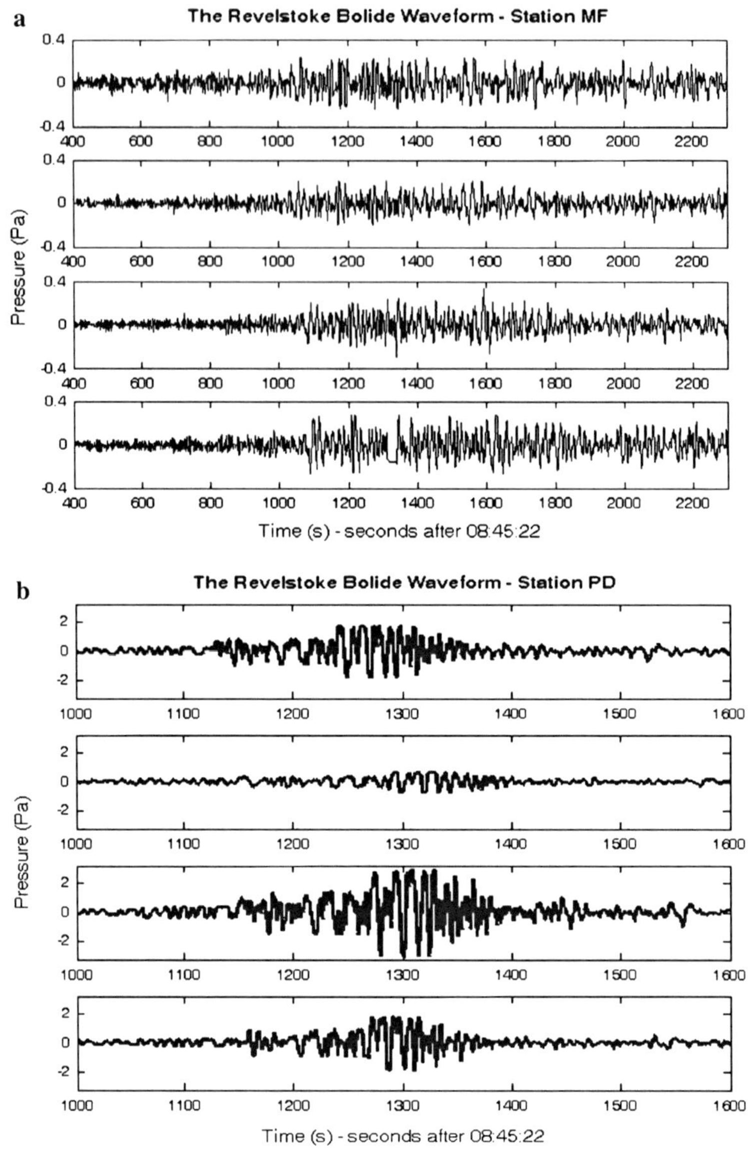

Fig. 1 (**a** and **b**) HF time series of amplitude (in Pa) for the Revelstoke airwave data for the Alaskan array-MF (**a**-top panel) and for the Greenland array-PD (**b**-bottom panel). The seismic/acoustic triangulated source location is: 51.1°N, 117.6°W

Given the Revelstoke signal variability (also observed at a range $\cong$ 1,550 km in Boulder, CO with $\Delta p \cong 0.80$ Pa—see ReVelle 1976), we suspect that focusing, source altitude and line source blast radius effects, or non-steady or range-dependent influences may have contributed to this variability.

We computed signal velocities and back azimuths (its origin time = 05:48 UT and the assumed arrival time in Fig. 1a and b was at 1,100 s).

(i) MF array—2497.7 km range: Signal velocity = 320 m/s, Back azimuth = 145.8°: 2 h, 10.28 min travel time
(ii) PD array—3497.1 km range: Signal velocity = 322 m/s; Back Azimuth = 196.8° for a 3 h, 0.867 min travel time

These velocities are too large for Stratospheric returns. Including errors, they could be Stratospheric or Lamb wave returns at PD (ReVelle 2007b).

3.2.2 *Preliminary Analysis of the August 3, 1963, S. African, P.E.I. Bolide Reconstruction of the Event*

We also analyzed the propagation for the S. African bolide. AGW/infrasonic waves originated from a remote region in the S. Atlantic and took many hours to reach these two arrays. In Fig. 2 (a and b) we plotted the signals recorded at array PB (at 13,824 km range) and at array JB (at 11,327 km range).

4 Preliminary Global Influx Rate Reanalysis

For Revelstoke we reduced the source energy from 26 to 1.4 kt based upon the nominal Edwards et al. (2006) solution for Revelstoke which determined its source energy to be from 0.61 to 3.2 kt. Similarly, for the S. African bolide we have reduced the source energy from 1,100 to 266 (±90 kt), again based upon the wave amplitude calibration work of Edwards et al. (2006) where the observed amplitudes from the digitized records were combined with known source-receiver ranges and averaged Stratospheric winds to arrive at these energy estimates. The errors reflect the standard deviation in the mean winds and the best-beam amplitudes. For these revised source energies, a preliminary revision to the global bolide influx rate using infrasonic methods alone including all AFTAC bolide data can be written (where N is the cumulative number of bolides per year over the entire earth of source energy, E_s and with r^2, the square of the resulting least-squares correlation coefficient for a constant power law curve fit):

$$N(\geq E_s) = 10.02 \cdot E_s^{-0.865}; \; r^2 = 0.93741 \qquad (3)$$

The new result above is not only *not* self-consistent with a predicted influx rate of 700–1,000 years for a 10 Mt energy release for Tunguska, but in addition for a source energy = 1.4 kt for Revelstoke, this unique event is predicted to occur >11 times each year over the earth, almost an order of magnitude above either the satellite estimated influx rate (Brown et al. 2002) or the lunar cratering influx rate of Werner et al (2002) at this energy. We consider this reoccurrence frequency for Revelstoke type events to be extremely unlikely. The reoccurrence timescale made using (3) for a 10 MT event is once every 287.53 years.

Revelstoke was so unique that it was studied for more than 2 years by US Geological Survey (USGS) researchers. The USGS detected the event seismically as well as infrasonically at ranges $\lesssim$1,000 km (Jordan and Bayer 1967; Bayer and Jordan 1967) in the Pacific Northwest. The overriding conclusion from the seismic body waves recorded as far away as 250 km from the event was that Revelstoke impacted the earth in a very remote region of some of the tallest mountains of the Canadian Rockies with an expected crater

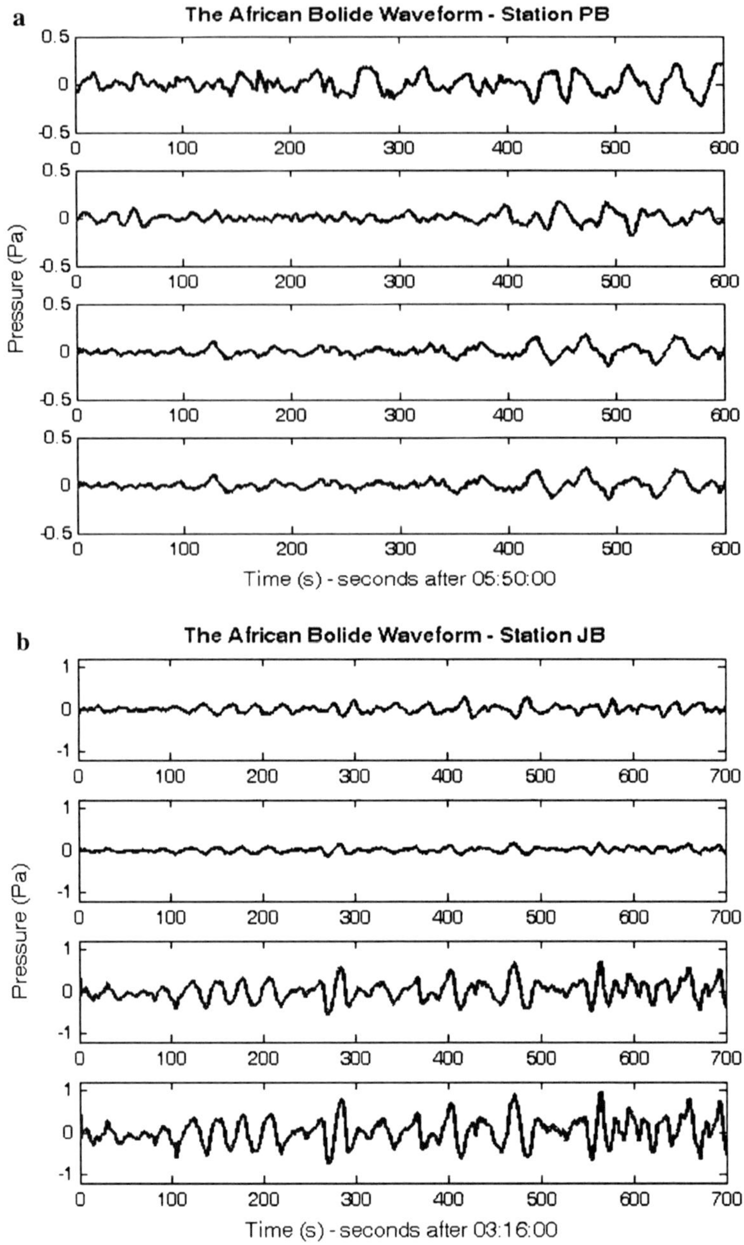

Fig. 2 (**a**) HF time series of amplitude (in Pa) for the S. African airwave data at the AFTAC European array-PB. (**b**) HF time series of amplitude (in Pa) for the S. African airwave data at AFTAC array-JB

diameter (using the deduced seismic body wave magnitude) of ~24.4 m. This impact is extremely unlikely to have occurred for source energies as small as 1.4 kt.

Finally, we further evaluate source energy issues by examining the source energies using observations solely of the wave period at maximum signal amplitude. Historically the wave period has been used as a good diagnostic for the bolide source energy since amplitudes have been observed to be highly variable due to a number of additional causes. In the statistical analyses of Edwards et al. (2006), horizontal mean winds in the Stratosphere were also included and this helped significantly in some cases to reduce the scatter of the statistical plots of amplitude versus range as a function of the bolide source energy. Nevertheless, since the atmospheric is a dynamic, time dependent medium, propagation effects will always cause uncertainty to remain that at times will be very significant for both wave period as well as wave amplitude, but in differing ways.

We assumed in this analysis that the wave period at maximum amplitude is increased by weak nonlinearity and atmospheric dispersive propagation and that each bolide was similarly affected by these processes. If the S. African bolide had a source energy $\cong$270 kt, we may ask what Revelstoke source energy is self-consistent with both the S. African bolide and with its own observed wave period at maximum amplitude. We already know that the wave period is $\propto$ to the blast radius, R_o and that the source kinetic energy is $\propto$ to R_o^3 (ReVelle 2004, 2007a). Given that the S. African bolide had a maximum period $\cong$48 s and that the Revelstoke bolide had a maximum period $\cong$16 s, we find that the Revelstoke source energy must be reduced by $\sim$27 times (3^3) or that its source energy should thus be at least $\sim$10 kt. This value is also fully consistent with our entry modeling work (ReVelle, 2007a, b).

Using 10 kt as the Revelstoke source energy, with all else the same as in Eq. 3, we determined an additional preliminary influx relation:

$$N(\geq E_s) = 10.21 \cdot E_s^{-0.843}; \; r^2 = 0.90447 \tag{4}$$

This relation predicts a Tunguska reoccurrence time interval – 229.75 years.

Further systematic reanalysis of all of the digitized data thus seems to be very important in order to assess the above uncertainties more carefully. Future efforts will be more thorough and will of course include an evaluation of the standard errors as well as an additional reevaluation of the influx rate with the most uncertain values (probably the largest and smallest source energies) having been removed. Until the latter effort is completed, the "true" influx probably lies somewhere between the previous estimates of Brown et al. (2002) and that deduced in Eq. 4 above.

5 Conclusions

We have just begun to systematically reanalyze the digital version of the historical AFTAC bolide infrasound database. At this time we have only reevaluated the source kinetic energies for two of the most energetic bolides within this dataset, namely the August 3, 1963 S. African bolide and the April 1, 1965 Revelstoke bolide, first using the wave amplitude methods of Edwards et al. (2006) and subsequently using the wave period at maximum signal amplitude as an energy diagnostic for these two events. We have also analyzed additional factors such as the plane wave back azimuth, arrival times and we have also computed the power spectra using discrete FFT algorithms for a number of additional AFTAC bolides. Finally using different values for the source energies of the S. African and the Revelstoke bolide, we have made two preliminary revisions of the global influx rate. Both revisions predict that the famous Tunguska bolide of June 30, 1908 should reoccur on a timescale of 200–300 years rather than the 700–1,000 years reoccurrence timescale reported in Brown et al. (2002). Our future analyses of these bolides will be reported on later. These analyses will include wind data from atmospheric global models such as those from the United Kingdom Meteorological Office (UKMO).

Acknowledgments The first author would like to acknowledge financial support from US DOE-HQ in NA-22. We are also greatly indebted to Ms. E. A. Sukara at UWO for carefully digitizing the analog records. One of us (D.O.R) would also like to thank Dr. Gary Geernaert, Head of the Los Alamos Institute of Geophysics and Planetary Physics (IGPP) for complete financial support. Without this financial help, this trip would not have been possible.

References

K.C. Bayer, J.N. Jordan, Seismic and acoustic waves from a meteor. J. Acoust. Soc. Am. **41**, 1580 (1967)

P.A. Bland, The impact rate on the Earth. Philos. Trans. Roy. Soc. **363**, 2793–2810 (2005)

P.G. Brown et al., The Southern Ontario Meteor Network (SOMN), Overview of Sensors, Analysis Techniques and Status, Meteoroids2007, Barcelona, this issue, in press (2007)

P.G. Brown, R.E. Spalding, D.O. ReVelle, E. Tagliaferri, S.P. Worden, The flux of small, near-Earth objects colliding with the Earth. Nature **420**, 314–316 (2002)

W.N. Edwards, P.G. Brown, D.O. ReVelle, Estimates of meteoroid kinetic energies from observations of infrasonic air waves. J.A.S.T.P. **68**, 1136–1160 (2006)

W.N. Edwards, Infrasonic Observations of Meteoroids: Preliminary Results from the Coordinated Optical-Radar-Infrasound Observing Campaign at SOMN, Meteoroids2007, Barcelona, this issue (2007), doi: 10.1007/s11038-007-9154-6

J.N. Jordan, K.C. Bayer, Exploding meteor located by seismographs and microbarographs. Earthquake Inform. Bull. **1**, 8–9 (1967)

J.L. Ortiz, F.J. Aceituno, J.A. Quesada, J. Aceituno, M. Fernandez, P. Santos-Sanz, J.M. Trigo-Rodriguez, J. Llorca, F.J. Martin-Torres, P. Montanes-Rodriguez, E. Palle, Detection of sporadic impact flashes on the moon: implications for the luminous efficiency of hypervelocity impacts and derived terrestrial impact rates. Icarus **184**, 319–326 (2006)

D.O. ReVelle, On meteor-generated infrasound. J. Geophys. Res. **81**, 1217–1230 (1976)

D.O. ReVelle, Interactions of large bodies with the Earth's atmosphere, in *I.A.U. Symposium*, No. 90, ed. by B.A. McIntosh, I. Halliday, *Solid Particles in the Solar System* (Ottawa, Canada, 1980), pp. 185–198

D.O. ReVelle, Historical detection of atmospheric impacts by large bolides using acoustic-gravity waves, near-Earth objects, U. N./Explorer's Club, N.Y.C Ann. N.Y. Acad. Sci. **822**, 284–302 (1997)

D.O. ReVelle, Global infrasonic monitoring of large bolides, in *Meteoroids2001* ed. by B. Warmbein, ESA SP-495, ESTEC (Noordwijk, The Netherlands, 2001), pp. 483–490

D.O. ReVelle, P.G. Brown, P. Spurny, Entry dynamics and acoustics/infrasonic/seismic analysis for the Neuschwanstein meteorite fall. M.A.P.S. **39**, 1605–1626 (2004)

D.O. ReVelle, Recent advances in bolide entry modeling—a bolide potpourri, earth, moon and planets, near-Earth objects, in *IAU Symposium*, Prague, Czech Republic, August, 2006, (2004) in press

D.O. ReVelle, NEO fireball diversity: energetics-based entry modeling and analysis techniques, in near-Earth objects, our celestial neighbors (IAUS 236)-opportunity and risk, ed. by A. Milani, G. Valsecchi, D. Vokrouhlicky (2007a) 524 pp.

D.O. ReVelle, Acoustic-gravity waves from bolide sources (Invited), Meteoroids2007 in Barcelona, Spain, Earth, Moon and Planets, this issue (2007b), doi:10.1007/s11038-007-9181-3

D.O. ReVelle, G.W. Wetherill, Terrestrial microbaro-graph "Airwave" recordings: the global influx rate of large meteoroids, Annual Report of the Director, Department of Terrestrial Magnetism (DTM), Carnegie Institution of Washington (CIW), Washington, D.C. (CIW Yearbook 77 for the period from July 1, 1977–June 30, 1978), pp. 490–493

S.C. Werner, A.W. Harris, G. Neukum, B.A. Ivanov, The near-Earth asteroid size-frequency distribution: a snapshot of the lunar impactor size-frequency distribution. Icarus **156**, 287–90 (2002)

Acoustic-Gravity Waves from Bolide Sources

Douglas O. ReVelle

Originally published in the journal Earth, Moon, and Planets, Volume 102, Nos 1–4.
DOI: 10.1007/s11038-007-9181-3

Abstract We have developed a new approach to modeling the acoustic-gravity wave (AGW) radiation from bolide sources. This first effort involves entry modeling of bolide sources that have available satellite data through procedures developed in ReVelle (Earth Moon Planets **95**, 441–476, 2004a; in: A. Milani, G. Valsecchi, D. Vokrouhlicky (eds) *NEO Fireball Diversity: Energetics-based Entry Modeling and Analysis Techniques, Near-earth Objects: Our Celestial Neighbors (IAU S236)*, 2007b). Results from the entry modeling are directly coupled to AGW production through line source blast wave theory for the initial wave amplitude and period at $x = 10$ (at 10 blast wave radii and perpendicular to the trajectory). The second effort involves the prediction of the formation and or dominance of the propagation of the atmospheric Lamb, edge-wave composite mode in a viscous fluid (Pierce, J. Acoust. Soc. Amer. **35**, 1798–1807, 1963) as a function of the source energy, horizontal range and source altitude using the Lamb wave frequency that was deduced directly during the entry modeling and that is used as a surrogate for the source energy. We have also determined that Lamb wave production by bolides at close range decreases dramatically as either the source energy decreases or the source altitude increases. Finally using procedures in Gill (*Atmospheric-Ocean Dynamics*, 1982) and in Tolstoy (*Wave Propagation*, 1973), we have analyzed two simple dispersion relationships and have calculated the expected dispersion for the Lamb edge-wave mode and for the excited, propagating internal acoustic waves. Finally, we have used the above formalism to fully evaluate these techniques for four large bolides, namely: the Tunguska bolide of June 30, 1908; the Revelstoke bolide of March 31, 1965; the Crete bolide of June 6, 2002 and the Antarctic bolide of September 3, 2004. Due to page limitations, we will only present results in detail for the Revelstoke bolide.

Keywords Bolides · Atmospheric acoustic-gravity waves · Revelstoke meteorite fall of March 31, 1965; Crete bolide of June 6, 2002; Antarctic bolide of September 3, 2004, Tunguska bolide of June 30, 1908 · Atmospheric dispersion of Lamb waves ·

D. O. ReVelle (✉)
Atmospheric, Climate and Environmental Dynamics Group, Earth and Environmental Sciences Division, Los Alamos National Laboratory, Los Alamos, NM 87545, USA
e-mail: revelle@lanl.gov

DOI: 10.1007/978-0-387-78419-9_49

Atmospheric dispersion of acoustic waves · Weak shock wave propagation in a nonisothermal model atmosphere · Formation and dominance of Lamb waves from explosions · Dirac delta function source · Heaviside step function source

1 Introduction and Overview

1.1 The Atmospheric Acoustic-Gravity Wave (AGW) Spectrum

The atmospheric acoustic-gravity wave spectrum in an isothermal, hydrostatic model atmosphere in middle latitudes extends from about 3 h to about 0.01 s in period. Within this regime there are two fundamental resonant periods, the acoustic cut-off frequency, ω_{ac} and the internal gravity wave cut-off frequency. The former is the low-frequency cut-off for internal acoustic waves in the fluid and the latter is the high frequency cut-off for internal gravity waves, the so-called Brunt–Vaisalla (or buoyancy) frequency. At these resonant periods the atmospheric response is expected to be enhanced compared to that at other frequencies within this band-pass. The Lamb edge-wave or fundamental atmospheric mode is a vertically evanescent wave (its wave energy rapidly decays away from the Earth's guiding lower boundary.) that exists at all AGW frequencies that is purely horizontal, but is the only type that exists between these two resonant frequencies. It acts as a propagating horizontal, longitudinal sound wave at frequencies higher than the acoustic cut-off and as a transverse internal gravity wave at frequencies below the Brunt–Vaisalla frequency, ω_{bv}. Within this spectrum numerous natural and manmade sources have been identified of which bolides are only a small part, but one which extends over a large range of frequencies because of the large range of source energies upon entry to the atmosphere (ReVelle 2004a). Infrasonic (or sub-audible) AGW waves extend from $\sim$20 Hz (near the lower threshold of human hearing) down to the acoustic cut-off frequency.

1.2 Waves from Bolides

From previous work, ReVelle (1997, 2004a), has shown the expected degree of diversity of AGW waves from bolide sources including internal acoustic and gravity waves as well as Lamb waves from very large low altitude sources. Source energies ranging from $\sim 10^{-5}$ kt to 1.0 MT ($4.185 \cdot 10^{12}$ Joules = 1 kt) have recently been analyzed (ReVelle 1997, 2001, 2007a). Historically AFTAC (The US Air Force Technical Applications Center, Patrick AFB, Florida) detected ten events from 1960–1974, with some events detected by multiple ground-based arrays almost 15,000 km from the bolide. More recently numerous bolide sources have also been detected by the IMS (International Monitoring System) infrasound arrays as well as at the Los Alamos National Laboratory (ReVelle 2001; 2004a, b; 2007a).

2 Summary of Modeling Approaches

2.1 Entry Modeling of Bolides

Recently ReVelle (2004a, 2007b) has produced entry models (that incorporate drag and deceleration, ablation, fragmentation, light and sound production, etc.) that have been extensively tested against observational material from numerous bolides with very good quantitative agreement having been achieved. Using this modeling procedure, we have produced detailed predictions of such quantities as the line source blast wave relaxation radius in order to specifically link the overall entry energetics analysis with the production of AGW by bolide sources. Later, we will summarize our solutions for the Revelstoke bolide.

2.2 Lamb Edge-Wave Modeling

Garrett (1969) and Pierce (1973) summarized work on the propagation and dispersion of the Lamb edge wave composite mode in the atmosphere, valid for small horizontal wavenumbers such that $\lambda_x/2\pi > \langle L \rangle$, the atmospheric dispersion coefficient. This dispersion is termed normal (as opposed to inverse dispersion which is more typical of acoustical waves as subsequently discussed below) since the lowest frequencies propagate with a phase velocity that exceeds that of progressively higher frequencies. The magnitude of $\langle L \rangle$ characterizes the degree of atmospheric dispersion. This ultimately characterizes the behavior of the Lamb wave in a nonisothermal, hydrostatic atmosphere. As $\langle L \rangle$ increases (decreases), the predicted phase velocity decreases (increases). Although $\langle L \rangle$ is a physical parameter that characterizes the atmospheric path over which the signals propagate, it is also a best fit modeling parameter that is used to produce agreement between an observed and modeled AGW signal as will be discussed below.

Calculations by Pierce (1973) indicate that normal dispersion is important for distances generally exceeding ~ 100 km from a source so that its regime of applicability is quickly evident during the propagation process, if the source energy is large and deposited at low altitudes.

2.3 AGW Modeling

Gill (1982) and Tolstoy (1973) summarized many of the features of AGW production and propagation in the atmosphere. Source models such as a *Dirac Delta* function or a *Heaviside* step function (whose horizontal derivative is the Delta function) or even a *Boxcar* function (which is two *Heaviside* step functions mathematically joined together) can all be used to evaluate the Lamb edge-wave pulse at great ranges. As discussed below the *Delta* function source produces a negative initial pulse while the *Heaviside* step function produces a positive initial pulse. Both types of such initial pulses have been observed.

For the original source pulse or the blast wave shape function at $x = 10$ (where $x \equiv r/R_o$, where r is range $\perp$ to the bolide trajectory and R_o is the blast radius) we have used the following blast wave signature as a function of the dimensionless argument α (with a small amount of frequency rounding added in order to simulate dissipative effects that have not yet been included in our analyses):

$$BW_{\text{sig}} = \Delta p_{\text{src}} \cdot (-0.620 \cdot \alpha^4 + 4.549 \cdot \alpha^3 - 10.475 \cdot \alpha^2 + 7.256 \cdot \alpha - 0.02202) \quad (1a)$$

$$\Delta p_{\text{src}} = 0.0575 \cdot p(z_{\text{src}}) \quad (1b)$$

where $\alpha = 2 \cdot \theta$ = twice the angular measure in radians from zero to π, $p(z_{\text{src}})$ = source pressure amplitude at $x = 10$ at the source height, z_{src}.

In order to utilize the above expression, it must have a nondimensional wave shape that can be normalized to unity. It is only approximately symmetric about the time axis so the resulting normalization is only approximate as well.

3 Lamb Edge-Wave Modeling and Predictions

3.1 Lamb Wave (LW) Formation and Dominance Distances

We have specifically evaluated Pierce's (1963) Lamb wave formation and dominance distances in a viscous, range-independent, windless and isothermal hydrostatic model atmosphere as a function of the source energy, horizontal range and source height. As a surrogate for the specifying the source energy we have used the Lamb wave frequency at the source. Since the R_1 parameter (Lamb wave dominance distance) does not exceed other necessary criterion devised by Pierce except at very great heights (above ~130 km), in a viscous fluid, we have assigned a multiple of the R_0 parameter (formation distance) as the relevant Lamb wave dominance distance. We have utilized a Rayleigh friction viscous decay modeling scheme for our calculations as originally suggested by Pierce. For Lamb wave periods shorter than ~400 s, the constant multiplier is typically two which we have used throughout for the multiplier criterion for Lamb wave dominance evaluations. The propagation possibilities can be summarized as:

Case 1 $R < [R_o, R_1]$: Weak shock and linear acoustic waves only

Case 2 $R \sim O[R_o, R_1]$: Weak shocks/linear and Lamb waves present

Case 3 $R > [R_o, R_1]$: Lamb waves dominate the initial response

The most relevant equations that were evaluated can be written as:

$$R_o = \{|\mu \cdot \beta/(B^2\omega)\} \cdot h = \text{LW formation distance} \tag{2a}$$

$$R_1 = \{|(\beta^2\mu^{-4}/(2\omega \cdot \pi^3))| \cdot h^2 \cdot \exp[2 \cdot B^2 \cdot h]\}^{1/3} = \text{LW dominance distance} \tag{2b}$$

where $h = z_s/H_p$ = dimensionless height of the source, $B^2 = (2-\gamma)/(2\gamma) \cong 0.2143$, $\gamma = 1.40$= specific heat ratio for air, $\mu^2 = \Omega^2 - A^2, \beta^2 = \Omega^2 - 1/4, A^2 = (\gamma-1)/\gamma^2 \cong 0.2041, i = (-1)^{1/2}$, $\Omega^2 = \omega(\omega + i\varepsilon)$ = scaled wave frequency^2 including viscous losses, ε = dimensionless Rayleigh friction coefficient = $\varepsilon^*/(c_s/H_p)$, ε^* = dimensional (linear) Rayleigh friction coefficient: s^{-1}, $\omega = \omega_o/(c_s/H_p)$ = non-dimensional (scaled) wave frequency, $c_s = (\gamma \cdot gH_p)^{1/2}$ = adiabatic, thermodynamic sound speed: m/s, H_p = atmospheric pressure scale height: km.

For sufficiently high acoustical frequencies, it was found that the Lamb wave formation/dominance distance can also be interpreted in terms of a propagation time or a local build-up time which can be categorized as a ratio of the altitude of the source to the horizontal wavelength of the Lamb wave. Thus, Lamb waves form most efficiently as a result of preferential constructive interference effects when the source altitude is sufficiently small with respect to the horizontal wavelength of the Lamb wave (ReVelle and Whitaker 1997). Thus, Lamb waves are expected to be delayed in their formation/dominance by either large source heights or at small source energies, i.e., progressively higher Lamb wave frequencies.

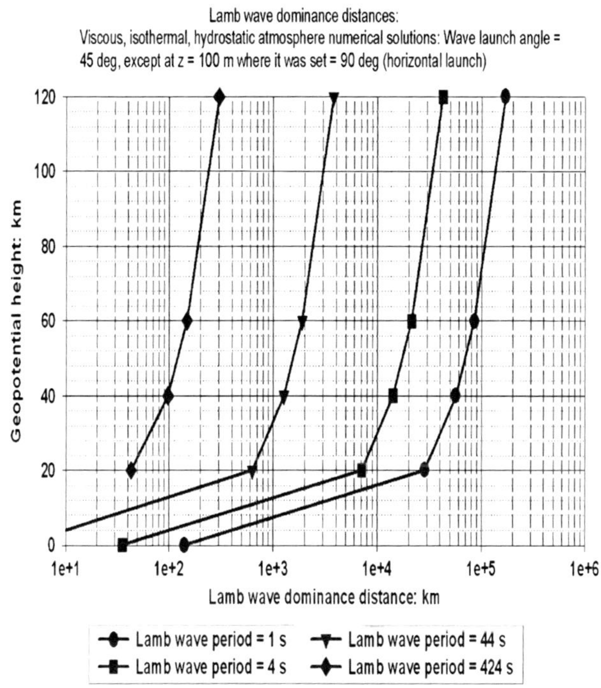

Fig. 1 Predicted Lamb wave dominance distance as a function of the Lamb wave period and of the geopotential source height

3.2 Applicability to Bolide Sources

We have examined Lamb wave formation/dominance distances in a viscous fluid with results plotted in Fig. 1 for all heights from the ground to 120 km as a function of the Lamb wave frequency (which itself is a function of the bolide source energy). Thus, unless source energies (or wave periods) are quite large, Lamb wave formation/dominance from small bolides is not likely (ReVelle 1996; ReVelle and Whitaker 1997). The current evaluations were accomplished in Cartesian coordinates so that for distances >> the earth's radius (~ 6,370 km), the current results are of course not very meaningful.

4 AGW Modeling and Predictions

4.1 Near-field, Weak Shock Behavior during Atmospheric Propagation from a Line Source Blast Wave

We start with the most general, one-dimensional weak shock relationship for a line source atmospheric explosion for weakly nonlinear propagating waves in the absence of significant wave energy dispersion or dissipation mechanisms (ReVelle 1976):

$$\Delta p \cong 0.2917 \cdot D_{\mathrm{ws}}(r) \cdot \{p_{\mathrm{o}} \cdot p(z)\}^{1/2} \cdot x^{-3/4}; \quad x > \sim 10^2 \tag{3}$$

where p_{o} = surface atmospheric pressure: Pa, $p(z)$ = air pressure at the source altitude: Pa, $\{p_{\mathrm{o}} \cdot p(z)\}^{1/2}$ = geometric mean pressure between the source and the observer (computed in a nonisothermal, hydrostatic model atmosphere), x = scaled total range from the line source explosion = r/R_o, r = total distance from the explosion: km, R_o = line source cylindrical blast wave relaxation radius: km, $D_{ws}(r)$ = weak shock dissipation function (assumed = 1, independent of wave frequency or altitude in this analysis).

The combination of parameters utilized in (3) above insures that the wave kinetic energy density is conserved during atmospheric propagation (see for example, ReVelle et al. (2004b) where this concept has been applied to our interpretation of the entry modeling and infrasound recordings from the Neuschwanstein meteorite fall in Germany.

4.2 AGW's: Two-Dimensional Waveguide Dispersion Relationships

Next we analyzed the mathematical relationship of the dispersive properties of two classes of atmospheric waves and proceeded to define the implications of these dispersive atmospheric wave properties. These relations determine how a brief initial pulse composed of only a few frequencies develops an extensive duration at progressively greater horizontal ranges. The dispersion relationship for internal propagating, high frequency, atmospheric acoustical waves in a nonisothermal model atmosphere is given by:

$$\omega = (c_s^2(z) \cdot k_x^2 + \omega_{ac}^2)^{1/2}; \quad k_y = 0 \tag{4a}$$

where ω = angular wave frequency: s^{-1}, c_s = adiabatic, thermodynamic sound speed in m/s as a function of height, k_x = x axis wavenumber (in the direction of the wave heading) in $m^{-1} = 2\pi/\lambda_x$, λ_x = x axis wavelength in m, k_y = y axis wavenumber = $2\pi/\lambda_y$ in m^{-1}; $\lambda_y \equiv \infty$ was assumed.

Similarly the dispersion relationship for Lamb waves in a nonisothermal atmosphere is given by (Garrett 1969; Pierce 1973), and valid to 3rd order in ε (In an isothermal atmosphere, the Lamb wave itself is non-dispersive):

$$\omega = c_s(z) \cdot k_x'; \quad k_x' = k_x \cdot \{1.0 - \langle L\rangle^2 \cdot k_x^2\} + O(\varepsilon^3) \tag{4b}$$

where $\langle L\rangle$ = mean atmospheric dispersion coefficient such that $k_x < \langle L\rangle$ or $\lambda_x/(2\pi) > \langle L\rangle, \langle L\rangle$ can take on values ranging from ~0.0010–10.0 km, ε = standard expansion parameter used for the power series representation of the angular frequency of the pulse such that $\varepsilon \ll 1$. Further details on $\langle L\rangle$ are provided in Table 1.

The first of these two relationships predicts *inverse* waveguide dispersion so that higher frequencies travel the fastest and are followed by progressively lower frequencies until the acoustic cut-off frequency of the idealized waveguide is approached from above, i.e., from higher frequencies down to the cut-off condition. The second predicts *normal* waveguide dispersion so that the natural evolution from a specific source function will produce a weakly dispersed pulse as a function of range (Gill 1982) with lower frequencies predicted to travel the fastest and arrive the soonest, etc.

It is very important to recognize that atmospheric Lamb waves can result from a number of other sources including the well-known process of meteorological geostrophic adjustment (Gill 1982). Thus, the presence of a Lamb Wave by itself is not necessarily indicative of a bolide entry since it has also been observed for a number of differing atmospheric circumstances. Its absence however, following a bolide entry observed at great range is usually indicative of either very low source energy or very high source altitude for the bolide, with both constraints as discussed earlier in this paper.

Table 1 Bolide entry modeling for the line source blast wave radius, wave period, deduced altitude of maximum conditions for the Lamb wave dominance distance, infrasonic IMS (Inter. Monitor. System) ranges, the horizontal wavelengths, max. atmospheric dispersion coefficients, mean atmospheric dispersion <*L*> values, etc

Bolide	Maximum blast wave radius: m	Maximum blast wave period: s	Height of maximum conditions: km	Lamb wave formation distance, R_0: km	Observed at the horizontal ranges: km, *A* = average range value	λ_x: km, observed wavelength at $c_s = 0.350$ km/s	$<L_{max}>$: km dispersion coefficient	$<L>$: km dispersion coefficient for testing
Tunguska bolide: 6/30/08	35,990 KE:z_{max} = 10.73 Mt	300.0 used (329.06 formally predicted)	19.44–19.50	At $\tau_L = 300.0$ s: $R_0 = 29.4$ km	Lenningrad, Russia: 3740, Great Britain: A: 5572	105.0 (for a 300 s period)	16.71	7.0 10.0
Revelstoke bolide: 3/31/65	655.7 KE:z_{max} = 5.28 kt	6.271	14.31	At $\tau_L = 6.271$ s: $R_0 = 3235.8$ km	NOAA: WPL Boulder, CO-1550; AFTAC: 2498, 3497. Also at numerous USGS arrays $<10^3$ km	5.425 (15.5 s period)	0.863	0.863 0.50
Crete bolide: 6/6/02	296.79 KE:z_{max} = 15.1 kt	2.570	32.70–32.92	At $\tau_L = 2.570$ s: $R_0 = 18,946.8$ km	I26: 1796; KNMI/DIA: DeBilt 2336; IRF: Lycksele, Sweden-3410	5.425 (15.5 s period)	0 .863	0.863 0.50
Antarctic bolide: 9/03/04	433.0 KE:z_{max} = 24.2 kt	3.974	28.38–28.44	At $\tau_L = 3.974$ s: $R_0 = 10, 608.4$ km	IMS: 1088, 3715, 5394, 7003, 12965	4.564 (13.04 s period)	0.726	0.726 0.50

4.3 Far-field Atmospheric Response

In what follows, we have used a one-dimensional *Dirac Delta* function with *Fourier's Integral Theorem* to represent a *propagating transient disturbance* (Tolstoy 1973; Gill 1982) in order to determine the wave solutions in low and high frequency AGW limits. We could also use a *Heaviside step function* or a *Boxcar function* for the source description however as noted earlier.

4.3.1 Low Frequency Limit (Lamb Edge-waves): Un-normalized Forms

Assuming that a Dirac Delta function is adequate to represent the bolide source and that the dispersion relationship, Eq. 4b, is adequate to describe low frequency, atmospheric acoustic-gravity waves, Pierce (1973), Gill (1982) and others, have shown that the predicted far field, low frequency wave shape is an Airy function which formally results from doing an integration at the source over all wavenumbers from $0 \rightarrow \infty$. The definition of an Airy function, *Ai(Z)* and the resulting un-normalized wave shape displacement parameter, $\eta(x, t)$ (dimensionless) and of the predicted pressure wave amplitude, $P(x, z, t)$ in Pa after η is normalized are given by:

$$Ai(Z) = (1/\pi) \cdot \int_{0}^{\infty} \cos(t^3/3 + Z \cdot t)\mathrm{d}t \tag{5a}$$

$$\eta(x,t) = 2 \cdot \{r^2/(3\langle L\rangle^2)\}^{-1/3} \cdot Ai\{(1 - t')/\{r^2/(3\langle L\rangle^2)\}^{1/3}\} \tag{5b}$$

$$P(x,z,t) = p(z) \cdot \eta(x,t)$$

where t' = dimensionless time ($= t/(r/c_{\text{eff}})$), $p(z)$ = pressure wave amplitude in Pa as a function of height in km, r = horizontal range in km, Z = argument of the Airy function integral, t = dummy integration variable, assuming a separation of variables type solution for linearized, i.e., vanishingly small amplitude, internal acoustic-gravity waves.

Pierce (1973) has also shown the relationship expected between the shape of the low frequency Lamb wave pulse as a function of the period of the source pulse and as a function of source energy and source altitude and the predicted dispersion timescale (obtained as a function of $\langle L\rangle$). Only for a weakly dispersed pulse at very large range will the above pulse shape be attainable. This weakly dispersed Airy pulse has the specific waveguide properties that low frequencies arrive first followed by higher frequencies.

We have not yet completely formulated the expected pulse behavior when the above situation has not been completely satisfied, but an approximate solution can readily be obtained by adding only a small percentage of the above solution to the weak shock solution given earlier in Eq. 2 and as will be discussed further in Eqs. 6a–6e below.

4.3.2 High Frequency Limit (Ducted Waves): Un-normalized Forms

Assuming once again that a Dirac Delta function is adequate to represent the bolide source, but now that the dispersion relationship, Eq. 4a, is adequate to describe high frequency, atmospheric acoustic waves, Tolstoy (1973) has shown that the predicted far field, high frequency wave shape is a Bessel function solution which again formally results from doing an integral over all wavenumbers from $0 \to \infty$. The formal mathematical definition in this case for a Bessel function $J_1(\zeta)$ is:

$$\text{If}\{(\omega_{ac}/c_{eff}) \cdot (c_{eff}^2 \cdot t^2 - R^2)^{1/2} \gg 1 : \zeta \equiv (\omega_{ac}/c_{eff}) \cdot (c_{eff}^2 \cdot t^2 - r^2)^{1/2} \tag{5c}$$

$$J_1(\zeta) = \{2c_{eff}/(\omega_{ac} \cdot \pi)\}^{1/2} \cdot (c_{eff}^2 \cdot t^2 - r^2)^{-1/4}\} \cdot \cos[\{(\omega_{ac}/c_{eff}) \cdot (c_{eff}^2 \cdot t^2 - r^2)^{1/2} - 3\pi/4] \tag{5d}$$

where c_{eff} = effective horizontal sound speed in km/s = f(T, M, u), T is air temperature in K, M is the mean molecular weight in kg/kmole and u is the horizontal wind speed in the direction of the wave motion in km/s, t = propagation time in s, r = horizontal range expressed in km.

This Bessel function solution has the specific waveguide properties that higher frequency waves arrive first followed by lower frequency waves.

Each of these functions also has to be normalized (see below) so that the resulting wave shape function is unitless with values between -1 and 1.

4.3.3 Delayed Response for Ducted Internal Atmospheric Waves

Assuming that the internal atmospheric acoustical waves ($\omega > \omega_{ac}$ solutions) originated at the same time as that of the Lamb wave from the bolide source, we have computed extended propagation paths through the Stratospheric (from the ground up to ~70 km) and Thermospheric (from the ground up to ~110 km) waveguides. This was accomplished by computing delayed internal wave arrival times compared to the Lamb wave (which travels strictly horizontally very near to the ground). These computed delays are consistent with the following atmospheric propagation mode and signal velocity types (≡ horizontal range/delay time) shown directly below:

(a) Downwind propagation—both Stratospheric and Thermospheric returns were allowed;
(b) Upwind propagation—only Thermospheric returns allowed (diffracted Stratospheric returns were neglected due to the very small expected amplitudes based upon previous detailed observations);
(c) Crosswind propagation—both Stratospheric as well as Thermospheric returns were allowed if the assumed crosswind angle was $< \sim 70°$.

Signal velocities (with c_{sig_S} for the Stratospheric waveguide, c_{sig_T} for the Thermospheric waveguide and c_{sig_C} for nearly-crosswind conditions) computed from our middle latitude model atmosphere (which we have not discussed at all due to imposed page length limits) have the range of values (depending on the angle between the wave heading vector and the prevailing, synoptic scale horizontal wind vector heading at various heights):

(i) $c_{sig\_S} = 0.285 - 0.312$ km/s: Downwind Regime;
(ii) $c_{sig\_T} = 0.212 - 0.2667$ km/s: Upwind Regime;
(iii) $c_{sig\_C} = 0.268 - 0.293$ km/s: Crosswind Regime.

An additional possible signal velocity return (with a velocity of ~0.31–0.32 km/s), produced by Tropospheric Polar jet stream refraction effects that are caused by the vertical shear of the prevailing horizontal winds is also computed in the model, but since it is not as commonly observed, we have not included it in our predictions (since it occurs only if the Tropospherical Jet Stream is physically present along the propagation path).

In addition, the number of equivalent, ray-mode theory wave "hops", the slant range from the bolide and total travel time for the primary ducted internal waves were computed and compared to horizontal range and travel times for the Lamb wave guided along the earth's surface.

4.4 Combined Near- and Far-field Pressure Wave Amplitude Response

Using a separation of variables approach, we constructed an amplitude prediction scheme as a function of the blast wave radius, the source height, the wave pulse shape, the wave kinetic energy density and the scaled range ($x \equiv r/R_o$) in the form (neglecting dissipation, ground reflection and penetration effects, etc.):

$$p_L(x,z,r,t) = p_{src}(R_o,z)\cdot\eta_L(x,t)\cdot Z_L(z)\cdot x_L(r) \quad (6a)$$

$$p_{iw}(x,z,r,t) = p_{src}(R_o,z)\cdot\eta_{iw}(x,t)\cdot Z_{iw}(z)\cdot x_{iw}(r) \quad (6b)$$

$$p_{ws}(x,z,r,t) = p_{src}(R_o,z)\cdot\eta_{ws}(x,t)\cdot Z_{ws}(z)\cdot x_{ws}(r) \quad (6c)$$

$$p_{tL}(x,z,r,t) = \Sigma p_i = p_L(x,z,r,t) + \sum p_{iw}(x,z,r,t); \quad R > \{R_0, R_1\} \quad (6d)$$

$$p_{tws}(x,z,r,t) = \Sigma p_i = p_{ws}(x,z,r,t) + \sum p_{iw}(x,z,r,t); \quad R < \{R_0, R_1\} \quad (6e)$$

where p_L = Lamb wave pressure amplitude: Pa, p_{iw} = internal wave pressure amplitude: Pa, p_{ws} = weak shock pressure amplitude: Pa, p_{tL} = total Lamb wave, low frequency regime, pressure amplitude: Pa, p_{tws} = total weak shock, high frequency regime, pressure amplitude: Pa, $p_{src}(R_o,z)$ = blast wave source pressure wave amplitude (in Pa) at $x = 10$ determined using bolide entry modeling and line source blast results in ReVelle (2007b) matching modeling against satellite luminosity data, $\eta_i(x,t)$ = normalized wave shape function (between -1 and 1): Unitless, $Z_i(z)$ = kinetic energy density conservation factor (inviscid fluid approx.): Unitless, $x_i(r)$ = dimensionless geometrical wave spreading function: Unitless. In a two-dimensional (2D) waveguide: $x_i(r) = \{R/R_o\}^{-\delta}$, δ = geometric wave amplitude spreading decay factor, $1/2 \le \delta \le 3/4$, depending on the propagation decay type i.e., for linear or weakly nonlinear amplitude decay, from a line source.

The summations of the internal waves above were performed over all ducted paths between source and observer. In addition, in the above we have computed *Z(z)* in a nonisothermal atmosphere where the linearized wave kinetic energy density at the source is matched to all other altitudes during the propagation so that $(1/2)\cdot\rho(z)\cdot\{u(z)'\}^2$ is a constant of the motion (where $u(z)'$ = perturbation wind due to the wave). This parameter is equivalent to the geometric mean pressure correction factor used above in (3) earlier and does not account for horizontal wind focusing/defocusing during propagation or for the effects of point and line caustics, etc.

In (6c) and (6e), we have also indicated the response for weak shocks from a line source explosion (Case 1 given earlier). This is modeled like the Lamb wave (without the dispersion) with both k_x and k_z small ($\cong 0$) with internal atmospheric waves assumed to be simultaneously present (Gill 1982).

5 Bolide Sources: Specific Applications

We present results for the Revelstoke bolide assuming that the maximum blast radius at the peak source altitude constitutes the bolide wave source. For conditions at higher (lower) altitudes, wave amplitudes will be lower (lower), due to smaller (higher) source pressures, smaller (smaller) blast wave radii and smaller (smaller) x_i values. Table 1 lists details for

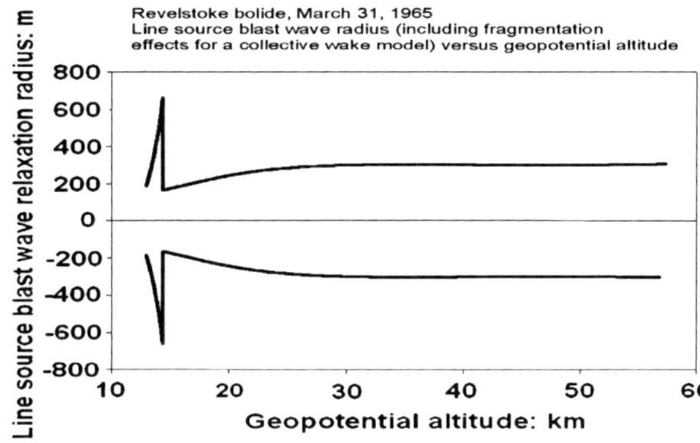

Fig. 2 Revelstoke: 3/31/1965, symmetric line source blast radius

all bolide sources studied using identical assumptions. The symmetric blast wave radii for Revelstoke at all heights are indicated in Fig. 2. The projection of this wave shape onto the ground is reminiscent of the observed butterfly "ballistic wave explosion" pattern of the fallen trees observed for the Tunguska bolide. The predicted Revelstoke AGW wave signatures are given in Figs. 3 and 4 for downwind (= 0°), upwind (= 180°) and crosswind conditions (= 90°) and for the nominal phase angle respectively at the Greenland AFTAC array (at ~3,497 km range). The observed Revelstoke amplitudes at great circle ranges in

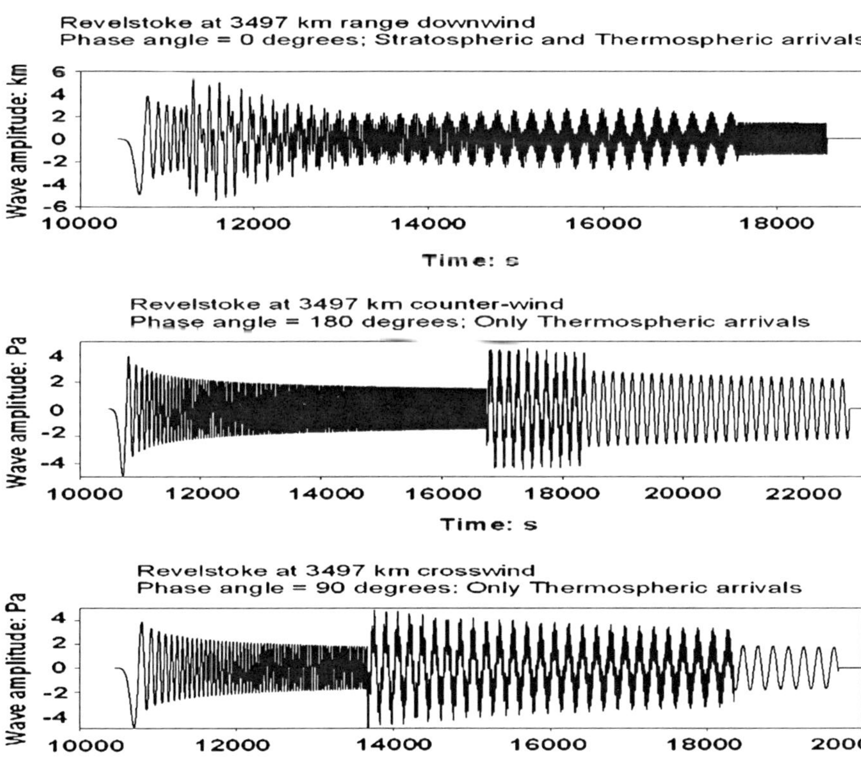

Fig. 3 Pressure waves from Revelstoke versus time (range decay constant = 0.50) for downwind, upwind and exactly crosswind conditions

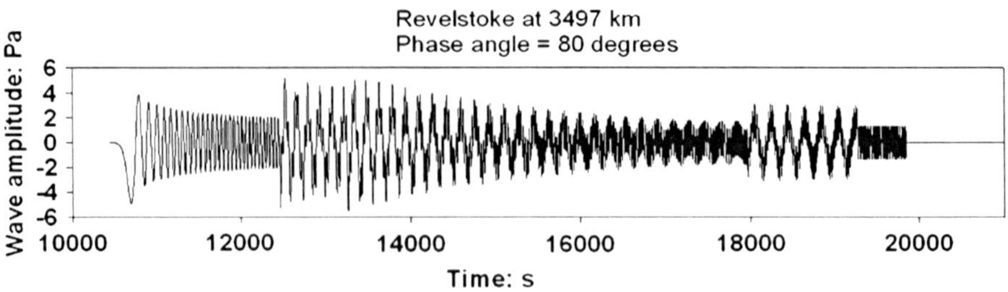

Fig. 4 Nominal Revelstoke predictions for a wind-wave vector phase angle = 80°

Alaska ($\sim$2,498 km), Greenland ($\sim$3,497 km) and at the former Wave Propagation Laboratory array at NOAA, Boulder, CO ($\sim$1,550 km) were $\sim$0.3, $\sim$2.0 and $\sim$0.8 Pa. In contrast, observed periods were $\sim 15.5 \pm 0.50$ s at all arrays. Our nominal waveform predictions for Revelstoke at the Greenland array are within a factor of two of the observed amplitude. From the large degree of amplitude variability, still other effects must be considered such as horizontal refraction, source altitude and blast radius effects, nonsteady flow effects, range-dependent atmospheric effects, etc.

Acknowledgements We appreciate ongoing financial support from the GNEM program at Los Alamos and at the US DOE-HQ in NA-22. We also acknowledge support from Los Alamos sources: Dr. Gary Geernaert at the Institute of Geophysics and Planetary Physics (IGPP) and Dr. David Lawrence at the ISR Division's Center for Space Science and Exploration who provided me with complete financial support during Meteoroids 2007.

References

C.G.R. Garrett, Atmospheric edge waves. Q. J. Roy. Met. Soc. **95**, 731–753 (1969)

A.E. Gill, *Atmospheric-Ocean Dynamics* (Academic Press Inc., Orlando,1982), 662 pp

A.D. Pierce, Propagation of acoustic-gravity waves from a small source above the ground in an isothermal atmosphere. J. Acoust. Soc. Amer. **35**, 1798–1807 (1963)

A.D. Pierce, Theory of Infrasound Generated by Explosions, Colloque Internat. sur les Infra-Sons, Centre Nation. Recherche Scientifique (CNRS), ed. by L. Pimonow, ls, quai, Anatole, France 75700 Paris, pp. 169–175 (1973)

D.O. ReVelle, On meteor generated infrasound. J. Geophys. Res. **81**, 1217–1230 (1976)

D.O. ReVelle, The development and propagation of Lamb waves from airborne explosive sources, LA-UR-96–881 (1996), 12 pp

D.O. ReVelle, R.W. Whitaker, Lamb waves from airborne explosive sources: viscous effects and comparisons to ducted acoustic arrivals. in *Proceedings of the 7th Symposium on long-range sound propagation*, eds. by D. Juve, H.E. Bass, K. Attenborough,Lyon, France, LA-UR-3594, 1997, 16 pp

D.O. ReVelle, Historical detection of atmospheric impacts by large bolides using acoustic-gravity waves, near-earth objects, ed. by J.L. Remo. Ann. N. Y. Acad. Sci. **822**, 284–302 (1997)

D.O. ReVelle, Global infrasonic monitoring of large bolides, Meteoroids 2001, ESA SP-495, ESTEC, Noordwijk, The Netherlands, ed. by B. Warmbein (2001), pp. 483–490

D.O. ReVelle, Recent advance in bolide entry modeling: a bolide potpourii. Earth Moon Planets **95**, 441–476 (2004a)

D.O. ReVelle, P.G. Brown, P. Spurny, Entry dynamics and acoustics/infrasonic/seismic analysis for the Neuschwanstein meteorite fall. Meteorit. Planet. Sci. **39**, 1–21 (2004b)

D.O. ReVelle, E. Sukara, W.N. Edwards, P.G. Brown, Reanalysis of the historic AFTAC bolide infrasound database, Earth, Moon and Planets, special issue on Meteoroids 2007, Barcelona, Spain (2007a, this issue). doi:10.1007/s11038-007-9173-3

D.O. ReVelle, *NEO Fireball Diversity: Energetics-based Entry Modeling and Analysis Techniques, Near-earth Objects: Our Celestial Neighbors (IAU S236)*, ed. by A. Milani, G. Valsecchi, D. Vokrouhlicky (Cambridge University Press, 2007b), 524 pp

I. Tolstoy, *Wave Propagation* (McGraw-Hill, New York 1973), 466 pp

Global Detection of Infrasonic Signals from Three Large Bolides

Stephen J. Arrowsmith · Doug ReVelle · Wayne Edwards · Peter Brown

Originally published in the journal Earth, Moon, and Planets, Volume 102, Nos 1–4.
DOI: 10.1007/s11038-007-9205-z

Abstract We present the infrasonic observations of three large bolides that were observed at numerous International Monitoring System (IMS) infrasound arrays on a global scale. First, a simple procedure for the global association of infrasound detections from large infrasound events is outlined. Infrasound signals are associated with large events based on arrival time, backazimuth and uniqueness at a given IMS array. Next, we apply the algorithm to three bolides and investigate some of the factors affecting the detectability of infrasound from large events. Our findings suggest that site-noise effects significantly degrade the capability of the IMS infrasound network, suggesting that more effort is required to reduce ambient site noise. These results have implications for the use of infrasound measurements (in particular those from IMS stations) as a tool for evaluating the global flux of near-Earth objects.

Keywords Bolide infrasound · Meteor detection · International Monitoring System

1 Introduction

Infrasound is acoustic energy that propagates at frequencies below the 20 Hz hearing threshold of the human ear. Unlike audible sound, infrasound can travel for long distances with relatively little attenuation. Large meteors generate infrasound as they enter Earths atmosphere (ReVelle 1976), which can propagate across the globe. The purpose of this paper is to outline a simple method for identifying infrasonic signals from large events on a global scale, and to apply the method to three superbolides. Following Docobo et al. (1998), a "superbolide" has a peak optical luminosity that exceeds −17 stellar magnitudes (referenced to an elevation of 100 km in the zenith). We also provide an assessment of the

S. J. Arrowsmith (✉) · D. ReVelle
Atmospheric, Climate and Environmental Dynamics Group, Los Alamos National Laboratory, Los Alamos, NM 87545, USA
e-mail: sarrowsmith@gmail.com

W. Edwards · P. Brown
Department of Physics and Astronomy, University of Western Ontario, London, ON, Canada

effect of site noise on the detectability of these large events. There have been relatively few studies of multi-station infrasound signals from large events on a global scale. Long before the advent of the present-day IMS, Wexler and Hass (1962) documented the global detection of infrasound signals from a large Soviet nuclear test. More recently, Brown et al. (2002) present the observations of infrasound at numerous IMS stations from two large bolides. Finally, Garces et al. (2005) and Le Pichon et al. (2006) document multi-station infrasonic signals from the 2004 Sumatra earthquake and tsunami and the 2005 Chilean earthquake respectively. In contrast with earlier studies, this study focuses on the more general problem of associating infrasound signals with global events, and on an assessment of factors that affect the global detectability of large events. We report on the global detection of three superbolides which entered Earths atmosphere on September 3rd, 2004 over Antarctica; on October 7th, 2004 over the Indian Ocean; and on December 9th, 2006 over North Africa.

2 Global Infrasound Event Association

The problem of associating an infrasound signal with a known event is non-trivial. The challenge can be phrased as follows: How can we confidently associate a large event (with

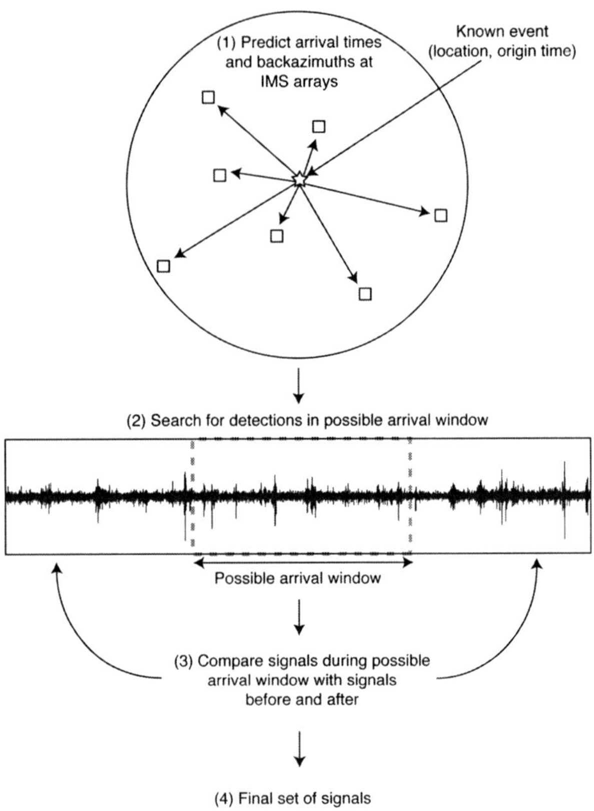

Fig. 1 Flowchart illustrating the event-based global association procedure (schematic diagram for illustration purposes only)

known event location and origin time) with an infrasound signal at a range of 100s to 10,000s of kilometers? Typically, event-based association utilizes the consistency of arrival-time and backazimuth at a given array with predicted values based on the known event location and origin time. However, standard detection algorithms can generate large numbers of coherent infrasonic signals with low signal-to-noise ratios (e.g., Brachet and Coyne 2006). This results in a significant risk of incorrectly associating an infrasonic signal with a given event. To mitigate against the possibility of mis-association, we introduce an additional requirement that a signal associated with a large bolide must be "unique" for a given array in a time window before and after the signal. Subsequently, we have developed a three-stage procedure for associating infrasound signals with large bolides (Fig. 1). First, for a known event location and origin time, we compute predicted arrival times and backazimuths at all IMS arrays located around the globe. Second, we apply an array-based signal detection algorithm to the data—the PMCC algorithm (Cansi 1995). The basis of the PMCC method, in common with other array-based detectors, is that signals are spatially coherent—arriving as plane waves, whereas noise is incoherent between the different elements in an array. By time-aligning arrivals at the different elements in an array, PMCC provides backazimuths and trace velocities for the incoming wave—useful information for event association. We search for PMCC detections in a time-window encompassing the range of all possible infrasonic arrivals (where the earliest possible signal propagates with a horizontal group velocity of 0.34 km/s, and the latest signal propagates with a group velocity of 0.22 km/s, Ceplecha et al. 1998). We then consider detections within an allowed backazimuth deviation of the true great-circle path, occuring within the predicted time window, to be "potential associations". The allowed backazimuth deviations are based on an empirical relation calculated by the U.S. Air Force for known explosions. Finally, we remove potential associations that are: (a) short-duration and narrow-band (and

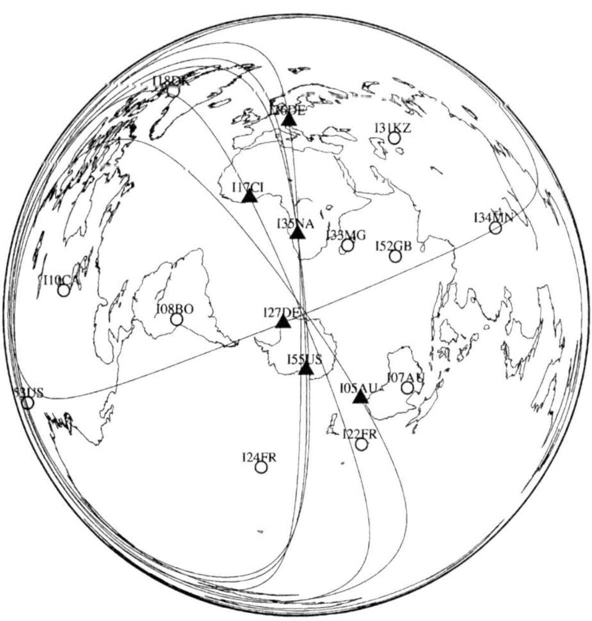

Fig. 2 Map showing stations that detected the September 3rd, 2004 Antarctic bolide (black triangles) and stations that did not (white circles). The event location is at the center of the map

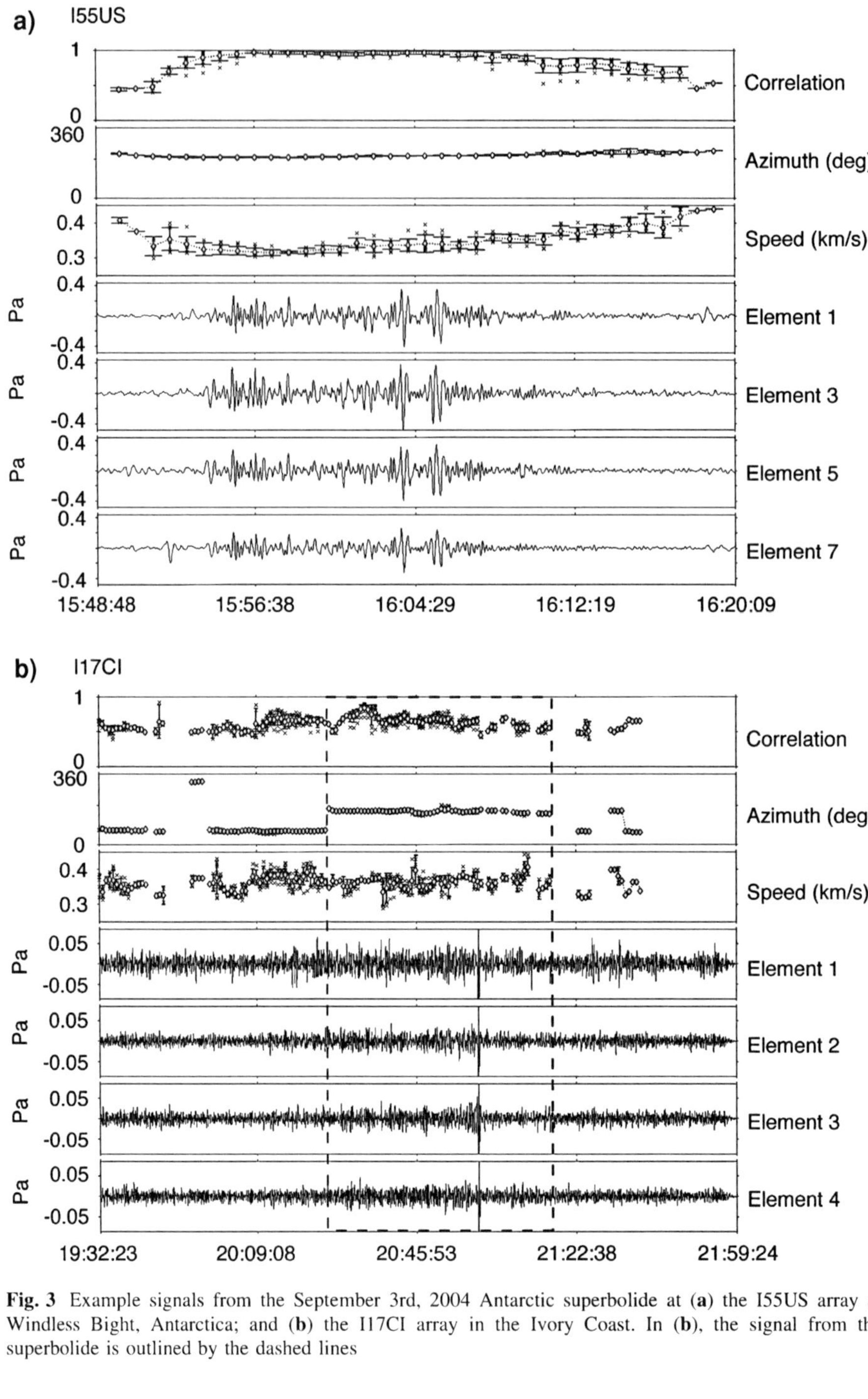

Fig. 3 Example signals from the September 3rd, 2004 Antarctic superbolide at (**a**) the I55US array at Windless Bight, Antarctica; and (**b**) the I17CI array in the Ivory Coast. In (**b**), the signal from the superbolide is outlined by the dashed lines

Table 1 Summary of associated detections at IMS infrasound arrays for the three superbolide events

Event	Array	Range (km)	Arrival time (UT)	Duration (s)	Peak amplitude (Pa)	Back-azimuth (°)
09/03/04	I27DE	1044	13:15:25	1300	0.25	92.0
09/03/04	I55US	3743	15:50:03	1565	0.44	208.4
09/03/04	I35NA	5390	17:20:45	1650	0.10	172.2
09/03/04	I05AU	7114	18:47:55	1900	0.06	204.8
09/03/04	I17CI	8423	20:26:15	2550	0.05	168.0
09/03/04	I26DE	12,918	00:27:45 (on 09/04)	2000	0.10	175.6
10/07/04	I52GB	2201	15:50:50	540	0.33	186.3
10/07/04	I32KE	4679	17:38:50	80	0.04	131.0
10/07/04	I55US	7204	20:01:15	1250	0.14	260.7
10/07/04	I17CI	9001	21:45:50	100	0.04	110.0
10/07/04	I26DE	10182	23:02:55	6150	0.10	127.6
10/07/04	I10CA	17241	06:17:55 (on 10/08)	1700	0.57	27.3
12/09/06	I26DE	2727	06:08:43	750	0.53	158.3
12/09/06	I35NA	5094	11:40:30	1360	0.47	12.2
12/09/06	I30JP	10320	15:13:30	70	0.07	301.2
12/09/06	I41PY	10632	16:05:20	110	0.40	63.2
12/09/06	I56US	10979	20:11:20	70	0.23	43.1
12/09/06	I04AU	11629	16:57:50	150	0.17	294.1

Note: The signal durations observed for these three superbolides do not increase as a function of range, as would be predicted theoretically. We speculate that the reason for this discrepancy is due to site noise

therefore unlikely to be associated with a large, distant source), and (b) similar to other detections obtained before and/or after the signal time window (Fig. 1). Test (b) reduces the likelihood of incorrectly associating an unrelated signal with the event of interest, since detections from coherent noise at a given station (which are typically highly repetitive) will be removed. The criteria used for measuring the similarity with other detections include backazimuth, frequency and trace velocity. The threshold constraints for matching two detections are set as follows: Maximum deviation in backazimuth = 5°, Maximum deviation in frequency = 1.2 Hz, Maximum deviation in trace velocity = 0.1 km/s. These constraints increase the likelihood that the signal is unique to a given event, i.e. it does not occur before or after the possible arrival window for an event of interest (Fig. 1).

3 Results and Conclusions

The event-based association algorithm described above is applied to three superbolides, which occurred from 2004 to 2006. Figure 2 shows the locations of arrays that detected one of the three events, a superbolide that entered over Antarctica on September 3rd, 2004 (for further details on this event, see Klekociuk et al. 2005). The event was detected by six IMS arrays. As shown in Fig. 2, the distribution of observations is strongly asymmetrical, suggesting that detectability is not a simple function of range. An example signal, recorded at the I55US infrasound array (which is located at Windless Bight in

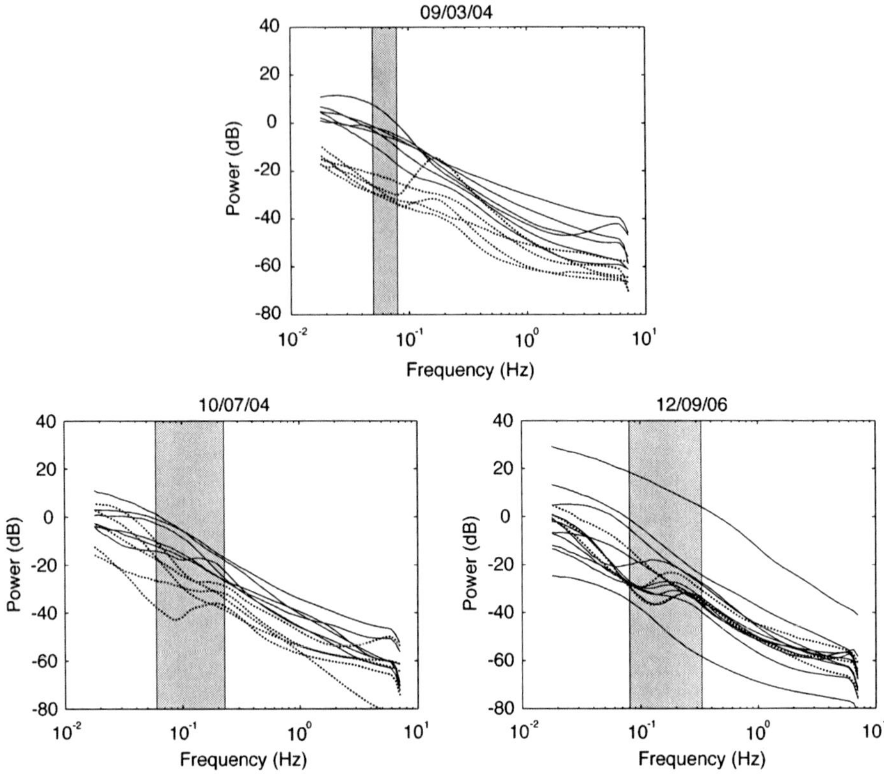

Fig. 4 Noise spectra for each of the three bolides. Solid black lines denote spectra for stations that detected the events and gray dashed lines denote spectra for stations that did not (for stations at ranges less than ~ 10,000 km). Regions shaded gray denote the frequency band over which detections were observed for each bolide

Antarctica), is shown in Fig. 3a. A long duration (~ 10 min long) signal is observed, with a backazimuth of 209 ± 4.8°. The relatively fixed backazimuth and trace velocity indicate the signal is from a single event. The clear signal correlation between the separate array elements can be observed. Signals at greater ranges are typically lower in amplitude and harder to pick out from the ambient noise, requiring array-processing techniques to extract their full extent. An example of such a signal from the September 3rd, 2004 Antarctic bolide is shown in Fig. 3b, which was recorded at the I17CI infrasound array in the Ivory Coast (see map in Fig. 2). Similar patterns of observations were observed for the two other events, with six detections, and an asymmetric distribution of observations observed in each case. A summary of the observations from all three bolides is provided in Table 1.

Whether or not a given station detects an event appears to be strongly governed by factors other than range. In particular, two factors that influence the detectability of infrasound from an event are: (1) ambient noise levels at a given station, and (2) propagation effects. In order to assess the importance of ambient noise levels on the signal detection, we have computed pre-event noise spectra for each station. Each power spectrum was computed for a beamformed waveform at each array using a time window of ~ 1 h (i.e., 2^{16} data points), and smoothed by averaging the power spectrum over

1/8 octave intervals. The resultant power spectra for stations at ranges <10,000 km are plotted in Fig. 4. The plot demonstrates that for two events there is a clear separation between noise-levels at stations that detected the event and stations that did not detect the events. In fact, for these two events the difference in noise levels between stations that detected the events, and stations that did not, is approximately 20 dB on average. This suggests that site-noise effects play a dominant role on the detectability of infrasound from such large events. We note that such a separation is not observed for more distant stations (>10,000 km), suggesting that propagation effects may dominate noise effects at larger ranges.

We have not performed detailed propagation modeling in this study but plan to do so for a future publication. As discussed previously, both site noise and propagation effects will affect the detectability of infrasound. However, our present results clearly suggest that site noise effects are a significant factor on the global detectability of the three superbolides.

Acknowledgements We gratefully appreciate financial support from the GNEM Program of the US DOE-HQ in NA-22. Waveform data were obtained from the SMDC monitoring website.

References

N. Brachet, J. Coyne, The Current Status of Infrasound Data Processing at the International Data Centre, Proceedings of the 28th Seismic Research Review, Orlando, Florida (2006)

P.G. Brown, R.W. Whitaker, D.O. ReVelle, Multi-station infrasonic observations of two large bolides: signal interpretation and implications for monitoring of atmospheric explosions. Geophys. Res. Lett. **29** (2002). doi:10.1029/2001GL013778

Y. Cansi, An automated seismic event processing for detection and location: The P.M.C.C. method. Geophys. Res. Lett. **22**, 1021–1024 (1995)

Z. Ceplecha, J. Borovicka, W.G. Elford, D.O. ReVelle, R.L. Hawkes, V. Porubcan, M. Simel, Meteor phenomena and bodies. Space Sci. Rev. **84**, 327–471 (1998)

J.A. Docobo, R.E. Spalding, Z. Ceplecha, F. Diaz-Fierros, V. Tamazian, Y. Onda, Investigation of a bright flying object over Northwest Spain, 1994 January 18. Meteorit Planet. Sci. **33**, 57–64 (1998)

M. Garces, P. Caron, C. Hetzer, A. Le Pichon, H. Bass, D. Drob, J. Bhattacharyya, Deep Infrasound radiated by the Sumatra earthquake and Tsunami. Eos **86**, 317–320 (2005)

A.R. Klekociuk, P.G. Brown, D.W. Pack, D.O. ReVelle, W.N. Edwards, R.E. Spalding, E. Tagliaferri, B.B. Yoo, J. Zagari, Meteoritic dust from the atmospheric disintegration of a large meteoroid. Nature **436**, 1132–1135 (2005)

A. Le Pichon, P. Mialle, J. Guilbert, J. Vergoz, Multistation infrasonic observations of the Chilean earthquake of 2005 June 13. Geophys. J. Int. **167**, 838–844 (2006)

D. ReVelle, On meteor-generated infrasound, J. Geophys. Res. **81**, 1217–1237 (1796)

H. Wexler, W.A. Hass, Global atmospheric pressure effects of the October 30, 1961. Explosion. J. Geophys. Res. **67**, 3875–3887 (1962)

Radio and Meteor Science Outcomes From Comparisons of Meteor Radar Observations at AMISR Poker Flat, Sondrestrom, and Arecibo

J. D. Mathews · S. J. Briczinski · D. D. Meisel · C. J. Heinselman

Originally published in the journal Earth, Moon, and Planets, Volume 102, Nos 1–4.
DOI: 10.1007/s11038-007-9168-0 © Springer Science+Business Media B.V. 2007

Abstract Radio science and meteor physics issues regarding meteor "head-echo" observations with high power, large aperture (HPLA) radars, include the frequency and latitude dependency of the observed meteor altitude, speed, and deceleration distributions. We address these issues via the first ever use and analysis of meteor observations from the Poker Flat AMISR (PFISR: 449.3 MHz), Sondrestrom (SRF: 1,290 MHz), and Arecibo (AO: 430 MHz) radars. The PFISR and SRF radars are located near the Arctic Circle while AO is in the tropics. The meteors observed at each radar were detected and analyzed using the same automated FFT periodic micrometeor searching algorithm. Meteor parameters (event altitude, velocity, and deceleration distributions) from all three facilities are compared revealing a clearly defined altitude "ceiling effect" in the 1,290 MHz results relative to the 430/449.3 MHz results. This effect is even more striking in that the Arecibo and PFISR distributions are similar even though the two radars are over 2,000 times different in sensitivity and at very different latitudes, thus providing the first statistical evidence that HPLA meteor radar observations are dominated by the incident wavelength, regardless of the other radar parameters. We also offer insights into the meteoroid fragmentation and "terminal" process.

Keywords Radar meteors · Headechoes · Meteor fragmentation · Radio science

J. D. Mathews (✉) · S. J. Briczinski
Communications and Space Sciences Laboratory (CSSL), The Pennsylvania State University, University Park, PA 16802-2707, USA
e-mail: JDMathews@psu.edu

D. D. Meisel
CSSL and Department of Physics and Astronomy, SUNY-Geneseo, Geneseo, NY 14454-1401, USA

C. J. Heinselman
SRI International, Menlo Park, CA 94025, USA

1 Introduction

Observations of sporadic radar meteors have been of increasing interest to the scientific community as the role of meteoroids in planetary astronomy, space weather, in the aeronomy of the meteor zone, and in various aspects of the plasma physics and radio science surrounding the meteoroid interaction with the atmosphere have become increasingly apparent (Janches et al. 2001; Mathews 2004; Mathews et al. 1997). Here we consider "head-echo" observations in which radar returns are from the distribution of plasma immediately surrounding the meteoroid and traveling with the meteoroid itself. The high power, large aperture (HPLA) radars at the 32 panel Advanced Modular Incoherent Scatter Radar at Poker Flat Alaska (PFISR-32), the Sondrestrom Research Facility (SRF) in Greenland, and at Arecibo Observatory (AO) all observe meteor head-echo scattering. We report on observations using these radars. Radar parameters are given in Table 1.

While several theories for the exact mechanism of the head-echo scattering have been proposed, all find that head-echo scattering is highly frequency dependent (Close et al. 2002; Mathews 2004). As such the ideal method for study of the scattering mechanism would employ a common-volume radar using multiple frequencies that could study individual meteor events. The only current radar with these capabilities (AO had this capability as described by Zhou et al. 1998) is ALTAIR (Kwajalein Atoll, Marshall Islands), which is largely inaccessible due to military applications although Close et al. (2002) and other papers including S. Close as an author, employ ALTAIR meteor observations in their results. In order to achieve similar ends, we employ observations at multiple locations but at nearly coincident time and examine the results statistically to provide insight into the meteor head-echo scattering mechanism.

In order to best compare meteor observations from different radars we employ observations from nearly the same calendar dates and local times and we apply the same automated FFT (Fast Fourier Transform) searching technique used at AO (Briczinski et al. 2006; Mathews et al. 2003) to nearly common observing strategies. The automated search routine provides minimal false positives (<1%) while still identifying events with

Table 1 Comparison of the operating parameters of the three radars used in this study

	PF AMISR-32	SRF	AO
Location	65.12° N, 147.47° W	66.99° N, 50.95° W	18.47° N, 66.73° W
Beam width	~2°	~1/2°	~1/6°
Transmit frequency (MHz)	449.3	1,290.0	430.0
Effective aperture (m^2)	180	800	73,000
System temperature (K)	135	110	80
Pulse length (μs)	50	82	45
Sampling rate (MHz)	1	1.25/2	1
Operating power (MW)	~0.5	~2.6	~1.6
Quality factor (MW-m^2/K)	0.667	18.9	1,460
Initial range (km)	84.0	75.5	75.0
Final range (km)	140.0	170.0	142.5
IPP (ms)	1	3.3	1
Observing time (h)	8	8	2
Meteors detected	443	271	2,486

signal-to-noise ratios (SNR) below unity thus maximizing the number of events positively identified.

The AO, SRF, and PFISR radars were chosen for our study for several reasons. (1) The PFISR and SRF radars can be configured with observation parameters similar to that at AO. (2) PFISR (449.3 MHz) and SRF (1,290 MHz) are widely different in frequency but have very similar latitudes (just below the Artic Circle), thus maximizing the frequency range of our results while minimizing possible spatial (latitude) dependency in meteor rates and sources. (3) AO (430 MHz) and PFISR (449.3 MHz) are very similar in frequency but vastly different in sensitivity thus likely involving different populations (mass and energy) of meteoroids. In this paper we compare the observed characteristics of radar meteors seen at the three radars. The results presented from (32 panel) PFISR and SRF are the first reports of altitude and Doppler measurements from campaigns at these facilities and are thus unique. For AMISR Poker Flat this is also the first meteor campaign at the facility. We also uniquely apply the same automated searching algorithm to all data sets to eliminate analysis disparities. In this paper we statistically compare the observed characteristics of radar meteors observed at the three radars to provide insight into the head-echo scattering mechanism and meteoroid properties.

2 Observations

Observed radar meteor parameters such as rates and speeds are seasonally varying (Janches and Chau 2005). Consequently, we conducted the meteor observations reported here within the same few weeks during the calendar year to minimize this source of variation. The zenith-pointed SRF 1,290 MHz radar system was used to observe meteors during the mornings of 31 July and 04 August 2005. Each observation window lasted approximately 4 h, 04.00–08.00 local time (LT). The PFISR-32 meteor observations occurred on 01 and 02 August 2006, and each observing window was 04.00–08.00 LT. In both instances this time period was chosen to be centered around local dawn at lower (non-polar) latitudes when the sporadic meteor event rate is a maximum (Mathews et al. 2001). The PFISR-32 beam was pointed 9° off of zenith at due north.

The AO micrometeors observations used here are for a 20 August 2004 2 h period (05.00–07.00 LT), again centered on local dawn, using the zenith pointed Arecibo 430 MHz radar system The properties of the three radars are summarized in Table 1 with particular emphasis on sensitivity.

At all of the facilities the received "head-echo" signals were Doppler shifted due to the 10–70 km/s meteoroid radial speed. The vector meteoroid encounter speed cannot be deduced as interferometric capabilities are currently unavailable at these three radars. Doppler speeds were obtained by fitting a complex exponential to the returned voltages, resulting in instantaneous (single-pulse) measurement errors on the order of 100 m/s (Briczinski et al. 2006).

Figure 1 gives the Range-Time-Intensity (RTI) and SNR (analogous to the optical "light curve") of events typical of the PFISR (left panel) and SRF (right panel) radars. AO meteor events are very similar in appearance to those at PFISR but sometimes different in character, as we will discuss. The Fig. 1 PFISR result shows a abrupt fragmentation event near 65 ms. Fragmentation and terminal—the meteor signal disappears in 1 ms—events are common in both PFISR and AO results. Fragmentation events consistent with two or three major fragments remaining—with each producing headechoes that produce distinctive interference patterns—are often observed at PFISR. The Fig. 1 SRF event represents

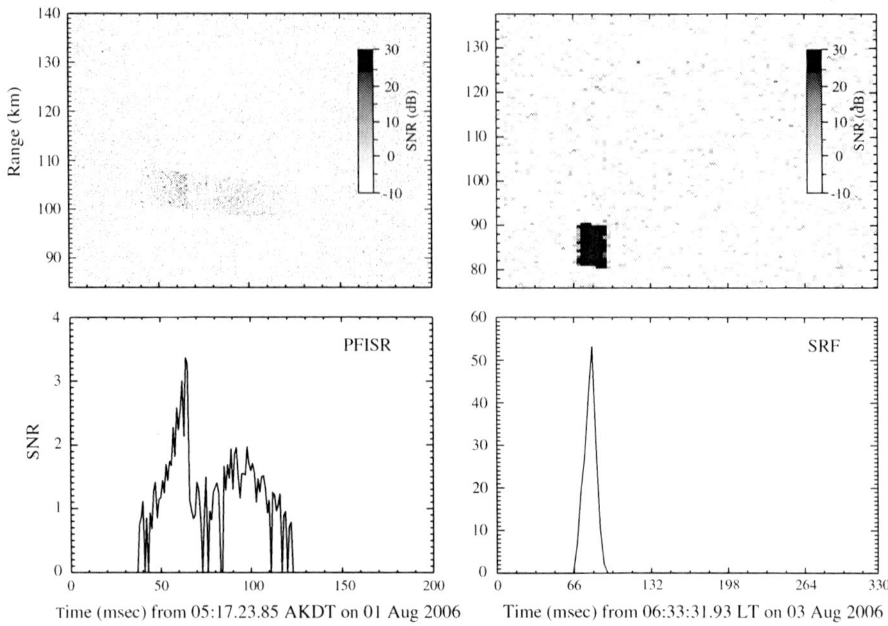

Fig. 1 RTI (Range-Time-Intensity; top) and SNR (Signal-to-Noise Ratio; bottom) versus time (analogous to light curves) plots showing meteor events typical to PFISR (left) and SRF (right). The PFISR meteor is also typical of AO meteors and shows a fragmentation event at ~65 ms. The abrupt decrease in signal strength at fragmentation is presumed to be due a now smaller meteoroid that generates less scattering plasma (Mathews 2004). The fragment is assumed to have terminally "flared" (see Fig. 6, Mathews 2004) as is often observed with much larger optical meteors. The strong SRF event occurs over ten radar pulses (IPPs) and is characteristic of nearly all SRF events

the vast majority of events observed with the 1,290 MHz radar. The SRF SNR (light) curve shows little if any structure and appears/disappears in one IPP—we give interpretation of this result in the conclusions. It is important to note that the fragmentation/terminal features are in no way associated with the respective antenna patterns. These features occur on time scales of order 1 ms or less during which even a 100 km/s meteoroid traveling horizontally would only traverse 100 m—a scale on which even the narrow Arecibo beam pattern does not change dramatically.

Figure 2 gives the altitude distributions of the meteor events observed at the three radars. Note that the AO (430 MHz)/PFISR (449.3 MHz) altitude distributions are quite similar with the AO distribution sharply centered at 105 km and the broader PFISR distribution centered at 100 km. The SRF distribution is centered 5–10 km below the other distributions at ~95 km and is asymmetric with a sharp onset at 100 km and a gradual decrease to 80 km. Also note that with common automated analysis software, the SRF detected just 60% of the number of meteor events seen with the much less sensitive PFISR system in the total of 8 h of observations. The much more sensitive AO system recorded ~4 times more events in 2 h as the combined PFISR/SRF systems.

The Fig. 3 speed distributions show that PFISR and SRF, at the same latitudes and local time, largely agree but with the SRF distribution peaking ~5 km/s faster. It is unclear if this small difference is significant to the processes discussed later. The AO distribution is fastest as it is closest to the apex of Earth's way (Janches and Chau 2005). We interpret

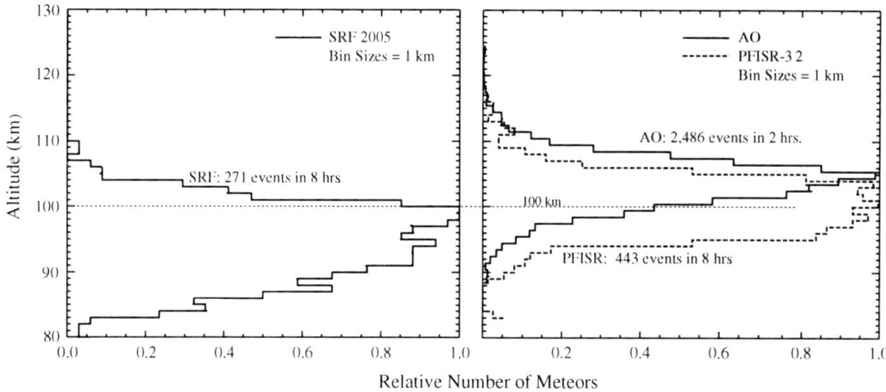

Fig. 2 The altitude distributions for the PFISR, SRF, and AO meteor events. Note the relative event numbers and that while the AO and PFISR results peak at nearly the same altitude, the SRF distribution peaks 5–10 km lower exhibiting the oft-noted upper-UHF "ceiling effect." Note that the PFISR distribution is ~3–5 km below the sharper AO distribution

these results as evidence that most of the high latitude events have a large across-the-beam component. Future interferometric capabilities at PFISR will allow this effect to be studied.

Figure 4 compares the decelerations and speed versus altitude and each other for the three radars. Note from all panels that high deceleration events are in the majority at SRF and are common at PFISR but are relatively rare at AO. The usual momentum equation (Mathews et al. 1997, 2001, 2003; Mathews 2004) interpretation of high meteor deceleration, especially deep in the atmosphere, is that the single-body meteoroids producing the observed ionization are very small. However from Table 1 it is clear that AO is ~2,100 times more sensitive than PFISR and ~77 times more sensitive than SRF while SRF is ~28 times more sensitive than PFISR. Additionally, as relative sensitivity decreases, the antenna beam width increases (except for transmitter power differences) so that the probability of seeing the lower-flux larger meteoroids increases. Note that the AO speed versus altitude distribution shows a clear tendency for higher speed events to be observed at higher altitudes. The PFISR results hint at the same outcome while the SRF results appear to be "flat" in this regard. We agree that the statistics are marginal for both SRF and PFISR—coordinated 24 h observations at both radars are planned.

3 Conclusions

AO (430 MHz) is ~2,100 times more sensitive than PFISR (449.3 MHz) and ~77 times more sensitive than SRF (1,290 MHz). Yet the event rate is lowest at SRF (~34 per hour) relative to PFISR (~55 per hour) and AO (~1,000 per hour) and the height distribution is also ~10 km below that observed with AO/PFISR. Both SRF and PFISR show a high proportion of high deceleration events relative to AO. We conclude that SRF sees a different class of events from AO with PFISR seeing both event classes.

The Fig. 1 "light curves" and Fig. 2 height distributions with different "ceilings" for SRF relative to AO/PFISR both point to different processes producing the large radar scattering cross-sections (RCS) necessary to observe these meteor events with the less sensitive SRF and PFISR radars. The observed high decelerations, especially at SRF, provide the final clue. These high decelerations would—under standard interpretation—

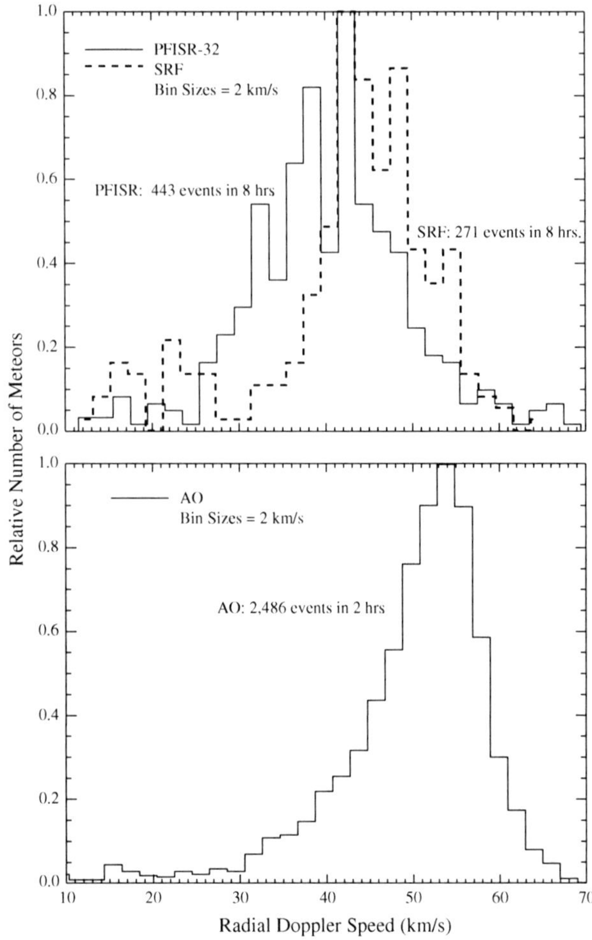

Fig. 3 The meteor radial speed distributions for the three radars

point to very small meteoroids. However, we surmise that the events seen at SRF and to somewhat lesser extent at PFISR are likely representative of a meteoroid fragmentation and/or terminal process yielding a highly confined distribution of small particles that decelerate rapidly while producing the RCS (from a compact, high-concentration plasma) necessary to appear as radar meteors in the—relative to AO—low sensitivity SRF and PFISR radars. This scenario also offers the explanation for the "height ceiling" effect noted in upper UHF radars (Westman et al. 2004; Pellinen-Wannberg 2004). That is, that relatively large meteoroids are invisible at higher altitudes because of the size relative to wavelength of the plasma scattering volume (Mathews 2004) but, upon fragmentation at lower altitudes, produces the observed radar meteor properties—in particular, high deceleration—with large, observable, RCS. Westman et al. (2004) report the existence of a UHF radar "ceiling" relative to the VHF radar. However, their results are difficult to interpret due to relatively small number of events at 224 MHz (VHF) relative to 931 MHz (UHF). Possible seasonal differences in the fluxes also cloud their results. Hunt et al. (2001) also demonstrate a weak ceiling effect when meteor results from the ALTAIR 160 and 422 MHz radars are compared. In both cases the deceleration distribution was not reported. Also note from Fig. 4 that the speed versus altitude correlation seen in the AO

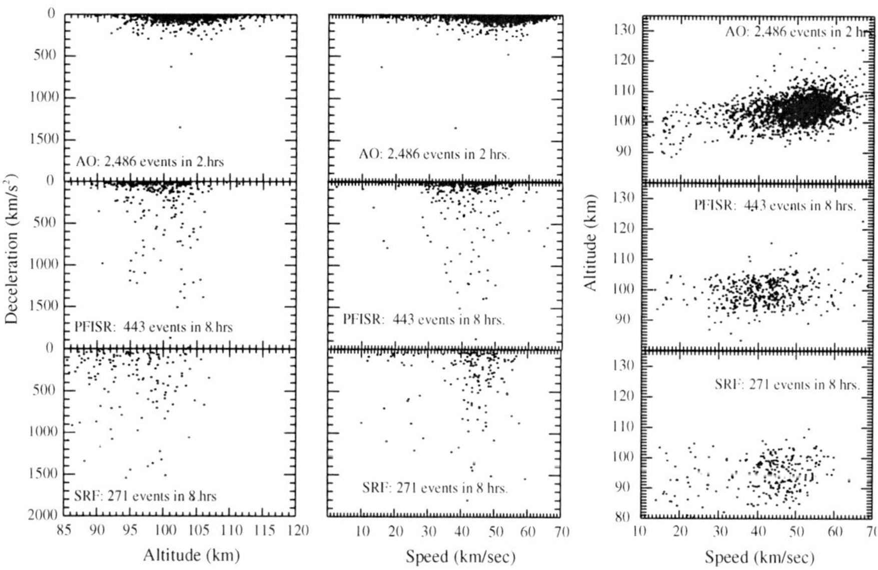

Fig. 4 Meteor deceleration and speed plotted versus altitude and each other for the three radars. Note that the SRF (1,290 MHz) results exhibit a high proportion of high deceleration events relative to AO and that the PFISR deceleration results lie between AO and SRF. If high decelerations were taken to represent, via the meteoroid momentum equation, a very small meteoroid, we would conclude that the SRF and PFISR radars proportionally "see" more smaller meteoroids than the much more sensitive AO radar

results and, perhaps, weakly in the PFISR results is not apparent in the SRF results. Future coordinated SRF/PFISR observational campaigns will address this issue.

It is the Fig. 4 deceleration results that provide the critical clue that forms the final basis for our conclusions. That is, we cannot argue that high deceleration events point to small single meteoroids, as this would imply that the less sensitive SRF and PFISR radars see smaller meteoroids than the very sensitive AO radar. Thus we conclude that fragmentation of the larger meteoroids that we expect to observe in the, relative to AO, wide-beam SRF/PFISR radars produces a radar scattering cross-section from a compact volume of plasma that is sufficient to be seen by these, relative to AO, low sensitivity radars. Recent meteor head- and trail-echo observations from Jicamarca offer further insights into meteoroid fragmentation processes (Malhotra et al. 2007). Further joint observations among the carefully cross-calibrated SRF, PFISR, AO, and—hopefully—the EISCAT radars should shed further "light" on these processes.

Acknowledgments This effort was supported under NSF Grants ATM 04-13009 and ITR/AP 04-27029 to the Pennsylvania State University. The Arecibo Observatory is part of the National Astronomy and Ionosphere Center, which is operated by Cornell University under cooperative agreement with the National Science Foundation. The Sondrestrom Research Facility and Poker Flat AMISR-32 radar are operated by SRI under cooperative agreement with the National Science Foundation.

References

S.J. Briczinski, C-H. Wen, J.D. Mathews, J.F. Doherty, Q-N. Zhou, Robust voltage fitting techniques for meteor Doppler speed determination. IEEE Trans. Geosci. Remote Sens. **44**, 3490–3496 (2006)

S. Close M. Oppenheim S. Hunt L. Dyrud, Scattering characteristics of high-resolution meteor head echoes detected at multiple frequencies. J. Geophys. Res. **107**(A10), 1295 (2002) doi: 1210.1029/2002JA00 9253

S. Hunt, S. Close, M. Oppenheim, L. Dyrud, Two-frequency meteor observations using the Advanced Research Project Agency Long Range Tracking and Instrumentation Radar (ALTAIR), in Proceedings of the Meteoroids 2001 Conference, vol. ESA SP-495, (Swedish Institute of Space Physics, Kiruna, Sweden, 2001) pp. 451–455

D. Janches, D.D. Meisel, J.D. Mathews, Orbital properties of the Arecibo micrometeoroids at Earth intersection. Icarus. **150**, 206–218 (2001)

D. Janches, J.L. Chau, Observed diurnal and seasonal behavior of the micrometeor flux using the Arecibo and Jicamarca radars. J. Atmosph. Solar-Terrestrial Phys. **67**, 1196–1210 (2005)

A. Malhotra, J.V. Urbina, J.D. Mathews, A radio science perspective on long duration meteor trails. J. Geophys. Res. (in press, 2007) doi: 10.1029/2007JA012576

J.D. Mathews, D.D. Meisel, K.P. Hunter, V.S. Getman, Q-H. Zhou, Very high resolution studies of micrometeors using the Arecibo 430 MHz radar. Icarus. **126**(1), 157–169 (1997)

J.D. Mathews, D. Janches, D.D. Meisel, Q-H. Zhou, The micrometeoroid mass flux into the upper atmosphere: Arecibo results and a comparison with prior estimates. Geophys. Res. Lett. **28**, 1929–1932 (2001)

J.D. Mathews, C.H. Wen, J.F. Doherty, S.J. Briczinski, D. Janches, D.D. Meisel, An update on UHF radar meteor observations and associated signal processing techniques at Arecibo Observatory. J Atmosph. Solar-Terrestrial Phys. **65**, 1139–1149 (2003)

J.D. Mathews, Radio science issues surrounding HF/VHF/UHF radar meteor studies. J. Atmosph. Solar-Terrestrial Phys. **66**, 285–299 (2004)

A. Pellinen-Wannberg, The EISCAT meteor-head method—a review and recent observations. Atmosph. Chem. Phys. **4**, 649–655 (2004)

A. Westman, G. Wannberg, A. Pellinen-Wannberg, Meteor head echo altitude distributions and the height cutoff effect studied with the EISCAT HPLA UHF and VHF radars. Ann. Geophys. **22**, 1575–1584 (2004)

Q-H. Zhou, P. Perillat, J.Y.N. Cho, J.D. Mathews, Simultaneous meteor echo observations by large aperture VHF and UHF radars. Radio Sci. **33**, 1641–1654 (1998)

Estimated Visual Magnitudes of the EISCAT UHF Meteors

Csilla Szasz · Johan Kero · Asta Pellinen-Wannberg · David D. Meisel · Gudmund Wannberg · Assar Westman

Originally published in the journal Earth, Moon, and Planets, Volume 102, Nos 1–4.
DOI: 10.1007/s11038-007-9206-y © Springer Science+Business Media B.V. 2007

Abstract We have investigated the conditions for simultaneous meteor observations with the EISCAT UHF radar system and telescopic optical devices. The observed characteristics of 410 meteors detected by all three UHF receivers are compared with model simulations and their luminosity is calculated as a part of a meteoroid ablation model using a fifth order Runge–Kutta numerical integration technique. The estimated absolute visual magnitudes are in the range of +9 to +5. The meteors should therefore be observable using intensified CCD or EMCCD (Electron Multiplying CCD) cameras with telephoto lenses. A possible setup of a coordinated radar and optical campaign is suggested.

Keywords Meteor · EISCAT · HPLA · Radar · Magnitude · Optical detection

Abbreviations

EISCAT	European incoherent scatter facility
HPLA	High power large aperture
EMCCD	Electron multiplying CCD

1 Introduction

Simultaneous high-resolution optical and radar observations of meteors are of great importance in the further understanding of the meteoroid-atmosphere interaction processes

C. Szasz (✉) · J. Kero
Swedish Institute of Space Physics, Kiruna, Sweden
e-mail: csilla@irf.se

A. Pellinen-Wannberg
Umeå University and Swedish Institute of Space Physics, Kiruna, Sweden

D. D. Meisel
SUNY Geneseo, Geneseo, NY, USA

G. Wannberg · A. Westman
EISCAT Scientific Association, Kiruna, Sweden

and the physics of the head echo. The head echoes observed with the meteor radar at the Springhill Meteor Observatory all had visual magnitudes in the range of +4 to −4 when detected optically (Jones and Webster 1991). Similar radar head echo and optical observations at High Power Large Aperture (HPLA) radar facilities have with one exception hitherto not succeeded due to the low light emission levels of typical HPLA meteors and problems with interference between transmitter equipment and collocated optical devices. One attempt is presented in Pellinen-Wannberg et al. (1998). Nishimura et al. (2001) have with the MU (Middle and Upper atmosphere) 46.5 MHz radar and one camera successfully recorded meteors down to magnitudes of +9.

The purpose of this study is to investigate the requisites and suitable conditions for simultaneous meteor observations with telescopic optical devices and the EISCAT UHF radar system. The radar operating frequency around 930 MHz and its three separate receivers at 200–400 km distance from each other would together with two optical devices provide excellent opportunities to compare the head echo scattering characteristics in the UHF band with optical emissions and investigate the accuracy of the two independent velocity and trajectory determination methods.

A short review of the EISCAT UHF system and the most recent meteor campaigns is given in Sect. 2. In Sects. 3 and 4 we present the procedure of estimating the visual magnitudes of the observed radar meteors by fitting their measured deceleration to a single object, numerical ablation model. Meteoroid ablation was described already by Öpik (1958) and further developed by Bronshten (1983) and Love and Brownlee (1991). We have in our calculations been guided by Rogers et al. (2005) who have combined the works mentioned and also added a sputtering model adopted from Tielens et al. (1994). We conclude that the absolute visual magnitudes of the EISCAT UHF meteors are in the range of +9 to +5 and present a possible setup of a coordinated radar and optical campaign in Sect. 5.

2 EISCAT UHF Observations

Four dedicated meteor experiments were run on the EISCAT UHF system between 2002 and 2005, as summarized in Table 1. Data was collected at vernal/autumnal equinox and summer/winter solstice.

The EISCAT 930 MHz UHF radar system comprises three 32 m paraboloids. A transmitter/receiver is located near Tromsø Norway, at 69.6° N, 19.2° E and two remote receivers are sited in Kiruna, Sweden, at 67.9° N, 20.4° E and Sodankylä, Finland, at 67.4° N, 26.6° E. All three antennae were pointed towards a common volume at a height of 96 km, the peak of the EISCAT UHF altitude distribution of detected meteors (Westman et al. 2004). The position of the common volume was 68.9° N and 21.9° E. The configuration used is of tetrahedron geometry as drawn in Fig. 1.

Table 1 Dates and times for meteor campaigns with the EISCAT UHF system

	Year	Start (UT)	Stop (UT)	No of events
Vernal equinox	2002	Mar 19, 12:00	Mar 20, 12:00	50
Summer solstice	2005	Jun 21, 14:00	Jun 22, 10:00	101
	2005	Jun 23, 10:00	Jun 23, 14:00	
Autumnal equinox	2005	Sep 21, 07:00	Sep 22, 07:00	194
Winter solstice	2004	Dec 21, 08:00	Dec 22, 08:00	65

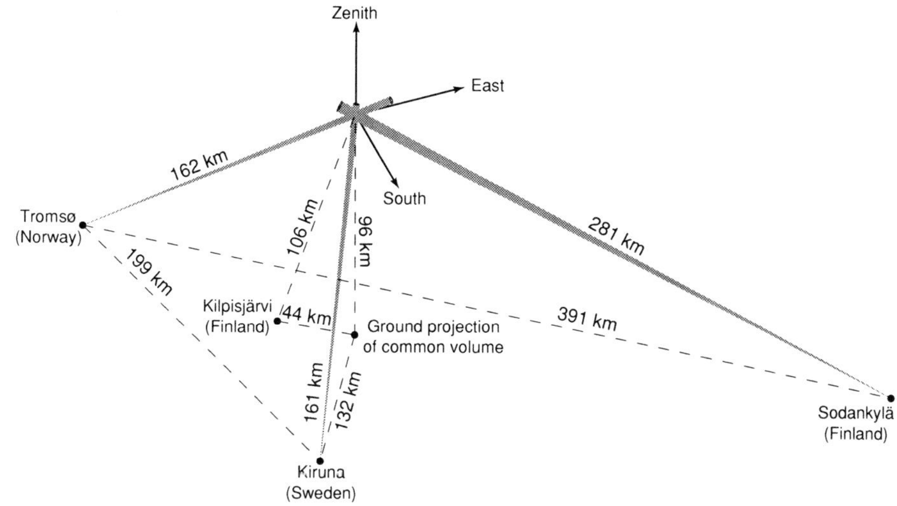

Fig. 1 Meteor observing geometry of the EISCAT UHF system with ranges from the transmitter/receiver and the two remote receivers to the common volume as well as ground distances between the sites. Distances are also indicated from Kiruna and Kilpisjärvi to the ground projection of the common volume

For meteoroids detected by all three receivers simultaneously, the precise geocentric velocity can be calculated. The velocity components measured by the remote receivers are pointing in the directions of the bisectors, defined in the plane spanned by each remote receiver's line of sight and the transmitter's line-of-sight. By dividing each velocity component along the bisectors and the Tromsø line-of-sight into orthogonal x, y and z components, three equations (one for each site) are obtained, with three unknowns (the three components of the velocity vector). The geocentric speed as a function of time (or position) of each meteoroid is then simply found by calculating the norm of the velocity vector for each interpulse period.

The results presented here are based on 410 tristatic meteor events which contain enough data points for line-of-sight velocity calculation to be compared to the Doppler velocity measurements.

3 Ablation Model

We have implemented a single-object ablation model to compare our observations with. The model is similar to Rogers et al. (2005) and references therein, originally based on Öpik (1958), Bronshten (1983) and Love and Brownlee (1991) with a sputtering model added described by Tielens et al. (1994). The input meteoroid parameters to the model are above-atmosphere velocity, mass, density and zenith distance. MSIS-E-90 (Hedin 1991) is used for atmospheric densities.

The radar cross section is estimated by assuming overdense scattering in the Rayleigh regime, and calculated in a similar fashion as described in Westman et al. (2004) and Close et al. (2002). Due to the hypersonic flow, we let the meteoric atoms constituting the mass loss in the ablation model expand radially outwards with a compressed mean free path adopted from Bronshten (1983). We assume that the head echo originates from an overdense region in the immediate vicinity of the meteoroid and use the primary ionization

Table 2 Values of parameters used in the ablation model

Parameter	Value	Unit
Meteoroid shape factor (spherical)	$\frac{\pi}{(4\pi/3)^{2/3}} \simeq 1.21$	
Specific heat of meteoroid	1200	J/K/kg
Clausius-Clapeyron coefficients	10.6, 13500–16120 K	
Emissivity	0.9	
Drag coefficient	1	
Heat transfer coefficient	0.2, 0.4, 0.6, 0.8, 1.0	
Latent heat of fusion + vaporization	$6.0 \cdot 10^6$	J/kg
Mean molecular mass of ablated vapour	56, 50, 20, 20	u
Meteoroid density	7800, 3300, 1000, 300	kg/m^3
Effective atmospheric temperature	280	K

coefficient of Jones (1997) and a spherical, collisionless expansion of the produced electrons to calculate the electron density. The size of the modelled Rayleigh target corresponds to the radial distance from the meteoroid at which the electron density equals the critical density of the UHF wave, about 10^{16}/m^3.

We have compared and fitted the precise particle deceleration and radar cross section obtained from the tristatic data to the ablation model by adjusting the input parameters propagated down through the atmosphere to our observation altitude using a fifth order Runge–Kutta numerical integration technique with a variable step size (Danby 1988). Four different densities, 0.3 g/cc for porous, 1 g/cc for cometary, 3.3 g/cc for asteroidal and 7.8 g/cc for iron material, were paired with mean molecular mass of ablated vapour of 20 u for graphite (both porous and cometary material), 50 u for silicon dioxide and 56 u for iron respectively (Tielens et al. 1994; Rogers et al. 2005). Every pair of density and molecular mass was propagated down through the atmosphere using every one of five different heat transfer coefficients, 0.2, 0.4, 0.6, 0.8 and 1. Each combination was fitted to the data by iteratively adjusting the input parameters and minimizing the least-square difference between model and measurements. Then the best of the fits was chosen and its input values used as estimates for the extra-atmospheric properties of our observed meteoroids. Other model parameters used are further described in Table 2.

The mass distribution found by this method is very similar to the one reported for the ALTAIR radar by Close et al. (2007).

4 Luminosity and Magnitude

Emitted power I [W] of each meteor is estimated as

$$I = \tau(v) \cdot \frac{\mathrm{d}m}{\mathrm{d}t} \cdot \frac{v^2}{2} \tag{1}$$

with the luminous efficiency $\tau(v)$ suggested by Hill et al. (2005). The mass loss due to ablation and sputtering, $\frac{\mathrm{d}m}{\mathrm{d}t}$, is estimated with the ablation model and v^2 is the measured meteoroid velocity. The maximum illuminance I_V [lux] in the visual band at a distance of R = 100 km is calculated as

$$I_v = \frac{683}{4\pi \cdot R^2} \cdot I. \tag{2}$$

The maximum absolute magnitude M_V in the visual band

$$M_V = -2.5 \cdot \log I_v - 14.2 \tag{3}$$

is plotted as a function of above-atmosphere meteoroid mass in Fig. 2a. Apparent magnitude m_V at the Kiruna receiver station (R_{kir} = 161 km)

$$m_V = M_V + 5 \cdot (\log R_{\mathrm{kir}} - 5) \tag{4}$$

is plotted as a function of above-atmosphere meteoroid velocity in Fig. 2b. The equations above can be found in almost any elementary textbook in astrophysics. The estimated absolute visual magnitudes are in the range of +9 to +5.

If a meteoroid has broken up into pieces before or during the observation, the measured deceleration represents the biggest remaining fragment (Ceplecha et al. 1998). The calculated luminosity is in this case an underestimation as the modelled ablation applies to this particular fragment only and not the sum of all fragments.

The uncertainties of the derived parameters introduced by errors in the measured radar cross-section and deceleration are small compared to the uncertainties introduced by the model assumptions. If we for example select a different meteoroid density than that determined as the most suitable one, the value of the above-atmospheric mass for a particular meteoroid may differ by an order of magnitude. The absolute visual magnitude generally changes with a value of less than one and the atmospheric entry velocity changes with a few percent.

Model assumptions as well as the role of fragmentation may be resolved by simultaneous optical and radar measurements. The accuracy of the present study is high enough for a feasibility study.

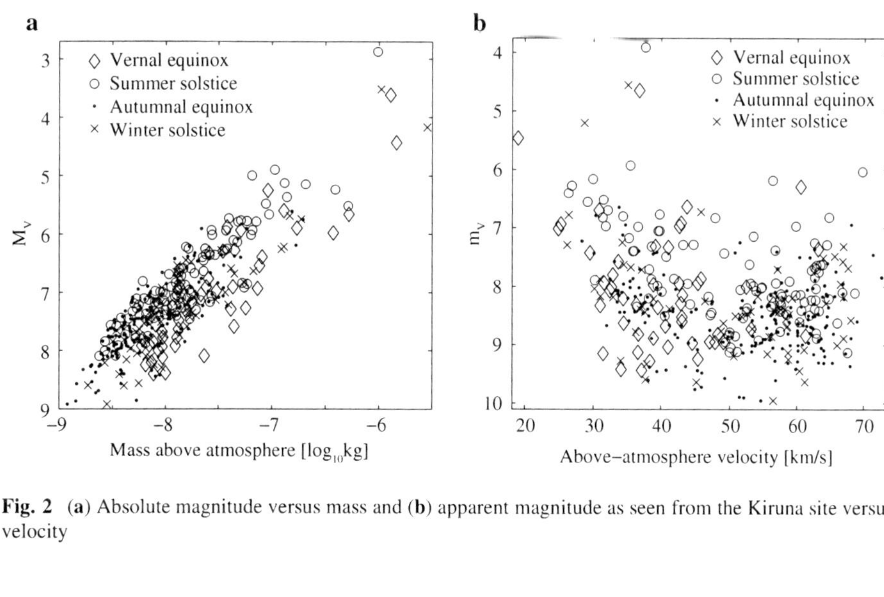

Fig. 2 (**a**) Absolute magnitude versus mass and (**b**) apparent magnitude as seen from the Kiruna site versus velocity

5 Conclusions and Future Work

The estimated visual magnitudes presented in this paper are comparable with the statistical estimations made by Pellinen-Wannberg et al. (1998) on a previous set of EISCAT meteor observations. According to both studies, the meteors should be observable using an intensified CCD or EMCCD camera with a telephoto lens.

The risk of interference from the high-power transmitter equipment makes the Tromsø site an inappropriate camera location. We propose to use two cameras, one collocated with the Kiruna receiver to enable direct comparisons between radar and optical observations, and a second one located in Kilpisjärvi, Finland, at 69.0° N and 20.9° E. A camera in Kilpisjärvi would provide a good complement to observations made in Kiruna. The elevation angle to the common volume is 65° and the azimuth makes an almost right angle with the Kiruna site azimuth (see Fig. 1).

As reported in Table 1, the highest meteor detection rates are found in autumn, when the apex source is circumpolar. The higher count rates are according to Fig. 2a–b caused by an excessive amount of faint meteors as compared to other seasons. As these faint meteors are most numerous in HPLA measurements, the optical equipment should be selected with an aim of observing them. The brightest meteors seem to be detected in summer, but the EISCAT UHF system being located above the Arctic Circle make optical observations from late spring to early autumn impossible. In late autumn, the night sky is dark enough for optical measurements and the outdoor temperature is tolerable for equipment and observers compared to winter conditions. Another factor to take into account is the geomagnetic activity as auroral emissions may outshine faint meteors. In summary, a coordinated measuring campaign as suggested in this study should be scheduled around or after autumnal equinox, in a period when the moon is close to new.

Acknowledgements We gratefully acknowledge the EISCAT staff for their assistance during the experiment. EISCAT is an international association supported by research organisations in China (CRIPR), Finland (SA), France (CNRS), Germany (DFG), Japan (NIPR and STEL), Norway (NFR), Sweden (VR) and the United Kingdom (STFC). Two of the authors (CS and JK) are financed by the Swedish National Graduate School of Space Technology.

References

V.A. Bronshten, *Physics of Meteoric Phenomena.* (D. Reidel Publishing Company, 1983)
Z. Ceplecha, J. Borovička, W.G. Elford, D.O. Revelle, R.L. Hawkes, V. Porubčan, M. Šimek, Space Sci. Rev. **84**, 327–471 (1998)
S. Close, M. Oppenheim, S. Hunt, L. Dyrud, J. Geophys. Res. (Space Phys.) **107**, 9–1 (2002)
S. Close, P. Brown, M. Campbell-Brown, M. Oppenheim, P. Colestock, Icarus. **186**, 547–556 (2007)
J.M.A. Danby, *Fundamentals of celestial mechanics*, 2nd. rev. and enlarged edn. (Willmann-Bell, 1988)
A.E. Hedin, J. Geophys. Res. **96**, 1159–1172 (1991)
K.A. Hill, L.A. Rogers, R.L. Hawkes, Astron. Astrophys. **444**, 615–624 (2005)
W. Jones, Mon. Not. R. Astron. Soc. **288**, 995–1003 (1997)
J. Jones, A.R. Webster, Planet. Space Sci. **39**, 873–878 (1991)
S.G. Love, D.E. Brownlee, Icarus. **89**, 26–43 (1991)
K. Nishimura, T. Sato, T. Nakamura, M. Ueda, IEICE Trans. Commun. **E84-C**(12), 1877–1884 (2001)
E.J. Öpik, *Physics of meteor flight in the atmosphere, No. 6 in Interscience tracts on physics and astronomy* (Interscience Publishers, Inc., 1958)
A. Pellinen-Wannberg, A. Westman, G. Wannberg, K. Kaila, Annales Geophysicae **16**, 1475–1485 (1998)
L.A. Rogers, K.A. Hill, R.L. Hawkes, Planet. Space Sci. **53**, 1341–1354 (2005)
A.G.G.M. Tielens, C.F. McKee, C.G. Seab, D.J. Hollenbach, Astrophys. J. **431**, 321–340 (1994)
A. Westman, G. Wannberg, A. Pellinen-Wannberg, Annales Geophysicae **22**, 1575–1584 (2004)

Improving the Accuracy of Meteoroid Mass Estimates from Head Echo Deceleration

Elizabeth Bass · Meers Oppenheim · Jorge Chau · Alice Olmstead

Originally published in the journal Earth, Moon, and Planets, Volume 102, Nos 1–4.
DOI: 10.1007/s11038-007-9202-2

Abstract This paper examines current techniques used to determine meteoroid mass from high-power, large aperture (HPLA) radar observations. We demonstrate why the standard approach of fitting a polynomial to velocity measurements gives inaccurate results by applying this technique to artificial datasets. We then suggest an alternate approach, fitting velocity data to an ablation model. Using data taken at the Jicamarca Radio Observatory in July 2005, we compare the results of both methods and demonstrate that fitting velocity data to an ablation model yields a reasonable result in some instances where alternate methods produce physically unrealistic mass estimates.

Keywords Meteors · Meteoroids · Radar

1 Introduction

Each year, over 10^7 kg of material enters the Earth's atmosphere, mostly composed of particles less than 1 mm wide (Love and Brownlee 1993; Ceplecha et al. 1998; Mathews et al. 2001; Janches et al. 2006). As these meteoroids pass through the atmosphere, they ablate atoms that ionize. HPLA radars can detect the surrounding plasma, called a head echo, at altitudes between 70 km and 140 km (Close 2004). Interferometric HPLA radars, such as the 50 MHz radar at JRO, accurately measure meteoroid trajectories and velocities. Meteoroid characteristics, such as mass, can be calculated from these measurements.

Using conservation of momentum, we find an expression for the meteoroid mass m:

$$\frac{m}{A} = -v\gamma\rho_{\text{air}} \sec\chi \left(\frac{dv}{dh}\right)^{-1} \tag{1}$$

E. Bass (✉) · M. Oppenheim · A. Olmstead
Astronomy Department, Boston University, 725 Commonwealth Avenue, Boston, MA 02215, USA
e-mail: enb@bu.edu

J. Chau
Instituto Geofísico del Perú, Radio Observatorio de Jicamarca, Apartado 13-0207, Lima 13, Peru

where A is cross sectional area, v is meteoroid velocity, ρ_{air} is air density, χ is zenith angle, dv/dh is the rate of change of the meteoroid's radial velocity with altitude, and γ is the drag coefficient. There are two challenges in applying this relationship. First, various parameters are not precisely known but can be estimated to within a factor of four. The second problem is determining dv/dh. Often, a function is fitted to the data, in range versus time or velocity versus time, in order to determine the deceleration. This function may be a polynomial, such as a line (Janches et al. 2000), or an exponential function (Close et al. 2005). In many cases, the mass calculation produces unphysical results, yielding masses that increase with time instead of decrease (Close et al. 2005).

In this paper, we demonstrate that fitting a polynomial to velocity measurements using artificial data generated by the ablation model outlined in Lebedinets et al. (1973) and Rogers et al. (2005) produces inaccurate results. In addition, we show that polynomial fits to real data yield different solutions depending on the order of the fit. Finally, we show that one obtains more physically plausible masses by fitting observational data to the ablation model.

2 Using Model Data to Calculate Mass

The model used to generate the artificial data involves a system of ordinary differential equations that relate the rate of change of meteoroid mass, velocity, and temperature using values from Rogers et al. (2005). This system is solved at each time step using a Runge-Kutta method. The solid lines in Fig. 1 indicate the resulting mass and velocity values.

To determine the range over which the meteor would be observed, a simple signal strength model was created, discussed in Close et al. (2004). Signal reflection was assumed when the plasma frequency of the head echo exceeded the radar frequency. We added noise to our theoretical signal and used the values as a way of weighting the ideal data when fitting a polynomial.

We made multiple polynomial fits to our ideal data in order to test the accuracy of a given fit. Meteoroid mass was then calculated from each fitted function. Figure 1 shows the fits and calculated masses. Low-order fits gave unphysical results. A mass that decreases

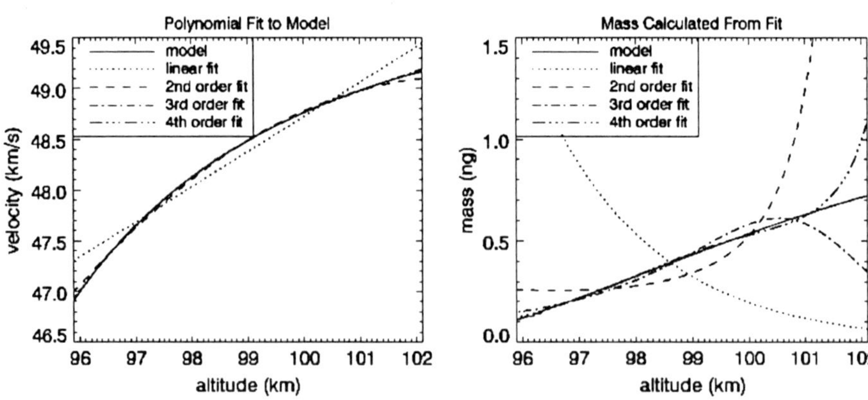

Fig. 1 Model data and polynomial fits (left) and mass calculations (right)

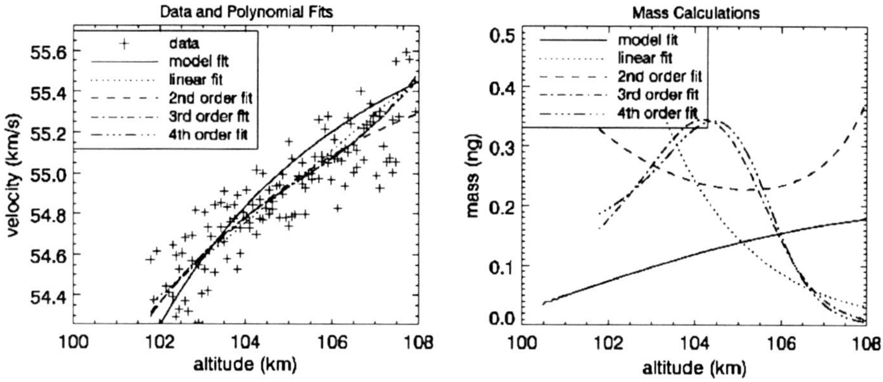

Fig. 2 Data with different fits (left) and the resulting mass calculations (right)

with altitude required a fourth order polynomial fit to the model data. Second and third order fits appear to correspond to the model data, but produce unreasonable mass estimates.

These results highlight the fact that an unphysical mass estimate does not necessarily mean there is a problem with the data. This shows that unphysical results could be obtained from head echo data because of poor velocity fitting, not the data itself. This conclusion becomes stronger when we evaluate the effects of polynomial fitting on real data in Sect. 3.

3 Fitting Data to Model

Another approach to calculating meteoroid mass is to fit the data to the ablation model itself, instead of fitting a function to velocity measurements.

To find the most accurate meteoroid characteristics, we run the simulator repeatedly until we find initial mass and velocity values that minimize the difference between the altitude and velocity measurements from our data and the model. Figure 2 compares the polynomial and model fitting results. An unphysical mass is found each time with a polynomial fit. While the maximum value in the third and fourth order fits is only a factor of ~ 4 greater than the model result, the increase and subsequent decrease in mass may lead an observer to discard the data. With higher order fits, noise has a greater effect, changing the velocity derivative and estimated mass. The model provides a reasonable mass estimate and a physically realistic mass along the entire trajectory, with the mass decreasing at all times.

4 Conclusions

As demonstrated in Sect. 2, Eq. 1 is highly sensitive to the velocity derivative, and small changes in a polynomial fit can greatly affect the resulting mass estimate. Fitting an ablation model to the data allows us to obtain a mass estimate for some cases where a polynomial fit produced unphysical results. We expect that this technique will help researchers make more accurate mass estimates and utilize a greater portion of their data.

Acknowledgements This work was supported by National Science Foundation grants TM-9986976, ATM-0332354, ATM-0334906, ATM-0432565, DGE-0221680, and DOE grant DE-FG02-06ER54887. The authors also thank the JRO (and IGP) staff for their assistance, especially F. Galindo for his help in processing the data.

References

Z. Ceplecha, J. Borovicka, W. Elford, D. Revelle, R. Hawkes, V. Porubcan, M. Simek, Meteor phenomena and bodies. Space Sci. Rev. **84**, 327–471 (1998)

S. Close, Theory and analysis of meteor head echoes and meteoroids using high-resolution multi-frequency radar data, Ph.D. Thesis, Boston University, (2004)

S. Close, M. Oppenheim, D. Durand, L. Dyrud, A new method for determining meteoroid mass from head echo data. J. Geophys. Res. 110, A09308 (2005). doi:10.1029/2004JA010950

D. Janches, J.D. Mathews, D.D. Meisel, Q.H. Zhou, Micrometeor observations using the Arecibo 430 MHz radar. Icarus **145**, 53–63 (2000)

D. Janches, C.J. Heinselman, J.L. Chau, A. Chandran, R. Woodman, Modeling the global micrometeor input function in the upper atmosphere observed by high power and large aperture radars. J. Geophys. Res. 111, A07317 (2006). doi:10.1029/2006JA011628

V.N. Lebedinets, A.V. Manochina, V.B. Shushkova, Interaction of the lower thermosphere with the solid component of the interplanetary medium. Planet Space Sci. **21**, 17–332 (1973)

S. Love, D. Brownlee, A direct measurement of the terrestrial mass accretion rate of cosmic dust. Science **262**, 550–553 (1993)

J.D. Mathews, D. Janches, D.D. Meisel, Q.H. Zhou, The micrometeoroid mass flux into the upper atmosphere: Arecibo results and a comparison with prior estimates. Geophys Res. Lett. **111**, 0A07317 (2001)

L.A. Rogers, K.A. Hill, R.L. Hawkes, Mass loss due to sputtering and thermal processes in meteoroid ablation. Planet Space Sci. **53**, 1341–1354 (2005)

Plasma and Electromagnetic Simulations of Meteor Head Echo Radar Reflections

Lars Dyrud · Derek Wilson · Steiner Boerve · Jan Trulsen · Hans Pecseli · Sigrid Close · Chen Chen · Yoonjae Lee

Originally published in the journal Earth, Moon, and Planets, Volume 102, Nos 1–4.
DOI: 10.1007/s11038-007-9189-8

Abstract Recently, meteor head echo detections from high powered large aperture radars (HPLA) have brought new measurements to bear on the study of sporadic interplanetary meteors. These same observations have demonstrated an ability to observe smaller meteoroids without some of the geometrical restrictions of specular radar techniques. Yet incorporating data from various radar reflection types and from different radars into a single consistent model has proven challenging. We believe this arises due to poorly understood radio scattering characteristics of the meteor plasma, especially in light of recent work showing that plasma turbulence and instability greatly influences meteor trail properties at every stage of evolution. In order to overcome some of the unknown relationships between meteoroid characteristics (such as mass and velocity) and the resulting head echo radar cross-sections (RCS), we present our results on meteor plasma simulations of head echo plasmas using particle in cell (PIC) ions, which show that electric fields strongly influence early stage meteor plasma evolution, by accelerating ions away from the meteoroid body at speeds as large as several kilometers per second. We also present the results of finite difference time domain electromagnetic simulations (FDTD), which can calculate the radar cross-section of the simulated meteor plasma electron distributions These simulations have shown that the radar cross-section depends in a complex manner on a number of parameters. In this paper we demonstrate that for a given head echo plasma the RCS as a function of radar frequency peaks at sqrt (2*peak plasma frequency) and then decays linearly on a dB scale with increasing radar frequency. We also demonstrate that for a fixed radar frequency, the RCS increases linearly on a dB scale with increasing head echo

L. Dyrud (✉) · D. Wilson · C. Chen · Y. Lee
Center for Remote Sensing Inc, Fairfax, VA, USA
e-mail: ldyrud@yahoo.com

S. Boerve
Norwegian Defense Research Establishment, Kjeller, Norway

J. Trulsen · H. Pecseli
University of Oslo, Oslo, Norway

S. Close
Las Alamos National Laboratory, Las Alamos, New Mexico

DOI: 10.1007/978-0-387-78419-9_54

plasma frequency. These simulations and resulting characterization of the head echo radar cross-section will both help relate HPLA radar observations to meteoroid properties and aid in determining a particular radar facility's ability to observe various meteoroid populations.

Keywords Meteors · Radar · Meteor head echoes

1 Introduction

Current estimates for the annual global meteor flux vary from 2,000 to 200,000 tons per year and estimates for the average velocity range between 10 and 60 km/s (Cziczo et al. 2001; Janches et al. 2000; Taylor 1995; Ceplecha et al. 1998; Mathews et al. 2001). Understanding the interplanetary meteoroid environment is important for several fields of study from solar system evolution, atmospheric physics, and most critically to manned and unmanned space flight. Yet, the basic properties of this global meteor flux, such as the average mass, velocity, and chemical composition remain poorly constrained (Mathews et al. 2001; Dyrud et al. 2004). Here we present an investigation aimed at improving our ability to characterize meteoroids via high powered large aperture radars (HPLA) radar observations of head echoes, and to more precisely characterize any bias or filter that a particular radar facility may have to a certain population of meteoroids, i.e., mass or velocity. We believe much of the mystery surrounding the basic parameters of a dominant source of the interplanetary meteor flux (mass ranges of $\sim$0.1–10^{-7} mg) exists for the following reasons; the unknown sampling characteristics of different radar meteor observation techniques, which are used to derive or constrain most models, and a need to relate meteor radar observables to the true meteoroid properties of interest. We believe this arises due to poorly understood radio scattering characteristics of the meteor plasma, especially in light of recent work showing that plasma turbulence and instability greatly influence meteor trail properties at every stage of evolution.

In this paper we demonstrate that plasma simulations and electromagnetic finite difference time domain electromagnetic simulations (FDTD) simulations can be utilized to provide detailed estimates of the radar scatter from meteors, specifically head echoes. Further, the work presented in this paper is motivated by the need for the most detailed understanding of head echo scattering processes in order to provide parameterizations for the modeling efforts of the global meteoroid flux from head echo observations by Janches et al. (2006); Fentzke and Janches (2007); and Plane (2004) and investigations into HPLA radar biases towards meteors of certain velocities and sizes as discussed within Close et al. (2007), and Janches et al. (2007). The introduction continues with a background scientific description to place this work in context.

For decades ground based meteor observations were typically made with photographic and TV cameras and specular meteor radars. Specular radars detect reflections from the trail of ionization formed perpendicular to the radar beam by a meteoroid during atmospheric entry. This specular condition requires that only trails formed perpendicular to the radar beam reflect strongly without destructive interference (Tayler 1995; Ceplecha et al. 1998). The resulting meteor parameters deduced from these radar observations are sensitive to the geometrical radio scattering requirements of this condition.

Over the past decade, two new types of radar meteor reflections have become known or widely used. These reflections are known as meteor head echoes and non-specular trails and are largely observed and studied with HPLA designed for incoherent scatter remote sensing of the ionosphere (Chapin and Kudeki 1994; Dyrud et al. 2002, Zhou et al. 2001). Examples of these two scattering mechanisms are shown in Fig. 1. This figure shows a meteor head echo followed by trail reflections, termed *non-specular*, which occur despite the fact that many trails are roughly aligned with the radar beam. While the head echo plasma is believed to be a cloud of electrons moving at the speed of the meteoroid, the non-specular trail echoes are attributed to coherent radio scatter from plasma turbulence–generated field aligned irregularities (FAI). Additionally, because these observations produce such detailed signatures, and seem to convey meteors entering anywhere within the radar beam, they show great promise as tools for deriving more complex parameters about meteoroids and the atmosphere they interact with.

We continue the introduction with some background explanation of the plasma processes expected to occur during meteoroid entry. Our current understanding of the physical processes occurring during the early stages of meteoroid atmospheric entry remains somewhat anecdotal and can be summarized as follows. As a meteor enters the Earth's atmosphere near 100 km altitude, the particle heats up and atoms begin boiling off the surface in a process known as ablation. Depending on energy, the ablated meteor atoms are ionized (freeing an electron from the atom, producing a positively charged ion and negatively charged electron) upon collision with an air molecule. These newly produced meteor ions cool after approximately 10 collisions, which takes between a fraction of a millisecond at 80 km and as long as one millisecond at 110 km (Jones 1995). This stage is depicted in cartoon form in panel (a) of Fig. 2. During this thermalization process, the plasma density near the meteoroid is very high, and it is assumed that head echo scattering occurs at this stage.

As we continue our description of the evolution of a meteor trail, the effects of plasma turbulence become all the more evident and important. Once the meteor plasma has cooled, the result is a large trail or column of enhanced ionization near 100 km altitude, which may extend between 10 and 20 km in length. It is during this stage of development that specular radar echoes from the trail commence. Our understanding of the next stages of evolution depicted in Fig. 2 result directly from super computer simulations of plasma instability and

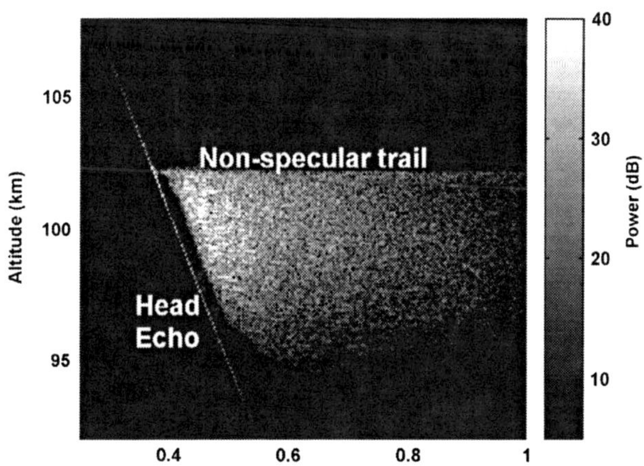

Fig. 1 Altitude-time-intensity image of a head and subsequent non-specular echoes over extended range from ALTAIR VHF Radar. The diagonal line to the left is called a head echo, while the echoes spread in range and time to the right are the non-specular trail. Figure reproduced from Close et al. (2002)

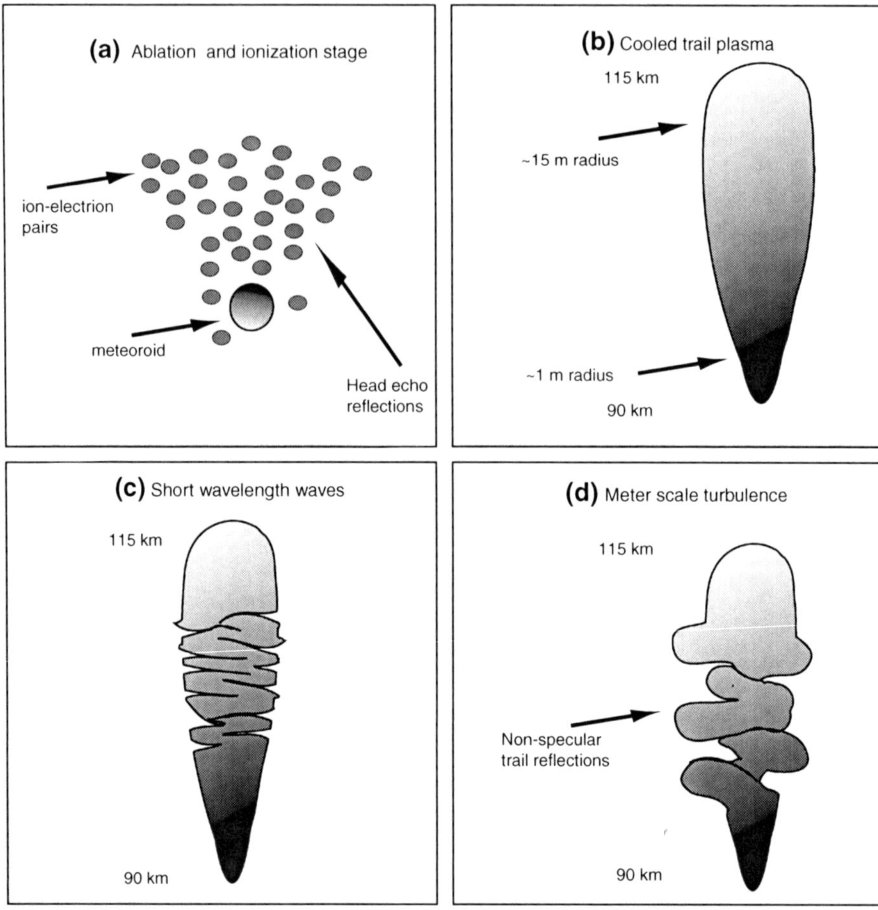

Fig. 2 Paneled cartoon depicting the four main stages of meteor trail evolution. The four stages ordered in increasing time are: (**a**) the ablation and ionization stage, head echo reflections are assumed to result from meteor plasma at this stage, (**b**) a cooled trail plasma column that has increasing radius at higher altitudes, (**c**) Farley-Buneman-gradient-drift (FBGD) waves at short wavelengths grow after only a few milliseconds after the stage (**b**), (**d**) The final stage depicts that the unstable portion of the meteor trail has become turbulent with structure at a broad range of wavelengths. Specular trail reflections are expected to occur throughout stages **b–d**, while non-specular trail reflections should only appear near stage d after sufficient field aligned irregularities (FAI) have formed

turbulence within meteor trails published in a number of papers (Dyrud et al. 2005; Dyrud et al. 2001, 2002; Oppenheim et al. 2000).

Regarding meteor trails: a tremendous amount of work has been done to characterize specular reflections of meteor trails (see Cervera and Elford (2004) and references therein), but none of these take into account the now known turbulence present in all meteor trails. An examination of the effect of plasma turbulence on specular trail observations represents a future direction of study that we expect to pursue, with some initial work presented by Dyrud et al. (2004). Regarding head echoes: very recent attempts have been made to model the electron distribution responsible for head echo reflection with certain degrees of success (Close et al. 2004; Pellinen-Wannberg 2004). Yet these analyses assumed that the meteor head echo plasma distribution is Gaussian in shape. In the following sections we use a particle in cell (PIC) plasma simulation to determine the plasma shape of a head echo, and FDTD EM simulations to simulate radar scatter from these plasmas.

2 Head Echo Plasma Simulation

Some results from our head echo plasma simulations are shown in Figs. 3 and 4. Figure 3 shows the simulation ion density as a function of two spatial dimensions, the x-axis is along the meteoroid path, and the y-axis is perpendicular to it. This plasma simulation uses a kinetic PIC treatment for ion physics by solving the complete Lorenz force equation for each ion (Birdsall and Langdon 1985). The electric field is then solved for iteratively assuming the electrons satisfy the standard non-linear Boltzman relation (Chen 1984). While the effects of the Earth's magnetic field electron motion should normally be considered for meteor and plasma physics near 100 km altitude, it has been ignored here, since the electrons have a significantly larger Larmour radius then the plasma density gradient length scale. This comparison indicates that the ambipolar electric field will dominate the short scale electron dynamics, and that magnetic field gyration can be ignored in the earliest timescales. Therefore the overall distribution of the plasma, once thermalized, near the tail end of Fig. 3, may not accurately represent the true physics of actual meteor plasma thermalization and expansion. However, the early stage plasma distribution near the meteoroid body should be accurately represented, and that is the focus of the research presented in this paper. The ions simulated had a mass of 50 AMU in order to represent an iron dominated meteor plasma, we also used an elastic collision cross-section of $2.61 \times 10^{-20}\,\mathrm{m}^2$, which was based upon numerical calculations conducted by H. R. Skullerud (Private Communication, see Skullerud et al. (1999) for more information regarding the numerical techniques). The background ion temperature, and electron temperature were taken as thermalized 257 K. It is possible that they electrons are hotter if

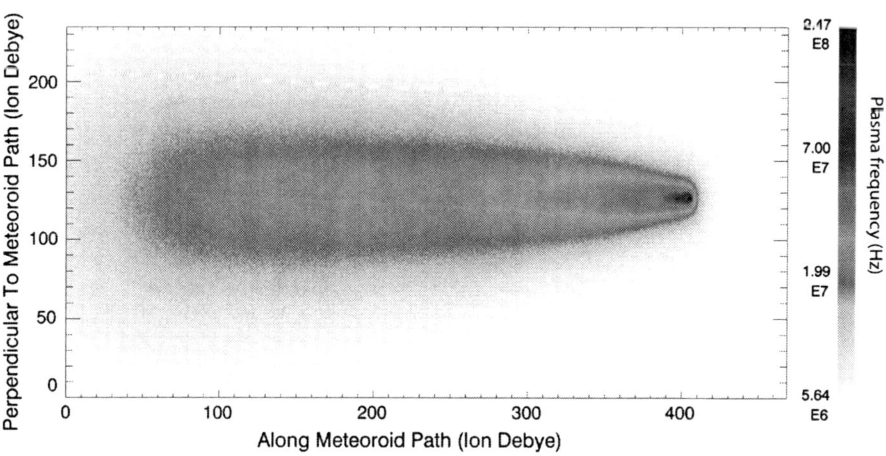

Fig. 3 Results from a meteor plasma simulation after the ablation stage. This figure shows the color representation of the plasma density surrounding a meteor. The meteor simulated here was producing 10^{12} ions per meter traveled, and was moving against the surrounding atmosphere at a rate of 40 km/s. The axes are in units of local Debye length; in the ionosphere near 100 km altitude Debye lengths are of the order of 1 cm

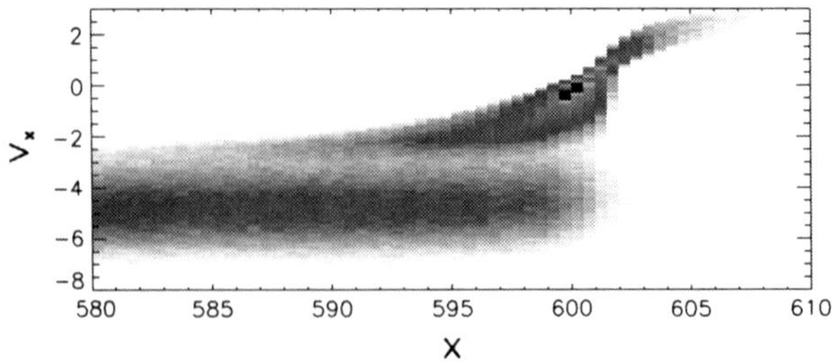

Fig. 4 Plot of ion phase space for the same simulation and time shown in Fig. 3: This figure shows the color representation of the plasma density surrounding a meteor. The meteor simulated here was producing 10^{12} ions per meter traveled, i.e., line density. The y-axis of this figure is in simulation units of velocity 1 unit = 8 km/s. The x-axis is in the same Debye length units as Fig. 3. The meteor ions are being produced at the x-coordinate of 600 and generated isotropically with a 2,000 K temperature. The air molecules then blow past the meteoroid at 40 km/s or in the dimensionless simulation units used in the figure, $v = -5$. The highest density portions are in black and was moving against the surrounding atmosphere at a rate of 40 km/s. The axes are in units of local Debye length; in the ionosphere near 100 km altitude Debye lengths are of the order of 1 cm

produced by impact ionization, however Murad et al. (2003) predicts a two-stage ionization, which would produce cooler electrons, and given the paucity of other theoretical work on meteoric ionization, we have adopted a thermalized temperature.

The electron density distribution from this plasma simulation is then input into an EM simulation to analyze the radio scattering properties of such an electron distribution. These EM simulations are discussed in the next section.

3 Finite Difference Time Domain (FDTD) Plasma Formulation

The analysis of electromagnetic fields generated in the scattering of waves by complex objects presents many difficulties, especially if such scatterers include scale sizes having characteristic dimensions comparable to the incident radiation wavelength, and exhibit dispersive characteristics such as in plasma. In many cases the only alternative to experimental measurements is the direct solution of Maxwell's equations by numerical methods. The FDTD method was first introduced by Yee (1966) and later developed by Taflove (1995) and others.

The standard FDTD formulation places a limitation such that the constitutive parameters must be specified as constants, i.e., μ, ε, and σ must be described by a single number. While this is true for free space, good conductors, and ideal dielectrics, it is only approximately true for most real materials. For some materials over a narrow band of frequencies, the approximation is excellent, while for other materials over a wider band of frequencies, it is not. For some materials, such as plasmas and ferrites, the permittivity may be zero or negative, so that the basic FDTD equations we have presented cannot be used at certain frequencies as some of the terms become singular. Thus special treatments are needed to use the FDTD method for simulation of dispersive materials and we briefly explain the FDTD algorithm to be used for simulation of dispersive plasma media. There are many different ways to simulate the electromagnetic interactions in plasma. Two major schemes that are used in FDTD simulations are the direct integration (DI) and the recursive convolution (RC) methods. In the former, Maxwell's equations are coupled to an auxiliary ordinary differential equation modeling the response of the current ($\boldsymbol{J}$) to the field ($\boldsymbol{E}$) and the latter is based on the time domain integral relating the flux (***D***) and the field (***E***). The DI method (Nickisch et al. 1992) was used in the simulation results shown in this paper.

For a complete treatment of the FDTD formulation in anisotropic plasma, the reader is referred to (Nickisch et al. 1992; Boardman 1982). For completeness, the set of field equations for a non-magnetized cold plasma that were used in our FDTD formulation are presented in Eqs. 1–3 (Burden 1985).

$$\nabla \times \vec{E} = -\mu_o \frac{\partial \vec{H}}{\partial t} \tag{1}$$

$$\nabla \times \vec{H} = \varepsilon_o \frac{\partial \vec{E}}{\partial t} + \vec{J} \tag{2}$$

$$\frac{\partial \vec{J}}{\partial t} + v\vec{J} = \varepsilon_o \omega_p^2 \vec{E} \tag{3}$$

where ***E*** and ***H*** are electric field (V/m) and magnetic field (A/m), respectively, ε_o and μ_o are electric permittivity (F/m) and magnetic permeability (H/m) of free space, respectively, ***J*** is plasma current density (A/m^2), ω_p is the plasma frequency (rad/s) and v is collision frequency (Hz).

Figure 4 shows an example result of the spatial distribution of the electric field output from one of our meteor FDTD simulations. We place the meteor plasma that was output from a numerical plasma simulation of a head echo (shown in Fig. 3), near the center of the EM simulation space, and then launch a broad-banded radar pulse towards the meteor and measure the reflected electric field. The comparison between the impinging and reflected electric field yields the radar cross-section, or RCS, for that particular object. Since FDTD simulations are conducted in the time domain and we use a broad-banded pulse, the post-processing of a single simulation outputs the type of information shown in Figs. 5 and 6, i.e., RCS as a function of impinging radar frequency.

We have simulated a 50 × 50 m box where the third dimension into the page is narrow and made effectively 2D by choosing a perfect electric conductor (PEC) boundary at the top and bottom of the box in the z direction. A grid spacing of 10 cm is achieved with a 500 × 500 grid, which limits the applicability of the results to about 300–350 MHz, due to discretization of the simulated fields. However, we have conducted several runs with 5 cm grid spacing and see no changes to the results shown here, and little deviation from the below demonstrated trends as the results are extend to higher radar frequencies. The polarization of the radio wave electric field is also linear in the into page direction. Figure 5 shows the electric field magnitude in color, the RF pulse, which is traversing upward and has passed the meteor head. Much weaker reflections from the meteor can be seen propagating downward. We continue with a description of some of the results of these FDTD simulations.

Figure 6 shows RCS as a function of radar frequency for four different meteor plasmas, which were derived using two different techniques. All of the lines shown are for plasma distributions with the same peak plasma frequency of 70 MHz. Plasma frequency is defined as $f_p = 9.0 * \text{sqrt}(n)$ where f is Hz and n is defined in units of m^{-3}. It is generally considered that plasma is overdense or behaves similar to a metal object for impinging EM waves below the plasma frequency and is evanescent for EM waves above that frequency.

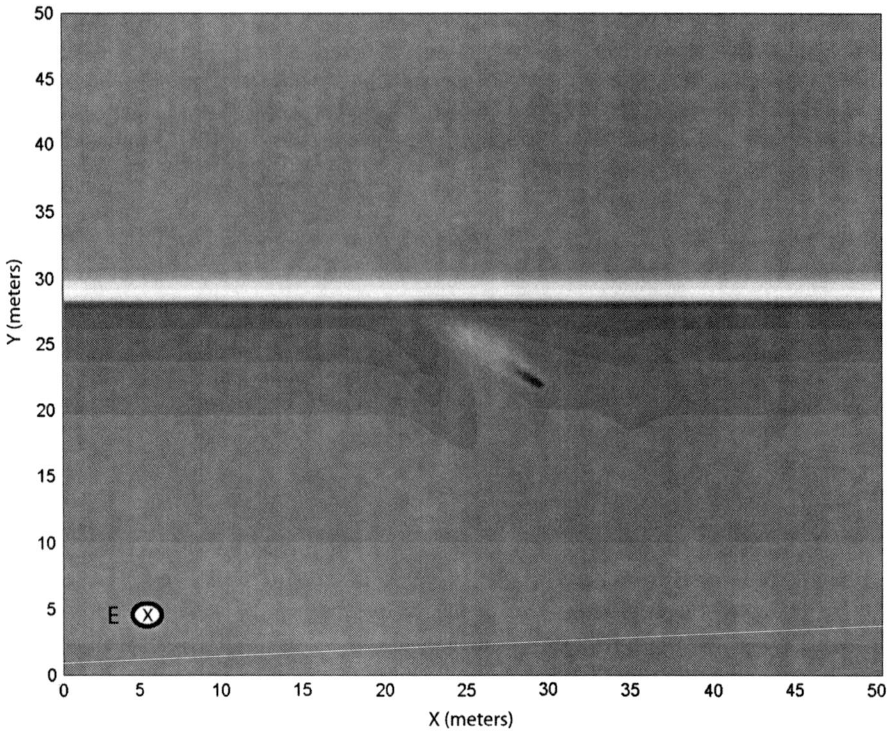

Fig. 5 Example results from one of the FDTD simulations conducted for analyzing the radar meteor scatter known as the meteor head echo. This plot shows the electric field in the simulation as a function of space, which is directed into the page. A broad-banded radar pulse was launched from the bottom of the page boundary and is seen propagating upward and passing the meteor head echo plasma. Faint reflections from the meteor are also seen propagating downwards. These reflections are then measured and compared with the input energy to determine the radar cross-section of this particular meteor as a function of input radar frequency

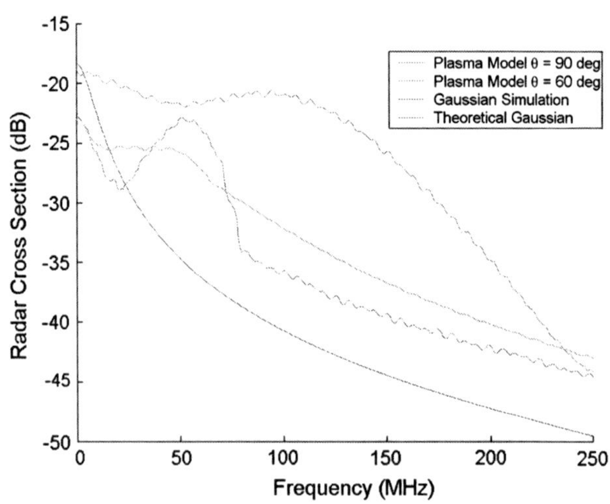

Fig. 6 Comparisons between analytical cross-section calculations, and simulations of Gaussian profiles and realistic plasma simulation produced profiles at two angles. All of the lines shown are results from a meteor plasma with similar peak plasma frequencies of 70 MHz

This plot makes clear that peak plasma density is not an ideal proxy for estimating head echo RCS or vice versa. Our explanation of this figure starts from the top line and ends with the bottom. The top line is one of our FDTD simulations for spherical-Gaussian head echo plasma. This simulation had the same peak density as all the other lines and a Gaussian sigma of 2 m, which gives it the same total number of electrons as the meteor shaped simulation. The next two lines are simulations of meteor shaped plasma at two different angles to the radar. The bottom line is an analytical calculation for the RCS of a Gaussian shaped meteor as a function of radar frequency from Close et al. (2004). So that the analytical solution would match the simulated RCS from the meteor distribution, we had to use a very small size of about 20 cm. This is similar in size to the very high density region near the tip shown in Fig. 3, but much smaller than the overall size of the meteor plasma that had a 3σ width (i.e., the distance transverse to the meteoroid direction where the head echo plasma effectively blends into the background ionospheric plasma) of 6 m across.

In order to provide a result one can be confident in, any good simulation should be compared with the existing analytical theory. Figure 6 presents comparisons with the analytical calculations of Close et al. (2004), and simulations of RCS as a function of a Gaussian shaped peak meteor density, and compared the results with the theoretical calculations of Close. What this plot shows is that the Close analytical model and the simulated head echo RCS possess the same slope as a function of radar frequency, but that the location of the knee in the Close model is predicted to be near the peak plasma frequency of the meteor head. Our simulations show that this knee occurs at approximately $f = \text{sqrt}(2) * (\text{peak plasma frequency})$.

4 Head Echo RCS as a Function of Peak Plasma Frequency

Here we use our simulation results to examine head echo RCS as a function of the peak plasma density in the head echo, which has been considered as a proxy for head echo RCS in a modeling effort of the interplanetary meteor flux by Janches et al. (2006). In this case we compare FDTD simulations with a fixed head echo size, and incident radar angle but scaled density. The angle chosen is for a meteoroid path that is straight down the beam, and the size of the head echo plasma had a 3σ width of 10 m, and is therefore similar in size to the meteoroid simulation shown in Fig. 3.

Both Figs. 7 and 8 show that for the same meteor size, RCS is proportional to peak plasma frequency. One notable feature shown in Fig. 7 is that for radar frequencies near the plasma frequency large fluctuations in RCS occur as a function of frequency, particularly in the 72 MHz line, Mie scattering resonances appear to influence the RCS with fluctuations near 8 dB deviation from the general smooth trend. Such rapid fluctuations may account for the occasional unusual head echo returns seen by some observers. It may be that deviations from smooth SNR as a function of altitude for head echoes may simple be a head echo that possesses a plasma density that has a plasma frequency near the radar frequency.

The head Echo RCS dependency on radar frequency and peak electron density is characterized by the following equation which was derived via a polynomial fit to the results of over 30 FDTD simulations of head echo plasmas with 3sgme sizes from 2 to 15 m.

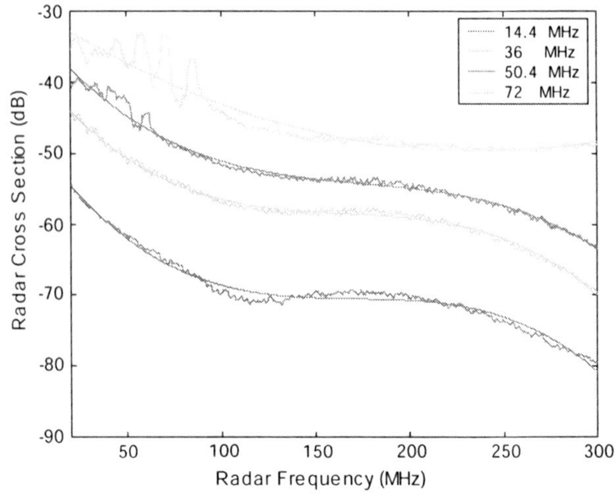

Fig. 7 Radar Cross-Section (RCS) versus Radar Frequency comparison at different meteor peak plasma frequencies with same meteor size. 3σ size = 10 m

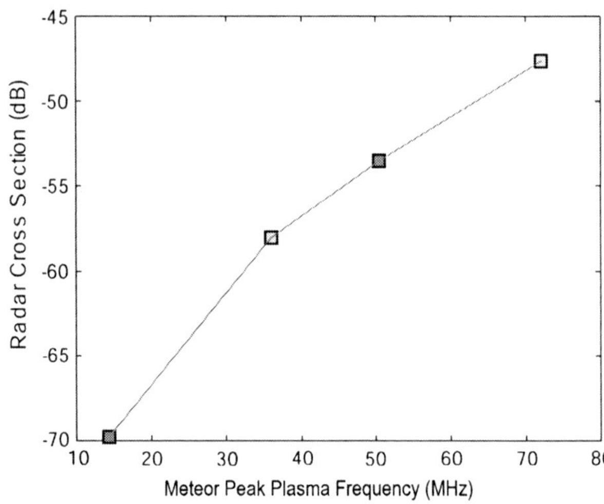

Fig. 8 Same as Fig. 7 but for a single radar frequency of 150 MHz, the results are a near linear trend in RCS as a function of peak plasma frequency

$$RCS(f) = -1.5 \times 10^{-7} \times f^3 + 2 \times 10^{-4} \times f^2 - 0.1 \times f + \alpha \quad (4)$$

where $\alpha \propto$ meteor peak plasma frequency, and α could be approximately expressed as:

$$\alpha(\lambda_p) = 0.38\lambda_p - \mathrm{C} \quad (5)$$

where λ_p is the peak plasma frequency defined in units of MHz, and C is a calibration constant in dB, for the plots shown here C = 74, but this number would be adjusted for comparisons with a particular observatory, or for calculations in dBsm.

5 Conclusions

This paper presents simulations that aim to improve our understanding of the radio scattering characteristics of the meteor plasma, especially in light of recent work showing that plasma turbulence and instability greatly influences meteor trail properties at every stage of evolution. We presented results on a meteor plasma simulation of head echo plasmas using PIC ions, which show that electric fields strongly influence early stage meteor plasma evolution, by accelerating ions away from the meteoroid body. This result should indicate the need for renewed modeling on the initial stages of meteor plasma expansion, since electric fields have never been considered in theoretical studies of what is known as the meteor trail initial radius (Jones 1995). We note here that these head echo plasma simulations are still in the preliminary stage, with significant work to be done on evaluating different parameter regimes. However we use it here because it represents the only estimate we have of the shape of the head echo scattering region and is likely to be significantly closer to reality then a spherical Gaussian distribution. Furthering the research on these early stage plasma simulations represents a future direction we plan to pursue.

This paper continued with the results of FDTD, which can calculate the radar cross-section of the simulated meteor plasmas by launching a simulated broad-banded EM pulse towards a simulated meteor plasma. These simulations have shown that the radar cross-section depends in a complex manner on a number of parameters. These include the angle between radar and meteor entry (as discussed in Dyrud et al. (2007), a large dependence on radar frequency, which shows that for a given meteor plasma size and density, the peak reflectivity for the meteor varies but is usually less then 100 MHz. Finally, we demonstrated that peak plasma frequency is not an ideal proxy for head echo RCS all by itself, but if the size of the head echo can be estimated, we have presented an empirical formulation for the variation of RCS as a function of changing peak electron density. Further, by conducting the FDTD simulations on plasma distributions form the plasma simulations we are able to study the effects of a number of processes that are not approachable through other means, such as non-Gaussian plasma distributions. In conclusion, we expect these results to be of use for those attempting to model the relationship between meteoroid parameters and radar observations of meteor head echoes.

References

C.K. Birdsall, A.B. Langdon, Plasma Physics via Computer Simulation. (McGraw Hill, New York, 1985)

A.D. Boardman, *Electromagnetic Surface Modes*. (John Wiley & Sons Ltd., 1982)

K.G. Burden, *The Propagation of Radio Waves*. (Cambridge University Press, 1985)

Z. Ceplecha, J. Borovicka, W.G. Elford, D.O. Revelle, R.L. Hawkes, V. Porubcan, M. Simek, Meteor phenomena and bodies. Space Sci. Rev. **84**, 327–471 (1998)

M.A. Cervera, W.G. Elford, The meteor radar response function: theory and application to narrow beam MST radar. Planet. Space Sci. **52**, 591–602 (2004)

E. Chapin, E. Kudeki, Radar interferometric imaging studies of long duration meteor echo observed at Jicamarca. J. Geophys. Res. **99**, 8937–8949 (1994)

F. Chen, *Introduction to Plasma Physics and Controlled Fusion*. vol. I, (Plenum Press, New York, 1984)

S. Close, M. Oppenheim, S. Hunt, L. Dyrud, Scattering characteristics of high-resolution meteor head echoes detected at multiple frequencies. J. Geophys. Res. (Space Physics) **107**(A10), 1295 (2002)

S. Close, M. Oppenheim, S. Hunt, A. Coster, A technique for calculating meteor plasma density and meteoroid mass from radar head echo scattering. Icarus **168**, 43–52 (2004)

S. Close, P. Brown, M. Campbell-Brown, M. Oppenheim, P. Colestock, Meteor head-echo radar data: mass-velocity selection effects. Icarus (2007). doi:10.1016/j.icarus.2006.09.07

D.J. Cziczo, D.S. Thomson, D.M. Murphy, Ablation, flux, and atmospheric implications of meteors inferred from stratospheric aerosol. Science **291**, 1772–1775 (2001)

L.P. Dyrud, M.M. Oppenheim, A.F. vom Endt, The anomalous diffusion of meteor trails. Geophys. Res. Lett. **28**, 2775–2778 (2001)

L.P. Dyrud, M.M. Oppenheim, A.F. vom Endt, Interpretation of non-specular radar meteor trails. Geophys. Res. Lett. **29** (2002)

L. Dyrud, K. Denney, J. Urbina, D. Janches, E. Kudeki, S. Franke, The meteor flux: it depends how you look. Earth, Moon, Planets, **95**, 89–100 (2004)

L.P. Dyrud, L. Ray, M. Oppenheim, S. Close, K. Denney, Modeling high-power large aperture radar meteor trails. J. Atmos. Terrestrial Phys. **67**(13), 1171–1177 (2005)

L. Dyrud et al., Plasma and electromagnetic wave simulations of meteors. J. Adv. Space Res. (2007). doi: 10.1016/j.asr.2007.03.048

J.T. Fentzke, D. Janches, A semi-emperical model of the contribution from sporadic meteoroid soirces on the meteor input function in the MLT observed at Arecibo. J. Geophys. Res. (Under Review) (2007)

D. Janches, J.D. Mathews, D.D. Meisel, Q.-H. Zhou, Micrometeor observations using the Arecibo 430 MHz radar. Icarus **145**, 53–63 (2000)

D. Janches, C.J. Heinselman, J.L. Chau, A. Chandran, R. Woodman, Modeling the global micrometeor input function in the upper atmosphere observed by high power and large aperture radars. J. Geophys. Res. (Space Physics) **111**, 7317 (2006)

D. Janches, S. Close, J.T. Fentzke, A comparison of detection sensitivity between ALTAIR and Arecibo meteor observations: Can high power and large aperture radars detect low velocity meteor head-echoes. (Accepted, Icarus, September 2007)

W. Jones, Theory of the initial radius of meteor trains. Mon. Not. R. Astron. Soc. **275**, 812–818 (1995)

J.D. Mathews, D. Janches, D.D. Meisel, Q.-H. Zhou, The micrometeoroid mass flux into the upper atmosphere: Arecibo results and a comparison with prior estimates. Geophys. Res. Lett. **28**, 1929–1932 (2001)

N.J. Nickish, P.M. Franke, *Finite-Difference Time-Domain Solution of Maxwell's Equations for the Dispersive Ionosphere.* (IEEE Antennas and Propagation Magazine, 1992), pp. 33–39

M.M. Oppenheim, A.F. vom Endt, L.P. Dyrud, Electrodynamics of meteor trail evolution in the equatorial E-region ionosphere. Geophys. Res. Lett. **27**, 3173–3176 (2000)

A. Pellinen-Wannberg, The EISCAT meteor-head method—a review recent observations. Atmos. Chem. Phys. **4**, 649–655 (2004)

J.M.C. Plane, A new time-resolved model for the mesospheric Na layer: constraints on the meteor input function. Atmos. Chem. Phys. **4**, 627–638 (2004)

H.R. Skullerud, T.H. Lovaas, K. Tsurugida, Diffusion and interaction potentials for K^+ ions in the noble gases. J. Phys. B At. Mol. Phys. **32**, 4509–4522 (1999)

A. Taflove, *Computational Electrodynamics—The Finite Difference Time Domain Method.* (Artech House, 1995)

A.D. Taylor, The harvard radio meteor project velocity distribution reappraised. Icarus **116**, 154–158 (1995)

K.S. Yee, Numerical solution of initial boundary value problems involving Maxwell's equations in isotropic media. IEEE Trans. Antenn. Propag. **AP-14**, 302–307 (1966)

Q.H. Zhou, J.D. Mathews, T. Nakumura, Implications of meteor observations by the MU radar. Geophys. Res. Lett. **28**, 1399 (2001)

A New Model for the Separation of Meteoroid Fragments in the Atmosphere

N. G. Barri

Originally published in the journal Earth, Moon, and Planets, Volume 102, Nos 1–4.
DOI: 10.1007/s11038-007-9204-0

Abstract This work is devoted to modeling of the transverse scattering of meteoroid fragments in the atmosphere by adopting supersonic gas dynamics around a system of bodies. Artem'eva and Shuvalov (1996, Shock Waves, 367) and Zhdan et al. (2004, Dokl. Phys., 315–317) found that the transverse force decreases with the increase of the distance between fragments, that is, fragments do not separate in a transverse direction under the action of constant repulsion force. This work on the decreasing transverse force uses the values of the transverse force coefficient by Zhdan et al. (2004, Dokl. Phys., 315–317) obtained from numerical modeling for spheres in a supersonic flow to derive the analytical solution of the dynamic equation for a fragment. The new model of layer-by-layer scattering of meteoroid fragments moving as a system of bodies is constructed on the basis of the analytical solutions derived in this work and the numerical data by Zhdan et al. (2005, Dokl. Phys., 514–518).

Keywords Meteoroid fragmentation · Transverse scattering modeling · Fragment separation · Fragment velocity · Morávka fireball

1 Introduction

Until recently, inaccurate models existed in literature to describe the separation dynamics of two fragments following the disruption of a body moving through the atmosphere with supersonic velocity. The notion of transverse scattering of meteoroid fragments was first introduced by Passey and Melosh (1980). They investigated the physics of meteoroid breakup in the atmosphere and its implications for the observed features of strewn fields. The effects that caused dispersion of meteoroid fragments are gravity, differential lift of the fragments, bow shock interactions just after breakup, centripetal separation by a rotating meteoroid, and crushing during deceleration. Passey and Melosh (1980) showed that the interactions of shock waves are a principal cause of the transverse dispersion of

N. G. Barri (✉)
Institute of Mechanics, Moscow State University, Vorob'evy gory, Moscow 119899, Russia
e-mail: n_barri@imec.msu.ru

fragments. These authors also investigated several known crater fields. Using numerical calculations Passey and Melosh (1980) reconstructed the events of meteorite fragment falls for each crater field. However these authors assumed transverse scattering velocity without consideration given to the fact that the interactions of shock waves will gradually decrease with the increase of the distance between fragments.

The present work will instead assume that the transverse force decreases with the increase of the distance between fragments.

2 Interactions of Two Fragments in a Supersonic Flow: A Literature Summary

Passey and Melosh (1980) defined the force acting on the fragments in the transverse direction F_r as the product of the total pressure and the effective cross-section area. The repulsion force was considered constant and that it had disappeared completely once the fragments had separated by a distance comparable to their size. The acceleration (a) for a fragment caused by this force is described as

$$a = \frac{F_r}{M} = \frac{\rho_a V_b^2 \pi R^2}{(4/3)\pi R^3 \rho_m} = \frac{3V_b^2 \rho_a}{4R\rho_m} \quad (1)$$

where M and R are the fragment mass and radius, V_b is the velocity at the instant of breakup, ρ_a is the local air density at the fragmentation altitude and ρ_m is the meteoroid density. Thus, in order to determine the transverse velocity of the fragments at the final moment of their interaction at constant acceleration Passey and Melosh (1980) used the following formula:

$$U_f = \left(\frac{3}{2} C V_b^2 \frac{\rho_a}{\rho_m}\right)^{1/2} \quad (2)$$

wherein the constant C was introduced under the assumption that the product CR defined a distance where interactions stopped. Values of the constant C range between 0.02 and 1.52. Passey and Melosh (1980) obtained this range by studying the cross-range spread of craters in known meteorite crater fields.

For the pressure force they used standard Eq. 3, which means Passey and Melosh (1980) accepted that the repulsion force coefficient C_r is constant and equal to 2. This work will argue that C_r is neither constant nor does it equal to 2.

$$F_r = \frac{1}{2} C_r \rho_a V^2 S \quad (3)$$

It follows from numerical modeling discussed by Artem'eva and Shuvalov (1996), Shuvalov et al. (2000), and by Zhdan et al. (2004) that the transverse force coefficient is a function of the distance between fragments. The results of numerical calculations on a flow over two semi-cylinders (Artem'eva and Shuvalov 1996; Shuvalov et al., 2000) and on a flow over two spheres (Zhdan et al. 2004) are presented in Fig. 1. One can see from Fig. 1 that values of the transverse force coefficient for spheres and for semi-cylinders are very close. The maximum of C_r in numerical experiments is approximately equal to 0.28 when the distance between the fragments is close to zero. The value of C_r monotonically decreases to zero for $h \approx 0.5$ where $2h$ is the non-dimensional distance between the fragments; as the unit length we take the radius R of a fragment.

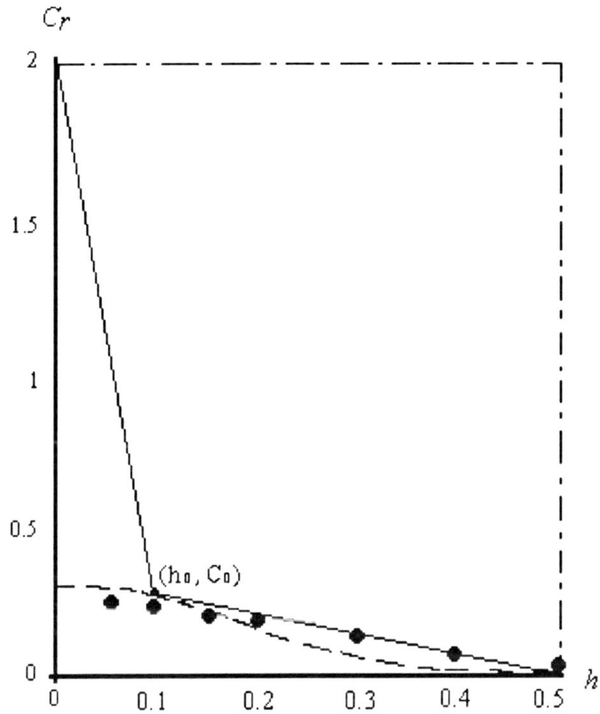

Fig. 1 The dependence of the repulsion force coefficient (C_r) on the distance (h) between fragments, viz. the value used by Passey and Melosh (1980) (dash-dotted line) and from numerical experimental data for spheres (Zhdan et al. 2004) (dashed line). The dots represent the numerical experimental data for semi-cylinders (Shuvalov et al. 2000). The solid line is the proposed approximation to numerical data taking into account the gas dynamics when for $h = 0$, C_r (max) should equal 2 (see Eq. 4)

Zhdan et al. (2004) calculated the values of the repulsion force coefficient for two fixed spheres at various distances between the nearest points of spheres. At $h = 0$ when the spheres touch, the numerical solution showed the value $C_r = 0.28$ (Zhdan et al. 2004). However during the initial time when the meteoroid is still intact as a single body, the pressure at the point of initial separation is equal to the total pressure and $F_r = \rho_a V_b^2 \pi R^2$, i.e., C_r should equal 2 for $h = 0$. In the case $h > 0$, the numerical experimental data can be used to describe the separating force as a function of the distance.

3 Interactions of Two Fragments in a Supersonic Flow: A New Approach

We propose the following approximation of the numerical experimental data. To be in agreement with the gas dynamics of fragment separation, it is necessary to make a correction in the data near $h = 0$. To further simplify our computations, the dependence $C_r(h)$ may be represented in the form of two intersecting straight lines at (h_0, C_0) (Fig. 1). The point (h_0, C_0) should be on the curve corresponding to the numerical experiments from Zhdan et al. (2004) and h_0 is chosen to be close to zero. The function

$$C_r(h) = \begin{cases} C_{r1}(h) = (h - h_0)\alpha + C_0, & 0 \le t \le t_0 \\ C_{r2}(h) = (h - h_0)\beta + C_0, & t_0 < t \le t_f \end{cases} \quad (4)$$

corresponds to the straight lines represented in Fig. 1, where (h_0, C_0) is the intersection point of the segments $C_{r1}(h)$ and $C_{r2}(h)$. This point corresponds to time t_0. Time t_f is the time corresponding to $h = 0.5$, distance at which the interactions among fragments stops. The coefficients α, and β determine the slopes of the straight lines.

In order to describe the transverse scattering of fragments we will make use of the dynamic equation in the following dimensionless form:

$$\frac{d^2h}{dt^2} = \frac{\rho_a}{\rho_m} C_r(h) \quad (5)$$

Here we take the radius R of a fragment as the unit length. As the time unit we take the $T = \sqrt{8/3}R/V_b$. Taking into account that the separating force coefficient is defined by Eq. 4, we solve this differential equation and obtain the following solution

$$h(t) = \begin{cases} h_1(t) = C_1 \cos mt + C_2 \sin mt - 2/\alpha, & 0 \le t \le t_0 \\ h_2(t) = C_3 \cos kt + C_4 \sin kt + 1/2, & t_0 < t \le t_f \end{cases} \quad (6)$$

where $m = \sqrt{p|\alpha|}, k = \sqrt{p|\beta|}$, $p = \rho_a/\rho_m$ and the constants C_1, C_2, C_3, C_4 that depend on C_0, h_0 are defined by the initial conditions $h_1(0) = 0, h_1'(0) = 0$, and $h_2(t_0) = h_0, h_2'(t_0) = h_1'(t_0)$.

Then we obtain the relationship between the transverse velocity and time, and the final transverse velocity of the fragments as $h_2'(t_f)$:

$$U_{f1} = \left(\frac{3}{2} C V_b^2 \frac{\rho_a}{\rho_m}\right)^{1/2} \left(\frac{|\beta|}{4C}(C_3^2 + C_4^2)\right)^{1/2} = U_f \cdot k \quad (7)$$

It is easy to check that U_{f1} differs from the velocity U_f (see Eq. 2) by the factor k, this factor tends to 1 as $C_0 \to 2$ and $h_0 \to 0.5$; if $C_0 = 0.27$ and $h_0 = 0.1$, then $k = 0.4$ (Fig. 2).

In this manner we obtained the analytical solution for the case of transverse scattering of two spheres under the action of a decreasing repulsion force. It turns out that the induced transverse velocity of a spherical fragment is much less than the values that were used in the papers by Passey and Melosh (1980) and is close to the values obtained by Artemieva and Shuvalov (1996) and by Shuvalov et al. (2000).

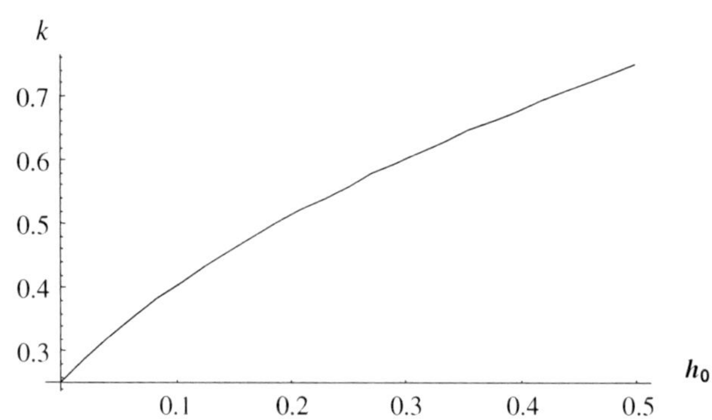

Fig. 2 The dependence of k on h_0 for $C_0 = 0.27$

4 New model of Meteoroid Fragments Separation

Various models for meteoroid fragmentation caused by aerodynamic strength failure were proposed previously by, among others, Baldwin and Sheaffer (1971), Grigoryan (1979), Hills and Goda (1993), Korobeynikov et al. (1994), Svetcov et al, (1995) and Ivanov and Ryzhanskii (1997). In many of these models the calculation of the trajectory of the fragment swarm is done by numerical integration using a formula for a single body with an increasing area for its effective cross-section (Grigoryan 1979; Hills and Goda 1993; Korobeynikov et al. 1994).

Other models consider the flight of individual fragments (Baldwin and Sheaffer 1971; Borovička et al. 1998; Artemieva and Shuvalov 2001).

The results obtained by Zhdan et al. (2005), Zhdan (2005) and Barri (2005) allow us to describe in more detail the transverse scattering of fragments caused by the interaction of shock waves and to construct a new model for meteoroid fragment scattering layer by layer. Zhdan et al. (2005) presented a numerical analysis of the problem for a finite number of spheres in a supersonic flow. The repulsion force coefficient for each sphere was calculated with respect to its relative position in the group. It was shown that the value of the repulsion force acting on the peripheral bodies of group is sufficiently larger than the value of the force acting on the internal bodies.

Basing on the results described above we propose a new model of layer-by-layer scattering of meteoroid fragments. The initially fragmented body will be interpreted as a compact collection of spherical fragments. We will consider two shapes for the initial body, i.e. either a cylindrical or a spherical shape (Fig. 3). In the case of a spherical body we will consider the layer to be the volume contained between two spheres of radius $R(2i - 1)$ and $R\,(2i + 1)$; $i \geq 1$, where i is layer's number. The centre of the sphere with radius R that coincides with the centre of spherical meteoroid will be the zero layer, $i = 0$. If a meteoroid has the cylindrical form, we will consider the layer as the volume contained between two cylinders with radii $R(2i - 1)$ and $R(2i + 1)$; $i \geq 1$, where i is layer's number. The cylinder with radius R will be a zero layer, $i = 0$.

In the proposed model the scattering of the meteoroid fragments has several stages whereby, at each stage, the interaction of a fragment of the outer layer and a fragment of the inner part was analyzed. The inner part is considered to be a single compact collection of spherical fragments. This interaction lasts until the distance between the outer layer and the inner, still intact, part reaches the radius of the inner part. Then, the next layer becomes an outer layer and it starts moving away from the main part of the meteoroid.

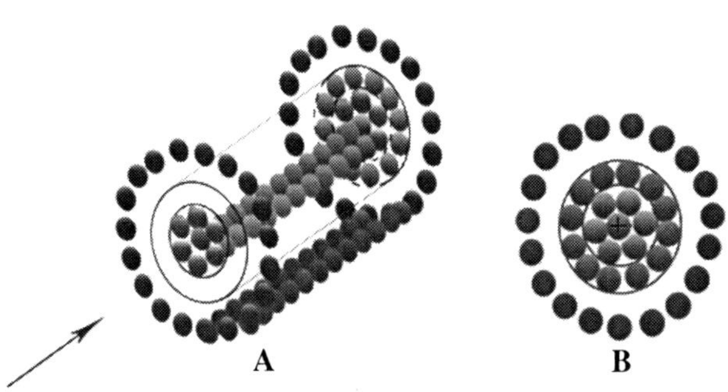

Fig. 3 Scheme of the new model of transverse scattering of meteoroid fragments for (**a**) a cylindrical meteoroid shape and (**b**) a spherical meteoroid

Table 1 The calculated durations of the scattering event based on the new proposed model

N	Radius of a fragment (cm)		Number of layers		Time of scattering (s) $A = 21$ km		Time of scattering (s) $A = 36$ km	
	C	S	C	S	C	S	C	S
10^3	22.8	20	7	7	0.021	0.017	0.058	0.046
10^5	4.93	4.3	29	30	0.022	0.018	0.059	0.049
10^7	1.06	0.92	131	135	0.022	0.018	0.059	0.049

The letters "C" and "S" denote the cylindrical and spherical cases, respectively. For the calculations we assumed that meteoroid separation started at two different altitude $A = 21$ km and $A = 36$ km

Given the number N of fragments, it is easy to determine their radius R by using the mass conservation law. We then determine the number of layers and number of fragments in each layer and find the size of the inner part and outer layer for each subsequent stage. In order to describe the transverse scattering of fragments we will use of the dynamic equation solutions of Eqs. 6 and 7 applying the corresponding relation for $C_r(h)$ provided by Zhdan (2005) numerical modeling on two spheres with different radii in a supersonic flow. In addition to the quotient of the radii, the coefficient $C_r(h)$ depends also on the distance between the spheres.

To construct the model, at each stage of the scattering process we will use of the corresponding relation $C_r(h)$ in respect to the ratio of the radii of the outer layer fragment and the inner part. Then we determine the duration of fragment separation to the distance when the interaction between all fragments stops; we add up the total of the interaction terms for all layers. Table 1 shows the results for a spherical body of radius 2 m, or an equivalent cylindrical body 4 m long and a 2-m radius, moving supersonically in the atmosphere at $V_e = 20$ km/s that were calculated applying the new proposed model.

5 Discussion and Conclusions

The total time of scattering by layers does not actually depend on the number of fragments. In addition, the time of scattering is an insignificant fraction of the total time a meteoroid travels through the atmosphere. Thus, meteoroid fragmentation and scattering of its fragments will be almost instantaneous and it reaches a state of independent movement for each fragment. The accuracy of modern meteor observations is not yet sufficient enough to detect the meteoroid scattering properties that were defined by the new numerical experiments presented here. The results of this new model of fragmentation behavior should be treated with reservation until it has been compared to the results of sufficient observational data on meteoroid fragmentation. Borovička and Kalenda (2003) made a detailed analysis of the fragmentation history of the Morávka fireball during its atmospheric entry. It will be very interesting to compare the methods of the work presented here with those applied to the Morávka fireball observations. We will explore how to best compare the observational and modeling data.

Acknowledgments I thank an anonymous reviewer but inparticular Dr. Olga Popova for constructive comments and suggestions. I also thank handling editor, Frans J.M. Rietmeijer, for many useful suggestions. Special thanks go to Professor Vladimir Stulov for valuable discussions on various aspects of this work. This work was supported by the Russian Foundation for Basic Research, project no. 07-01-00009.

References

N.A. Artem'eva, V.V. Shuvalov, Interaction of shock waves during the passage of a disrupted meteoroid through the atmosphere. Shock Waves **5**, 359–367 (1996)

N.A. Artemieva, V.V. Shuvalov, Fragmented meteoroid movement through the planetary atmosphere. J. Geophys. Res. **106**, 3297–3310 (2001)

B. Baldwin, Y. Sheaffer, Ablation and breakup of large meteoroids during atmospheric entry. J. Geophys. Res. **76**(19), 4653–4668 (1971)

N.G. Barri, A model for the separation of fragments of a destroyed meteoroid. Moscow Univ. Mech. Bull. **60**(4), 20–22 (2005)

J. Borovička, P. Kalenda, The Morávka meteorite fall: 4. Meteoroid dynamics and fragmentation in the atmosphere. Meteorit Planet Sci **38**(7), 1023–1043 (2003)

J. Borovičcka, O. P. Popova, I. V. Nemtchinov, P. Spurný, Z. Ceplecha, Bolides produced by impacts of large meteoroids into the Earth's atmosphere: comparison of theory with observations. I. Benešov bolide dynamics and fragmentation. Astron. Astrophys. **334**, 713–728 (1998)

S.S. Grigoryan, About the movement and destruction of meteoroids in the planet's atmospheres (in Russian). Space Explor. **17**(6), 875–893 (1979)

J.G. Hills, P. Goda, Fragmentation of small asteroids in the atmosphere. Astron. J. **105**, 1114–1144 (1993)

A.G. Ivanov, V.A. Ryzhanskii, Fragmentation of a small celestial body on its interaction with the atmosphere of a planet. Dokl. Phys. **42**(3), 139–143 (1997)

V.P. Korobeynikov, V.I. Vlasov, D.B. Volkov, Modeling of space bodies distraction during the movement in the planet's atmospheres (in Russian). Math. Model. **6**(8), 61–75 (1994)

Q.R. Passey, H.J. Melosh, Effects of atmospheric breakup on crater field formation. Icarus **42**(2), 211–233 (1980)

V.V. Shuvalov, N.A. Artemyeva, I.A. Trubetskaya, A modeling of a movement of a destroyed meteoroid with taking into account of evaporation. Solar Syst. Res. **34**(1), 49–60 (2000)

V.V. Svetcov, I.V. Nemtchinov, A.V. Teterev, Disintegration of large meteoroids in Earth's atmosphere: theoretical models. Icarus **116**, 131–153 (1995)

I.A. Zhdan, The aerodynamic resistance of the bodies system in a supersonic flow (abstract) (in Russian). Lomonosov conference, Moscow State University, Moscow, 88 p (2005)

I.A. Zhdan, V.P. Stulov, P.V. Stulov, Aerodynamic interaction of two bodies in a supersonic flow. Dokl. Phys. **49**(5), 315–317 (2004)

I.A. Zhdan, V.P. Stulov, P.V. Stulov, 3D configurations of broken body fragments in a supersonic flow. Dokl. Phys. **50**(10), 514–518 (2005)

Radar Backscatter from Underdense Meteors and Diffusion Rates

Werner Singer · Ralph Latteck · Luis Federico Millan · Nick J. Mitchell · Jens Fiedler

Originally published in the journal Earth, Moon, and Planets, Volume 102, Nos 1–4.
DOI: 10.1007/s11038-007-9220-0

Abstract Many meteoroids burn up between about 120 km and 70 km, deposit metals and dust and form ionized trails which are detected by radars. Model studies about the influence of neutral or positively charged background dust on the ambipolar diffusion indicate that significant smaller decay times should be observed for weak meteor echoes compared to strong meteor echoes which can affect the estimation of temperatures. The variation of meteor decay times in dependence on echo strength, height, and season was studied using radar observations at 69° N, 22° S, and 67° S. Significantly reduced decay times were found for weak echoes below about 88 km at low latitudes throughout the year, and at high latitudes with the exception of summer. In summer at high latitudes, decreasing decay times of weak and strong meteors are observed at altitudes below about 85 km during the appearance of noctilucent clouds. The impact of reduced decay times on the estimation of neutral temperatures from decay times is discussed.

Keywords Diffusion · Mesopause · Meteor · Meteor radar · Temperature

1 Introduction

Many meteoroids burn up between about 120 km and 70 km and form ionized trails. The meteor ablation is an important source for the metal atoms of the upper atmosphere and for condensation nuclei (meteor dust particles), the existence of which is required for the formation of noctilucent clouds (NLC) in the polar mesopause region. The ionized trails are detected by radars and radar measurements have allowed determining characteristics of

W. Singer (✉) · R. Latteck · J. Fiedler
Leibniz Institute of Atmospheric Physics, Schlossstr. 6, 18225 Kuhlungsborn, Germany
e-mail: singer@iap-kborn.de

L. F. Millan
Chalmers University of Technology, Gothenborg, Sweden

N. J. Mitchell
University of Bath, Bath, UK

both the meteor trails and the atmosphere around the trails. Radar studies of underdense trails are used to infer atmospheric temperatures from the diffusion rate of the trail which expands firstly by molecular diffusion (Chilson et al. 1996; Hocking 1999). Typical decay times for underdense meteors vary between 0.015 and 0.3 s for radar frequencies in the range 30–55 MHz.

The influence of a possible background of nanometer sized neutral or positively charged dust on the diffusion of meteor trails has been studied by Havnes and Sigernes (2005) to see if the decay time of underdense trails can be affected by a part of trail electrons being absorbed by dust as the trail expands. The results showed a decrease of the decay times of up to 10 per cent in relation with the dust free case. The largest effect would be observed for meteors with low electron line densities (weak meteor echoes). In situ observations by sounding rockets detected neutral or charged particles of comparable size and charge (Gelinas et al. 1998; Lynch et al. 2005; Rapp et al. 2005). In this study we use meteor observations to investigate systematically the decay times at different heights, time of year, and geographic latitude in dependence on echo strength.

2 Radar Experiment and Observations

The meteor radars used in this study are interferometric SKiYMET radars of nearly identical hardware and meteor detection software (Hocking et al. 2001). The radars used are located at Andenes (Norway, 69.3° N, 16.0° E), Learmonth (Australia, 22.2° S, 114.1° E), and Rothera (Antarctica, 67.5° S, 68.0° W) and were operated at frequencies 32.55, 35.24, and 32.5 MHz with peak powers of 12, 6, and 12 kW. Details of the Andenes meteor radar are given by Singer et al. (2004a).

We analysed data from January 2005 until March 2006 to cover a complete summer season at northern and southern polar latitudes. For the arctic summer 2005 lidar observations of NLC were available. We rely on reliable data and selected unambiguous detections of meteors with signal-to-noise ratios better than 6 dB and located at zenith angles between 10° and 60°. The meteors are separated into weak and strong echoes according to their electron line densities, q, estimated from the observed peak amplitudes according to McKinley (1961, p. 189)

$$q^2 = P_R R^3 / (2.5 \cdot 10^{-32} P_T G^2 \lambda^3) \tag{1}$$

where R is the range of the meteor, P_R is the received echo power, P_T is the transmitted power (including feeding loss, $P_T = 9{,}500$ W at Andenes), G is the antenna directivity ($G = 5.4$ for Andenes), and λ is the radar wave length. The echo power has been calibrated for the Andenes meteor radar by feeding defined fractions of the transmitted pulse delayed by 100 μs into the receiver instead of the antenna signal (Latteck et al. 2005, 2007) which results in.

$$P_R = A^2 \cdot 4 \cdot 10^{-21} \tag{2}$$

where A is the amplitude in digitiser units (P_R/W). The same relation has been assumed for the Learmonth and Rothera radar as the three systems apply the same receivers and data acquisition hardware. The differences of the decay times between weak and strong meteors maximise for the following electron line densities—weak/strong meteors: 8.0E11–1.7E12 m^{-1}/1.7E12–7.0E12 m^{-1} (Andenes, Rothera); 1.0E12–2.0E12 m^{-1}/2.0E12–7.0E12 m^{-1} (Learmonth). The data were binned into height ranges of 3 km centred on 82,

85, 88, 91, and 94 km. Our study has been confined to altitudes below 95 km as above this altitude anomalous diffusion can influence the observations (Ceplecha et al. 1998; Dyrud et al. 2001).

3 Results

Figure 1 shows the seasonal variation of the decay times at 69° N and 22° S determined for weak and strong meteors at 82 km. Mean values and error bars are estimated from 5-d composite days shifted by 5 days with about 1,000–3,000 meteors at 82 km and about 2,000–9,000 meteors at 91 km. The error bars represent a significance level of 90% but most of the values reach significance levels up to 99%. The significance of the estimated mean decay times for weak and strong meteors was tested with help of the double sided significance levels of the t-distribution using the mean values, variances and meteor counts (Taubenheim 1969, p. 97). At all latitudes the decay times of weak meteors are reduced by up to about 20% compared to strong echoes. A strong seasonal variation appeared at high latitudes where the lowest decay times occurred in summer. The difference between the decay times of weak and strong meteors decreased with increasing altitudes and disappeared typically around 91 km. Height-dependent decay times for weak and strong meteors are presented in Fig. 2 for winter and summer conditions at 69° N, 22° S, and 67° S. The data from 67° S are presented with a shift of 6 months to allow an easy comparison of the seasonal variations on both hemispheres. The decay times showed the expected decrease with altitude in winter at all latitudes but in summer only at low latitudes. At high latitudes a remarkable decrease of the decay times of weak and strong meteors with decreasing altitude was found below 85 km. In summer the polar mesosphere is characterised by neutral temperatures below 140 K which allow the formation of icy dust particles at mesopause altitudes ($\sim$90 km) with a size of 10–15 nm, the size is increasing during sedimentation up to about 90 nm before the particles evaporate around 80 km due to increasing temperature. These icy particles are responsible for the occurrence of noctilucent clouds which are visible for lidars due to Mie scattering from the 20–90 nm sized icy particles (Baumgarten et al. 2007; Fiedler et al. 2005).

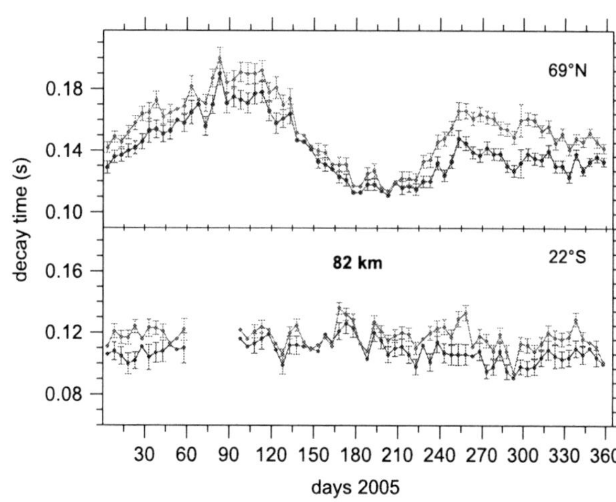

Fig. 1 Seasonal variation of the meteor decay times at 82 km observed by meteor radars at 69° N and 22° S. Decay times of strong meteor echoes are shown in red, decay times of weak echoes in blue (see text for details about the relation between electron line densities and weak/strong meteors)

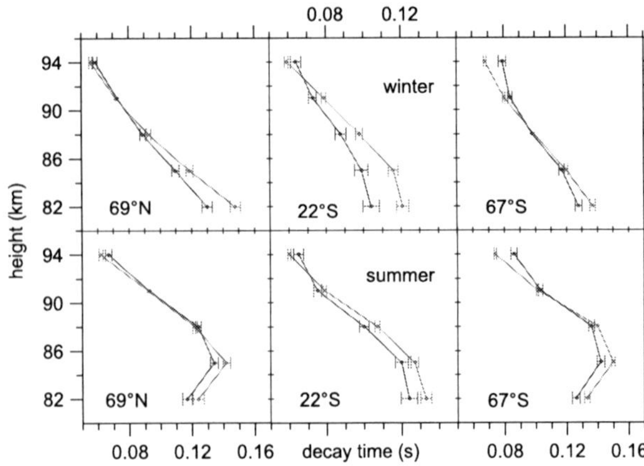

Fig. 2 Height variation of meteor decay times at 69° N, 22° S, and 67° S for winter and summer. Decay times of strong meteor echoes are shown in red, decay times of weak echoes in blue (winter: 22 November–01 December 2005; summer: 25 June–04 July 2005)

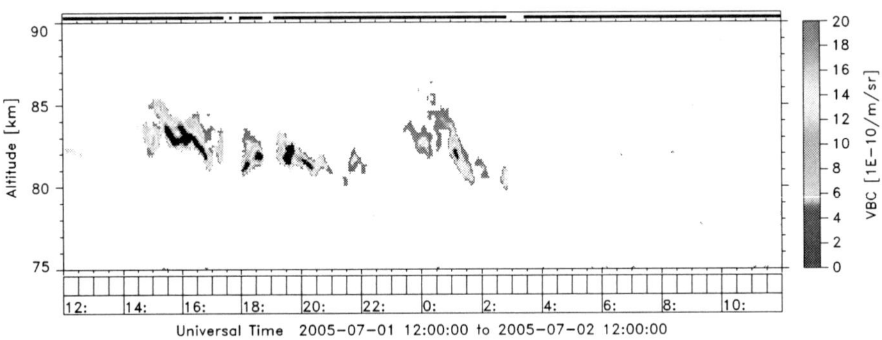

Fig. 3 Noctilucent clouds observed with the ALOMAR RMR lidar at 532 nm on 1st to 2nd July 2005 (the black bar at the top indicates the times of lidar operation)

Six days of continuous observations with the ALOMAR Rayleigh-Mie-Raman lidar at Andenes in July 2005 provided the capability to study meteor decay times in presence of NLC and at times where NLC did not appear. The NLC layers were located between about 80 km and 85 km with a peak altitude around 83 km (Fig. 3) in total for about 48 h. Meteors were selected for this period and for 54 h when NLC were not observed. The meteor decay times of weak meteors are reduced in the whole height range up to 94 km with the largest deviation around the NLC peak height (left panel of Fig. 4) compared with the separation into weak and strong trails (right panel of Fig. 4). In case of presence of the larger size NLC particles the decaytimes of strong echoes are reduced too at altitudes below 85 km.

4 Discussion and Summary

Reduced decay times (increased diffusion rates) were found for weak meteor echoes compared to strong echoes at low and polar latitudes at altitudes below 88 km all-the-year. The effect increases with decreasing height. This behaviour is in general agreement with model studies by Havnes and Sigernes (2005) showing reduced decay times in presence of neutral or positively charged dust due to absorption of trail electrons. Anomalous diffusion

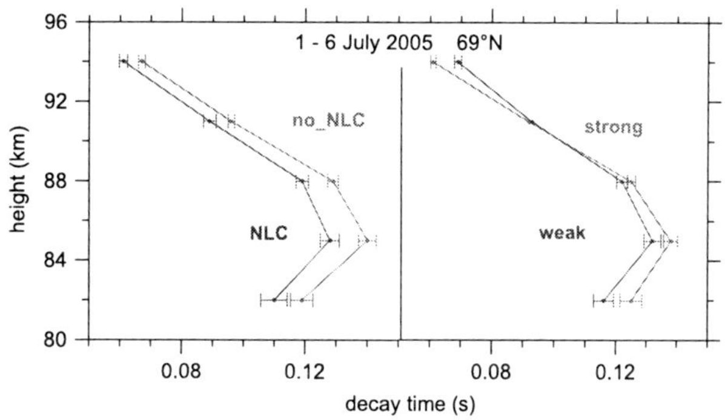

Fig. 4 Height variation of meteor decay times at 69° N for the period 1–6 July 2005 with continuous lidar measurements to observe NLC. Left panel: decay times for periods when NLC were present (NLC) and when none NLC (no_NLC) were observed. Right panel: decay times sorted into weak and strong meteors

which dominates the trail expansion above 95 km (Dyrud et al. 2001) can be ruled out with great probability at heights below 91 km. Hocking (2005) found that the influence is at best weak for altitudes above 93 km when daily averages are used. In addition, a second but unexpected behaviour was found at high latitudes in summer with the reduction of all decay times at the lowest altitudes. This behaviour is probably related to the presence of larger icy particles in the lower part of the cold summer mesopause region where the largest reduction occurred around the peak altitude of the NLC layers.

The generally reduced decay times in summer will affect the estimation of neutral temperatures from meteor decay times using the temperature gradient method (Hocking 1999) as well as the pressure model technique (e.g., Holdsworth et al. 2006). The consequence for the temperature gradient method is discussed in more detail. The method relies on the slope of the graph height versus $\log_{10}(1/\text{decay time})$ and an empirical model of the vertical temperature gradient at the peak of the layer (for details see Hocking et al. 2004; Singer et al. 2004b). A height dependent change of the decay time will change the slope and the estimated temperature. To evaluate the bias the seasonal variation of the temperature at 69° N was estimated for time bins of 72 h using all meteors without any selection and using alone strong meteors with electron line densities between 1.7E12 and 7.0E12 electrons/m. The results are shown in Fig. 5 and a mean bias to lower temperatures

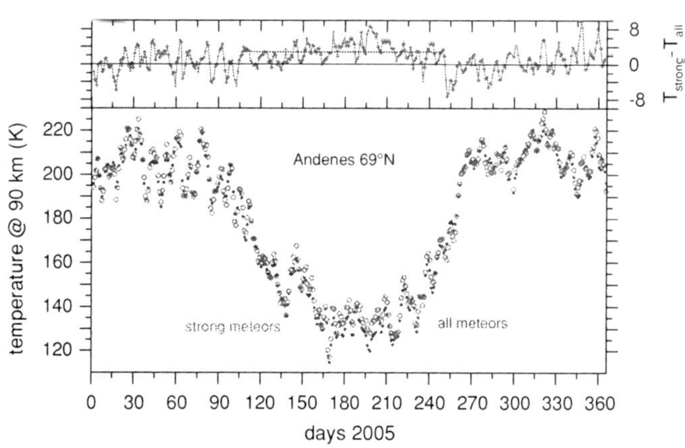

Fig. 5 Seasonal variation of the mesopause temperature at 69° N in 2005. The temperatures were estimated from meteor decay times of all (•) and of strong (o) meteor echoes using the temperature gradient method. The temperature difference between estimates from strong and all meteor echoes shows a mean bias of 2.8 K in summer

in the order of 3 ± 2 K is evident under summer conditions. This bias is probably underestimated as the slope of the graph height versus $\log_{10}$(1/decay time) was derived from the decay times of strong meteors at altitudes 80–98 km which are also reduced at the lowest altitudes in summer. The bias at low latitudes is in general within the measurements error. A more detailed study of the summer temperatures will be part of a separate paper.

The study of meteor decay times (diffusion rates) as function of meteor echo strength at high and low latitudes showed that (1) the decay times of weak underdense meteor echoes are reduced below about 88 km throughout the year with the largest reduction at 82 km. This behaviour is in agreement with model studies about the absorption of trail electrons by nanometer sized neutral or positively charged background dust which results in an enhanced diffusion rate. (2) At high northern and southern latitudes the anomalous effect of increasing diffusion with decreasing altitude has been observed for weak and strong meteor echoes below about 85 km in summer. This behaviour seems to be related to the presence of larger size icy particles in the cold summer mesopause region during the appearance of NLC. This reduction of the summer decay times at polar latitudes causes a bias to lower temperatures if neutral temperatures are estimated from meteor decay times.

Acknowledgements We appreciate the excellent support by IPS Radio and Space Services and Genesis Software Pty Ltd to operate the meteor radar at Learmonth. The radar experiment at Andenes received funding from the EU 6th framework programme project ALOMAR eARI.

References

G. Baumgarten, J. Fiedler, G. von Cossart, The size of noctilucent cloud particles above ALOMAR (69N, 16E): optical modelling and method description. J. Adv. Space Res. (2007). doi:10.1016/j.asr.200701.018

Z. Ceplecha, J. Borovicka, W.G. Elford, D.O. Revelle, R.L. Hawkes, V. Porubcan, M. Simek, Meteor phenomena and bodies. Space Sci. Rev. **84**, 327–471 (1998)

P.B. Chilson, P. Czechowsky, G. Schmidt, A comparison of ambipolar diffusion coefficients in meteor trains using VHF radar and UV lidar. Geophys. Res. Lett. **23**, 2845–2748 (1996)

L.P. Dyrud, M.M. Oppenheim, A.F. vom Endt, The anomalous diffusion of meteor trails. Geophys. Res. Lett. **28**, 2775–2778 (2001)

J. Fiedler, G. Baumgarten, G. von Cossart, Mean diurnal variations of noctilucent clouds during 7 years of lidar observations at ALOMAR. Ann. Geophys. **23**, 1175–1181 (2005)

L.J. Gelinas, K.A. Lynch, M.C. Kelley, S. Collins, S. Baker, Q. Zhou, J.S. Friedman, First observation of meteoritic charged dust in the tropical mesosphere. Geophys. Res. Lett. **25**, 4047–4050 (1998)

O. Havnes, F. Sigernes, On the influence of background dust on radar scattering from meteor trails. J. Atmos. Solar-Terr. Phys. **67**, 659–664 (2005)

W.K. Hocking, Temperatures using radar-meteor decay times. Geophys. Res. Lett. **26**, 3297–3300 (1999)

W.K. Hocking, Experimental radar studies of anisotropic diffusion of high altitude meteor trails. Earth Moon Planets (2005). doi:10.1007/s11038-005-3446-5

W.K. Hocking, B. Fuller, Vandepeer, real-time determination of meteor-related parameters utilizing modern digital technology, B. J. Atmos. Solar-Terr. Phys. **63**, 155–169 (2001)

W.K. Hocking, W. Singer, J. Bremer, N.J. Mitchell, P. Batista, B. Clemesha, M. Donner, Meteor radar temperatures at multiple sites derived with SkiYMET radars and compared to OH, rocket and lidar measurements. J. Atmos. Solar-Terr. Phys. **66**, 585–593 (2004)

R. Latteck, W. Singer, S. Kirkwood, L. O. Jönsson, H. Eriksson, Observation of mesosphere summer echoes with calibrated VHF radars at latitudes between 54°N and 69°N in summer 2004. in *Proc. of the 17th ESA Symposium on European Rocket and Balloon Programmes and Related Research, Sandefjord, Norway, 30 May–2 June 2005, ESA SP-590* (2005), pp. 121–126

R. Latteck, W. Singer, R. J. Morris, D. A. Holdsworth, D. J. Murphy, Observation of polar mesosphere summer echoes with calibrated VHF radars at 69° in the Northern and Southern Hemisphere. Geophys. Res. Lett. **34**, L14805 (2007). doi:10.1029/2007GL030032

K.A. Lynch, L.J. Gelinas, M.C. Kelley, S. Collins, M. Widholm, D. Rau, E. MacDonald, Y. Liu, J. Ulwick, P. Mace, Multiple sounding rocket observations of charged dust in the polar winter mesosphere. J. Geophys. Res. **72**, A03302 (2005). doi:10.1029/2004JA010502

D.W.R. McKinley, *Meteor Sience and Engineering* (McGraw-Hill, New York, 1961), p. 309

M. Rapp, J. Hedin, I. Strelnikova, M. Friedrich, J. Gumbel, F.-J. Lübken, Observations of positively charged nanoparticles in the nighttime polar mesosphere, Geophys. Res. Lett. **32**, L23821 (2005). doi:10.1029/2005GL024676

W. Singer, J. Weiß U. von Zahn, Diurnal and annual variations of meteor rates at the arctic circle. Atmos. Chem. Phys. **4**, 1355–1363 (2004a)

W. Singer, J. Bremer, J. Weiss, W.K. Hocking, J. Höffner, M. Donner, P. Espy, J. Atmos. Solar-Terr. Phys. **66**, 607–616 (2004b)

J. Taubenheim, Statistische Auswertung geophysikalischer und meteorologischer Daten (Akademische Verlagsgesellschaft Geest und Portig K.-G., Leipzig, 1969)

Quantitative Comparison of a New Ab Initio Micrometeor Ablation Model with an Observationally Verifiable Standard Model

David D. Meisel · Csilla Szasz · Johan Kero

Originally published in the journal Earth, Moon, and Planets, Volume 102, Nos 1–4.
DOI: 10.1007/s11038-007-9222-y

Abstract The Arecibo UHF radar is able to detect the head-echos of micron-sized meteoroids up to velocities of 75 km/s over a height range of 80–140 km. Because of their small size there are many uncertainties involved in calculating their above atmosphere properties as needed for orbit determination. An ab initio model of meteor ablation has been devised that should work over the mass range 10^{-16} kg to 10^{-7} kg, but the faint end of this range cannot be observed by any other method and so direct verification is not possible. On the other hand, the EISCAT UHF radar system detects micrometeors in the high mass part of this range and its observations can be fit to a "standard" ablation model and calibrated to optical observations (Szasz et al. 2007). In this paper, we present a preliminary comparison of the two models, one observationally confirmable. Among the features of the ab initio model that are different from the "standard" model are: (1) uses the experimentally based low pressure vaporization theory of O'Hanlon (A users's guide to vacuum technology, 2003) for ablation, (2) uses velocity dependent functions fit from experimental data on heat transfer, luminosity and ionization efficiencies measured by Friichtenicht and Becker (NASA Special Publication 319: 53, 1973) for micron sized particles, (3) assumes a density and temperature dependence of the micrometeoroids and ablation product specific heats, (4) assumes a density and size dependent value for the thermal emissivity and (5) uses a unified synthesis of experimental data for the most important meteoroid elements and their oxides through least square fits (as functions of temperature, density, and/or melting point) of the tables of thermodynamic parameters given in Weast (CRC Handbook of Physics and Chemistry, 1984), Gray (American Institute of Physics Handbook, 1972), and Cox (Allen's Astrophysical Quantities 2000). This utilization of mostly experimentally determined data is the main reason for calling this an ab initio model and is made necessary by the fact that individual average meteoroid mass densities are now derivable from Arecibo observations.

D. D. Meisel
SUNY Geneseo, Geneseo, NY, USA

C. Szasz (✉) · J. Kero
Swedish Institute of Space Physics, Kiruna, Sweden
e-mail: csilla@irf.se

Keywords Ablation · Meteor · HPLA · Radar · AO

1 Introduction

It is essential for accurate determination of meteoroid orbits that extra-atmospheric values of mass, radius, density, and velocity be obtained. The problem is not trivial and there is a tendency in modern studies of meteors to avoid these great difficulties by either trying to obtain general analytical function corrections or even worse by not making any corrections at all! The correction problem is particularly acute for the highest velocity meteors that are candidates for extra-solar origin, but it is precisely this mass range where direct model verification by an independent observation method cannot be done for such small particles. Thus we have attempted a model-to-model comparison as an alternative calibration.

General problems of determining the correct ablation model have been extensively reviewed by Popova (2004) and will not be repeated here. Given the plethera of available ablation models, it may be asked why devise another? The main reason for reconsidering the modeling of micrometeoroids is observational. First, the smallest particles detected by the Arecibo UHF radar can often occur at lower altitudes than standard models predict. Secondly, for the highest speed micrometeoroids, standard model ablation temperatures are often much higher than melting and boiling points of most plausible meteoric materials would allow for evaporation from solid objects. Third, most models are not in a density dependent form suitable for direct interpretation of the Arecibo data.

The "standard" model to be used for comparison is described by Szasz et al. (2007) and will not be described again here. Suffice is to say that the type and form of differential equation (drag, mass loss, thermal balance, and path length), integration scheme (Fifth-order Runge-Kutta with step-size adjustment (Danby 1988), sputtering model (Tielens et al. 1994), and atmospheric model (MSIS-E-90: http://www.modelweb.gsfc.nasa.gov/models/msis.html) are identical. The initial heights, temperatures, masses, radii, density, geographic locations, inclination of the meteor path to the normal to the atmosphere, and dates were kept identical. Fragmentation is ignored in both models.

The basic physics of meteoroid entry modeling is described briefly and qualitatively as follows. As shown by Love and Brownlee (1991), the entry of very slow (<20 km/s) and small particles heat up very little and hence their motion is dominated by the classical drag equation that decelerates them until they slow down to terminal velocity. Thus the first differential equation is a momentum loss formula assuming a velocity squared drag term.

At higher velocities (>20 km/s), the drag heats the particle and the impinging air. The air and the meteoroid share the loss of kinetic energy so a second differential equation is added that describes the net heating of the meteoroid and in particular its change of temperature with time. The energy gain is modeled by integration of the drag equation to give a velocity cubed dependence.

As the temperature rises toward the meteoroid melting/boiling point, mass loss by thermal ablation occurs. Mass loss by sputtering does not usually show an experimental temperature dependence at low bombarding particle densities and so is only included in the mass loss equation.

The date chosen was September 22, 2005. The equations referenced to a curved Earth were used to get the heights above the geoid for the MSIS-E-90 model. The initial meteorid temperature assumed was 290 K at 500 km. The geographic location was Kiruna, Sweden. The models were run with identical initial conditions: (1) four values of mass, 3.3×10^{-8} kg,

1.4×10^{-11} kg, 5.2×10^{-13} kg, and 1.3×10^{-15} kg that are derived from four separate values each of radius and density (1, 5, 10, and 100 μm paired with 0.3, 1, 3.3, and 7.8 g/cc respectively, (2) the zenith distance of the normal to the atmosphere, 0° or 45° respectively, and (3) the above atmosphere velocity was 20, 40, and 60 km/s respectively.

2 The Differences Between the ab initio and Standard Models

The differences between the ab initio model and "standard" model are mainly in the use in the ab initio model of least squares fits of laboratory data rather than observational parameters derived from "large" mass meteor data. In particular, the ab initio model:

(1) uses laboratory vaporization data as a function of density as incorporated in the theory of O'Hanlon (2003) for ablation instead of the traditional Clausius-Clapyron approach;
(2) uses the experimental heat transfer, luminous efficiency, and ionization efficiency measured as functions of velocity by Friichtenicht and Becker (1973) for small particles (see comment below);
(3) assumes a mass density and temperature dependence of the micrometeoroid and ablation product specific heats;
(4) assumes a mass density and size dependent value for the thermal emissivity of the meteoroid as appropriate for small particles;
(5) uses a unified synthesis of experimental data for the most important meteoroid elements and their oxides through least square fits (as functions of temperature, density, and/or melting point) derived from the tables of thermodynamic parameters given in Weast (1984), Gray (1972), and Cox (2000).

A rather confusing point in small particle modeling concerns the role of the heat transfer coefficient (Popova 2004). In the free molecular flow regime, the transfer coefficient is supposed to be taken as unity and this was done in the standard model computations used here. Yet Friichtenicht and Becker (1973) found a velocity dependent experimental heat transfer coefficient in extensive accelerator experiments even for micron sized particles presumed to be in free molecular flow. It might be assumed that this was simply an effect of increasing ablation efficiency combined with radiative cooling at higher particle velocities. If that were the case there would be no reason to take either of these effects into account explicitly in the energy balance equation. However, we interpret the variable heat transfer coefficient of Friichtenicht and Becker as simply being a shift out of strict free molecular flow conditions resulting from intense ablation as discussed by Popova (2004). Thus in the ab initio model, the terms in question are retained in the energy balance equation but taking into account the likelihood that the average specific heat of the ablated atoms will depend on the composition of the meteoroid and in the context of the ab initio model, on the assumed mean mass density of the meteoroid. It should be noted that along with the adoption of the Friichtenicht and Becker heat transfer coefficient, the ab initio model uses the experimental results on the luminous and ionization efficiencies as well thus avoiding the need to consider in detail electron, ion, and photon production mechanisms utilized in the standard model. We will not discuss a comparison of these pending the outcome of the proposed photometric calibration of objects detected with the EISCAT radar (Szasz et al. (2007).

3 Comparisons, Summary, and Conclusions

In general, inclinations of the path simply raise the altitude where things happen by 3–5 km regardless of model and are not presented in more detail here. In Fig. 1, we give

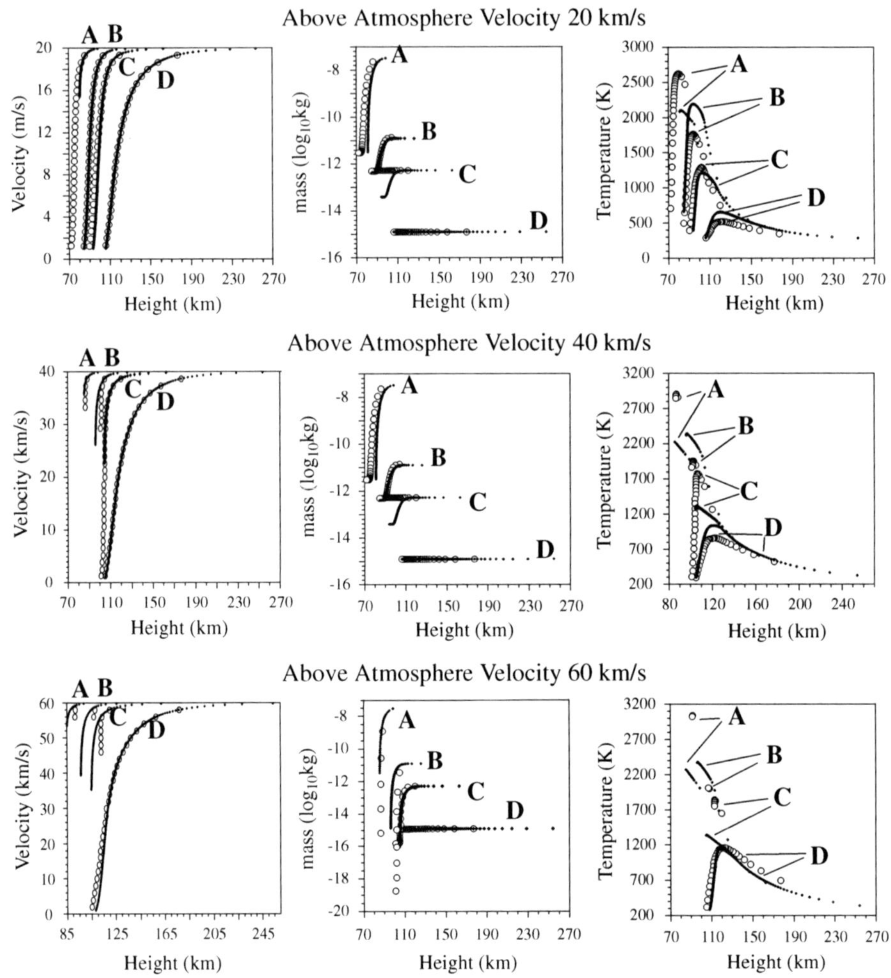

Fig. 1 Summary results for vertical incidence are shown for the sets of runs in the two models described. Three separate subgraphs are given for each above atmosphere velocity. The leftmost column is velocity (km/s) versus height (km), the middle column is for residual mass ($\log_{10}$ kg) versus height (km), and the rightmost one is temperature (K) versus height (km). The standard model points are plotted as open circles while the ab initio model points are plotted as small filled symbols. Results for three above atmosphere velocities are shown with 20 km/s at the top, 40 km/s in the middle, and 60 km/s at the bottom. Each subgraph shows the results for four masses $A = 3.3 \times 10^{-8}$ kg, $B = 1.4 \times 10^{-11}$ kg, $C = 5.2 \times 10^{-13}$ kg, and $D = 1.3 \times 10^{-15}$ kg. These correspond roughly to compositional classes iron, chondrite, average cometary, and porous cometary respectively. The temperature graphs are the most discordant, but the corresponding point sequences have been identified with two "arrows". The difference in the lengths of the sequences shown occur because the conditions for numerical integration termination are a bit different between the modeling programs and not for any significant numerical computation difference between the two computers used

results of calculations of vertical incidence paths. Three quantities—velocity (km/s), (remaining) mass, and meteoroid temperature—are plotted versus height with four masses ($A = 3.3 \times 10^{-8}$ kg, $B = 1.4 \times 10^{-11}$, $C = 5.2 \times 10{-}13$ kg, $D = 1.3 \times 10^{-15}$ kg) per graph and three different velocities (20, 40, and 60 km/s). The four masses have radii 100, 10, 5 and 1 μm appropriate to the assumed densities of 7.8, 3.3, 1.0, and 0.3 g/cc respectively. These densities correspond roughly to iron, chondritic, average cometary, and porous cometary compositions respectively. In the graphs, the ab initio results are plotted as sequences of small filled symbols while the standard model results are plotted as larger open circles.

Briefly the results in the graphs can be summarized as follows:

(1) The temperature versus height results do not agree regardless of velocity. This is doubtless due to the great differences in the treatment of the heat balance equation between the two models. At low velocities, the peak temperatures of either model do occur at comparable heights, but not so at the highest initial velocity.
(2) The velocity versus height relationships show good agreement at low initial velocities with a tendency of the standard model particles to decelerate faster than ab intio model particles, particularly at high initial velocity.
(3) The ablation (as indicated by the decreasing mass) versus height generally shows reasonable agreement in trends except for the C mass at low initial velocity where the standard model particles penetrates deeply while the ab initio particles are nearly completely ablated.

From this initial and preliminary comparison of the two models, it appears that the ab initio model will provide a useful extension of meteor ablation theory to the majority of the Arecibo detected micrometeoroids. We conclude that further two model comparisons will be useful to understand any comprehensive comparison between the EISCAT and Arecibo radar data.

References

A.N. Cox (ed.), *Allen's Astrophysical Quantities*, 4th edn. (AIP and Springer, 2000)

J.M.A. Danby, *Fundamentals of Celestial Mechanics*, 2nd. rev. and enlarged edn. (Willmann-Bell, 1988)

J.F. Friichtenicht, D.G. Becker, Determination of meteor parameters using laboratory simulation techniques. NASA Special Publication **319**, 53 (1973)

D.E. Gray (ed.), *American Institute of Physics Handbook*, 3rd edn. (McGraw-Hill, 1972)

S.G. Love, D.E. Brownlee, Heating and thermal transformation of micrometeoroids entering the Earth's atmosphere. Icarus **89**, 26–43 (1991)

J. O'Hanlon, *A Users's Guide to Vacuum Technology*. (John Wiley and Sons, New York, 2003)

O. Popova, Meteoroid ablation models. Earth Moon Planet. **95**, 303–319 (2004)

C. Szasz, J. Kero, D.D. Meisel, A. Pellinen-Wannberg, G. Wannberg, A. Westman, Estimated visual magnitudes of the EISCAT UHF meteors. Earth Moon Planet. (2007). doi:10.1007/s11038-007-9206-y

A.G.G.M. Tielens, C.F. McKee, C.G. Seab, D.J. Hollenbach, The physics of grain–grain collisions and gas-grain sputtering in interstellar shocks. Astrophys. J. **431**, 321–340 (1994)

R.C. Weast (ed.), *CRC Handbook of Physics and Chemistry*, 64th edn. (CRC Publishing, 1984)

Chapter 4. Meteoroid Parent Bodies and Impact Hazard

Meteoroids, Meteors, and the Near-Earth Object Impact Hazard

Clark R. Chapman

Originally published in the journal Earth, Moon, and Planets, Volume 102, Nos 1–4.
DOI: 10.1007/s11038-007-9219-6

Abstract In considering the modern-day hazard from infalling near-Earth asteroids and comets, the focus has shifted toward the smallest, most frequent impacts that can do damage on the ground, like the 1908 Tunguska aerial burst. There is considerable uncertainty about the potential for damage by objects in the range 20 to 100 m diameter. Since smaller, less dangerous, meter-sized meteoroids are part of a continuum of small interplanetary bodies, derived by a collisional cascade and Yarkovsky spin-up, research on such phenomena by meteor scientists can shed light on a vital question that will soon have great practical relevance as new telescopic searches for near-Earth asteroids come on line: what is the threshold size between harmless high-altitude airbursts and impacts that can cause lethal damage on the ground?

Keywords Asteroid · Impact hazard · Bolides · Meteoroids · Tunguska · Spaceguard Survey · Near-Earth asteroids

1 Introduction

The Earth has not only been bombarded by asteroids, comets, and their smaller pieces—meteoroids—over its history but continues to be struck today. During the last quarter-century, awareness has increased of the natural hazard posed by such cosmic projectiles. For a comprehensive review of the impact hazard, see Chapman (2004). The basic magnitude of the threat, in terms of time-averaged human fatalities in industrialized countries, is similar to that of individual kinds of natural disasters, such as hurricanes. However, as described by Chapman and Morrison (1994), by far the greatest fraction of the hazard resides in impacts by asteroids or comets larger than about 2 km diameter, where there is a significant risk of a sudden global climate crisis that could cause hundreds of millions of people or more to starve. Chapman and Morrison (1994) estimated the chances that an individual would die by near-Earth object (NEO) impact as 1-in-25,000. [NEOs are both comets and near-Earth asteroids

Based on an Invited Talk at the "Meteoroids 2007" conference in Barcelona, Spain, 15 June 2007.

C. R. Chapman (✉)
Southwest Research Institute, Suite 300, 1050 Walnut Street, Boulder, CO 80302, USA
e-mail: cchapman@boulder.swri.edu

DOI: 10.1007/978-0-387-78419-9_58

(NEAs) whose orbits pass close to or cross the Earth's orbit; comets are believed to be responsible for a very small fraction of the overall hazard.]

During the past decade, the Spaceguard Survey (http://impact.arc.nasa.gov/intro.cfm) has detected about three-quarters of NEAs >1 km diameter, none of which will impact Earth during the next century, resulting in decreased chances for a large impact during our lifetimes. Moreover, bias correction and other analyses of these telescopic surveys suggest fewer numbers of 100-m-scale impactors than had been previously estimated (more on this below). The overall result is that the chance of dying from an NEA impact has been reduced by at least an order-of-magnitude. Harris (pers. comm. 2007) now considers the chances of dying by impact to be only 1-in-720,000, in the range of death by fireworks accidents or amusement park rides.

Most of the risk reduction has resulted from finding those NEAs that would kill millions or billions of people, but which strike only every million years or less frequently, and showing that they will not collide with Earth during the next century. Most of the remaining hazard resides in much smaller NEAs, 50–300 m in diameter, which strike much more frequently and for which there is a substantial chance (exceeding 1%) of one striking during our lifetimes. It is the cross-over region between these smallest-but-still-dangerous impactors and the still smaller but brilliant bolides, caused by meteoroids meters to a few tens of meters in size and studied by meteor researchers, that I emphasize in this article.

There have been reports of doubtful credibility from antiquity, as well as more recent anecdotes, of death by meteorite falls. While such an accident is certainly possible, there has been no confirmed, credible report of a human being dying from a meteorite strike. A human being was injured by a meteorite in 1954 and a dog was reportedly killed by a fragment of the Martian meteorite Nakhla in 1911, though this has been questioned. Confirmed strikes on automobiles and roof-tops reflect the greater cross-sectional areas presented by these larger, common targets. The Science Definition Team (2003) study suggests that much of the remaining impact hazard resides in Tunguska-scale events that would plausibly kill hundreds to thousands of people. Tunguska was the impact that happened one century ago in Siberia, with an estimated yield of $\sim$15 MT (megatons of TNT equivalent); I discuss Tunguska in greater detail below. Also very dangerous, and little addressed so far by the telescopic surveys, are somewhat larger NEAs 200–500 m in diameter, which could cause a tsunami rivaling or exceeding the Indian Ocean tsunami of 2004. However, it is plausible that there would be adequate warning for most people to evacuate, restricting much of the damage to infrastructure only.

2 Emphasis on More Frequent but Smaller Impacts

Consideration of the dangers of impacts by relatively small NEAs now dominates discussion of the impact hazard. This is partly because the original Spaceguard Survey, designed to address primarily NEAs >1 km diameter, is approaching completion of its ten-year goal to find 90% of such very large NEAs. And it is partly because both the social sciences and practical politics teaches us that people are more concerned about potential catastrophes that are more likely to affect themselves or their children or grandchildren, as distinct from extremely rare and unlikely catastrophes, even if the latter are much more lethal. Especially given the fact that an NEA impact can be prevented, by means of a space mission that would "nudge" the NEA away from its impact trajectory, it is relevant in practical terms to consider how we might address a potentially lethal NEA impact that has a few percent chance of happening this century...or how we must deal with the even more likely very-near-misses, predictions of dangerous impacts with temporarily high probabilities, or megaton-scale impacts that may explode too high to be dangerous but which could frighten people or even be mistaken for a nuclear attack.

Consider the case of Apophis (see also Sansaturio and Arratia 2007). For a few days around Christmas 2004, this 250–300 m NEA was given an official probability (by the Jet Propulsion Laboratory Sentry system and by the Univ. of Pisa NEODyS system) of about 3% of impacting Earth on 13 April 2029. The places on Earth that were at risk of being struck were central Europe, the Middle East, and populous regions in Asia such as the Ganges river valley. About a month later, radar echoes received by the Arecibo radar refined knowledge of Apophis' position and removed any chance of collision in 2029, although Apophis will still pass below the geosynchronous artificial satellites and will be visible to the unaided eye as a 3rd magnitude star rapidly crossing the sky. (There remains a 1-in-45,000 chance that Apophis will pass through a resonant-return "keyhole" in 2029, so that it impacts Earth on 13 April 2036.) News about this 3% possibility of an impact that would have destroyed a whole country or caused a tsunami rivaling the 2004 Indian Ocean disaster was pushed aside by the holiday and then by news of the *actual* Indian Ocean tsunami. Still, future NEO scares are certain to happen during the next decades, even though an actual impact is quite unlikely.

It appears probable that a new survey (informally called the Spaceguard Two Survey, as a follow-on to the original survey) will commence shortly, with the goal of finding 90% of NEAs >140 m diameter within the next 15 years or so. A U.S. law passed by Congress and signed by the President in late 2005 mandates that NASA conduct such a survey, although in a March 2007 report to Congress (NASA 2007) NASA claimed that it lacked funds to carry out such a survey. The large, wide-field telescope projects required for such a survey, however, are already well underway, with some under construction, and one telescope—the first of four PanSTARRs instruments—about to become operational. It is plausible, even without substantial support from NASA, that the observing programs of these telescopes will conduct a non-optimized search for small NEAs and largely meet the Spaceguard Two goals by around 2025.

3 The NEO Size Distribution

The size distribution of projectiles striking the Earth has traditionally been divided into at least two size ranges: (a) asteroids generally >100 m diameter that can be readily discovered by Earth-based telescopes and (b) meteoroids generally <few meters in size and interplanetary dust, whose flux is estimated by various techniques in meteor science and by dust counters on spacecraft (cf. Zolensky et al. 2006). The gap between the brightest bolides and the faintest NEAs has narrowed in recent years, due in part to the comprehensive analysis of infrasound and downward-looking satellite data (cf. Brown et al. 2002) and by the ongoing telescopic surveys (Fig. 1; Harris 2008). There is also new data on the frequency of very small impacts by meteorite-sized bodies on the lunar surface (Ortiz et al. 2006). The data are least certain near 10 m diameter, due to low-number statistics and because of uncertain systematic errors in bias corrections, luminous efficiencies, etc.

Until recently, it has been assumed that the size distribution is roughly linear on a log–log plot; i.e. that a power-law with a constant index fits the data over a wide span of sizes. Brown et al. (2002) concluded that a constant index fits data from 5 to 200 m, and the Science Definition Team (2003) concluded that a constant index adequately (though

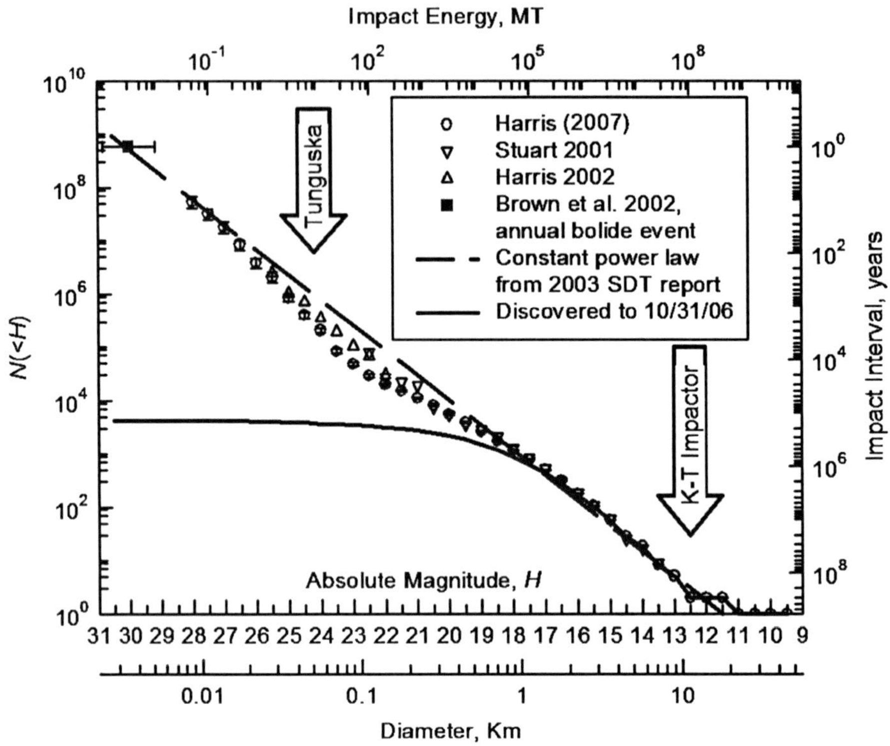

Fig. 1 This graph shows various estimates of the size vs. impact frequency of NEAs, including the most recent estimate of Harris (2008). Equivalent astronomical absolute magnitude and impact energy in megatons are shown. The solid curve shows the number actually known as of late 2006. Reproduced courtesy of Alan W. Harris (who retains the copyright)

not perfectly) fits data between about 3 and 10 km. The two indices are not exactly the same, however, indicating that the slope steepens somewhat toward smaller sizes; indeed, it steepens yet again at the smallest sizes (interplanetary dust; cf. Zolensky et al. 2006). In fact, it has long been known from the lunar cratering record (cf. Chapman and Haefner 1967) that the size-distribution of the projectile flux must be somewhat "wavy", at least averaged over the billions of years that the lunar surface has served as an impact counter. It has recently been re-recognized (cf. McEwen and Bierhaus 2006) that the dramatic steepening of the lunar crater size distribution at sizes <2 km diameter (made by projectiles <100 m diameter) is augmented significantly by secondary craters, as originally proposed by Shoemaker (1965); nevertheless, it is apparent that the NEA/meteoroid size distribution also steepens somewhat over this range.

Harris (2008) has recently emphasized that the previously uncertain deficit (relative to the constant power-law) of NEAs between 20 and 500 m diameter seems to be real and somewhat larger than previously estimated. The deficit is fully a factor of 3 below the power-law near 100 m diameter. This result contributes to a reduced hazard from Tunguska-sized impacts, as noted earlier. On the other hand, the reduction in frequency of 10 MT events to about 1 every 3,000 years seems to be increasingly incompatible with the fact that Tunguska itself struck only 100 years ago. (Indeed it struck on land and comparable explosions over the ocean might not have been recognized until recent decades; so there could have been more than one Tunguska-scale impact in the past couple hundred years.)

4 Tunguska and the Transition from Harmless Bolides to Dangerous Impacts

Most interpretations of phenomena associated with Tunguska place the energy in the 10–40 MT range, favoring lower values around 15 MT, making it about a 1-in-4,000 year event. However, Boslough and Crawford (1997) argue that the Tunguska devastation might have been caused by a 3 MT explosion, or about a 1-in-700 year event according to Fig. 1. Boslough (2007) has recently discussed reasons why the damage from such small impactors might be amplified beyond what previous calculations have shown. Nevertheless, to be a once-in-200 year event, an impact <1 MT is required, which would be caused by an impacting projectile (at typical velocities and with stony composition) only 20 to 25 m across (Fig. 1). Most calculations suggest that bodies that small explode far too high in the atmosphere to cause significant damage on the ground, let alone the devastation over 1,000s of sq. km represented by Tunguska. (One co-author of the Science Definition Team (2003) report has said that he would run *toward* such an impact, to observe it, rather than run away to escape what he believes would be a negligible chance of damage on the ground.) With a contrary perspective, Bland and Artemieva (2003) present calculations showing that atmospheric fragmentation of incoming small NEAs is more effective than previously modeled; they argue, opposite to Boslough, that ground damage is greatly reduced from previous estimates.

Of course, small metallic objects do penetrate the atmosphere, with only modest fragmentation, resulting in nearly all of the known impact craters on Earth <1 km diameter (e.g. the Henbury crater cluster in Australia). But only a few percent of NEAs, and a few percent of meteoroids striking the top of the Earth's atmosphere, are believed to have metallic strengths. If the fraction of metallic meteoroids several meters to several tens of meters in size is larger than has been estimated, then surface destruction would be greater than is currently being inferred from the NEA size distribution.

Currently, analyses of mortality from the impact hazard (e.g. Science Definition Team 2003) assume negligible effects from NEAs <50 m diameter. If dangerous effects, including mortality, were caused by NEAs only half as big (25 m), then dangerous events would occur almost a factor of ten more often. To put the issue in a practical context, what should the response be of national and international emergency management officials to a prediction that a 35 m NEA will strike a populated country a decade in the future? Following current interpretations, we would simply tell people near ground-zero to stay inside and not look directly at the high-altitude explosion. But if objects of that size could cause Tunguska-like damage, we might not only evacuate people for 100 km surrounding ground-zero but we would certainly consider a space mission to move or blow-up the threatening NEA. A major issue that must be thoroughly evaluated with high priority concerns the damage, on the Earth's surface, caused by explosions of various sizes at different altitudes in the Earth's atmosphere. Much of what we currently believe is based on bomb tests from nearly half-a-century ago (Glasstone and Dolan 1977). We now have tools to examine these matters that do not involve actual explosions in the real world. We must better understand the range of possible effects of large atmospheric explosions on the environment, on artificial structures, and on human beings.

It is because of all of these uncertainties that it is vital to learn about the numbers and natures of objects in the upper end of the range of bolides, studied chiefly by military assets, and in the lower end of the range of NEAs accessible to telescopes (optical, infrared, and radar). Since these transitional NEAs are part of a continuum—a collisional cascade spreading over a large range of sizes—the attributes of meter-sized objects are related to the larger, tens-of-meters sized objects in the transition region. Not only are many of the meter-size projectiles fragments of these larger objects, but their numbers—both in near-Earth space and in the main asteroid belt from which most of them come—govern the rates of catastrophic disruption and hence the numbers and impact frequencies of these larger objects. Hence the properties and numbers of the largest bolides studied by meteor specialists are directly relevant to establishing the hazardous effects of threshold-sized NEAs.

5 Related Issues and Conclusion

It is worth considering now what the implications will be of the Spaceguard Two Survey, described above, once it gets underway in the next few years and by the time it is concluded in the mid 2020s. The discovery rate for 10 m NEAs will go up by more than a thousand times! By the end of the search, even though it is focused on bodies >140 m, the search will have found more than half of the 50 m (Tunguska-sized?) NEAs. We will then be tracking 1–2 million 30 m bodies; even though impact damage *may* be small or negligible, any threatening NEA of such size will command attention. By the end of the survey, we should know the orbits of a quarter-million meteoroids 5 m in size: think of the implications for meteoroid researchers! At a minimum, we will be able to check and correct for the existing uncertainties in luminous efficiencies, bias corrections, etc. that affect both the telescopic data and interpretations of actual bolides, because the issue of small-number statistics will have been erased.

We should re-think the issue of the danger of meteorite falls—those events like Sikhote-Alin that shower the landscape with "rocks falling from the skies." In the past, the hazard from creation of such strewn fields has been negligible because of the much lower population density of human beings during past decades, centuries, and millennia. The population density of the world is now about seven times what it was in 1,800, when meteorites were first recognized as being rocks from interplanetary space. And it averaged about four times less than that during the previous millennium. So the chances of human fatalities from meteorite falls during the 21st century are greater than the cumulative chances during all of recorded human history. In fact, if Bland and Artemieva (2003) are correct in asserting that NEAs up to 200 m in size are generally fragmented in the atmosphere, then falls of tens-of-centimeter to ten-meter scale meteorites may be more common than previously estimated.

Our picture of the physical nature of small NEAs is rapidly changing. It is now realized that about a fifth of NEAs are binaries, or asteroids with satellites, and another fifth are probable contact binaries. How these statistics vary with size is uncertain. Almost all NEAs >200 m in diameter are believed to be "rubble piles", composed of multiple pieces held together loosely by gravity. Since Yarkovsky forces tend to spin up small NEAs, they are likely to disaggregate into their constituent pieces when their spin periods become shorter than about 2 h. The spins of most NEAs <100 m in size are so fast that they must be coherent monoliths, with appreciable tensile strength; perhaps many of them were previously constituents of rubble piles that have since come apart.

One dramatic case of a rapidly spinning NEA, exchanging mass with its nearby satellite, is 1999 KW4 (Scheeres et al. 2006). This developing picture of the physical nature of small NEAs is in rapid flux, and puzzles remain. For example, the astronomical evidence suggests that nearly all binary NEAs are either in contact or separated by just several radii from the larger body. Yet the frequency of double craters on the Earth and Venus (Cook et al. 2003) imply that about 15% of NEAs are much more widely separated binaries (the common close binaries would form a single crater). Obviously, observations pertaining to the properties of projectiles that cause large bolides, which may be fragments of either catastrophic collisional disruption of larger bodies *or* the disintegration of rubble piles under Yarkovsky spin-up, can potentially help us understand the processes that shape the physical properties of NEAs and meteorites. Are meteoroids that cause bolides also binary objects? Are apparently paired meteors telling us something about the forces affecting meteoroids as they approach the Earth? Is there correspondence between inferred properties of bolide-producing meteoroids and the astronomical evidence that about half of NEAs are either C-type asteroids or dormant comets, which are increasingly suspected of having very low bulk densities?

Finally, although the specialties of meteor science, meteoritics, and asteroid astronomy lie within the physical sciences, we must remember that these scientific specialties—like no others within astronomy besides research on the Sun and its influence on Earth—have very practical implications. Social science research has already predicted that the NEA hazard may have societal and political consequences beyond its "objective" impact (Slovic 2007). And the realities of the news media and political decision-making processes force us to acknowledge that an NEA impact (predicted or actual) with lethal consequences similar to the effects of a large earthquake, flood, or typhoon may stimulate psychological and political reactions far out-of-proportion to the fact that such an impact is roughly two-orders-of-magnitude less likely to happen than one of these familiar natural disasters. And even smaller, more frequent events—which are sure to happen during the next few decades—could have unfortunate consequences; for example, if an unusually large (but not unexpected) airburst were to occur over a war zone and be misinterpreted, thus triggering a nuclear response. The chances for such misinterpretation have been reduced since the scenario was first raised (Shoemaker 1983). But with the imminent frequent discovery of innumerable small NEAs during the next decade, as Spaceguard Two comes on line, it becomes vital to understand those smallest meteoroid impacts that can be lethal *or* that may be *perceived* as significantly threatening or dangerous.

References

P.A. Bland, N.A. Artemieva, Nature **424**, 288–291 (2003)

M. Boslough, presentation at 2007 planetary defense conference (2007), available via http://www.aero.org/conferences/planetarydefense/2007papers/S4-2-Boslough-Abstract.pdf. Accessed 14 Sept 2007

M.B.E. Boslough, D.A. Crawford, in *Near-Earth Objects: The United Nations International Conference*, vol. 822, ed. by J.L. Remo (Annals N.Y. Acad. Sci., 1997), pp. 236–282

P. Brown, R.E. Spalding, D.O. ReVelle, E. Tagliaferri, S.P. Worden, Nature **420**, 294–296 (2002)

C.R. Chapman, Earth Planet. Sci. Lett. **222**, 1–15 (2004)

C.R. Chapman, R.R. Haefner, J. Geophys. Res. **72**, 549–557 (1967)

C.R. Chapman, D. Morrison, Nature **367**, 33–40 (1994)

C.M. Cook, H.J. Melosh, W.F. Bottke, Icarus **165**, 90–100 (2003)

S. Glasstone, P.J. Dolan, *The Effects of Nuclear Weapons*, 3rd edn. (U.S. Govt. Printing Office, Washington, 1977), 644 pp, available via http://www.princeton.edu/~globsec/publications/effects/effects5.pdf. Accessed 14 Sept 2007

A.W. Harris, submitted to Icarus (2008)

A.S. McEwen, E.B. Bierhaus, Ann. Rev. Earth Planet. Sci. **34**, 535–567 (2006)

NASA (National Aeronautics and Space Administration), Near-Earth object survey and deflection, analysis of alternatives: report to congress, 28 pp (2003), available via http://www.nasa.gov/pdf/171331main_NEO_report_march07.pdf. Accessed 14 Sept 2007

J.L. Ortiz, F.J. Aceituno, J.A. Quesada, J. Aceituno, M. Fernández, P. Santo-Sanz, J.M. Trigo-Rodríguez, J. Llorca, F.J. Martín-Torres, P. Montañés-Rodríguez, E. Pallé, Icarus **184**, 319–326 (2006)
E. Sansaturio, O. Arratia, Earth Moon Planet., this issue (2007). doi:10.1007/s11038-007-9165-3
D.J. Scheeres et al., Dynamical configuration of binary near-Earth asteroid (66391) 1999 KW4. Science **314**, 1280–1283 (2006)
Science Definition Team, Study to determine the feasibility of extending the search for near-Earth objects to smaller limiting diameters, NASA Office of Space Science, Solar System Exploration Div., Washington, 154 pp, (2003), available via http://neo.jpl.nasa.gov/neo/neoreport030825.pdf. Accessed 14 Sept 2007
E.M. Shoemaker, in *Ranger VII, Part II, Experimenters' Analyses and Interpretations* (Tech. Rept. No. 32-700, Jet Propulsion Laboratory, Pasadena, 1965), pp. 75–134
E.M. Shoemaker, NASA workshop, collision of asteroids and comets with Earth: physical and human consequences (Snowmass, Colorado, 13–16 July 1981) (1983)
P. Slovic, in *Comet/Asteroid Impacts and Human Society*, eds. P. Bobrowsky, H. Rickman (Springer, Berlin, 2007), pp. 369–382
M. Zolensky, P. Bland, P. Brown, I. Halliday, in *Meteorites and the Early Solar System II*, eds. D.S. Lauretta, H.Y. McSween (Univ. Ariz. Press, Tucson, 2006), pp. 859–888

Apophis: the Story Behind the Scenes

María Eugenia Sansaturio · Oscar Arratia

Originally published in the journal Earth, Moon, and Planets, Volume 102, Nos 1–4.
DOI: 10.1007/s11038-007-9165-3

Abstract On December 20, 2004 the Minor Planet Center issued the Minor Planet Electronic Circular (MPEC) 2004-Y25 announcing the discovery of a new Near Earth Asteroid (NEA) with designation 2004 MN_4. Only two days later, when the Christmas holidays were about to begin, it was already apparent that this asteroid, currently known as Apophis, would be notorious: our close-approach monitoring system, CLOMON2, was already showing a Virtual Impactor (VI) in 2029 reaching the level 2 in the Torino Scale, the first asteroid to reach that level since our monitoring system had been operational. However, this was just the beginning of what it was to come in the subsequent days. In this paper we will give an overview of the NEODyS-CLOMON2 system and provide the details on how Apophis' collision scenario evolved, the way NEODyS' team handled it and the crazy 2004' Christmas holidays we had due to this unexpected guest.

Keywords Apophis · Near Earth Asteroids · Impact risk · Impact probability · Virtual asteroids · Virtual impactors · NEODyS

1 Introduction

The collision between the Earth and an asteroid can be described as an *extreme event*. On one hand, it is *extremely rare*: although tons of material enter the Earth's atmosphere on a daily basis, humankind has not yet witnessed the impact of a body belonging to the asteroid class. On the other hand, the effects of such an impact could be *extremely catastrophic*: the energy released in the process ranges from tens of Megatons (for 50-m bodies) to millions of Megatons (for bodies with several Kilometers in diameter), reaching anyway global consequences for asteroids with diameter equal or greater than 1 km.

M. E. Sansaturio · O. Arratia (✉)
E.T.S. de Ingenieros Industriales, University of Valladolid, Paseo del Cauce, Valladolid 47011, Spain
e-mail: oscarr@eis.uva.es

M. E. Sansaturio
e-mail: meusan@eis.uva.es

DOI: 10.1007/978-0-387-78419-9_59

The extreme character of the impacts between the Earth and the asteroids makes it difficult for the general public to understand the true nature of the problem we are facing. The fact that this type of collisions are *rare* does not mean that they are *impossible*. In fact, there are evidences both indirect (craters on the surfaces of the rocky celestial bodies, including the Earth) and direct (collision of the comet Shomaker–Levy 9 with Jupiter in 1994) showing that this kind of episodes have occurred in the past and will take place again in the future.

Among the population of asteroids only those with orbits close to that of the Earth, which are known as Near Earth Asteroids (NEAs), represent a real risk. There is a number of different measures that can be taken to mitigate this hazard. The most basic measures have a preventative nature and aim at cataloging the whole NEA population. A complete catalog of accurate orbits would allow us to know, well in advance, the asteroids that are on a collision course. The second type of measures, more selective and targeted for a particular threatening asteroid, include different deflection techniques to avoid its collision with the Earth. Finally, if the previous actions fail, it is still possible to implement a third kind of measures, such as population evacuations before the impact or construction of refuges near the shock area, to mitigate the effects of a certain collision. It is worth noting that the success of the measures of the second and third kind heavily relies on the time interval ranging from the publication of a certain impact to the impact itself. Obviously, those predictions can only be made when there is enough information available on the asteroids, hence the importance of the cataloging.

This paper is focused on the real case provided by the asteroid initially known as 2004 MN_4 and later numbered (99942) and named *Apophis*. According to recent measurements, this object has a diameter around 250 m and an estimated mass of 2.1×10^{10} kg, which is derived from an assumed density. The collision problem posed by this asteroid has not been fully solved yet. However, the largest threat raised by Apophis, a possible collision in the year 2029, was completely removed by using the first preventative method mentioned above. The collaborative work to rule out that impact provides a great example of the good practices needed to handle a delicate situation in which the information that becomes public must be carefully presented and commented to avoid unnecessary alarms.

The paper is organized as follows: in Sect. 2 we provide an introductory description of the NEODyS/CLOMON2 system. The risks scales used in assessing the hazard posed by an asteroid potentially colliding with the Earth are presented in Sect. 3. In the next Section we review the collision scenario for Apophis during the Christmas holidays in 2004 from the perspective of the NEODyS team. The last Section collects some concluding remarks that can be extracted from this experience.

2 The NEODyS/CLOMON2 System

Let us start by presenting a brief tour through the NEODyS/CLOMON2 system to help understand the context in which the computations for 2004 MN_4 were done.

NEODyS (Chesley and Milani 1999), which is an acronym for *Near Earth Objects Dynamic Site*, is a database containing all the orbital information concerning the currently known NEAs and was created in 1999, almost simultaneously with the NASA NEO Program Office at the Jet Propulsion Laboratory (JPL). It is accessible on the Internet at the following web addresses:

- http://newton.dm.unipi.it/cgi-bin/neodys/neoibo
- http://unicorn.eis.uva.es/cgi-bin/neodys/neoibo

The first corresponds to a server physically located in Italy, at the University of Pisa, while the second one is a duplicate maintained at the University of Valladolid in Spain. Both together make NEODyS a very reliable service, almost completely independent from possible computer failures, network disruptions or other technical events, due to its location in different parts of the world.

The system input is taken from the Minor Planet Electronic Circulars (MPECs), sent via e-mail by the Minor Planet Center (MPC). The data is processed by means of different perl and shell scripts and fortran programs (OrbFit[1]). Its main tasks are scheduled as follows:

- 10:00 am (CET): Update of the data, orbits and priority list,
- 12:00 am (CET): Update of the ephemerides,
- 13:00 am (CET): Impact monitoring.

Most of the system runs automatically, although some parts still require human intervention.

Through the web interface of the NEODyS system it is possible to access a large amount of information such as the list of all known NEAs, the list of observatory codes and their statistics, as well as a selected collection of related sites. Concerning a particular NEA, you will find all the orbital information including orbital elements with uncertainty, information about the orbital fit and a list of close approaches to the terrestrial planets and Jupiter, as well as to Ceres, Pallas and Vesta, from 1950 to 2100. It is important to bear in mind that all this information refers to the nominal solution, that is the best fit in the sense of leasts squares. On the other hand, NEODyS offers interesting services from the observer point of view: prediction of ephemerides and observations or tools for searching the database. However, what makes NEODyS different from other systems is what is known as *Risk Page*[2]. This service lists all the objects for which it has not yet been possible to discard the possibility of impacting the Earth until 2080. In reality, its content is not but the final result of the impact monitoring program CLOMON2, which was the last feature added to NEODyS at the end of 1999.

The CLOMON2 (Milani et al. 2005a) program consists of three stages:

1. Generation of multiple Virtual Asteroids (VAs) (Milani et al. 2005b), in our case 2401, that sample the confidence region.
2. Propagation of these VAs until 2080.
3. Analysis of the close approaches of all these propagated orbits.

On average, the run of CLOMON2 takes about 2–3 h per object, although for some complicated cases it can even last more than one day. The risk table posted on the web is, in fact, the output of the third stage and the meaning of the different columns can be found in the *help* available in the same page. Currently, CLOMON2 and Sentry[3] are the only two impact monitoring systems that operate in a systematic and automatic way on a daily basis. Regarding automation, it is important to note that the NEODyS/CLOMON2 system does not post any risk table automatically, but there is always human intervention. On the contrary, JPL/Sentry posts automatically the risk tables containing low risk Virtual Impactors (VIs), that is with Palermo Scale (PS) below –2 and zero Torino Scale (TS), while

---

[1] http://newton.dm.unipi.it/neodys/astinfo/orbfit/

[2] http://newton.dm.unipi.it/cgi-bin/neodys/neoibo?riskpage:0;main and http://unicorn.eis.uva.es/cgi-bin/neodys/neoibo?riskpage:0;main

[3] http://neo.jpl.nasa.gov/risk/

human intervention is required for the others. This is so because, according to the IAU standard, we have to undergo through the Technical Verification Procedure for every event with PS above -2 and we also consult for whatever is TS $>$ 0. Note that this verification is always slowed down by the local time differences.

3 Risk Scales

The risk scales just mentioned try to quantify in a simple way the risk associated with the possible collision of an asteroid with the Earth and its purpose is to serve as a mean of communication for the astronomers and the general public when evaluating the seriousness of the potential collisions.

In general terms, we can state that when evaluating the risk of an impact at least three factors have to be considered:

- on one hand it is obvious that the risk has to be an increasing function of the probability of the impact, in such a way that less probable impacts lead to low risks and viceversa.
- On the other hand, a risk scale must be sensitive to the energy released in the collision. This quantity is directly related to the size of the asteroid and its relative velocity with respect to our planet.
- Finally, we cannot neglect the time left till the predicted possible impact since, for instance, a well in advance prediction would allow us to adopt measures aimed at the mitigation of the possible effects of the threat.

The first risk scale to be introduced was the Torino Scale (Binzel 2000). It uses numbers from 0 to 10 in combination with colors and words to classify the impact risks (Fig. 1). This scale is discrete and the fact that it takes into account the third factor only in a binary way[4] makes it a bit inconvenient, so that a new scale was proposed: the Palermo Scale (Chesley et al. 2002). This scale is continuous, includes the three forementioned factors and is mainly used among professionals. Basically, the PS compares the destructive effect of a certain impact with that of the whole population of asteroids, both known and unknown.

Most of the asteroids posted in the Risk Pages of NEODyS and JPL are TS = 0 and PS < -2. The first notorious case was 2002 NT_7, since although it was TS = 1—but very close to the TS = 2 region, on July 23, 2002—it was the first asteroid to reach a positive PS value. 2002 NT_7 held this PS record until our protagonist 2004 MN_4 came into scene beating any previous record, both in the Torino and the Palermo scales.

4 The Apophis Case

On December 20, 2004 the MPC issued MPEC-Y25 (Gilmore et al. 2004) containing observations over three consecutive nights of a new *Potentially Hazardous Asteroid* (PHA). Figure 2 shows the relative position of this asteroid and the Earth when Gordon Garradd observed it from Siding Spring on December 18. However, it was not the first time this object had been observed. Indeed, it had been discovered on June 19, 2004 by Roy

[4] If the possible impact is to happen within 100 years, then the factor is unity; if it will happen later than 100 years from now, then it is zero

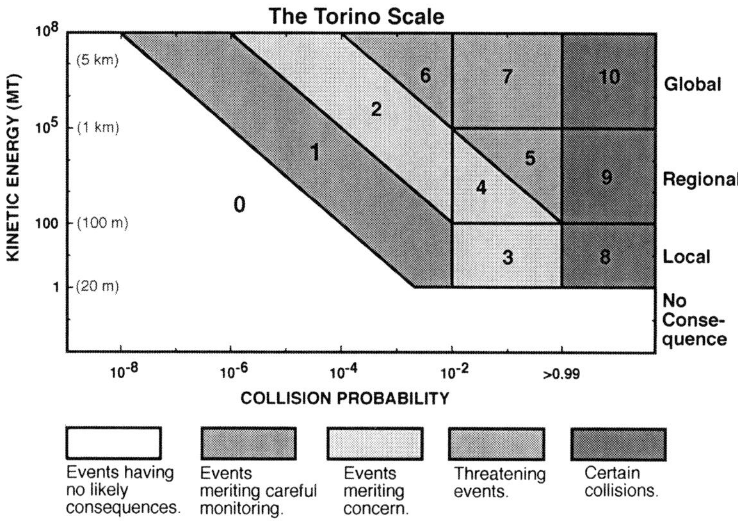

Fig. 1 A simple graphical tool to compute the number in the Torino Scale and hence the associated color and word to classify the risk from white (no risk) to red (certain impact) passing through green, yellow and orange

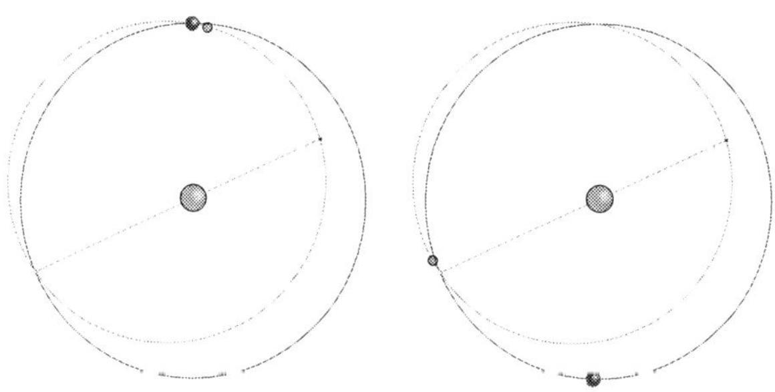

Fig. 2 Position of the Earth (dark sphere) and Apophis (light sphere) on their respective orbits when the asteroid was discovered on June 19, 2004 (right) and rediscovered on December 18, 2004 (left). In addition to the orbits, the plots also show the ascending (white) and descending (black) nodes of Apophis joined by the nodal line

Tucker, David Tholen and Fabrizio Bernardi from Kitt Peak (see also Fig. 2), who observed it over two consecutive nights by using an instrument never previously used to find asteroids.

As a result, the processing of the astrometry and photometry had some problems related to distortions in the large CCD mosaic and clock errors. Then, when the asteroid was serendipitously recovered by G. Garradd it was soon found to be the same as 2004 MN_4, but the fit was very poor for the June data as shown in Fig. 3.

When CLOMON2 ran on this data it found a VI in 2029 with a few 10^{-4} Impact Probability (IP) and TS = 1. In the afternoon of December 20 CET, we discussed the results with the JPL people. Both monitors were showing the VI in 2029 in a reasonable, though not perfect, agreement. In view of the poor fit (Fig. 3), we considered the possibility of impact in 2029 a dubious result and decided not to post it in our Risk Pages, waiting for new and/or corrected data, which we expected soon, since Steve Chesley,

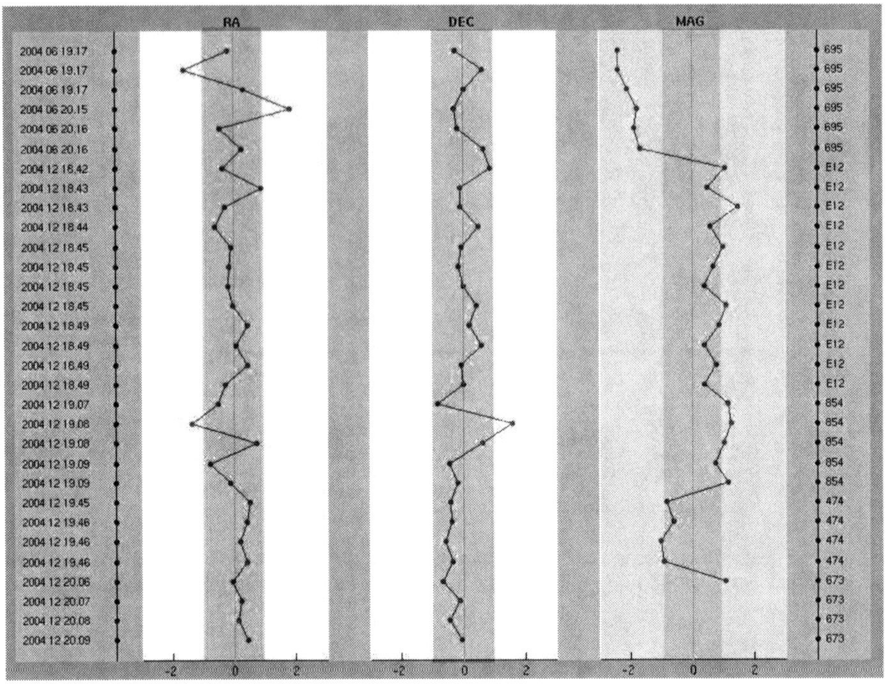

Fig. 3 Depiction of the residuals in right ascension (RA), declination (DEC) and magnitude (MAG) corresponding to the data available on December 20. In the case of RA and DEC the grey strip has a width of 2 arcsec, while for MAG the width is 2 magnitude units

Sentry's lead member, had contacted D. Tholen concerning the problematic June data and asked for new observations to a group of observers.

This call for observations had no response on December 21, but the following day four new observations from E12 were published and we received the remeasurements of the June 19 data. The improvement for these observations was clear (Fig. 4) but the fit of the "uncorrected" June 20 data became even worse. When we processed all the available data, CLOMON2 got even more worrying results than the previous day, the 2029 VI reaching TS = 2 and PS = +0.72. After consultation with the Sentry team, we decided to wait for the second half of Tholen's work to be completed.

On December 23, Tholen obtained accurate remeasurements for all the June Kitt Peak observations (Fig. 4), sent them to us and we processed them. The results of CLOMON2 were a little less bad than the previous day, but still at the TS = 2 and PS $>$ 0 level. The agreement between CLOMON2 and Sentry was good to the point that there was no way to cast into doubt the existence of the 2029 VI, although the quality and time distribution of the data was not at all what we would have liked. At that point it was decided to post the results in the Risk Pages simultaneously and including a note in each of the two web sites stressing that the situation was bound to change as new data became available.

At the end of the day, a few dozen additional observations came directly to us from Robert McNaught. They were incorporated to the orbital solution and, in the evening (CET), we posted an update of the results.

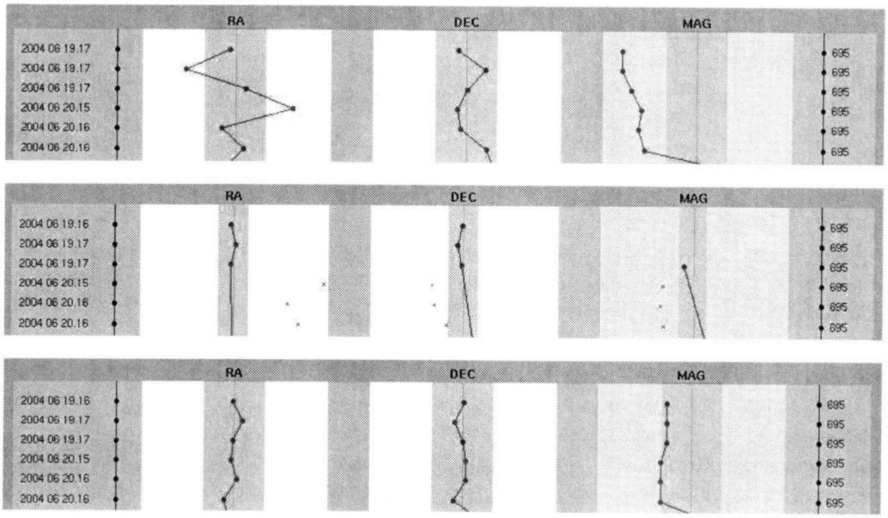

Fig. 4 Comparison of the residuals of the discovery observations of Apophis. The fit on December 20 is at the top. The middle plot represents the fit after correcting the observations of the first night. The fit at the bottom includes also the correction in the data of the second night. Note that the scale measuring the residuals has been shrunk in the second case to show the markers correspoding to outliers

On Christmas' Eve, the new available observations gave rise to new outputs of CLOMON2 with unprecedented TS = 4 and PS > +1. We posted the results near 20:00 CET and continued to make a call for observations.

During the following two days (25–26) new observations started to flow in. The MPC issued four special MPECs (three on the 25th and one on the 26th) for 2004 MN_4. Due to the seriousness of the case, we realized we needed to apply some weighting (Carpino et al. 2003) to the observations due to the large number of observations coming from some observatories. M. E. Sansaturio carried out those discussions with Chesley through e-mail, despite the holidays and the local time difference.

There is an important aspect to comment at this point: during those days the impact monitoring tasks involved not just running the existing computer programs and doing simple discussions on the residuals in order to get the best orbital solution. On the contrary, the NEODyS team was furiously changing all parts of the software which were not performing as expected because of the new features of the case. Certainly, 2004 MN_4 has proven to be a challenge for both monitoring systems.

CLOMON2 had a mistake in the definition of the Target Plane (TP) coordinates, irrelevant but for exceptionally low velocity at infinity and exceptionally high IP. Sentry also had some problems, but they appeared in late January 2005, when the first radar measurements were got, also because of some new features of the 2004 MN_4 encounter in 2029. Thus, the software was being changed on the fly and under a great preasure.

Finally, it is also relevant here to remember that, on December 26, a dramatic tsunami took place in Asia. This and the fact that we were immerse in the core of the Christmas holidays help to understand why this outstanding case hardly had media coverage.

On December 27, we got the maximum IP ever: the infamous 1 in 38 chances of impact. The high IP can be easily explained by looking at the situation in the Target Plane (Fig. 5).

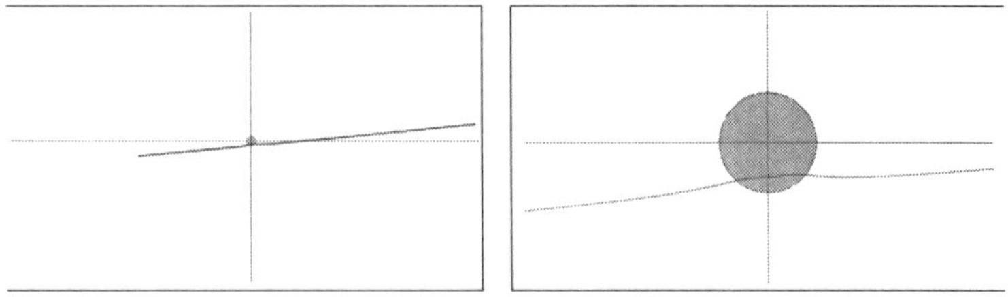

Fig. 5 The picture on the left shows the trail left on the TP by the set of VAs that generates the 2029 VI. The picture on the right shows a detail of the trail in the neighborhood of the Earth. In both plots the circle represents the impact cross section of the Earth once the gravitational focusing has been taken into account

On one hand, there is almost no dispersion of the Line Of Variations prior to the impact. The close approach shows all the 2401 VAs almost perfectly aligned. On the other hand, there are many VAs within a distance of 0.7 Earth radii.

This was the situation at 14:00 CET when CLOMON2 finished its run. However, the MPC issued 4 new special MPECs. In particular, MPEC-Y70 (Gleason et al. 2004) contained pre-discovery observations for 2004 MN_4 from observatory code 691 (Spacewatch), which extended the arc back to March 2004. In addition, almost at the same time, Tholen sent time corrections to the June 19 observations. We processed this new data and started a new run of CLOMON2 near 23:00 CET. The run finished at 1:37 December 28, the most outstanding result being that the 2029 VI had been ruled out, which brought an end to the "2004 MN_4 crisis".

Nevertheless, there is something to be noted about the March Spacewatch observations. In late January 2005, when the first radar data points were got for Apophis, it was apparent that these observations were not compatible with the radar data, the miss distance being adjusted by a factor two closer to Earth. According to the information provided by the Spacewatch team at the time, these images were found even though the ephemeris prediction plus trailing losses placed the asteroid well below the detection threshold for the frames. Chesley contacted the Spacewatch team to ask them to revisit the measurements, which they did and even had an external astrometric reduction (by the MPC staff Tim Spahr) done for verification. The corrected March observations provided by Spahr were then properly weighted and incorporated in the orbital solution.

However, it is important to point out that the problems exhibited by the March observations were only detected a posteriori when the orbital solution included the radar data points. Contrary to what happened with the discovery observations from Kitt Peak, for which the poor fit was evident from the very first day, the March observations showed no clear indication of their low quality in the fit. Thus, no action to "improve" them was taken when they were made available by the MPC. In any case, the corresponding computations lead to the conclusion that the radar observations obtained at the end of January 2005 would have ruled out the 2029 VI, irrespective of the March data weighting.

Finally, apart from the already discarded 2029 VI, there were some other low risk VIs out of which the one in 2036 has persisted till nowadays with a current impact probability of one in 45,000. If not before, it is expected (Chesley 2006) that new radar measurements in 2013 will exclude the whole set of VIs associated to this asteroid.

Table 1 Evolution of the 2029 VI of Apophis

Run #	CET	IP (%)	TS	PS
1	Dec 20 19:18	0.07	1	– 0.52
2	Dec 22 22:35	0.97	2	+0.72
3	Dec 23 18:43	0.6–1 in 170	2	+0.42
4	Dec 23 21:53	0.8–1 in 125	2	+0.65
5	Dec24 19:15	1.6–1 in 60	4	+1.01
6	Dec 25 15:25	2.3–1 in 43	4	+1.03
7	Dec25 21:55	2.2–1 in 45	4	+1.02
8	Dec 26 16:04	2.2–1 in 45	4	+0.98
9	Dec 27 14:00	2.6–1 in 38	4	+1.07
10	Dec 28 01:37	0	0	–

5 Conclusions

Table 1 provides a brief summary of the story of the 2029 VI of Apophis, showing the evolution of the risk published on the NEODyS Risk Page from December 20 to December 27, 2004. It exhibits the usual pattern followed by a large VI that is not a real impactor: tipically, obtaining the very first new observations reduces the size of the uncertainty region, although the VI is still there and therefore the IP grows. When the number (and the accuracy) of the observations is sufficiently large the confidence region becomes so small that the VI is excluded, finally dropping the IP to zero.

From a scientific point of view the Apophis case was really extraordinary: in only one week it generated more than 200 observations, more than 20 runs of CLOMON2 and hundreds of technical e-mails. To discard the possibility of the 2029 impact it was necesary to coordinate the specialized work of a lot of people, including both professionals and amateurs, distributed all around the globe. This is an excelent example of the good results that can be achieved with an efficient collaboration among the scientific community.

Finally, we want to stress that a great work has been done to assess the hazard posed by the NEA population and nowadays we know that it is really possible to put it under control. Also the public opinion is starting to be aware of the problem. It is now the responsibility of the governments to do the economic efforts to provide the scientific community with the necessary material means to complete the task.

Acknowledgements This work has been partially supported by: (a) the Spanish *Ministerio de Ciencia y Tecnología* and the European funds FEDER through the grant AYA2007-64592 and (b) the *Junta de Castilla y León* through the grant VA060A07. The authors would like to acknowledge the other people who were most involved in the work on Apophis during the days from Dec 20 to 27, 2004: A. Milani, G.B. Valssecchi and G. Tommei, in Italy; S.R. Chesley, D. Tholen and F. Bernardi in U.S.A.; G. Garrad and R. McNaught in Australia, as well as all the amateur observers, who greatly contributed to gather observations of Apophis and the MPC staff, who issued so many MPECs during that week. Without their contributions, the work described in this paper would not have been possible. Finally, we thank the referees (Clark Chapman and Andrea Milani) for their constructive comments.

References

R.P. Binzel, The torino impact hazard scale. Planet. Space Sci. **48**, 297 (2000)
M. Carpino et al., Error statistics of asteroid optical astrometric observations. Icarus **166**, 248 (2003)

S.R. Chesley, Potential impact detection for near-earth asteroids: the case of 99942 Apophis (2004 MN_4). In: *Asteroids, Cometes, Meteors—Proceedings IAU Symposium No. 229*, ed. by D. Lazzaro, S. Ferraz-Mello, J. A. Fernández (Cambridge University Press, Cambridge New York Port Melbourne Madrid Cape Town 2006) pp. 215–228

S.R. Chesley, A. Milani, NEODyS: an online information system for near-Earth objects. AAS/Division for Planetary Sciences Meeting Abstracts, vol. 31 (1999), pp. 28–34

S.R. Chesley et al., Quantifying the risk posed by potential earth impacts. Icarus **159**, 423 (2002)

A.C. Gilmore et al., 2004 MN4. Minor Planet Electr. Circ. **Y25**, 12 (2004)

A.E. Gleason et al., 2004 MN4. Minor Planet Electr. Circ. **Y70**, 70 (2004)

A. Milani et al., Nonlinear impact monitoring: line of variation searches for impactors. Icarus **173**, 362 (2005a)

A. Milani et al., Multiple solutions for asteroid orbits: Computational procedure and applications, A&A **431**, 729 (2005b)

What was the Volatile Composition of the Planetesimals that Formed the Earth?

Joseph A. Nuth III

Originally published in the journal Earth, Moon, and Planets, Volume 102, Nos 1–4.
DOI: 10.1007/s11038-007 9208-9 © US Government 2007

Abstract Is there an asteroid type or meteorite class that best exemplifies the materials that went into the Earth? Carbonaceous chondrites were once the objects of choice, and in the minds of many this choice is still valid. However, the origin of primitive chondritic meteorites is unclear. At the extremes they could either be fragments of very small parent bodies that never became hot enough to undergo geochemical modification other than mild lithification, or remnants of the uppermost layers of a body that had undergone a significant degree of internal differentiation, while the top layers remained cool due to radiative heat loss or loss of volatiles to space. This latter case is problematic if one considers these objects as precursors to the Earth since the timescale for the evolution of such a small body could be longer than the timescale for the accretion of the Earth. Large-scale circulation of materials in the primitive solar nebula could greatly increase the diversity of materials near 1 AU while also making the entire inner solar system both more homogeneous and much wetter than previously expected. The total mass of the nebula is an important, but poorly constrained factor controlling the growth of planetesimals. There is also a selection effect that dominates our sampling of the planetesimals that may have existed 4.5 billion years ago; namely, small fragile bodies are more likely to be lost from the system or ground down by collisions between small bodies, yet these are precisely those that may have dominated the population from which the Earth accreted. The composition of these aggregates could have played a very important role in the early chemical evolution of the Earth. In particular, the Earth may have been much wetter and richer in hydrocarbons and other reducing materials than previously suspected.

Keywords Accretion · Earth · Planetesimals · Water · Organics · Oceans · Volatiles

---

J. A. Nuth III (✉)
Astrochemistry Laboratory, Code 691 NASA's Goddard Space Flight Center,
Greenbelt, MD 20771, USA
e-mail: Joseph.A.Nuth@nasa.gov

1 Introduction

There are many reasons to worry about the chemical composition of the building blocks from which the terrestrial planets were assembled. The standard model is to postulate that the terrestrial planets formed from carbonaceous chondrites or even enstatite chondrites, yet this view has been questioned (Drake and Righter 2002; Righter et al. 2006; Righter 2007) on grounds that no known meteorite type appears to satisfy the observed compositional and isotopic constraints posed by the Earth's primitive upper mantle. From a geological perspective the chemical compositions of the Earth's core and lower mantle are important factors in understanding the workings of the deep interior of our planet. From an astrobiology perspective, the quantity of organic molecules and water that was accreted into the early Earth could have been a controlling factor in the origin of life, as these will have a major effect on the evolution of the oceans and on the oxidation state of the terrestrial atmosphere. These same considerations apply to the reconstruction of the early histories and evolution of Venus and Mars as well as to the potential that life might evolve in other protoplanetary systems. The origin of the terrestrial oceans is still a matter of debate, especially the relative contribution of water (Drake 2005) that may have outgassed from the Earth's interior compared to the quantity that may have been contributed by comets or even asteroids as a late arriving planetary veneer.

Unfortunately gathering an unbiased sample of the materials that accreted to form the Earth may be nearly impossible. The surface of the Earth has certainly been affected by planet building processes, including core formation, volcanic eruptions and chemical weathering to the point that nothing remains of the original building blocks. The great impact that formed the Moon certainly would have erased virtually any trace of the original material left at the Earth's surface. We also know that the surface of the Earth was changed significantly during the Late Heavy Bombardment (Levison et al. 2001, 2006) that peaked approximately 3.9 billion years ago and emplaced more than 20,000 craters with diameters greater than 20 km and more than 40 impact basins with diameters greater than 1000 km on the surface of the Moon. Because of the greater gravitational cross section of the Earth one should expect much larger fluxes of impactors on Earth than on the Moon. These planetesimals impacting the Earth and the moon evolved for more than 600 million years. Over this time period considerable chemical change could have occurred, especially to the concentrations of the more volatile elements.

In what follows I will examine factors responsible for the evolution of the planetesimals from which the Earth formed, and the likely timescales for their evolution based on a range of reasonable assumptions. I will also examine a generalized scenario for planetesimal accretion based on the comet accretion scenario of Weidenschilling (1997) and will then look to the current meteorite population as well as to their parent bodies to examine any bias(es) that might preclude the use of this data set as a starting point in modeling the formation and evolution of the Earth. Finally I will present the most likely candidate for a generalized model for the average primitive body aggregated into the growing Earth and discuss the implications of this choice for its early history, especially the oxidation state of the early atmosphere, the origin of the oceans and the conditions leading to the origin of life.

2 Thermally Driven Evolution of Planetesimals

There are many factors that influence the thermal evolution of a planetesimal ranging from the relative concentrations of radioactive elements and water, to proximity to the heat and magnetic field of the sun, to the sheer size of the body itself. I will briefly examine each of these factors below and will conclude that if all other factors are roughly equal, then a larger (~ 1000 km) body (protoplanet) will evolve much more rapidly than a smaller (~ 10 km) body (planetesimal).

2.1 Radioactive Heating

Although potassium, thorium and uranium decay are now responsible for heating the interior of the Earth, the short lived radioactive elements ^{26}Al and ^{60}Fe most likely provided the energy that drove geological processing of the earliest stages of planetary history (Chabot and Haack 2006). While ^{60}Fe can only be produced in a supernova explosion, ^{26}Al and other short-lived elements can be made in the winds of massive stars as well as by spallation reactions due to irradiation of heavier elements by protons and alpha particles produced in an intense, active, early sun (Gounelle et al. 2001). There is considerable controversy over the sun's birthplace. Was the sun formed in a quiescent environment such as is found in the modern Taurus Auriga star-forming region (Hartmann et al. 1998; 2001), or as was more recently suggested did the sun form in the presence of one or more massive stars such as is seen in the Orion Nebula (Hester et al. 2004)?

In the traditional view the solar nebula would be reasonably homogeneous and all planetesimals would contain roughly the same proportion of all of the radioactive elements. In such a star-forming region the concentration of short-lived spallation-produced nuclides might have increased sporadically due to solar flares, but most of this material would remain concentrated near the protostar with limited but steady, outward transport (Boss 2004; Ciesla 2007). In such a scenario the earliest formed planetesimals could experience slightly less radioactive heating than later-formed bodies since they would contain fewer radioactive nuclei and might therefore evolve a bit more slowly.

In a star formation region such as the Orion Nebula, the nebula could have been seeded with ^{26}Al over time by nearby stellar winds before being injected with ^{60}Fe due to one or more supernova explosions (Wadhwa et al. 2006). In regions of massive star formation where short-lived radioactive elements could be injected well after some planetesimals had already formed, those formed after the nebula was seeded with short-lived radioactive nuclei could evolve much more rapidly than the earliest formed bodies. A complicating factor is the distribution of the radioactive elements, that is, if early formed planetesimals accreted large fractions of calcium-aluminum rich inclusions (CAIs), and thus more ^{26}Al, then even small (5 km radius) bodies could rapidly (10^6 years) melt (Das and Srinivasan 2007).

2.2 Water

The temperature in the solar nebula decreased with increasing heliocentric distance, which causes a snow line where water condenses to ice at some distance from the sun. Most planetesimals that begin to accrete outside the snowline will contain some water as ice grains become trapped together with silicates and other more refractory materials within these kilometer-scale bodies. Only those bodies that accreted completely inside the snowline might be devoid of ice or water. Such planetesimals could still contain substantial quantities of hydrated silicates and it is possible that dust grains inside the snow line could carry substantial quantities of adsorbed water (Stimpfl et al. 2006a, b). There is likely to be a gradient in water content in planetesimals orbiting in the region of the nebula where the

Earth formed (Ciesla and Cuzzi 2005, 2006) ranging from some small percentage of hydrous silicates nearest the sun to substantial quantities of internal ice in the outermost regions of the asteroid belt. If the orbital eccentricities of these bodies are excited by the formation of a giant planet (Wetherill 1985, 1996) this gradient could become disturbed due to the penetration of ice-bearing planetesimals deep into the innermost nebula. Yet it is unlikely that this gradient could be completely erased. If all other factors are equal, lower water content should lead to more rapid heating of a primitive planetesimal, since the same amount of radioactive decay will lead to higher internal temperatures without the moderating effects of melting rocks (dissolved water lowers the melting point of magmas), subliming ices or the escape of steam from the hot interior.

2.3 Heating Mechanisms

2.3.1 Solar Heating

The protosun heats the innermost regions of the nebular accretion disk to sufficiently high temperatures (>1500 K) to evaporate silicates and other refractory grains (Woolum and Cassen 1999). Directly heated regions will naturally conduct some fraction of this energy into the interior of the disk where it will add to the gravitational energy dissipated during infall from the molecular cloud. This heating leads to a compositional gradient of small dust grains in the nebula ranging from highly refractory calcium and aluminum bearing minerals nearest the sun, to silicates, and even ice-coated silicates and ices farther out in the disk. One might assume that the larger planetesimals (>10 km) would be effectively immune from solar heating, after the dissipation of the gaseous disk, except for the net energy conducted from their surfaces due to the emission of infrared radiation or the conduction from the tenuous residual nebular gas.

2.3.2 Magnetic Induction Heating

A more viable way to heat planetesimals is magnetic induction. The large magnetic flares observed by the Chandra Mission (Feigelson et al. 2006) and especially by the COUP project (Feigelson et al. 2005) imply the presence of a very strong magnetic field associated with young stars. If a planetesimal is conductive, then the Hall effect will heat it to some degree; e.g., moving a conductor in the presence of a magnetic field induces electrical current flow and the resistance of the medium leads to power dissipation—and heating—within the planetesimal (Sonett and Colburn 1968; Sonett et al. 1968). Nearly all primitive meteorites and chondritic aggregate interplanetary dust particles (for a petrologic review see Rietmeijer 1998) contain metals and sulfides, which are very good natural conductors. Melting ice could lead to the dissolution of some mineral species that would greatly enhance the electrical conductivity of some meteorite parent bodies and leave as evidence veins of aqueous minerals, mostly various salts, that chemically precipitated from hot evaporating fluids: Such reactions could also destroy metal and sulfide particles, leading to local decreases in heating. In bodies free of ice, heating of carbonaceous grain coatings could lead to the deposition of graphitic material at grain boundaries, possibly forming large conductive networks that would also lead to more efficient magnetic induction heating of the interior of the body. However, such a mechanism generally favors heating the outer portions of most asteroids.

2.4 Planetesimal Mass

Despite the potential importance of the factors discussed above, the most important single consideration controlling the loss of volatiles from a planetesimal is undoubtedly its size. A larger body will contain more radioactive elements per unit of surface area, will release more energy from each accretional impactor, will have a larger gravitational cross-section, thus attracting impactors at a higher rate, and will also release internal energy more slowly to space due to a smaller surface to volume ratio. Once the planetesimal grows sufficiently large (>1000 km), geological forces such as core-mantle formation based on density will release gravitational energy to drive volatile loss. From this point of view, the larger a planetesimal grows, the faster it changes from an undifferentiated planetesimal to a fully differentiated planet (Kleine et al. 2002). Therefore an important question to consider is "Do all planetesimals accrete and grow at the same rate?"

3 Planetesimal Accretion

One of the most reasonable models of planetesimal accretion, starting from dust grains, was published by Weidenschilling (1997) and based on the assumption of a minimum mass solar nebula. The minimum mass nebula is the lowest mass from which the solar system could possibly form and it is highly probable that our nebula was more massive (Weidenschilling 1997). In this model individual dust and ice grains aggregate slowly because they are closely tied to the ambient gas, and grain diffusion via Brownian motion is the dominant cause of collisions among these grains. As the aggregates increase in size, and depending on their fractal dimension (Meakin and Donn 1988), they slowly begin to decouple from the gas and gently spiral inward, collecting more grains and aggregates via increasingly energetic collisions. In this model comets begin to accrete at ~100–200 AU and complete their growth to ~10–15 km scale bodies near 5–10 AU. At this size they orbit independent of the ambient solar nebula gas.

Nebular mass (and the higher pressure of gas in the nebula) affects the scale of the zone of growth for a planetesimal, the timescale for planetesimal growth and determines the smallest planetesimal size of a given density that will orbit the sun unaffected by the drag of the ambient gas. Yet a higher nebular mass does not change the general picture described above. At higher nebular pressures, growth occurs faster, the distance traveled during the growth process is smaller and the final planetesimal will be larger, denser, or both. Yet the vast majority of planetesimals could start accumulating well outside the snowline (Lunine 2006; Ciesla and Charnley 2006) and should thus contain some quantity of water and organics. Planetesimals in a massive nebula would not travel as far and might be devoid of water if they started aggregating well within the snowline. Although planetesimals in a higher mass nebula would accrete grains and collide with similar sized aggregates at slightly higher velocities, they would still be only moderately compacted bodies that had yet to experience significant internal changes via typical geological processes (Donn 1991, and references therein).

While the vast majority of planetesimals remain small (kilometer scale) a few stochastically "runaway cores" quickly grow to planetary scale bodies many hundreds to thousands of kilometers in radius (Wetherill and Stewart 1989, 1993). These runaway cores grow at the expense of the existing planetesimal population; e.g. they grow by accretion of unprocessed planetesimals, mostly by virtue of their ever-increasing gravitational cross-sections. As they grow larger, they evolve faster; differentiating a core, mantle and crust, outgassing volatiles to form an atmosphere and possibly a hydrosphere as at least some of these volatiles should condense around the largest of these bodies.

4 Implications for the Chemical Composition of the Earth

If mass is the single most important factor controlling the heating of a planetesimal, and if models of planetary formation are correct about the prevalence of runaway accretion in aggregation of the terrestrial planets, then the Earth accreted largely from planetesimals that had just barely begun to get warm due to radioactive heating and had no chance to evolve into the "dead" parent bodies of modern meteorites (asteroids). Some questions logically arise. What differences exist between the initial unaltered planetesimals that accreted into the runaway Earth and the rest of that same population of planetesimals that evolved due to internal heating and volatile loss for several hundred million years? What differences exist between a population of initially unaltered planetesimals that had evolved for several hundred million years and the modern asteroid population from which our meteorite collection is derived?

4.1 Changes on Half-Billion Year Timescales

As small planetesimals slowly heat due to the decay of short lived radioactive elements (e.g. La Tourette and Wasserburg 1998), water ice and clathrates melt or sublime, lost to the vacuum of a nebula that has already lost most of its gas. Depending on the ratio of rock to ice, and the composition and character of the dust, heating could leave planetesimals almost completely hydrated, including intercalated water in clay minerals. At lower water to rock ratios (less than 1–1), all internal water could be lost with enough radioactive energy left over to melt large fractions of the interior. At high water to rock ratios some small degree of gravitational and collisional settling and consolidation occurs, as water is lost from the interior, yet blocky aggregates could become cemented together due to hydrous alteration. Such processes could leave large voids in the interior of the planetesimal. At water to rock ratios so low that the higher radiation concentration causes silicates to melt, the surface tension of the melt will lead to compaction and, upon cooling, the planetesimal will be solid throughout. This leads to chemical and textural differences in planetesimals on timescales of roughly 500 million years. Depending on the concentration of radioactive elements, planetesimal size and the radius of its orbit, most ten kilometer scale planetesimals should stop increasing in temperature within much less than 500 million years (Ghosh et al. 2006). Further change in such small bodies is driven more by physical processes.

Planetary bodies—the products of runaway accretion—evolve on faster timescales than the ten kilometer scale planetesimals from which they are derived. The earth was fully-grown within 10 million years and core formation occurred within 30 million years of nebular collapse (Jacobsen 2003; Nichols 2006). Total melting within even water-free 10-kilometer scale planetesimals may not have occurred for several tens of million years (Huss et al. 2006), though much shorter timescales are possible if the concentration of ^{26}Al is enhanced (Das and Srinivasan 2007). Planetesimals accreting into growing planetary embryos were much less evolved than the embryos themselves.

The accretion of planetary bodies is much more energetic than similar processes in kilometer scale aggregates, especially when two large bodies collide, such as during the moon forming event (Cameron and Benz 1991; Canup and Asphaug 2001; Canup 2004).

Such processes can result in the loss of all volatiles from the surfaces of both bodies and to degassing of the mantle of the largest body. It is likely that the moon formed from material that had been dispersed as small particles around the Earth at very high temperatures, and was thus completely degassed (Jacobsen 2005). The fraction of the proto-Earth that may have been degassed during this collision has yet to be well determined, though mantle depth magmas can hold up to 3% water by mass (Righter 2007).

4.2 Changes on Longer Timescales

Roughly 3.9 billion years ago a Lunar Cataclysm or Late-Heavy Bombardment was responsible for creating several impact basins and more than 20,000 craters on the moon (Hartmann et al. 2000; Levison et al. 2001, 2006). The much higher mass of the Earth, and the greater gravitational cross section for impactors due to gravitational focusing, would bring a much higher impactor flux to the surface of the Earth than was experienced by the Moon. If the Late Heavy Bombardment was triggered by the rearrangement of the outer solar system (Gomes et al. 2004, 2005) then impactors would be a mixture of unprocessed comets and evolved asteroids, neither of which would be representative of the building blocks of the Earth. Comets rarely migrated to the inner solar system during the planetary accretion phase, being effectively stopped by Jupiter, Saturn, Uranus and Neptune. The asteroids had already evolved according to their size, internal complement of radioactive elements and proximity to the sun; they were considerably different from the primitive, volatile-rich aggregates available less than 30 million years after nebular collapse.

The current asteroid belt has evolved considerably over the last 3.5–4.0 billion years of relative stability throughout the rest of the system. Change is driven primarily by resonant orbital interactions with giant planets, especially Jupiter. Resonant interactions increase orbital eccentricity causing asteroids to interact with others in nearby orbits. Even slight changes in stable systems can lead to catastrophic collisions that disrupt kilometer scale objects. It is easier to disrupt a fluffy, loosely bound, aggregate of hydrous silicate than to break stony iron rocks. All asteroids are gradually worn away, but the volatile rich, loosely aggregated, less cohesive bodies are destroyed more rapidly.

Meteorites in today's collection arrived on Earth within the last few hundred thousand years, after having spent a few tens to hundreds of millions of years in slowly evolving orbits after ejection from parent asteroids via collisions. Our collection is a sample of the modern asteroid population, and may not be complete, let alone representative. The most volatile rich and least structurally sound asteroids were long ago ground into dust and lost from the system. Those that remain are much more cohesive. There are also severe selection effects in the terrestrial atmosphere where ram pressure disrupts low sheer strength objects (Chyba et al. 1994).

Unless a volatile rich bolide is seen to fall (such as Tagish Lake), it may be overlooked as a meteorite. Even if such an object were collected in the field, unless special precautions were taken, volatiles could be lost due to melting and sublimation (e.g. Brown et al. 2000). The true fraction of volatile rich asteroids is therefore likely to be under-represented in our meteorite collection.

4.3 Chemical Implications

From the arguments discussed above, no modern meteorite type is likely to be representative of the primary building blocks of the Earth (Drake and Righter 2002). While the refractory content of CI chondrites provides a reasonable starting point to estimate the chemical composition of the core, mantle and crust, even such a primitive specimen will not contain sufficient water and volatile carbon to match the majority of planetesimals available 4.5 billion years ago. Thermally driven processing of modern asteroids and loss of the most fragile and primitive specimens to collisional disruption over the lifetime of the solar system severely biases our perceptions of the volatile content of the bodies that went into the Earth. Even the mild heating experienced by ordinary chondrites of metamorphic grade 3.6 has been shown to result in loss of all pre-solar carbonaceous grains (Huss 1990; Huss and Lewis 1994, 1995) due to reactions between diamond, graphite and silicon carbide and the oxide components that dominate meteorites. If such robust forms of carbon can be destroyed then more reactive materials, possibly in closer contact with oxide grain surfaces (e.g. organic grain coatings) would also be lost at low temperatures (400 K) on billion year timescales.

The Earth accreted from much wetter, more carbon-rich planetesimals than any existing in our modern meteorite collections. During the Moon forming event internal water may have been lost from a proto-Earth but it is not clear that all water would have been lost from the deep mantle despite the catastrophic nature of such a collision. Water could have maintained oxidized iron in the Earth's crust and upper mantle, releasing large quantities of hydrogen that would migrate into the primitive and very hot, SiO-rich atmosphere. Hydrocarbons trapped in the terrestrial interior could have reacted with silicates to produce large amounts of CO, or might have escaped directly into the super heated atmosphere where reduced species would be stabilized by large quantities of hydrogen escaping from the interior. Escaping CO and hydrogen could have reacted at high pressures and temperatures on grain surfaces via Fischer-Tropsch type reactions to form simple hydrocarbons. Escaping nitrogen could also react to form ammonia via the analogous Haber-Bosch process. The net consequence of such a scenario is formation of a methane-, ammonia-, and water-rich atmosphere as postulated by Miller (1953) and Miller and Urey (1959) in experiments to understand chemical evolution leading to the origin of life on Earth. This atmosphere could persist as long as hydrogen, CO and hydrocarbons were escaping from the terrestrial interior. Loss of hydrogen from the top of the atmosphere combined with the attainment of a quasi steady state rate of exchange between the atmosphere, crust and upper mantle would gradually lead to the evolution of a less reducing system.

There has been speculation that comets may have provided sufficient water to account for the terrestrial oceans. This assumes that the Earth formed from CI chondrites. If the Earth accreted from primitive bodies that were wetter and more hydrocarbon rich than CIs, then this presents an interesting question, If a large impactor had *not* collided with the proto-Earth, thus drying out a considerable fraction of the material that became its crust and upper mantle, would the surface of the Earth have been completely covered by oceans? Would dry land exist on Earth today? Consequences for the natural history of the Earth and for the evolution of communicative civilizations on our planet are significant.

5 Summary

Models for the accretion of planetesimals tens of kilometers in diameter predict that they will grow from a wide feeding zone that could span the snowline in the primitive solar nebula. Our understanding of the evolution of isolated planetesimals of this size predicts that timescales on the order of tens to hundreds of millions of years are required to produce asteroids from which modern meteorites are derived, unless they contain enhanced levels of ^{26}Al. In contrast, the Earth accreted very rapidly, on the order of from ten to twenty million years maximum, as constrained from the time inferred for terrestrial core formation. The Earth must have therefore accreted from extremely primitive small bodies, many of which could have contained significant fractions of water and organics.

Heating of these primitive planetesimals leads to loss of water and volatile organics within the first few hundred million years. Remnants of the most volatile rich bodies would evolve into loosely bound aggregates of clay minerals and could contain large voids. Collisional evolution of the asteroid population over the last 3–4 billion years selectively destroys fragile bodies while preserving those that evolve into rocks. Atmospheric ram pressures experienced when bolides enter the atmosphere easily disrupt fragile meteoroids with high concentrations of organic materials, ices, or both. Their fragments are both unlikely to survive long on the surface of the Earth, or be identified as meteorites. We are forced to conclude that not only does our modern meteorite collection not contain samples of the primitive planetesimals that accreted to form the Earth, since they long ago evolved to much drier and less organic rich materials, but that it is unlikely that even the desiccated remnants of this population have been collected.

An exception could be the very primitive Orgueil carbonaceous chondrite meteorite that was a witnessed fall (Gounelle et al. 2006). Specimens that were subsequently recovered were still warm and could be cut with a knife and, when sharpened, could be used like a pencil, suggesting that many of its most volatile components were indeed lost during the bolide phase and during its very short terrestrial residence. Therefore no meteorite type adequately represents the chemical composition of the planetesimals that accreted to form the Earth. The intensifying research efforts on all aspects of meteors by traditional ground-based observers plus quantitative chemical data from spectral observations may lead to understanding of the fundamental differences between the meteorite database and the chemical properties of less-evolved material from different reservoirs in the solar nebula.

Acknowledgements I would like to thank NASA's Cosmochemistry Program for its generous support of my research program. I would also like to acknowledge the many helpful comments from Drs. Rhian Jones and Kevin Righter that greatly improved this manuscript, even though following all of their suggestions would have nearly doubled its original length.

References

A.P. Boss, Evolution of the solar nebula. VI. Mixing and transport of isotopic heterogeneity. Astrophys. J. **616**, 1265–1277 (2004)

P.G. Brown, A.R. Hildebrand, M.E. Zolensky, M. Grady et al., The fall, recovery, orbit, and composition of the Tagish Lake meteorite: a new type of carbonaceous chondrite. Science **290**, 320–325 (2000)

A.G.W. Cameron, W. Benz, The origin of the moon and the single impact hypothesis. IV. Icarus **92**, 204–216 (1991)

R.M. Canup, Simulations of a late lunar forming impact. Icarus **168**, 433–456 (2004)

R.M. Canup, E. Asphaug, Origin of the Moon in a giant impact near the end of the Earth's formation. Nature **412**, 708–712 (2001)

N.L. Chabot H. Haack, Evolution of asteroidal cores, in *Meteorites and the Early Solar System II*, ed. by D.S. Lauretta, H.Y. McSween (University of Arizona Press, Tucson, 2006), pp. 747–741

C.F. Chyba, T.C. Owen, W.-H. Ip, Impact delivery of volatiles and organic molecules to earth, in *Hazards Due to Comets and Asteroids*, ed. by T. Gehrels (University of Arizona Press, Tucson, 1994), pp. 9–58

F.J. Ciesla, Outward transport of high temperature materials around the midplane of the solar nebula. Science **318**, 613–615 (2007)

F.J. Ciesla, S.B. Charnley, The physics and chemistry of nebular evolution, in *Meteorites and the Early Solar System II*, ed by D.S. Lauretta, H.Y. McSween (University of Arizona Press, Tucson, 2006), pp. 209–230

F.J. Ciesla, J.N. Cuzzi, The distribution of water in a viscous protoplanetary disk (abstract), in *Lunar and Planetary Science*, 36, #1479 (CD-ROM) (Lunar and Planetary Institute, Houston, Texas, 2005)
F.J. Ciesla, J.N. Cuzzi, The evolution of the water distribution in a viscous protoplanetary disk. Icarus **181**, 178–204 (2006)
A. Das, G. Srinivisan, Rapid melting of planetesimals due to radioactive decay of Al-26: a case study of planetary bodies with variable aluminum abundance (abstract). Lunar Planetary Science Conference 38, #2370, Lunar and Planetary Institute, Houston, Texas, 2007
B. Donn, The accumulation and structure of comets, in *Comets in the Post Halley Era*, ed by R.L. Newburn, M. Neugebauer, J. Rahe (Kluwer, Dordrecht, The Netherlands, 1991), pp. 335–359
M.J. Drake, Origin of water in the terrestrial planets. Meteorit. Planet. Sci. **40**, 519–527 (2005)
M.J. Drake, K. Righter, Determining the composition of the earth. Nature **416**, 39–44 (2002)
E.D. Feigelson, K. Getman, L. Townsley, G. Garmire, T. Preibisch, N. Grosso, T. Montmerle, A. Muench, M. McCaughrean, Global X-Ray properties of the Orion Nebula region. Astrophys. J. Suppl. **160**, 379–389 (2005)
E. Feigelson, L. Townsley, M. Gudel, K. Stassun, X-ray properties of young stars and stellar clusters, in *Protostars and Planets V*, ed by B. Reipurth, D. Jewett, K. Keil (University of Arizona Press, Tucson, 2006), pp. 313–328
R.S. Gomes, A. Morbidelli, H.F. Levison, Planetary migration in a planetesimal disk: why did Neptune stop at 30 AU? Icarus **170**, 492–507 (2004)
R.S. Gomes, H.F. Levison, K. Tsiganis, A. Morbidelli, Origin of the cataclysmic Late Heavy Bombardment period of the terrestrial planets. Nature **435**, 466–469 (2005)
A. Ghosh, S.J. Weidenschilling, H.Y. McSween, A. Rubin, Asteroidal heating and thermal stratification of the asteroid belt, in *Meteorites and the Early Solar System II* ed by D.S. Lauretta, H.Y. McSween (University of Arizona Press, Tucson, 2006), pp. 555–566
M. Gounelle, F.H. Shu, H. Shang, A.E. Glassgold, E.K. Rehm, T. Lee, Extinct radioactivities and protosolar cosmic-rays: Self-shielding and light elements. Astrophys. J. **548**, 1051–1070 (2001)
M. Gounelle, O. Spurný, P.A. Bland, The orbit and atmospheric trajectory of the Orgueil meteorite from historical records. Meteorit. Planet. Sci. **41**, 135–150 (2006)
L. Hartmann, N. Calvet, E. Gullbring, P. D'Alessio, Accretion and the evolution of T Tauri disks. Astrophys. J. **495**, 385–400 (1998)
L. Hartmann, J. Ballesteros-Paredes, E.A. Bergin, Rapid formation of molecular clouds and stars in the solar neighborhood. Astrophys. J. **562**, 852–868 (2001)
W.K. Hartmann, G. Ryder, L. Dones, D. Grinspoon, The time-dependent intense bombardment of the primordial Earth/Moon system, in *Origin of the Earth and Moon*, ed by R.M. Canup, K. Righter (University of Arizona Press, Tucson, 2000), pp. 493–512
J.J. Hester, S.J. Desch, K.R. Healy, L.A. Leshin, The cradle of the solar system. Science **304**, 1116–1117 (2004)
G.R. Huss, Ubiquitous interstellar diamond and SiC in primitive chondrites: abundances reflect metamorphism. Nature **347**, 159–162 (1990)
G.R. Huss, R.S. Lewis, Noble gases in presolar diamonds II: component abundances reflect thermal processing. Meteoritics **29**, 811–829 (1994)
G.R. Huss, R.S. Lewis, Presolar diamond, SiC, and graphite in primitive chondrites: abundances as a function of meteorite class and petrologic type. Geochim. Cosmochim. Acta. **59**, 115–160 (1995)
G. R. Huss, A. E. Rubin, J.N. Grossman, Thermal metamorphism in chondrites, in *Meteorites and the Early Solar System II*, ed by D.S. Lauretta, H.Y. McSween (University of Arizona Press, Tucson 2006), pp. 567–586
S. Jacobsen, How old is planet earth? Science **300**, 1513–1514 (2003)
S.B. Jacobsen, The Hf-W isotopic system and the origin of the Earth and Moon. Annu. Rev. Earth Planet. Sci. **33**, 531–570 (2005)
T. Kleine, C. Münker, K. Metzger, H. Palme, Rapid accretion and early core formation on asteroids and the terrestrial planets from Hf-W chronometry. Nature **418**, 952–955 (2002)
T. La Tourette, G.J. Wasserburg, Mg diffusion in anorthite: implications for the formation of early solar system planetesimals. Earth Planet. Sci. Lett. **158**, 91–108 (1998)
H.F. Levison, L. Dones, C.R. Chapman, S.A. Stern, M.J. Duncan, K. Zahnle, Could the lunar "Late Heavy Bombardment" have been triggered by the formation of Uranus and Neptune? Icarus **151**, 286–306 (2001)
H.F. Levison, A. Morbidelli, R. Gomes, D. Backman, Planet migration in planetesimal disks, in *Protostars and Planets V*, ed by B. Reipurth, D. Jewett, K. Keil (University of Arizona Press, Tucson, 2006), pp. 669–684

J.I. Lunine, Origin of water ice in the solar system, in *Meteorites and the Early Solar System II*, ed by D.S. Lauretta, H.Y. McSween (University of Arizona Press, Tucson, 2006), pp. 309–219

P. Meakin, B. Donn, Aerodynamic properties of fractal grains: implications for the primordial solar nebula. Astrophys. J. **329**, L39–L41 (1988)

S.L. Miller, A production of amino acids under possible primitive Earth conditions. Science **117**, 528–529 (1953)

S.L. Miller, H.C. Urey, Organic compound synthesis on the primitive earth. Science **130**, 245–251 (1959)

R.H. Nichols, Chronological constraints on planetesimal accretion, in *Meteorites and the Early Solar System II*, ed by D.S. Lauretta, H.Y. McSween (University of Arizona Press, Tucson, 2006), pp. 463–472

F.J.M. Rietmeijer, Interplanetary dust particles, in *Planetary Materials, Reviews in Mineralogy*, vol. 36, ed by J.J. Papike (Mineralogical Society of America, Chantilly, Virginia, 1998), pp. 2-1–2-95

K. Righter, Not so rare Earth? New developments in understanding the origin of the Earth and Moon. Chemie der Erde. **67**, 179–200 (2007)

K. Righter, M.J. Drake, E. Scott, Compositional relationships between meteorites and terrestrial planets, in *Meteorites and the Early Solar System II*, ed by D.S. Lauretta, H.Y. McSween (University of Arizona Press, Tucson 2006), pp. 803–828

C.P. Sonett, D.S. Colburn, The principle of solar wind induced planetary dynamos. Phys. Earth Planet. Inter. **1**, 326–346 (1968)

C.P. Sonett, D.S. Colburn, K. Schwartz, Electrical heating of meteorite parent bodies and planets by dynamo induction from a premain sequence T Tauri "solar wind". Nature **219**, 924–926 (1968)

M. Stimpfl, M.J. Drake, N.H. de Leeuw, P. Deymeier, A.M. Walker, Effect of composition on adsorption of water on perfect olivine (abstract). Geochim. Cosmochim. Acta. **70**, A615 (2006a)

M. Stimpfl, A.M. Walker, M.J. Drake, N.H. de Leeuw, P. Deymier, An angstrom-sized window on the origin of water in the inner solar system: atomistic simulation of adsorption of water on olivine. J. Cryst. Growth. **294**, 83–95 (2006b)

M. Wadhwa, Y. Amelin, A.M. Davis, G.W. Lugmair, B. Meyer, M. Gounelle, S.J. Desch, From dust to planetesimals: implications for the solar protoplanetary disk from short-lived radionuclides, in *Protostars and Planets V*, ed by B. Reipurth, D. Jewett, K. Keil (University of Arizona Press, Tucson, 2006), pp. 835–848

S.R. Weidenschilling, The origin of comets in the solar nebula: A unified model. Icarus **127**, 290–306 (1997)

G.W. Wetherill, Occurrence of giant impacts during the growth of the terrestrial planets Science **228**, 877–879 (1985)

G.W. Wetherill, The formation and habitability of extra-solar planets. Icarus **119**, 219–238 (1996)

G.W. Wetherill, G.R. Stewart, Accumulation of a swarm of small planetesimals. Icarus **77**, 330–357 (1989)

G.W. Wetherill, G.R. Stewart, Formation of planetary embryos—effects of fragmentation, low relative velocity, and independent variation of eccentricity and inclination. Icarus **106**, 190 (1993)

D.S. Woolum, P. Cassen, Astronomical constraints on nebular temperatures: implications for planetesimal formation. Meteorit. Planet. Sci. **34**, 897–907 (1999)

Physical, Chemical, and Mineralogical Properties of Comet 81P/Wild 2 Particles Collected by Stardust

George James Flynn

Originally published in the journal Earth, Moon, and Planets, Volume 102, Nos 1–4.
DOI: 10.1007/s11038-007-9214-y © Springer Science+Business Media B.V. 2007

Abstract NASA's Stardust spacecraft collected dust from the coma of Comet 81P/Wild 2 by impact into aerogel capture cells or into Al-foils. The first direct, laboratory measurement of the physical, chemical, and mineralogical properties of cometary dust grains ranging from $<10^{-15}$ to $\sim 10^{-4}$ g were made on this dust. Deposition of material along the entry tracks in aerogel and the presence of compound craters in the Al-foils both indicate that many of the Wild 2 particles in the size range sampled by Stardust are weakly bound aggregates of a diverse range of minerals. Mineralogical characterization of fragments extracted from tracks indicates that most tracks were dominated by olivine, low-Ca pyroxene, or Fe-sulfides, although one track was dominated by refractory minerals similar to Ca–Al inclusions in primitive meteorites. Minor mineral phases, including Cu–Fe-sulfide, Fe–Zn-sulfide, carbonate and metal oxides, were found along some tracks. The high degree of variability of the element/Fe ratios for S, Ca, Ti, Cr, Mn, Ni, Cu, Zn, and Ga among the 23 tracks from aerogel capture cells analyzed during Stardust Preliminary Examination is consistent with the mineralogical variability. This indicates Wild 2 particles have widely varying compositions at the largest size analyzed (>10 μm). Because Stardust collected particles from several jets, sampling material from different regions of the interior of Wild 2, these particles are expected to be representative of the non-volatile component of the comet over the size range sampled. Thus, the stream of particles associated with Comet Wild 2 contains individual grains of diverse elemental and mineralogical compositions, some rich in Fe and S, some in Mg, and others in Ca and Al. The mean refractory element abundance pattern in the Wild 2 particles that were examined is consistent with the CI meteorite pattern for Mg, Si, Cr, Fe, and Ni to 35%, and for Ca, Ti and Mn to 60%, but S/Si and Fe/Si both show a statistically significant depletion from the CI values and the moderately volatile elements Cu, Zn, Ga are enriched relative to CI. This elemental abundance pattern is similar to that in anhydrous, porous interplanetary dust particles (IDPs), suggesting that, if Wild 2 dust preserves the original composition of the Solar Nebula, the anhydrous, porous IDPs, not the CI meteorites, may best reflect the Solar

G. J. Flynn (✉)
Department of Physics, State University of New York – Plattsburgh, 101 Broad St., Plattsburgh, NY 12901, USA
e-mail: george.Flynn@plattsburgh.edu

Nebula abundances. This might be tested by elemental composition measurements on cometary meteors.

Keywords CI meteorites · Wild 2 · Comets · Interplanetary dust particles · IDPs · Solar Nebula · Meteor streams

1 Introduction

Particles emitted by comets are believed to be the same dust that accreted, along with ices, as the Solar System was forming, ~4.57 billion years ago. This dust has been held in "cold storage," experiencing minimal thermal or aqueous processing, in the comets since their formation. Thus, comets are generally thought to have preserved a record of the processes and conditions in the Solar Nebula at the time of Solar System formation. This record is not available in bodies such as meteorites that experienced significant aqueous or thermal metamorphism since their formation. Thus the detailed characterization of cometary dust can contribute to the understanding of the processes and conditions in the Solar Nebula at the time of dust formation (Rietmeijer 2005).

In 1986 two VEGA spacecraft and the Giotto spacecraft flew through the dust coma of Comet 1P/Halley. Instruments on these spacecraft provided information on the size-frequency distribution and elemental composition of small particles, generally from 5×10^{-12} to 5×10^{-17} g. The elemental composition of an estimated 0.5 ng of material was determined by the VEGA instruments (Fomenkova et al. 1992). The properties of the grains analyzed at comet Halley are summarized by Jessberger et al. (1988), Mukhin et al. (1991), and Schulze et al. (1997).

NASA's Stardust spacecraft flew through the dust coma of Comet 81P/Wild 2 on January 2, 2004, collecting samples of Wild 2 dust by impact at ~6.1 km/s into low-density silica aerogel that was specially fabricated for the mission and into Al-foil (Tsou et al. 2004; Brownlee et al. 2006). About 3×10^{-4} g of Wild 2 material is estimated to have impacted the Stardust collector during the encounter (Hörz et al. 2006). These Wild 2 samples were delivered to Earth on January 15, 2006.

A small fraction of the Wild 2 samples collected by Stardust were studied during the Preliminary Examination. Results were reported by six teams of investigators who focused on craters (Hörz et al. 2006), organics (Sandford et al. 2006), isotopes (McKeegan et al. 2006), infrared spectroscopy (Keller et al. 2006), elemental composition (Flynn et al. 2006), and mineralogy/petrology (Zolensky et al. 2006). Rietmeijer (2008) has compared results obtained on the Wild 2 particles with the results obtained on meteors.

The dust trail from Wild 2 does not intercept the Earth, so Wild 2 does not produce a meteor stream in the Earth's atmosphere. Nonetheless, the physical, mineralogical, and chemical properties of the Wild 2 particles collected by the Stardust spacecraft should provide indications of the properties of the particles in meteor streams from other comets. Conversely, the analysis of cometary meteors can indicate if the properties of Wild 2 particles are typical for cometary material, and can resolve some of the questions raised by the Wild 2 analyses.

2 Wild 2 Particles Collected by Stardust

Long-exposure images taken by the Stardust spacecraft during the encounter (Fig. 1) show that Wild 2 is an active comet, with at least 20 areas on the surface emitting highly

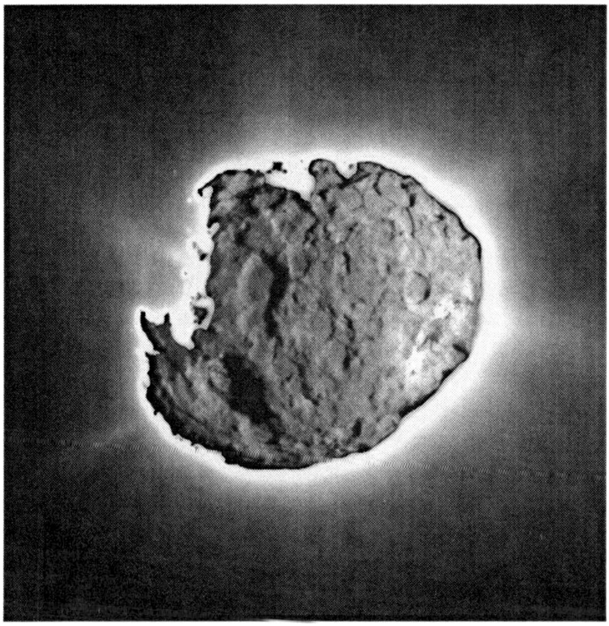

Fig. 1 Composite image of Wild 2, with a short exposure showing the surface detail overlain on a long exposure taken just 10 s later showing the jets taken during the close-approach phase of Stardust's January 2, 2004 flyby (NASA photo)

collimated jets of gas and dust (Brownlee et al. 2004). These jets are believed to originate in pockets of gas and dust in the interior of the comet. The Dust Flux Monitor Instrument, an active dust counter carried on Stardust, indicated that the spacecraft passed through several intense swarms of particles (Tuzzolino et al. 2004), some of which Sekanina et al. (2004) associated with specific jets seen in the images. These results indicate that Wild 2 particles collected in the aerogel and Al-foil flown on Stardust spacecraft sample several different source regions in the interior of the comet. Thus, the collected particles are believed to be representative of the non-volatile component of Wild 2 over the size range that was sampled during the encounter.

During the Preliminary Examination, the Elemental Composition team mapped the distributions of several major and minor elements, including S, Ca, Ti, Cr, Mn, Fe, Ni, Cu, Zn, and Ga along 23 tracks from the Stardust aerogel capture cells using Synchrotron X-Ray Microprobes (SXRMs) and determined the elemental compositions of residues deposited in 7 craters in Stardust Al-foils using Energy Dispersive X-ray Analysis in Scanning Electron Microscopes and Time of Flight-Secondary Ion Mass Spectrometry. The procedures and results are described in detail by Flynn et al. (2006). The Mineralogy/Petrology team determined the mineralogies of fragments extracted from 52 tracks from the Stardust aerogel capture cells using Synchrotron X-Ray Diffraction, Transmission Electron Microscopy, and X-Ray Absorption Near-Edge Structure spectroscopy, described by Zolensky et al. (2006). The Cratering team inferred the mineralogies of particles in craters based on residue compositions, described by Hörz et al. (2006). Although only a small fraction of the total collected mass of Wild 2 material was characterized during the Preliminary Examination, important insights into the physical, chemical, and mineralogical properties of Wild 2 particles were obtained.

3 Particle Sizes

The Stardust aerogel capture cells have a total collection area of 1,039 cm^2 and the Al-foils have a total collection area of 152 cm^2 (Hörz et al. 2006). Particles striking the foils produced elongated cavities called "tracks" as they decelerated in the aerogel. Particles impacting onto the Al-foil produced craters, many of which contained residue from the impactor.

The 6.1 km/s collection velocity can be duplicated in laboratory light-gas guns. A variety of projectiles were shot into Stardust flight-spare aerogel to determine the relationship between track dimensions in the aerogel and incident particle masses and into Al-foil to determine the relationship between crater dimensions and incident particle masses. These calibration experiments indicate that the smallest craters in the Stardust Al-foils were produced by Wild 2 particles having masses less than 10^{-15} g, while the largest track observed in the Stardust aerogel corresponds to an incident particle having a mass of $\sim 10^{-4}$ g (Hörz et al. 2006).

The mass–frequency distribution of the particles that impacted the Stardust collectors, derived from the plot of the cumulative mass–frequency distribution given by Hörz et al. (2006), is shown in Fig. 2. The Wild 2 coma, as sampled by the Stardust spacecraft, is dominated by the largest particles that were collected ($>10^{-5}$ g), and most of the mass collected by Stardust at Wild 2 was in a few particles, each >100 μm in size (Hörz et al. 2006). A similar mass–frequency distribution was reported for particles impacting the Long Duration Exposure Facility (LDEF) (Love and Brownlee 1993). For LDEF impacts, the mass flux increased from the smallest mass measured (10^{-9} g) up to a particle mass of $\sim 10^{-5}$ g then decreased for the largest mass they observed, $\sim 10^{-4}$ g (Love and Brownlee 1993).

4 Physical Properties

Control experiments demonstrate that if a single crystalline grain larger than a few micrometers in size is captured in aerogel at ~6 km/s, the resulting damage track is conical, with most of the mass of the incident particle remaining as a single, relatively unaltered particle at the end of the track. The SXRM mapping of the Wild 2 tracks extracted from Stardust aerogel capture cells showed that most of the 23 tracks mapped

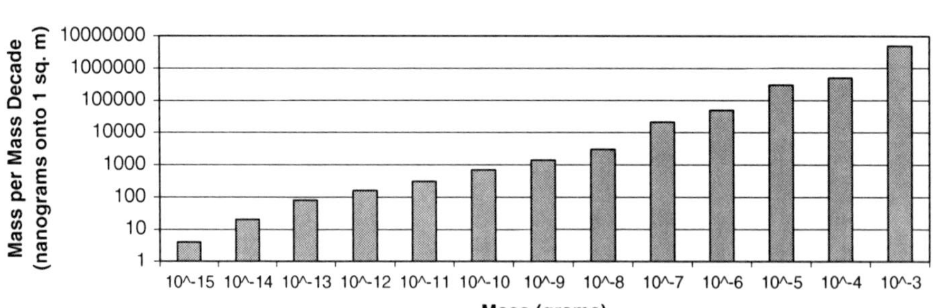

Fig. 2 The mass per mass decade incident on the Stardust collector, scaled to a 1 m^2 area, derived from Fig. 4 of Hörz et al. (2006). The sharp increase in the mass contributed by the largest particles indicates that the five particles >100 μm contain most of the mass colleted at Wild 2, and that, unless they have exceptional compositions, the smallest particles cannot significantly perturb the mean composition

during the Preliminary Examination have large amounts of material distributed along the walls of the track. Some of the tracks have large bulbs, which may result from severe fragmentation and deceleration of many tiny fragments that separated from the incident particle immediately after penetration of the aerogel (Trigo-Rodríguez 2008), with multiple tracks emerging from the bottom of the bulb.

As a result of fragmentation, a single particle incident on the aerogel frequently resulted in a multitude of fragments distributed along the damage track. The nomenclature convention established during the Stardust Preliminary Examination uses "particle" to denote the entire object that hit the collector and "fragment" to denote pieces of that object distributed along the track or in the crater. The one exception is that the phrase "terminal particle" is used to denote the fragment at the end of the longest section of the track.

Figure 3 shows one example of the Fe distribution along a conical Wild 2 track, determined by SXRM mapping. Similar measurements on other Wild 2 tracks indicate that the Fe mass of the terminal particle varies from a high of 80% of the total Fe detected in the map of the entire volume of the track to 0% of the total for two tracks that had no detectable terminal particles (Flynn et al. 2006).

The material distributed along the track walls varied dramatically in composition from one spot to the next, demonstrating that the incident particle was weak enough to break up during capture, and consisted of discrete grains of diverse composition and mineralogy. Extraction of fragments from the aerogel demonstrates that Wild 2 also contains discrete mineral grains, mainly olivine, pyroxene, and Fe-sulfide, some of which exceed 10 μm in size. One track contained fragments similar to the Ca–Al-rich inclusions found in CV carbonaceous chondrites (e.g., Allende).

Many of the craters have "compound structures," characterized by irregular outlines and overlapping depressions, which result from impactors with heterogeneous mass distributions (Hörz et al. 2006). Composition measurements on the residue in craters

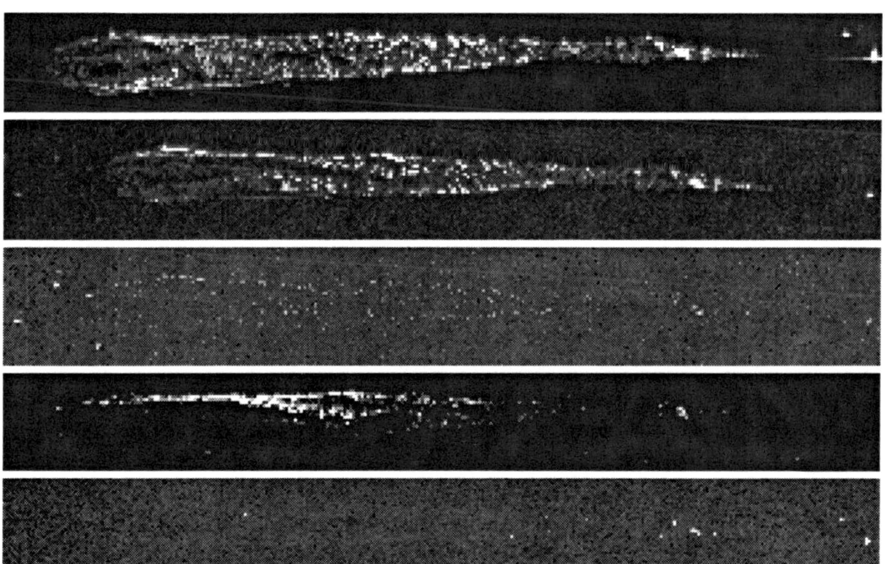

Fig. 3 X-ray fluorescence intensity maps of Fe (top), Ni, Cu, Zn, and Cr (bottom) along Wild 2 Track 19, an ~860 μm long track produced by an ~3 μm incident particle

indicates that poly-mineralic impactors dominate over mono-mineralic impactors even down to the smallest craters, less than 100 nm in size (Hörz et al. 2006). However, bowl-shaped craters, consistent with a single, homogeneous impactor are also seen on the Stardust Al-foils.

Both the crater and track morphologies indicate that many of the Wild 2 grains collected by the Stardust spacecraft are weakly bound aggregates of fine-grained, sub-micron material and larger crystalline grains, some larger than 10 μm. Because the Stardust aerogel capture cells are composed of amorphous silica, the amorphous silica content of the incident particles was not determined during the Preliminary Examination, although careful petrographic and compositional analysis may allow cometary glass to be distinguished from melted aerogel in future work.

5 Heterogeneity of the Wild 2 Samples

Because individual minerals differ significantly in elemental composition, compositional heterogeneity can be used as an indicator of the size-scale of the individual mineral grains of a sample. Maps showing the distributions several major and minor elements along tracks in the aerogel were obtained during the Preliminary Examination (Flynn et al. 2006).

One example of these analysis, for Wild 2 Track 19, a cone-shaped track having a length of 860 μm that was produced by an incident particle of $\sim$2–3 μm in size, is shown in Fig. 3. The Fe distribution map for Track 19 indicates that Fe was deposited along the wall of the entry track as well as in the terminal particle. Nickel and Cu show a similar distribution. But, Zn was deposited along only part of the track wall, suggesting that one side of the initial particle hosted a Zn-rich phase that abraded or vaporized as the particle decelerated in the aerogel. The terminal particle has a very low Zn content. Chromium is localized in a few discrete fragments along the track.

The 20 most intense Fe spots found in the Track 19 map were analyzed individually, for longer times, to obtain high-quality X-ray fluorescence spectra. Because the elemental composition of the CI carbonaceous chondrite meteorites is believed to represent the bulk composition of the Solar System (Anders and Grevesse 1989), the Wild 2 composition results are compared to the CI composition. The CI- and Fe-normalized S, Ca, Cr, Mn, Ni, Cu, Zn, Ga and Ge abundances for these fragments, shown in Fig. 4, demonstrate that these 20 fragments from a single Wild 2 particle exhibit element/Fe ratios that differ from one another by 2 orders-of-magnitude or more. This indicates that the incident particle deposited material of diverse compositions, presumably representing a diversity of minerals, along the track.

The Mineralogy/Petrology team extracted Wild 2 material, sometimes consisting of mixtures of grains and compacted or melted aerogel, from 52 tracks, some of which were tracks previously mapped by the Elemental Composition team. They studied in detail 26 tracks that were chosen at random from those of average length. Eight of these tracks were dominated by olivine, 7 by low-Ca pyroxene, 3 by roughly equal amounts of olivine and pyroxene, 5 by Fe-sulfides, one by Na-silicate minerals, and one by refractory minerals similar to Ca–Al inclusions in primitive meteorites like the Allende CV3 carbonaceous chondrite (Zolensky et al. 2006). They found crystalline grains distributed along the length of the track, not just in the terminal particles. Many tracks contained minor mineral phases as well, including Cu–Fe-sulfide, Fe–Zn-sulfide, and possible K-feldspars (Zolensky et al. 2006), consistent with the high degree of elemental heterogeneity mapped along some tracks.

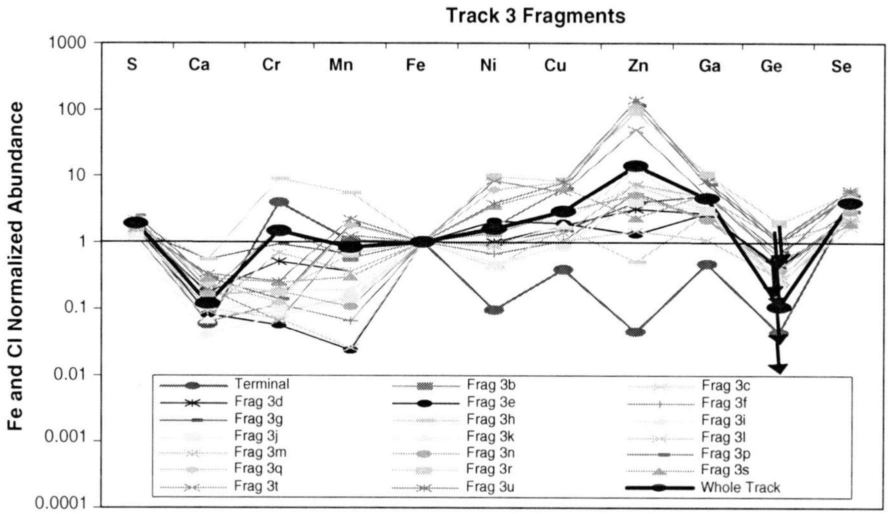

Fig. 4 CI normalized element/Fe ratios for 20 spots along Wild 2 Track 19, an 860 μm long, probably the result of capture of an initial ~2–3 μm particle. The element distribution along this track is extremely heterogeneous, with the 20 spot analyses showing variations of 2 orders-of-magnitude or more for most of the element/Fe ratios. This indicates that Wild 2 dust includes a significant fine-grained component. The elemental composition of the terminal particle (blue line) differs significantly from the average composition of this whole track (black line) with the low content of Ni and the moderately volatile elements in the terminal particle suggesting it is an anhydrous silicate

Even more striking is the difference in elemental composition from track to track, shown in Fig. 5. Only Fe or Fe and Ni were detected in two of the 23 tracks examined during the Stardust Preliminary Examination. But S, Ca, Cr, Mn, Fe, Ni, Cu, Zn, and Ga abundances were obtained on most of the remaining 21 tracks. Figure 4 shows the diversity in compositions for these elements in the 21 whole tracks examined during Preliminary Examination. Many elements show three or four order-of-magnitude variation in element/Fe ratios from one track to another. Even among the 6 largest particles, all estimated to be >10 μm in size based on their measured Fe contents, the element/Fe ratios show order-of-magnitude variations for all of the elements measured (Fig. 4). This indicates particles from Wild 2 are elementally heterogeneous at the largest size scale measured (>10 μm). Nonetheless, crater composition measurements indicate that composite particles dominate at the smallest size (~100 nm) (Horz et al. 2006). Thus, Wild 2 contains much sub-micron material along with larger mineral grains, >10 μm (Zolensky et al. 2006).

Most striking is the difference between the elemental composition of the material distributed along the whole volume of the track including the terminal particle and the composition of just the terminal particle (Fig. 4). Most of the 21 tracks that had detectable terminal particles show differences of an order-of-magnitude or more for several of the element/Fe ratios between the whole track and the terminal particle. The terminal particle frequently has a significantly lower Ni/Fe ratio than the whole track, consistent with observations by the Mineralogy/Petrology Preliminary Examination team that the terminal particles are frequently anhydrous silicates (olivine or pyroxene), which usually have low Ni/Fe. Much of the finer-grained material deposited along the track has a composition closer to the CI composition, suggesting that Wild 2 particles in the size range examined consist of large mineral grains embedded in a fine-grained matrix.

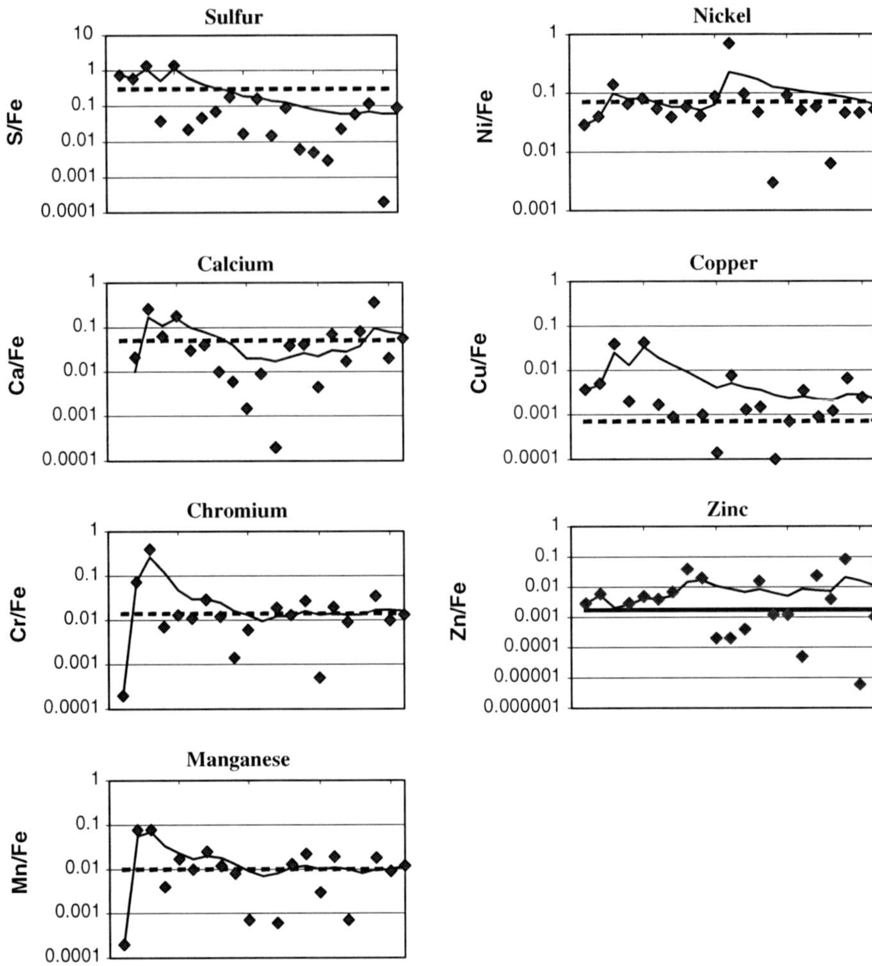

Fig. 5 Each plot shows the whole track element/Fe ratio (symbols) and the Cumulative Element/Fe ratio—the sum of the Element ratio to the Sum of the Fe for all particles up to that Fe-mass (solid line), with increasing Fe mass to the right. The CI Composition (Lodders 2003) is shown by the dashed line. Scatter in the whole track element/Fe ratios provides insight into the size of the largest minerals analyzed Good convergence in the Cumulative Element/Fe ratio suggests enough tracks were examined to provide a valid mean composition

6 Mean Elemental Composition of the Wild 2 Particles

The Si and O contained in particles that produced tracks in the aerogel could not be assessed during the Preliminary Examination because these analyses were performed on whole tracks in keystones of aerogel (described by Westphal et al. 2004) and Si and O are major elements in the aerogel. The aerogel also contains trace quantities of many other elements (Tsou et al. 2004) and these elements are frequently distributed inhomogeneously (Flynn et al. 2006), precluding quantitative background subtraction. Magnesium and Al could not be quantitatively determined for particles captured in the aerogel because the thickness and density of the overlying aerogel was not known sufficiently well to allow correction for the absorption of the low-energy fluorescence X-rays from Mg and Al. Quantitative analysis of S in tracks is difficult because the S fluorescence X-rays can experience significant attenuation by the aerogel. For elements heavier than Ca these absorption corrections are small enough that the uncertainty in thickness is only a minor effect.

The Wild 2 particles may show compositional variation with size, with FeS particles being prominent in the smaller craters (Hörz et al. 2006), but a lower fraction of the tracks were dominated by FeS (Flynn et al. 2006). Compositional variation was also seen in the particles analyzed at Comet Halley, with many of the smallest grains being Mg-rich and Si-poor, while the larger grains have a CI-like Mg/Si ratio (Mukhin et al. 1991). Because the composition varies with particle size, it is necessary to mass-weight the grains in determining the bulk composition. If all particles were given an equal weight in the average, the contribution from the smaller particles would be overestimated (Fomenkova et al. 1992). The "whole track" content of S, Ca, Ti, Cr, Mn, Fe, Ni, Cu, Zn, and Ga was determined by summing the mass of that element detected in each of the 23 Wild 2 tracks analyzed during the Preliminary Examination (see Flynn et al. 2006), a process that mass-weights the average.

Although minerals may be altered during aerogel capture, only the most volatile elements are likely to be redistributed far from the track. Track associated organic matter was detected as far as several track diameters from the track, demonstrating redistribution of a volatile organic species (Sandford et al. 2006). Thus, the whole track elemental composition is the elemental composition of the incident particle, except possibly for the most volatile elements (e.g., S).

The mean composition of the 23 tracks analyzed during the Preliminary Examination is quite well-determined. However, the large, non-normal distribution of element abundances in the 23 whole track compositions, shown in Fig. 5, made the assessment in the uncertainty of the bulk composition difficult. This dispersion in compositions dominated over the analytical uncertainties in determining the uncertainty in the mean composition of Wild 2 particles in the range of sizes that was analyzed (Flynn et al. 2006). To assess the uncertainty in the mean composition a statistical approach was adopted during the Preliminary Examination (see Flynn et al. 2006). If the 23 tracks studied by the Elemental Composition Preliminary Examination team are a random sample of Wild 2 particles in the size range studied, then the 1σ uncertainty range given for each element/Fe ratio, given in Table 1 as the mean composition + or − the modeled 1σ variation, is expected to reflect the range of mean compositions of sets of 23 other randomly selected Wild 2 particles in the same size range 68% of the time. The modeled 2σ uncertainty ranges, corresponding to the 95% confidence interval for the mean composition, are shown in Figs. 3 and 4 of Flynn et al. (2006).

Impact residue, which is abundant in all large craters in the Stardust Al-foils that were examined during Preliminary Examination, provides chemical composition data complementary to that obtained on the tracks and allows direct measurement of element to Si ratios. The elemental composition of the residue in 7 craters in the Al-foils was measured by SEM-EDX, and five of these craters were also measured by ToF-SIMS. Aluminum could not be determined for the crater residue because of the fluorescence from the underlying Al-foil, and the presence of minor quantities of other elements in the Al-foil compromised their detection in the residues. Table 1 shows the mean contents of elements detected in residue of 5 or more craters. The 1σ uncertainty was determined by the same technique as for the tracks, although the validity of this statistical technique for such a small data set is less certain.

Table 1 Mean CI-Normalized Composition of Wild 2 Particles

Element/Fe	Mean of 23 whole tracks (1σ uncertainty range)	Mean of 7 crater residues (1σ uncertainty range)	Rescaled 23 track mean (1σ uncertainty range)
Mg		1.13 (+0.22, −0.05)	
S	0.17 (+0.12, −0.06)	0.13 (+0.40, −0.06)	0.13 (+0.12, −0.06)
Ca	1.25 (+0.47, −0.43)	0.51 (+0.12, 0.05)	0.94 (+0.47, −0.43)
Ti	0.42 (+1.74, −0.23)		0.32 (+1.74, −0.23)
Cr	1.30 (+0.24, −0.31)	0.31 (+0.33, −0.04)	0.98 (+0.24, −0.31)
Mn	1.32 (+0.32, −0.37)		0.99 (+0.32, −0.37)
Fe	1[a]	0.75 (+0.05, −0.40)	0.75[b]
Ni	0.82 (+0.37, −0.24)		0.62 (+0.37, −0.24)
Cu	2.06 (+1.14, −0.89)		1.55 (+1.14, −0.89)
Zn	4.60 (+6.30, −3.10)		3.45 (+6.30, −3.10)
Ga	10.0 (+8.9, −7.5)		7.5 (+8.9, −7.5)

[a] Fe fixed at 1 × CI

[b] Fe adjusted to the value 0.75 × CI determined from crater analysis

Since Fe is the only major element easily quantifiable in the SXRM analysis of whole tracks in aerogel keystones, the 23 track mean composition was CI- and Fe-normalized in Table 1. However, Si was easily detected in the craters, and the Fe/Si ratio was measured to be ~0.75, although one large crater dominated the element abundance determination in crater residues (Flynn et al. 2006). To permit direct comparison of the mean compositions of the craters and the tracks, Table 1 also shows the track data rescaled to a CI-normalized Fe abundance of 0.75. With this rescaling, two of the three elements determined in both tracks and craters, S and Ca, have overlapping 1σ confidence ranges, but a discrepancy remains for Cr.

The similarity of the CI meteorite composition to the composition of the Solar photosphere for elements that are well-determined in the Solar photosphere led to the suggestion that the CI meteorite composition is the mean composition of the Solar System (Anders and Grevese 1989). The mean composition of the Wild 2 particles is consistent with the CI meteorite composition for the refractory elements Mg, Si, Cr, Ni, Ca, Ti and Mn, but both S and Fe appear to be depleted (at the 2σ confidence limit) compared to the CI composition and the moderately volatile elements, Cu, Zn, and Ga, are enriched relative to the CI composition. The ~10 μm size anhydrous, porous interplanetary dust particles (IDPs) (described in Bradley et al. (1988) and Rietmeijer (1998)) collected from the Earth's stratosphere, some of which have been suggested to be derived from comets based on atmospheric entry speeds inferred from heating during atmospheric deceleration (Brownlee et al. 1993), show an element abundance pattern similar to these Wild 2 grains. These anhydrous IDPs are generally chondritic, but both S (0.8 × CI) and Fe (0.78 × CI) are depleted from CI (Schramm et al. 1989) and the moderately volatile elements Cu, Zn, and Ga are enriched by factors of 2 or 3 compared to CI (Flynn et al. 1996).

7 Conclusions

The slope of the cumulative size frequency distribution of the particles in the Wild 2 coma is similar to that measured in the Halley coma, but the smallest particles (~10^{-15} g) seem to be less abundant in Wild 2 than in Halley (Hörz et al. 2006). The Wild 2 particles collected by the Stardust spacecraft were generally weakly bound aggregates of sub-micron grains and, frequently, larger individual mineral grains. The Preliminary Examination results demonstrate, for the first time by the direct sampling of a comet, that Wild 2 contains some mineral grains (olivine, pyroxene, and Fe-sulfide) >10 μm in size as well as assemblages of refractory minerals similar to the Ca–Al inclusions found in some primitive chondrites. Some of the anhydrous silicates are Mg-rich, Fe-poor minerals, while most of the sulfides are Fe-rich, and the Ca–Al material is Ca-rich but poor in Fe. Thus the stream of particles associated with Comet Wild 2 contains individual grains of diverse compositions, some that are rich in Fe and S, some rich in Mg, and others rich in Ca.

The Stardust Preliminary examination did not establish an upper limit on the grain size of individual minerals in Wild 2, since significant compositional heterogeneity was observed from one track to the next, up to the largest size studied (>10 μm). Examination of the largest Wild 2 particles collected by Stardust may establish this limit for Wild 2. The measurement of compositional heterogeneity as a function of size for the objects in the meteor streams associated with other comets could establish maximum grain size limits for the parent bodies of these meteor streams. Determining the maximum grain size in cometary bodies is critical in establishing the minimum sample size that it is desirable to collect on future comet sample-return missions.

The 23 tracks and 7 craters analyzed during the Preliminary Examination are estimated to have been produced by Wild 2 particles that had a total mass of about ~300 ng, about one-thousandth of the total mass collected by the Stardust spacecraft. The bulk refractory element abundance pattern in the Wild 2 particles is consistent with the CI meteorite pattern for Mg, Si, Cr, Fe, and Ni to 35%, and for Ca, Ti and Mn to 60%, but the Fe/Si shows a statistically significant depletion from the CI value (Flynn et al. 2006). The moderately volatile elements Cu, Zn, Ga, are enriched compared to CI, while S appears to be depleted from CI, although the effects of S mobilization during capture and the fluorescence absorption correction due to overlying aerogel are still being assessed (Flynn et al. 2006). Both the enrichment in moderately volatile elements and the depletion in S and Fe have previously been reported in the fine-grained, anhydrous IDPs (Flynn et al. 1996; Schramm et al. 1989), and an Fe depletion was reported from the Giotto analysis of dust from comet Halley (Jessberger et al. 1988). Elemental analysis of the larger particles that Stardust collected from the Wild 2 coma, which contain most of the collected mass, should significantly reduce the uncertainty in the mean composition.

The elemental composition determined from the Wild 2 samples is consistent with that obtained for Halley dust, but about two orders-of-magnitude more mass was analyzed during the Stardust Preliminary Examination than by the impact ionization mass spectrometers on the Giotto and VEGA spacecraft. The Stardust results extend the measurement of comet composition to include several moderately volatile minor elements. The major differences between the Wild 2 results and the CI meteorite composition are for elements whose abundances are not well-determined in the Solar photosphere (Anders and Grevesse 1989), suggesting that CI may not represent the mean Solar System composition (Flynn et al. 2006). Precise measurement of the abundances of elements that show differences between the CI meteorite composition and the Wild 2 or anhydrous IDP compositions, particularly S, Ca, Fe and moderately volatile minor elements, for other comets or for the particles in their meteor streams will aid in determining if the CI meteorite composition best represents the bulk composition of the Solar Nebula.

Acknowledgements The Stardust mission was the 4th mission in the NASA Discovery Program. This work was supported by NASA Cosmochemistry Grant NNG06GG13G.

References

E. Anders, N. Grevesse, Abundances of the elements: meteoritic and solar. Geochim. Cosmochim. Acta **53**, 197–214 (1989)

J.P. Bradley, S.A. Sandford, R.M. Walker, Interplanetary dust particles, in *Meteorites and the Early Solar System*, ed. by J. Kerridge, M.S. Matthews (University of Arizona Press, 1988), pp. 861–898

D.E. Brownlee, D.J. Joswiak, S.G. Love, A.O. Nier, D.J. Schlütter, J.P. Bradley, Identification of cometary and asteroidal particles in stratospheric IDP collections. Lunar Planet. Sci. **XXIV**, 205–206 (1993)

D.E. Brownlee, et al. Surface of a Young Jupiter Family Comet 81P/Wild 2: view from the Stardust spacecraft. Science **304**, 1764–1769 (2004)

D.E. Brownlee, et al. Comet 81P/Wild2 under a microscope. Science **314**, 1711–1716 (2006)

G.J. Flynn, S. Bajt, S.R. Sutton, M.E. Zolensky, K.L. Thomas, L.P. Keller, The abundance pattern of elements having low nebular condensation temperatures in interplanetary dust particles: evidence for a new chemical type of chondritic material, in *Physics; Chemistry; and Dynamics of Interplanetary Dust*, vol. 104, ed. by B.A.S. Gustafson, M.S. Hanner (Astronomical Society of the Pacific Conference Series, 1996), pp. 291–297

G.J. Flynn, et al. Elemental compositions of Comet 81P/Wild 2 samples collected by Stardust. Science **314**, 1731–1735 (2006)

M.N. Fomenkova, J.F. Kerridge, K. Marti, L.-A. McFadden, Compositional trends in rock-forming elements of Comet Halley dust. Science **258**, 266–269 (1992)

F. Hörz, et al. Impact features on Stardust: implications for Comet 81P/Wild 2 dust. Science **314**, 1716–1719 (2006)

E.K. Jessberger, A. Christoforidis, J. Kissel, Aspects of the major element composition of Halley's dust. Nature **332**, 691–695 (1988)

L.P. Keller, et al. Infrared spectroscopy of Comet 81P/Wild 2 samples returned by Stardust. Science **314**, 1728–1731 (2006)

K. Lodders, Solar system abundances and condensation temperatures of the elements. Astrophys. J. **591**, 1220–1247 (2003)

S.G. Love, D.E. Brownlee, A direct measurement of the terrestrial mass accretion rate of cosmic dust. Science **262**, 550–553 (1993)

K.D. McKeegan, et al. Isotopic compositions of cometary matter returned by Stardust. Science **314**, 1724–1728 (2006)

L. Mukhin, G. Dolnikov, E. Evlanov, M. Fomenkova, O. Prilutsky, R. Sagdeev, Re-evaluation of the chemistry of dust grains in the coma of Comet Halley. Nature **350**, 480–481 (1991)

F.J.M. Rietmeijer, Interplanetary dust particles, in *Planetary Materials, Reviews in Mineralogy*, vol. 36, ed. by J.J. Papike (Mineralogical Society of America, Chantilly, Virginia, 1998), pp. 2-1–2-95

F.J.M. Rietmeijer, Interplanetary dust and carbonaceous meteorites: constraints on porosity, mineralogy and chemistry of meteors from rubble-pile planetesimals, in *Modern Meteor Science, An interdisciplinary view*, ed. by R. Hawkes, I. Mann, P. Brown (Springer, 2005), pp. 321–338

F.J.M. Rietmeijer, Natural variation in comet aggregate meteoroid composition. Earth Moon Planets (2008, this issue)

S.A. Sandford, et al. Organics captured from Comet 81P/Wild 2 by the Stardust spacecraft. Science **314**, 1720–1724 (2006)

L.S. Schramm, D.E. Brownlee, M.M. Wheelock, Major element composition of stratospheric micrometeorites. Meteoritics **24**, 99–112 (1989)

H. Schulze, J. Kissel, E.K. Jessberger, K. Elmar, Chemistry and mineralogy of Comet Halley's dust, in *From Stardust to Planetesimals*, vol. 122, ed. by Y.J. Pendleton, A.G.G.M. Tielens (ASP Conference Series, 1997), pp. 397–414

Z. Sekanina, et al. Modeling of the nucleus and jets of Comet 81P/Wild 2 based on the Stardust encounter data. Science **304**, 1769–1774 (2004)

J.M. Trigo-Rodríguez, G. Dominguez, M.J. Burchell, F. Hörz, J. Llorca, P. Tsou, W.W. Anderson, Towards an understanding of hypervelocity impact into aerogel. Meteorit. Planet. Sci. (2008, in press)

P. Tsou, D.E. Brownlee, J.D. Anderson, S. Bhaskaran, A.R. Cheuvront, B.C. Clark, T. Duxbury, T. Economou, S.F. Green, M.S. Hanner, F. Hörz, J. Kissel, J.A.M. McDonnell, R.L. Newburn, R.E. Ryan,

S.A. Sandford, Z. Sekanina, A.J. Tuzzolino, J.M. Vellinga, M.E. Zolensky, Stardust encounters comet 81P/Wild 2. J. Geophys. Res. **109**(E12), E12S01 (2004)

A. Tuzzolino, et al. Dust measurements in the coma of Comet 81P/Wild 2 by the dust flux monitor instrument. Science **304**, 1776–1780 (2004)

A.J. Westphal, C. Snead, A. Butterworth, G. Graham, J.P. Bradley, S. Bajt, P.G. Grant, G. Bench, S. Brennan, P. Pianetta, Aerogel keystones: extraction of complete hypervelocity impact events from aerogel collectors. Meteorit. Planet. Sci. **39**, 1375–1386 (2004)

M.E. Zolensky, et al. Mineralogy and petrology of Comet 91P/Wild 2 nucleus samples. Science **314**, 1735–1739 (2006)

Natural Variations in Comet-Aggregate Meteoroid Compositions

Frans J. M. Rietmeijer

Originally published in the journal Earth, Moon, and Planets, Volume 102, Nos 1–4.
DOI: 10.1007/s11038-007-9164-4

Abstract Bulk compositions of aggregate meteoroids made of the originally accreted dust with its highly varied in mineral content and chemistry and considerable grain size variations do not have a chondritic bulk composition. Deviations from CI element abundances reflect indigenous variations within and among comet nuclei. These unmodified meteoroids that are heterogeneous in all their properties are fundamentally different from meteoroids with a CI bulk composition that are fine-grained, equigranular materials and chemically and mineralogically homogeneous. Collection and data reduction bias exists but the compositions of individual fast meteors are entirely constrained by the measured main component meteor abundances.

Keywords Aqueous alteration · CI composition · Comet nucleus · Comet dust · Chemical abundances · Differential ablation · Interplanetary dust particles · Halley · Main meteor component · Meteoroids · Minerals · Leonids · Perseids · Solar abundances · Wild 2

1 Introduction

The elemental abundances that are well determined in the solar photosphere are similar to those measured in carbonaceous chondrite meteorites, except for highly volatile elements (Anders and Grevesse 1989). It was postulated that all matter in the solar system had condensed from a solar nebula gas of a chondritic (CI; after the Ivuna carbonaceous chondrite meteorite) composition (Suess 1969). In broad terms, solid grains condensed and evolved to large (several millimeters) minerals that accreted and were nuclei for condensation of volatile gasses (e.g. H_2O) and then formed the first planetesimals (future comet nuclei and primitive asteroids). Once such bodies had grown into protoplanets with internal heat-producing sources, the accreted solar nebula dust was modified by thermal

F. J. M. Rietmeijer (✉)
Department of Earth and Planetary Sciences, MSC03-2040, 1-University of New Mexico, Albuquerque, NM 87131-0001, USA
e-mail: fransjmr@unm.edu

metamorphism or aqueous alteration. Both processes might have acted at different times, and repeatedly in a protoplanet (Brearley and Jones 1998) wherein ultimately no more traces of solar nebula dust remained and chemical heterogeneity was erased.

Isotopic anomalies in some meteorite grains that pointed to extrasolar origins showed that the nebula within Jupiter's orbit was never entirely dust-free of refractory, high-temperature minerals. Inwards of Jupiter's orbit in the region generally known as the inner solar nebula the temperatures were only high enough to evaporate most other presolar dust. Recently the sharp distinction between comets and asteroids was further eroded with the finding of a new population of comets residing in the main asteroid belt (Hsieh and Jewitt 2006). As these icy bodies formed in the regions wherein they currently reside (Hsieh and Jewitt 2006), it supports a thermal gradient in the solar nebula wherein the temperatures decreased gradually to ensure preservation of increasingly more volatile solids in a transition zone wherein for example original D/H ratios of organic matter in meteorites were preserved intact. It also implies that the 'snow-line' that marks the boundary of the solar nebula region where nebular temperatures were low enough to condense water ice that is still found in comet nuclei and in the outermost P- and D-asteroids. In this region that we now know was probably within Jupiter's orbit, the original molecular cloud dust was unaffected by evaporation, which makes comet nuclei attractive targets to study the transition from pre-solar to dust formed and processed in the inner solar nebula. Comet 81P/Wild 2 showed that some of this nebula dust was transported in the sun's bipolar outflows to above the mid-plane where it could reach the Kuiper belt (Brownlee et al. 2006; Zolensky et al. 2006). Comet nuclei, or at least Jupiter-Family (J-F) comets, might be aggregates of molecular cloud dust, inner solar nebula dust, and processed protoplanet dust liberated by collisions among asteroids, comet nuclei and asteroid debris into comet nuclei (Cintala 1981). In this paper I will address several questions: (1) Should all comet meteoroids have a CI composition? (2) When yes, what are the physiochemical implications? (3) When no, what is the extent of compositional variations? (4) How critical is differential ablation for comet meteoroid compositions? and (5) Is chemical data reduction of pristine aggregated comet meteoroids biased?

First we should ask what kind of material represents the CI composition? The laboratory measurements of this composition were made on fully hydrated meteorite samples. Some had veins filled with salt minerals. It is not known how much material was leached from, or added to, the sample in hand when it was still in the parent body. The CI meteorites are typically 'clumps of clay' of several centimeters wherein aqueous alteration had erased all chemical and mineralogical heterogeneity and the original grain size range and distributions. The layer silicates that replaced the original minerals have a limited and uniform size range <625 nm. The layer silicates in the Orgeuil meteorite are <200 nm in size with a modal value between 50 nm and 75 nm (Mackinnon and Kaser 1988). Hence, the flat patterns of normalized element distributions such as the Orgueil CI meteorite that was recently linked to a J-F, or maybe a Halley-type, comet (Gounelle et al. 2006). The flat distribution pattern indicates a texturally and chemically homogenous, processed (that is, hydrated) comet meteoroid of CI-like material strength (cf. Borovička 2007).

2 Aggregates

Interplanetary dust particles (IDPs) are typically ~ 10 μm in size. They range from completely anhydrous aggregates to particles with a fully hydrated matrix. With continuing dust evolution cluster IDPs formed that are highly porous aggregates of several distinct

~10-μm, dust types (Rietmeijer and Nuth 2004): (1) aggregate IDPs, (2) mostly Mg-rich, Mg-Fe-Ca-silicate dust, (3) Fe–Ni–S and Fe–Ni metal dust and (4) composite refractory particles (Rietmeijer 2002a, 2005). Aggregate and cluster IDPs survived atmospheric deceleration in part because their material strength (Flynn 2005) was greater than for similar aggregates when freshly ejected from an active comet nucleus prior to on-orbit processing that increased the mechanical strength of such aggregate meteoroids (Borovička 2007). The particles ejected from comets Halley and Wild 2 showed rampant fragmentation (Tuzzolino et al. 2004) and the first laboratory results showed that "The bulk of the comet 81P/Wild 2 (hereafter Wild 2) samples returned to Earth by the Stardust spacecraft appear to be weakly constructed mixtures of nanometer-scale grains, with occasional much larger (over 1 μm) ferromagnesian silicates, Fe-Ni sulfides, Fe-Ni metal, and accessory phases" (Zolensky et al. 2006).

The much larger grains included terminal grains, ~5–20 μm, at the very end of a deceleration track carved by an impacted Wild 2 particle, viz. (1) Mg-rich olivine (Fig. 1a), (2) Mg-rich, low-Ca-pyroxene, (3) composite grains of about equal amounts of these silicate minerals, (4) mainly Fe,Ni-sulfide grains (Fig. 1b), (5) Fe,Ni metal grain, (6) a polymineralic refractory grain, (7) a composite grain of Na, K-bearing silicates (Zolensky et al. 2006), and (8) an 8-μm grain that was a large FeS crystal, a small low-Ca pyroxene and a patch of fine-grained chondritic aggregate material (Brownlee et al. 2006). The latter resembled the sulfide IDPs that were part of cluster IDPs (Rietmeijer 2004).

Aggregate and cluster IDPs, and the Wild 2 particles showed that (1) a chondritic bulk composition of unmodified aggregates is achieved by accreting many dust grains, most of which do not have a chondritic composition, and (2) the dust grain sizes ranged from ~100 nm up to 20–50 μm independent of composition. They revealed a feature unique to unmodified, typically anhydrous, aggregates namely they reflect the types and sizes of dust that were available in their dynamic regions of dust accretion. The results were unpredictable mixtures but fortunately the initial dust had a simple mineralogy dominated by "amorphous silicate" grains, Mg–Fe-silicates and Fe–Ni-sulfides that only became more complex during dust evolution to grains with increasing size (Rietmeijer 2002a, 2005). The

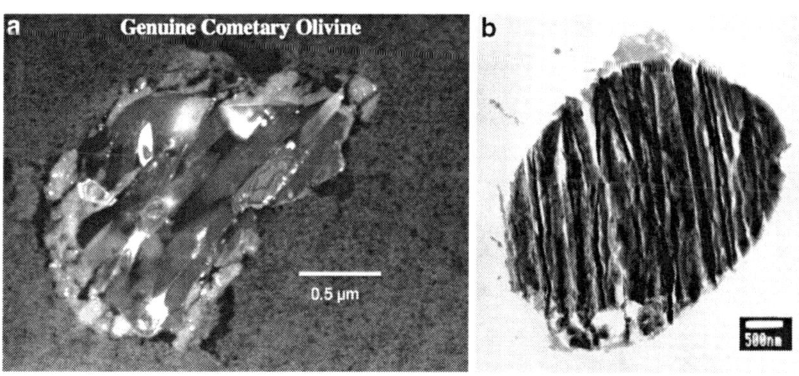

Fig. 1 High resolution TEM images of two terminal Wild 2 grains, (**a**) a Mg-rich olivine crystal with a narrow rim of Si-rich glass and Fe–Ni–S inclusions of modified comet nanograins in melted aerogel, and (**b**) a pyrrhotite crystal. The shattered mineral appearances resulted from ultramicrotome sample preparation but it expresses structural weakening caused by the ultrarapid heating and similarly fast cooling during 6.1 km/s hypervelocity impact capture. In both images the background is an embedding material. Photo credits LLNL/JPL/NASA JSC

earliest aggregates were fine-grained with the compositional variations of comet Halley dust and aggregate IDP matrix. Later-formed aggregates had a wide size range for chemically complex grains such as the Wild 2 polymineralic aggregates of "hundreds of micrometers in size" (Hörz et al. 2006). The extended accretion time allowed aggregation of a larger sample of dust populations that were evolving in a turbulent solar nebula.

3 Bias in Data Reduction

This effect is unique to unmodified aggregates. Normalizing the element abundances of collected extraterrestrial materials and from meteor spectral analyses to a CI standard is helpful for comparative purposes. It involved the selection of an internal (to each sample) standard that is typically Si, Mg or Fe and then the CI ratio for same elements. The bulk chemical compositions of such dust aggregates are essentially a mixture of silicate (olivine, pyroxene) grains and FeS grains (Rietmeijer 2002a). Iron is mostly in massive sulfides and less abundant Fe,Ni-metal grains. Magnesium typically resides in silicates but Fe-bearing silicates, $Fe/(Fe + Mg) < 0.5$, occurred in aggregate and cluster IDPs (Rietmeijer 1998) and comet Wild 2 (Zolensky et al. 2006); both with constituents of widely ranging grain sizes and thus fundamentally different from CI "clay".

3.1 Equigranular Aggregates

Comet Halley dust was an aggregate of nanograins comparable to the matrix of aggregate IDPs with ferromagnesiosilica grains and a few (sub-)micron FeS grains (Rietmeijer 1998, 2005). Its Fe-normalized element abundances were not CI-like (Jessberger et al. 1988) but showed the ragged pattern of unprocessed dust (Fig. 2). The comet contained surprisingly few FeS grains (Jessberger et al. 1988). To show the effect on the Fe-normalized abundances, I artificially increased the 'amount of FeS' (or larger grains) in comet Halley and its element distribution pattern drops below the CI line (Fig. 2) and, for example, the comet appears to be less Na-rich but comparable to aggregate IDPs with micrometer FeS grains.

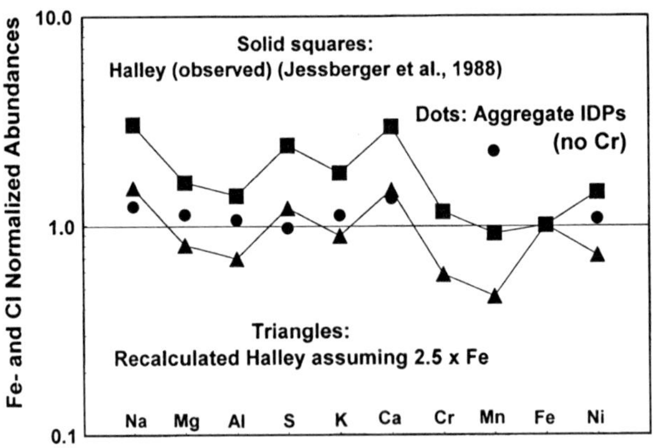

Fig. 2 Fe- and CI-normalized measured abundances of comet Halley dust (Jessberger et al. 1988; solid squares) compared to the same data but with an assumed twice-higher iron content (solid triangles) and for comparison chondritic porous aggregate IDPs (Rietmeijer 2002a; dots)

Doubling the amount of Mg-silicates had a similar but smaller effect. The example showed how selecting the normalizing element might affect the outcome. The cause was a collection bias when only very few sulfide grains ejected were small enough to enter the size/mass shielded on-board mass spectrometers. Bias from sample size on the bulk composition is probably not an issue for cluster IDPs and comet Wild 2 (Flynn et al. 2006). It remains under investigation (Flynn 2005) but mixing ~10-μm aggregate IDPs and similarly large silicate and sulfide IDPs might be a reasonable good approximation of their parent body composition (Flynn et al. 2007).

3.2 Inequigranular Aggregates

Comet Wild 2 compositions were obtained form (1) crater residues on the Al- collector frame, (2) entire deceleration tracks in aerogel, and (3) Si-rich glass grains from track walls (Fig. 3). These very different patterns are for particles and grains from the same comet. The CI-like pattern represents the Fe–Ni–S (a few to ~100 nm) and Mg-rich silicate (~500 nm) grains scattered in the Si-rich glass of quenched aerogel melts. Their precursors were nanometer grains in the weakly constructed comet dust aggregates. Although unprocessed, i.e. anhydrous, their small equigranular grains effectively mimicked this physical property of homogenized CI "clays" for the elements shown.

The grain compositions in the tracks and the impact crater residues defined the element abundances for the large comet particles of grains with considerable size ranges and mineral heterogeneity. There is no evidence for aqueous alteration of comet Wild 2 particles (Zolensky et al. 2006). That is, Wild 2 particles were anhydrous aggregates that were heterogeneous in texture, mineralogy and chemistry. Hence, the irregular normalized element distribution pattern.

Five out of 26 tracks analyzed had a terminal grain that was mainly FeS or Fe,Ni-metal (Zolensky et al., 2006). The Fe,Ni nanograins in the Si-rich glass contained 5–53 at% Ni (assuming no Ni is in sulfides) and the metal/sulfide ratios had one maximum at 78/22 (Leroux et al. 2007). Most iron of the Wild 2 bulk composition was present in these

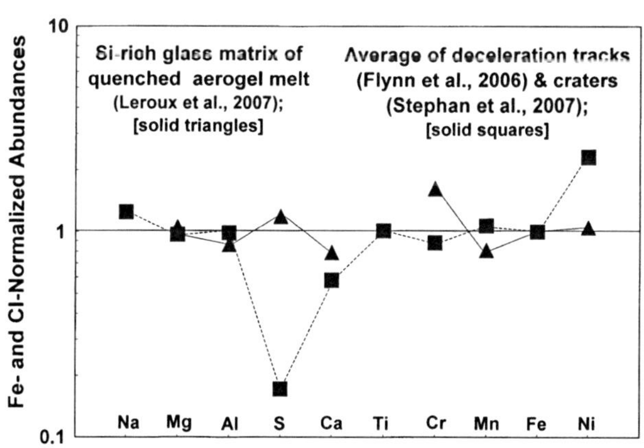

Fig. 3 The average Fe- and CI-normalized abundances of comet Wild 2 particles calculated by the author from the original sources of the impact crater residue (Stephan et al. 2007) and entire deceleration track (Flynn et al. 2006) data (solid squares), and Si-rich glass grains (Leroux et al. 2007; solid triangles). The normalized Ca values range from 1.25 (tracks) to 0.51 (average crater residues)

Fe-sulfides and Ni-rich, Fe,Ni metal with sizes from a few nanometers to several microns. It is likely that these host phases biased its Fe-normalized element abundances (Fig. 3) opposite to the Fe-poor comet Halley dust (Fig. 2). It does beg the question how much indigenous chemical variability exists in comet nuclei that are natural rubble piles of modified (hydrated) and unmodified (anhydrous) dust aggregates (Rietmeijer 2000, 2005).

4 Chemical Variability in Comet Meteoroids

The Fe- and CI-normalized average element abundances for Leonid, Perseid and a few very fast sporadic meteors (Trigo-Rodriguez et al. 2003; Borovička 2005; Jenniskens 2007) display a ragged pattern (Fig. 4). The error bar-like features indicate the observed element range (Table 1). All elements, except Ca, had a Gaussian distribution that is apparently independent of the main and second component regimes. The bi-modal Ca distribution shows its sensitivity to the thermal regime (Borovička 2005). These meteoroids from mostly two Halley-type comets were Na-rich as Rietmeijer (1999) had suggested for comet dust.

The ragged pattern supports that the meteoroids were unprocessed, anhydrous aggregates. These element distributions are remarkably similar for J-F comet Wild 2 and the Halley-type comets Tempel-Tuttle and Swift-Tuttle (Fig. 5). Despite its ~ 1 micron alkali-bearing silicates the sodium content of Wild 2 is lower than for the meteoroids. Perhaps its high iron content shifted the pattern downward. If so, the corollary would be that the meteoroids (Fig. 5) did not contain no or few large Fe,Ni sulfide and/or metal grains. The simplest explanation accepts the observed Wild 2 results as the new standard for bulk comet dust compositions. With this in mind, the question of indigenous chemical variations in comet nuclei could be assessed from the meteor data that are probably small parts of these nuclei.

Surely there are uncertainties in spectroscopic meteor measurements and analyses. Such uncertainties might vary among different observatories but they ought to be reasonably

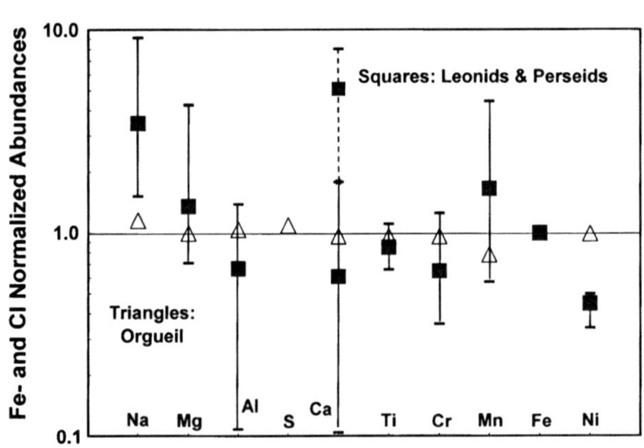

Fig. 4 Calculated average Fe- and CI-normalized abundances and ranges for Leonid, Perseid, and a few very fast sporadic meteors (filled squares) from Trigo-Rodriguez et al. (2003), Borovička (2005) and Jenniskens (2007). Calcium shows a bimodal pattern of the main component (solid line) and 2nd component (dashed line) regimes. The Ti and Ni data are subject to considerable uncertainty (Trigo-Rodriguez et al. 2003). The typical flat CI distribution pattern for the hydrated Orgueil meteorite (Wiik 1969) (triangles) is shown for comparison

Table 1 Fe- and CI-normalized, main component, element abundances for Leonid and Perseid meteors and two fast sporadic meteors calculated by the author from data by Trigo-Rodriguez et al. (2003), Borovička (2005) and Jenniskens (2007)

Element	Mean	Standard deviation	Range	*N*
Na	4.46	2.82	1.52–9.11	12
Mg	2.12	1.28	0.56–4.43	14
Al	0.55	0.32	0.11–1.08	8
Ca (Main)	0.48	0.23	0.10–0.96	12
Ca (2nd)	6.20	2.59	3.20–8.00	3
Ti	0.85	0.19	0.66–1.11	4
Cr	0.65	0.30	0.33–1.25	11
Mn	1.35	1.28	0.49–4.43	12
Fe	1			15
Ni	0.45	0.09	0.34–0.50	3

The Mg data include the main and second component abundances. The different population numbers (*N*) indicate that many elements were only measured in few meteors. Each population is a normal distribution; for Ca two populations are shown corresponding to the main and 2nd components

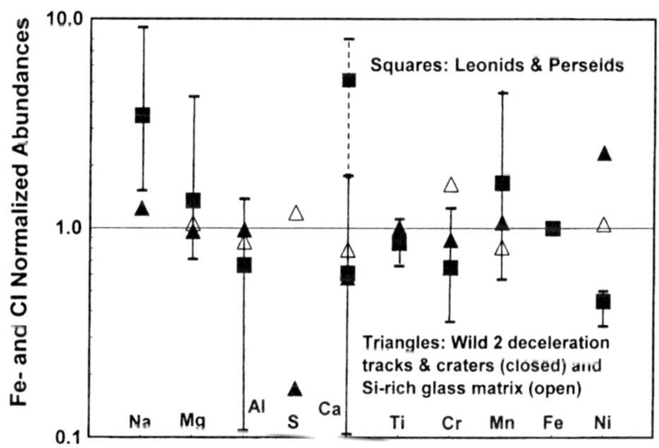

Fig. 5 The measured comet Wild 2 dust abundances (Flynn et al. 2006; Stephan et al. 2007) (solid triangles) and Si-rich glass (Leroux et al. 2007) (open triangles) compared to the calculated average Fe- and CI-normalized abundances and ranges for Leonid, Perseid, and a few very fast sporadic meteors (Trigo-Rodriguez et al. 2003; Borovička 2005; Jenniskens 2007)

small. Instead of being 'error bar'-like features (Fig. 4), it seems reasonable to postulate that the individual meteor measurements (Fig. 6) show the natural, indigenous variations in the meteoroid sources. It is significant that the variations measured in the laboratory or in situ (comet Halley) plot within the element ranges defined by the meteoroids.

How variable and non-CI-like would be allowable in a comet meteoroid to remain acceptable as a comet particle? The chemical variations (Fig. 6) seem too large to be solely accounted for by bias from wide grain size distributions, high Fe–Ni sulfide and metal grains (in case of Fe-normalized data), or large (10–20 μm) Mg-silicate grains (Rietmeijer 2002a, 2005; Rietmeijer and Nuth 2004). The orbital parameters or association with an annual shower and the occasional storm are the key meteoroid parameter to make a link to its source. Multiple meteors from a single source might produce both ragged and flat element distribution patterns in which case the parent comet would be a mixed rubble pile, it might show variable degrees of on-orbit processing as a function of meteor stream age (cf. Borovička 2007), or both.

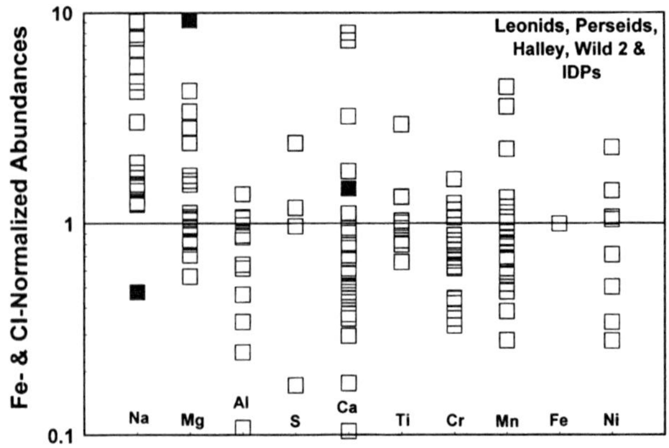

Fig. 6 A compilation of the Fe- and CI-normalized abundances in these meteoroids (Fig. 4) (open squares) augmented with a Leonid fireball (Abe et al. 2005) (solid squares) that yielded the lowest Na value, comet Halley (Fig. 2), comet Wild 2 (Fig. 5), and aggregate IDPs (Rietmeijer 2002a)

5 Differential Ablation

Rare refractory IDPs exist (Zolensky 1987). Ca–Al-rich inclusions ranging from ∼100 μm to ∼10 mm are common in carbonaceous chondrite meteorites (see Rietmeijer and Nuth 2000). Yet, comet meteors are naturally low in Al, Ca and Ti (Fig. 6). It is significant that the Wild 2 composition was low in these elements despite a terminal grain of several microns that was entirely made of refractory minerals (Zolensky et al. 2006). The Wild 2, Leonid and Perseid meteoroid compositions are indistinguishable (Fig. 5). At least for fast comet meteors the bulk composition is accurately constrained by elements released in the main component.

Indisputably the 2nd component revealed an additional Ca-reservoir. As kinetic constraints delay evaporation of 'large' grains over finer grains, Ca release in the main and 2nd component regimes might indicate the presence of refractory dust with significantly different sizes, or different refractory minerals of similar grain size. There are very few minerals that could yield a strong 2nd component Ca signal and little else (Rietmeijer 2002b), which poses a bit of a problem. Calcium carbonates have low boiling points (<1400 K) but perhaps ablation of a cm-sized carbonate grain could be delayed long enough for major Ca-release to appear in the 2nd component. In this case, there should be the concomitant 'C' and 'CO' species emission, perhaps associated with a "humped" light curve (Murray et al. 2000). High-Ca refractory grains in Ca–Al-rich inclusions (Brearley and Jones 1998), IDPs and in Wild 2 have very low iron contents. Thus, any Fe-normalized Ca abundances from the 2nd component would plot well above the CI line thereby overemphasizing its importance in terms of its real abundance.

6 Discussion

The main component Fe-normalized abundances in Perseid meteors ranged from 0.2 to 0.45 for Fe = 0.85 in CI (Borovička 2005) to 0.71–0.98 (av. 0.85) (Trigo-Rodriguez et al. 2003). As the CI abundances were obtained on fully hydrated materials, it could follow that the meteoroids of 0.2–29 g (photographic mass) studied by Trigo-Rodriguez et al. (2003) were small hydrated objects. If so, they should have released hydrogen. Absent this release, the CI composition could be interpreted as compact, equigranular anhydrous comet debris. Borovička (2005) reported a range of 0.01–1 kg for Leonid and Perseid meteoroids. The upper limit could be a porous aggregate meteoroid $\sim$27 cm in diameter, assuming a density of 0.1 g/cm^3. Its composition and texture might resemble comet Halley's dust and its Fe-normalized element abundances would appear to be >CI. Rietmeijer and Nuth (2004) had postulated that newly processed and accreted Fe–Ni-sulfide, Fe–Ni-metal, and Mg-rich silicate grains became parts of ever-larger aggregates as hierarchical dust accretion continued. In this scenario, increasingly more iron became concentrated in fewer large Fe-rich grains. Thus, unique to aggregate meteoroids, the fragmentation of structurally weak comet particles determined meteoroid properties that could impose a bias in the normalization of their element abundances. Magnesium in such meteoroids followed a similar scenario.

What size (mass) of a cometary meteoroid would be representative of the bulk composition of its parent? The minute mass of comet Halley dust (Fomenkova et al. 1992) after the severe fragmentation of the ejected dust prior to collection was most likely insufficient even when the comet is a homogeneous dirty snowball (Sagdeev et al. 1986) of nanometer-scale silicates (cf. Rietmeijer 2002a, b). The collected sample of Wild 2 dust is much larger and efforts are under way (Flynn 2007) to assess how with this larger sample we can be confident that it represents the comet's bulk composition.

Meteor light curves might contain information to trace the hierarchical dust accretion history of comet meteoroids, if not in an absolute way, then by comparative studies of the light curves, fragmentation behavior and deceleration for meteor from the same source and the relative onsets of element releases. Normalization bias could be partially responsible for the ranges in elemental abundance variations (Fig. 6) but it can be used constructively to indicate that the normalizing element is located in massive grains that make up an aggregate meteoroid. This exercise is currently based on small-number statistics and more quantitative meteoroid compositions are required to verify the observations in this paper.

7 Conclusions

Comet aggregate meteoroids have unique morphological structural, mineralogical and chemical properties because parent body processes had not modified their sources. Sampling and normalization bias occurs but the wide range in element abundances of comet meteoroids reveals the extent of indigenous variability in dust textures, mineral types and grain sizes among and within the comet nuclei of fast meteoroids. These meteoroid bulk compositions are well constrained by the main component releases. Differential ablation of probably relatively large refractory grains in a fast comet aggregate meteoroid has no significant impact on the bulk composition from the main component. The tacit corollary of this study is that a perfect CI bulk composition will be restricted to equigranular meteoroids because they were hydrated in their parent body (e.g. Orgueil meteorite) or thermally (e.g. Wild 2 grains in quenched aerogel melt). More studies of all aspects of the physical interactions of meteoroids from the same source during deceleration are necessary at this time when data are already forthcoming from the STARDUST mission (Brownlee et al. 2006), and later the ROSETTA mission (Colangeli et al. 2004). There is no plausible logical reason that each fast comet meteor should fit the CI-composition. Adherence to this compositional constraint will prevent us to appreciate indigenous chemical and mineralogical variability within a comet nucleus and among the comets that produce fast meteors.

Acknowledgments I am supported by grants NNG04GM48A (NASA Goddard Space Flight Center), NNG05GM84G (NASA Headquarters STARDUST participating Scientist Program, and NNX07AM65G (SSAP). Two anonymous reviewers made helpful suggestions.

References

S. Abe, N. Ebizuka, H. Yano, J.-I. Watanabe, J. Borovocka, Detection of the N_2^+ first negative system in a bright Leonid fireball. Astrophys. J. **618**, L141–L144 (2005)

E. Anders, N. Grevesse, Abundances of the elements: meteoritic and solar. Geochim. Cosmochim. Acta. **53**, 197–214 (1989)

J. Borovička, Elemental abundances in Leonid and Perseid meteoroids, in *Modern Meteor Science, An interdisciplinary view*, ed. by R. Hawkes, I. Mann, P. Brown (Springer, 2005), pp. 245–253

J. Borovička, Properties of meteoroids from different classes of parent bodies, in *Near Earth Objects, Our Celestial Neighbors: Opportunity and Risk, Proc. IAU Symposium. 236* ed. by G.B. Valsecchi, D. Vokrouhlický (Cambridge, Cambridge University Press, 2007), pp. 107–120

A.J. Brearley, R.H. Jones, Chondritic meteorites, in *Reviews in Mineralogy*, vol. 36, ed. by J.J. Papike (Mineralogical Society of America, Chantilly, Virginia, 1998), pp. 3–1 - 3–398

D.E. Brownlee et al., Comet 81P/Wild 2 under a microscope. Science **314**, 1711–1716 (2006)

M.J. Cintala, Meteoroid impact into short-period comet nuclei. Nature **291**, 134–136 (1981)

L. Colangeli, E. Mazzotta Epifani, P. Palumbo (eds.), *The New ROSETTA Targets—Observation, Simulations and Instrument Performances*. (Astrophys. Space Sci. Library, Kluwer Academic Publishers, 307 p., 2004)

G.J. Flynn, Physical properties of meteorites and interplanetary dust particles: Clues to the properties of meteors and their parent bodies, in *Modern Meteor Science, An interdisciplinary view*, ed. by R. Hawkes, I. Mann, P. Brown (Springer, 2005) pp. 361–374

G.J. Flynn, Physical, chemical, and mineralogical properties of comet 91P/Wild 2 particles collected by Stardust (2007, this issue)

G.J. Flynn et al., Elemental compositions of comet 81P/Wild 2 samples collected by Stardust. Science **314**, 1731–1735 (2006)

G.J. Flynn, A. Lanzirotti, S.R. Sutton, I. Sitnitsky, Chemical compositions of five large cluster IDPs (abstract). Lunar Planet. Sci. 38, #2290, Lunar and Planetary Institute, Houston Texas (2007)

M.N. Fomenkova, J.F. Kerridge, K. Marti, L.-A. McFadden, Compositional trends in rock-forming elements of comet Halley dust. Science **258**, 266–269 (1992)

M. Gounelle, O. Spurný, P.A. Bland, The orbit and atmospheric trajectory of the Orgueil meteorite from historical records. Meteoriti. Planet. Sci. **41**, 135–150 (2006)

F. Hörz et al., Impact features on Stardust: implications for comet 81P/Wild 2 dust. Science **314**, 1716–1719 (2006)

H.H. Hsieh, D. Jewitt, A population of comets in the main asteroid belt. Science **312**, 561–563 (2006)

P. Jenniskens, Quantitative meteor spectroscopy: elemental abundances. Adv. Space Res. **39**, 491–512 (2007)

E.K. Jessberger, A. Christoforidis, J. Kissel, Aspects of major element composition of Halley's dust. Nature 332, 691–695 (1988)

H. Leroux, F.J.M. Rietmeijer, M.A. Velbel, A.J. Brearley, D. Jacob, F. Langenhorst, J.C. Bridges, T.J. Zega, R.M. Stroud, P. Cordier, R.P. Harvey, M. Lee, M. Gounelle, M.E. Zolensky, A TEM study of thermally modified comet 81P/Wild 2 dust particles by interactions with the aerogel matrix during the Stardust capture process. Meteorit. Planet. Sci. (2007, in press)

I.D.R. Mackinnon, S.A. Kaser, The clay-size fraction of CI chondrites Alais and Orgueil: an AEM study (abstract). Lunar Planet. Sci. **19**, 708–709 (1988), Lunar and Planetary Institute, Houston Texas

I.S. Murray, M. Beech, M.J. Taylor, R.L. Hawkes, Comparison of 1998 and 1999 Leonid light curve morphology and meteoroid structure. Earth Moon Planets **82–83**, 351–367 (2000)

F.J.M. Rietmeijer, Interplanetary dust particles, in *Revs. in Mineralogy*, ed. by J.J. Papike, vol. 36 (Mineralogical Society of America, Chantilly, Virginia, 1998), pp. 2-1–2-95

F.J.M. Rietmeijer, Sodium tails of comets: Na/O and Na/Si abundances in interplanetary dust particles. Astrophys. J. **514**, L125–L127 (1999)

F.J.M. Rietmeijer, Interrelationships among meteoric metals, meteors, interplanetary dust, micrometeorites, and meteorites. Meteorit. Planet. Sci. **35**, 1025–1041 (2000)

F.J.M. Rietmeijer, The earliest chemical dust evolution in the solar nebula. Chemie der Erde **62**, 1–45 (2002a)

F.J.M. Rietmeijer, Shower meteoroids: constraints from interplanetary dust particles and Leonid meteors. Earth Moon Planets **88**, 35–58 (2002b)

F.J.M. Rietmeijer, Dynamic pyrometamorphism during atmospheric entry of large (~ 10 micron) pyrrhotite fragments from cluster IDPs. Meteorit. Planet. Sci. **39**, 1869–1887 (2004)

F.J.M. Rietmeijer, Interplanetary dust and carbonaceous meteorites: constraints on porosity, mineralogy and chemistry of meteors from rubble-pile planetesimals, in *Modern Meteor Science, An Interdisciplinary View*, ed. by R. Hawkes, I. Mann, P. Brown (Springer, 2005), pp. 321–338

F.J.M. Rietmeijer, J.A. Nuth III, Collected extraterrestrial materials: constraints on meteor and fireball compositions, in *Leonid Storm Research* ed. by P. Jenniskens, F.J.M. Rietmeijer, N. Brosch, M. Fonda (Kluwer Acad. Publ., Dordrecht, The Netherlands, 2000), pp. 325–350

F.J.M. Rietmeijer, J.A. Nuth III, Grain sizes of ejected comet dust: Condensed dust analogs, interplanetary dust particles and meteors, in *The New ROSETTA Targets—Observation, simulations and instrument performances, Astrophys. Space Sci. Library*, ed. by L. Colangeli, E. Mazzotta Epifani, P. Palumbo (Kluwer Academic Publishers, 2004), pp. 97–110

R.Z. Sagdeev, J. Blamont, A.A. Galeev, V.I. Moroz, V.D. Shapiro, V.I. Shevchenko, K. Szegő, VEGA spacecraft encounters with comet Halley. Nature **321**, 259–262 (1986)

T. Stephan, D. Rost, E.P. Vicenci, E.S. Bullock, G.J. MacPherson, A.J. Westphal, C.J. Snead, G.J. Flynn, S.A. Sandford, M.E. Zolensky, TOF-SIMS analysis of cometary matter in Stardust aerogel track. Meteorit. Planet. Sci. (2007, in press)

H. Suess, I. Nuclear abundance rules and the composition of meteorites, in *Meteorite Research*, ed by P.M. Millman, Meteorite Research (Springer-Verlag New York Inc./New York-D. Reidel Publishing Company/Dordrecht-Holland), pp. 3–6

J.M. Trigo-Rodriguez, J. Llorca, J. Borovička, J. Fabregat, Chemical abundances determined from meteor spectra: I. Ratios of the main chemical elements. Meteorit. Planet. Sci. **38**, 1283–1294 (2003)

A.J. Tuzzolino, T.E. Economou, B. Clark, P. Tsou, D.E. Brownlee, S.F. Green, J.A.M. McDonnell, N. McBride, M.T.S.H. Colwell, Dust measurements in the coma of comet 81P/Wild 2 by the Dust Flux Monitor Experiment. Science **304**, 1776–1780 (2004)

H.B. Wiik, On regular discontinuities in the composition of meteorites. Commentationes Physico-Mathematicae **34**, 1350145 (1969)

M.E. Zolensky, Refractory interplanetary dust particles. Science **237**, 1466–1468 (1987)

M.E. Zolensky et al., Mineralogy and petrology of comet 81P/Wild 2 nucleus samples. Science **314**, 1735–1739 (2006)

Carbon in Meteoroids: Wild 2 Dust Analyses, IDPs and Cometary Dust Analogues

Alessandra Rotundi · Frans J. M. Rietmeijer

Originally published in the journal Earth, Moon, and Planets, Volume 102, Nos 1–4.
DOI: 10.1007/s11038-007-9218-7

Abstract Assuming that similar organic components as in comet 81P/Wild 2 are present in incoming meteoroids, we try to anticipate the observable signatures they would produce for meteor detection techniques. In this analysis we consider the elemental and organic components in cometary aggregate interplanetary dust particles and laboratory analyses of inter- and circumstellar carbon dust analogues. On the basis of our analysis we submit that (semi) quantitative measurements of H, N and C produced during meteor ablation will open an entire new aspect to using meteoroids as tracers of these volatile element abundances in active comets and their contributions to the mesospheric metal layers.

Keywords Comets · 81P/Wild 2 · Meteoroids · Interplanetary dust particles · Cometary dust analogues · Carbon · Organics

1 Introduction

The laboratory analyses of meteorites, in particular those associated with a recorded fireball event, micrometeorites and interplanetary dust particles (IDPs) are the more direct sources of information to constrain physical and chemical characteristics of meteoroids (Rietmeijer 2000). These materials that reached the Earth's surface or were collected in the lower stratosphere are biased because of their physical properties, *e.g.* material strength (Flynn 2004; Borovička 2007; Trigo-Rodríguez and Llorca 2006, 2007) and orbital properties (Ceplecha et al. 1998; Borovička 2005). Meteors include Zodiacal cloud debris released from asteroids and active comets, which are the most prolific producers of small < cm-sized interplanetary matter, and, mostly comet, debris organized in meteoroid

A. Rotundi (✉)
Dip. Di Scienze Applicate, Università di Napoli Parthenope, Centro Direzionale di Napoli, Isola C4, 80143 Napoli, Italy
e-mail: rotundi@uniparthenope.it

F. J. M. Rietmeijer
Department of Earth and Planetary Sciences, MSC03-2040, University of New Mexico, Albuquerque, NM 87131, USA

DOI: 10.1007/978-0-387-78419-9_63

streams (Jenniskens 2006). TV-camera observations, photographic spectroscopic techniques and spectral line identification of meteors over the last 30 years have revealed considerable insights into the chemical and physical properties of meteoroids (Jenniskens et al. 2000; Hawkes et al. 2005). More recently, quantitative chemical analyses from high-resolution spectroscopy of meteors became available (e.g. Borovička 2005) and on the physical processes, e.g. ablation and sputtering, during meteoroids atmospheric entry (Popova 2005; Popova et al. 2007). Meteor studies, and laboratory analyses (Nuth et al. 2002; Rotundi et al. 2002) of comet and asteroid dust analogues can provide specific chemical, mineralogical and textural information on comets and asteroids but nothing is better than laboratory analyses of dust collected from a known active comet. Thus, the Stardust mission to comet 81P/Wild 2 (Brownlee et al. 2004, 2006 will become a calibration point for meteor data analyses. Specifically, the comet carbon-bearing components (cf. Sandford et al. 2006) will provide a clear definition of what we would expect to observe with regard to these components during the physical interactions of incoming meteoroids, ranging with the Earth atmosphere. They will help (1) to optimise techniques for meteoroid observations of the volatile components and (2) to explore for new observable parameters.

Inorganic and organic carbons and hydrocarbons abound in comets and many asteroids as evidenced by astronomical observations (Ehrenfreund and Charnley 2000; Kelley et al. 2006; Bockelée-Morvan et al. 2000), in meteorites (a.o. Nakamura-Messenger et al. 2006; Garvie and Buseck 2006; Pizzarello et al. 2001; Busemann et al. 2006; Kerridge and Matthews 1988) and IDPs (Rietmeijer 1998; Flynn et al. 2003, 2004). Ablation protected the interior of meteorites beneath the fusion crust but atmospheric entry heating of micrometeorites (Genge 2008) and IDPs (Rietmeijer 1998) can be severe enough to modify indigenous volatile components. Carbonaceous materials in carbon-rich IDPs were fused into contiguous patches wherein less volatile silicate and sulphide minerals were embedded (Rietmeijer 1992; Thomas et al. 1993). Inorganic and organic carbons and hydrocarbons present in meteoroids are expected to be modified due to interactions with the atmosphere, if not compromised by thermal fractionation.

The physical interactions of meteoroids with the atmosphere include (1) very high-altitude sputtering above ~130 km altitude (Koten et al. 2006; Hill et al. 2004; Rogers et al. 2005; Spurný et al. 2000) and (2) ablation that is dominated by two distinct thermal regimes at ~5000 K, the main component, and at ~10,000 K, the second component (Borovička 2005). These processes will be most efficient for cometary meteors such as the Leonids and Perseids (Borovička 2005) that are both associated with Halley-type comets. Comet Halley itself was the target of the Vega and Giotto missions during the 1985/86-perihelion passage (Grewing et al. 1987). The carbon and organics in this comet that were inferred from the on-board mass spectrometers show a very complex and diverse CHON speciation (Kissel and Krueger 1987; Fomenkova 1999; Fomenkova et al. 1994). Still today, because these data could not give information on the comet grain structures, the interrelationship of CHON species with the other comet dust is unproven. Is comet Halley dust made of mixtures of ice and interstellar grains forming core-mantle grains (Greenberg 1998), or not, although in their analysis Kissel and Krueger (1987) preferred this interpretation.

The "humped" light curves among Leonid meteors were interpreted as evidence for compound meteoroids that had a strong and a structurally weak component (Murray et al. 2000). Using what was known about aggregate IDPs, Rietmeijer (2002) suggested that a compound meteoroid could be a micrometer-size sulphide, or a silicate, grain with attached fine-grained aggregate material. Scaling up to cluster IDPs the massive and aggregate grains could be ~10 μm in diameter (Rietmeijer and Nuth 2004). A compound grain of

$\sim 8\ \mu$ is present in comet 81P/Wild 2. It consists of a large sulphide, a small silicate grain and a chondritic aggregate (Brownlee et al. 2006) wherein carbon material had survived (Matrajt et al. 2007a, b). Brownlee et al. (2006) suggested that survival of the fragile aggregate material was possible because of the entry geometry whereby it was shielded in the wake of the structurally stronger mineral grains. Shielding effects could also occur in meteoroids during atmospheric entry but at this time we will not pursue this unique behaviour however exciting it might be for volatile species survival.

Meteors are a source of organics deposited into the atmosphere during ablation (Jenniskens 2001). Here then we will discuss the laboratory measurements of organic and inorganic carbon species in particles that were ejected from the nucleus of comet Wild 2. On the basis of these, still limited, data we will explore what their behaviour might have been, during the physical interactions with the Earth's atmosphere, if they had been present in a cometary meteoroid, and how, if at all, they could leave observable meteor signatures.

2 The nature of Stardust samples

The Stardust mission captured thousands of particles, 3-2,000 μm in diameter (Tuzzolino et al. 2004) in the coma of comet 81P/Wild at a distance of 235 km ± 1 km from the nucleus. The comet particles impacted under-dense silica (SiO_2) aerogel tiles and Al foil at 6.1 km/s. The particle deceleration tracks in aerogel showed distinct morphologies that are a first order indication of the (bulk) particle structure, e.g. a structurally-coherent or an aggregate particle (Hörz et al. 2006). From systematic analyses of track morphology and particles extracted from a still limited number of tracks Zolensky et al. (2006) concluded that "The bulk of the comet 81P/Wild 2 (hereafter Wild 2) samples returned to Earth by the Stardust spacecraft appear to be weakly constructed mixtures of nanometer-scale grains, with occasional much larger (over 1 μm) ferromagnesian silicates, Fe-Ni sulfides, Fe-Ni metal, and accessory phases". Bulbous tracks were most likely produced by aggregate particles that according to Zolensky et al. (2006) were "disaggregated into individual components, with the larger, denser components penetrating more deeply into the aerogel, making thin tracks with terminal grains". The cause of the bulbous cavity is uncertain. It could be some type of "explosive" event caused by ultra-rapid vaporisation of indigenous comet volatile phases (Sandford et al. 2006), silica aerogel (Rietmeijer et al. 2007), or both. Dominguez et al. (2007) argued that for the fragmentation and subsequent deceleration of sub-micron grains in the particle. The first mechanism is supported by (1) the formation of ~ 100 nm iron-silicide spheres near the penetration hole of track #35 that involved silica vapour (Rietmeijer et al. 2007), (2) distributions of <100 nm Fe, Ni-sulphides, sulphur 'hot spots' and melted aerogel silica glass with numerous dispersed nanometer grains distant from track walls (Leroux et al. 2007; Zolensky et al. 2006) and (3) aliphatic C–H carbon that was spatially associated in a few tracks but distributed as much as several track diameters from the track centre (Sandford et al. 2006). The extent of silica evaporation during track formation, or perhaps only near the particle penetration hole, is not yet quantified but there is no evidence yet that Wild 2 particles were volatile rich and/or particularly carbon-rich (Burchell et al. 2007). There is evidence for apparent survival of carbon and/or hydrocarbon materials under unusual conditions, volatilization and vaporization and condensation (re-deposition) of modified materials. More laboratory analyses of the samples are required to establish that evaporation of volatile phase in comet particles was responsible for the bulbous part formation (Burchell et al. 2007), and if so, it places kinetic constraints on the survival of carbon and/or hydrocarbon materials in the tracks.

3 Organics in Wild2 and in IDPs

The spectra acquired by the mass spectrometer during the flyby of Wild 2 registered nitrogen-containing species (Kissel et al. 2004) that were confirmed by subsequent laboratory analyses. Some of the organics present in the Wild 2 grains detected by laboratory analyses are summarised below. The abundances, functionality, relative C, H, N and O abundances and compositions of organics were heterogeneous among grains extracted from the same track, i.e. within the same comet particle (Rotundi et al. 2007; Sandford et al. 2006). Rotundi et al. (2007) found that the grains included highly disordered amorphous carbons that are probably aromatic hydrocarbons and C-H chain molecules. Spatially correlated C, N, and S distributions and low C/N ratios are consistent with the presence of volatile organic molecules, as e.g. HCN.

The C-bearing component in sixteen grains (9–15 μm in size), eleven of which were extracted from the bulbous part of track #35, remained apparently unmodified during deceleration. They show the typical D (disordered) and G (graphite) Raman band parameters of highly disordered carbonaceous material and are similar to those measured in IDPs (Rotundi et al. 2007). These light element compounds in IDPs remain poorly characterized but they include "light hydrocarbons", organic carbons, amorphous and poorly graphitized carbons, not all necessarily organic in nature (Rietmeijer 1998; Flynn et al. 2003).

One almost pure C (minor O) grain showed the Raman band parameters for extremely disordered carbon similar to amorphous carbon produced by irradiation of carbonaceous materials (Ferini et al. 2004). This grain was also enriched in deuterium (McKeegan et al. 2006). Its presence is prima facie evidence that at least some fraction of indigenous Wild 2 carbonaceous materials did survive the deceleration process unmodified.

Four of the eleven grains extracted from the bulb of track #35 and analysed by Raman spectroscopy were also analysed by infrared micro-spectroscopy. They all show the C–H aliphatic band (Rotundi et al. 2007). One of these four grains shows a particularly intense O-H band suggesting a possible presence of a volatile component.

The cometary grains CH_2/CH_3 infrared (3.4 μm) bands ratios (Table 1) are on average higher than those calculated for IDPs (Matrajt et al. 2004) suggesting relatively long or less branched, aliphatic chains. This spectral feature was not detected in bulk IDPs but it appeared after crushing the IDPs showing organic matter was not present at the IDP surface (Flynn et al. 2004).

Some fraction of organics was mobilized, another apparently survived. What ratio of CHON-like and almost pure carbon materials survived the collection process, either unmodified or slightly modified, or were evaporated is still unknown. The surviving comet

Table 1 Organic matter and pure carbons in IDPs and in Wild 2 grains

	CH_2/CH_3	Pure carbon	Organic Carbons (chain and rings)	Carbon content
IDPs	3.4–5.5	Amorphous, Poorly-graphitised	C–O, C–H, C–N, PAHs	12 (wt%) av. on >100 IDPs
Wild 2	5.7–9.6	Amorphous	C–O, C–H, C–N, CN^-, CH^+, PAHs	10 (wt%) av. on 4 grains of 1 Wild2 particle

IDPs: Schramm et al. 1989; Thomas et al. 1994; Flynn et al. 2004; Matrajt et al. 2004
Wild 2: Sandford et al. 2006; Rotundi et al. 2007; Kissel et al. 2004

materials provide an observational basis to assess how they might behave during atmospheric interactions if were they incorporated in a decelerating cometary meteoroid.

4 Laboratory Carbon Analogue Studies

4.1 Carbon Condensation

Meteor ablation reaches temperatures that would allow elemental carbon evaporation (2,000–3,000 K). There could be two different effects associated with the behaviour of amorphous pure carbon compounds and organics in meteoroids: (1) introduction of processed C-bearing species to the atmosphere and (2) condensation of pure carbon and C–H-species in the meteor's wake or a persistent train. Condensation will critically depend on vapour density and prevailing kinetic conditions as shown by laboratory condensation experiments that produced (1) chain-like aggregates of amorphous carbon soot grains (fullerenes and soot), (2) multi-walled fullerenic carbon onions and hollow tubes, (3) amorphous carbon, and (4) poorly graphitized and graphitic carbons (Rotundi et al. 1998, 2006; Rietmeijer 2006; Rietmeijer et al. 2004). It raises the possibility that similar 10–100 nm size carbons could form in dense carbon-rich vapours generated in the wake of carbon-rich cometary meteoroids. The C_{60} molecule could be "seeds" for soot formation (Rotundi et al. 2006) but it is doubtful there will be sufficient time for significant and detectable soot formation during meteor ablation. Further considering the difficulties involved in making unique fullerene identifications from UV spectral analyses (the 217.5 nm-feature) (see chapters in Rietmeijer 2006) there is little hope that fullerenes and soot in the atmosphere above $\sim$80 km altitude can be linked to meteor activity, except perhaps by direct capture.

4.2 Carbon Spectral Signatures

Amorphous carbons only show distinct characteristic spectral features in the infrared (IR), the ultraviolet (UV) and far UV ranges carbon as a function of hydrogen content and structure of the carbon grains (Fig. 1). The shape and intensity of the typical 3.4 μm IR feature of hydrogenated carbon materials, that is due to symmetric and asymmetric C–H vibrations in the methyl (CH_3) and methylene (CH_2) functional groups, varies systematically as a function of hydrogen content (Mennella et al. 2002). Laboratory experiments on submicron hydrogenated amorphous carbon grains that were subjected to thermal annealing showed that the characteristic 217.5 nm UV carbon spectral feature shifted as a function of the hydrogen content of carbon samples (Mennella et al. 1995). This feature became more pronounced and shifted towards longer wavelengths as the annealing temperature increased. This UV feature is absent for amorphous carbon with a high percentage of hydrogen. During annealing it emerges and increases in strength, as hydrogen is lost. After 3 h at 250°C the peak position is at 194 nm. It is at 259 nm after 3 h at 800°C.

4.3 Carbon Detection in Meteors

Temperatures in meteors define two regimes at $\sim$5000 K and $\sim$10,000 K. Both regimes are typically achieved in high-velocity meteoroids (e.g. Leonid and Pereid meteors); in

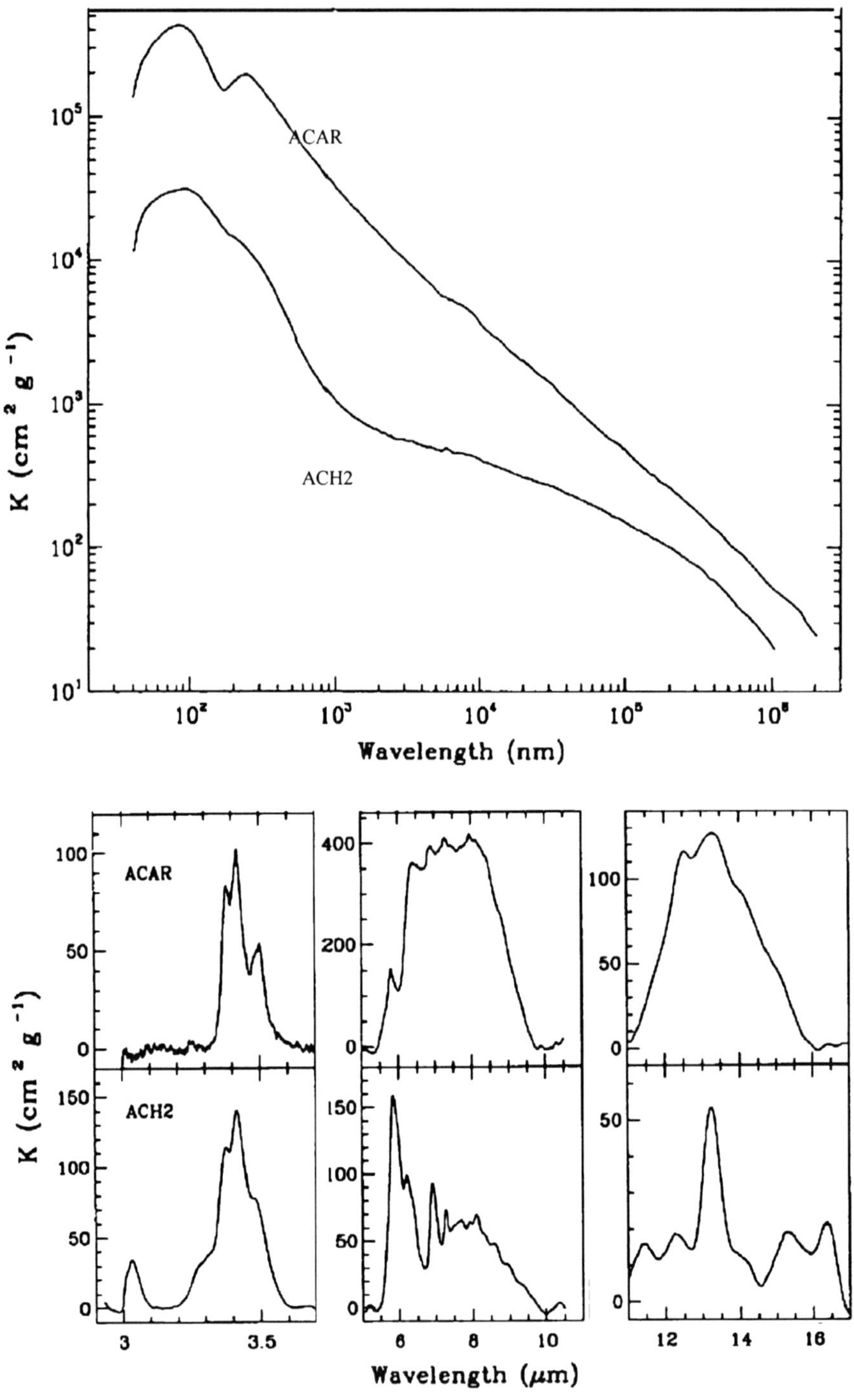

Fig. 1 The amorphous carbon spectra of produced by arc discharge in an argon atmosphere (ACAR) and in a hydrogen atmosphere (ACH2) from the far UV to the far IR (top); enlarged spectra in the IR range (bottom). Adapted from Colangeli et al. (1995)

low-velocity meteoroids the high temperature regime is not reached (Borovička 1994). The lower regime is comparable to temperatures experienced by Wild 2 grains decelerating in aerogel (cf. Rietmeijer et al. 2007). It is possible that much higher temperatures were reached but it is doubtful they reached as high as 10,000 K, which was based on modelling studies of aerogel with a higher density than used for Stardust (Anderson and Ahrens 1994).

When we assume that pure carbon compounds and organics similar to those in Wild 2 are present in decelerating cometary meteoroids with entry velocities in the range of ~25 to ~70 km/s (cf. Rietmeijer 2000) what would be the meteor ablation signatures to search for, assuming detections could be made in the appropriate spectral ranges. Meteor temperatures can be measured as a function of time together with the abundances of chemical elements being released (Trigo-Rodríguez 2002; Trigo-Rodríguez et al. 2003). We show the example of magnesium in a Perseid meteoroid (Fig. 2). It suffices to say that temperature data will be potentially available to trace volatile element releases from C–H–O–N compounds. We do not understand the nature of the correlations and anti-correlations between the Mg-abundances and temperature as a function of time but they contain a hint of a kinetic effects controlling ablation.

The duration of an impact event in Stardust aerogel is measured in nano-to milliseconds (Brownlee et al. 2006). The duration of a typical visual fast meteor between altitudes 120 km to 90 km is 0.5 to 1 second. The point being that time and temperature are measurable quantities that would allow kinetic effects to be explored on comparable scales.

The O and N rich Wild 2 organics offer opportunities as both O and N are readily seen in meteor spectra (Borovička 2005). Their exact origins in the upper atmosphere are not

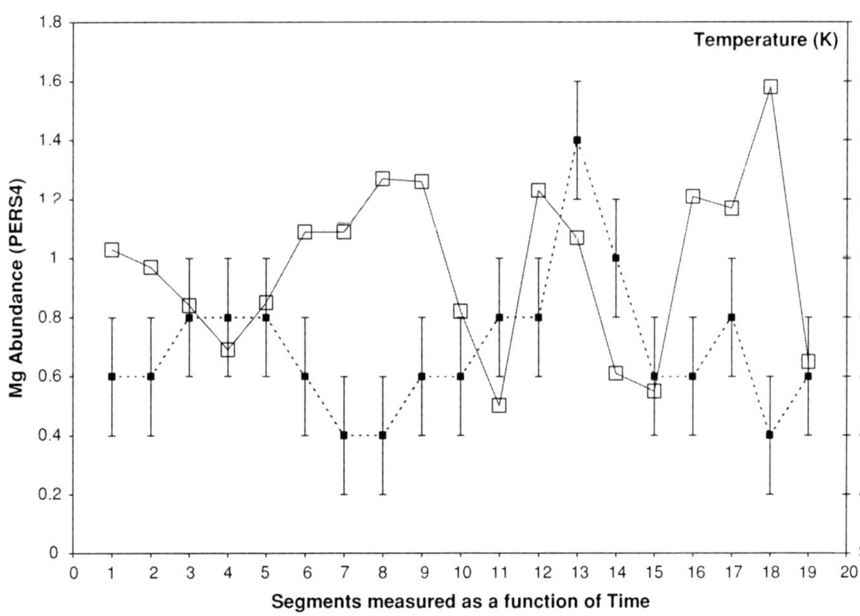

Fig. 2 Magnesium abundances (open squares and solid lines) in a Perseid meteor (PERS4) as a function of temperature (K) (solid squares and dashed lines) identified on the right–hand side of the diagram and both plotted as a function of time (are decreasing altitude) indicated by the numerical values of the measured segments from 1 to 19. The present authors produced this diagram from data by Trigo-Rodríguez (2002)

clear with regard to the fraction of indigenous meteoroid concentrations and excitation of N and O species in the atmosphere (a.o. Jenniskens et al. 2004; Abe et al. 2005). The Wild 2 oxygen and nitrogen data could be benchmark values that, in conjunction with their atmospheric mixing ratios, might provide an opportunity to detect these elements in cometary shower meteors, or to provide lower concentration limits when in excess of the chondritic abundances of these elements.

If, as aggregate IDP data suggest, carbon in comet dust is much higher than the chondritic carbon abundance, there should be potentially detectable signals of neutral and ionised C-species, carbon vapour condensates, particularly fullerenes, or both in meteors, which will be a challenge to spectroscope observers.

Hydrogen (Jenniskens and Mandell 2004; Abe et al. 2007; Borovička and Jenniskens 2000) and OH (Abe et al. 2002; Jenniskens and Laux 2004) are detectable in meteor spectra and Wild 2 proves it is present in comets. Wild 2 hydrogen abundances are still only semi-quantitatively constrained. They could eventually become "typical comet H values" that when found to be systematically lower in cometary meteoroids might hold clues to on-orbit dust aging in a manner similar to the Na-abundances that Trigo-Rodríguez et al. (2004) and Borovička (2005, 2007) used to trace on-orbit meteoroid chemical modification.

5 Conclusions

The elemental carbon and organic contents of cometary meteoroids are not firmly established which is a matter of detection limits and capabilities of the currently used instruments. We can say that compound meteoroids consisting of strong and a structurally weak component that produce 'humped' light curves are among cometary debris, witness the one terminal grain that was part of a Wild 2 particle. They consist of a massive and an aggregate component that includes light element phases. Amorphous elemental carbons and organic materials in cometary meteoroids, that are similar to those identified in comet Wild 2 and aggregate IDPs, will interact with the upper atmosphere. The measured C, H, O and N abundances in Wild 2 set some constraints on detections of these elements during the meteor phase. We are cautiously optimistic that with the advances in modern meteor science these elements will be detected and measured in meteors. They will provide data on their origins, abundances and aging processes in cometary meteoroids.

Acknowledgments We thank George Flynn for a very thoughtful and constructive review. AR was supported by MIUR–PRIN and ASI grant. FJMR was supported by grant NNX07AM65G through the NASA Stardust Analyses Program.

References

S. Abe, H. Yano, N. Ebizuka,T. Kasuga, J.-I. Watanbe, M. Sugimoto, N. Fujino, T. Fuse, R. Ogasawara, First results of OH emission from meteor and after glow: Search for organics in cometary meteoroids, in *Asteroids, Comets, Meteors-ACM 2002*, ed. by B. Warmbein (ESA SP-500, ESA Publications Division, The Netherlands, 2002), pp. 213–216

S. Abe, N. Ebizuka, H. Yano, J.-I. Watanabe, J. Borovička, Detection of the N_2^+ first negative system in a bright Leonid fireball. Astrophys. J. **618**, L141–L144 (2005)

S. Abe, N. Ebizuka, H. Yano, J.-I. Watanabe, J. Borovička, Search for OH(A X) and detection of N_2^+ (B X) in ultraviolet meteor spectrum. Adv. Space Res. **39**, 538–543 (2007)

W. Anderson, T.J. Ahrens, Physics of interplanetary dust capture via impact into organic foam. J. Geophys. Res. E99, 2063–2071 (1994)

J. Borovička, Two components in meteor spectra. Planet. Space Sci. **42**, 145–150 (1994)

J. Borovička, Physical and chemical properties of meteoroids as deduced from observations, in *Proc. International Astron. Union*, vol. 1, ed. by D.L. Sylvio-Ferraz, M. Angel, F. Julio (Cambridge University Press, Cambridge, 2005), pp. 249–271

J. Borovička, Properties of meteoroids from different classes of parent bodies, in *Proc. IAU Symposium 236, Near Earth Objects, Our Celestial Neighbors: Opportunity and Risk*, ed. by G.B. Valsecchi, D. Vokrouhlický. (Cambridge University Press, Cambridge, 2007), pp. 107–120

J. Borovička, P. Jenniskens, Time resolved spectroscopy of a Leonid fireball afterglow. Earth Moon Planet. **82/83**, 399–428 (2000)

D. Bockelée-Morvan, D.C. Lis, J.E. Wink et al., New molecules found in comet C/1995 O1 (Hale-Bopp). Investigating the link between cometary and interstellar material. Astron. Astrophys. **353**, 1101 (2000)

D.E. Brownlee et al., Surface of young Jupiter family comet 81P/Wild 2: View from the Stardust spacecraft. Science **304**, 1764–1769 (2004)

D.E. Brownlee et al., Comet 81P/Wild 2 under a microscope. Science **314**, 1711–1716 (2006)

M.J. Burchell, S.A.J. Fairey, P. Wozniakiewicz, D.E. Brownlee, F. Hörz, A.T. Kearsley, T.H. See, P. Tsou, A. Westphal, S.F. Green, J.M. Trigo-Rodríguez, G. Dominguez, Characteristics of cometary dust tracks in Stardust aerogel and laboratory calibrations, Meteorit. Planet. Sci. (in press, 2007)

H. Busemann, A.F. Young, C.M.O'D. Alexander, P. Hoppe, S. Mukhopadhyay, L.R. Nittler, Interstellar chemistry recorded in organic matter from primitive meteorites. Science **312**, 727–730 (2006)

Z. Ceplecha, J. Borovička, W.G. Elford, D.O. ReVelle, R.L. Hawkes, V. Porubcan, M. Simek, Meteor phenomena and bodies. Space Sci. Revs. **84**, 327–471 (1998)

L. Colangeli, V. Mennella, P. Palumbo, A. Rotundi, E. Bussoletti, Mass extinction coefficients of various submicron amorphous carbon grains: Tabulated values from 40 nm to 2 mm. Astron. Astrophys. Suppl. Ser. **113**, 561–577 (1995)

G. Dominguez, J.M. Trigo-Rodriguez, M.J. Burchell, F. Hörz, J. Llorca, P. Tsou, B. Anderson, Analysis of factors contributing to bulbous capture tracks from cometary dust particles in the Stardust aerogel collector (abstract). Meteorit. Planet. Sci. **42**(Suppl), A37 (2007)

P. Ehrenfreund, S.B. Charnley, Organic molecules in the interstellar medium, comets, and meteorites: A voyage from dark clouds to the early Earth. Annu. Rev. Astron. Astrophys. **38**, 427–483 (2000)

G. Ferini, G.A. Baratta, M.E. Palumbo, A Raman study of ion irradiated icy mixtures. Astron. Astrophys. **414**, 757–766 (2004)

G.J. Flynn, Physical properties of meteorites and interplanetary dust particles: clues to the properties of the meteors and their parent bodies. Earth Moon Planet. **95**, 361–374 (2004)

G.J. Flynn, L.P. Keller, M. Feser, S. Wirick, C. Jacobsen, The origin of organic matter in the solar system: Evidence from the interplanetary dust particles. Geochim. Cosmochim. Acta. **67**, 4791–4806 (2003)

G.J. Flynn, L.P. Keller, C. Jacobsen, S. Wirick, An assessment of the amount and types of organic matter contributed to the Earth by interplanetary dust. Adv. Space Res. **33**, 57–66 (2004)

M.N. Fomenkova, On the organic refractory component of cometary dust. Space Sci. Rev. **90**, 109–114 (1999)

M.N. Fomenkova, S. Chang, L.M. Mukhin, Carbonaceous components in the comet Halley dust. Geochim. Cosmochim. Acta. 58, 4503–4512 (1994)

M.J. Genge, Micrometeorites and their implications for meteors. Earth Moon Planet., this volume (2008)

L.A.J. Garvie, P.R. Buseck, Carbonaceous materials in the acid residue from the Orgueil carbonaceous chondrite meteorite. Meteorit. Planet. Sci. **41**, 633–642 (2006)

M. Grewing, F. Praderie, R. Reinhard. (eds.), Exploration of Halley's comet (Springer, Heidelberg, 1987), 984 p

J.M. Greenberg, Making a comet nucleus. Astron. Astrophys. **330**, 375–380 (1998)

Hawkes, R, Mann, I, Brown, P. (eds.), Modern Meteor Science. An interdisciplinary view (Springer, Dordrecht, 2005), 732 p

K.A. Hill, L.A. Rogers, R.L. Hawkes, Sputtering and high altitude meteors. Earth Moon Planet. **95**, 403–412 (2004)

F. Hörz et al., Impact features on Stardust: Implications for comet 81 P/Wild 2 dust. Science **314**, 1716–1719 (2006)

P. Jenniskens, Meteors as a delivery vehicle for organic matter to the early Earth, in *Proc. Meteoroids 2001 Conf*, ed. B. Warmbein (ESA SP-495, ESA Publications Division, Noordwijk, 2001), pp. 247–254

P. Jenniskens, *Meteor Showers and their Parent Comets*. (Cambridge University Press, Cambridge, 2006), p.790

P. Jenniskens, C.O. Laux, Search for the OH (X2Pi) Meinel band emission in meteors as a tracer of mineral water in comets: Detection of N_2^+ (A-X). Astrobiol. **4**, 109–121 (2004)

P. Jenniskens, A.M. Mandell, Hydrogen emission in meteors as a potential marker for the exogenous delivery of organics and water. Astrobiol. **4**, 123–121 (2004)

P. Jenniskens, F.J.M. Rietmeijer, N. Brosch, M. Fonda (eds.), *Leonid Storm Research* (Kluwer Academic Publishers, Dordrecht, 2000), 606 p

P. Jenniskens, E.L. Schaller, C.O. Laux, M.A. Wilson, G. Schmidt, R.L. Rairden, Meteors do not break exogenous organic molecules into high yields of diatomics. Astrobiol. **4**, 67–79 (2004)

M.S. Kelley, C.E. Woodward, D.E. Harker, D.H. Wooden, R.D. Gehrz, H. Campins, M.S. Hanner, S.M. Lederer, D.J. Osip, J. Pittichova, E. Polomski, A Spitzer study of comets 2P/Encke, 67P/Churyumov-Gerasimenko, and C/2001 HT50 (Linear-Neat). Astrophys. J. **651**, 1256–1271 (2006)

J.F. Kerridge, M.S. Matthews (eds.), *Meteorites and the Early Solar System* (The University of Arizona Press, Tucson, Arizona, 1988) 1269 p

J. Kissel, F.R. Krueger, The organic component in dust from comet Halley as measured by the PUMA mass spectrometer on board Vega 1. Nature **326**, 755–760 (1987)

J. Kissel, F.R. Krueger, J. Silén, B.C. Clark, The cometary and interstellar dist analyzer at 81P/comet Wild 2. Science **304**, 1774–1776 (2004)

P. Koten, P. Spurny, J. Borovička, S. Evans, A. Elliott, H. Betlem, R. Štork, K. Jobse, The beginning heights and light curves of high-altitude meteors. Meteorit. Planet. Sci. 41, 1305–1320 (2006)

H. Leroux, F.J.M. Rietmeijer, M.A. Velbel, A.J. Brearley, D. Jacob, F. Langenhorst, J.C. Bridges, T.J. Zega, R.M. Stroud, P. Cordier, R.P. Harvey, M. Lee, M. Gounelle, M.E. Zolensky, A TEM study of thermally modified comet 81P/Wild 2 dust particles by interactions with the aerogel matrix during the Stardust capture process. Meteorit. Planet. Sci. (in press, 2007)

G. Matrajt, J. Borg, P.I. Raynal, Z. Djouadi, L. d'Hendecourt, G. Flynn, D. Deboffle, FTIR and Raman analyses of the Tagish Lake meteorite: Relationship with the aliphatic hydrocarbons observed in the diffuse interstellar medium. Astron. Astrophys. 416, 983–990 (2004)

G. Matrajt, S. Wirick, M. Ito, S. Messenger, D. Brownlee, D. Joswiak, Carbon investigation of Stardust particles: A TEM, NanoSIMS and XANES study, Lunar Planet Sci., **38**, LPI Contribution #1338, Lunar and Planetary Institute, Houston, Texas (2004a)

G. Matrajt, S. Wirick, S. Messenger, M. Ito, D. Joswiak, D. Brownlee, A TEM, C-XANES and NanoSIMS investigation of a fragment from the Stardust track ADA (abstract). Meteorit. Planet. Sci. **42**(Suppl), A99 (2007b)

T. McKeegan et al., Isotopic compositions of cometary matter returned by Stardust. Science **314**, 1724–1728 (2006)

V. Mennella, L. Colangeli, A. Blanco, E. Bussoaletti, S. Fonti, P. Palumbo, H.C. Mertins, A dehydrogenation study of cosmic carbon analogue grains. Astrophys. J. **444**, 288–292 (1995)

V. Mennella, J.R. Brucato, L. Colangeli, P. Palumbo, CH bond formation in carbon grains by exposure to atomic hydrogen: the evolution of the carrier of the interstellar 3.4 μm band. Astrophys. J. **569**, 531–540 (2002)

I.S. Murray, M. Beech, M.J. Taylor, R.L. Hawkes, Comparison of 1998 and 1999 Leonid light curve morphology and meteoroid structure. Earth Moon Planet. **82/83**, 351–367 (2000)

K. Nakamura-Messenger, S. Messenger, L.P. Keller, S.J. Clemett, M.E. Zolensky, Organic globules in the Tagish Lake meteorite: Remnants of the protosolar disk. Science **314**, 1439–1442 (2006)

J.A. Nuth, F.J.M. Rietmeijer, H.G.M. Hill, Condensation processes in astrophysical environments: The composition and structure of cometary grains. Meteorit. Planet. Sci. **37**, 1579–1590 (2002)

S. Pizzarello, Y. Huang, L. Becker et al., The organic content of the Tagish Lake meteorite. Science **293**, 2236–2239 (2001)

O. Popova, Meteoroid ablation models. Earth Moon Planet. **95**, 303–319 (2005)

O. Popova, A.S. Strelkov, S.N. Sidneva, Sputtering of fast meteoroids' surface. Adv. Space Res. 39, 567–573 (2007)

F.J.M. Rietmeijer, Mineralogy of primitive chondritic protoplanets in the early solar system, in *Trends in Mineralogy*, vol. 1 (Council Scientific Research Integration, India, 1992), pp. 23–41

F.J.M. Rietmeijer, Interplanetary dust particles, in *Planetary Materials, Reviews in Mineralogy*, vol. 36, ed. by J.J. Papike. (Mineralogical Society of America, Chantilly, Virginia, 1998), 2-1–2-95

F.J.M. Rietmeijer, Interrelationships among meteoritic metals, meteors, interplanetary dust, micrometeorites, and meteorites. Meteorit. Planet. Sci. **35**, 1025–1041 (2000)

F.J.M. Rietmeijer, Shower meteoroids: Constraints from interplanetary dust particles and Leonid meteors. Earth Moon Planet. **88**, 35–58 (2002)

F.J.M. Rietmeijer (eds.), *Natural Fullerenes and related structures of elemental carbon* (Springer, Dordrecht, 2006), 295 p. + xii

F.J.M. Rietmeijer, J.A. Nuth III, Grain sizes of ejected comet dust: Condensed dust analogs, interplanetary dust particles and meteors, in *The new ROSETTA targets–Observation, simulations and instrument performances* eds. by L. Colangeli, E. Mazzotta Epifani, P. Palumbo (Astrophys. Space Sci. Library, Kluwer Academic Publishers, 2004), pp. 97–110

F.J.M. Rietmeijer, A. Rotundi, D. Heymann, C_{60} and giant fullerenes in soot condensed in vapors with variable C/H_2 ratio. Fullerenes Nanotubes Carbon Nanostruc. **12**, 659–680 (2004)

F.J.M. Rietmeijer, T. Nakamura, A. Tsuchiyama, K. Uesugi, T. Nakano, H. Leroux, Origin and formation of iron-silicide phases in the aerogel of the Stardust mission. Meteorit. Planet. Sci. (in press, 2007)

L.A. Rogers, K.A. Hill, R.L. Hawkes, Mass loss due to sputtering and thermal processes in meteoroid ablation. Planet. Space Sci. **53**, 1341–1354 (2005)

A. Rotundi, F.J.M. Rietmeijer, L. Colangeli, V. Mennella, P. Palumbo, E. Bussoletti, Identification of carbon forms in soot materials of astrophysical interest. Astron. Astrophys. **329**, 1087–1096 (1998)

A. Rotundi, J.R. Brucato, L. Colangeli, G. Ferrini, V. Mennella, E. Palomba, P. Palumbo, Production, processing and characterisation techniques for cosmic dust analogues. Meteorit. Planet. Sci. **37**, 1623–1635 (2002)

A. Rotundi, F.J.M. Rietmeijer, J. Borg, Natural C60 and large fullerenes: A matter of detection, in *Natural Fullerenes and Related Structures of Elemental Carbon*, ed. by F.J.M. Rietmeijer (Springer Kluwer, 2006), pp. 71–94

A. Rotundi, G.A. Baratta, J. Borg, J.R. Brucato, H. Busemann, L. Colangeli, L. d'Hendecourt, Z. Djouadi, G. Ferrini, I.A. Franchi, M. Fries, F. Grossemy, L.P. Keller, V. Mennella, K. Nakamura, L.R. Nittler, M.E. Palumbo, S.A. Sandford, A. Steele, B. Wopenka, Combined micro-IR and micro-Raman analyses of comet 81P/Wild 2 particles collected by Stardust. Meteorit. Planet. Sci. (in press, 2007)

S.A. Sandford et al., Organics captured from comet 81P/Wild 2 by the Stardust spacecraft. Science **314**, 1720–1724 (2006)

L.S. Schramm, D.E. Brownlee, M.M. Wheelock, Major element composition of stratospheric micrometeorites. Meteorit. **24**, 99–112 (1989)

P. Spurný, H. Betlem, K. Jobse, P. Koten, J. van't Leven, New type of radiation of bright Leonid meteors above 130 km. Meteorit. Planet. Sci. **35**, 1109–1115 (2000)

K.L. Thomas, G.E. Blanford, L.P. Keller, W. Klöck, D.S. McKay, Carbon abundance and silicate mineralogy of anhydrous interplanetary dust particles. Geochim. Cosmochim. Acta. **57**, 1551–1566 (1993)

K.L. Thomas, L.P. Keller, G.E. Blanford, D.S. McKay, Quantitative analyses of carbon in anhydrous and hydrated interplanetary dust particles, in *Analysis of Interplanetary Dust*, eds. by M.E. Zolensky, T.L. Wilson, F.J.M. Rietmeijer, G.J. Flynn. AIP Conf. Proc, vol. 310 (American Institute of Physics Press, New York, 1994), pp. 165–174

J.M. Trigo-Rodríguez, Spectroscopic analysis of cometary and asteroidal fragments during their entry into the terrestrial atmosphere (in Spanish). PhD. Thesis, University of Valencia, Spain (2002)

J.M. Trigo-Rodríguez, J. Llorca, The strength of cometary meteoroids: Clues to the structure and evolution of comets. Mon. Not. R. Astron. Soc. 372, 655–660 (2006)

J.M. Trigo-Rodríguez, J. Llorca, Erratum: The strength of cometary meteoroids: clues to the structure and evolution of comets. Mon. Not. R. Astron. Soc. (2007). doi:10.1111/j.13652966.2006.11310.x

J.M. Trigo-Rodríguez, J. Llorca, J. Fabregat, Chemical abundances determined from meteor spectra: II. Evidence for enlarged sodium abundances in meteoroids. Mon. Not. R. Astron. Soc. **348**, 802–810 (2004)

J.M. Trigo-Rodríguez, J. Llorca, J. Borovička, J. Fabregat, Chemical abundances determined from meteor spectra: I. Ratios of the main chemical elements. Meteorit. Planet. Sci. **38**, 1283–1294 (2003)

A.J. Tuzzolino, et al., Dust measurements in the coma of comet 81P/Wild 2 by the Dust Flux Monitor Instrument. Science **304**, 1776–1780 (2004)

M.E. Zolensky et al., Mineralogy and petrology of comet 81P/Wild 2 nucleus samples. Science **314**, 1735–1739 (2006)

Analysis of a Low Density Meteoroid with Enhanced Sodium

Jiří Borovička · Pavel Koten · Pavel Spurný · Rostislav Štork

Originally published in the journal Earth, Moon, and Planets, Volume 102, Nos 1–4.
DOI: 10.1007/s11038-007-9194-y

Abstract We present an analysis of sporadic meteor number 07406018, observed by image intensified video cameras at two stations, which showed a pronounced deceleration along its trajectory. We have applied the erosion model to analyze simultaneously the deceleration and light curve. We have found that the meteoroid had a low density of about 500 kg m^{-3}, consistent with its cometary orbit. The meteoroid structure was, nevertheless, markedly different from the Draconid meteoroids, studied recently with the same model. The size of the constituent grains was larger and the erosion energy was higher than in Draconids. The meteor spectrum was also different from Draconid spectra and showed very bright Na lines. The meteoroid composition was probably different from normal cometary composition.

Keywords Meteors · Meteoroids · Comets · Composition

1 Introduction

In our recent work (Borovička et al. 2007), presented also at the Meteoroid 2007 conference, we have developed a new model, called the erosion model, to analyze the simultaneously observed decelerations and light curves of Draconid meteors. Draconid meteoroids were found to be porous aggregates of constituent grains. The typical grain size varied from meteoroid to meteoroid but was always smaller than 100 microns (assuming grain density of 3,000 kg m^{-3}). The typical bulk density of the meteoroids was 300 kg m^{-3}, corresponding to the porosity of 90%. The grains started to separate from the surface of the meteoroid after the surface received the energy of 10^6 J m^{-2}. We call this phase erosion. The energy of erosion was 15–30 times lower than the energy of vaporization.

Draconids are not the only meteors in our database of video meteors that show significant deceleration. Here we present an analysis of the sporadic meteor numbered 07406018. The initial velocity of this meteor was 19.6 km s^{-1}, i.e. by only 4 km s^{-1} lower

J. Borovička (✉) · P. Koten · P. Spurný · R. Štork
Astronomical Institute of the Academy of Sciences, Fričova 298, 25165 Ondřejov, Czech Republic
e-mail: borovic@asu.cas.cz

DOI: 10.1007/978-0-387-78419-9_64

than the Draconid velocity. Also the trajectory slope and meteoroid mass was similar to the brightest Draconid observed. Nevertheless, the atmospheric behavior of meteor 07406018 was clearly different since it penetrated to much lower height.

The spectrum of meteor 07406018 was also different from Draconid spectra. It showed very bright sodium line. Sodium is an interesting chemical element which can be easily observed in meteors. Our survey of video spectra of faint sporadic meteors (Borovička et al. 2005) revealed three populations of meteoroids without sodium; but also meteoroids with significantly enhanced sodium. To a lesser extend, sodium abundance was found to be enhanced also in bright shower meteors (Trigo-Rodríguez et al. 2004; Borovička 2005; Jenniskens 2007).

2 Basic Data

Meteor 07406018 was observed on April 6, 2007, 21:30:53 UT from the Ondřejov (14°46′49″ E, 49°54′37″ N) and Kunžak (15°12′03″ E, 49°06′27″ N) observatories in the Czech Republic. Both stations used S-VHS video cameras with the Mullard XX1332 second generation image intensifiers and the lens Arsat 1.4/50 mm. The field of view was 54°. In Kunžak, the first half of the meteor occurred out of the field of view. Nevertheless, the meteor trajectory could be computed well. In Ondřejov, another camera of the same type equipped with the transmission grating with 600 grooves/mm recorded the spectrum of the whole meteor.

The data on meteor trajectory and orbit are given in Table 1. The meteor was first observed at a height of 100 km. Ceplecha (1988) presented a classification, called K_B criterion, of physical properties meteors according to beginning height and velocity. According to this classification, the meteor was clearly of cometary origin with K_B = 6.62. In this respect it is quite comparable with Draconids (see Koten et al. 2007).

The orbit of the meteor had a low inclination of 2.8° and a relatively large semimajor axis 4.8 AU. According to the Tisserand parameter with respect to the Jupiter (T_J = 2.18), the orbit can be classified as of Jupiter-family-comet type.

3 Deceleration and Light Curve

Meteor deceleration is demonstrated in Fig. 1. From meteor beginning down to the height of about 80 km, the positional data are consistent with constant meteor velocity. Further down, however, the meteor lag increased progressively. At the end, the meteor was 3.5 km higher than it would be if the velocity remained constant. The velocity decreased to 11 km s^{-1} at the height of 65 km.

Meteor light curve is shown in Fig. 2. The meteor was relatively bright, reaching the absolute magnitude of almost −2. During the maximum, the meteor entered the non-linear part of the image intensifier response, so the magnitude measurements around the maximum are less certain (we estimate the error to 0.3 mag). In any case, the light curve was rather smooth with the maximum in the second half of the trajectory. The F parameter (Fleming et al. 1993; Kotel et al. 2007) is F = 0.65. The light curve is similar to the theoretical light curve of single ablating body (Beech and Hargrove 2005), so we first tried to model the meteor without any fragmentation (the single-body model).

The single-body model was relatively successful. The fits are shown with the thin line in Figs. 1 and 2. The equations used for our meteor modeling are given in Borovička et al.

Table 1 Basic data on trajectory and orbit (J2000.0) of meteor 07406018

Meteor number		07406018
Date		April 6, 2007
Time of meteor beginning		21:30:53 UT
Apparent radiant (°)	α	172.56 ± 0.13
	δ	14.07 ± 0.13
Initial velocity (km/s)	v_∞	19.60 ± 0.20
Beginning height (km)	h_B	100.0 ± 0.2
End height (km)	h_E	64.6 ± 0.2
Coordinates of the beginning (°)	λ_B	14.3044 ± 0.0010
	φ_B	48.8481 ± 0.0016
Coordinates of the end (°)	λ_E	14.2970 ± 0.0004
	φ_E	49.0673 ± 0.0015
Trajectory length (km)	L	43.1 ± 0.4
Slope relative to vertical (°)	z_R	35.0 ± 0.1
Meteor duration (s)	D	2.4
Maximum magnitude	M_{max}	−1.9 ± 0.3
K_B criterion	K_B	6.62
Geocentric radiant (°)	α_G	171.51 ± 0.14
	δ_G	10.67 ± 0.16
Geocentric velocity (km/s)	v_G	16.15 ± 0.24
Semimajor axis (AU)	a	4.8 ± 0.4
Eccentricity	e	0.820 ± 0.015
Perihelion distance (AU)	q	0.872 ± 0.002
Aphelion distance (AU)	Q	8.8 ± 0.8
Argument of perihelion (°)	ω	224.48 ± 0.15
Longitude of ascending node (°)	Ω	16.6100 ± 0.0001
Inclination (°)	i	2.77 ± 0.09
Tisserand parameter	T_J	2.18

(2007). The single-body fit provided the initial meteoroid mass, $m_\infty = 4.4$ g, the ablation coefficient, $\sigma = 0.037\ \mathrm{s^2\ km^{-2}}$, and the shape-density coefficient, $K = 0.027\ \mathrm{m^2\ kg^{-2/3}}$. The shape-density coefficient is large and means low meteoroid density. Single bodies with lower shape-density coefficient would show deceleration only at the end of the trajectory and the maximum of the light curve would occur later than in 07406018. If the meteoroid was spherical, its density was 300 kg $\mathrm{m^{-3}}$. Only a very flat shape (diameter to height = 5:1) would yield the density 1,000 kg $\mathrm{m^{-3}}$.

We consider it unlikely that low density body would suffer no fragmentation. Moreover, the light curve fit is not perfect. In particular, the descend branch is steeper than observed. We have therefore analyzed the meteor with the erosion model developed for Draconids (Borovička et al. 2007). The fits are shown with the thick line in Figs. 1 and 2. The light curve was fitted much better than with the single-body model. The resulting parameters of the fit and the inferred properties of the meteoroid are given in Table 2. The quantities which are not dependent on meteoroid mass are compared to Draconids.

Since the fit was not obtained by a rigorous mathematical procedure but by a trial-and-error method, it is not possible to provide standard deviations of the parameters.

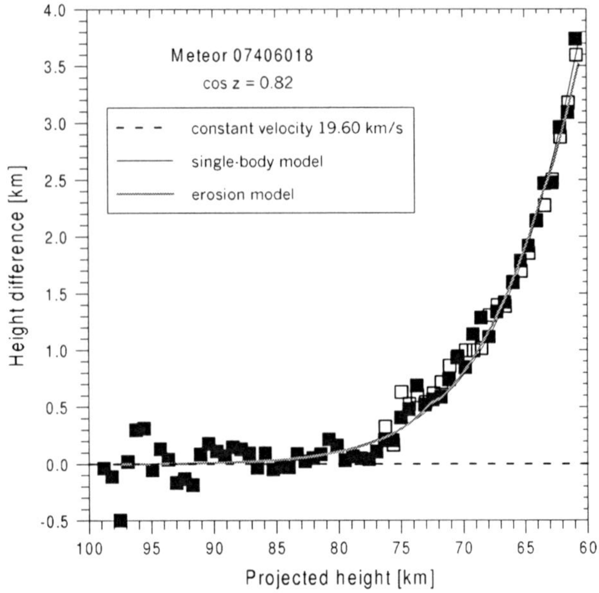

Fig. 1 Deceleration curve of meteor 07406018. Meteor height expected for the given time and constant meteor velocity of 19.60 km s^{-1} is given on the horizontal axis. The difference between the observed height and the expected height is given on the vertical axis. Filled squares are measurements from Ondřejov, empty squares come from Kunžak. The (almost identical) fits by the single-body model and the erosion model are shown

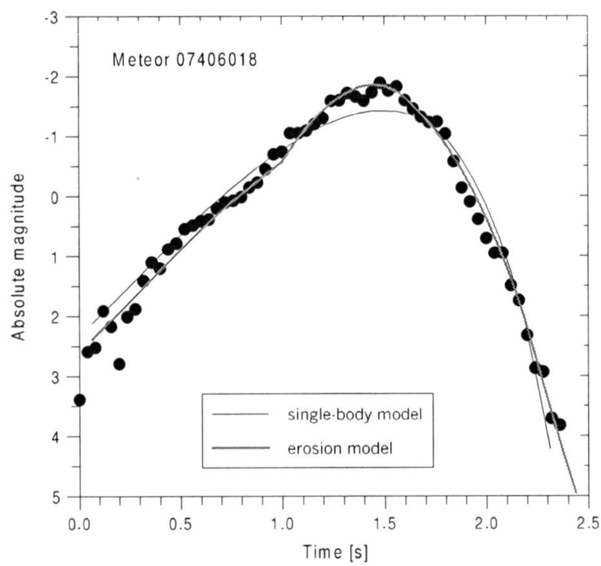

Fig. 2 Light curve of meteor 07406018. The circles are measurements from Ondřejov. Time zero corresponds to meteor height 98.8 km. The fits by the single-body model and the erosion model are shown

Nevertheless, a range of acceptable solutions was found in which higher values of σ and η coefficients are compensated by larger grain masses and earlier erosion start, and vice versa. The possible range of parameters is given in Table 2 as well. Note, however, that this the range inside the model. All computations used fixed values of external parameters. The drag coefficient, Γ, and the heat transfer coefficient, Λ, were assumed to be equal to unity. The classical meteor luminous efficiency function (Ceplecha and McCrosky 1976)

Table 2 Properties of meteoroid 07406018 according to the erosion model and comparison with Draconids

		Meteoroid 07406018[a]	Average Draconid[b]
Initial mass (kg)	m_∞	5.5×10^{-3} *5.4–5.9 × 10⁻³*	
Initial shape-density coefficient ($m^2\ kg^{-2/3}$)	K_∞	0.027 *0.024–0.032*	0.040
Initial size (mm)	$S_\infty^{1/2}$	29 *27–32*	
Grain mass upper limit (kg)	m_u	8×10^{-6} *7–12 × 10⁻⁶*	7×10^{-10}
Grain mass lower limit (kg)	m_l	5×10^{-7} *2–10 × 10⁻⁷*	2×10^{-11}
Grain mass distribution index	s	1.7 *1.2–2.2*	2.4
Grain size upper limit (μm)		1500 *1,450–1,700*	70
Grain size lower limit (μm)		600 *400–800*	40
Number of grains in the largest mass bin	n_0	50 *20–60*	
Total number of grains	N	2,000 *1,000–6,000*	
Ablation coefficient ($s^2\ km^{-2}$)	σ	0.025 *0.020–0.030*	0.024
Erosion coefficient ($s^2\ km^{-2}$)	η	0.10 *0.07–0.15*	0.50
Ablation to erosion energy ratio	η/σ	4 *3.5–5*	20
Height of erosion start (km)	h_{es}	82.8 *81.5–85*	100.6
Height of erosion end (km)	h_{ee}	73.2 *71.5–75*	
Energy received before erosion start:			
per unit cross-section ($J\ m^{-2}$)	E_S	1.5×10^7 *1.1–1.8 × 10⁷*	1.4×10^6
per unit mass ($J\ kg^{-1}$)	E_V	2.3×10^6 *1.4–3.2 × 10⁶*	

[a] The numbers in italics show the possible range of the parameters within the erosion model

[b] Borovička et al. (2007)

was used. The density of the grains was assumed 3,000 kg m^{-3}. As explained in Borovička et al. (2007), the data do not allow to infer grain density. If the grain density was lower, the number of grains and the grain mass must be scaled accordingly.

The erosion model gave the same shape-density coefficient as the single-body model. The low density of the meteoroid was therefore confirmed. The meteoroid was composed of large, about millimeter-sized, grains, according to the erosion model. This is in contrast with Draconids, where the grains were more than one order of magnitude smaller. The ablation coefficient was the same as for Draconids. The erosion coefficient was, however, much lower. In 07406018, the energy necessary to separate the mass in form of grains was only four times lower than the energy of complete ablation (vaporization), and five times higher than in Draconids. The energy necessary for the start of erosion was by one order of magnitude higher. Part of this difference could be, nevertheless, caused by a lower heat transfer, since 07406018 penetrated to denser atmosphere.

4 Spectrum

The spectrum (SX 498) of meteor 07406018 is shown in Fig. 3. The spectrum is different from typical meteor spectra, including Draconids (see Borovička et al. 2007). Between 500–600 nm, the spectrum shows typical emissions of Mg, Fe, and Na but with unusual dominance of the Na line. In the infrared part, the lines of Na at 819 nm and K at 770 nm, which are rarely seen in meteor spectra, are present. There might be a contribution of the O line at 777 nm, which is very bright in fast meteors, but we conclude that K is the main contributor (see the abundance analysis below). The other K line at 766 nm was blocked by atmospheric absorption (the O_2 band). The spectrum did not change significantly along the meteor trajectory.

The spectrum also contains a bright continuum. The presence of continua in meteor video spectra is common and usually they can be fitted by the Planck function for temperature 4,500 K (Borovička et al. 1999, 2005). In SX 498 the best fit was obtained for the temperature of 2,800 K only. The continuum accounted for 70% of meteor light between 450 and 850 nm. The origin of the continuum is not clear.

The dominance of the Na line might be connected with the low velocity of the meteor. Figure 4 (modified from Borovička et al. 2005) shows that the Na/Mg intensity ratio really increases with decreasing meteor velocity. This trend probably reflects a lower temperature of the radiating plasma in slow meteors. Nevertheless, the Na/Mg ratio in SX 498 is well above the average for the given velocity and falls within the region classified by Borovička

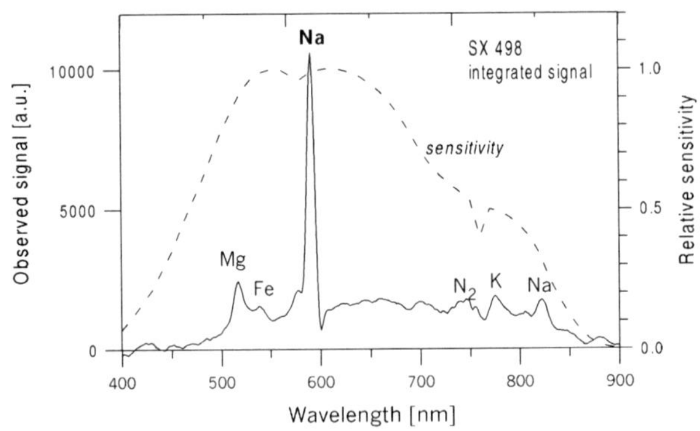

Fig. 3 Spectrum of meteor 07406018. The sum of spectra extracted from individual video frames is shown. Important emissions are identified. The minimum near 600 nm is an artifact. The plotted intensities are uncalibrated. The dashed line shows relative sensitivity of the system derived from stellar spectra observation

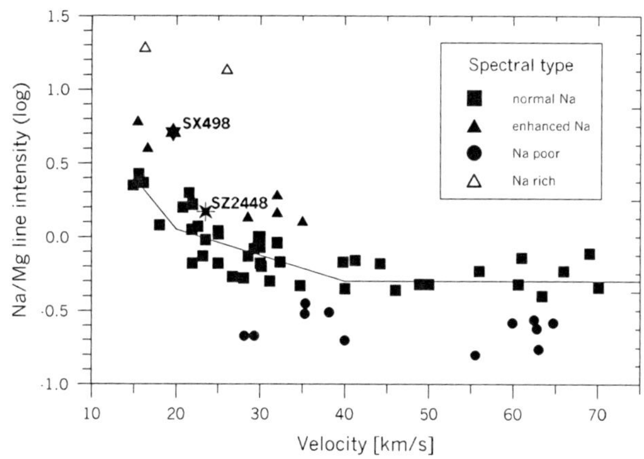

Fig. 4 Observed intensity ratio of the Na line (589 nm) to the Mg line (518 nm) as a function of meteor velocity, according to Borovička et al. (2005). The spectra have been classified into four types. The trend for normal spectra is shown by the solid line. The Draconid spectrum SZ 2448 and the spectrum SX 498 of the 07406018 meteor are included for comparison

et al. (2005) as Enhanced-Na meteoroids. These meteoroids are supposed to have higher Na content than corresponds to the normal Solar System abundances (occurring in CI chondrites and presumably also in comets). Note that meteors with even large Na line dominance exist and have been called Na-rich (Fig. 4).

The abundances in the radiating plasma of the 07406018 meteor cannot be unambiguously determined because only few lines are present in the spectrum. The intensity of the second Na line at 819 nm shows that the plasma was optically thick. The intensity ratio of both Na lines depends on temperature, T, and column density of the Na I atoms, $N_{\mathrm{Na\,I}}$ (see Borovička 1993, for the theory). We have computed the column densities and the abundances of observable elements (after correcting for ionization) for several assumed temperatures, from 4,500 K (the usual meteor plasma temperature) to 2,800 K. The results are shown in Table 3. The upper limit of Li abundance was computed from the absence of the Li line at 671 nm.

We can see that for any temperature, the Fe/Na, Fe/Mg, and Li/Na abundance ratios are lower than chondritic. The radiating plasma, however, may not reflect the meteoroid composition. But even if we consider possible incomplete evaporation of the meteoroid, the observed Fe–Mg–Na abundances cannot be in accordance with chondritic composition of the meteoroid. Incomplete evaporation can cause Na overabundance in the vapor phase but Fe/Mg would be higher than chondritic (for vaporization temperatures T_{vap} < 4,500 K; see Schaefer and Fegley 2005).

We consider plasma temperature $T \approx 3{,}500$ K as the most probable. In that case, K/Na was nearly chondritic in the radiating plasma, Fe/Na was 40 times lower and Mg/Na was 15 times lower than chondritic.

5 Discussion

We have obtained complex data on an interesting sporadic meteor 07406018. The data include trajectory, orbit, deceleration, light curve, and spectrum. We have shown that the meteoroid had a low bulk density, 300 kg m^{-3} if it had spherical shape, somewhat more if

Table 3 Abundances of atoms and elements in the radiating plasma of meteor 07406018 computed from the observed line intensities for several assumed plasma temperatures

T (K)	$N_{\mathrm{Na\,I}}$ (m^{-2})	Fe I/Na I	Mg I/Na I	K I/Na I	Fe/Na	Mg/Na	K/Na	Li/Na (10^{-4})
4,500	2.5×10^{17}	15	14	0.015	0.08	0.12	0.12	<5.2
4,000	4×10^{17}	16	25	0.010	0.13	0.24	0.10	<2.9
3,500	7×10^{17}	16	55	0.004	0.37	1.3	0.056	<1.3
3,000	1.5×10^{18}	15	155	0.0012	2.6	27	0.022	<0.74
2,800	2.2×10^{18}	13	280	0.0008	5.0	110	0.014	<0.50
CI					15.0	18.1	0.062	9.65

Abundances in CI chondrites (Lodders 2003) are given for comparison

it was flat. The density about 500 kg m^{-3} is the most probable. Though the data could be roughly explained by a homogenous single-body meteoroid, a better fit was obtained for a meteoroid composed from a number of grains which were gradually released between the heights 83–73 km. The low meteoroid density can be then naturally explained by a high porosity. The value of the porosity and the size of the grains depends on the density of the grains, which could not be determined from the data. Assuming grain density 3,000 kg m^{-3}, there were about 2,000 grains typically 1 mm in size (0.6–1.5 mm). The meteoroid porosity would be about 80%. Considering the suspected unusual chemical composition, the grain density could be much lower. If the grain density was 1,200 kg m^{-3}, for example, there were 300 grains, 1.5–4 mm in size, and the meteoroid porosity was 60%.

In comparison with Draconids, which belong to the most fragile material entering the Earth's atmosphere, the 07406018 meteoroid was considerably stronger. The energy per unit cross-section necessary to start the grain release was 10 times higher. The energy necessary for releasing unit mass in form of fragments was five times higher. On the other hand, the ablation energy was the same. The Draconids are composed from small (<0.1 mm) silicate grains and have porosities of 90% (Borovička et al. 2007). The 07406018 meteoroid was composed from much larger grains, probably made of lighter elements, and sticking better together. Alternatively, if the concept of the glue holding the grains together is valid (Hawkes and Jones 1975), the difference may be caused by different properties of the glue.

According to the spectrum, 07406018 was classified as an Enhanced-Na meteor. Such meteors are characterized by brighter Na line than majority of meteors of similar velocities (Borovička et al. 2005). In case of 07406018, two Na lines were visible and some analysis of chemical composition of the radiating plasma was possible. Unless there were some strong non-equilibrium effects, which are not present in majority of other meteors, the Na–Mg–Fe abundance ratios are not consistent with chondritic composition of the meteoroid. The meteoroid was rich in alkali metals—with K/Na ratio nearly chondritic—and poor in silicates (represented by Mg) and especially in heavy metals (represented by Fe). Unfortunately, we have no information on other important elements.

According to the orbit, 07406018 was classified as cometary. The low bulk density is consistent with this classification. The non-chondritic composition, however, contradicts the view of comets as pristine objects. Some other cometary meteoroids were, nevertheless, classified as of the Enhanced-Na type (Borovička et al. 2005). We suspect that there may be cm-sized inhomogeneities in comets. On the other hand, considering the low inclination of the 07406018 orbit, other origin of the meteoroid may be also possible, e.g. within the outer asteroid belt, Trojan region, or Jovian satellite system.

Acknowledgements This work was supported from grant no. 205/05/0543 from GAČR and grant no. B300030502 from GA AV ČR.

References

M. Beech, M. Hargrove, Classical meteor light curve morphology. Earth Moon Planets **95**, 389–394 (2005)

J. Borovička, A fireball spectrum analysis. Astron. Astrophys. **279**, 627–645 (1993)

J. Borovička, Elemental abundances in Leonid and Perseid meteoroids. Earth Moon Planets **95**, 245–253 (2005)

J. Borovička, J. Štork, J. Boček, First results from video spectroscopy of 1998 Leonid meteors. Meteorit. Planet. Sci. **34**, 987–994 (1999)

J. Borovička, P. Koten, P. Spurný, J. Boček, R. Štork, A survey of meteor spectra and orbits: evidence for three populations of Na-free meteoroids. Icarus **174**, 15–30 (2005)

J. Borovička, P. Spurný, P. Koten, Atmospheric deceleration and light curves of Draconid meteors and implications for the structure of cometary dust. Astron. Astrophys. **473**, 661–672 (2007)

Z. Ceplecha, Earth's influx of different populations of sporadic meteoroids from photographic and television data. Bull. Astron. Inst. Czech. **39**, 221–236 (1988)

Z. Ceplecha, R. E. McCrosky, Fireball end heights—A diagnostic for the structure of meteoric material. J. Geophys. Res. **81**, 6257–6275 (1976)

F.E.B. Fleming, R.L. Hawkes, J. Jones, Light curves of faint television meteors. in *Meteoroids and Their Parent Bodies*, ed. by J. Štohl, I. P. Willimas (Bratislava, 1993), pp. 261–264

R.L. Hawkes, J. Jones, A quantitative model for the ablation of dustball meteors. Mon. Not. R. Astron. Soc. **173**, 339–356 (1975)

P. Jenniskens, Quantitative meteor spectroscopy: Elemental abundances. Adv. Space Res. **39**, 491–512 (2007)

P. Koten, J. Borovička, P. Spurný, R. Štork, Optical observations of enhanced activity of the 2005 Draconid meteor shower. Astron. Astrophys. **466**, 729–735 (2007)

K. Lodders, Solar system abundances and condensation temperatures of the elements. Astrophys. J. **591**, 1220–1247 (2003)

L. Schaefer, B. Fegley Jr., Application of an equilibrium vaporization model to the ablation of chondritic and achondritic meteoroids. Earth Moon Planets **95**, 413–423 (2005)

J.M. Trigo-Rodríguez, J. Llorca, J. Fabregat, Chemical abundances determined from meteor spectra—II. Evidence for enelarged sodium abundances in meteoroids. Mon. Not. R. Astron. Soc. **348**, 802–810 (2004)

NEOCAM: The Near Earth Object Chemical Analysis Mission

Joseph A. Nuth III · John L. Lowrance · George R. Carruthers

Originally published in the journal Earth, Moon, and Planets, Volume 102, Nos 1–4.
DOI: 10.1007/s11038-007-9178-y

Abstract The prime measurement objective of the Near Earth Object Chemical Analysis Mission (NEOCAM) is to obtain the ultraviolet spectra of meteors entering the terrestrial atmosphere from ~ 125 to 300 nm in meteor showers. All of the spectra will be collected using a slitless ultraviolet spectrometer in Earth orbit. Analysis of these spectra will reveal the degree of chemical diversity in the meteors, as observed in a single meteor shower. Such meteors are traceable to a specific parent body and we know exactly when the meteoroids in a particular shower were released from that parent body (Asher, in: Arlt (ed.) Proc. International Meteor Conference, 2000; Lyytinen and van Flandern, Earth Moon Planets 82–83:149–166, 2000). By observing multiple apparitions of meteor showers we can therefore obtain quasi-stratigraphic information on an individual comet or asteroid. We might also be able to measure systematic effects of chemical weathering in meteoroids from specific parent bodies by looking for correlations in the depletions of the more volatile elements as a function of space exposure (Borovička et al., Icarus 174:15–30, 2005). By observing the relation between meteor entry characteristics (such as the rate of deceleration or breakup) and chemistry we can determine if our meteorite collection is deficient in the most volatile-rich samples. Finally, we can obtain a direct measurement of metal deposition into the terrestrial stratosphere that may act to catalyze atmospheric chemical reactions.

Keywords Meteor · Spectra, ultraviolet · Meteor shower · Space observations

---

J. A. Nuth III (✉)
Astrochemistry Laboratory, Code 691, NASA's Goddard Space Flight Center, Greenbelt, MD 20771, USA
e-mail: Joseph.A.Nuth@NASA.gov

J. L. Lowrance
Princeton Scientific Instruments, Inc., 7 Deer Park Drive, Monmouth Junction, NJ 08852, USA

G. R. Carruthers
Ultraviolet Measurements Section, Space Science Division, Code 7645, Naval Research Laboratory, Washington, DC 20375-5320, USA

1 Introduction

We are currently in an era of unprecedented opportunity and can propose to fly spacecraft to nearly any object in the solar system. Missions to an asteroid (NEAR, Hyabusa, Dawn) and a comet (STARDUST, Deep Impact) have been launched and several more missions to such primitive bodies are in construction or have been proposed (Contour, OSIRIS). However, in spite of the wealth of knowledge to be returned by these spacecraft on the chemical composition and physical structure of individual primitive bodies in our solar system, our real knowledge of the structure and composition of comets and asteroids will remain far from complete. By definition, each primitive body is a nearly pristine specimen of the planetesimals that formed in our planetary system. They have not been processed to attain equilibrium or to homogenize the random nature of the nebular accretion process into a uniform, predictable geochemical system. In a sense, these bodies represent grab bag samples of the solar nebula, e.g. 81P/comet Wild 2 (Brownlee et al. 2006), that is itself a highly dynamic system, the composition of which will be time dependent at any given locale (see Fig. 1). Characterization of a representative sample of large numbers of comets or asteroids via individual space missions is not yet an option.

The composition of these small bodies can tell us much about the chemistry of the solar nebula and the distribution of the biogenic elements prior to the Origin of Life on the Earth. In fact, the distribution of these elements in primitive comets and asteroids can help to discriminate between a terrestrial setting for the chemical evolution of complex, pre-biotic molecules (e.g. Miller 1953; Miller and Urey 1959) and an exogenic origin where such molecules were delivered by asteroids or comets to the primitive Earth (e.g. Oro 1961). Much of what is now known about the structure and composition of both comets and asteroids has been gleaned from painstaking telescopic observations (Gehrels 1994; Festou et al. 2004) and by the careful analysis of meteorites, micrometeorites, and interplanetary dust particles (IDPs) (Kerridge and Matthews 1988; Papike 1998; Brownlee et al. 1997; Engrand et al. 1999; Taylor et al. 2000; Rietmeijer 2002). We know that both comets and

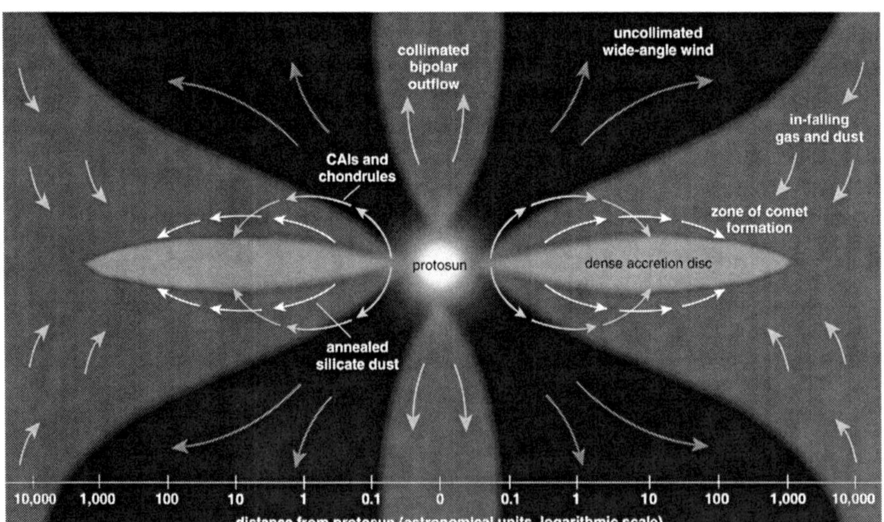

Fig. 1 Schematic diagram showing various levels of mixing in the primitive Solar Nebula (Nuth 2001)

asteroids display an enormous diversity in their chemical and physical properties, yet individual missions to a representative sample of either population are beyond our present capability. The missions that are planned will provide extremely important information on the least-altered remnants of the primitive solar nebula and valuable data on the correlation between telescopic and in-situ observations of primitive bodies. However, there is much more information available—at relatively little cost—on the chemical composition and physical properties of a very large number of individual comets and asteroids that cross the orbit of the earth. In a very real sense, samples of these parent bodies come to us on a regular basis (Asher 2000; Lyytinen and van Flandern 2000): we just need the tools to analyze them.

Meteor showers and sporadic meteors (including IDPs) bring as much as 20,000 tons of extraterrestrial material to the earth in any given year (Brownlee, personal communication). Debris from comets, asteroids and even planetary sources has rained down upon the earth since its formation and the average composition of the meteoritic component—including IDPs—is relatively well understood (Sears and Dodd 1988; Kallemeyn 1988; Papike 1998). However, speculation concerning the rest of the infalling material ranges from the ridiculous to the sublime: an ocean of water from micro-cometesimals over the last 4.5 billion years (Frank et al. 1986; Frank and Craven 1988), to "Diseases from Space" (Hoyle and Wickramasinge 1989). Comets and meteorites may once have delivered nutrients essential for chemical evolution to the primitive Earth (Marcus and Olsen 1991) and some remnant of this organic flux may still fall today. Spectral studies of incoming meteors have shown them to be non-uniform in composition, origin and entry characteristics (Harvey 1971, 1974; Millman 1976; Russell 1981; Jenniskens et al. 2000a, b; Hawkes et al. 2005; Borovička 2005, 2006, 2007).

Unfortunately, ground-based studies of meteoric spectra contain only limited quantitative compositional information since the visible spectra of meteors are dominated by a very large number of iron lines while the atmosphere absorbs their ultraviolet emission. Space-based observations of the ultraviolet spectra of incoming meteors are largely free of iron lines and may reveal data on the abundances of many other elements including carbon, magnesium, calcium, nitrogen, silicon, phosphorus, oxygen, hydrogen and sulfur.

In what follows we will briefly describe a space flight mission concept designed to obtain quantitative ultraviolet spectra of meteors, especially those in meteor showers, using a slitless ultraviolet meteor spectrometer (SUMS) originally designed as an attached payload for the International Space Station (Nuth et al. 1999).

2 Ultraviolet Observations of Meteors

Meisel (1976) estimated the fluxes from incoming meteors as a function of their entry velocity based on theoretical arguments and predicted that they should be detectable from orbit. The instrument that we will discuss below has been optimized to detect this signal and test this theoretical prediction. Dr. George Carruthers and his team at the Naval Research Laboratory actually observed an incoming Leonid meteor in the ultraviolet. This amazing observation was made with the Global Imaging Monitor of the Ionosphere (GIMI) instrument on board the Advanced Research and Global Observation Satellite (ARGOS) operated by US DoD's Space Test Program and launched on February 23, 1999. The GIMI Image of the meteor was obtained on November 18, 1999 from an altitude of 833 km using an exposure time of 110 s. If the bolide were traveling at more than the ~60 km/s or so typical for Leonid meteors, the total exposure time (110 s) for the image is on the order of

30–50 times longer than the time the meteor spent in the field of view. This obviously increases the noise enormously. The GIMI Instrument was not designed to detect meteors, but did manage to detect one. This image proves conclusively that incoming meteors do produce ultraviolet radiation upon atmospheric entry.

A second US DoD mission (the Mid-Course Space Experiment Satellite—MSX) actually obtained the spectrum of an incoming Leonid meteor over the wavelength range between 110 nm to 900 nm (Romick et al. 2000; Jenniskens et al. 2000a, b; Carbary et al. 2003, 2004). Unfortunately, the instruments on this mission also used relatively long exposure times ($\sim$0.5 s) relative to the time required by the meteor to traverse an individual pixel in the CCD detector as seen through the slit of their spectrometers. Even without efforts to optimize the signal-to-noise at the detector, this experiment detected strong lines from Fe, Ca, Mg, O and N as well as molecular lines from NO, O_2, N_2, CO and OH. The most intense emissions were seen between 230 and 290 nm. No effort has yet been made to analyze the spectrum obtained by MSX.

One potential problem is the separation of spectral lines from individual meteoric components from the spectrum of the surrounding, highly excited atmosphere. This should not really be a problem for the UV lines of metallic elements such as sodium, potassium, magnesium and calcium, or even for phosphorus, sulfur and SiO. However, since we are also very interested in the determination of the ratio of the more volatile, biogenic elements such as carbon, oxygen and nitrogen to the refractory components of meteors, any level of atmospheric emission may be problematic. It may be impossible to directly separate the meteoric nitrogen or oxygen signal from the atmospheric nitrogen or oxygen (noise) emission. However, if a decelerating meteor were to show emission from only carbon, hydrogen, nitrogen and oxygen, then we could certainly infer the entry of a very volatile-rich body. Similarly, weak metallic lines combined with strong volatile emission may provide at least a qualitative, or even semi-quantitative measure of the volatile content of individual meteors that could be usefully correlated with their entry characteristics and orbital data. Finally, comparison of large numbers of spectra, obtained at a range of entry velocities and originating from a large number of parent bodies can be used to put limits on the atmospheric contribution to the observed UV emission through modeling studies. The models already constructed by Meisel (1976) could be tested against a large data set obtained via the SUMS instrument to reduce the level of uncertainty in factors such as the emission efficiency of individual lines as a function of entry velocity. Another factor that would contribute to quantifying such data is the simultaneous detection of several ionization states of an individual element or of several vibrational states of a molecular species as a function of entry velocity. The instrument described below has been designed to separate the spectral lines of volatile constituents from those of more refractory elements.

The scientific payoff from observations of entering meteors is highly dependent on the time available to collect the observations. Long-term observations of numerous meteor showers over many years will yield much more information than the mere detection of an incoming meteoroid during a shower, even if the spectrum of the meteoroid is obtained at high resolution.

3 Summary of the Conceptual Instrument Design

The radiation from a meteor "burning" in the atmosphere originates from a narrow $\sim$1 m wide track and is observed as a point source moving at a velocity of 10–70 km/s, 80–120 km above the Earth's surface. The meteor's radiation at wavelengths less than ~310 nm is absorbed by the Earth's atmosphere before reaching the ground. It has been estimated by Meisel (1976) that the meteor velocity must exceed 40 km/s to generate UV radiation and the UV is associated primarily with the gas very near the body of the meteor. The number of meteoroids entering the atmosphere per unit time increases as their size decreases and the brightness in the visible is essentially an exponential function of the rate, i.e. the fainter meteors are much more numerous than the brighter ones. It is reasonable to expect the brightness in the UV to follow the same trend.

The faintest meteor that can be detected depends, in the limit, on the quantum (Shot) noise in the background presented by the Earth's atmosphere. This is the same problem as detecting faint stars against the night sky background. The higher the angular resolution of the instrument, the smaller the background signal per pixel, and the lower the noise associated with the background. Making the exposure time short can further minimize the background signal for meteoric spectra. The optimum exposure is one that matches the time the meteor spends in traversing a pixel. This maximizes the contrast between the meteor and the sky background. The meteor to sky background contrast can be further improved by controlling the spectral transmission of the optics and the spectral sensitivity of the image sensor. It is important to make the camera "solar blind" to eliminate scattered sunlight and light emanating from the surface of the earth such as fires, city lights, etc. Even so, the UV signature of the atmosphere is much brighter in the day (Fig. 2) than at night (Fig. 3). At night Lyman-α (121.6 nm) emission from hydrogen atoms dominate the background due to solar Lyman-α scattered at very high altitudes by Earth's exospheric hydrogen geocorona. However, this wavelength is not transmitted by the optical system of the proposed instrument and hence provides no significant interference.

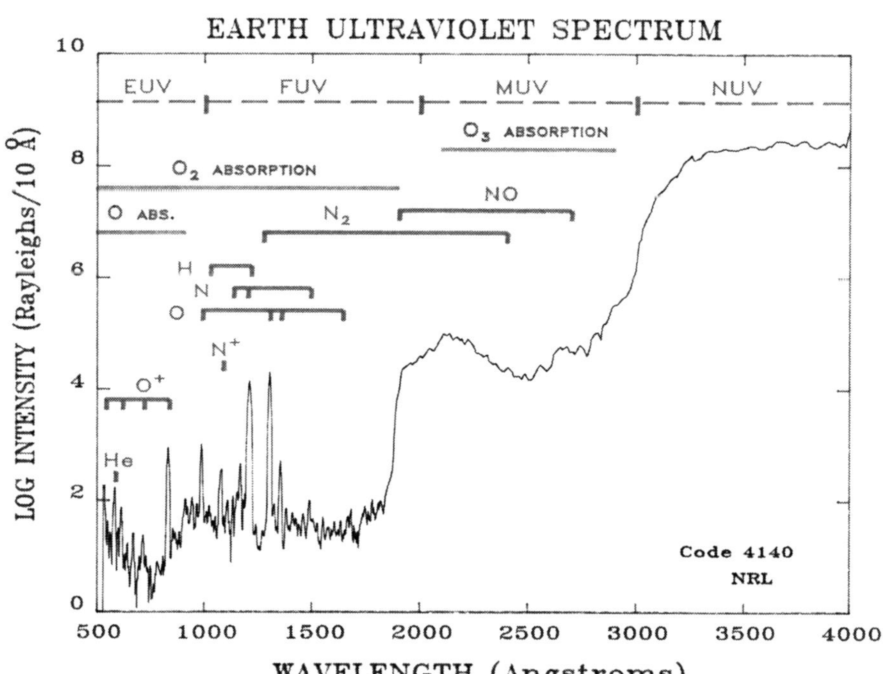

Fig. 2 The ultraviolet spectrum of the terrestrial atmosphere during the day (Carruthers, personal communication)

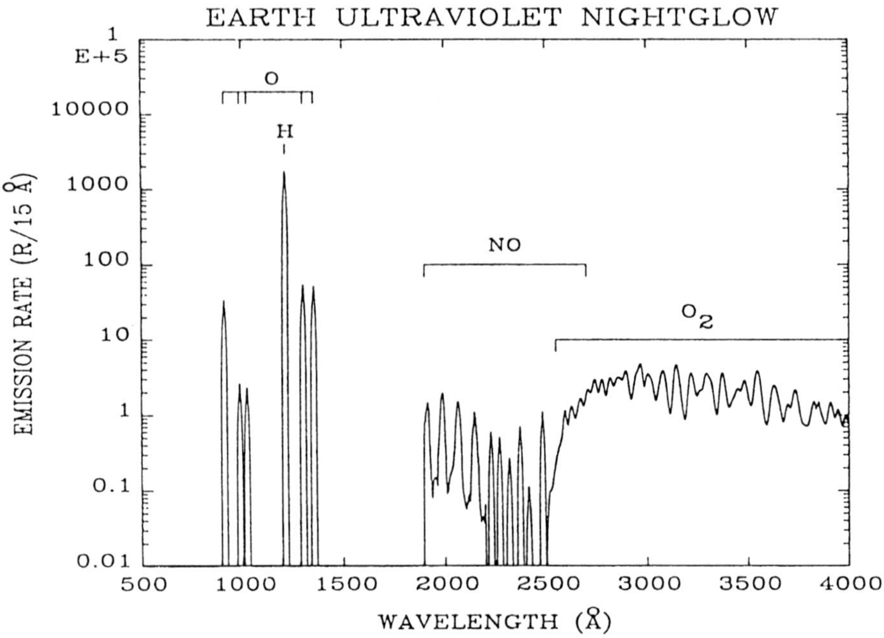

Fig. 3 The ultraviolet spectrum of the terrestrial atmosphere at night (Carruthers, personal communication)

In order to determine the chemical constituents of the meteor, the spectral emission lines in the radiation must be detected. As discussed earlier, there is a distinct advantage in making the spectral measurement in the UV. The highest contrast between the emission lines and the continuum/background is achieved by making the spectral dispersion such that the width of the individual emission lines is approximately one pixel wide. The spectrographic camera's field of view should also be wide enough to cover the entire track of a given meteor. A wide field of view also maximizes the number of meteors that are detected. In addition, since sporadic meteors are infrequent and unpredictable, there is a need for real-time data reduction in order to make the data storage requirement tractable if data is obtained at high speed.

The Slitless UV Spectrometer, shown schematically in Fig. 4, consists of three main subsystems. The first is an f/1.5, 300 mm focal length Schmidt telescope with a 9.4-degree field of view, a spectral dispersion element and a corrector plate. The second component is a solar blind image sensor consisting of an image intensifier and fast frame rate CCD camera with a multi-port readout to achieve a frame rate of 380 frames/s. The third subsystem is an electronics package containing circuits to operate the image sensor, digital electronics to implement real-time image analysis to detect the random and infrequent meteors (compared to the 380 frame per second camera speed) and digital data archiving electronics to store the meteoric UV spectra, as they are collected during the mission.

The Schmidt type telescope optics are preceded by a spectral dispersing element: currently baselined as a 600 line/mm grating mosaic. A field-flattening lens matches the convex focal plane of the Schmidt telescope to the focal surface of an image intensifier having a solar blind photo cathode to limit sensitivity to the middle and far ultraviolet and that is insensitive to visible light from the Earth or to scattered visible sunlight. The visible wavelength output image from the image intensifier tube is optically coupled to a CCD

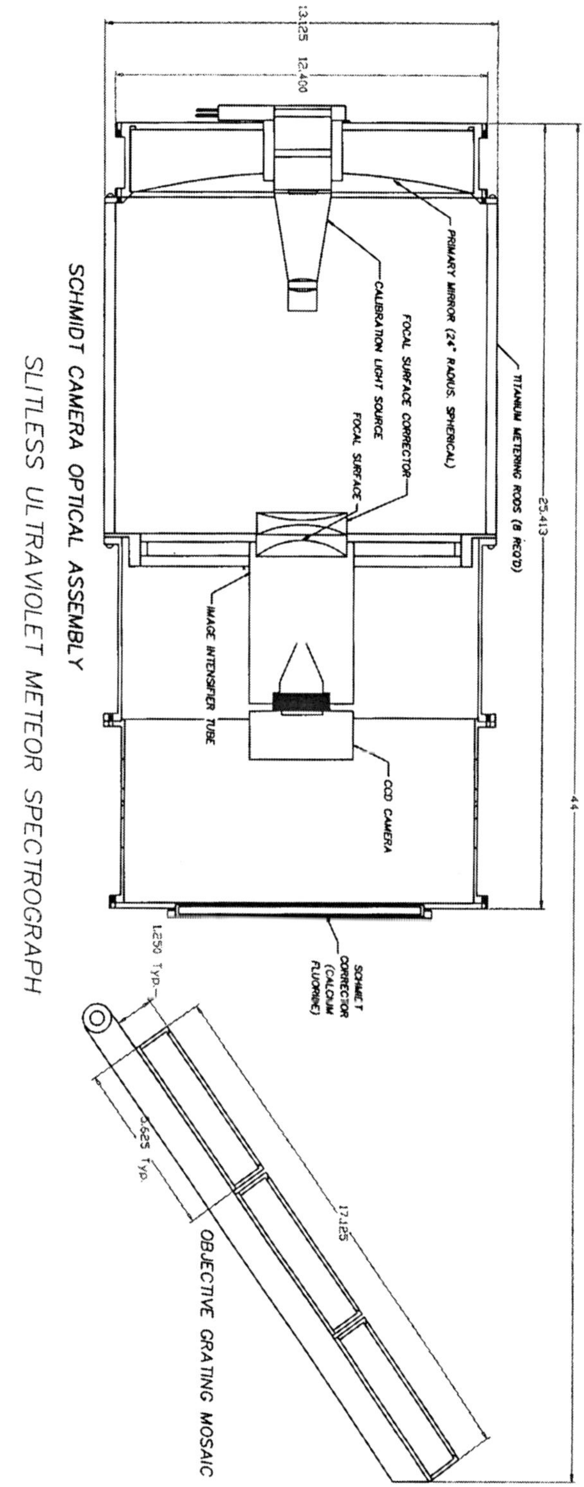

Fig. 4 Schematic diagram of the baseline slitless ultraviolet meteor spectrometer (SUMS)

type solid-state image sensor operating at fast frame rate (380 frames/s). An internal calibration source will be installed to allow in-flight testing of the CCD camera by projection of test patterns at several ultraviolet wavelengths. The spectral dispersion of the grating is aligned with the CCD. In this way the spectrum of a meteor will generally fall roughly along a horizontal line rather than at some arbitrary angle to the raster lines of the CCD.

4 SUMS Observations

The very high-frame rate operation of a 512 × 512 CCD array quickly yields an unmanageable quantity of raw data: impractical to store for the expected several-day periods between scheduled downlinks. We have therefore devised a meteor detection algorithm that will independently operate in all 16 parallel-output data streams from the CCD array. The algorithm searches for rapid changes in the intensity of individual pixels from one frame to the next: short-term memory is used to store a frame and compare it to the next one on a continuous basis. If no differences are found above a programmable threshold value then the previous frame is overwritten by the next readout and a new comparison is made. If a single pixel exceeds the threshold, then the previous frame (no signal), the current frame and all subsequent (changing) frames of the 16 parallel ports are stored in long-term memory together with appropriate timing data from which the spacecraft's position can be derived. This record continues until no detectable differences are seen in the readout of any of the 16 output ports, plus a few additional frames recorded to extract the instantaneous background. In order to ensure that we set the detection thresholds to appropriate levels we will also mount a normal video camera to monitor the same optical path observed by the SUMS instrument, though probably we will use a much wider field of view. If we see meteors in the video camera that should have been observed by the SUMS instrument, but were not, we will set the detection threshold to lower levels.

While the basic design scheme of the SUMS instrument is to image a lined grating and capture the resultant spectral information, it is possible to do nearly the same thing in a transmission mode. In this case, the instrument would be pointed down at the atmosphere from orbit and the meteors would be viewed through a transmission grating. It would also be advantageous in this configuration to place a UV transmitting wide-angle lens in front of the transmission grating to collect meteor spectra from horizon to horizon. Whereas in the reflection mode, meteor spectra originate from a small cone directly beneath the SUMS instrument, in the transmission mode, the distance to the meteor is no longer certain. Close, faint meteors could yield the same signal as more distant but brighter bolides. Therefore, when the SUMS instrument is used in reflection we can obtain quantitative meteor spectra with precise data on the absolute line intensities. This data will be very useful to calibrate models of the plasma emission from such meteoroids. On the other hand, when used in transmission mode, the SUMS instrument will obtain relative compositions of the brighter bolides in any given shower, as these bright meteors will be most readily observed from larger distances. In the ideal scenario a meteor shower will be observed from orbit using both configurations of the SUMS instrument with simultaneous observations from the ground. In this mode the distance to the incoming bolides can be calculated based on the timing and trajectory information available from the satellite record correlated to the observations from the ground. Therefore both observing modes of the SUMS instrument could yield quantitative spectral data.

In order to facilitate coordination with ground-based observational campaigns, it is our intention to post the raw spectral data to the mission website as soon as it is obtained. We will also post all spectral calibration data as it is obtained. Reduced spectral datasets will be prepared and posted as soon as is practical, depending on the level of support available. All of this data will be freely available to the scientific community.

5 Summary

NEOCAM: the Near Earth Object Chemical Analysis Mission is a free flying spacecraft that will look down on the terrestrial atmosphere in order to observe the trails of incoming meteors using one or more SUMS. SUMS is an intensified, fast frame rate, solar blind, CCD camera that in its baseline design will use a 12″ f/1.5 telescope to image a grating that views the atmosphere in order to capture the spectra of meteors over the range from 123 nm to 300 nm. The primary objective of the mission is to obtain quantitative spectra from a large number of meteors in a number of meteor showers over several successive apparitions in order to derive the chemical composition of the parent body responsible for each shower, the chemical heterogeneity of the meteors released at any one time from these parent bodies, and the overall homogeneity of the parent body by comparison of the compositional distribution of meteors released at intervals of many years. Secondary objectives for NEOCAM include studies of the potential correlation of the composition of meteors with internal strength, as observed by monitoring the deceleration and possible breakup of the bolide during atmospheric entry. In addition, a second spectrometer configuration can be used to obtain the spectra of brighter meteors from as far away as the horizon in order to characterize their relative chemical abundances and to obtain a measure of the ratio of meteors derived from asteroids and comets in the population of random meteors as well as to capture the widest possible range of spectral variation in the bolides in an individual meteor shower.

Acknowledgments Thanks to Dr. Martin Beech for a helpful review and to Dr. Frans Rietmeijer for some much needed editorial help, especially suggestions for additional references.

References

D.J. Asher, in *Leonid Dust Trail Theories*, ed. by R. Arlt. Proc. International Meteor Conference, Frasso Sabino, 1999 (IMO, 2000), pp. 5–21

J. Borovička, Elemental abundances in Leonid and Perseid meteoroids, in *Modern Meteor Science, An Interdisciplinary View*, ed. by R. Hawkes, I. Mann, P. Brown (Springer, 2005), pp. 245–253

J. Borovička, in *Physical and Chemical Properties of Meteoroids as Deduced from Observations*, ed. by D. Lazarro, J.A. Ferraz-Mello, J.A. Fernandez. Proc. IAU Symp. Asteroids, comets, meteors, 2006, vol. 229 (International Astronomical Union, 2006), pp. 249–271

J. Borovička, in *Properties of Meteoroids from Different Classes of Parent Bodies*, ed. by G.B. Valsecchi, D. Vokrouhlický. Proc. IAU Symposium. Near Earth objects, our celestial neighbors: opportunity and risk, vol. 236 (Cambridge University Press, Cambridge, 2007), pp. 107–120

J. Borovička, P. Koten, P. Spurný, J. Boček, R. Štork, A survey of meteor spectra, orbits: evidence fort three populations of Na-free meteoroids. Icarus **174**, 15–30 (2005)

D.E. Brownlee, et al., Comet 81P/Wild 2 under a microscope. Science **314**, 1711–1716 (2006)

D.E. Brownlee, B. Bates, L. Schramm, The elemental composition of stony cosmic spherules. Meteorit. Planet Sci. **32**, 157–175 (1997)

J.F. Carbary, D. Morrison, G.J. Romick, J.-H. Yee, Leonid meteor spectrum from 110 to 860 nm. Icarus **161**, 223–234 (2003)

J.F. Carbary, D. Morrison, G.J. Romick, J.H. Yee, Spectrum of a Leonid meteor from 110 to 860 nm. Adv. Space Res. **33**, 1455–1458 (2004)

C. Engrand, E. DeLoule, F. Robert, M. Maurette, G. Kurat, Extraterrestrial water in micrometeorites and cosmic spherules from Antarctic: an ion microprobe study. Meteorit. Planet Sci. **34**, 773–786 (1999)

M.C. Festou, H.U. Keller, H.A. Weaver (eds.), Comets II (The University of Arizona Press, Tucson, Arizona, 2004), p. 780

L.A. Frank, J.D. Craven, Imaging results from Dynamics Explorer 1. Rev. Geophys. **26**, 249–283 (1988)

L.A. Frank, J.B. Sigwarth, J.D. Craven, Comment on the paper 'On the influx of small comets into the earth's upper atmosphere. II – Interpretation'. Geophys. Res. Lett. **13**, 701; Authors' reply, 703–704 (1986)

T. Gehrels (ed.), Hazards Due to Comets and Asteroids (The University of Arizona Press, Tucson, Arizona, 1994), p. 1300

G.A. Harvey, Calcium H-line and K-line anomaly in meteor spectra. Astrophys. J. **165**, 669–671 (1971)

G.A. Harvey, Strongly differentiated material in high-inclination and retrograde orbits. Astrophys. J. **79**, 333–336 (1974)

R. Hawkes, I. Mann, Brown P. (eds.), Modern Meteor Science: An Interdisciplinary View (Springer, the Netherlands, 2005), p. 732

F. Hoyle, C. Wickramasinge, *Evolution From Space* (Paladin Press, London, 1989)

P. Jenniskens, D. Nugent, E. Tedesco, J. Murthy, 1997 Leonid shower from space Earth Moon Planets **82–83**, 305–312 (2000a)

P. Jenniskens, F.J.M. Rietmeijer, N. Brosch, M. Fonda (eds.), Leonid Storm Research (Kluwer Academic Publishers, Dordrecht, the Netherlands, 2000b), p. 606

G. Kallemeyn, Elemental variations in bulk chondrites: a brief review, in *Meteorites and the Early Solar System*, ed. by J.F. Kerridge, M.S. Matthews (University of Arizona Press, Tucson, Arizona, 1988), pp. 390–393

J.F. Kerridge, M.S. Matthews (eds.), Meteorites and the Early Solar System (University Arizona Press, Tucson, Arizona, 1988), p. 11269

E.J. Lyytinen, T. van Flandern, Predicting the strength of the Leonid outbursts. Earth Moon Planets **82–83**, 149–166 (2000)

J.N. Marcus, M.A. Olsen, Biological implications of organic compounds in comets, in *Comets in the Post-Halley Era*, ed. by R.L. Newburn, M. Neugebauer, J. Rahe (Kluwer, Dordrecht, 1991), pp. 439–462

D.D. Meisel, A study of meteor spectroscopy and physics from Earth-orbit: a preliminary survey into ultraviolet meteor spectra, NASA CR-2664 (1976)

S.L. Miller, A production of amino acids under possible primitive earth conditions. Science **117**, 528–529 (1953)

S.L. Miller, H.C. Urey, Organic compound synthesis on the primitive Earth. Science **130**, 245–251 (1959)

P.M. Millman, Quantratid meteors from 41,000 feet. Sky Telescope **51**, 225–228 (1976)

J.A. Nuth, How were the comets made? Am. Sci. **89**, 230–237 (2001)

J.A. Nuth, J.L. Lowrance, G. Renda, G.R. Carruthers, A fast UV slitless spectrometer for meteor research, Proc. 44th Annual SPIE Meeting, Denver, CO July 19–21. SPIE **3818**, 108–124 (1999)

J. Oro, Comets and the formation of biochemical compounds on the primitive Earth. Nature **190**, 389–391 (1961)

J.J. Papike (ed.), Planetary Materials, Revs. Mineralogy 36 (Mineralogical Society of America, Chantilly, Virginia, 1998), p. 1052

F.J.M. Rietmeijer, The earliest chemical dust evolution in the solar nebula. Chemie der Erde **62**, 1–45 (2002)

G. Romick, D. Morrison, P. McEvaddy, J. Yee, D. Ossing, A. Bowman, R. Reinders, G. Baer, L. Paxton, D. Murdock, C. Meng, Observations of Leonid meteor spectra from 110 nm to 900 nm from the MSX Satellite, Abstract P61A-10, 2000 Spring Meeting American Geophysical Union (2000)

J.A. Russell, Spectral-height relations in Perseid meteors. Astrophys. J. **243**, 317–321 (1981)

D. Sears, R. Dodd, Overview and classification of meteorites. in *Meteorites and the Early Solar System, 3–31*, ed. by J.F. Kerridge, M.S. Matthews (University of Arizona Press, Tucson, Arizona, 1988)

S. Taylor, J.H. Lever, R.P. Harvey, Numbers, types, and compositions of an unbiased collection of cosmic spherules. Meteorit. Planet Sci. **35**, 651–666 (2000)

Mostly Dormant Comets and their Disintegration into Meteoroid Streams: A Review

Peter Jenniskens

Originally published in the journal Earth, Moon, and Planets, Volume 102, Nos 1–4.
DOI: 10.1007/s11038-007-9169-z

Abstract The history of associating meteor showers with asteroidal-looking objects is long, dating to before the 1983 discovery that 3200 Phaethon moves among the Geminids. Only since the more recent recognition that 2003 EH1 moves among the Quadrantids are we certain that dormant comets are associated with meteoroid streams. Since that time, many orphan streams have found parent bodies among the newly discovered Near Earth Objects. The seven established associations pertain mostly to showers in eccentric or highly inclined orbits. At least 35 other objects are tentatively linked to streams in less inclined orbits that are more difficult to distinguish from those of asteroids. There is mounting evidence that the streams originated from discrete breakup events, rather than long episodes of gradual water vapor outgassing. If all these associations can be confirmed, they represent a significant fraction of all dormant comets that are in near-Earth orbits, suggesting that dormant comets break at least as frequently as the lifetime of the streams. I find that most pertain to NEOs that have not yet fully decoupled from Jupiter. The picture that is emerging is one of rapid disintegration of comets after being captured by Jupiter, and consequently, that objects such as 3200 Phaethon most likely originated from among the most primitive asteroids in the main belt, instead. They too decay mostly by disintegration into comet fragments and meteoroid streams. The disintegration of dormant comets is likely the main source of our meteor showers and the main supply of dust to the zodiacal cloud.

Keywords Meteor shower · Meteoroid stream · Comet · Asteroid · Near-Earth Object · Minor planet · Comet-asteroid transition object · Interplanetary dust · Comet fragmentation · Zodiacal cloud

---

Editorial handling: Frans Rietmeijer.

P. Jenniskens (✉)
Carl Sagan Center, SETI Institute, 515 N. Whisman Road, Mountain View, CA 94043, USA
e-mail: pjenniskens@mail.arc.nasa.gov

1 Introduction

The topics of this review are the minor planets in short orbital periods (<20 years) that are potential parent bodies of our ecliptic (and toroidal) meteor showers. They derive from two source regions: the asteroid belt between Mars and Jupiter and the Kuiper Belt beyond Neptune, specifically its Scattered Disc. Some of the planetesimals that were formed in the region between Jupiter and Neptune can now be found in the Oort cloud and are the source of our Long-Period and Halley-type comets.

The International Astronomical Union has not yet adopted clear definitions on what the terms "asteroid" or "comet", or even "minor planet" refer to. All these objects fall under the category "*small solar system bodies*", including meteoroids.

The name "*asteroid*" is used by some to imply all minor planets that appear star-like. Others restrict the name to planetesimals that were formed in the region between Mars and Jupiter. I will adopt an even stricter definition: those that were formed in the region between Mars and Jupiter and also have lost most of their unbound water. Most remaining water in asteroids was incorporated into mineral structures, mostly forming clays, resulting in strong rocky materials. Asteroids are the source of our meteorites, material strong enough to survive the impact in Earth's atmosphere. Most are suspected to originate from the inner parts of the asteroid belt, where S-type asteroids are common.

The name "*comet*" is used by some to imply all minor planets that have a fuzzy halo or tail. Others restrict the name to planetesimals formed from dust grains coated with a layer of water ice. Such grains did not exist in the inner solar system, where small rocky planets were formed. Water ice was present in the region where rapid dust accumulation resulted in the growth of Jupiter, our most massive planet. All planetesimals formed in the neighbourhood of Jupiter and outwards are comets. Because of the presence of volatile ices (and abundant organic molecules), the dust of comets tends to be very fragile as soon as the ice evaporates. A loose agglomerate of dust grains remains, such as collected from comet 81P/Wild 2 by NASA's Stardust mission (Brownlee et al. 2006; Zolensky et al. 2006).

It was recently realised that the border between "asteroids" and "comets" may well be diffuse and is somewhere in the asteroid belt. Some very *primitive asteroids* could still contain water ice and result in comet-like activity following a collision or when perturbed inwards. Several main belt asteroids have been discovered that showed brief cometary activity (Hsieh and Jewitt 2006). These objects are strictly comets, in my definition and that of Hsieh and Jewitt, but they are comets from a third source region: just inside the orbit of Jupiter.

Near-Earth Objects (NEO) are asteroids and comets whose orbits have a perihelion distance $q < 1.3$ AU, which can bring them close to Earth's orbit. Asteroids are perturbed into near-Earth orbits through the action of the ν_6 secular resonance (line of apsides of the asteroids move at the same rate as that of Saturn) on the inside of the asteroid belt, and numerous mean motion resonances with Jupiter throughout the asteroid belt, notably the 3:1 mean motion resonance. Once a large or small asteroid attains an orbit that resonates with that of Jupiter, it will quickly change eccentricity and the perihelion distance will decline. The aphelion of the orbit stays in the asteroid belt (2.5–4 AU). Asteroid belt comets would also be perturbed by mean motion resonances, perhaps most notably by the 2:1 resonance.

Comets are perturbed into near-Earth orbits through the action of resonances with Neptune in the scattered disc of the Kuiper Belt. They gradually evolve into orbits with a perihelion inside that of Neptune (when they are called "Centaurs") and can then be captured by Uranus, Saturn, and finally Jupiter. When they loose momentum in a close encounter, they end up having an aphelion close to the orbit of the planet and a perihelion much further inwards. Jupiter Family Comets have an aphelion near the orbit of Jupiter, until they are decoupled from Jupiter in a close encounter with one of the terrestrial planets, which can lower the aphelion distance (Everhart 1972).

2 A Brief History of Associating Asteroids to Meteor Showers

The first fifteen asteroid-looking NEO were discovered in 1898 (433 Eros), 1911 (719 Albert), 1918 (887 Alinda), 1929 (1627 Ivar), 1924 (1036 Ganymed), 1932 (1862 Apollo, the first Earth-crossing asteroid) and (1221 Amor), 1936 (2101 Adonis), 1937 (69230 Hermes), 1947 (2201 Oljato), 1948 (1685 Toro) and (1863 Antinous), 1949 (1566 Icarus), 1950 (29075), and 1951 (1620 Geographos).

When Whipple (1938) calculated the first meteoroid orbits from his multi-station photographic project, he pointed out that several of his ecliptic short period orbits were similar to those of Apollo type asteroids Apollo, Adonis (which he called Anteros), and Hermes. They had a similar low inclination (1.9–6°) and one had a semi-major axis of only 1.91 AU, shorter than that of Jupiter Family Comets. Whipple's 1936 1937 results for the orbit of the Geminids, with $\langle a \rangle = 1.396$ and $\langle e \rangle = 0.900$ (Lovell 1954), implied a stream that was unique compared to orbits of comets or asteroids known at that time, but not unlike the orbit of Icarus discovered shortly thereafter.

Cuno Hoffmeister (1948) first recognized the complex of ecliptic showers and pointed out that this could well be part of the system of minor planets. He noticed a resemblance between the orbit of Adonis and that of his Scorpiid-Sagittariid Complex, but he found a discouraging difference of some 25° between the longitude of perihelion of the asteroid and that of the middle of the meteor complex. He also noticed a similarity between his Piscids Complex and asteroid Hermes, and his Virginids Complex and Apollo. Plavec (1953, 1954) investigated the evolution of these proposed shower and asteroid associations, but did not confirm their generic relationship.

Later, Sekanina (1976) reinforced the orbital similarity of Adonis with six streams detected in the Harvard Meteor Project radar data, radiating from Sagittarius, Aquarius, and Capricorn, but he could not prove the evolutionary relationship more quantitatively. Besides Adonis, he pointed to the possible existence of associations of meteor streams with minor planets 433 Eros (Sekanina's xi Cygnids), 1627 Ivar (his August mu Draconids), 1566 Icarus (#171 Daytime Arietids and his Taurids-Arietids), 1862 Apollo (#66 Northern omega Scorpiids), 69230 Hermes (#156 N. Daytime May Arietids and #234 October epsilon Piscids), 1620 Geographos (#39 N. alpha-Leonids), 1685 Toro (his January Aquariids), 1950 DA (#39 N. alpha-Leonids and #133 April Psi Ursae-Majorids), 1959 LM (6 toroidal streams in June), 4788 P-L (his Canes Venaticids), 1973 NA (#187 psi Cassiopeiids), and 1973 EC (his kappa Geminids and his lambda Aurigids).

None of these proposed associations were particularly convincing. The tool used to make such associations was the Dissimilarity criterion (D criterion) introduced by Southworth and Hawkins (1963) and several varieties since. It makes a comparison between two sets of orbits and quantifies how much they differ. With many ecliptic streams much dispersed and very badly described by observations, it was not too difficult to find potential parent bodies.

Sekanina also identified as associated with meteor showers a series of deeply penetrating fireballs of the Prairy Network and European Fireball Network, the type of fireballs that are suspected meteorite droppers. This would suggest that asteroids, like comets, travel with a stream of debris. These early results suggested that *all* near-Earth objects with a node near Earth orbit had an associated meteoroid streams, irrespective of taxonomy.

The problem with the proposed meteoroid streams from an asteroidal origin are the high cosmic ray exposure ages of meteorites. They measure the time since the meteoroid was away from a larger parent body that generated neutrons by cosmic ray impacts, either sitting on the surface or being part of the parent body. Those timescales range typically from 1 to 30 million years. Over such long timescales, the streams will loose cohesion, disperse widely, and can get separated from the parent asteroid (Levin 1956). An asteroidal meteoroid stream is possible only if the meteoroid subsequently breaks in a collision, or otherwise, not long before some of the fragments hit Earth (Pauls and Gladmann 2005). The search for asteroidal streams was pursued by A. K. Terentjeva (1968, 1989), who has published many possible associations from fireball orbit surveys. Table 9 in Jenniskens (2006) gives a list of the more likely potential asteroidal meteoroid streams, after separating out the deeply penetrating fireballs from others, most of which are only pairs or triplets of similar orbits. All of those proposed streams need confirmation, before it can be certain that these are meteoroids from the same parent bodies. As it stands, no asteroidal meteoroid streams are established.

3 A Brief History of Association with Dormant Comets

For years, the relationship between asteroids, Jupiter family comets, and meteoroids was widely debated, but little was known about the dynamical processes that determined their interrelationships. Following Whipple's (1950) formulation of a comet model, where the sublimation of water vapor caused the comet to accelerate and loose mass by ejecting ice and dust in space, it was realised that comets can get "defunct" or dormant after having exhausted their gas reserves (Samoilova-Yakhontova 1950; Öpik 1963). There were in fact examples, such as comet 28P/Neujmin 1, which was stellar in appearance only two weeks after its perihelion passage during discovery in 1913. Only shortly afterwards a faint coma and tail were detected. With an orbital period of 18 yrs, this was not likely an asteroid. Later, Öpik (1968) argued from the lack of a known mechanism at the time to turn circular asteroidal orbits into eccentric orbits, that many of the Apollo asteroids had to be dormant comet nuclei, and argued that these comets were the source of meteorites. Kresák (1987) pointed to evidence of dormant phases in the aging of periodic comets from missed comet apparitions. In the mid 1980s, the list of asteroid-looking objects moving on cometary orbits increased significantly (Kresák and Stohl 1989).

Since that time, the orbital dynamics is better understood and we now guess that only 5–9% of NEO are dormant comets. Weissman et al. (2002) and Binzel and Lupisho (2006) have given reviews of the physical characteristics of such objects and the dynamical studies that estimate their abundance. These objects are now typically recognized by their Jupiter-Family-Comet orbits (Tisserand parameter between 2 and 3) and dark albedo (A'Hearn 1985). Another way of recognizing dormant comets among the population of Near Earth Objects (NEOs) is the presence of a meteoroid stream from past cometary activity.

3.1 The Association with Meteor Showers

In 1983, a fast moving object was discovered in IRAS observations of the sky at mid-Infrared wavelengths and Whipple (1983) realised that this asteroidal-looking object 3200 Phaethon moved among the Geminids. Phaethon shows no cometary activity (McFadden et al. 1985; Hsieh and Jewitt 2005) and has such a high Tisserand invariant that it was suspected to be an asteroid, perhaps just an interloper, or an object that generated meteoroids from a collision with a small main belt asteroid near aphelion (Hunt et al. 1985). Gustafson (1989), however, demonstrated that the Geminids were generated at perihelion, not at aphelion. He suspected activity over an extended period of time, from a now dormant comet. Phaethon was thought to be the rocky core of a de-volatilized comet, the missing link showing that comets can evolve into Apollo-type asteroids. In the words of Hughes (1985): "*The discovery of the asteroid-like object 1983 TB in the Geminid stream has strengthened the possibility that some comets can either choke themselves to death by forming a thick crust, or have a core of volatile-free material that remains after the majority of the gas and dust has escaped.*"

Earlier, Whipple and Hamid (1952) had discussed the Taurid stream in connection with 2P/Encke and concluded that other objects than this comet had to contribute to the formation of the stream. Napier and Clube (1979) and Napier (1983) proposed that minor planets *2201 (Oljato), 2212 (Hephaistos), 5025 P-L, 1979 XB, 1982 TA, 1984 KB, 1987 SB, 1991 TB* and others were such comet fragments, now dormant, together forming a massive Taurid Complex (Clube and Napier 1984; Bailey et al. 1986; Clube 1986, 1987; Olsson-Steel 1987, 1988; Steel et al. 1991; Asher et al. 1992; Porubçan and Kornos 2002). The main premise of a progressively disintegrating comet has held up, but the original comet was not quite as big as needed to justify their hypothesis of frequent past terrestrial catastrophic events. Nearly all proposed parent bodies have since been dismissed as asteroids, based on taxonomy (Jenniskens 2006). Most are O or S-type asteroids that became NEO through the ν_6 secular resonance mechanism that is also responsible for some of our meteorites.

The discovery of Phaethon and the possible existence of a Taurid Complex resulted in a new search for associations of asteroid-looking objects and meteor showers. Olsson-Steel (1987, 1988) linked *1566 (Icarus)* to the Daytime Arietids, now thought to be associated with the Marsden Sungrazers instead. He, too, pursued the idea that some of the NEO could be dormant comets and therefore associated with meteoroid streams from past activity.

Following on this work, Hasegawa et al. (1992) published a series of theoretical radiants and considered orbits of NEO up to the end of 1989. Drummond (1982, 1991) has compared the orbital elements of 139 NEO to meteoroid streams up to 1990 KA. He also compared the orbits of meteorite falls to those of minor planets, and like Halliday et al. (1990), identified four possible streams among meteoroid dropping fireballs. Kostolansky (1998) searched for asteroid parent bodies for 4409 photographed meteor orbits. Babadzhanov (1998) investigated the orbital evolution of candidate Taurid complex bodies over long enough periods of time to complete a nutation cycle and identified observed meteoroid streams at the four possible nodes for all objects. None of the proposed associations have been confirmed (but see Beech 2006).

4 The New Era: Comet Disintegration as the Major Source of Dust

The massive disruption of comets was recognized as a possible source of meteoroids, but such disruptions were deemed too rare among active Jupiter Family Comets to be a significant source of our meteor showers (Hughes 1985). The state of affairs before 2003 was best expressed by Hughes saying: "*There is no reason why the parent comet should undergo perturbations of a similar magnitude so even though they started in the same place, the stream and comet can quickly separate as time passes. This is probably the only satisfactory explanation as to why two out of three of the streams in Cook's* (1973) *list do not have recognizable parents.*" The parent bodies were somewhere, but they had now evolved beyond their streams. "*It seems*", according to Hughes, "*that in the large majority of cases comets decay gently.*"

In my 2006 book "Meteor Showers and their Parent Comets", I have argued that all of that changed in 2003, when it was discovered that *2003 EH1* moves among the highly inclined Quadrantid meteoroid stream, with only a 1 in 2 million chance of being a coincidental interloper (Jenniskens 2003, 2004). The association has since been studied by Williams et al. (2004), who confirmed that a comet observed in 1491 (C/1490 Y1) could well be the moment of breakup that generated the Quadrantid stream. Alternatively, Wiegert and Brown (2005a) have calculated backward in time the orbits of photographed Quadrantids, to conclude that the stream may be as young as two hundred years. Note, however, that the dispersion in the backward integrated orbits rapidly increases only when integrated to before 1490.

Over such a short timescale, the dispersion of dust can be simulated in numerical modeling, and from the distribution of nodes and the activity of the shower in Earth's path, a mass can be calculated for the whole stream. That mass (Table 1) is of order 1×10^{13} kg, needing a thousand years to generate during normal comet activity.

There is mounting evidence, in my opinion, that in fact most of our streams originate from discrete breakup events, rather than long episodes of gradual water vapor outgassing. Jenniskens and Lyytinen (2005) demonstrated that *2003 WY25* can be a fragments of an 1819 (or shortly before) breakup of *D/1819 W1 (Blanpain)* and that the dust of such a breakup would have evolved into Earth's path to create the 1956 Phoenicids. 2003 WY25 has since been found to be weakly active at perihelion (Jewitt 2006). Watanabe et al. (2006), too, recognized that the 1956 Phoenicids could have been the product of a breakup in 1819.

Table 1 Mass estimates of remaining comet fragments and their meteoroid stream, in units of 1 billion kg, after Jenniskens (2006)

Parent	Mass	Breakup	Product	Mass	Notes
3D/Biela	~14,000	AD 1842/43	Dormant comet	?	Remaining fragments
			Main dust mass	?	Dust during fragmentation
			Andromedids	33	1846/52 dust only
D/Blanpain	<5,600	AD 1819	2003 WY25	30	
			Other fragments	?	
			Phoenicids	100	
C/1490 Y1	~50,000	~AD 1490	2003 EH1	16,000	
			Quadrantids	10,000	
Unknown progenitor	–	>AD 1059	Marsden group	~10,000?	
			Daytime Arietids	~8,000	
Unknown progenitor	–	~AD 1030	3200 Phaethon	69,000	
			Geminids	28,000	
Unknown progenitor	–	~ AD 600	2004 TG10	360	
			N. Taurids	10,300	
Unknown progenitor	–	~AD 10	169P/NEAT	17,000	
			Alpha Capricornids	5,200	

After this, it was found that *2002 EX12* moves among the alpha-Capricornids (Wiegert and Brown 2005b; Jenniskens 2006). This minor planet is now better known by the name 169P/NEAT, after it was found that the object was weakly active at perihelion (Jäger and Hale 2005), after (!) we associated the object with the alpha-Capricornids. Again, the comet was only weakly active at perihelion, not active enough to account for the massive stream. The proposed shower formation age (~ AD 10) is still very uncertain.

The best documented case of comet fragmentation is that of the Sungrazers. Among the various sungrazer comet groups are the Marsden and Kracht Sungrazers which move in prograde orbits. These are small comet fragments that are detected only because they pass close to the Sun during perihelion, at which time they brighten from backscattered sunlight and pass the field of view of spaceborne Sun observatories. They are observed too briefly for a good orbit determination. Seargent (2002) first recognized the similarity in orbital elements between the Daytime Arietids and Marsden sungrazers and when it was recognized that some sungrazer comets return on a short-period comet orbit, the orbit could be greatly improved and the association was established. Sekanina and Chodas (2005) have argued that the Daytime Arietids and delta-Aquariids were created after 1059 AD. Shortly before that time, the Machholz family progenitor broke and a train of comet fragments had a close encounter with Jupiter in 1059 AD, which accelerated its evolution along the nutation cycle. The meteoroid streams were created by subsequent disintegration of some of these fragments.

There is, however, an interesting discrepancy in orbital period. Most Machholz family objects (9P/Machholz 1, Marsden and Kracht Sungrazers, Delta Aquariids, even 2003 EH1 and the Quadrantids) have a semi-major axis of about 3.1 AU. The Daytime Arietids have a semi-major axis of only 1.5 AU, half this value (Campbell-Brown 2005). The reason for this discrepancy may hold clues to understanding why some of our meteor streams have a relatively short semi-major axis. Perhaps the progenitor had a close encounter with Earth before (or during?) breakup.

Further evidence of frequent comet disintegrations has come from the confirmation that, in addition to 2P/Encke, there are other comet fragments among the Taurid showers (Table 2), several objects now being discovered that are a much better match to the Taurid showers than any of the objects proposed before (Porubçan et al. 2005; Jenniskens 2006).

Finally, Ohtsuka et al. (2005) recognized that 2005 UD moves among the Daytime Sextantids. Indeed, 2005 UD and 3200 Phaethon appear to have originated from a common ancestor, with 2005 UD over time evolving into an orbit not unlike that of Phaethon today. Recently, Jewitt and Hsieh (2006) found that 2005 UD is smaller than Phaethon (1.3 ± 0.1 km), but has the same bluish color, albedo = 0.11, and similar rotation period (5.249 h), consistent with both objects originating from one parent object. The Geminids are thought to have originated from Phaethon (or more precisely from a parent body that left Phaethon and the Geminids as products) at about 1030 AD (Jenniskens 2006). The whole complex of comet fragments broke at an earlier time.

The type of disintegration is not unlike that of the recent 1995 breakup of 73P/Schwassmann-Wachmann 3, which will cause a shower of tau-Herculids in 2022 (Lüthen et al. 2001). About as much mass is released in the form of dust and small fragments than the remaining mass (Table 1). That said, the mechanism of fragmentation may well be very different.

Few of these associations have been studied in detail thus far, but those that have point at a recent formation history of our meteor showers. All strong showers identified containing now dormant or weakly active comets have fragmented in the last 2,000 years (Table 1).

Table 2 (Mostly) dormant Jupiter family comets and their established meteor showers

Name	T_J	Tax.	D_{SH}	D_B	IAU#	Shower	H_N	H_{10}	D (km)
Phaethon family (Geminid Complex)									
1983 TB (Phaethon)	4.51	B	0.05	0.17	#4	GEM	14.6	–	5.1
2005 UD	4.51	B	0.14	0.53	#221	DSX	17.5	–	–
Machholz family (Daytime Arietid Complex)									
96P/Machholz	1.94	–	–	–	?#185	DBA	17.0	13.1	6.4
2003 EH1	2.07	–	0.07	0.62	#10	QUA	16.7	–	–
Marsden sungrazers	1.92	–	0.08	0.96	#171	ARI	20.0	–	Small
Kracht sungrazers	1.97	–	0.53	0.42	#5	SDA	20.0	–	Small
Encke family (Taurid Complex)									
2P/Encke	3.03	–	0.23	1.38	#28	SOA	17.3	13.3	4.8
2004 TG10	2.99	–	0.06	0.43	#17	NTA	19.5	–	–
2003 WP21	3.09	–	0.13	0.73	#2	NTA#9	21.4	–	–
2002 XM35	2.96	–	0.05	0.40	#256	ORN	23.0	–	–

T_J = Tisserand parameter with respect to Jupiter

Tax. = Taxonomy classification

D_{SH} = Dissimilarity criterion, measure of how good NEO orbit matches to mean stream orbit

D_B = Dissimilarity criterion based on three invariants in secular orbital evolution

IAU# = IAU shower number (Task Group on Meteor Shower Nomenclature working list)

Shower = IAU shower code (Task Group on Meteor Shower Nomenclature working list)

H_N = Nuclear magnitude at distance of 1 AU from Earth and Sun

H_{10} = Comet magnitude at distance of 1 AU from Earth and Sun

D = Diameter in km

5 Streams from (Weakly) Active Jupiter Family Comets: Also From Disintegration?

I would like to add here that comet disintegration may even play a role in creating meteoroid streams from active Jupiter Family Comets, perhaps dominating the mass loss from the normal water vapor outgassing as envisioned by Whipple (1951). From the streams listed in Table 3 (in that sequence), the situation is as follows:

Comet 26P/Grigg-Skjellerup was visited by Giotto after its Halley flyby and scattered light was observed from a cloud of particles, which suggested to McBride et al. (1997) that a larger fragment had come off.

Comet 21P/Giaconini-Zinner is an active comet and the Draconids are thought to have originated from normal comet outgassing. However, the meteor magnitude distribution in the stream is high, indicative of agressive disintegration of the dust after ejection. This implies evaporation from a water-rich layer, possibly freshly exposed in a comet breakup.

The Andromedids of comet 3D/Biela were created in a breakup in 1842/43 AD. In a recent paper, Jenniskens and Vaubaillon (2007) investigated the cause of the 1872 and 1885 Andromedid storms and concluded that the dust encountered was that generated during the continued fragmentation in the 1846 and 1852 returns. The mass generated in the 1842/43 breakup did not meet with Earth orbit. Normal activity from prior years did not result in meteor showers, but there were no favorable dust trail crossings.

Sykes and Walker (1992) found that the IRAS dust trail of comet 7P/Pons-Winnecke could have been created in about 1 orbit in normal comet activity and it is not clear if this

Table 3 (Weakly active) Jupiter family comets and their meteor showers

Name	T_J	Tax.	D_{SH}	D_B	IAU#	shower	H_N	H_{10}	D (km)
Established streams									
26P/Grigg-Skjellerup	2.81	–	0.01	0.02	#137	PPU	12.5	12.5	2.6
169P/NEAT	2.89	–	0.01	0.10	#1	CAP	18.0	18.0	–
21P/Giacobini-Zinner	2.47	–	0.02	0.20	#9	DRA	14.0	8.9	2.0
2003 WY25	2.82	–	0.14	0.39	#254	PHO	20.9	20.9	0.4
3D/Biela	2.53	–	0.17	0.45	#18	AND	–	7.1	–
73P/Sch.-Wach.	2.78	–	0.16	0.85	#61	TAH	15.0	12.0	–
7P/Pons-Winnecke	2.68	–	0.13	0.98	#170	JBO	16.0	11.5	–
D/Haneda-Campos	2.76	–	0.23	1.14	#233	OCC	–	11.9	–
Not (yet) established streams									
6P/d'Arrest	2.71	–	0.15	1.14	#73	ZDR	–	6.0	3.2
45P/Honda-Mrkos-Paj.	2.58	–	0.14	1.16	#199	ADC	18.0	14.0	1.6
P/2005 JQ5 (Catalina)	2.96	–	0.23	1.16	#66	NSC	–	18.5	–
D/Helfenzrieder	2.70	–	0.21	1.28	#11	EVI	–	4.5	–

Shower IAU number, code, and reference orbit refer to orbit and activity period information in Jenniskens (2006). D_{SH} and D_B are dissimilarity criteria defined in the text. H_N and H_{10} are the absolute magnitude of the comet nucleus and comet in active state, respectively

T_J = Tisserand parameter with respect to Jupiter

Tax. = Taxonomy classification

D_{SH} = Dissimilarity criterion, measure of how good NEO orbit matches to mean stream orbit

D_B = Dissimilarity criterion based on three invariants in secular orbital evolution

IAU# = IAU shower number (Task Group on Meteor Shower Nomenclature working list)

Shower = IAU shower code (Task Group on Meteor Shower Nomenclature working list)

H_N = Nuclear magnitude at distance of 1 AU from Earth and Sun

H_{10} = Comet magnitude at distance of 1 AU from Earth and Sun

D = Diameter in km

activity is normal. The comet is known for periodic outbursts of June Bootids from orphan trails, now much different from that of the comet orbit.

D/1978 R1 (Haneda-Campos) is now lost and probably also a dormant comet. It was responsible for a brief, but strong, shower in October, the #233 October Capricornids, with considerable mass (Wood 1988). The comet itself was only seen in 1978 and has not yet reappeared as a dormant comet.

6 Not So Well Established Associations

Given that all strong streams seem to have such remnant fragments, it is likely that many more associations will be recognized. Until now, however, a very small number have been associated with meteoroid streams, which is somewhat surprising given the number of known NEO. As of January 1, 2007, 701 Near-Earth Asteroids greater than 1 km in size have been identified as well as 64 Near Earth Comets (JPL website). In total, about 4407 NEOs have been discovered, 822 of which are potentially hazardous. This number is expected to increase in the near future. The estimated population larger than 1 km is about 1,100, while the population greater than 140 m in size is about 100,000 objects.

Reason for that low number may be the difficulty of establishing an association with a NEO in a low-inclination orbit of moderate eccentricity. The likelihood of chance associations increases dramatically with the increase in the population density of NEO in a, e, i space. The problem with lower inclination streams is that many more potential candidate parent bodies exist, as demonstrated by many proposed identifications before 2003 that did not pan out.

Ways to decrease the likelihood of chance association are: 1) to better describe the meteoroid stream so that dynamical studies are possible that trace the stream back to its point of origin; and 2) to discover that the proposed parent body has features expected for a (mostly) dormant comet nucleus, such as weak activity at perihelion, a dark nucleus (e.g., A'Hearn 1985; Dandy et al. 2003; Binzel and Lupishko 2006), or be dynamically related to Jupiter Family Comets (e.g., Bottke et al. 2002).

Fortunately, we can now make a strong argument that the association of parent bodies and their streams ought to be relatively tight. In search of other such associations, we are looking for (remnants of) parent bodies that can have created meteoroid streams within the past nutation cycle of the secular orbital evolution (one rotation of the nodal line relative to the line of apsides, which takes typically less than 4000 years). Some of these objects can now pass more than 0.2 AU from Earth's orbit. 2003 EH1, for example, passes at 0.21 AU due to periodic perturbations by Jupiter at aphelion. In the same way, 2002 EX12 does not pass close to Earth's orbit, but is found along the evolving orbit of the alpha-Capricornid shower, just slightly further along the nutation cycle than the meteoroids we recognize at Earth.

Table 4 lists associations made in this manner by searching for theoretical radiants close to those of observed meteor showers, and then testing how dissimilar the orbits are relative to those expected from the nutation cycle. The table gives the Tisserand parameter with respect to Jupiter and a dissimilarity criterion derived from invariants of secular perturbation (D_B).

Obrubov (1991) and Babadzhanov (1989) have first used invariants in secular perturbation theory derived by Lidov (1961, 1962) to search for asteroid–comet associations along the nutation cycle. Their first invariant is derived from the constant energy and momentum (related to the Tisserand parameter with respect to Jupiter). Their second invariant is a consequence of a perturbing function in the elliptical twice-averaged three-body problem being constant, as derived by Lidov (1961, 1962). Their third invariant is that of the longitude of perihelion, which moves much slower than the nodal line. In a recent paper, Jenniskens (2007) defined a dissimilarity criterion (D_B) based on these invariants. Table 4 (update from those given in Jenniskens 2006, until January 1, 2007) is ordered according to this dissimilarity criterion. Associations with $D_B < 1.0$ are thought to be siblings. Associations with $D_B = 1.0$–1.5 are thought to be aunts and uncles, like the Machholz family comet showers. Hence, the most likely associations are those with $D_B < 1.0$. The listed associations are less likely going down the list. Jenniskens (2007) also gives other criteria to evaluate the likelihood of association, including one proportional to the population density of NEO in a, e, i space.

7 Rapid Evolution of Jupiter Family Comets

Decoupling occurs due to a series of encounters with Earth and Venus. According to Wetherill (1991), shortly after decoupling, Jupiter Family comets have orbits with $a = 2.1$–2.5 AU, $Q < 4.35$ AU, and q mostly just outside Earth's orbit. Those with $q < 1$

Table 4 Less certain associations with not so well established showers

Name	T_J	Tax	D_{SH}	D_B	IAU#	Shower	H_N	D (km)	Ref. Orbit	I.D.
Jupiter family comets										
2006 UF17	2.91	–	0.11	0.31	#11	EVI	21.55	0.3	PJ ch28	PJ
2006 CS	2.44	–	0.10	0.38	#130	DME	16.56	3.2	PJ	PJ
2004 BZ74	2.37	–	0.12	0.44	#55	ASC	18.4	1.3	PG	PJ
1998 SH2	2.93	–	0.05	0.58	#21	AVB	20.8	0.4	L71B	PJ/PWK
2005 UR	2.92	–	0.08	0.65	#25	NOA	21.60	0.3	T89 (59)	PJ
2005 EM169	2.81	–	0.15	0.71	#129	QPE	24.67	0.1	NL (61.3.1)	PJ
6344 P-L	2.95	–	0.08	0.70	#236	GPS	20.38	0.5	T89 (54)	PJ
2002 FC	2.94	–	0.10	0.86	#134	NGV	18.82	1.1	T89 (22N)	PJ
1999 RD32	2.87	–	0.14	0.89	#112	NDL	16.32	3.6	L72B	PJ
1986 JK (Hypnos)	2.93	C	0.17	0.95	#63	COR	18.3	1.4	H48	O87
2002 GZ8	2.97	–	0.15	1.02	#260	GTI	18.15	1.5	PG	PJ
2002 KG4	2.77	–	0.21	1.06	June	Camelop.	20.85	0.4	ZS	PJ
1973 NA (5496)	2.53	–	0.28	1.18	#187	PCA	15.30	5.8	ZS	PSV92
2006 TA8	2.75	–	0.23	1.32	#235	LCY	20.92	0.4	T89 (49c)	PJ
2001 ME1	2.67	P	0.28	1.34	#167	NSS	16.60	3.1	L71B	PJ
2001 YB5	2.89	–	0.18	1.48	#97	SCC	20.62	0.5	T89 (6a)	PJ
2004 NL8	2.99	–	0.20	1.49	#117	DCQ	17.12	2.5	ZS	PJ
Asteroid-like orbits										
2004 GC19	3.49	–	0.03	0.13	#135	SGV	24.05	0.1	T89 (22S)	PJ
2004 TB18	3.89	–	0.05	0.31	#92	UER	17.54	2.0	PJ	PJ
2004 HW	3.04	–	0.06	0.35	#63	COR	17.1	2.5	H48	PJ
1999 RM45	3.95	–	0.05	0.38	#121	NHY	19.33	0.9	PJ	PJ
2002 SY50	3.87	–	0.13	0.47	#154	DEA	17.57	2.0	ZS	JFV
2003 CR20	3.32	–	0.08	0.52	#124	SVI	18.61	1.2	ZS73	PJ
2006 AR3	3.17	–	0.10	0.59	#76	KAQ	20.39	0.5	PG	PJ
2003 BD44	3.62	–	0.13	0.56	#135	SGV	16.62	3.1	PG	PJ/PKW
2005 CA	3.03	–	0.11	0.77	#235	LCY	15.33	5.7	PG	PJ
2002 GM5	3.36	–	0.12	0.78	#136	SLE	21.44	0.3	PG	PJ
2006 JV26	3.33	–	0.08	0.89	#139	GLI	25.19	0.1	ZS73	PJ
2003 QC10	4.48	–	0.09	0.89	#155	NMA	17.83	1.8	ZS	PJ
1995 EK1	3.12	–	0.14	0.95	#136	SLE	17.54	2.0	PG	K98
2005 NZ6	3.43	–	0.20	0.97	#144	APS	17.40	2.2	KL (4)	PJ
2003 YM137	3.03	–	0.21	0.99	#125	SAL	18.72	1.2	T89 (16)	PJ
1999 FN53	3.96	–	0.14	1.00	May	Ursids	18.39	1.4	ZS	PKW
2004 YD5	3.11	–	0.21	1.17	#167	NSS	29.26	0.01	ZS73	PJ
107P/Wilson-Harr.	3.08	CF	0.13	1.42	#218	GSA	16.0	4.0	T89 (48)	PJ

For each group, the likelihood decreases going down the list

D (km) = diameter based on nuclear magnitude and adopted albedo = 0.04

Ref Orbit: See references in Jenniskens (2006), Table 7. Examples are: L71B = Lindblad (1971); PG = Porubçan and Gavajdova (1994); PJ = Jenniskens (2006); ZS = Sekanina (1973, 1976)

I. D.: References for who made the identification. JVF: Jopek et al. (1999); K98 = Kostolansky 1998; O87 = Olsson-Steel (1987); PJ = Jenniskens (2006); PKW = Porubçan et al. (2005); PSV92 = Porubçan et al. (1992); PWK = Porubçan et al. (2004); W = Whipple (1940)

AU are distributed mostly between 0.5 and 0.9 AU. Many of our potential parent bodies are objects with semi-major axis above a = 2.5 AU. These are Jupiter Family Comets that are not yet fully decoupled from Jupiter.

The typical lifetime for decoupling is 100,000 to 1,000,000 years (Wetherill 1991). This is much longer than the lifetime of an active comet ($\sim$12,000 years according to Levison and Duncan 1997) and the typical nutation cycle ($\sim$4,000 years). Hence, most not-yet decoupled Jupiter-family comets are expected to be dormant. This is consistent with finding many dormant objects in this transition regime.

Our lists include 24 candidate dormant (or weakly active) comets that appear to be Jupiter Family comets (T_J = 2–3). This is a significant fraction of all such objects, estimated at 123 ± 41 by Binzel and Lupishko (2006). If all these objects are confirmed parent bodies, then this would imply that these objects break on a time scale equal or less that of the rate of meteoroid streams evolving into Earth orbit (<2,000 years).

In each fragmentation, about half of the mass of the comet is lost in the form of meteoroids (Table 1). A typical comet would evolve from a diameter of D = 3 km to D < 0.5 km in only eight disruptions. For the lifetime of Jupiter family comets of 12,000 years, this would imply a period between disruptions of about 1500 years, in good agreement. If the comet would stay at its most active over that period of time, the same amount of mass would be lost, but that is clearly not the case. Hence, fragmentation is the main mass loss mechanism for dormant comets in the inner solar system.

Already 20 years ago, Kresák and Kresákova (1987) concluded that the visible release of dust from Jupiter Family Comets was insufficient to maintain the zodiacal cloud in equilibrium. They were first to suggest that "*the progressive decay of the dark matter, including extinct cometary nuclei, their fragments, and products of asteroidal collisions, represents the dominant source of replenishment of the interplanetary dust complex.*"

8 Origin of Phaethon

This young dynamical lifetime of Jupiter Family Comets creates a problem in explaining the origin of 3200 Phaethon and 2005 UD, which currently move in an asteroid-like orbit (T_J >> 3). Close encounters with the terrestial planets are needed, but those are infrequent. It is possible that their predecessor originated from among the most primitive asteroids in the (outer) asteroid belt.

As noticed in the past, quite a number of our meteor showers have a small semi-major axis, and may be related to such outer belt comet-asteroid transition objects (Table 4). However, many of these streams are in doubt and need to be established first.

9 Further Work

Before any of the associations listed in Table 4 can be established, we need to be certain that proposed streams exist, as in being streams of meteoroids from the same parent body. Observational programs are encouraged that can help confirm the existence of the streams.

To help confirm the association of a NEO with a given meteoroid stream, dynamical studies are needed that trace the meteoroid stream and the proposed parent body to the time of fragmentation. The dispersion of dust reveals the age of a stream. Orbits more precise than those derived by radar are needed for such studies. This calls for a significant push to better characterize the known meteoroid streams by photographic, digital CCD, and intensified video techniques.

The likelihood of an association can also be increased from studies of the proposed parent body. Table 4 can serve as a list of priority for taxonomic studies of NEOs, to address some of the other criteria that could help identify dormant comets (Binzel and Lupisho 2006): they should have low geometric albedo ($<$0.075), taxonomic classes D, P, or C (possibly F or B), and rotation rates lower on average than the mean rate of asteroidal NEOs.

As a more general course of action, the fragmentation mechanisms needs to be better understood (Hughes 1990; Gronkowski 2007). There may be more than one. The streams themselves may give information about their cause. In the case of the Andromedids, for example, most mass in the resulting meteoroid stream is in the form of small particles, presumably because rapid evaporation of residual ices in the comet boulders broke the larger meteoroids. This could also be why few meteoroid streams are known for the prevalence of boulders. In the case of comet 2P/Encke, on the other hand, the gas drag limit would predict no larger than $\sim$kilogram sized meteoroids, but the largest observed Taurids are at least two orders of magnitude in mass bigger. These meteoroids could originate from parts of the comet that had already lost much of their volatiles. Other such cases might be found in a search for meteoroid stream association in the population of tens to hundreds of kilogram objects.

Acknowledgments This paper was greatly improved by helpful comments from editor Frans Rietmeijer, as well as from Peter Brown and an anonymous reviewer. I thank NASA's Planetary Astronomy program and the NASA Goddard Space Flight Center's IRAD program for partial support of this research effort.

References

M.F. A'Hearn, Are cometary nuclei like asteroids? in *Asteroids, Comets, Meteors II*. ed. by C.-I. Lagerkvist, B.A. Lindblad, H. Lundstedt, H. Rickman (Uppsala University, Uppsala, 1985), pp. 187–190

D.J. Asher, S.V.M. Clube, D.I. Steel, The Taurid complex asteroids. in *Meteoroids and their parent bodies*, ed. by J. Stohl, I. P. Williams (Astronomical Institute, Slovak Academy of Sciences, Bratislava, 1992), pp. 93–96

P.B. Babadzhanov, Formation of twin meteor showers. in *Asteroids, Comets, Meteors III*, ed. by C.-I. Lagerkvist, H. Rickman, B.A. Lindblad, M. Lindgren (Uppsalla University, Uppsala, 1989), pp. 497–503

P.B. Babadzhanov, Near-Earth asteroids associated with meteor showers. in *Meteoroids 1998*, ed. by W.J. Baggaley, V. Porubçan (Astronomical Institute, Slovak Acad. Sci., Bratislava, 1998), pp. 185–190

M.E. Bailey, S.V.M. Clube, W.M. Napier, The origin of comets. Vistas Astron. **29**, 53–112 (1986)

M. Beech, Canadian fireball activity from 1962 to 1989. JIMO **34**, 104–110 (2006)

R.P. Binzel, D.F. Lupishko, Properties of the Near-Earth Object population: the ACM 2005 view. in *Asteroids, Comets, Meteors*. IAU Symposium 229. ed. by L. Daniela, M. Sylvio Ferraz, F.J. Angel (Cambridge University Press, Cambridge, 2006), pp. 207–214

W.F. Bottke, A. Morbidelli, R. Jedicke, J.-M. Petit, H.F. Levison, P. Michel, T.S. Metcalfe, Debiased orbital and absolute magnitude distribution of the near-Earth objects. Icarus **156**, 399–433 (2002)

D. Brownlee et al., Comet 81P/Wild 2 under a microscope. Science **314**, 1711–1716 (2006)

M. Campbell-Brown, Arietid meteor orbits measurements. Earth Moon Planets **95**, 279–287 (2005)

S.V. Clube, Giant comets or ordinary comets: parent bodies or planetetesimals. *Proceedings of 20th ESLAB Symposium on the Exploration of Halley's Comet. Vol. 2: dust and nucleus*. (Heidelberg, ESA SP-250, 1986), pp. 403–408

S.V.M. Clube, The origin of dust in the solar system. Phil. Trans. R. Soc. London **323**, 421–436 (1987)

S.V.M. Clube, W.M. Napier, The microstructure of terrestrial catastrophism. MNRAS **211**, 953–968 (1984)

A.F. Cook, A working list of meteor streams. in *Evolutionary and physical properties of meteoroids*. ed. by C.L. Hemenway, P.M. Millman, A.F. Cook (NASA SP-319, 1973), p. 183

C.L. Dandy, A. Fitzsimmons, S.J. Collander-Brown, Optical colors of 56 near-Earth objects: trends with size and orbit. Icarus **163**, 363–373 (2003)
J.D. Drummond, Theoretical meteor radiants of Apollo, Amor, and Aten asteroids. Icarus **49**, 143–153 (1982)
J.D. Drummond, Earth-approaching asteroid streams. Icarus **89**, 14–25 (1991)
E. Everhart, The origin of short-period comets. Astrophys. Lett. **10**, 131–135 (1972)
P. Gronkowski, The search for a cometary outbursts mechanism: a comparison of various theories. Astron. Nachr. **328**, 126–136 (2007)
B.Å.S. Gustafson, Geminid meteoroids traced to cometary activity on Phaethon. Astron. Astrophys. **225**, 533–540 (1989)
I.A. Halliday, A.T. Blackwell, A.A. Griffin, Evidence for the existence of groups of meteorite-producing asteroidal fragments Meteoritics **25**, 93–99 (1990)
I. Hasegawa, Y. Ueyama, K. Ohtsuka, Predictions of the meteor radiant point associated with an earth-approaching minor planet. PASJ **44**, 45–54 (1992)
C. Hoffmeister, *Meteorströme*. (Verlag Johann Ambrosine Bart, Leipzig, 1948)
H.H. Hsieh, D. Jewitt, Search for activity in 3200 Phaethon. Ap. J. **624**, 1093–1096 (2005)
H.H. Hsieh, D. Jewitt, A population of comets in the main asteroid belt. Science **312**, 561–563 (2006)
D.W. Hughes, The relationship between comets and meteoroid streams. in *Asteroids, Comets, Meteors II*. ed. by C.I. Lagervkist, B.A. Lindblad, H. Lundstedt, H. Rickman (Astronomical Observatory, Uppsala, 1985), pp. 503–519
D.W. Hughes, Cometary outbursts—A review. QJRAS **31**, 69–94 (1990)
J. Hunt, K. Fox, I.P. Williams, Asteroidal origin for the Geminid meteor stream. in *Asteroids, Comets, Meteors II*. C.-I. Lagerkvist, B.A. Lindblad, H. Lundstedt, H. Rickman (Astronomical Observatory, Uppsala, 1985), pp. 549–553
M. Jäger, A. Hale, Comet 169P/NEAT. IAU Circular 8600. in ed. by D.W.E. Green (Central Bureau for Astronomical Telegrams, Smithsonian Astrophysical Observatory, Cambridge, MA, 2005)
P. Jenniskens, 2003 EH_1 and the Quadrantids. IAU Circular 8252. ed. by D.W.E. Green (Central Bureau for Astronomical Telegrams, Smithsonian Astrophysical Observatory, Cambridge, MA, 2003)
P. Jenniskens, 2003 EH_1 is the parent of the Quadrantids. Astron. J. **127**, 3018–3022 (2004)
P. Jenniskens, *Meteor showers and their parent comets*. (Cambridge University Press, Cambridge, U.K., 2006) 790 pp
P. Jenniskens, Meteoroid streams that trace to candidate dormant comets. Icarus (2007) (in press)
P. Jenniskens, E. Lyytinen, Meteor showers from broken comets: D/1819 W1 (Blanpain), 2003 WY25, and the phoenicids. Astron. J. **130**, 1286–1290 (2005)
P. Jenniskens, J. Vaubaillon, 3D/Biela and the Andromedids: fragmenting versus sublimating comets. Astron. J. **134**, 1037–1045 (2007)
D. Jewitt, Comet D/1819 W1 (Blanpain): not dead yet. Astron. J. **131**, 2327–2331 (2006)
D. Jewitt, H. Hsieh, Physical observations of 2005 UD: a mini-Phaethon. Astron. J. **132**, 1624–1629 (2006)
T.J. Jopek, G.B. Valsecchi, Cl. Froeschlé, Meteoroid stream identification: a new approach—II. Application to 865 photographic meteor orbits. MNRAS **304**, 751–758 (1999)
E. Kostolansky, On a search for meteoroid streams of asteroidal origin among photographic meteor orbits. in *Meteoroids 1998*, ed. by W.J. Baggaley, V. Porubçan (Astron. Inst., Slovak Acad. Sci., Bratislava, 1999, 1998), pp. 191–194
L. Kresák, Dormant phases in the aging of periodic comets. Astron. Astrophys. **187**, 906–908 (1987)
L. Kresák, M. Kresáková, The contribution of periodic comets to the zodiacal cloud. *Proceedings of the 10th European Regional Astronomy Meeting of the IAU, Vol. 2*. (Czechoslovak Academy of Sciences, Ondrejov, Czechoslovakia, 1987), pp. 265–270
L. Kresák, J. Stohl, Generic relationships between comets, asteroids, and meteors. in *Asteroids, Comets, Meteors III*. ed. by C.-I. Lagerkvist, H. Rickman, B.A. Lindblad, M. Lindgren (Uppsala University, Uppsala, 1989), pp. 379–388
B.Y. Levin, Fiziceskaja teorija meteorov i meteornoe vescestvo v solnecnoj sisteme. Akademizdat, Moskva (1956) (in Russian)
H.F. Levison, M.J. Duncan, From the Kuiper belt to Jupiter-family comets: the spatial distribution of ecliptic comets. Icarus **127**, 13–32 (1997)
M.L. Lidov, Orbital evolution of the artificial satellites of planets under the action of gravitational perturbations (in Russian). Iskusstvennie sputniki Zemli (Artificial Earth's satellites) **8**, 5–45 (1961)
M.L. Lidov, The evolution of orbits of artificial satellites of planets under the action of gravitational perturbations of external bodies. Planet. Space Sci. **9**, 719–759 (1962)
B.A. Lindblad, 2. A computerized stream search among 2401 photographic meteor orbits. Smithson. Contr. Astrophys. **12**, 14–24 (1971)
A.C.B. Lovell, *Meteor Astronomy*. (Clarendon Press, Oxford, 1954), p. 317

H. Lüthen, R. Alt, M. Jäger, The disintegratiing comet 73P/Schwassmann-Wachmann 3 and its meteors. JIMO **29**, 15–28 (2001)
N. McBride, S.F. Green, A.C. Levasseur-Regourd, B. Goidet-Devel, J.-B. Renard, The inner dust coma of comet 26P/Grigg-Skjellerup: multiple jets and nucleus fragments? MNRAS **289**, 535–553 (1997)
L.A. McFadden, M. F. A'Hearn, R.L. Millis, G.E. Danielson, A search for cometary emissions and meteor debris associated with (3200) 1983 TB. PASP **97**, 899–900 (1985)
W.M. Napier, The orbital evolution of short period comets. in *Asteroids, Comets, Meteors*, ed. by C.-I. Lagerkvist, H. Rickman (Uppsala Observatory, Uppsala, 1983), pp. 391–395
W.M. Napier, S.V.M. Clube, A theory of terrestrial catastrophism. Nature **282**, 455–459 (1979)
Yu.V. Obrubov, Complexes of minor bodies in the solar system (in Russian). Astronomicheskii Zhurnal **68**, 1063–1073 (1991)
K. Ohtsuka, T. Sekiguchi, D. Kinoshita, J.-I. Watanabe, 2005 UD and the Daytime Sextantids. *CBET 283*. ed. by D.W.E. Green (Central Bureau for Astronomical Telegrams, Smithsonian Astrophysical Observatory, Cambridge, MA, 2005)
D. Olsson-Steel, Theoretical meteor radiants of Earth-approaching asteroids and comets. Australian J. Astron. **2**, 21–35 (1987)
D. Olsson-Steel, Identification of meteoroid streams from Apollo asteroids in the Adelaide radar orbit surveys. Icarus **75**, 64–96 (1988)
E. Öpik, Survival of cometary nuclei and the asteroids. Adv. Astron. Astrophys. **2**, 219–262 (1963)
E. Öpik, The cometary origin of meteorites. Irish Astron. J. **8**, 185–208 (1968)
A. Pauls, B. Gladmann, Decoherence time scales for "meteoroid streams". Meteoritics Planet. Sci. **40**, 1241–1256 (2005)
M. Plavec, On the relations between minor planets and meteor streams. Bull Astron. Inst. Czech. **4**, 195–195 (1953)
M. Plavec, On the relations between minor planets and meteor streams. Bull. Astron. Inst. Czechosl. **5**, 38–42 (1954)
V. Porubçan, M. Gavajdová, A search for fireball streams among photographic meteors. Planet. Space Sci. **42**, 151–155 (1994)
V. Porbuçan, L. Kornos, The taurid meteor shower. in *Proceedings of ACM 2002*. ed. by B. Warmbein (ESA Publication Division, Noordwijk, ESA SP-50, 2002), pp. 177–180
V. Porubçan, L. Kornos, I.P. Williams, The Taurid complex meteor showers and asteroids. Contr. Astron. Obs. Skalnaté Pleso. **36**, 103–117 (2005)
V. Porubçan, J. Stohl, R. Vana, On associations of Apollo asteroids with meteor streams. in *Asteroids, Comets, Meteors 1991*. (Lunar and Planetary Institute, Houston, 1992), pp. 473–476
V. Porubçan, I.P. Williams, L. Kornos, Associations between asteroids and meteoroid streams. Earth, Moon Planets **95**, 697–712 (2004)
N.S. Samoilova-Yakhontova, The minor planets (in Russian). Uspekhi Astronomicheskikh Nauk **5**, 136 (1950)
D.A.J. Seargent, Comets C/1999 173, 2002 E1 (SOHO). MPEC 2002-E18. ed. by B.G. Marsden (Central Bureau for Astronomical Telegrams, Smithsonian Astrophysical Observatory, Cambridge, MA, 2002)
Z. Sekanina, Statistical model of meteor streams. III. Stream search among 19 303 radio meteors. Icarus **18**, 253–284 (1973)
Z. Sekanina, Statistical model of meteor streams. IV. A study of radio streams from the synoptic year. Icarus **27**, 265–321 (1976)
Z. Sekanina, P.W. Chodas, Origin of the Marsden and Kracht groups of sunskirting comets. I. Association with Comet 96P/Machholz and its interplanetary complex. Astron. J. Supp. Ser. **161**, 551–586 (2005)
R.B. Southworth, G.S. Hawkins, Statistics of meteor streams. Smithson. Contr. Astrophys. **7**, 261–285 (1963)
D.I. Steel, D.J. Asher, S.V.M. Clube, The structure and evolution of the Taurid complex. MNRAS **251**, 632–648 (1991)
M.V. Sykes, R.G. Walker, Cometary dust trails. I—Survey. Icarus **95**, 180–210 (1992)
A.K. Terentjeva, Investigation of minor meteor streams. in *Physics and Dynamics of Meteors*. Ed. by L. Kresák, P. Millman (Reidel, Dordrecht, 1968), pp. 408–427
A.K. Terentjeva, Fireball streams. in *Asteroids, Comets, Meteors III*. ed. by C.I. Lagerkvist, H. Rickman, B.A. Lindblad (Astronomical Observatory, Uppsala, 1989), pp. 579–784
J.-I. Watanabe, M. Sato, T. Kasuga, Pursuing a historical meteor shower. Astron. Herald. **99**, 629–636 (2006)
P.R. Weissman, W.F. Bottke, H.F. Levison, Evolution of comets into asteroids. in *Asteroids III*. ed. by W.F. Bottke, P. Paolicchi, R.P. Binzel, A. Cellino (The University of Arizona Press, Tucson, Arizona, 2002), pp. 669–686

G.W. Wetherill, End products of cometary evolution—cometary origin of Earth-crossing bodies of asteroidal appearance. in *Comets in the post Halley era, Vol. 1*. (Kluwer, Dordrecht, 1991), pp. 537–556
F.L. Whipple, Photographic meteor studies I. Proc. Amer. Phil. Soc. **79**, 499–548 (1938)
F.L. Whipple, Photographic meteor studies. III. The Taurid shower. Proc. Am. Phil. Soc. **83**, 711–745 (1940)
F.L. Whipple, A comet model. I. The acceleration of comet Encke. Astroph. J **111**, 375–394 (1950)
F.L. Whipple, A comet model. II. Physical relations for comets and meteors. Astrophys. J. **113**, 464–474 (1951)
F.L. Whipple, 1983 TB and the Geminid meteors. IAUC 3881, ed. by B.G. Marsden (Central Bureau for Astronomical Telegrams, Smithsonian Astrophysical Observatory, Cambridge, MA, 1983)
F.L. Whipple, S.E. Hamid, On the origin of the Taurid meteor streams. Helwan Obs. Bull. **41**, 1–30 (1952)
P. Wiegert, P. Brown, The Quadrantid meteoroid complex. Icarus **179**, 139–157 (2005a)
P. Wiegert, P. Brown, The problem of linking minor meteor showers to their parent bodies: initial considerations. Earth Moon Planets **95**, 19–25 (2005b)
I.P. Williams, G.O. Ryabova, A.P. Baturin, A.M. Chernitsov, The parent of the Quadrantid meteoroid stream and asteroid 2003 EH1. MNRAS **355**, 1171–1181 (2004)
J. Wood, De october Capricorniden: een nieuwe zwerm (in Dutch). Radiant, J. DMS **10**, 71–73 (1988), (also: NAPO-MS Bull. 200)
M.E. Zolenski et al., Miineralogy and petrology of comet Wild 2 nucleus samples. Science **314**, 1735–1739 (2006)

Large Dust Grains Around Cometary Nuclei

A. Molina · F. Moreno · F. J. Jiménez-Fernández

Originally published in the journal Earth, Moon, and Planets, Volume 102, Nos 1–4.
DOI: 10.1007/s11038-007-9158-2

Abstract Large amounts of particles ejected from the nucleus surface are present in the vicinity of the cometary nuclei when comets are near the Sun (at heliocentric distances ≤2 AU). The largest dust grains ejected may constitute a hazard for spatial vehicles. We tried to obtain the bounded orbits of those particles and to investigate their stability along several orbital periods. The model includes the solar and the cometary gravitational forces and the solar radiation pressure force. The nucleus is assumed to be spherical. The dust grains are also assumed to be spherical, and radially ejected. We include the effects of centrifugal forces owing to the comet rotation. An expression for the most heavy particles that can be lifted is proposed. Using the usual values adopted for the case of Halley's comet, the largest grains that can be lifted have a diameter about 5 cm, and the term due to the rotation is negligible. However, that term increases the obtained value for the maximum diameter of the lifted grain in a significant amount when the rotation period is of the order of a few hours.

Keywords Comets: general · Comets: individual · Interplanetary medium · Meteoroids

1 Introduction

Particles of different sizes (in the millimetric to decimetric size range) around asteroids and comets have been reported for a long time. These particles may constitute a hazard for spacecrafts visiting the neighbours of the comets and asteroids. Although it might be thought that this risk does not exist if the comet-to-sun direction is avoided, the danger

A. Molina (✉)
Departamento de Física Aplicada, Facultad de Ciencias, Universidad de Granada, Avenida Severo Ochoa s/n, Granada 18071, Spain
e-mail: amolina@iaa.es

A. Molina · F. Moreno · F. J. Jiménez-Fernández
Instituto de Astrofísica de Andalucía, CSIC, Camino Bajo de Huétor 50, Granada 18008, Spain

J.M. Trigo-Rodriguez et al. (eds.), *Advances in Meteoroid and Meteor Science*.
DOI: 10.1007/978-0-387-78419-9_67

remains if large grains are orbiting around the nucleus. Suisei and Giotto were hit by millimetric grains when both spacecraft visited the Comet Halley in 1986. In the case of Suisei, that hit was over 150.000 km distance and the two particles were several milligrams and probably nearly millimetric in size, and in the Giotto impact the grains were larger, about forty milligrams (see, for example, Campbell et al. (1989) and references therein). The danger is much larger if the spatial vehicle is planned to make a close approach to the nucleus of the comet, as planned for Rosetta mission, which is intended to visit the Comet 67P/Churyumov-Gerasimenko in 2014. The interest has increased after the detection of large grains (>2 cm) in the coma of comet LINEAR from radar observations (Nolan et al. 2006).

The purpose of this work is to show the appropriated expressions to study the dynamical properties of the ejected cometary nuclei dust particles, including those terms coming from the comet rotation. Specifically, a determination of the largest size grain that can be lifted from the nucleus surface is made. The numerical integration of the differential equations of the motion will allow us to obtain the orbits of the dust particle around the nuclei of the comets.

2 Equation of Motion

We write $\sum \vec{F}_i = m_{\mathrm{d}}\vec{a}$, where $\vec{F}_i$ is each force term applied over the mass m_{d} of the dust particle which suffers an acceleration $\vec{a}$. We consider the following forces (all units are in the International System of units):

(a) Drag force, $\vec{F}_{\mathrm{D}}$. This force is due to the gas drag, and it can be derived from Navier–Stokes equations after suitable assumptions. We consider radially-symmetric outgassing. If the gas velocity is assumed to be constant and the dust velocity is considered much lower than the gas velocity, drag force is $\vec{F}_{\mathrm{D}} = (1/32)C_{\mathrm{D}}d^2\dot{m}_{\mathrm{g}}v_{\mathrm{g}}r_{\mathrm{d}}^{-3}\vec{r}_{\mathrm{d}}$, which is a particular case of that one given by Wallis (1982), where C_{D} is the drag coefficient, d the diameter of the grain, $\dot{m}_{\mathrm{g}}$ the gas loss rate, v_{g} the gas velocity, and r_{d} the dust grain to the nucleus centre distance.

(b) Gravitational force, $\vec{F}_{\mathrm{G}}$. This force can be written as $\vec{F}_{\mathrm{G}} = -m_{\mathrm{d}}\left[M_{\mathrm{c}}Gr_{\mathrm{d}}^{-3}\vec{r}_{\mathrm{d}} + M_{\mathrm{S}}Gr_{\mathrm{S}}^{-3}\vec{r}_{\mathrm{S}}\right]$, being m_{d} the dust grain mass, M_{c} the comet mass, M_{S} the Solar mass, $\vec{r}_{\mathrm{s}}$ the vector joining the centre of Sun to the dust grain and G the gravitational constant.

(c) Solar radiation pressure force, $\vec{F}_{\mathrm{rad}}$. Introducing the dimensionless parameter β as the ratio of the radiation pressure to the Sun gravitational force, the radiation pressure becomes $\vec{F}_{\mathrm{rad}} = m_{\mathrm{d}}M_{\mathrm{S}}G\beta r_{\mathrm{S}}^{-3}\vec{r}_{\mathrm{S}}$.

(d) Inertial forces, $\vec{F}_{\mathrm{I}}$. In this work, we used a nucleus-attached reference system with origin at the centre of the comet. Obviously, this frame is non inertial and then we must consider two inertial forces. One of them is due to the gravitational comet attraction by the Sun and the other is due to the rotation of the comet (spin). The final expression of this force is $\vec{F}_{\mathrm{I}} = GM_{\mathrm{S}}m_{\mathrm{d}}r_{\mathrm{c}}^{-3}\vec{r}_{\mathrm{c}} + m_{\mathrm{d}}\left\{\vec{\Omega} \times (\vec{r}_{\mathrm{d}} \times \vec{\Omega}) + 2\vec{v}_{\mathrm{d}} \times \vec{\Omega}\right\}$, where $\vec{\Omega}$ is the nucleus angular speed and $\vec{v}_{\mathrm{d}}$ is the dust particle speed. Here, $\vec{\Omega}$ is considered constant with time and so no angular acceleration term is included.

Now we have every F_i to include in $\sum \vec{F}_i = m_{\mathrm{d}}\vec{a}$. Due to $\vec{r}_{\mathrm{s}} = \vec{r}_{\mathrm{c}} + \vec{r}_{\mathrm{d}}$, we can rewrite the previous expression with only $\vec{r}_{\mathrm{c}}$ and $\vec{r}_{\mathrm{d}}$, expanding r_{s}^{-3} in a Taylor series and assuming that $(r_{\mathrm{d}}/r_{\mathrm{c}})^n = 0$ for $n > 1$ (Richter et al. 1995). Then, we divide every term by $m_{\mathrm{d}} = (4/3)\pi(d/2)^3\rho_{\mathrm{d}}$ and use β ($\beta = 1 - \mu$) to obtain:

$$\frac{d^2\vec{r}_d}{dt^2} = \left(\frac{3C_D d^2 \dot{m}_g v_g}{16\pi C_p Q_p}\beta - GM_c\right)\frac{\vec{r}_d}{r_d^3} + \beta M_S G\frac{\vec{r}_c}{r_c^3} + \frac{\mu M_S G}{r_c^3}\left(\frac{3\vec{r}_d\vec{r}_c}{r_c^2}\vec{r}_c - \vec{r}_d\right) \\ + \Omega^2 \cdot \vec{r}_d - (\vec{\Omega}\cdot\vec{r}_d)\cdot\vec{\Omega} + 2\vec{v}_d \times \vec{\Omega} \quad (1)$$

which is similar to that by Richter et al. (1995 and that by Fulle (1997), but including new terms due to the rotation of the comet. These new terms, which only have to be taking into account before the particles are ejected, are not necessarily negligible, and thus, for example, the term $\Omega^2 \cdot \vec{r}$ can be 50% of that due to the comet gravitation for a spin period of 5 h.

3 Dust Grain Ejection

As noted by Crifo et al. (2005), dust grains will lift from the nucleus surface if $\sum \vec{F}_i \cdot \hat{n} > 0$, being $\hat{n}$ an unitary normal vector to the surface. In the limit, that scalar product will be equal to 0 for the largest grain that can be lifted from the nucleus surface. Thus, multiplying both sides of the Eq. 1 by $\hat{n}$, rearranging and simplifying, and considering that and the second and the third terms of the Eq. 1 do not contribute to the Eq. 2 because $R << r_c$, we obtain:

$$d_{max} = \frac{1}{\rho_d g_{eff}}\frac{3C_D \dot{m}_g v_g}{16\pi R^2} \quad (2)$$

where $g_{eff} = g - \Omega^2 R\sin^2\varphi + 2(\vec{v}_d \times \vec{\Omega})\cdot\hat{n}$, g is the gravity of the comet, φ is the angle between $\vec{\Omega}$ and $\hat{n}$, R is the radius of the cometary nucleus, and d_{max} is the maximum diameter of the grain that can be lifted. Since Whipple (1951) similar formulae have been reported (see, for example, Ma et al. 2002). Assuming C_D = 2 (sphere), $\rho_d = 10^3$ kg m^{-3} to the density, 5×10^3 m for R and 1.5×10^7 kg m s^{-1}, appropriated values for comet Halley (Krankowsky 1986), we obtain a result of 5 cm for the diameter of the largest grain lifted from the nucleus surface neglecting the nucleus rotation. When the rotation is included, the maximum size increases to 6.25 cm at the nucleus equator, remaining 5 cm size at the poles. We have assumed an isotropic outgassing, but if the gas ejection is confined to certain active areas, the maximum diameter could be much larger.

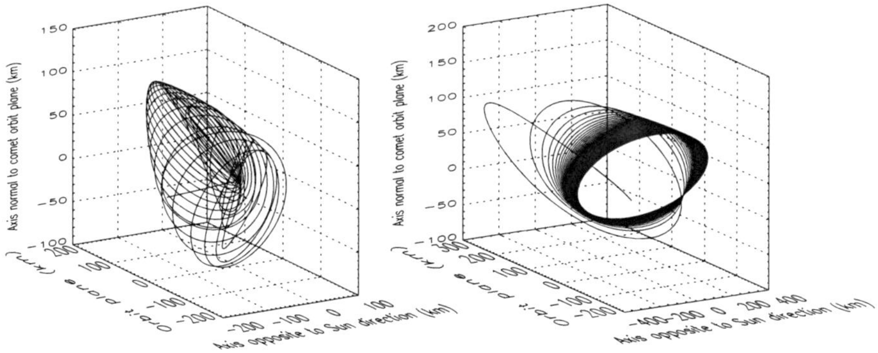

Fig. 1 Orbit simulations: (a) Dust particle with $\beta = 5.94 \times 10^{-6}$, released from the nucleus surface of 49/Wirtanen at latitude 25° and longitude +7.7°. (b) Dust particle with $\beta = 2.98 \times 10^{-7}$, released from the nucleus surface of 1P/Halley at latitude 18° and longitude +4°

4 Orbit Simulations

As stated before, our purpose is to obtain the orbits of dust particles released from the nucleus surface, and to investigate the circumstances under which the orbits become bounded for a considerable fraction of the comet orbital period. Our model, in which we plan to add the corresponding inertial terms due to comet spin, reproduces fairly well the calculations previously made by Fulle (1997). In Fig. 1, we can see dust particles trajectories, where the origin is the comet nucleus in both cases. The two orbits shown are typical cases of orbital stability, which are obtained for very specific values of the physical parameters involved, as detailed in the caption.

We will extend our study to other comets in the near future.

Acknowledgements This work was supported by contract AYA2004-03250, PNE2006-02934 and FEDER funds.

References

D.B. Campbell et al. ApJ **338**, 1094 (1989)
J.F. Crifo et al. Icarus **176**, 192 (2005)
M. Fulle, Astron. Astrophys. **325**, 1237 (1997)
D. Krankowsky et al. Nature **321**, 326 (1986)
Y. Ma et al. Mon. Not. R. Astron. Soc. **337**, 1081 (2002)
M.C. Nolan et al. Icarus **432**, 181 (2006)
K. Richter et al. Icarus **114**, 355 (1995)
M.K. Wallis, in *Comets*, ed. by L.L. Wilkening (University of Arizona Press, Tucson, 1982), p. 357
F.L. Whipple, ApJ **113**, 464 (1951)

Micrometeorites and Their Implications for Meteors

Matthew J. Genge

Originally published in the journal Earth, Moon, and Planets, Volume 102, Nos 1–4.
DOI: 10.1007/s11038-007-9185-z

Abstract Micrometeorites (MMs) are extraterrestrial dust particles, in the size range 25–400 μm, recovered from the Earth's surface. They have experienced a wide range of heating during atmospheric entry from completely molten spherules to particles heated to temperatures <300°C that have retained low temperature minerals. The majority of MMs have mineralogies, textures and compositions that strongly resemble components from chondritic meteorites suggesting these correspond to sporadic, low geocentric velocity meteors. Changes in MMs due to entry heating, however, have implications for meteoric processes in general that may allow the observed behaviour of meteors to be directly related to the material properties of their meteoroids.

Keywords Micrometeorites · Meteors · Asteroid · Comet

1 Introduction

Micrometeorites (MMs) are that fraction of the extraterrestrial dust flux that survives atmospheric entry to be recovered from the Earth's surface. These particles are mostly in the size range 50–1,000 μm and unlike the larger meteorites (>1 cm), recovered from the Earth's surface, and smaller interplanetary dust particles (<30 μm), collected in the stratosphere, are likely to include the surviving remnants of meteors. Micrometeorites can, therefore, provide valuable constraints on processes operating during atmospheric entry of small meteoroids that are particularly applicable to the interpretation of meteor phenomena.

Micrometeorites have been collected in regions of the Earth's surface where the abundance of terrestrial particles is low: (1) deep sea sediments (Brownlee 1985),

M. J. Genge
Impact and Astromaterials Research Centre, Department of Earth Science and Engineering, Imperial College London, Exhibition Road, London SW7 2AZ, UK

M. J. Genge (✉)
Department of Mineralogy, The Natural History Museum, Cromwell Road, London SW7 2BT, UK
e-mail: m.genge@imperial.ac.uk

DOI: 10.1007/978-0-387-78419-9_68

(2) glacial lakes in Greenland (Maurette et al. 1986), and (3) Antarctic blue ice and snow (Maurette et al. 1991; Taylor et al. 1998). Large numbers of relatively pristine particles have been collected by melting and filtering of Antarctic ice and snow. These Antarctic MMs are mostly >50 μm since terrestrial dust is very abundant at small sizes.

Extraterrestrial dust is also collected in the stratosphere by NASA ER-2 aircraft. These samples are known as interplanetary dust particles (IDPs) and are mostly smaller than 30 μm. These materials include particles that are very different from MMs collected from the Earth's surface (Rietmeijer 1998). The implications of these materials for meteors has been considered by Rietmeijer (2000), the nature of the particles will not, therefore, be described in detail in this paper.

Previous studies of MMs have focused primarily on their nature and identity of their parent bodies, often through mineralogical, chemical and isotopic comparisons with meteorites (Genge et al. 1997; Kurat et al. 1994). The current paper examines the implications of MMs for the atmospheric entry of micrometeoroids and meteor phenomena.

2 Identification and Analysis of Micrometeorites

Definitive evidence for extraterrestrial origin of MMs has been made on the basis of cosmogenic nuclei and noble gas measurements that demonstrate exposure to solar radiation in interplanetary space as dust particles (Olinger 1990; Raisbeck and Yiou 1989). The identification of dust particles as MMs, however, can be made on one or more of a number of other non-isotopic criteria (Genge et al. 2008). Features that strongly suggest an extraterrestrial origin are any of: (1) the presence of a partial or complete shell of magnetite around MMs, which is thought to arising from entry heating (Toppani and Libourel 2003; Toppani et al. 2001), (2) the presence of Ni-bearing iron metal and/or sulphides, and (3) a solar bulk composition for major and minor rock forming elements. Features that are less determinative are: (1) high CaO, Cr_2O_3 olivines and very FeO-poor olivines that are exceedingly rare in terrestrial rocks (Brearley and Jones 1998), (2) vesicular surface melt layers (Genge 2006), and (3) spherical particle morphologies.

Separation of collected MMs is achieved by manual picking under a binocular microscope, principally on the basis of the presence of magnetite rims, indicated by a black colour, and spherical or lobate particle shapes, which may indicate melting. The extraterrestrial nature of particles is confirmed by analytical scanning electron microscope observations on polished samples.

3 Micrometeorite Types

Two main groups can be identified on the basis of surviving pre-atmospheric textures: (1) fine-grained MMs (FgMMs), which are dominated by a fine-grained porous groundmass of micron-sized mineral grains, and (2) coarse-grained MMs (CgMMs), which are dominated by anhydrous silicates with grain-sizes larger than several microns, often with glassy mesostasis. Heating during atmospheric entry, however, complicates the classification of particles because it results in significant changes in the primary mineralogy, texture and even the compositions of particles. Micrometeorites are, therefore, also divided into several groups depending on the extent of thermal modification during atmospheric entry. The proportion of melted, partially melted and unmelted MMs varies with particle size (Maurette et al. 1991; Taylor et al. 2000). For sizes >100 μm melted MMs make up

70–90% of particles (Maurette et al. 1991; Taylor et al. 2000), for those 50–100 μm in size melted and partially melted MMs make up ~50% of particles (Genge et al. 1997). The most important MM types are described briefly below. Rare groups of particle, such as refractory unmelted MMs will not be described in detail here. A full description of MM types is given in Genge et al. (2008).

3.1 Melted Micrometeorites: Cosmic Spherules

Melted micrometeorites are known as cosmic spherules (CSs) and have experienced large degrees of fusion of primary phases during atmospheric entry and thus behave as low viscosity melts that form molten droplets during atmospheric entry (Fig. 1a–f). The maxima in the extraterrestrial mass flux at ~200 μm (Love and Brownlee 1993) strongly implies that the majority of CSs formed by melting of dust particles rather than as ablation droplets from larger meteoroids.

Cosmic spherules show considerable diversity in textures, compositions and mineralogy and are sub-divided into several chemical and textural groups. The basic chemical sub-types of CSs, which are also reflected in their principle mineralogy, are the iron-rich spherules (I-type), a glass with magnetite (G-type) group (Blanchard et al. 1980) and silicate-type (S-type) CSs. I-type and G-type spherules comprise only a few percent of MMs and consist principally of the iron oxides, magnetite and wüstite.

The silicate S-type make up 97% cosmic spherules (Taylor et al. 2000). Most have broadly chondritic compositions (Brownlee et al. 1997), notable exceptions are the CAT spherules that have Mg/Si ratios >1.7 and are highly enriched in Ca, Al and Ti (Taylor et al. 2000). S-type spherules can be sub-divided into several sub-classes depending on their quench textures, which are thought to reflect their peak atmospheric temperatures (Taylor and Brownlee 1991). These are: (1) CAT spherules that have high Mg/Si ratios, Ca, Al and Ti abundances. These particles are thought to have been partially evaporated during entry heating. (2) Glass spherules (Fig. 1a) which consist entirely of glass, and are thought to have formed at the high peak temperatures. Glass spherules sometimes contain large vesicles. (3) Cryptocrystalline (CC) spherules (Fig. 1b), which contain sub-micron crystallites and can have significant sub-micron magnetite. Their textures are often dominated by elongate olivine dendrite crystals that radiate from the surface of the spherule. (4) Barred olivine (BO) spherules (Fig. 1c), which are dominated by parallel growth olivine, which occurs as parallel bars, within a glassy mesostasis that often contains magnetite. Some BO spherules exhibit FeNi metal beads that have sometimes oxidised to form iron-oxide located at one end of the particle. Barred olivine spherules are thought to form at lower peak temperatures than CC spherules (Taylor et al. 2000). (5) Porphyritic olivine (PO) spherules (Fig. 1e, f), which are dominated by olivine mi-crophenocrysts with equant, euhedral or skeletal morphologies within a glassy mesostasis, usually with accessory magnetite and/or chromite. Relict unmelted olivine (and less frequently pyroxene) are common in PO spherules. PO spherules experienced the lowest peak temperatures of any cosmic spherules, many are highly vesicular and are likely to be gradational to partially melted MMs. Some PO spherules also contain areas dominated by Fe–Ni-metal and/or Ni-bearing sulphides (Fig. 1e, f). These are thought to form as immiscible metallic liquids during heating and are often found at the margins of spherules suggesting they were in the processes of separating during cooling (Genge and Grady 1998).

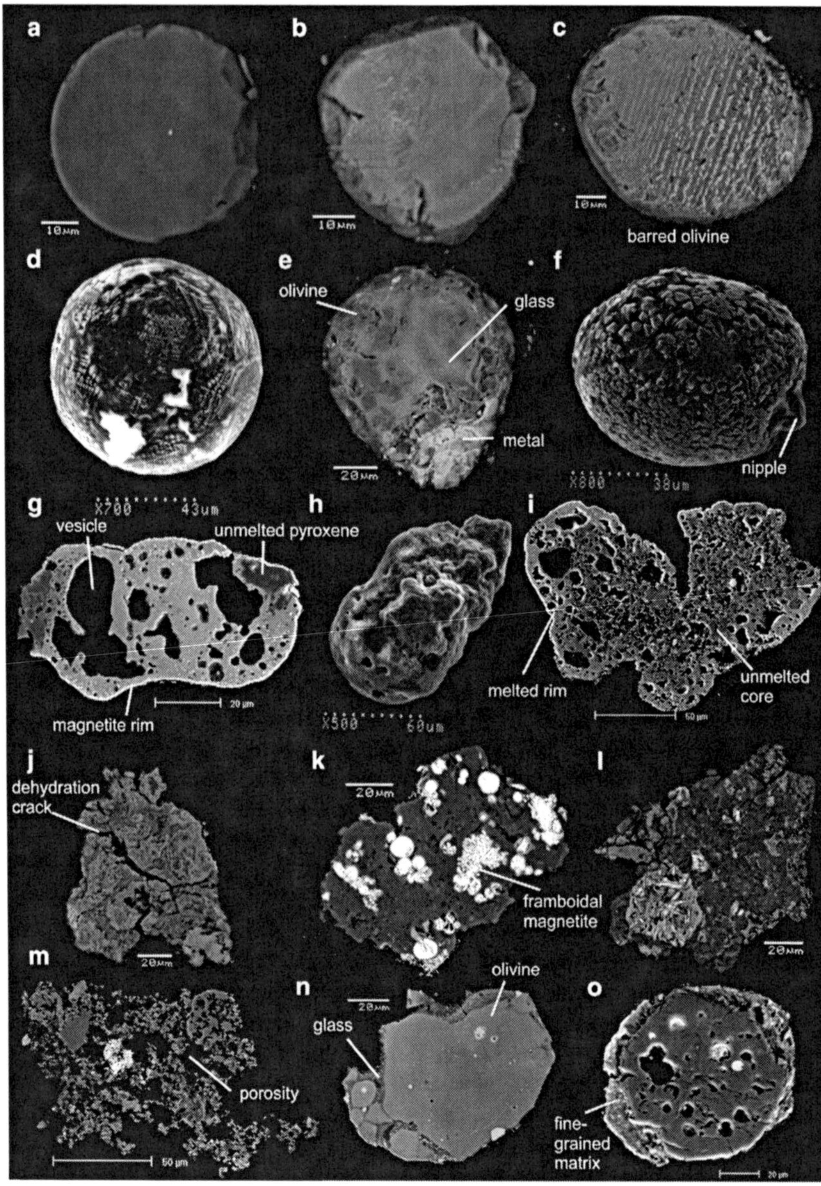

Fig. 1 Scanning electron microscope images of micrometeorites. The backscattered electron images (BEIs) are of polished particles and show interior textures. Contrast in these images relates to atomic mass with bright objects usually dominated by Fe. Particles d, f and h are secondary electron images of the external surfaces of particles. The MM types are (a) glassy cosmic spherule (CS), (b) cryptocrystalline CS, (c) barred olivine CS, (d–f) porphyritic olivine CS, where spherule e contains a metal droplet, and spherule f has a metal separation nipple, (g–i) scoriaceous MMs, where g has relict unmelted crystals, h shows bursting of vesicles on the exterior surface, and i has a large unmelted core, (j) altered fine-grained MM (FgMM) containing dehydration cracks, (k) C1 FgMM, (l) C2 FgMM, (m) C3 FgMM, (n) coarse-grained MM, and (o) composite MM with both coarse-grained and fine-grained portions

3.2 Partially Melted MMs: Scoriaceous Fine-grained MMs

Scoriaceous micrometeorites (ScMMs) are irregular, but smooth, highly vesicular particles (Fig. 1g–i) that have external rims of magnetite. Scoriaceous micrometeorites are dominated by a small olivine crystals, usually with crystal sizes <1 μm, within an interstitial silicate glass phase indicating significant melting, however, ScMMs commonly contain areas of relict fine-grained matrix similar to that of many unmelted MMs (Fig. 1i). Many ScMMs are, therefore, partially melted, consisting of an igneous rim surrounding an unmelted core. These are, therefore, a gradational group to unmelted particles.

The most common relict grains in ScMMs are Mg-rich pyroxene and Mg-rich olivine (Fig. 1g). Vesicle abundances in ScMMs are frequently high and sometimes exceed 50% by volume. Occasionally ScMMs contain Fe–Ni metal/oxides or Ni-bearing sulphides and like those of CSs these may have formed as immiscible metallic/sulphide liquids or by incomplete melting.

3.3 Unmelted MMs

Three varieties of unmelted MMs are identified: (1) fine-grained MMs (FgMMs), (2) coarse-grained MMs (cgMMs) and (3) refractory MMs. Fine-grained MMs are those dominated by a fine-grained porous groundmass of micron-sized mineral grains and are similar to the fine-grained matrices of chondritic meteorites (Fig. 1j–m). Like these materials they have broadly solar compositions (Rietmeijer 2000), mostly in the range of CI, CM and CR chondrite matrices, for most major and minor elements (Genge et al. 1997; Kurat et al. 1994).

The pre-atmospheric mineralogy of fine-grained matrix is thought to have been dominated by phyllosilicates, however, these are rarely preserved. Most FgMMs are dominated by amorphous silicate grains, or sub-micron olivine and pyroxene within glass, both thought to have formed by thermal decomposition of phyllosilicate during entry heating. Dehydration of phyllosilicates is associated with a decrease in volume. Dehydration cracks are thus common in FgMMs. Some heated particles also have melted rims that resemble the matrices of ScMMs (Fig. 1j).

Where phyllosilicates are preserved transmission electron microscope (TEM) and X-ray diffraction studies show they are dominated by smectite (Genge et al. 2001; Gounelle et al. 2002; Nakamura et al. 2001; Noguchi and Nakamura 2000; Noguchi et al. 2002) although serpentine has also been identified (Genge et al. 2001). Despite the thermal modification of particles many, nevertheless, retain their primary textures.

Three broad sub-groups of fine-grained MM are identified: (1) C1 fgMMs that are dominated by compact, chemically homogeneous phyllosilicate-dominated matrix (Fig. 1k) The particles lack isolated olivine and pyroxene, and contain framboidal magnetite clusters. Texturally they resemble the CI chondrites. (2) C2 fgMMs that are dominated by compact, chemically heteorogeneous phyllosilicate-dominated matrix (Fig. 1l). These particles contain isolated olivine and pyroxene grains <10 μm in size and include the mineral tochilinite. Texturally these particles resemble CM2 chondrites. (3) C3 fgMMs, which are porous, chemical heterogeneous particles dominated by micron-sized olivine and pyroxene (Fig. 1m). These particles contain larger (<10 μm) isolated olivine and pyroxene grains and have no direct meteorite analogs. Primary differences from chondrites, for example, in the high pyroxene to olivine ratios of MMs and the exact mineralogy of matrix are also evident (Maurette et al. 1991).

Unmelted coarse-grained MMs (CgMMs) usually have igneous textures and are dominated by olivine, pyroxene crystals within glassy mesostasis (Fig. 1n). They often contain either metal and sulphide, or iron oxides. The textures of CgMMs fall largely into four sub-groups: (1) porphyritic particles, dominated by crystals within a glassy mesostasis, (2) granular particles, dominated by olivine and pyroxene with little glass, (3) radiating pyroxene particles, dominated by elongate dendritic crystals of pyroxene that radiate across the particle, and (4) barred olivine particles, containing bar-like crystals of olivine within pyroxene or glass. All these textural groups are similar to those of chondrules, mm-sized igneous objects within the chondritic meteorites. The minor element compositions of olivine and pyroxene within these particles are likewise consistent with those of chondrules from carbonaceous and ordinary chondrites (Genge et al. 2005).

Refractory MMs are rare particles containing Ca–Al–Ti rich minerals, such as hibonite, perovskite, melilite and spinel, which are associated with CAIs from chondritic meteorites. The majority of these particles contain isolated grains of refractory minerals within a FgMM, however, particles dominated by refractory minerals have been observed.

Composite unmelted MMs have also been observed that contain portions characteristic of both CgMMs and FgMMs (Genge 2006). Such particles usually consist of a coarse-grained, igneous-textured object dominated by anhydrous silicates and glassy mesostasis, surrounded by a partial rim of fine-grained matrix (Fig. 1o). They indicate that CgMMs and FgMMs can be derived from the same parent bodies and that CgMMs are present as small objects similar to chondrules.

4 Implications for Meteors

4.1 Survival of Micrometeoroids

Ceplecha et al. (1998) suggest that typical meteors are associated with micrometeoroids 0.05 mm–20 cm and at least a proportion of micrometeorites, therefore, may be the surviving remnants of meteors. Meteoroid velocity is also, however, a crucial parameter since observation of meteoric phenomena requires evaporation of the meteoroid by heating by incident air molecules to form a plasma within the meteor trail. High velocity meteoroids which experience significant evaporation are more likely to be observable but are less likely to be preserved as micrometeorites. Ionisation of gas species is particularly important since the light produced by optical meteors largely originates from the de-excitation of metals by radiation and observation by radar depends on the density of free electrons in the meteor trail. Ionisation temperatures are >3,000 K (e.g. Ceplecha et al. 1998), however, evaporation rates of silicate liquids become extremely high at temperatures >2,200 K with the result that the surface temperatures of 100 μm droplets are unlikely to exceed this limit (Schaefer and Fegley 2004). Observations of meteors produced by meteoroids of this size, therefore, suggest that gas temperature and meteoroid surface temperature are not the same, presumably due to excitation of gas species by direct collections with incident air molecules. Micrometeoroids which lose sufficient mass through evaporation potentially could be observed as meteors.

Entry heating models indicate that CSs, in particular large mm-sized spherules, can lose 90% of their mass through evaporation during atmospheric entry (Love and Brownlee 1991) and potentially represent the surviving meteoroids of meteors. The compositions of most CSs, however, are solar (e.g. Genge et al. 1997) and do not support changes due to differential ablation that would be expected during significant evaporative mass loss. Only

CAT spherules, which are enriched in refractory elements such as Al and Ca, are consistent with such extreme evaporative fractionation (Taylor et al. 2000).

Micrometeorites are most likely to survive atmospheric entry unmelted at low entry velocities (Love and Brownlee 1991) implying these are derived from low geocentric velocity sporadic sources. The mineralogy, textures and compositions of unmelted MMs support this inference since they are similar to chondritic meteorites and thus imply an asteroidal source (e.g. Genge et al. 1997, 2008). The occurrence of abundant phyllosilicates in MMs in particular differs from prevailing models of cometary nuclei, and the results from comet Wild-2 particles (Brownlee et al. 2006). The presence of phyllosilicates in comets, however, remains controversial (Gounelle et al. 2006; Lisse et al. 2006; Rietmeijer 1998). Low geocentric velocity cometary dust particles may, nevertheless, exist particularly amongst highly porous FgMMs.

Anhydrous IDPs, which are aggregates of sub-micron silicate grains contained within carbonaceous material, have been considered to represent cometary materials, a view that is broadly consistent with the early results of the Stardust Mission (Brownlee et al. 2006). Refractory Ca–Ti-rich particles and silicate igneous objects, reminiscent of microchondrules, however, have been discovered amongst Stardust samples (Brownlee et al. 2006) implying that comet Wild-2 shares some mineralogical and textural features with MMs and chondritic meteorites.

Micrometeorites are, therefore, directly analogous to the meteoroids of sporadic meteoroids derived from asteroidal sources. They may also provide constraints on the nature of cometary meteoroids that produce meteor showers, given the uncertainty in the nature of these materials. A proportion of CSs may also be derived directly from cometary sources since entry heating models suggest these can survive atmospheric entry at high velocities of >30 km s^{-1} (Love and Brownlee 1991). It is, however, unlikely that meteoroids from meteor showers are common amongst MMs due to their high entry velocities.

4.2 Entry Heating Phenomena of Meteoroids

Although unmelted MMs are samples of relatively low geocentric velocity dust particles the mineralogical and physical changes they experience during atmospheric entry heating have applications to meteor phenomena in general, including those from high entry velocity streams, since the meteoroids of all meteors experience a degree of pre-heating during non-luminous flight that can modify their physical and chemical properties. Such changes allow definite predictions to be made that potentially can relate the material properties of micrometeoroids to their deceleration and evaporation rates.

The onset of melting is likely to have a significant effect on the atmospheric entry heating of micrometeoroids. Partially melted MMs indicate that phyllosilicate-bearing particles experience surface melting in which high temperature gradients (>600°C) are supported by endothermic dehydration reactions (Genge 2006). Continued heating of such micrometeoroids leads to progressive fusion of the solid core of the particle. Due to the latent heat of fusion and the high thermal conductivity of silicate melts the surface melt layer will remain at the melting temperature of the fine-grained matrix ($\sim$1,350°C) until the solid core has been consumed. The surface temperature of micrometeoroids, and thus their evaporation rates, will, therefore, remain constant over a portion of the entry heating. Isothermal surface temperatures apply only to particles containing volatile-bearing phases such as phyllosilicate.

Melting of micrometeoroids during entry heating also results in changes in the density and volume of particles that influences their deceleration. These changes depend on the material properties of the micrometeoroid. Compact particles, such as cgMMs, experience a decrease in density during melting from that of the crystalline solid aggregate (>3.0 g cm^{-3}) to a partially melted material as glass begins to melt. On fusion their densities will decrease smoothly with temperature as they partially melt towards that of a ferromagnesian melt of approximately 2.7 g cm^{-3}.

The change in density of FgMMs on melting is likely to depend on the abundance of vesicles formed by the exsolution of gases from volatile components. Unmelted compact FgMMs have densities similar to CI chondrites of ~2.1 g cm^{-3} (Britt and Consolmagno 2003), however, high porous particles may have densities as low as 1.0 g cm^{-3}. The abundance of vesicles in ScMMs often approaches 50% by volume, suggesting the density of the molten particle will evolve towards 1.3 g cm^{-3}. The significant decrease in the density of compact phyllosilicate-bearing particles is associated with an increase in the volume of the particle and thus will produce a pronounced increase in the deceleration of the micrometeoroid. With increasing temperature the decrease in melt viscosity will allow escape of vesicles from the molten micrometeoroid, as indicated by the decrease in vesicle abundances from ScMMs to CSs. Particle density will, therefore, increase with heating towards that of the melt. The presence of unmelted olivine and pyroxene will, however, result in slightly higher densities.

Within cometary micrometeoroids, vesicularity is also likely to play a significant role. If these objects are similar to anhydrous IDPs, they may have densities as low as 0.6 g cm^{-3} (Flynn and Sutton 1991). The density change on melting is likely to be sensitive to the degassing prior to fusion. If the particle does not significantly degas it is likely to become a highly vesicular foam on melting (e.g. Rietmeijer 1996) and thus retain its low density. If it does degas prior to melting there will be a significant increase in density and decrease in volume.

Molecular gas species generated during degassing of micrometeoroids will vary with particle type. Phyllosilicates will release water vapour during dehydration, carbonaceous materials will generate CO_2 as a result of pyrolysis, and sulphides are likely to generate both SO_2 and H_2S on decomposition. Highly volatile components are likely to be released during pre-heating of meteors, however, the presence of vesicles within cosmic spherules suggests that some may be retained into luminous flight. Emission from molecular gas species has yet to be detected in meteors.

4.3 Fragmentation of Micrometeoroids

Meteor fragmentation events are a function of the mechanical properties of micrometeoroids. The nature of MMs allows fragmentation events to be interpreted in terms of the material properties of micrometeoroids. The break-up of solid meteoroids is likely to occur during the pre-heating of meteors prior to luminous flight and result in a burst of closely related meteors.

Fragmentation of phyllosilicate-bearing particles is likely to occur during dehydration of the fine-grained matrix due to the formation of dehydration cracks. These represent planes of mechanical weakness and are likely to dictate the size of secondary particles during fragmentation events. Examination of MMs suggests that fragmentation due to dehydration cracks will lead to 2–5 fragments with diameters ~0.25–0.5 times that of the original particle, and a second population of particles with diameters <0.1× that of the original particle.

Cometary, IDP-like micrometeoroids, are also likely to experience decomposition of their volatile carbonaceous components prior to melting which may likewise lead to fragmentation. Mechanical disaggregation of the carbonaceous "glue" of such particles is likely to result in liberation of the sub-micron sized silicate grains contained within the meteoroid. Sudden fragmentation to sub-micron-grains is, therefore, likely to be a feature of IDP-like micrometeoroids, although similar fragmentation may also occur for the most porous and fragile FgMM-like particles, albeit to micron-sized mineral grains.

Compact igneous particles, similar to chondritic cgMMs, that may largely represent pieces of chondrule, are mechanically strong objects and unlikely to fragment during deceleration. Composite particles, that include both fine-grained portions and coarse-grained portions, however, may break-up into a single large fragment and a range of micron-sized grains.

Fragmentation of molten micrometeoroids is problematic since continuous ablation due to the removal of surface melt is only expected for larger meteoroids in the slip-flow regime. Most micrometeroids <400 μm have a size much less than the mean-free path of atmospheric species and thus decelerate in the free molecular flow regime in which there is no shear component over the surface of the particle. The development of instabilities in droplet shapes has been suggested as a potential mechanism for the fragmentation of molten particles, however, for small droplets surface tension is likely to strongly resist the development of instabilities (Bronshten 1983). The regular shapes of CSs suggest that such instabilities are rare.

The presence of metal and iron-sulphide droplets within some CSs provides one mechanism by which fragmentation of molten particles may occur. Metal and iron sulphide are generated either by non-equilibrium melting of large pre-existing mineral grains or through redox reactions during melting. Metal generation through redox reactions should be relatively common within carbon-rich micrometeoroids, similar to fgMMs and IDPs, since on melting carbon reacts with oxygen within the melt to form CO_2 leading to reduction of Fe^{2+} in the melt to metallic Fe^0 (Genge and Grady 1998). Amongst MMs there is evidence from the compositions of spherules that metal separation during to deceleration is common.

Once generated the higher density of the metallic liquid than the silicate melt leads to migration of the metallic liquid to the leading surface of the meteoroid, to form a surface protrusion, which in many cases will then separate entirely. Only metal droplets that failed to separate are preserved amongst MMs. Metal separation will result in fragmentation into two droplets, a high density ($\sim$8 g cm^{-3}), small droplet, and a lower density, larger silicate droplet. Due to the differences in size and density the metal droplet will experience less deceleration after separation and thus will rapidly move ahead of the silicate sphere. The exact relative acceleration will depend on the sulphide to metal content of the droplet, and the degree of oxidation of the metal, since these will influence the density of the iron-rich liquid. The occurrence of metal separation events, which can be identified by the relative deceleration of the produced meteoroids, implies the precursor meteoroid either contained metal grains, or was carbon-rich, prior to melting. Meteors with iron dominated spectra (Type Z meteors; e.g. Ceplecha et al. 1998), may, therefore, represent separated metal droplets rather than primary compositional variations in the precursor meteoroids.

5 Conclusions

The nature of unmelted MMs suggest that the majority of these particles represent asteroidal materials and thus probably correspond to low geocentric velocity sporadic meteors rather than meteor streams. The nature of melted MMs, nevertheless, has implications for the atmospheric entry behaviour of meteors in general. Specifically: (1) melted rims on phyllosilicate-bearing MMs indicates these particles have thermal gradients and isothermal surface temperatures, (2) highly vesicular partially melted particles indicate large changes in meteoroid volume and density on melting, (3) dehydration cracks in phyllosilicate-bearing MMs provide a mechanism for fragmentation that is distinct from anhydrous IDP-like particles, and (4) metal separation from CSs provides a mechanism for fragmentation of molten micrometeoroids.

Acknowledgements Susan Taylor and Mike Zolensky are thanked for their reviews which were helpful and positive. Frans Rietmeijer is in particular thanked for his many useful suggestions.

References

M.B. Blanchard, D.E. Brownlee, T.E. Bunch, P.W. Hodge, F.T. Kyte, Meteoroid ablation spheres from deep sea sediments. Earth Planet. Sci. Lett. **46**, 178–190 (1980)

A.J. Brearley, R.H. Jones, Chondritic meteorites, in *Revs. in Mineralogy*, vol. 36, ed. by J.J. Papike (Mineralogical Society of America, Chantilly, Virginia, 1998), pp. 3-1–3-301

D.T. Britt, G.J. Consolmagno, Stony meteorite porosities and densities: a review of the data through 2001: Meteorit. Planet. Sci. **38**, 1161–1180 (2003)

V.A. Bronshten, *Physics of Meteoric Phenomena* (D. Reidel, Dordrecht, Holland, 1983), p. 450

D.E. Brownlee, Cosmic dust: collection and research. Ann. Rev. Earth Planet. Sci. **13**, 147–173 (1985)

D.E. Brownlee, B. Bates, L.S. Schramm, The elemental composition of stony cosmic spherules. Meteorit. Planet. Sci. **32**, 157–175 (1997)

D.E. Brownlee et al., Comet 81P/Wild 2 under a microscope. Science **314**, 1711–1716 (2006)

Z. Ceplecha, J. Borovička, W.G. Elford, D.O. Revelle, R.L. Hawkes, V. Porubčan, M. Simek, Meteor phenomena and bodies. Space Sci. Rev. **84**, 327–471 (1998)

G.J. Flynn, S.R. Sutton, Cosmic dust particle densities – evidence for two populations of stony micrometeorites (abstract), Lunar and Planetary Science Conference, 31, pp. 541–547, Lunar and Planetary Institute, Houston, Texas, 1991

M.J. Genge, Igneous rims on micrometeorites. Geochim. Cosmochim. Acta. **70**, 2603–2621 (2006)

M.J. Genge, J.P. Bradley, C. Engrand, M. Gounelle, R.P. Harvey, M.M. Grady, The petrology of fine-grained micrometeorites: evidence for the diversity of primitive asteroids (abstract). Lunar and Planetary Science Conference, 32, #1546, Lunar and Planetary Institute, Houston, Texas, 2001

M.J. Genge, M.M. Grady, Melted micrometeorites from Antarctic ice with evidence for the separation of immiscible Fe–Ni–S liquids during entry heating. Meteorit. Planet. Sci. **33**, 425–434 (1998)

M.J. Genge, M.M. Grady, R. Hutchison, The textures and compositions of fine-grained Antarctic micrometeorites – implications for comparisons with meteorites. Meteorit. Planet. Sci. **61**, 5149–5162 (1997)

M.J. Genge, A. Gileski, M.M. Grady, Chondrules in Antarctic micrometeorites. Meteorit. Planet. Sci. **40**, 225–238 (2005)

M.J. Genge, C. Engrand, M. Gounelle, S. Taylor, The classification of micrometeorites. Meteorit. Planet. Sci. (2008, in press)

M. Gounelle, B. Devouard, C. Engrand, M.J. Genge, A. Toppani, H. Leroux, TEM study of Antarctic micrometeorites: a preliminary report (abstract). Meteorit. Planet. Sci., **37**, A55 (2002)

M. Gounelle, P. Spurný, P.A. Bland, The orbit and atmospheric trajectory of the Orgueil meteorite from historical records. Meteorit. Planet. Sci. **41**, 135–150 (2006)

G. Kurat, C. Koeberl, T. Presper, B. Franz, M. Maurette, Petrology and geochemistry of Antarctic micrometeorites. Meteorit. Planet. Sci. **58**, 3879–3904 (1994)

C.M. Lisse, J. VanCleve, A.C. Adams, M.F. A'Hearn, Y.R. Fernandez, T.L. Farnham, L. Armus, C.J. Grillmair, J. Ingalls, M.J.S. Belton, O. Groussin, L.A. McFadden, K.J. Meech, P.H. Schultz, B.C. Clark,

L.M. Feaga, J.M. Sunshine, Spitzer spectral observations of deep impact ejecta. Science **313**, 635–640 (2006)
S.G. Love, D.E. Brownlee, Heating and thermal transformation of micrometeoroids entering the Earth's atmosphere. Icarus **89**, 26–43 (1991)
S.G. Love, D.E. Brownlee, A direct measurement of the terrestrial mass accretion rate of cosmic dust. Science **262**, 550–553 (1993)
M. Maurette, C. Hammer, D.E. Brownlee, N. Reeh, H.H. Thomsen, Placers of cosmic dust in the blue ice lakes of Greenland. Science **233**, 869–872 (1986)
M. Maurette, C. Olinger, M.C. Michel-Levy, G. Kurat, M. Pourchet, F. Brandstätter, M. Bourot-Denise, A collection of diverse micrometeorites recovered from 100 tonnes of Antarctic blue ice. Nature **351**, 44–47 (1991)
T. Nakamura, T. Noguchi, T. Yada, Y. Nakamuta, N. Takaoka, Bulk mineralogy of individual micrometeorites determined by X-ray diffraction analysis and transmission electron microscopy. Meteorit. Planet. Sci. **65**, 4385–4397 (2001)
T. Noguchi, T. Nakamura, Mineralogy of Antarctic micrometeorites recovered from the Dome Fuji Station, 24th Symposium on Antarctic Meteorites (abstract), NIPR Symposium. Antarctic Meteorit. Res. **13**, 285–301 (2000)
T. Noguchi, T. Nakamura, W. Nozaki, Mineralogy of phyllosilicate-rich micrometeorites and comparison with Tagish Lake and Sayama meteorites. Earth Planet. Sci. Lett. **202**, 229–246 (2002)
C.T. Olinger, Isotopic measurements of solar noble gases in individual micrometeorites from Greenland and Antarctica, Ph.D. Thesis, Washington Univ. Seattle, 1990
G.M. Raisbeck, F. Yiou, Cosmic ray exposure ages of cosmic spherules (abstract). Meteoritics **24**, 318 (1989)
F.J.M. Rietmeijer, The ultrafine mineralogy of a molten interplanetary dust particle as an example of the quench regime of atmospheric entry heating. Meteorit. Planet. Sci. **31**, 237–242 (1996)
F.J.M. Rietmeijer, Interplanetary dust particles, in *Revs. in Mineralogy*, vol. 36, ed. by J.J. Papike (Mineralogical Society of America, Chantilly, Virginia, 1998), pp. 2-1–2-95
F.J.M. Rietmeijer, Interrelationships among meteoric metals, meteors, interplanetary dust, micrometeorites, and meteorites. Meteorit. Planet. Sci. **35**, 1025–1041 (2000)
L. Schaefer, B. Fegley Jr., Application of an equilibrium vaporization model to the ablation of chondritic and achondritic meteoroids. Earth Moon Planets **95**, 413–423 (2004)
S. Taylor, D.E. Brownlee, Cosmic spherules in the geological record. Meteoritics **26**, 203–211 (1991)
S. Taylor, J.H. Lever, R.P. Harvey, Accretion rate of cosmic spherules measured at the South Pole. Nature **392**, 899–903 (1998)
S. Taylor, J.H. Lever, R.P. Harvey, Numbers, types and compositions of an unbiased collection of cosmic spherules. Meteorit. Planet. Sci. **55**, 651–666 (2000)
A. Toppani, G. Libourel, Factors controlling compositions of cosmic spinels: application to atmospheric entry conditions of meteoritic materials. Meteorit. Planet. Sci. **67**, 4621–4638 (2003)
A. Toppani, G. Libourel, C. Engrand, M. Maurette, Experimental simulation of atmospheric entry of micrometeorites. Meteorit. Planet. Sci. **36**, 1377–1396 (2001)

March 1, 2005 Daylight Fireball Over Galicia (NW of Spain) and Minho (N. Portugal)

José Angel Docobo · Josep Maria Trigo-Rodríguez · Jiri Borovicka · Vakhtang S. Tamazian · Vera Assis Fernandes · Jordi Llorca

Originally published in the journal Earth, Moon, and Planets, Volume 102, Nos 1–4.
DOI: 10.1007/s11038-007-9191-1 © Springer Science+Business Media B.V. 2007

Abstract A daylight bolide was observed over Galicia (NW Spain) and Minho (N. Portugal) on March 1, 2005 at 15 h10 min ± 3 min UTC. We interviewed 23 eyewitnesses of the event in order to obtain the azimuth, altitude, and slope of the fireball's trajectory. Reports suggest an atmospheric ending height below 20 km, indicating that meteorite survival was likely. From the reconstructed trajectory and the fireball's duration, we obtained the approximate heliocentric orbits for the meteoroid. Assuming an entry velocity higher than 20 km s^{-1} which is consistent with its estimated duration, the meteoroid originated in the asteroid belt.

Keywords Meteors · Meteoroids

J. A. Docobo (✉) · V. S. Tamazian
Observatorio Astronómico Ramón Mª Aller, Universidade de Santiago de Compostela, Santiago de Compostela 15782, Spain
e-mail: oadoco@usc.es

J. M. Trigo-Rodríguez
Institut de Ciències de l'Espai-CSIC, Bellaterra, Barcelona, Spain

J. M. Trigo-Rodríguez
Institut d'Estudis Espacials de Catalunya, Barcelona, Spain

J. Borovicka
Astronomical Institute of Academy of Sciences, Ondrejov Observatory, Czech Republic

V. A. Fernandes
Universidade de Coimbra, Coimbra, Portugal

V. A. Fernandes
University of Manchester, Manchester, UK

J. Llorca
Institut de Tècniques Energètiques, Universitat Politècnica de Catalunya, Barcelona, Spain

1 Introduction

Study of daylight superbolides (fireballs with brightness greater than −17 visual magnitude) from the ground can provide important information about interplanetary bodies intercepting the Earth's orbit such as their size, velocity and the impact energy released when they collide with the atmosphere.

Accurate reconstruction of the original orbit in the solar system of a meteorite has been obtained on only nine occasions (Ceplecha 1961; McCrosky et al. 1971; Halliday et al. 1978; Brown et al. 1994, 1996, 2002, 2004; Docobo and Ceplecha 1999; Borovička et al. 1999, 2003; Spurný et al. 2003; Trigo-Rodríguez et al. 2006). In addition, 14 less accurate orbits have been derived from visual observations (La Paz 1949; Krinov 1960; Folinsbee et al. 1969; Levin et al. 1976; Ballabh et al. 1978; Halliday and McIntosh 1990; Jenniskens et al. 1992).

On March 1, 2005, the sky was completely clear over the Iberian Peninsula. At 16 h10 min local time (15 h10 min UTC; hereinafter given in UTC) a call was received at the Astronomical Observatory 'Ramon Maria Aller' of the University of Santiago de Compostela (OARMA) from M. Alvarez, an inhabitant of a small village close to the town of Padrón near Santiago de Compostela. The eyewitness claimed to have observed, while driving, an extremely bright object crossing the sky at high velocity. It was a few minutes after 15 h00 min on a sunny afternoon. From his experience in observing meteors and fireballs at night, he was sure that he saw a celestial object and decided to alert OARMA. At the present time, we have personally interviewed 22 witnesses in Spain and one witness in Portugal, which was the direction in which the observed object was apparently moving.

We present here a case study of a bright daylight fireball (hereafter, DF) because it could have been a meteorite-dropping event, although the compiled data based on the witness accounts are not enough to obtain its precise orbit. We believe that bright fireball events should be studied in detail around the world in order to collect information on the origin of the relatively rare meteorite-dropping bolides. This work is in the line of research conducted by the Spanish Meteor and Fireball Network (SPMN).

2 Observational Data

Unfortunately, this event was not recorded photographically or digitally. As precise visual measurements using a theodolite had been very useful on other occasions (Docobo et al. 1998; 1999), we visited each of the 23 witnesses in order to reconstruct the trajectory on the basis of sightings made at these exact witness locations. Each of the 23 observers provided us with (1) the approximate time and duration (Δt) of the event, (2) the position of the beginning point (BP) and end point (EP) of the apparent trajectory, and (3) the inclination (i) relative to the vertical. The fireball trajectory was very long and sometimes the EP indicated was, in reality, the point where it was occulted by some terrestrial object such as a house or a tree. We clarified this key point with each eyewitness to avoid mistakes in the computation of the fireball's ending height. All observations listed in Table 1 are arranged in descending order of the observer latitudes.

The errors reported in azimuth (A) and elevation (h) were estimated from the witness statements. These errors are similar to those obtained in the previous works where measurements have been obtained by the use of theodolites. The azimuth was measured from the North. The apparent slope of the DF was requested in order to estimate the radiant point by using the Jenniskens et al. (1992) approach.

Table 1 Locations (listed numerically) and geographical coordinates of each witness reporting a visual sighting of the DF and the calculated azimuth (A) and elevation of the beginning and end points for the event and the apparent inclination of the trajectory (i) and the reported estimates of the event duration (Δt) in seconds

Location	Geographical coordinates		Beginning point (BP)		End point (EP)		i	Δt
#	Φ N	λ W	A (°)	h (°)	A (°)	h (°)	(°)	(s)
1.	43°30′18″	7°16′10″	173°	16° ± 1	189°	14° ± 1	70°	2
2.	43.10.52	7.53.23	160 ± 2	25 ± 5	170 ± 2	13.3	40	1.5
3.	43.09.56	7.59.53	160 ± 2	9 ± 1	182 ± 3	4 ± 0.5	60	2.5
4.	42.56.10	8.20.25	108 ± 2	20 ± 1	160.5 ± 2	12 ± 0.5	80	3
5.	42.48.15	8.26.30	55 ± 2	17 ± 1	*160*	*0*	*70*	*4*
6.	42.46.01	8.38.50	103.5	25	114.5 ± 2	15	55	2
7.	42.45.04	8.20.40	137 ± 3	30 ± 2	152.5	8	50	2
8.	42.43.00	8.27.00	100 ± 2	22 ± 1	163 ± 2	8 ± 0.5	70	3
9.	42.42.30	8.14.23	158	15.5	168.5	8.1	60	2
10.	42.41.05	8.28.27	93.5 ± 2	22.5 ± 1	161.5 ± 2	8 ± 1	75	3
11.	42.38.16	8.45.44	130 ± 2	13 ± 1	147 ± 7	10.3 ± 1	75	1.5
12.	42.23.51	7.52.38	157	41 ± 2	177	15	40	3.5
13	42.19.51	7.51.35	25 ± 5	24.5 ± 1	183 ± 1	13.5 ± 1	25	5
14.	42.18.08	8.08.00	160 ± 2	20 ± 3	*173 ± 3*	*12 ± 0.5*	65	3
15.	42.17.55	8.41.05	105 ± 5	20 ± 2	135 ± 5	10 ± 2	60	2
16.	42.16.59	8.41.00	96	25 ± 2	125 ± 5	11.7 ± 0.5	45	1.5
17.	42.15.18	8.13.00	145 ± 5	18 ± 1	173 ± 3	12.5	45	2
18.	42.07.01	8.43.28	116 ± 3	17.5 ± 0.5	133 ± 3	12 ± 2	60	1
19.	42.06.58	8.49.06	105 ± 3	16.5 ± 1	121.5 ± 3	11.5 ± 0.5	50	2
20.	42.06.15	8.36.25	97	15 ± 1	121.5 ± 1.5	12.5 ± 1	50	2.5
21.	42.05.53	8.44.02	88 ± 3	14 ± 0.5	130	8.5	60	2
22.	42.05.17	8.19.02	*45*	*40*	145.5 ± 3	13.5	30	5.5
23.	41.54.13	8.04.51	298	6.5	286 ± 1	5	70	<1

Uncertain or low precision measurements are italicized

After a careful evaluation of all sightings, we concluded that the event took place at 15 h10 min ± 3 min UTC. At that time, the Sun was rather high in the sky, with horizontal coordinates for Santiago de Compostela of $A = 223°$, $h = 30°$. This circumstance acted as a selection effect and explains why almost all witnesses saw the event when looking towards the East or Southeast.

One witness heard a buzzing sound, immediately opened a window, and was able to observe the object's path across the sky. Another witness claimed to have seen two smaller objects that followed the principal one along its path. A witness just within the city of Ourense observed an apparent arc of about 160° with a maximum elevation of 45–50°, which allowed this person to see the object both during its approach as well as when it moved away from his position. In fact, he saw the fireball from behind as it was disappearing. Twelve witnesses provided hand drawings that graphically depicted this phenomenon. Some reported seeing color in the fireball's head.

The National Geographical Institute informed us that, on that day at 15 h13 min 25 s UTC, they recorded a seismic event at the station of Mazaricos ($\varphi = 42.949$,

$\lambda = 8.9765$ W, Alt = 405 m). With the present trajectory data and uncertainty regarding the time of appearance, we cannot guarantee that this seismic event was related to this fireball, but we think that it deserves to be mentioned.

3 Calculation of the Trajectory

Using the method described by Borovička (1990), the best solution for the event corresponds to the beginning point at $\lambda = -6.^\circ 96$, $\varphi = 43^\circ.72$, H = 81 km and the end point at $\lambda = -7^\circ.97$, $\varphi = 41^\circ.54$ and H = 16 km. The error of the end point is on the order of 5 km although a solution with the end point at 10 km to the West is also possible. The corresponding apparent radiant has $A = 198 \pm 6^\circ$ (counted from the South) and $z = 77 \pm 5^\circ$, that generate $\alpha = 164 \pm 11^\circ$ and $\delta = +57 \pm 4^\circ$. The calculated trajectory is plotted on a map of the area (Fig. 1). Table 2 lists the numerical values of possible orbital solutions. Notice that the node difference arises from the transformation of the longitude of the node to the standard epoch J2000.0. The transformation depends on the inclination of the orbit that is different for each solution.

From estimates of the duration of the event, the entry velocity seems to have been more than 20 km s^{-1}. We selected seven geometrically-best observations (numbers 2, 7, 11, 12, 15, 16, 19 in Table 1) and computed the velocity on the basis of the length of the observed trajectory and estimated duration provided by each of the above mentioned witnesses. The velocity estimates vary widely from 12 to 31 km s^{-1}. The arithmetic mean is 21 km s^{-1} and the median is 20 km s^{-1}. These estimates generally apply to the part of the trajectory in the height range 30–21 km. The initial velocity is expected to be higher because of fireball deceleration at lower heights. Nevertheless, keeping in mind the scatter of the duration estimates, we still consider an initial velocity of 20 km s^{-1}, or even less, as possible.

The brightness of the fireball is difficult to judge but since so many people saw it in broad daylight, we estimate that the magnitude was -18 or brighter. With this data, a very

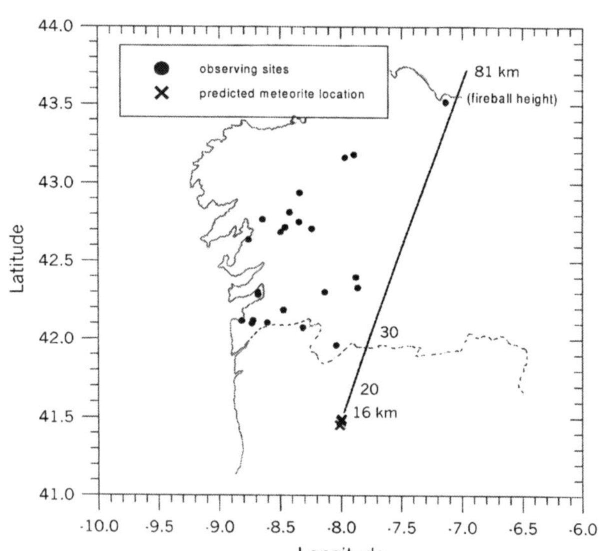

Fig. 1 The dots mark the locations of the reporting witnesses. The fireball trajectory and the computed fireball height from the reconstructed atmospheric trajectory are also given

Table 2 Possible orbital solutions (J2000.0) for different assumed initial velocities

V_{inf} (km s^{-1})	e	a	Q	i	Ω	ω
24	0.81	4.5	0.85	23°	341°.068	228°
21	0.62	2.2	0.85	20	341.067	232
18	0.44	1.51	0.85	16	341.067	239
15	0.27	1.15	0.84	11	341.067	254

rough estimate of meteoroid initial mass is 2×10^4 kg. Assuming an initial velocity of 22 km s^{-1}, for example, and an apparent ablation coefficient of 0.015 s^2 km^{-2}, the fireball would reach the maximum absolute magnitude of −19 at the height of 29 km. The velocity would be 20 km s^{-1} at 30 km and 11 km s^{-1} at 20 km. The average velocity between the heights 30 and 21 km would be 17 km s^{-1}, which lies within the range of uncertainty of the visual estimates. At the height of 16 km, the fireball brightness would drop below magnitude −10. This idealized picture, however, ignores meteoroid fragmentation which is common in well-documented meteorite falls (Borovička and Kalenda 2003) and which at least one observer noticed in the present case.

4 The Meteorite Landing Area

With the EP lying below 20 km, a meteorite fall is likely. The nominal landing point of a 15 kg meteorite is at longitude −8 h01 min and latitude 41° 45′. This is a reasonable value for the ending mass according to our model. In any case, the exact mass is unknown and smaller meteorites would lie closer to the trajectory end. Wind was not considered. The error in the prediction of the meteorite fall location may be as large as 10 km.

According to our calculations, the most probable area of meteorite fall is situated in an area between Cabeceiras de Basto-Faia and Modim de Basto, very close to Guimarães (Portugal). In the case of significant fragmentation the areas of Varcea Cova or even Calvos (more to the North) should also be investigated. The first attempt of a joint Spanish–Portuguese expedition to search for meteorites in this area took place in May 2006. There were many difficulties mainly due to the dense vegetation and mountainous character of the region.

5 Conclusions

A bright daylight fireball was observed moving from Galicia (Spain) to Minho (Portugal) on March 1, 2005 at about 15 h10 ± 3 min UTC. We interviewed 23 eyewitnesses of this event. By measuring with a theodolite from all observation points we have reconstructed the approximate atmospheric trajectory and probable heliocentric orbit of this potentially meteorite-dropping fireball. So far, no meteorites have been recovered. This study illustrates that reconstructions of bright bolides based on accurately measured azimuth and altitude is possible when using data obtained from casual eyewitnesses. Of course, the success depends on how well the eyewitnesses can remember the fireball path. It is important to obtain sufficiently large statistical sample of data to be able to exclude inaccurate or false reports. In the present case we finally used the data from 17 eyewitnesses, at least partially. Seven reports proved to be particularly good with both indicated points lying within 3.5° of the resulting trajectory. These reports belong to the best we encountered during similar studies (Borovička et al. 1999; Jenniskens et al. 1992).

Acknowledgments We thank all of the eyewitnesses for their valuable collaboration and everyone who helped in the search for possible meteorite fragments. The authors thank the anonymous referee and R. Spalding for useful comments. Our special thanks to F. Rietmeijer for many constructive suggestions that helped to improve the paper. J. A. Docobo and V. Tamazian thank the Vicerreitoria de Investigación e Innovación of Universidade de Santiago de Compostela for providing financial support. J. M. Trigo-Rodriguez thanks the Spanish Ministerio de Educación y Ciencia for a Juan de la Cierva grant.

References

G.M. Ballabh, A. Bhatnagar, N. Bhandari, The orbit of the Dhajala meteorite. Icarus **33**, 361–367 (1978)

J. Borovička, The comparison of two methods of determining meteor trajectories from photographs. Bull. Astron. Inst. Czechosl. **41**, 391–396 (1990)

J. Borovička, P. Kalenda, The Morávka meteorite fall: 4. Meteoroid dynamics and fragmentation in the atmosphere. Meteorit. Planet. Sci. **38**, 1023–1043 (2003)

J. Borovička, M.C. Pineda de Carías, A. Ocampo, E. Tagliaferri, R. E. Spalding, About a big fireball seen in Honduras, in *Meteoroids*, 1998, ed. by W.J. Baggaley, V. Porubčan (Astron. Inst., Slovak Acad. Sci, Bratislava, 1999), pp. 139–142

J. Borovička, P. Spurný, P. Kalenda, E. Tagliaferri, The morávka meteorite fall: I. Description of the events and determination of the fireball trajectory and orbit from video records. Meteorit. Planet. Sci. **38**, 975–987 (2003)

P.G. Brown, Z. Ceplecha, R.L. Hawkes, G. Wetherill, M. Beech, K. Mossman, The orbit and atmospheric trajectory of the peekskill meteorite from video records. Nature **367**, 624–626 (1994)

P. Brown, A.R. Hildebrand, D.W.E. Green, D. Page, C. Jacobs, D. ReVelle, E. Tagliaferri, J. Wacker, B. Wetmiller, The fall of the St-Robert meteorite. Meteorit. Planet. Sci. **31**, 502–517 (1996)

P.G. Brown, D.O. ReVelle, E. Tagliaferri, A.R. Hildebrand, An entry model for the tagish lake fireball using seismic, satellite and infrasound records. Meteorit. Planet. Sci.. **37**, 661–675 (2002)

P.G. Brown, D. Pack, W.N. Edwards, D.O. ReVelle, B.B. Yoo, R.E. Spalding, E. Tagliaferri, The orbit, atmospheric dynamics, and initial mass of the park forest meteorite. Meteorit. Planet. Sci. **39**, 1781–1796 (2004)

Z. Ceplecha, Multiple fall of příbram meteorites photographed. Bull. Astron. Inst. Czechosl. **12**, 21–47 (1961)

J. A. Docobo, R. E. Spalding, Z. Ceplecha, F. Diaz-Fierros, V. Tamazian, Y. Onda, Investigation of a bright flying object over Northwest Spain, 1994 January 18. Meteorit. Planet. Sci. **33**, 57–64 (1998)

J. A. Docobo, Z. Ceplecha, Video record (CD copy attached) of the Spain bolide of June 14, 1996: The atmospheric trajectory and orbit. Astron. Astrophys. Suppl. Ser. **138**, 1–9 (1999)

R.E. Folinsbee, L.A. Bayrock, G.L. Cumming, D.G.W. Smith, Vilna meteorite-camera, visual, seismic and analytic records. J. Roy. Astron. Soc. Canada **72**, 61–64 (1969)

I. Halliday, A.T. Blackwell, A.A. Griffin, The innisfree meteorite and the Canadian camera network. J. Roy. Astron. Soc. Canada **72**, 15–39 (1978)

I. Halliday, B. McIntosh, Orbit of the murchison meteorite. Meteoritics **25**, 339–340 (1990)

P. Jenniskens, J. Borovicka, H. Betlem, C. Ter Kuile, F. Bettonvil, D. Heinlein, Orbits of meteorite producing fireballs: the glanerbrug—a case study. Astron. Astrophys. **255**, 373–376 (1992)

E.L. Krinov, *Principles of Meteorites* (Pergamon Press, New York, 1960), pp. 652–654

L. La Paz, The achondritic shower of February 18, 1948. Publ. Astron. Soc. Pac. **61**, 63–73 (1949)

B.J. Levin, A.N. Simonenko, E. Anders, Farmington meteorite: a fragment of an apollo asteroid. Icarus **28**, 307–324 (1976)

R.E. McCrosky, A. Posen, G. Schwartz, C.-Y. Shao, Lost city meteorite: its recovery and a comparison with other fireballs. J. Geophys. Res. **76**, 4090–4108 (1971)

P. Spurný, J. Oberst, D. Heinlein, Photographic observations of neuschwanstein, a second meteorite from the orbit of the příbram chondrite. Nature **423**, 151–153 (2003)

J.M. Trigo-Rodríguez, J. Borovička, P. Spurný, J.L. Ortiz, J.A. Docobo, A.J. Castro-Tirado, J. Llorca, The villalbeto de la Peña meteorite fall: II. Determination of atmospheric trajectory and orbit. Meteorit. Planet. Sci. **41**, 505–517 (2006)

Mineralogy of HED Meteorites Using the Modified Gaussian Model

Lina Canas · René Duffard · Teresa Seixas

Originally published in the journal Earth, Moon, and Planets, Volume 102, Nos 1–4.
DOI: 10.1007/s11038-007-9177-z

Abstract The correlation between specific meteorites and asteroids is a long-standing problem. The best-known correlation seems to be the HED–Vesta, although several problems still remain to be solved. We report the spectral reflectance analysis (0.4–2.5 μm) of a set of HED meteorites, taken from the RELAB database and three V-type asteroids, taken from MIT-UH-IRTF Joint Campaign for NEO Reconnaissance. We used the Modified Gaussian Model to fit the spectra to a series of overlapping, modified Gaussian absorptions. The fitted individual bands are validated against established laboratory calibrations. With spectral resolution extending to the near-infrared, we are able to resolve the presence of both high-calcium pyroxene (HCP) and low-calcium pyroxene (LCP) and, thus, use the HCP/(HCP + LCP) ratios to remotely trace igneous processing on the parent asteroids. A search of this mineral provides a useful probe of differentiation. The high HCP/(HCP + LCP) ratios found require extensive differentiation of these asteroids and/or their primordial parent body. The degree of melting obtained for the eucrites, using the former ratio, is comparable with that obtained for all V-type asteroids here analyzed, suggesting a comparable geologic history.

Keywords Vesta · HEDs · Meteorites · Asteroids · Modified Gaussian Model

1 Introduction

Meteorites are extraterrestrial samples from small bodies of the solar system—the asteroids. Given their ancient radiometric ages indicating that their ages are very close to that of

L. Canas (✉) · T. Seixas
Faculdade de Ciências da Universidade do Porto, Porto, Portugal
e-mail: lina.canas@multimeios.pt

T. Seixas
e-mail: tmseixas@fc.up.pt

R. Duffard
Instituto Astrofisica de Andalucia, Granada 18008, Spain
e-mail: duffard@iaa.es

the solar system itself, studies of these samples will provide important key information of their initial composition and subsequent processes of chemical evolution and the thermal and physical processes that affected their parent asteroids. From such studies we gain an understanding of the stages of protoplanet formation and the subsequent evolution of smaller bodies (Pieters and McFadden 1994).

One of the most interesting bodies among the asteroid population is 4 Vesta. Combining remote sensing, modeling and density estimates indicate that is a differentiated object with a crust-mantle structure.

The asteroid 4 Vesta, is believed to be the parent body of the howardite-eucrite-diogenite (HED) suite of differentiated meteorites (McCord et al. 1970; Consolmagno and Drake 1977) that are available for laboratory analyses. These meteorite samples are spectroscopically similar to the Vesta V-type asteroids (Duffard et al. 2004). These meteorites that are differentiated basalts (eucrites), pyroxenites (diogenites), and breccia mixtures of mainly these two types (howardites) have experienced several changes including their impact-related ejection from Vesta and subsequent on-orbit collisions and dynamical evolutions.

The V-type asteroids show visible at infrared (IR) wavelengths (0.4–2.5 μm) mineral reflection spectra that have absorption features typical of crystalline structure for chemically, stoichiometric minerals.

Using carefully selected mineral analogs that are commensurate with the bulk compositions of the HED meteorites, a IR remote sensing database can be developed to estimate the mineralogy exposed at terrestrial and extraterrestrial rock surfaces, such as Vesta.

Using high signal-to-noise data and high spectral resolution data it is now possible to resolve the presence of both high calcium pyroxenes and low calcium pyroxenes and use HCP/(LCP + HCP) ratios to remotely trace the nature of igneous processes on asteroids, such as Vesta (Sunshine et al. 2004).

2 Method

The Modified Gaussian Model (MGM) arose from the necessity of a more general fitting method of analysis that could resolve and distinguish individual spectral absorption features and representing them with discrete mathematical distributions. The use of this quantitative correlation is only dependent of the spectrum itself. MGM method supplies an objective and consistent tool to examine the individual absorption features of a spectrum (Sunshine and Pieters 1990).

Given a set of different MGM input parameters (eight individual absorption band centers, widths and strength) and operating the first step of the fitting process, we must meet strict sequential steps in order to reach physically coherent results. The calibration procedures, to which all samples analyzed in the present study were submitted, can be summarized as follows. Given the input parameters we first proceed to an initial fit. Once the fit-result is obtained we proceed to check the bandwidth calibration and assure it is within the tabulated values (Sunshine and Pieters 1993), all the values of bandwidth to respect of band centers found fall within a similar and expected region. Band center calibration follows and again, comparing the results with the tabulated values, we then see if the LCP/HCP ratio band near 1 and 2 μm region is acceptable, all results obtained fall within the expected area of the calibration data (Adams 1974). If all these conditions are met we achieved the final result, individual absorption bands, which combined can describe the composed absorption bands, if not then a change in the band parameters (centers, widths, strength) is needed and we restart the process. An important addition to this procedure is to always mind the fit of the residual and keep its structures to a minimum.

3 Results

In the present work a set of ten eucrites taken from the RELAB database MB-TXH-066, MB-TXH-069-A, MB-TXH-070-A, MT-TXH-043-A, MT-TXH-059, MP-TXH-054-A, MP-TXH-076-A, MP-TXH-087-A, MP-TXH-042-A, MB-TXH-084-A and had their spectra fitted as described above. Also a set of three V-type asteroids 3908, 4055, 6611were submitted to the same described procedure.

Given the number of fitted eucrites, here we only present the obtained plot, in Fig. 1, of one of their fitted spectra, MB-TXH-070-A, as an example of the spectra achieved. We find that with the residual line kept to a minimum of structures we can fit the two main absorption features near 1 and 2 μm.

In the present work we used the model parameters of absorption bands in the MGM fit listed (Sunshine and Pieters 1993) as input. First we started with orthopyroxene (LCP) input parameters only (six absorption band centers, widths and strength) and observed a non-conformity in the obtained results, the individual absorption bands weren't enough to explain the overall spectra. Therefore we did not manage to attain all the given constraints, that is, we could not explain the full width of the second combined band. We needed to address the problem including additional individual bands to obtain consistent results, to fit the composite spectra of our samples. A second analysis was performed with changes in the input parameters to account for a possible percentage of clinopyroxenes (HCP) (75/25 LCP/HCP) in the analyzed samples eucrites and V-type asteroids.

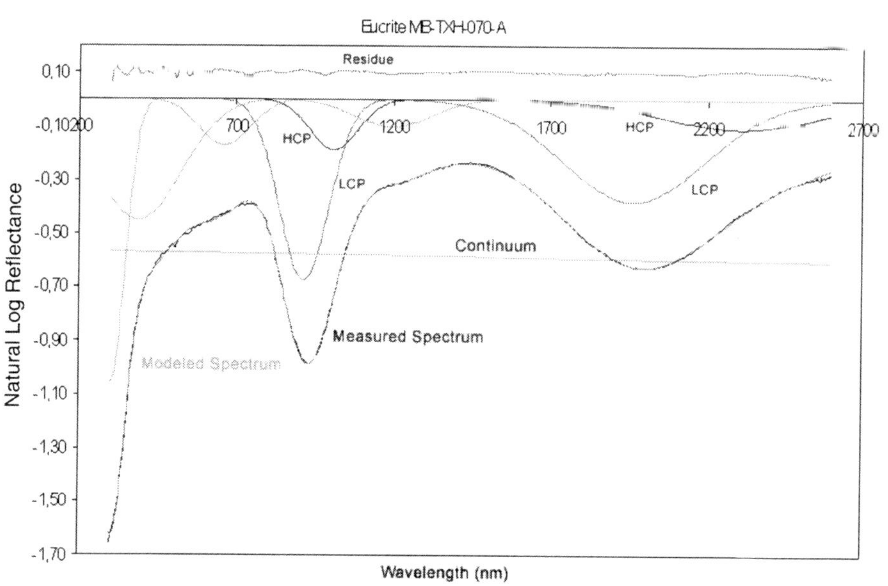

Fig. 1 A example of an MGM fit of the MB-TXH-070-A eucrite

Sunshine and Pieters (1993) derived a relationship from spectra of a set of powders of known proportions of high- and low-calcium pyroxenes and this relationship can be used to separate meteorite classes that have undergone various degrees of igneous processing (Sunshine et al. 2004). We proceeded to the comparison of our results, plotting them against these tabulated values of reference. The ratios of LCP/HCP bands (which are a measurement of the ratio of the band strengths) from our fitted samples were used to perform a logarithmic transformation that allows us to determine HCP/(HCP + LCP) ratio as seen in Fig. 2.

Based on the pre elaborated systematic variation in the relative strength of pyroxene absorption function of the HCP/(HCP + LCP) ratios which are also, in it's turn is related to melting percentage and using calculations of melting with the MELTS program (Ghiroso and Sack 1995; Asimov and Ghiroso 1998) we determined the correspondent values of degree of melting of our samples as seen in Fig. 3. In this manner, we calculated the corresponding degrees of melting of our meteorite and asteroid samples.

4 Discussion and Conclusions

In this study we analyzed a set of ten eucrites and three type V asteroids with MGM, which has been shown to accurately model the shape of isolated absorptions and thus provides a high degree of confidence in resolving overlapping absorption bands. Eucrites are enriched in high-calcium pyroxene, consistent with their origin by crystallization of partial melts. In order to achieve some degree of confidence in the sensitivity of the MGM program we started with the referred composition of orthopyroxenes and found, as expected, that the fitted spectra could not be reproduced by such a combination of that individual bands. Calculations suggest that the HCP/(HCP + LCP) ratio is a sensitive indicator of the degree of partial melting of a chondritic precursor and could be an important tool for deciphering the igneous history of differentiated asteroids (Sunshine et al. 2004).

Analyzing the data obtained for the percentage of melting of the eucrite samples we found that the degree of melting varied from ~21 to 24% corresponding to an HCP/(LCP + HCP) ratio of 0.46–0.55. For the results obtained of the three V-type asteroids we found ~ 18–19% of melting based on HCP/(LCP + HCP) ratios ranging from 0.59 to 0.66.

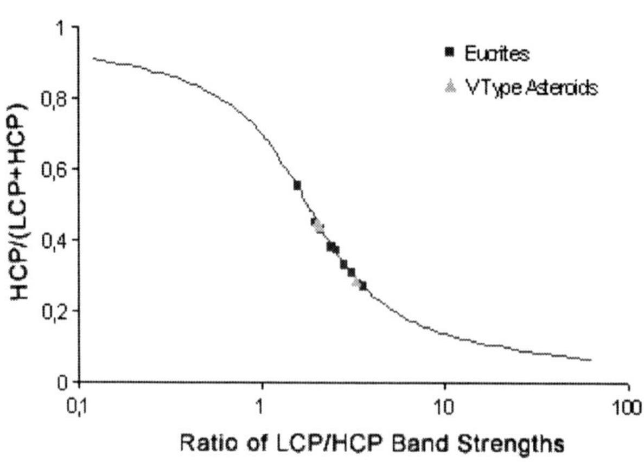

Fig. 2 Ratio of band intensity, CBRS, as function of relative quantities of HCP. Based on systematic variation in the relative strength of pyroxene absorption as a function of HCP/(HCP + LCP)

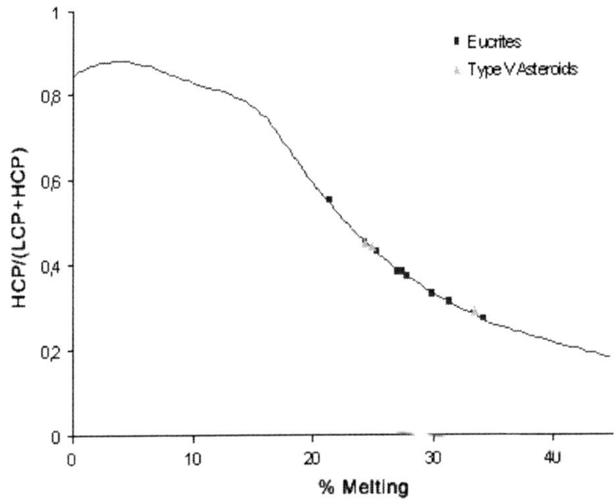

Fig. 3 HCP/(HCP + LCP) ratios function of percent melting. Based on the calculations of melting percentage using the MELTS program

As a whole, the parent body of the HED meteorites would exhibit lower values, consistent with the inclusion of some percentage of orthopyroxene-rich, augite-poor diogenitic material with HCP/(HCP + LCP) ratios approaching zero (Sunshine et al. 2004).

These HCP/(HCP + LCP) ratios suggest extensive differentiation of their parent body Vesta or other Vestoids. The degree of melting obtained for the eucrites using the HCP/(HCP + LCP) ratio can be comparable to the V-type asteroids analyzed and suggest a geological comparable history.

Theoretical predictions expecting asteroid and meteorite differentiation to produce changes in high-calcium pyroxene abundances is clear, the strongest evidence still lies in the study and analysis of both meteorites and remote sensing observed asteroid spectra which experienced various degrees of differentiation and its study stands as an important path to follow in present and future researches.

Acknowledgments Part of the data utilized in this work were obtained as part of the MIT-UH-IRTF Joint Campaign for NEO Reconnaissance. The IRTF is operated by the University of Hawaii under Cooperative Agreement no. NCC 5-538 with the National Aeronautics and Space Administration, Office of Space Science, Planetary Astronomy Program. The MIT component of this work is supported by the National Science Foundation under Grant No. 0506716.

References

J.B. Adams, Visible and near-infrared diffuse reflectance spectra of pyroxenes as applied to remeote sensing of solid objects in the solar system. J. Geophys. Res. **79**(32), 4829–4836 (1974)

P.D. Asimov, M.S. Ghiroso, Algorithmic modifications extending MELTS to calculate sub-solidus phase relations. Am. Mineral. **83**, 1127 (1998)

G.J. Consolmagno, M.J. Drake, Composition and evolution of the eucrite parent body—evidence from rare earth elements. Geochim. Cosmochim. Acta **41**, 1271–1282 (1977)

R. Duffard, D. Lazzaro, J. Licandro, M.C. De Sanctis, M.T. Capria, J.M. Carvano, Mineralogical characterization of some basaltic asteroids in the neighborhood of (4) Vesta: first results. Icarus. **171**, 120–132 (2004)

M.S. Ghiroso, R.O. Sack, Chemical mass transfer in magmatic processes IV. A revised and internally consistent thermodynamic model for the interpolation and extrapolation of liquid-solid equilibria in magmatic systems at elevated temperatures and pressures. Contrib. Mineral. Petr. **119**, 197 (1995)

K. Keil, Geological history of asteroid 4 Vesta: the "smallest terrestrial planet". in *Asteroids III*, ed. by W.F. Bottke Jr., A. Cellino, P. Paolicchi, R.P. Binzel (University of Arizona Press, Tucson, 2002), pp. 653–667

T.B. McCord, J.B. Adams, T.V. Johnson, Asteroid Vesta: spectral reflectivity and compositional implications. Science. **168**, 1445–1447 (1970)

C.M. Pieters, L.A. McFadden, Meteorite and Asteroid reflectance spectroscopy: clues to early solar system processes. Ann. Rev. Earth Planet. Sci. **22**, 457–497 (1994)

J. Sunshine, C. Pieters, S. Pratt, Deconvolution of mineral absorption bands: an improved approach. J. Geophys. Res. **95**(B5), 6955–6966 (1990)

J.M. Sunshine, C.M. Pieters, Estimating modal abundances from the spectra of natural and laboratory pyroxene mixtures using the modified Gaussian model. J. Geophys. Res. **98**(E5), 9075–9087 (1993)

J.M. Sunshine, S.J. Bus, T.J. McCoy, T.H. Burbine, C.M. Corrigan, R.P. Binzel, High-calcium pyroxene as an indicator of igneous differentiation in asteroids and meteorites. Meteoritics & Planet. Sci. **39**, 1343–1357 (2004)

Measurement of Ejecta from Normal Incident Hypervelocity Impact on Lunar Regolith Simulant

David L. Edwards · William Cooke · Danielle E. Moser · Wesley Swift

Originally published in the journal Earth, Moon, and Planets, Volume 102, Nos 1–4.
DOI: 10.1007/s11038-007-9198-7

Abstract The National Aeronautics and Space Administration (NASA) continues to make progress toward long-term lunar habitation. Critical to the design of a lunar habitat is an understanding of the lunar surface environment. A subject for further definition is the lunar impact ejecta environment. The document NASA SP-8013 was developed for the Apollo program and is the latest definition of the ejecta environment. There is concern that NASA SP-8013 may over-estimate the lunar ejecta environment. NASA's Meteoroid Environment Office (MEO) has initiated several tasks to improve the accuracy of our understanding of the lunar surface ejecta environment.

This paper reports the results of experiments on projectile impact into powered pumice targets, simulating unconsolidated lunar regolith. The Ames Vertical Gun Range (AVGR) was used to accelerate spherical Pyrex projectiles of 0.29*g* to velocities ranging between 2.5 and 5.18 km/s. Impact on the pumice target occurred at normal incidence. The ejected particles were detected by thin aluminum foil targets placed around the pumice target in a 0.5 Torr vacuum. A simplistic technique to characterize the ejected particles was formulated. Improvements to this technique will be discussed for implementation in future tests.

Keywords Lunar · Meteoroid impact · Ejecta distribution · Impact testing

---

D. L. Edwards (✉) · W. Cooke
National Aeronautics and Space Administration (NASA)/George C. Marshall Space Flight Center (MSFC)/Natural Environments Branch/ EV13 MSFC, Huntsville, AL 35812, USA
e-mail: David.L.Edwards@nasa.gov

W. Cooke
e-mail: William.Cooke@nasa.gov

D. E. Moser
Morgan, A Stanley Company MSFC, Huntsville, AL 35812, USA
e-mail: Danielle.E.Moser@nasa.gov

W. Swift
Raytheon MSFC, Huntsville, AL 35812, USA
e-mail: Wesley.Swift@nasa.gov

1 Introduction

The series of tests discussed in this paper grew out of two focus areas, related to exploration of the lunar surface. The primary goal of this series of tests was to calibrate ground-based cameras utilized to observe and record meteoroid impacts on the lunar surface. The other focus was the need to increase our understanding of the ejecta environment on the lunar surface. NASA's Meteoroid Environment Office (MEO) offered the opportunity to gather ejecta distribution data during a series of test shots using the Ames Vertical Gun Range (AVGR). The AVGR is a 0.30 caliber light gas gun that can launch projectiles to velocities ranging from 0.5 to nearly 7 km/s. A very unique feature of the AVGR is the ability to vary the gun's angle of elevation with respect to the target.

The angle of elevation of the gun can be varied in 15° increments from 0° to 90°, thus permitting oblique angles of impact. Impact events can be recorded with a variety of high-speed imaging options (Dino 2007).

NASA SP-8013 describes the ejecta environment subsequent to meteoroid impact (NASA SP-8013 1969). The graph shown in Fig. 1 indicates approximately a four order of magnitude increase in the number of ejecta particles of mass, *m*, from the baseline primary meteoroid impact flux of mass, *m*, as defined by the Grün model (W. Cooke, Unpublished presentation of lunar impact ejecta analysis). The design of a long-term habitation structure to survive the ejecta environment described in NASA SP-8013 would require excessive mass, making it difficult and potentially cost prohibitive to launch and deliver the structure to the lunar surface. A better understanding of the lunar ejecta environment is required to optimize the lunar habitat design.

2 Experiment

The purpose of this series of experiments was to collect information enabling the characterization of the ejecta angular distribution resulting from a hypervelocity impact into simulated regolith. These ejecta characterization experiments were secondary experiments in the AVGR, and as such were dependent upon the test parameters required by the primary

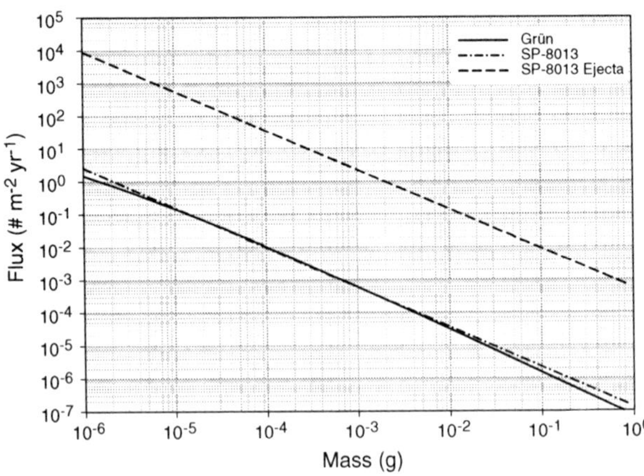

Fig. 1 Estimation of the lunar ejecta flux compared to the prediction of primary impact flux from the Grün model

experiment: the calibration of video cameras from the primary impact flash. Therefore, one of the constraints of the ejecta characterization experiment was that the ejecta experiment could not influence, bias, perturb, or otherwise contaminate the calibration of the video camera. The experiment set-up, shown in Fig. 2, consisted of placing sheets of aluminum foil in two specific locations, identified as position "D" and position "B" around the periphery of the AVGR vacuum chamber.

The 0.17 mm thick foils were positioned such that the ejecta would impact the foils at near-normal incident angles. In each case, the projectile was fired vertically (90°) into a target of fine grain, less than 60 μm diameter, pumice. The projectile in each case was 6.35 mm diameter, 0.29*g* Pyrex sphere. The projectile velocities were 2.5, 3.78, and 5.18 km/s for the three test shots.

3 Results

The impact test performed at 2.5 km/s did not produce ejecta that penetrated the foil detectors. Projectile impact velocities of 3.78 and 5.18 km/s produced primary ejecta that penetrated the foil targets. The foil targets were divided into 1° wide horizontal bands and the number of penetrations in each band was counted. The results from counting penetration in foils from positions D and B are shown in Fig. 3 and indicate a preferred ejecta angle of 38–40° with respect to the horizontal.

4 Summary and Discussion

The Ames Vertical Gun Range (AVGR) was used to accelerate spherical Pyrex projectiles of 0.29*g* to velocities ranging between 2.5 and 5.18 km/s. Impact on the powered pumice target occurred at normal incidence.

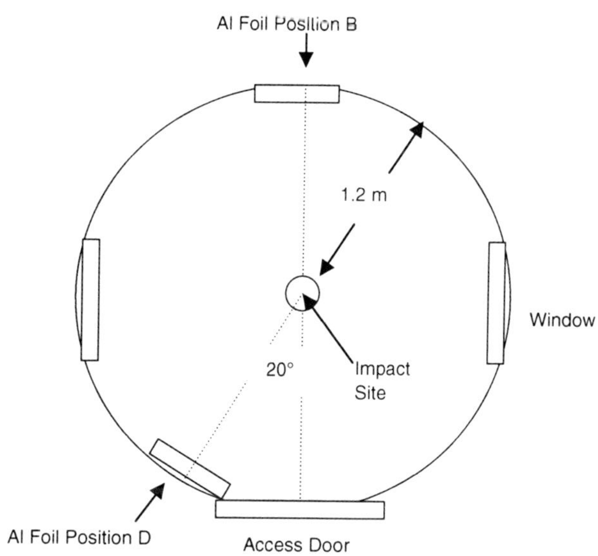

Fig. 2 Top view schematic of the test chamber—specifically indicating the positions of the aluminum foil detectors D and B

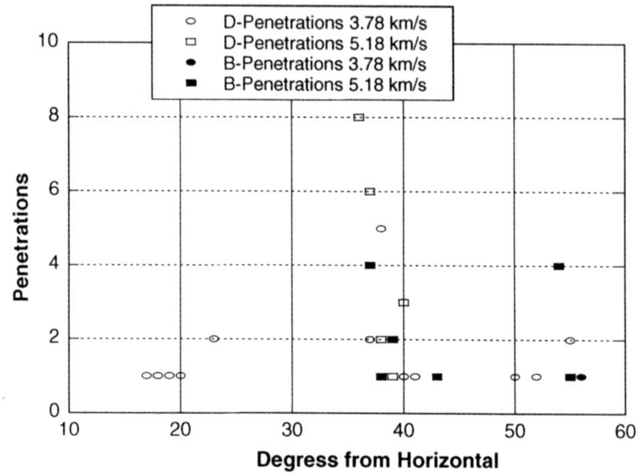

Fig. 3 Angular distributions of ejecta with sufficient velocity to penetrate the aluminum foil detectors

The ejected particles were detected by thin aluminum foil targets placed around the fine grain, 60 μm diameter, pumice target in a 0.5 Torr vacuum. The results presented in this paper indicate that a peak ejection angle for penetrating ejecta is approximately 38° off the horizontal. Previous work by Yamamoto resulted in a peak ejection angle of approximately 30° (Yamamoto 2002, p. 92). Yamamoto et al. used a "staple-shaped" copper projectile with impact velocities ranging from 243 to 272 m/s and impacted a target consisting of soda-lime particles with a nominal diameter of 220 μm. Cintala et al. performed a series of impact tests using spherical aluminum particles accelerated to velocities ranging from 0.8 to 1.92 km/s (Cintala 1999). The incident projectiles had a nominal diameter of 4.76 mm and impacted coarse-grained sand with grain sizes ranging from 1 to 3 mm. Cintala provides extensive detail for characterizing the ejecta angular and size distributions and recorded ejecta angles ranging between 38° and 55°. Cintala, Yamamoto, and this work used varying techniques to determine ejecta distributions, with Cintala and Yamamoto also providing ejecta velocities.

This experiment made no attempt to measure ejection velocities. Speculation on anticipated ejecta velocities was aided by referencing Yamamoto's work, which states "In the case of the vertical impact of the projectile, most ejecta have velocities lower than 24% of the projectile speed."(Yamamoto 2002, p. 87) If Yamamoto's 24% prediction can serve as a guide for the higher velocities in this experiment, then using 24% of the incident projectile speeds of 2.5, 3.78, and 5.18 km/s provides upper threshold ejecta speeds of approximately 600, 907, and 1,243 m/s, respectively.

Using the penetration equation given in NASA SP-8013 (p. 7) describing threshold penetrations of "single thin ductile metal plates," we get a penetration threshold velocity for the aluminum foil of approximately 988 m/s. This equation is:

$$t = K_1 \rho^{1/6} m^{0.352} V^{0.875} \tag{1}$$

where, t is the thickness of the foil penetrated, K_1 is a constant, ρ is the mass density of the ejecta (1.3 gm/cm^3 for pumice), m is the mass of the ejecta particle, and V is the ejecta velocity. This calculation assumes all ejecta particles are 60 μm diameter pumice. If Yamamoto's 24% prediction holds true for this body of work, then there is an explanation of why the impact at 2.5 km/s did not produce ejecta that penetrated the aluminum foil and why the 3.78 and 5.18 km/s impacts did produce foil penetrating ejecta. This explanation needs to be confirmed by a future experiment.

The lunar meteoroid ejecta environment needs to be better characterized. This work, integrated with other activities being initiated by NASA's Constellation program, can lead to a more accurate understanding of the ejecta environment. To build on this existing database, additional tests at the AVGR are scheduled. These tests will include varying the regolith target to begin to understand the differences in mare and highland regolith by using foil detectors of various thicknesses to gain information on ejecta velocity distributions. Detectors will also be placed at various distances from the impact site to better characterize the ejecta dispersions.

References

J. Dino (ed.) http://www.nasa.gov/centers/ames/research/technology-onepagers/range-complex.html cited 5 May 2007 (2007)

M.J. Cintala, L. Berthoud, F. Hörz, Ejection-velocity distributions from impacts into coarse-grained sand. Meteorit. Planet. Sci. **34**, 605–623 (1999)

NASA SP-8013 Meteoroid environment model-1969 [Near earth to lunar Surface], March 1969

S. Yamamoto, Measurement of impact ejecta from regolith targets in oblique impacts. Icarus **158**, 87–97 (2002)

Understanding the WMAP Results: Low-Order Multipoles and Dust in the Vicinity of the Solar System

Valeri Dikarev · Oliver Preuß · Sami Solanki · Harald Krüger · Alexander Krivov

Originally published in the journal Earth, Moon, and Planets, Volume 102, Nos 1–4.
DOI: 10.1007/s11038-007-9172-4

Abstract Analyses of the cosmic microwave background (CMB) radiation maps produced by the Wilkinson Microwave Anisotropy Probe (WMAP) have revealed anomalies not predicted by the standard cosmological theory. It has been suggested that a dust cloud in the vicinity of the Solar system may be the cause for these anomalies. In this paper, the thermal emission by particles from two known interplanetary meteoroid complexes is tested against the CMB maps. Conclusions are drawn based on the geometry of cloud projections onto the WMAP sky whether these clouds are likely to explain the observed anomaly. The smooth background Zodiacal cloud and one of the Taurid meteor complex branches do not explain the WMAP anomaly.

Keywords CMB · WMAP · Solar system · Dust

1 Introduction

The cosmic microwave background (CMB) radiation was released 14×10^9 years ago, 380,000 years after the Big Bang, i.e. when the Universe became transparent. Ever cooling since then, the CMB has reached the average temperature of 2.725 K today. Fluctuations of the CMB around this temperature, of the order of 100 μK, carry information about matter and gravitational potential distribution at the earliest time of the Universe. Thus cosmological theories can be tested against maps of CMB.

V. Dikarev (✉) · O. Preuß · S. Solanki · H. Krüger
MPI für Sonnensystemforschung, 37191 Katlenburg-Lindau, Germany
e-mail: dikarev@mps.mpg.de

V. Dikarev
MPI für Kernphysik, 69117 Heidelberg, Germany

V. Dikarev
Astronomical Institute of St. Petersburg University, St. Petersburg, Russia

A. Krivov
AIU FSU Jena, Jena, Germany

DOI: 10.1007/978-0-387-78419-9_72

The Wilkinson Microwave Anisotropy Probe (WMAP, see Bennett et al. 2003) has surveyed the microwave sky at a so far unprecedented accuracy and resolution, allowing one to make maps of CMB fluctuations and analyse them (Bennet et al. 2003; Hinshaw et al. 2007). As the fluctuations are random, and the sky is a single implementation, statistical properties of the fluctuations are determined. The entire map is expanded into spherical harmonic functions, and cosmological theories are applied to predict the power of the multipoles obtained from the data. While the standard inflationary cosmology is confirmed by the WMAP data at a very high confidence level for nearly all multipole orders, significant contradictions have been found as well. In particular, contrary to predicted independence, the quadrupole and octopole moments are mutually aligned to a very unlikely degree. Moreover, they are aligned with the Solar-system geometry at a confidence level that is again difficult to ignore (see Fig. 1).

It has been suggested that a dust cloud in the vicinity of the Solar system could be the reason for the anomaly (Copi et al. 2006; de Oliveira-Costa and Tegmark 2006; Schwarz et al. 2004; Starkman and Schwarz 2005). Calculations of the thermal emission spectra of interplanetary meteoroids performed in Dikarev et al. (2007) have shown that mm-sized carbonaceous as well as cm-sized silicate particles are characterised by a flat absorption efficiency throughout the infrared and millimetre wavelengths, whereas the absorption efficiency of small dust grains of a few tens micrometres in size and below, decreases steeply with the wavelength λ increase, close to the emissivity law λ^{-2} usually applied to describe spectra of fine grains. In order to provide the microwave emission sufficient to cause the

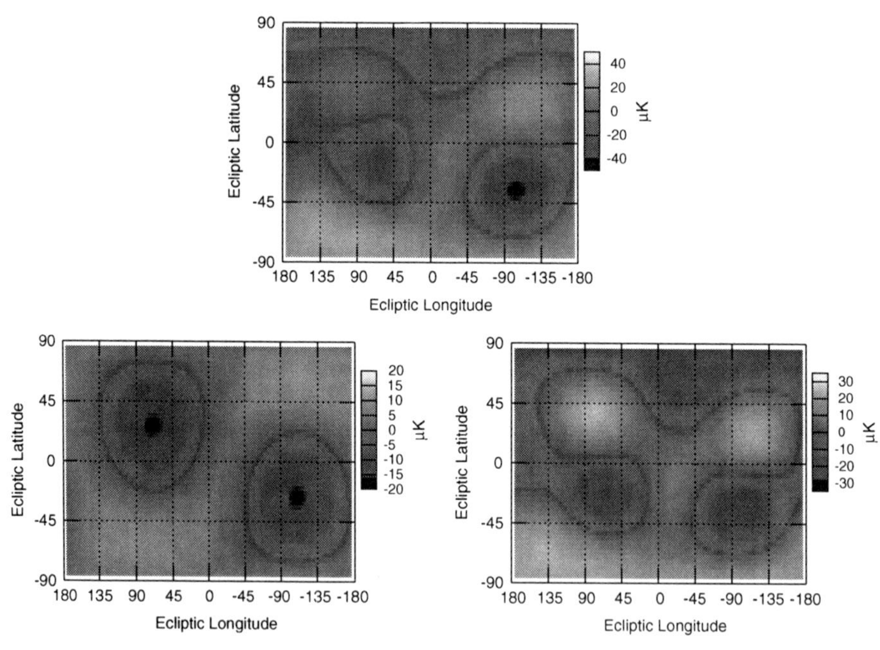

Fig. 1 Quadrupole (bottom left) and octopole (bottom right) from the three-year temperature map (Hinshaw et al. 2007), and their superposition (top) in the ecliptic coordinate system. A first look at the octopole component suggests an unusual planarity compared to what is expected from an isotropic random field with no preferred axis. Also the quadrupole and octopole show a striking correlation with the ecliptic plane which clearly separates three extrema in the north and three in the south. In addition, the superposition map reveals a north-south ecliptic asymmetry where the extrema in the north are visibly weaker than those in the south

WMAP anomaly, one would therefore need a cloud of big particles with a significantly smaller visual and infrared optical depth than under the hypothesis of a cloud of tiny grains, possibly explaining why this cloud has remained unknown. The large meteoroids, however, have been observed only scarcely, almost entirely as meteors in Earth's atmosphere. Their number densities in the Solar system remain largely unconstrained. Therefore, we constructed relative microwave brightness maps of several known dust clouds, assuming that the big particles may be more abundant in these clouds than it was previously thought. In this paper, we report two of this series of tests of existing dust clouds against the WMAP data that we have carried out in order to find the mysterious cause of the WMAP anomaly.

2 Projecting Dust Clouds onto the WMAP Sky

The full sky maps made by the WMAP are dominated by the CMB fluctuations, which are Gaussian random, and appear to be too noisy to put any firm constraints on the dust cloud that may be responsible for the quadrupole and octopole anomalies. Therefore, instead of a comparison between the full maps and dust cloud images, we make a comparison of the quadrupole and octopole moments of CMB fluctuations, i.e. where the hypothetical dust cloud is best revealed, and the corresponding momenta of the multipole expansions of real dust cloud images.

The expansion is implemented as an integration over the entire celestial sphere:

$$a_{l,n} = \int Y^*_{l,n}(\Omega) f(\Omega) \mathrm{d}\Omega, \tag{1}$$

where $Y_{l,n}$ is a spherical harmonic function and $a_{l,n}$ is the corresponding expansion coefficient, f is the function expanded, $l = 1,\ldots,\infty$, and $n = \quad l,\ldots,l$. The quadrupole and octopole momenta are indexed by $l = 2$ and 3, respectively.

The expanded function f is defined so as to simulate the dust cloud number density, temperature and thermal emission, and the WMAP observation strategy. For each line of sight, the integral

$$I = \int_0^\infty B[\lambda, T(r)]\, C_{\mathrm{abs}}[x(l), y(l), z(l)]\, \mathrm{d}l \tag{2}$$

is evaluated, where B is the blackbody emission power at the wavelength λ, C_{abs} is the absorption cross-section of dust per unit volume, assumed to be proportional to the number density of dust, l is the position on the line of sight, x, y, and z are its heliocentric coordinates. The temperature T of dust is adopted from the model (Kelsall et al. 1998)

$$T(r) = 286\ \mathrm{K} \times (1\ \mathrm{AU}/r)^{0.467}. \tag{3}$$

The WMAP observation strategy is approximately described as viewing from the moving Earth everywhere within the range of solar elongations from 90° to 135°, with a uniform coverage inside this range (c.f. Bennet et al. 2003).

3 Dusty Multipoles

The Zodiacal cloud is a flat cloud nearly symmetric about a plane slightly inclined with respect to the ecliptic plane. If its inclination were zero, one would not expect it to cause the anomalies reported by the WMAP. However, the inclination is about 2°, and the half latitudinal width at half maximum of the number density is about 15°. Therefore the motion of the Earth (and WMAP) with respect to the cloud's symmetry plane causes e.g. polar brightness variations of 20 to 30% due to the column density variations (Kelsall et al. 1998)! Also, the Earth's heliocentric distance oscillations lead to the temperature variations of dust on the same rest-frame line of sight. It seems to be worthwhile to check if the motion of the WMAP observatory with respect to this cloud can explain the anomalous multipoles.

The major component of the model (Kelsall et al. 1998), the smooth background cloud, is implemented. Even though the relevance of that model fitted to the infrared data to the interplanetary dust seen in the microwaves is doubtful, there is an argument in support of its use in rough first-step estimates. To put it simply, the model is not worse than any other number density representation in resembling the most general, large-scale properties of the zodiacal dust cloud, i.e. a flat, azimuthally symmetric dust complex inclined with respect to the ecliptic plane.

The smooth background cloud emission maps are plotted in Fig. 2. The top row consists of two maps of the total emission, before the expansion into multipoles, in arbitrary units, at 1 mm (left) and 5 mm (right) wavelengths. Substantially different in normalisations, their relative distributions are also different, since at 5 mm more remote and colder dust is seen, which extends further in the ecliptic plane, while the number density drops down exponentially in the vertical direction.

To some extent in accord with expectations, the octopole moment of the smooth background cloud does reveal an interesting structure with maxima and minima following anti-symmetrically along the ecliptic longitude. The quadrupole, however, is closely confined to the ecliptic plane. Moreover, the octopole magnitude with respect to that of the quadrupole is negligible: less than 1%, contrary to the WMAP results which suggest similar quadrupole and octopole magnitudes.

In a 'gedanken experiment', we tried to enhance the anticipated effects of the cloud's inclination and Earth's orbital eccentricity by artificially increasing both of them. This leads indeed to more equal quadrupole and octopole magnitudes, with a structure reminiscent of the WMAP multipoles, although not identical. As neither cloud's inclination nor Earth's eccentricity can be modified, the next idea is to find an existing meteoroid complex in the Solar system that is on an eccentric and inclined orbit.

One possible candidate for such a complex is the Taurid meteor complex. It was interesting to simulate as well, since some believed (Štohl and Porubčan 1991; Whipple and Hamid 1952) it is related to comet 2P/Encke and may have been produced in a recent disruption of a large comet, perhaps by collision with an asteroid. This can make the complex potentially a large modern reservoir of big meteoroids being collisionally transformed into small dust.

The orbit elements of one of the Taurid's radiants are adopted, the semimajor axis 2.2 AU, eccentricity 0.8, inclination 3.6°, perihelion arguments 294°, and node longitude 225° (Štohl and Porubčan 1991). Figure 3 shows how the Taurid meteor stream would look as viewed from the Sun. The longitude of pericentre is indeed near 180° ecliptic longitude where the figure shows an enlargement of the stream thickness due to its proximity, while the surface brightness peaks near zero longitude where cold dust dwells near the aphelion. Figure 4 shows the emission from the Taurid meteor complex as seen by WMAP. The bizarre appearance of the stream in the top left panel of Fig. 4 is due to the motion of the Earth and WMAP observation strategy.

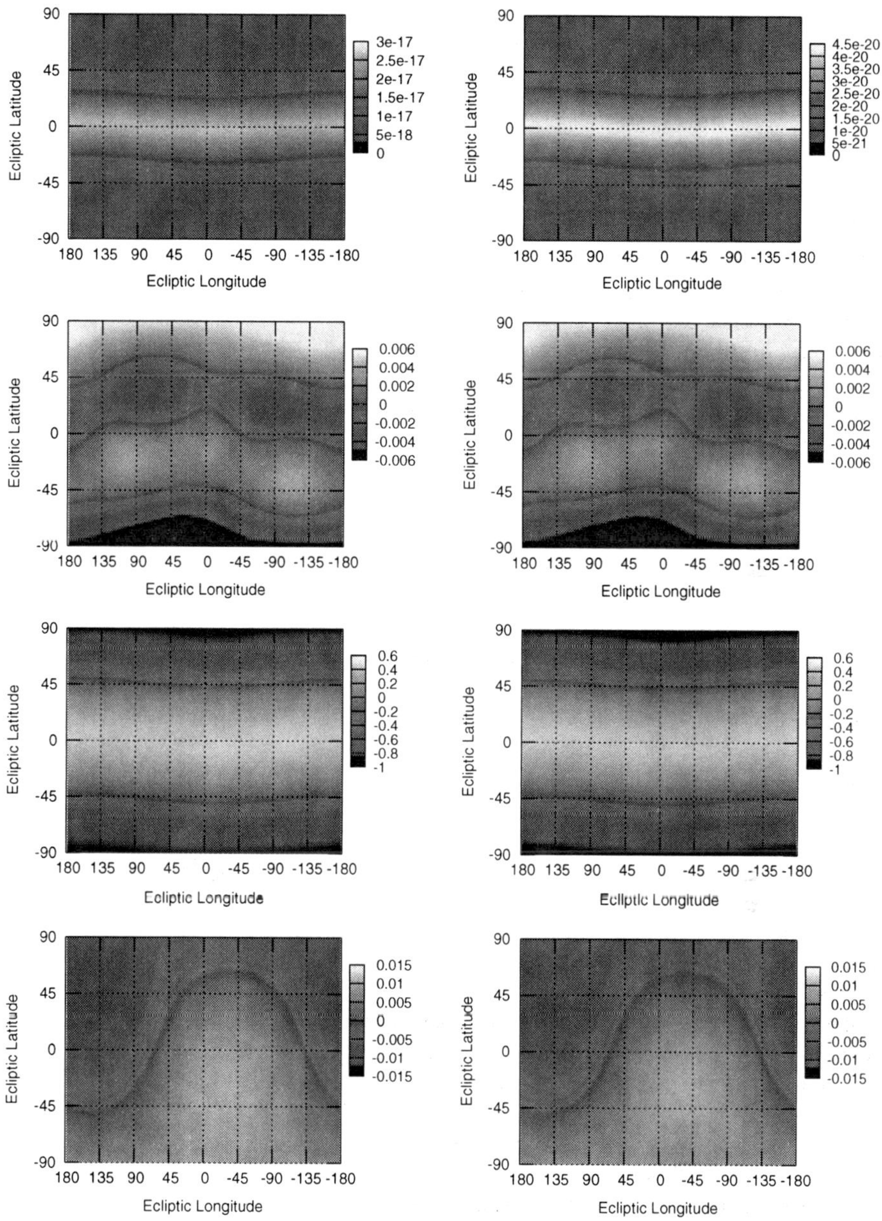

Fig. 2 Maps of the dust emission due to the smooth background zodiacal cloud of Kelsall et al. (1998). Left column: 1 mm wavelength, right column: 5 mm. From top to bottom: the total emission in arbitrary units, then the octopole, quadrupole and dipole in units of the monopole

Interestingly, one can see indeed comparable magnitudes of the quadrupole and octopole, probably because all the low-order multipoles are equally bad at reproducing the low-scale features. Yet their alignment is not obviously close to the WMAP puzzling results. Moreover, as most of the multipole power comes from a very small spot on the sky, and if

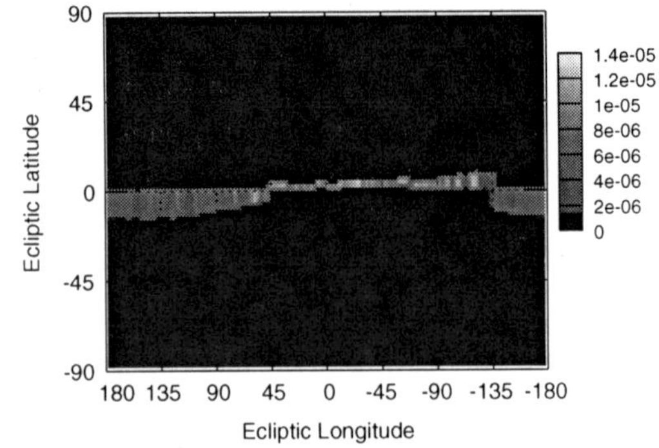

Fig. 3 Maps of the thermal emission by meteoroids of the Taurid meteor complex at the wavelength of 1 mm, as it would appear to an observer at the Sun, arbitrary units

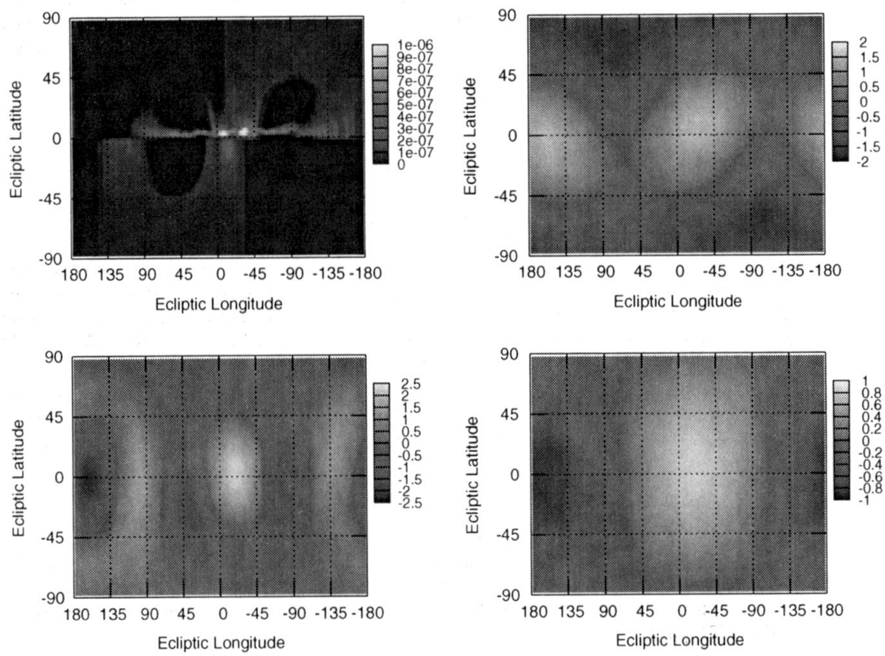

Fig. 4 Maps of the thermal emission by the meteoroids of the Taurid meteor complex at the wavelength of 1 mm. Top left: the total emission, before the expansion into multipoles, in arbitrary units; bottom left: the octopole; top right: the quadrupole; bottom right: the dipole; the multipoles are in the units of monopole

indeed this single bright spot were responsible for the alignment of low-l multipoles, then it would certainly be recognized on the full sky maps, even before the multipole expansion procedure was applied. This is not the case, however, so this hypothesis can also be rejected.

4 Conclusion

We have reported an ongoing search for the reason of the WMAP anomaly. An unexpected alignment of the quadrupole and octopole moments of the expansion of the map of CMB fluctuations mutually and with the Solar-system geometry was interpreted as an indication of the presence of a dust cloud that radiates in the microwaves. In this paper, the smooth background component of the Zodiacal cloud and the Taurid meteor complex are both tested for compatibility with the CMB quadrupole and octopole. They are both rejected based on different argumentation. We plan to consider more known and hypothetical dust clouds, bound to and neighbouring the Solar system, in our subsequent publications (Dikarev et al. 2007).

References

C.L. Bennett, M. Bay, M. Halpern, G. Hinshaw, C. Jackson, N. Jarosik, A. Kogut, M. Limon, S.S. Meyer, L. Page, D.N. Spergel, G.S. Tucker, D.T. Wilkinson, E. Wollack, E.L. Wright, The microwave anisotropy probe mission. Astrophys. J. **583**, 1–23 (2003)

C.L. Bennett, M. Halpern, G. Hinshaw, N. Jarosik, A. Kogut, M. Limon, S.S. Meyer, L. Page, D.N. Spergel, G.S. Tucker, E. Wollack, E.L. Wright, C. Barnes, M.R. Greason, R.S. Hill, E. Komatsu, M.R. Nolta, N. Odegard, H.V. Peiris, L. Verde, J.L. Weiland, First-year Wilkinson microwave anisotropy probe (WMAP) observations: preliminary maps and basic results. Astrophys. J. Suppl. **148**, 1–27 (2003). doi: 10.1086/377253, arXiv:astro-ph/0302207

C.J. Copi, D. Huterer, D.J. Schwarz, G.D. Starkman, On the large-angle anomalies of the microwave sky. Mon. Not. R. Astron. Soc. **367**, 79–102 (2006). doi: 10.1111/j.1365-2966.2005.09980.x, arXiv:astro-ph/0508047

A. de Oliveira-Costa, M. Tegmark, CMB multipole measurements in the presence of foregrounds. Phys. Rev. D **74**(2) (2006), 023,005–+, arXiv:astro-ph/0603369

V. Dikarev, O. Preuß, S. Solanki, H. Krüger, A. Krivov, The local dust foregrounds in the microwave sky: I. Thermal emission spectra. Astrophys. J. submitted (2007)

G. Hinshaw, M.R. Nolta, C.L. Bennett, R. Bean, O. Doré, M.R. Greason, M. Halpern, R.S. Hill, N. Jarosik, A. Kogut, E. Komatsu, M. Limon, N. Odegard, S.S. Meyer, L. Page, H.V. Peiris, D.N. Spergel, G.S. Tucker, L. Verde, J.L. Weiland, E. Wollack, E.L. Wright, Three-year Wilkinson microwave anisotropy probe (WMAP) observations: temperature analysis. Astrophys. J. Suppl. **170**, 288–334 (2007). doi: 10.1086/513698, arXiv:astro-ph/0603451

T. Kelsall, J.L. Weiland, B.A. Franz, W.T. Reach, R.G. Arendt, E. Dwek, H.T. Freudenreich, M.G. Hauser, S.H. Moseley, N.P. Odegard, R.F. Silverberg, E.L. Wright, The COBE diffuse infrared background experiment search for the cosmic infrared background. II. Model of the interplanetary dust cloud. Astrophys. J. **508**, 44–73 (1998)

D.J. Schwarz, G.D. Starkman, D. Huterer, C.J. Copi, Is the low-ℓ microwave background cosmic? Phys. Rev. Lett. **93**(22) (2004), 221,301–+. doi: 10.1103/PhysRevLett.93.221301, astro-ph/0403353

G.D. Starkman, D.J. Schwarz, Is the universe out of tune? Sci. Am. **291**, 36–43 (2005)

J. Štohl, V. Porubčan, Dinamical aspects of the Taurid meteor complex, in *Chaos, Resonance and Collective Dynamical Phenomena in the Solar System*, ed. by S. Ferraz-Mello (1991), pp. 315–324

F.L. Whipple, S.E. Hamid, On the origin of the Taurid meteor streams, in *Helman Obs. Bull.* 41, Harward Reprint 361 (1952), pp. 1–30

Breinigsville, PA USA
02 September 2009
223408BV00005B/23/P

9 780387 784182